MW00837558

Advances in Electromagnetics Empowered by Artificial Intelligence and Deep Learning

IEEE Press
445 Hoes Lane
Piscataway, NJ 08854

Advances in Electromagnetics Empowered by Artificial Intelligence and Deep Learning

Edited by

Sawyer D. Campbell and Douglas H. Werner
Department of Electrical Engineering
The Pennsylvania State University
University Park, Pennsylvania, USA

IEEE Press Series on Electromagnetic Wave Theory

Published by John Wiley & Sons, Inc., Hoboken, New Jersey.

Published simultaneously in Canada.

Library of Congress Cataloging-in-Publication Data Applied for:
Hardback ISBN: 9781119853893

Cover Image and Design: Wiley

Set in 9.5/12.5pt STIXTwoText by Straive, Chennai, India

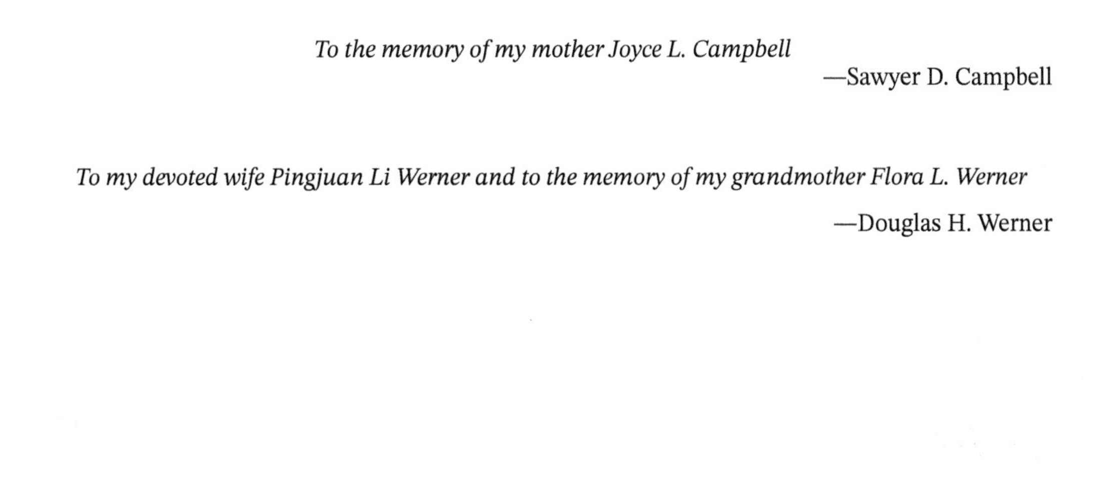

To the memory of my mother Joyce L. Campbell

—Sawyer D. Campbell

To my devoted wife Pingjuan Li Werner and to the memory of my grandmother Flora L. Werner

—Douglas H. Werner

Contents

About the Editors

Sawyer D. Campbell is an Associate Research Professor in Electrical Engineering and associate director of the Computational Electromagnetics and Antennas Research Laboratory (CEARL), as well as a faculty member of the Materials Research Institute (MRI), at The Pennsylvania State University. He has published over 150 technical papers and proceedings articles and is the author of two books and five book chapters. He is a Senior Member of the Institute of Electrical and Electronics Engineers (IEEE), OPTICA, and SPIE and Life Member of the Applied Computational Electromagnetics Society (ACES). He is the past Chair and current Vice Chair/Treasurer of the IEEE Central Pennsylvania Section.

Douglas H. Werner holds the John L. and Genevieve H. McCain Chair Professorship in Electrical Engineering and is the director of the Computational Electromagnetics and Antennas Research Laboratory (CEARL), as well as a faculty member of the Materials Research Institute (MRI), at The Pennsylvania State University. Prof. Werner has received numerous awards and recognitions for his work in the areas of electromagnetics and optics. He holds 20 patents, has published over 1000 technical papers and proceedings articles, and is the author of 7 books and 35 book chapters. He is a Fellow of the Institute of Electrical and Electronics Engineers (IEEE), the Institute of Engineering and Technology (IET), Optica, the International Society for Optics and Photonics (SPIE), the Applied Computational Electromagnetics Society (ACES), the Progress In Electromagnetics Research (PIER) Electromagnetics Academy, and the National Academy of Inventors (NAI).

List of Contributors

Sensong An
Department of Electrical & Computer Engineering
University of Massachusetts Lowell
Lowell, MA
USA

and

Department of Materials Science & Engineering
Massachusetts Institute of Technology
Cambridge, MA
USA

N. Anselmi
ELEDIA Research Center (ELEDIA@UniTN – University of Trento)
DICAM – Department of Civil, Environmental, and Mechanical Engineering
Trento
Italy

Wenshan Cai
School of Electrical and Computer Engineering
Georgia Institute of Technology
Atlanta, GA
USA

Sawyer D. Campbell
The Pennsylvania State University
University Park, PA
USA

Yu Cao
School of Electrical, Computer and Energy Engineering
Arizona State University
Tempe, AZ
USA

Nagadastagiri Reddy Challapalle
School of Electrical Engineering and Computer Science
The Pennsylvania State University
University Park, PA
USA

Christos Christodoulou
Department of Electrical and Computer Engineering
The University of New Mexico
Albuquerque, NM
USA

Xiaocong Du
School of Electrical, Computer and Energy Engineering
Arizona State University
Tempe, AZ
USA

Ahmet M. Elbir
Interdisciplinary Centre for Security
Reliability and Trust (SnT)
University of Luxembourg
Luxembourg

Jonathan A. Fan
Department of Electrical Engineering
Stanford University
Stanford, CA
USA

Feng Feng
School of Microelectronics
Tianjin University
Tianjin
China

Clayton Fowler
Department of Electrical & Computer Engineering
University of Massachusetts Lowell
Lowell, MA
USA

Isha Garg
Elmore School of Electrical and Computer Engineering
Purdue University
West Lafayette, IN
USA

Arjun Gupta
Facebook
Menlo Park, CA
USA

Rong-Han Hong
Institute of Electromagnetics and Acoustics
Xiamen University
Xiamen
China

Wei Hong
State Key Laboratory of Millimeter Waves
School of Information Science and Engineering
Southeast University
Nanjing, Jiangsu Province
China

and

Department of New Communications
Purple Mountain Laboratories
Nanjing, Jiangsu Province
China

Hao-Jie Hu
Institute of Electromagnetics and Acoustics
Xiamen University
Xiamen
China

Ronald P. Jenkins
The Pennsylvania State University
University Park, PA
USA

Jing Jin
College of Physical Science and Technology
Central China Normal University
Wuhan
China

Youngeun Kim
School of Engineering & Applied Science
Yale University
New Haven, CT
USA

Youngwook Kim
Electronic Engineering
Sogang University
Seoul
South Korea

Slawomir Koziel
Faculty of Electronics, Telecommunications and Informatics
Gdansk University of Technology
Gdansk
Poland

and

Engineering Optimization & Modeling Center
Reykjavik University
Reykjavik
Iceland

Gokul Krishnan
School of Electrical, Computer and Energy Engineering
Arizona State University
Tempe, AZ
USA

Mrinal Kumar
Department of Mechanical and Aerospace Engineering
The Ohio State University
Columbus, OH
USA

Chonghan Lee
School of Electrical Engineering and Computer Science
The Pennsylvania State University
University Park, PA
USA

Yuhang Li
School of Engineering & Applied Science
Yale University
New Haven, CT
USA

Qing Liu
Institute of Electromagnetics and Acoustics
Xiamen University
Xiamen
China

Zhaocheng Liu
School of Electrical and Computer Engineering
Georgia Institute of Technology
Atlanta, GA
USA

Robert Lupoiu
Department of Electrical Engineering
Stanford University
Stanford, CA
USA

Jordan M. Malof
Department of Electrical and Computer Engineering
Duke University
Durham, NC
USA

Manel Martínez-Ramón
Department of Electrical and Computer Engineering
The University of New Mexico
Albuquerque, NM
USA

A. Massa
ELEDIA Research Center (ELEDIA@UniTN – University of Trento)
DICAM – Department of Civil, Environmental, and Mechanical Engineering
Trento
Italy

and

ELEDIA Research Center
(ELEDIA@TSINGHUA – Tsinghua University)
Haidian, Beijing
China

and

ELEDIA Research Center (ELEDIA@UESTC – UESTC)
School of Electronic Science and Engineering
University of Electronic Science and Technology of China
Chengdu
China

and

School of Electrical Engineering
Tel Aviv University
Tel Aviv
Israel

and

ELEDIA Research Center (ELEDIA@UIC – University of Illinois Chicago)
Chicago, IL
USA

Kumar Vijay Mishra
Computational and Information Sciences Directorate (CISD)
United States DEVCOM Army Research Laboratory
Adelphi, MD
USA

Weicong Na
Faculty of Information Technology
Beijing University of Technology
Beijing
China

Vijaykrishnan Narayanan
School of Electrical Engineering and Computer Science
The Pennsylvania State University
University Park, PA
USA

Indranil Nayak
ElectroScience Laboratory and Department of Electrical and Computer Engineering
The Ohio State University
Columbus, OH
USA

G. Oliveri
ELEDIA Research Center (ELEDIA@UniTN – University of Trento)
DICAM – Department of Civil, Environmental, and Mechanical Engineering
Trento
Italy

Willie J. Padilla
Department of Electrical and Computer Engineering
Duke University
Durham, NC
USA

Priyadarshini Panda
School of Engineering & Applied Science
Yale University
New Haven, CT
USA

Anna Pietrenko-Dabrowska
Faculty of Electronics, Telecommunications and Informatics
Gdansk University of Technology
Gdansk
Poland

L. Poli
ELEDIA Research Center (ELEDIA@UniTN – University of Trento)
DICAM – Department of Civil, Environmental, and Mechanical Engineering
Trento
Italy

A. Polo
ELEDIA Research Center (ELEDIA@UniTN – University of Trento)
DICAM – Department of Civil, Environmental, and Mechanical Engineering
Trento
Italy

Simiao Ren
Department of Electrical and Computer Engineering
Duke University
Durham, NC
USA

P. Rocca
ELEDIA Research Center (ELEDIA@UniTN – University of Trento)
DICAM – Department of Civil, Environmental, and Mechanical Engineering
Trento
Italy

and

ELEDIA Research Center
(ELEDIA@XIDIAN – Xidian University)
Xi'an, Shaanxi Province
China

José Luis Rojo Álvarez
Departamento de Teoría de la señal y Comunicaciones y Sistemas Telemáticos y Computación
Universidad rey Juan Carlos
Fuenlabrada, Madrid
Spain

Kaushik Roy
Elmore School of Electrical and Computer Engineering
Purdue University
West Lafayette, IN
USA

M. Salucci
ELEDIA Research Center (ELEDIA@UniTN – University of Trento)
DICAM – Department of Civil, Environmental, and Mechanical Engineering
Trento
Italy

Wei Shao
School of Physics, University of Electronic Science and Technology of China
Institute of Applied Physics
Chengdu
China

Jingbo Sun
School of Electrical, Computer and Energy Engineering
Arizona State University
Tempe, AZ
USA

Fernando L. Teixeira
ElectroScience Laboratory and Department of Electrical and Computer Engineering
The Ohio State University
Columbus, OH
USA

Yeshwanth Venkatesha
School of Engineering & Applied Science
Yale University
New Haven, CT
USA

Bing-Zhong Wang
School of Physics, University of Electronic Science and Technology of China
Institute of Applied Physics
Chengdu
China

Haiming Wang
State Key Laboratory of Millimeter Waves
School of Information Science and Engineering
Southeast University
Nanjing, Jiangsu Province
China

and

Department of New Communications
Purple Mountain Laboratories
Nanjing, Jiangsu Province
China

Zhenyu Wang
School of Electrical, Computer and Energy Engineering
Arizona State University
Tempe, AZ
USA

Douglas H. Werner
The Pennsylvania State University
University Park, PA
USA

Qi Wu
State Key Laboratory of Millimeter Waves
School of Information Science and Engineering
Southeast University
Nanjing, Jiangsu Province
China

and

Department of New Communications
Purple Mountain Laboratories
Nanjing, Jiangsu Province
China

Li-Ye Xiao
Department of Electronic Science
Xiamen University, Institute of
Electromagnetics and Acoustics
Xiamen
China

Amir I. Zaghloul
Bradley Department of Electrical and
Computer Engineering
Virginia Tech
Blacksburg, VA
USA

Hualiang Zhang
Department of Electrical & Computer
Engineering
University of Massachusetts Lowell
Lowell, MA
USA

Qi-Jun Zhang
Department of Electronics
Carleton University
Ottawa, ON
Canada

Bowen Zheng
Department of Electrical & Computer
Engineering
University of Massachusetts Lowell
Lowell, MA
USA

Yi Zheng
School of Electrical Engineering and Computer
Science
The Pennsylvania State University
University Park, PA
USA

Dayu Zhu
School of Electrical and Computer Engineering
Georgia Institute of Technology
Atlanta, GA
USA

Preface

The subject of this book is the application of the rapidly growing areas of artificial intelligence (AI) and deep learning (DL) in electromagnetics (EMs). AI and DL have the potential to disrupt the state-of-the-art in a number of research disciplines within the greater electromagnetics, optics, and photonics fields, particularly in the areas of inverse-modeling and inverse-design. While a number of high-profile papers have been published in these areas in the last few years, many researchers and engineers have yet to explore AI and DL solutions for their problems of interest. Nevertheless, the use of AI and DL within electromagnetics and other technical areas is only set to grow as more scientists and engineers learn about how to apply these techniques to their research. To this end, we organized this book to serve both as an introduction to the basics of AI and DL as well as to present cutting-edge research advances in applications of AI and DL in radio-frequency (RF) and optical modeling, simulation, and inverse-design. This book provides a comprehensive treatment of the field on subjects ranging from fundamental theoretical principles and new technological developments to state-of-the-art device design, as well as examples encompassing a wide range of related sub-areas. The content of the book covers all-dielectric and metallo-dielectric optical metasurface deep-learning-accelerated inverse-design, deep neural networks for inverse scattering and the inverse design of artificial electromagnetic materials, applications of deep learning for advanced antenna and array design, reduced-order model development, and other related topics.

This volume seeks to address questions such as "What is deep learning?," "How does one train a deep neural network,?" "How does one apply AI/DL to electromagnetics, optics, scattering, and propagation problems?," and "What is the current state-of-the-art in applied AI/DL in electromagnetics?" The first chapters of the book provide a comprehensive overview of the fundamental concepts and taxonomy of artificial intelligence, neural networks, and deep learning in order to provide the reader with a firm foundation on which to stand before exploring the more technical application areas presented in the remaining chapters. Throughout this volume, theoretical discussions are complemented by a broad range of design examples and numerical studies. We hope that this book will be an indispensable resource for graduate students, researchers, and professionals in the greater electromagnetics, antennas, photonics, and optical communities.

This book comprises a total of 17 invited chapters contributed from leading experts in the fields of AI, DL, computer science, optics, photonics, and electromagnetics. A brief summary of each chapter is provided as follows.

Chapter 1 introduces the fundamentals of neural networks and a taxonomy of terms, concepts, and language that is commonly used in AI and DL works. Moreover, the chapter contains a discussion of model development and how backpropagation is used to train complex network architectures. Chapter 2 provides a survey of recent advancements in AI and DL in the areas of supervised and unsupervised learning, physics-inspired machine learning models, among others as well as a discussion of the various types of hardware that is used to efficiently train neural networks. Chapter 3 focuses on the use of machine learning and surrogate models within the system-by-design paradigm for the efficient optimization-driven solution of complex electromagnetic design problems such as reflectarrays and metamaterial lenses. Chapter 4 introduces both the fundamentals and advanced formulations of artificial neural network (ANN) techniques for knowledge-based parametric electromagnetic (EM) modeling and optimization of microwave components. Chapter 5 presents two semi-supervised learning schemes to model microwave passive components for antenna and array modeling and optimization, and an autoencoder neural network used to reduce time-domain simulation data dimensionality. Chapter 6 introduces generative machine learning for photonic design which enables users to provide a desired transmittance profile to a trained deep neural network which then produces the structure which yields the desired spectra; a true inverse-design scheme. Chapter 7 discusses emergent concepts at the interface of the data sciences and conventional computational electromagnetics (CEM) algorithms (e.g. those based on finite differences, finite elements, and the method of moments). Chapter 8 combines DL with multiobjective optimization to examine the tradeoffs between performance and fabrication process uncertainties of nanofabricated optical metasurfaces with the goal of pushing optical metasurface fabrication toward wafer-scale. Chapter 9 explores machine learning (ML)/DL techniques to reduce the computational cost associated with the inverse-design of reconfigurable intelligent surfaces (RISs) which offer the potential for adaptable wireless channels and smart radio environments. Chapter 10 presents a selection of neural network architectures for Huygens' metasurface design (e.g. fully connected neural networks, convolutional neural networks, recurrent neural networks, and generative adversarial networks) while discussing neuromorphic photonics wherein meta-atoms can be used to physically construct neural networks for optical computing. Chapter 11 examines the use of deep neural networks in the design synthesis of artificial electromagnetic materials. For both forward and inverse design paradigms, the major fundamental challenges of design within that paradigm, and how deep neural networks have recently been used to overcome these challenges are presented. Chapter 12 introduces the framework of machine learning-assisted optimization (MLAO) and discusses its application to antenna and antenna array design as a way to overcome the limitations of traditional design methodologies. Chapter 13 summarizes the basics of uniform and non-uniform array processing using kernel learning methods which are naturally well adapted to the signal processing nature of antenna arrays. Chapter 14 describes a procedure for improved-efficacy electromagnetic-driven global optimization of high-frequency structures by exploiting response feature technology along with inverse surrogates to permit rapid determination of the parameter space components while rendering a high-quality starting point, which requires only further local refinement. Chapter 15 introduces four DL techniques to reduce the computational burden of high contrast inverse scattering of electrically large structures. These techniques can accelerate the process of reconstructing model parameters such as permittivity, conductivity, and permeability of unknown objects located inside an inaccessible region by analyzing the scattered fields from a domain of interest. Chapter 16 describes various applications of DL in the classification of radar images such as micro-Doppler spectrograms, range-Doppler diagrams, and synthetic aperture radar images for applications including human motion classification, hand gesture recognition, drone detection, vehicle detection, ship detection, and more. Finally, Chapter 17 explores the use of Koopman autoencoders for producing reduced-order models that mitigate the computational burden of traditional electromagnetic particle-in-cell algorithms, which are used to simulate kinetic plasmas due to their ability to accurately capture complicated transient nonlinear phenomena.

We owe a great debt to all of the authors of each of the 17 chapters for their wonderful contributions to this book, which we believe will provide readers with a timely and invaluable reference to the current state-of-the-art in applied AI and DL in electromagnetics. We would also like to express our gratitude to the Wiley/IEEE Press staff for their assistance and patience throughout the entire process of realizing this book – without their help, none of this would be possible.

June 2023

Sawyer D. Campbell and Douglas H. Werner
Department of Electrical Engineering
The Pennsylvania State University
University Park, Pennsylvania, USA

Section I

Introduction to AI-Based Regression and Classification

1

Introduction to Neural Networks

Isha Garg and Kaushik Roy

Elmore School of Electrical and Computer Engineering, Purdue University, West Lafayette, IN, USA

The availability of compute power and abundance of data has resulted in the tremendous success of deep learning algorithms. Neural Networks often outperform their human counterparts in a variety of tasks, ranging from image classification to sentiment analysis of text. Neural networks can even play video games [1] and generate artwork [2]. In this chapter we introduce the basic concepts needed to understand neural networks. The simplest way to think of how these networks learn is to think of how a human learns to play a sport, let's say tennis. When someone who has never played tennis is put on a court, it takes them just a few volleys to figure out how to respond to an incoming shot. It might take a long time to get good at a sport, but it is quite magical that only through some trial and error, we can learn how to swing a racquet to a tennis ball heading our way. If we were to write a mathematical model to calculate the angle of swing, it would require many complicated variables such as the wind velocity, the incoming angle, the height from the ground, etc. Yet, we just learn from real-life examples that swinging a particular way has a particular impact, without knowing these variables. This implicit learning from examples is what sets machine learning models apart from their rule-based computational counterparts, such as a calculator. These models learn an implicit structure of the data they see without any explicit definitions on what to look for. The learning is guided by a lot of examples available with ground truth, and the models learn what is needed from these datapoints. Not only do they learn the datapoints they have seen, they are also able to generalize to unseen examples. For instance, we can train a model to differentiate between cats and dogs with, say, 100 examples. Now when we show them new examples of cats that were not present in the training set, they are still able to classify them correctly. There are enormous applications of the field, and a lot of ever-evolving subfields. In Section 1.1, we introduce some basic taxonomy of concepts that will help us understand the basics of neural networks.

1.1 Taxonomy

1.1.1 Supervised Versus Unsupervised Learning

In the unsupervised learning scenario, datapoints are present without labels. The aim is to learn an internal latent representation of data that catches repeated patterns and can make some decisions based on it. By latent representations, we mean an unexposed representation of data that is no longer in the original format, such as pixels of images. The hope is that repetition magnifies

Advances in Electromagnetics Empowered by Artificial Intelligence and Deep Learning, First Edition.
Edited by Sawyer D. Campbell and Douglas H. Werner.

the significant aspects of images, or aids in learning compressed internal representations that can remove noisy artifacts or just learn lower dimensional representations for compressed storage, such as in AutoEncoders [3]. An advantage of having a meaningful latent space is that it can be used to generate new data. Most concepts covered in this chapter pertain to discriminatory models for regression or classification. However, in models such as Variational AutoEncoders [4], the latent space can be perturbed in order to create new data that is not present in the dataset. These models are called generative models. Unsupervised learning problems are harder and an active area of research, since it removes the need to label data. In this chapter, we will stick to supervised learning problems of the discriminatory kind.

1.1.2 Regression Versus Classification

The kind of output expected from a supervised learning discriminatory task dictates whether the problem is one of regression or classification. In regression, the output is a continuous value, such as predicting the price of a house. In classification, the task corresponds to figuring out which of the predefined classes an input belongs to. For example, determining whether an image is of a cat or a dog is a two-class classification problem. In this chapter, we will give examples of both regression and classification tasks.

1.1.3 Training, Validation, and Test Sets

An important concept in deep learning is that of inference versus training. Training is the method of determining the parameters of a model. Inference is making a prediction on an input once the training is complete. When we have a dataset, we can have many different resulting models and predictions for the same query, depending on initialization and the choice of some tunable parameters, which are called hyperparameters. This means we need a way of testing different models. We cannot test on the same data as that used for training, since models with enough complexity tend to memorize training data, leading to the problem of overfitting. Overfitting can be mitigated by using regularization techniques, discussed later. To avoid memorizing data, there is a need to partition the entire set of data into a training and a testing set. We choose a certain value of the hyperparameters and train on the training dataset, get a trained model, and then test it on the testing dataset to get the final reported accuracy. However, this is not the best practice. Let's say a particular choice of hyperparameters did not yield good testing accuracy, and hence we alter these hyperparameters and retrain the model until we get the best testing accuracy. In this process, we are overfitting to the testing set, because we, as the hyperparameter tuners, are exposed to the testing accuracy. This is not a fair generalized testing scenario. Thus, we need a third partition of the data, called the validation set, upon which the hyperparameters should be tuned. The testing set is shown only once to the final chosen model, and the accuracy obtained on that is reported as the models final accuracy. A commonly used split percentage for the dataset is 70%-15%-15% for training, validation, and testing, respectively.

The rest of this chapter is organized as follows. We first introduce linear regression models in Section 1.2 and then extend them to logistic classification in Section 1.3. These sections make up the base for the neurons that serve as building blocks for neural networks. In each section, we introduce the corresponding objective functions and training methodologies. Changes to the objective function to tackle overfitting are discussed in Section 1.4. We then discuss stacking neurons into layers and layers into fully connected neural networks in Section 1.5. We introduce more complex layers that make up convolutional neural networks in Section 1.5 and conclude the chapter in Section 1.6.

1.2 Linear Regression

In this section, we consider a supervised regression problem and explore a simple technique that makes the assumption of linearity in modeling. The canonical example often used to explain this is that of predicting the price of a house. Let's say that you are a realtor with a lot of experience. You have sold a lot of houses and maintain a neat log of all their details. Now you get a new house to sell and need to price it based on your experience with all the previous houses. If you were going to do this manually, you might consider all your records and look for a house that is "similar" to the one that you have sold and give an estimate close to that. This is similar to what a nearest neighbor algorithm would do. But we are going to go a step further and use a linear model to fit a high dimensional curve as best as we can to the data of the old houses, and then determine where a new house would perform according to this model.

Being a diligent realtor, you noted all the features you thought were relevant to the price of a house, such as the length of the house, the width of the house, the number of rooms, and the zip code. Let's say, you have D such features, and N such houses in your log. One can imagine each house as a point in D dimensional space. We have N such points, and the problem of regression essentially reduces to the best curve we can fit to this data. Since we are assuming a linear model, we will try and fit a line to this data. The hypothesis underlying the model can be denoted as $h(\theta)$, with θ being the parameters to be learned. The equation for this model for a single datapoint, x, thus becomes:

$$\hat{y} = h(\theta, \mathbf{x}) = \theta_0 + \theta_1 x_1 + \theta_2 x_2 + \cdots + \theta_D x_D \tag{1.1}$$

We can write this in abridged form using matrix or vector multiplication. Each house is represented as a D-dimensional vector, and all the N houses together can be concatenated into the columns of an $D \times N$ matrix, $\mathbf{X}$. Since θ_0 does not multiply with any input, it is known as the bias, denoted by b. The remaining parameters are denoted by the matrix $\boldsymbol{\theta}$.

$$\hat{y} = h(\theta, \mathbf{X}) = \theta^{\mathrm{T}}\mathbf{X} + \mathbf{b} \tag{1.2}$$

We use parameters and weights interchangeably to refer to $\boldsymbol{\theta}$. For the single point case, $\boldsymbol{\theta}$ and $\mathbf{X}$ are D-dimensional vectors and b and $\hat{y}$ are scalar values. For multiple points, $\mathbf{X} \in \mathbf{R}^{\mathbf{D}\times\mathbf{N}}$, $\theta \in \mathbf{R}^{\mathbf{D}\times\mathbf{1}}$, b is a scalar repeated for each sample, and the output is a value for each sample, i.e. $\hat{\mathbf{y}} \in \mathbf{R}^N$. Often, for ease of notation, b is absorbed into $\boldsymbol{\theta}$, with a corresponding 1 appended to each datapoint in $\mathbf{X}$ so that the equation is simplified to $\hat{\mathbf{y}} = \theta^T\mathbf{X}$. $\boldsymbol{\theta}$ is the matrix that takes inputs from D-dimensional input space to the output space, which in this case is one-dimensional, the price of the house. This equation is shown pictorially as a single node, also called a neuron, in Figure 1.1.

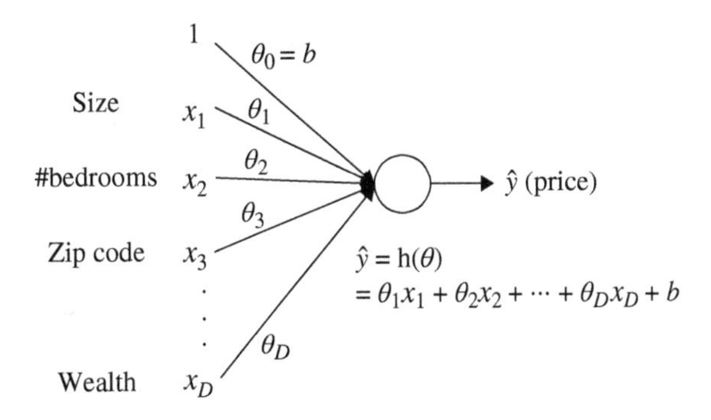

Figure 1.1 A simple linear model that takes as input a D dimensional feature vector and predicts a single dimensional output, in this case, price. The linear hypothesis is shown.

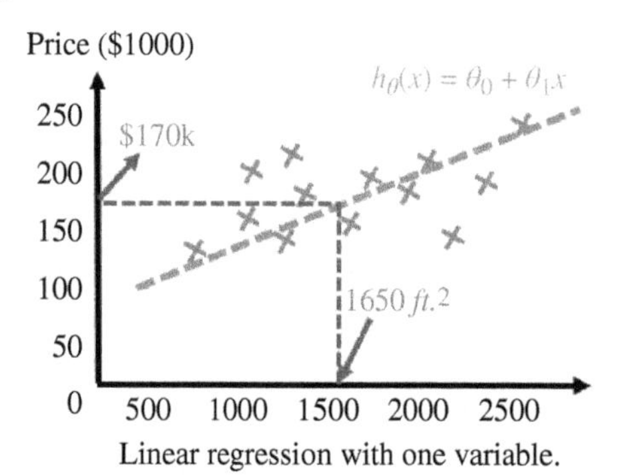

Figure 1.2 Linear Regression with one variable for visualization purposes. The task of regression can be viewed as fitting the best curve to the data. Since we assume linear models, we show the best fit line. Inference of a single point is shown; when a new data point comes in, we can predict its price by seeing where it lies on the line.

Since D dimensional space is really hard to imagine and pictorially represent, for the sake of visualization, we assume $D = 1$ in Figure 1.2. This means that we have only one feature: that of, say, square feet area. We can imagine all the points laid out on the 2D space with the X-axis being the area feature value, and the Y-axis representing the price that the house sold for. The true data distribution might not be linear, and hence the datapoints might not match the best fit line as shown in the figure. The mismatch between these points from the line is called the training error, also referred to as cost or loss of the model. The error between the testing points and their true value is correspondingly referred to as testing error. Note that during inference, the parameters are held constant and Eq. (1.2) can be written as a function of $\mathbf{X}$ alone. However, we need to train this model, i.e. we need to learn the parameters $\boldsymbol{\theta}$ such that the predicted output matches the ground truth. In order to find such θ, we employ objective functions which minimize the expected empirical training error.

1.2.1 Objective Functions

An objective function, as apparent from the name, is a mathematical formulation of what we want to achieve with our model. The objective function is also called the loss function or the cost function, since it encapsulates the costs or losses incurred by the model. In the linear regression model shown above, we want the prediction to align with the ground truth price of the house. In the case of classification, the objective would be to minimize the number of misclassifications.

In the example of predicting the house price, a possible objective function is the distance between the best fit curve and the ground truth. The distance is often measured in terms of norm. The Lp-norm of an n-dimensional vector is defined as:

$$||x||_p = (|x_1|^p + |x_2|^p + \cdots |x_n|^p)^{1/p} \tag{1.3}$$

Commonly used norms are $p = 1$ and $p = 2$, which translate to Manhattan (L1) and Euclidean (L2) distance, respectively. Let's assume our objective is to minimize the L2 distance between the predicted house price $\hat{y}_i$ and the ground truth value y_i for all the N samples of houses in our log. The datapoint specific cost is averaged into the overall cost, depicted by $J(\theta)$.

$$\hat{y}_i = \theta^T x_i + b \tag{1.4}$$

$$J(\theta) = \frac{1}{n}\sum_{i=1}^{N} ||\hat{y}_i - y_i||_2^2 \tag{1.5}$$

$$\text{where for any vector a, } ||\mathbf{a}||_2 = \sqrt{\sum_j a_j^2} \tag{1.6}$$

where datapoints are indexed by the subscript i. Note that the cost is a scalar value. If the cost is high, the model does not fit the data well. In Section 1.2.2, we discuss how to find the parameters to obtain the best fit to the training data, or minimize the cost function, $J(\boldsymbol{\theta})$. This process is called optimization and the commonly used method for performing this optimization is Stochastic Gradient Descent (SGD).

1.2.2 Stochastic Gradient Descent

In Section 1.2.1, we introduced the cost function that captures the task we want our model to perform. We now discuss how this cost function is used in training, to find the right parameters for the task. Most cost functions used in neural networks are much more complicated and non-convex, and we used Stochastic Gradient Descent to minimize them. Hence, even though the cost function we discussed for linear regression is convex and can be minimized analytically, we explore how to utilize gradient descent to minimize it. The analogy used to understand gradient descent is usually one of a hiker finding themselves blindfolded on a hill, and trying to find their way to the bottom of the hill. In this analogy, the hill refers to the landscape of the loss function such that the height corresponds to the cost. At the bottom of the hill lies the minima, corresponding to the optimized set of weights that minimize the cost function and achieve the objective. The hiker in question, which represents the set of current weights, desires to move down the hill iteratively, until they reach the minima or close enough to it. A two-dimensional loss landscape (with two weights) is shown in Figure 1.3.

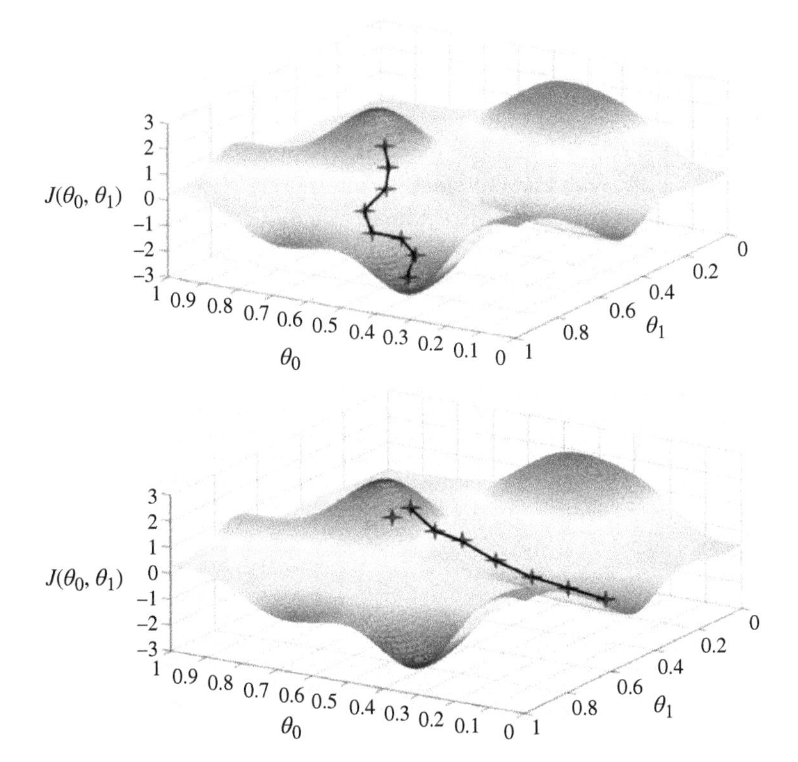

Figure 1.3 A non-convex loss landscape with multiple different local minimas. Different initializations and hyperparameters can result in convergence to different minimas.

The hiker has two immediate questions to answer: in which direction should they step, and how long should the step be (or equivalently, how many steps) in that direction. The latter is called the learning rate, denoted by α. Right away, it is easy to see that too small a step will take too long to get to the minima, and if too large, it is possible to completely miss the minima and instead diverge away as shown in Figure 1.4. In addition, the landscape may not be convex and too small a learning rate can get stuck in local minimas, which can be hard to get out of. The learning rate is a hyperparameter, and is often decayed over the course of learning to smaller values to trade off the number of iterations and the closeness to minima. The final minima that the algorithm converges to is also dependent on the initialization, and hence the entire process is stochastic in nature, and we can get many different sets of weights (corresponding to different local minimas) upon training the same model multiple times, as shown in Figure 1.3.

Now, let's consider the direction of descent. The quickest way to the bottom is via the direction of steepest descent, which is the negative of the gradient at that point. Let's take a deeper look at the gradient, also known as the derivative. Assume x is single dimensional, and the loss is always a scalar value, and hence $f(x)$ is a function from $\mathbf{R} \rightarrow \mathbf{R}$. The derivative of $f(x)$ with respect to x:

$$\frac{df(x)}{dx} = \lim_{h \to 0} \frac{f(x+h) - f(x)}{h} \tag{1.7}$$

It essentially captures the sensitivity of the function to a small change in the value of x. When x is multi-dimensional, f maps from $\mathbf{R}^d \rightarrow \mathbf{R}$. In this case, we use partial derivatives, that show the sensitivity of f($\mathbf{x}$) with respect to each dimension of $\mathbf{x}$ (x_i), denoted by $\frac{\partial f(\mathbf{x})}{\partial x_i}$. This computation, however, is difficult to approximate as the change needs to be infinitesimal to calculate correctly. The final expression for the weight update in each iteration is given by:

$$\theta^{(i+1)} = \theta^{(i)} - \alpha \frac{\partial L}{\partial \theta^{(i)}} \tag{1.8}$$

Henceforth, for clarification, the iterations will be shown as a superscript, and each sample is shown as subscript. Let's make this clearer with an example. Let's return to the case of the linear model shown in Equation (1.2), used for regression to predict house prices. Let's assume the loss function is a simple L2 loss, shown below.

$$\hat{y}_i = \theta^{(i)^T} \mathbf{x_i} + b \tag{1.9}$$

$$J = \frac{1}{2N} \sum_{i=1}^{N} \|\hat{y}_i - y_i\|_2^2 \tag{1.10}$$

In each iteration, we take the derivative of the total loss with respect to the weights, and take a direction in the negative of the derivative, scaled by the learning rate α. The iterative weight update can be calculated using the chain rule of derivatives, shown below:

$$\frac{\partial f}{\partial x} = \frac{\partial f}{\partial g} \frac{\partial g}{\partial x} \tag{1.11}$$

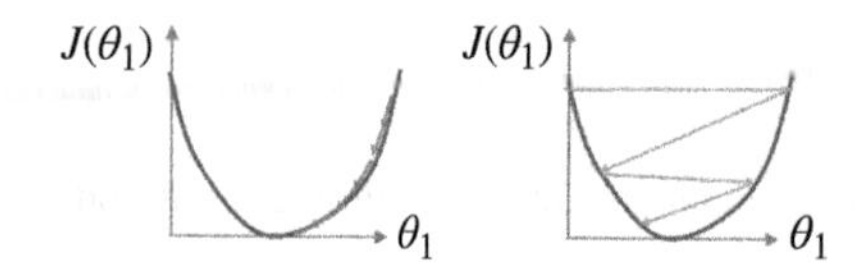

Figure 1.4 A pictorial representation of the role of learning rate in convergence to a minima. Too small a step takes a long time to reach the minima, and too big a step can diverge and miss the minima altogether.

We put all this together to get one iteration of weight update for the linear regression case with L2 loss:

$$\frac{\partial J}{\partial \theta^{(i)}} = \frac{\partial J}{\partial \hat{y}_i}\frac{\partial \hat{y}_i}{\partial \theta^{(i)}} = \frac{1}{N}\sum_{i=1}^{N}(\hat{y}_i - y_i)\mathbf{x_i} \tag{1.12}$$

$$\theta^{(i+1)} = \theta^{(i)} - \alpha\frac{1}{N}\sum_{i=1}^{N}(\hat{y}_i - y_i)\mathbf{x_i} \tag{1.13}$$

Here, N is the size of the entire training dataset. In practical scenarios, N is usually large, which means that we cannot take a step until we have parsed all the data. In practical settings, we partition the training dataset into minibatches of size 32, 64, 128, or 256 and take one step per minibatch. This is called Batch Gradient Descent and helps us converge to the minima faster. The extreme case of updating after every single data point, i.e. minibatch size = 1, is called Stochastic Gradient Descent. This is because the underlying operations of batched gradient descent are matrix multiplications, which can be implemented by general purpose hardwares, such as GPUs, very efficiently. The minibatch size is upper-bounded by the GPU memory, since the GPU has to hold the entire matrix in its memory to perform the matrix multiplication. We generally refer to updates with minibatches as SGD in the literature, and optimize the batchsize as a hyperparameter. Ruder [5] provides a more detailed discussion. Advanced versions of SGD introduce concepts such as momentum to recall past gradients, and automatically tuned parameter-specific learning rates, such as in Adagrad [6].

The standard practice also involves normalizing the input data, so that the values of all features have similar impact. For the housing price example, the number of rooms will often be values between 1 and 9 and area might be values in the hundreds to thousands of square feet. If the raw values are used in the hypothesis, the area will dominate the house price, or the weight assigned to the number of rooms would have to be really large to have a similar impact as area. To avoid this, the features are first normalized to values between 0 and 1, and then centered by subtracting the mean. They are also then scaled by dividing by the variance, q, a process referred to as mean-std normalization. In practical scenarios, the input data matrix can be very high dimensional in terms of features. This adds significant computational expense to the learning procedure. In addition, a lot of features are correlated to each other and do not offer additional information for learning. For example, if the features are in terms of length, width, and area, just the area is enough to capture the concept of size. To counter this, a standard practice is feature selection, by say, Principal Component Analysis [7], which will remove redundant features.

1.3 Logistic Classification

In Section 1.2, we considered a regression problem, that of predicting the price of a house, which is a real valued output. Now, we look at a classification problem, with $C = 2$ classes. Let's consider a tumor classification example. Given some input feature of a tumor, one classification task could be to predict whether the tumor is malignant ($y = 1$) or benign ($y = 0$). Let's assume the input data, as before, is D-dimensional. Our model is thus a mapping from $\mathbf{R}^D \rightarrow \mathbf{R}^C$, where each of the C parameters of the output corresponds to a score of the input belonging to that class. Let's assume that the same linear model still applies,

$$h(\theta, \mathbf{x_i}) = \theta^T\mathbf{x_i} + \mathbf{b} \qquad i = 1, 2 \ldots\ldots, N \tag{1.14}$$

where, as before, N is the number of datapoints, $\mathbf{x} \in \mathbf{R}^D$ is the input, and the output $\hat{\mathbf{y}} \in \mathbf{R}^C$. The predicted output, $\hat{\mathbf{y}}_\mathbf{i}$, is distinct from the ground truth label, $\mathbf{y_i}$. The weight matrix, θ, is a $\mathbf{D} \times \mathbf{C}$

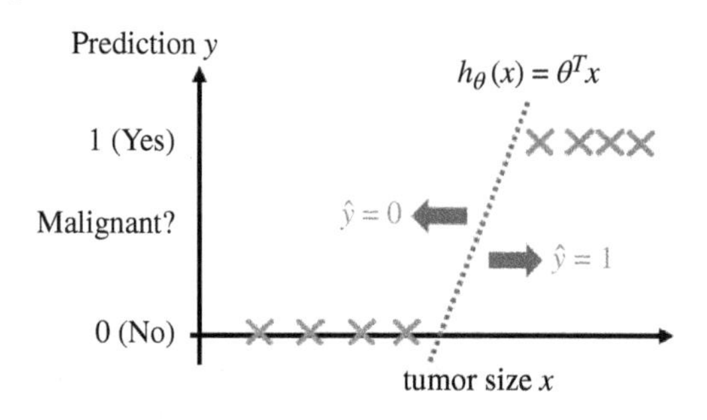

Figure 1.5 The linear classification model: data with ground truth 0 shown in light gray, and for ground truth 1 shown in dark gray. In the classification case, the learned line, shown as the dotted gray line, serves as a boundary, delineating regions for each class.

matrix that takes us from the input space of data to the output space of class scores. Now, instead of drawing the best-fit line, the task is to find the best linear boundary delineating space for each class as shown in Figure 1.5. Additionally, we have a vector of biases, $\mathbf{b} \in \mathbf{R^c}$, one for each class. The predicted output $\mathbf{y_i}$ consists of C values and can be interpreted as the score of each class. Hence a simple classification decision can be based on thresholding. If the output is greater than 0.5, predict class 1; otherwise, predict class 0, as shown below:

$$\hat{y}_i = \begin{cases} 0 & \text{if } h(\theta, \mathbf{x_i}) < 0.5 \\ 1 & \text{if } h(\theta, \mathbf{x_i}) \geq 0.5 \end{cases}$$

The single node discussed until now, also known as a neuron, is the form of the earliest perceptron, introduced in 1958 by Frank Rosenblatt in [8]. We mentioned that the output is interpreted as the scores allotted to each class. A better way to understand the output is if we interpreted them as probabilities. To do this, we have to normalize the scores to lie between 0 and 1. There are two popular ways of doing this, one via the sigmoid function, which turns linear regression/classification into logistic regression/classification, and another via the softmax function. The sigmoid function for a scalar value x is shown below, with the corresponding graph plotted in Figure 1.6.

$$\sigma(x) = \frac{e^x}{1 + e^x} = \frac{1}{1 + e^{-x}} \tag{1.15}$$

The sigmoid function acts independently on each output score, and squashes it to a value between 0 and 1. While it ensures that each score lies between 0 and 1, it does not ensure that they sum to 1. Hence, it is not strictly a probability metric. However, it comes in handy in case of classification problems where the labels are not mutually exclusive, i.e. multiple classes can be correct for the same sample. The pre-normalized outputs are referred to as logits, often denoted by $\mathbf{z}$. The corresponding changed hypothesis function now becomes

$$\mathbf{z_i} = \theta^T \mathbf{x_i} \tag{1.16}$$

$$h(\theta, \mathbf{x_i}) = \sigma(\mathbf{z_i}) \tag{1.17}$$

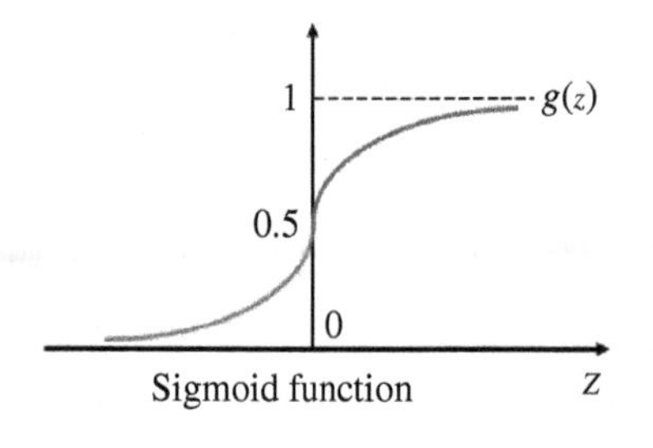

Figure 1.6 The graph of the sigmoid function that squashes outputs into a range of 0 to 1.

The softmax function is an extension of the sigmoid. It normalizes the score of each class after taking the scores of other classes into account. It ensures that the resulting scores sum to 1, so each value can be interpreted as the confidence of belonging to that class. It is useful when the labels are exclusive, i.e. only one class can be present at a time. The equation for softmax is shown below for a vector $\mathbf{x}$ of dimension D:

$$softmax(\mathbf{x}) = \frac{e^{\mathbf{x}}}{\sum_{j=1}^{D} e^{x_j}} \tag{1.18}$$

where x_j corresponds to the jth element in the vector $\mathbf{x}$. Similar to the case with sigmoid, the new hypothesis function now becomes:

$$\mathbf{z_i} = \theta^T \mathbf{x_i} \tag{1.19}$$

$$h(\theta, \mathbf{x_i}) = \mathbf{\hat{y}_i} = softmax(\mathbf{z_i}) \tag{1.20}$$

We no longer need to threshold, since the output is directly the score of the class, and the class with the maximum probability can be predicted as the classification output. Softmax is a commonly used last layer for typical classification problems with more complex models as well. The probability outputs available from softmax are often used in an information-theoretic objective function, called the cross entropy loss function. Since the outputs function as probabilities, a way of measuring the distance between them is the Kullback–Liebler (KL) divergence [9], and is closely related to cross entropy. For two distributions $p(x)$ and $q(x)$, where p is considered the true distribution, and q is the distribution that approximates p, the cross entropy is defined by:

$$H(p, q) = -\sum_{x} p(x) \log q(x) \tag{1.21}$$

In the case of a C class classification, we want to measure the error between the true distribution y_i, which is a just point mass on the correct class and zero elsewhere, and the predicted distribution $\mathbf{\hat{y}_i}$. Hence the loss for the datapoint becomes:

$$l_i = -\log \hat{y}_{i,y_i} \tag{1.22}$$

where $\hat{y}_{i,y_i}$ is the predicted score of the ground truth class for sample i. To optimize this objective function, we employ gradient descent in the direction of the derivative of this cost with respect to the parameters, with a tunable learning rate, similar to logistic regression example.

1.4 Regularization

Let's return to the example of predicting house prices. In Section 1.2, we tried to fit a linear line to the data. It is possible that the best fit of the line may not fit the data well. It could be because the underlying distribution was not linear or the linear model did not have the sufficient complexity to fit the data. This problem is referred to as underfitting, also known as a high bias problem. One way to fix this is to use a more complex model, such as a polynomial of degree 2. Let's say in the example of house price prediction, we only had length and breadth of the house of features. Having second-degree polynomials would allow us to multiply them and have area (*length* × *breadth*) as well as one of the features, which might give us a better fit. Taking this further, we can fit an increasingly higher degree polynomial to our data, and it will in most cases end up fitting our training data near perfectly. However, this near-perfect fit of training data to the complex hypothesis means that

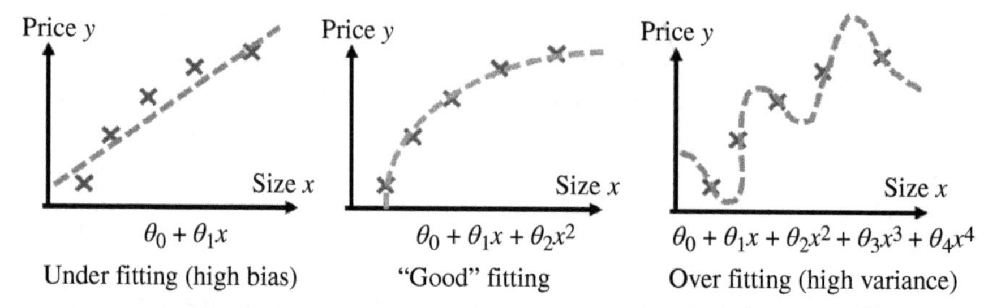

Figure 1.7 Different complexity models can fit the data to different degrees. Too simple a model cannot explain the data well and underfits. Too complex a model memorizes the data and overfits the training set.

the model is memorizing the data, in which case it would not generalize well to the testing or the validation set. This is shown as overfitting in Figure 1.7, also referred to as a high variance problem.

The cost functions used in practical scenarios are highly non-convex, which means there are many local minimas and non-unique sets of weights that the optimization process can converge to, as shown in Figure 1.3. We can encode preferences for certain kinds of weights by expanding the cost function. Since complex models often overfit the data, hurting their generalization performance, we wish to encode a preference for simpler models. By simpler models, we mean those in which no one weight or parameter has the capability to largely effect the cost by itself. A simple way to achieve this is to minimize the Lp norm of all the weights [10]. This means that as any weight grows large, a large value is added to the cost function, which is not preferred since the objective is to minimize this cost function. Thus, the minimization procedure would naturally prefer smaller weights. A common form is to add the L2 norm of weights as a regularization objective to the original cost function, as shown below.

$$L = \frac{1}{2N}\left[\sum_i l_i + \lambda \|\theta\|_2^2\right] \tag{1.23}$$

where λ is the tradeoff parameter that decides the strength of regularization. Note that the regularization function is independent of the data samples, and is just a function of the weights. Let's derive the weight update rule in case of the linear regression problem discussed earlier.

$$\hat{y}_i = \theta^{(i)^T}\mathbf{x_i} \tag{1.24}$$

$$J = \frac{1}{2N}\left[\sum_{i=1}^{N} \|\hat{y}_i - y_i\|_2^2 + \lambda \|\theta^{(i)}\|_2^2\right] \tag{1.25}$$

$$\frac{\partial J}{\partial \theta^{(i)}} = \frac{1}{N}\left[\sum_{i=1}^{N} (\hat{y}_i - y_i)\mathbf{x_i} + \lambda \theta^{(i)}\right] \tag{1.26}$$

The corresponding weight update rule is given by:

$$\theta^{(i+1)} = \theta^{(i)} - \alpha \frac{\partial J}{\partial \theta^{(i)}} \tag{1.27}$$

$$\theta^{(i+1)} = \theta^{(i)}\left[1 - \alpha \frac{\lambda}{N}\right] - \alpha \left[\frac{1}{N}\sum_{i=1}^{N} (\hat{y}_i - y_i)\mathbf{x_i}\right] \tag{1.28}$$

1.5 Neural Networks

In Section 1.4, we introduced polynomial functions as an alternative to linear functions in order to better fit more complex data. However, data can be high dimensional and polynomial models suffer from the curse of dimensionality. Let's return to the house prediction example one last time, and assume we have $D = 100$ features in our input data. Let's take a very reasonable hypothesis, that of polynomial functions with a degree of 2. We now have $O(D^2)$ combinations in the input, in this case, 5000. Hence, the model will also have that many extra parameters. It is common to have millions of parameters in the input, such as images that are made of $D = 224 \times 224 \times 3$ pixels, and this is significantly computationally prohibitive. If we expand the hypothesis to a kth order polynomial, the features would grow by $O(D^k)$. We would also need correspondingly larger number of data samples that can be quite expensive to obtain. Clearly, we need different model structures to process such complex data.

Neural networks (NNs) were introduced to counter this explosion in input dimensionality, yet enable rich and complex learning. They stack together multiple perceptrons (or neurons) in a layer together to aid in complex data mapping. In order to aid non-linear mapping, multiple layers are stacked together, with non-linearities in the middle. Training NNs is a simple extension of the chain rule of gradient descent, a process known as backpropagation. NNs have shown astounding amounts of success in all forms of data ranging from images, texts, audio to medical data. We will now discuss their structure and training in detail via the example of training on images. We borrow a lot of the concepts from linear models to build our way up to the final neural network structure.

The basic neuron of the NNs remains the same as in the linear regression model from the house prediction model, shown in Figure 1.8. The input, $\mathbf{x}$, multiplies with a weight represented by an edge in the figure and denoted as before by $\boldsymbol{\theta}$. The output of the neuron is $\boldsymbol{\theta}^T\mathbf{x}$.

Many of these neurons are stacked together in one layer, with each input connecting to each neuron for now. This connectivity pattern results in what is referred to as a fully connected layer, as all neurons are connected to all inputs. We will explore more connectivity patterns when we discuss convolutional neural networks. Let's assume we have s_1 such inputs and the data, as before, is D dimensional. Therefore, the weight matrix dimensions change, $\boldsymbol{\theta} \in \mathbf{R}^{D \times s_1}$. Let's call the output of each neuron as its activation, since it represents how active that neuron is, represented by a_i, $i = 1, 2 \dots s_1$. We can represent all activations as a vector $T^{(j)} \in \mathbf{R}_1^N$, corresponding to a the jth layer.

Till now, we described the first layer, i.e. $j = 1$. We can stack together multiple layers, as shown in Figure 1.9. The layers $T_j, j = 1, 2$ are called the hidden layer representations of the neural networks. In each layer j, neuron i produces activation $a_i^{(j)}$. Each layer gets a bias, with the corresponding input set to 1, represented by the subscript 0. Each layer's weight matrix, connecting layer j to $j + 1$

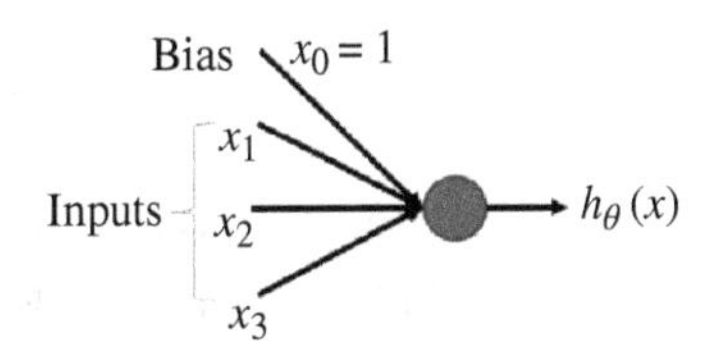

Figure 1.8 Each neuron in a Neural Network essentially performs regression.

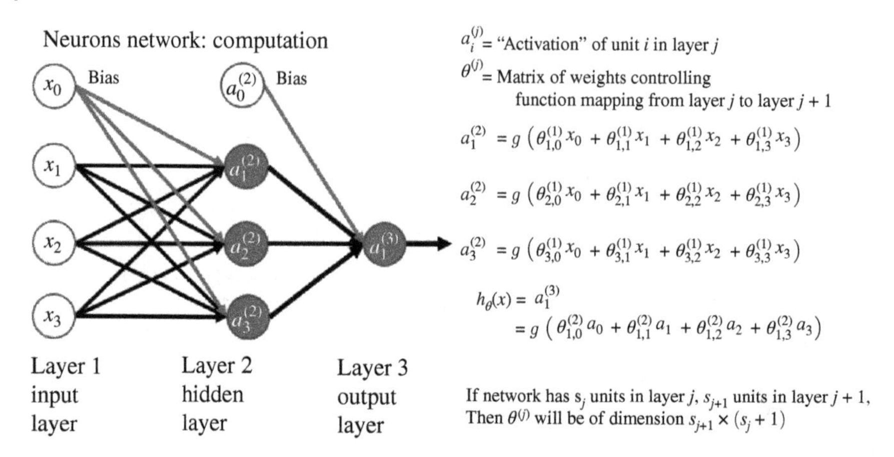

Figure 1.9 A three-layer neural network with the corresponding activation shown.

is denoted by $\theta^{(j)} \in \mathbf{R}^{s_j \times s_{j+1}}$. The equation for the two hidden-layered NN shown in Figure 1.9, is expressed by:

$$\mathbf{T}^{(1)} = \theta^{(1)T}\mathbf{x} \tag{1.29}$$

$$\mathbf{T}^{(2)} = \theta^{(2)T}\mathbf{T}^{(1)} \tag{1.30}$$

$$= (\theta^{(2)T} \cdot \theta^{(1)T})\mathbf{x} \tag{1.31}$$

$$= \theta^{(3)T}x \tag{1.32}$$

$$\text{where } \theta^{(3)} = \theta^{(1)}\theta^{(2)} \tag{1.33}$$

Equation (1.32) implies that stacking two layers with s_1 and s_2 adds no extra complexity than a single layer with just s_2 neurons. This is due to linearity of matrix multiplications: sequential multiplication with two matrices can be denoted as multiplication with a different matrix. We avoid this by adding a non-linearity after each neuron. As discussed earlier, this linearity could be sigmoid, or the more commonly used ReLU, a Rectified Linear Unit. ReLU returns 0 for all inputs less than 0, and passes the input unaltered after 0, and became the default choice for non-linearity after its success in [11]. Let's represent the choice of non-linearity by g(.), and Equations (1.29)–(1.33) are updated as follows:

$$\mathbf{T}^{(1)} = g(\theta^{(1)T}\mathbf{x}) \tag{1.34}$$

$$\mathbf{T}^{(2)} = g(\theta^{(2)T}\mathbf{T}^{(1)}) \tag{1.35}$$

$$= g(\theta^{(2)T}g(\theta^{(1)T}\mathbf{x})) \tag{1.36}$$

$$\neq g(\theta^{(3)T})\mathbf{x}, \text{ for some } \theta^{(3)} \tag{1.37}$$

The forward pass of the input through all layers, generating activations at each neuron and representations at each layer, right up to the cost is called a forward pass or forward propagation. Let's explore this with the multi-class classification example shown in Figure 1.10. In this case, we get an input image and have to predict which class the image belongs to: pedestrian, car, motorcycle, or dog. Earlier we had binary classification, and we only need a single neuron to perform that, since we can threshold on its output. But if we have C classes, we need C neurons in the last layer, corresponding to the C scores for each class. The ground truth labels are now interpreted as one-hot vectors, i.e. $\mathbf{y_i} \in \mathbf{R}^\mathbf{C}$, where all elements of $\mathbf{y_i}$ are 0 except the true label, which is a 1 as shown in the figure. Each neuron still performs a one versus all binary classification. For example, the last

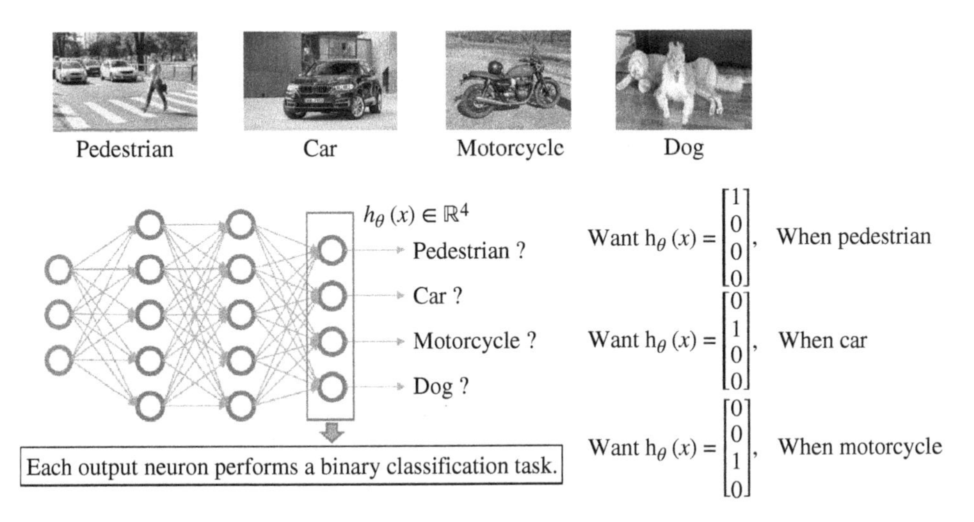

Figure 1.10 Using a NN to perform a 4 class classification task. Source: Sergey Ryzhov/Adobe Stock; Moose/Adobe Stock; brudertack69/Adobe Stock.

neuron in the last layer corresponding to truck class essentially predicts the confidence of the image containing a truck versus not containing a truck. In this form, all the layers of the NN until the last layer perform some sort of feature extraction. This feature vector corresponding to each input is fed into the last layer, which serves as the classifier.

Let's look at the cost function for this example. Let's assume there are L layers in the NN. Each layer has $s_l \;\; l = 1, 2, \ldots L$ neurons. There are N training samples shown as pairs of input and ground truth labels: $(\mathbf{x_i}, \mathbf{y_i}), \;\; i = 1, 2, \ldots. N$. The previous equations showed how to forward propagate the input to all layers. The activation of layer l is represented by $\mathbf{T^{(l)}}$. The last layer logits are therefore $\mathbf{T^{(L)}}$. Similar to logistic classification, the logits represent unnormalized scores for classes and will be passed through a softmax function for a cross entropy objective function.

$$\mathbf{\hat{y}_i} = softmax(\mathbf{T_i^{(L)}}) \tag{1.38}$$

$$J(\theta) = \sum_{c=1}^{C} \mathbf{y_i} \log \boldsymbol{\hat{y}}_{\mathbf{i,c}} \tag{1.39}$$

$$= -\log \hat{y}_{i,y_i} \tag{1.40}$$

The ground truth label $\mathbf{y_i}$ is one-hot encoded; hence, it only has one non-zero element corresponding to the ground truth class, getting rid of the sum in equation 1.39. The regularization term, if included, would be the L2 norm of all the weights for all layers in the neural network.

We now show how to train all the layers simultaneously. We discussed how the forward pass generates activations at all the intermediate layers, and the loss to be optimized as the objective function. The gradient of the loss for optimization is first calculated at the last layer which has direct access to the cost function, and then flows backward to each parameter using the chain rule. This process is called a backward pass or backward propagation. The weight update rule remains the same for all neurons as earlier with their respective gradients. Let's look at the gradient for the weights in layer j, represented by $\theta^{(j)}$. For ease of notation, we return to the case of three layers and revisit the corresponding forward pass that was described in equations 1.36 for an input $\mathbf{x_i}$.

$$\mathbf{T_i^{(1)}} = g(\theta^{(1)T}\mathbf{x_i}) \tag{1.41}$$

$$\mathbf{T_i^{(2)}} = g(\theta^{(2)T}\mathbf{T_i^{(1)}}) \tag{1.42}$$

$$\hat{\mathbf{y}}_\mathbf{i} = softmax(\mathbf{T}_\mathbf{i}^{(2)}) \tag{1.43}$$

$$J(\boldsymbol{\theta}) = -\log \hat{y}_{i,y_i} \tag{1.44}$$

For each of these equations, we can write the gradient rule and then string them together to get the required gradients.

$$\frac{\partial J}{\partial \theta^{(1)}} = \frac{\partial J}{\partial \hat{\mathbf{y}}_\mathbf{i}} \times \frac{\partial \hat{\mathbf{y}}_\mathbf{i}}{\partial \mathbf{T}_\mathbf{i}^{(2)}} \times \frac{\partial \mathbf{T}_\mathbf{i}^{(2)}}{\partial \mathbf{T}_\mathbf{i}^{(1)}} \times \frac{\partial \mathbf{T}_\mathbf{i}^{(1)}}{\partial \theta^{(1)}} \tag{1.45}$$

This means that NNs avoid the curse of dimensionality in input features and are able to be trained using SGD. Stacking many such layers allows us to achieve more complex learning. This stacking is what gave rise to the term "deep learning." In practice, the softwares used for training NNs use automatic differentiation, a powerful procedure that can calculate gradients quickly. Calculation of gradients is basically matrix–vector or matrix–matrix multiplications, something GPUs are really good at. Combining this computational power with the advent of big data has made NNs very powerful. A particular kind of NN, called the Convolutional Neural Network, allows us to take this even further, and we will discuss that in detail next.

1.6 Convolutional Neural Networks

The NNs introduced in section 1.5 had all layers fully connected. Since these networks regularly deal with high dimensional data, the weight matrices for the fully connected layers can grow quite large. To counter this, we utilize a special class of NNs, called Convolutional Neural Nets, or CNNs, which are particularly useful for extracting features from image and audio data. Additionally, each pixel in the image domain does not form a feature by itself. Quite often a group of neighboring pixels form features relevant to concepts in images that might help make classification decisions, for examples. This informs the connectivity pattern shift from fully connected into convolutional styles

Convolutional neural networks were first introduced in [12] for document character recognition. This network, called Le-Net, performed really well on a handwritten digit dataset, known as MNIST [13]. Since then, there have been many complex networks and datasets introduced. The dataset used for characterizing real-life images is made of millions of $224 \times 224 \times 3$ resolution images, known as ImageNet [14]. The state-of-the-art performance on this dataset is held by powerful networks like ResNets [15] and Vision Transformers [16]. These networks are trained and follow the same inference methodology as NNs, discussed in Section 1.5. However, they differ in the structure of layers. An example of one such CNN structure is shown in Figure 1.11. The first layer in this example is a convolutional layer, made up of convolutional kernels that act on the input. The output of this layer is referred to as feature maps. Often, convolutional layers are followed by subsampling layers to reduce the dimensionality of the maps. The `[conv-maxpool]` structure repeats for a while, ending with one or more fully connected layers. The last layer is the fully connected classifier, which acts upon the extracted features to make the classification decision.

An important concept is that of the receptive field. The receptive field of a kernel is the size of input it accesses. Hence, for the first layer, it is directly the size of the kernel. But as shown in Figure 1.11, each layer takes as input the output of the previous layer, and hence each layer's receptive fields translate backward into larger areas of input. This implies that the receptive field grows as we go deeper into the network. In Section 1.6.1, we discuss some of the layers that make up CNNs, and their functionality.

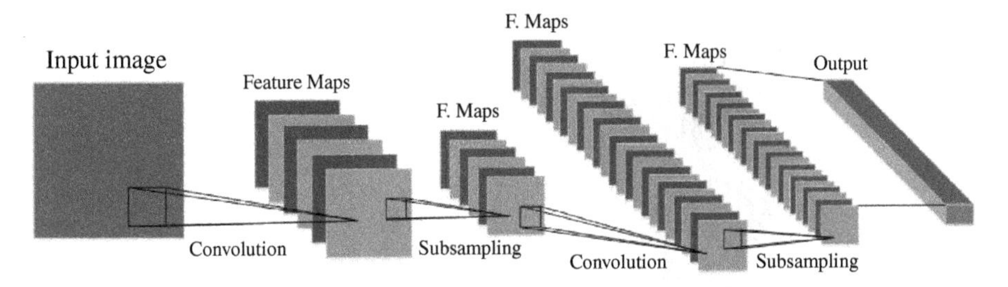

Figure 1.11 A simple CNN structure made of convolutional, subsampling and fully connected layers.

1.6.1 Convolutional Layers

Convolutional layers derive their name from the convolutional operation in signal processing, with the exception that the kernel is not flipped as it passes over the input (and hence, it is actually performing correlation instead of convolution). Let's first consider the simplified one-dimensional convolution operation that occurs between a kernel and a single channel input. A convolutional kernel acts on a patch of the input the same size as the kernel. This is shown in Figure 1.12. The input image, I, and the kernel, K, both have a single channel in this example. Let's assume we have a kernel of size 3×3, shown using binary values. The full input is larger sized, but only a 3×3 patch is taken, as highlighted. A dot product is performed between the image patch and the kernel, which is just element-wise multiplication as expanded above the arrow. The resulting value of 2 is the first pixel of the output feature map. The next patch that convolves with the kernel happens when the highlighted 3×3 slides over to the right. The amount of pixels it moves to the right is known as stride. A stride of 1 is shown in the figure. The same dot product repeats on the slided patch to give the next output of 1. This 3×3 window slides over the entire image, giving rise to the entire feature map shown on the right in Figure 1.12.

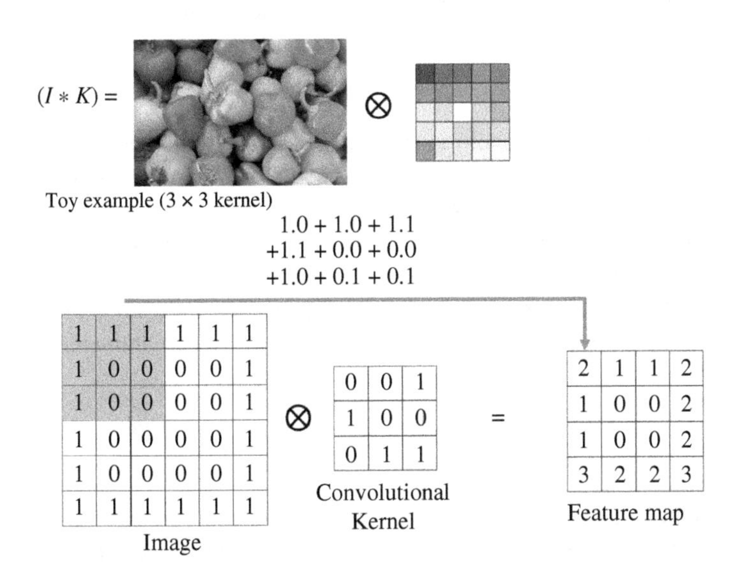

Figure 1.12 The operations involved in convolution are shown. The highlighted part of the input is the patch the kernel convolves with to give a single output in the feature map. This window then slides over the entire image resulting in the whole feature map. Source: DAVID CARREON/Adobe Stock.

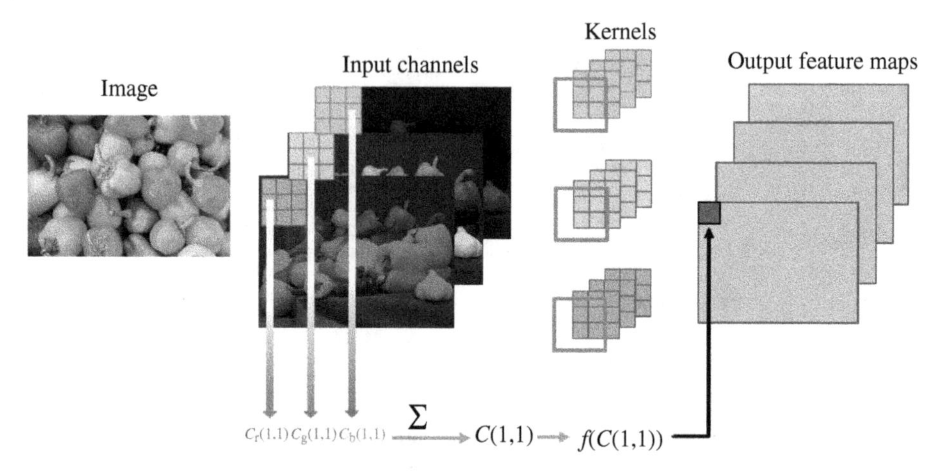

Figure 1.13 One input patch of the image of the same size as the kernel convolves with all channels of one kernel to give rise to a single pixel in the output feature map. Source: DAVID CARREON/Adobe Stock.

Now let us look at what happens with multi-channeled inputs and multiple kernels. Figure 1.13 shows the first convolutional layer. In this case the input, **X**, is an RGB image, and hence, has 3 channels, 1 for each color. Let's assume the image height and width are H_i and W_i respectively. Hence $X \in \mathbf{R}^{H_i \times W_i \times 3}$. The kernels are usually square, let's say of size $K_i \times K_i \times 3$. In Figure 1.13, the image is shown expanded into its 3 different channels. Correspondingly 4 different kernels are shown stacked one upon the other, with the 3 channels for each kernel expanded vertically. The first kernel, with its 3 channels, convolves with a similarly sized 3×3 patch of the input, and the resulting convolution outputs one pixel. As shown in the image, each channel of the kernel convolves with the corresponding channel of the input. The kernel then slides over by the value of stride to the next patch in the input, creating the next pixel in the feature map. This computation can be performed in parallel for each kernel, with different kernels adding pixels to the depth of the feature map. Hence, if there are M kernels in the layers ($M = 4$ in this example), the output feature map will have a depth of M. This convolution process repeats for later layers, where the input is now the feature maps from previous layers.

Ultimately when the learning is complete, the convolutional kernels learn features relevant to recognizing objects in images. The early layer filters respond to edges and color blobs, but later layers build upon these simpler features and learn more complex features, responding to concepts like faces and patterns. The intuition behind sliding the same kernel across the image is that in images, a feature could be located at any part of the image, such as the center or the bottom left or top right. This connectivity pattern encodes a preference for feature location invariance into the CNN. The size and number of convolutional filters, along with the stride are hyperparameters, along with the total number of layers.

1.6.2 Pooling Layers

Pooling layers, also known as subsampling layers, are useful to reduce the size of the feature maps resulting from convolutions. There are two typical pooling types: max pooling and average pooling. Let's say we have pooling kernels of size 2×2. Similar to the convolutional kernels, pooling kernels will slide over the image in blocks of size of size 2×2, with the prescribed stride. Each block will result in one output. For max pool the output would be the maximum of the 4 values in the 2×2

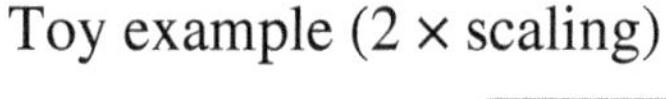

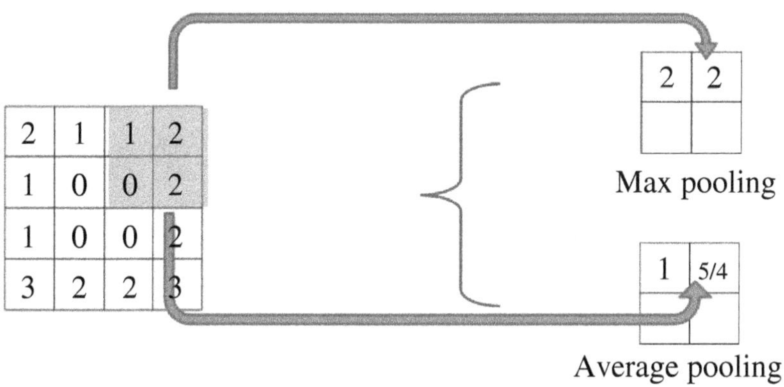

Figure 1.14 The output of max and average pooling applied to blocks of size 2 × 2.

block, and for average pooling, it would be the average of all 4 values, as shown in Figure 1.14. This subsampling introduces some robustness or slight invariance to the feature extraction as well: if a feature is detected at a particular location or a slightly shifted location (with a receptive field of the surrounding 3 values), the output is the maximum or average detection in that region.

1.6.3 Highway Connections

The first CNNs were made of repeating blocks of convolutional layers followed by pooling layers. However, many such blocks were needed to learn complex features. Going too deep results in the problem of vanishing gradients. As gradients backpropagate from the classifier backwards through all the layers, they tend to shrink in magnitude such that the earlier layers get very small gradients and hence, may not be able to learn much. To solve this problem, ResNet [15]-style architectures were introduced that utilize shortcut, highway, or residual connections between the outputs of multiple layers. The output of normal convolution and the highway convolution get concatenated. Hence, when the gradient flows backward, there is a strong gradient flowing back to the earlier layers via the means of these highway connections.

1.6.4 Recurrent Layers

Until now, we have only introduced feedforward layers, layers that only pass an input in one direction. Sometimes, it is useful for layers to have recurrent connections [17], that is layers that are also connected to themselves. This helps a lot in sequential learning, where any output is dependent on the previous outputs, such as in text or video frames. Recurrent networks were introduced to hold some memory of previous inputs by introducing self-connected layers. To train them, however, backpropagation had to unroll the network "in time," resulting in a very large network. This unrolling causes RNNs to suffer from the same problem of vanishing gradients that very deep networks suffered from without highway connections. To counter such problems, Long Short Term Memory Networks (LSTMs) [18] were introduced, which have proven successful at sequence learning tasks. An emerging paradigm of networks that incorporate inherent recurrence are networks that mimic the spike-based learning that occurs in mammalian brains, known as spiking neural networks (SNNs) [19, 20]. They operate on spikes, which can be thought of events that occur when something changes. They accumulate spikes over time to make any inference and use integrate and fire neurons. Each neuron activates spikes when the accumulated spikes crosses a threshold. These are particularly useful in the case of event-driven sensors that naturally emit data as a time series of spikes. However, they can be used with static data such as images as well, by encoding the pixel intensity in say, the number of spikes over a certain time or the time between subsequent spikes.

There have been many more networks that have grown to billions of parameters in size. They are being utilized for a plethora of tasks, outperforming humans at quite a few of them. A lot of ongoing research focuses on making networks more accurate, making training faster, introducing more learning complexity and generalizability across a range of tasks and newer application domains. We encourage readers to seek out some interesting state-of-the-art challenges or domains of interest and explore state-of-the-art methods in those areas.

1.7 Conclusion

In this chapter, we endeavored to introduce some concepts of machine learning that help us build an intuition for understanding the nuts and bolts of neural networks. We introduced neural networks and convolutional neural networks that have shown tremendous success in many deep learning applications. We showed objective functions that encapsulate our learning goals and how backpropagation can be used to train these networks to achieve these objective. This is a fast-evolving field with applications in nearly every domain. We encourage readers to utilize this as a base and find their application of choice and dive into how deep learning can be utilized to revolutionize that area.

References

1 Mnih, V., Kavukcuoglu, K., Silver, D. et al. (2013). Playing Atari with deep reinforcement learning. *CoRR*, abs/1312.5602. http://arxiv.org/abs/1312.5602.

2 Ramesh, A., Dhariwal, P., Nichol, A. et al. (2022). Hierarchical text-conditional image generation with clip latents. https://arxiv.org/abs/2204.06125.

3 Goodfellow, I., Bengio, Y., and Courville, A. (2016). *Deep Learning*. MIT Press. http://www.deeplearningbook.org.

4 Kingma, D.P. and Welling, M. (2013). Auto-encoding variational bayes. https://arxiv.org/abs/1312.6114.

5 Ruder, S. (2016). An overview of gradient descent optimization algorithms. *arXiv preprint arXiv:1609.04747*.

6 Duchi, J., Hazan, E., and Singer, Y. (2011). Adaptive subgradient methods for online learning and stochastic optimization. *Journal of Machine Learning Research* 12 (7): 2121–2159.

7 Karl Pearson, F.R.S. (1901). LIII. On lines and planes of closest fit to systems of points in space. *The London, Edinburgh, and Dublin Philosophical Magazine and Journal of Science* 2 (11): 559–572. https://doi.org/10.1080/14786440109462720.

8 Rosenblatt, F. (1958). The perceptron: a probabilistic model for information storage and organization in the brain. *Psychological Review* 65 (6): 386.

9 Kullback, S. and Leibler, R.A. (1951). On information and sufficiency. *The Annals of Mathematical Statistics* 22 (1): 79–86.

10 Ng, A.Y. (2004). Feature selection, L^1 vs. L^2 regularization, and rotational invariance. *Proceedings of the 21st International Conference on Machine Learning*, p. 78.

11 Nair, V. and Hinton, G.E. (2010). Rectified linear units improve restricted Boltzmann machines. *ICML*.

12 LeCun, Y., Bottou, L., Bengio, Y., and Haffner, P. (1998). Gradient-based learning applied to document recognition. *Proceedings of the IEEE* 86 (11): 2278–2324.

13 Deng, L. (2012). The MNIST database of handwritten digit images for machine learning research. *IEEE Signal Processing Magazine* 29 (6): 141–142.

14 Deng, J., Dong, W., Socher, R. et al. (2009). ImageNet: a large-scale hierarchical image database. *2009 IEEE Conference on Computer Vision and Pattern Recognition*, pp. 248–255. IEEE.

15 He, K., Zhang, X., Ren, S., and Sun, J. (2015). Deep residual learning for image recognition. *CoRR*, abs/1512.03385. http://arxiv.org/abs/1512.03385.

16 Dosovitskiy, A., Beyer, L., Kolesnikov, A. et al. (2020). An image is worth 16x16 words: transformers for image recognition at scale. *CoRR*, abs/2010.11929. https://arxiv.org/abs/2010.11929.

17 Rumelhart, D.E., Hinton, G.E., and Williams, R.J. (1986). Learning internal representations by error propagation. Parallel Distributed Processing: Explorations in the Microstructure of Cognition, Vol. 1: Foundations MIT Press Cambridge, MA, USA. 318–362.

18 Hochreiter, S. and Schmidhuber, J. (1997). Long short-term memory. *Neural Computation* 9 (8): 1735–1780. https://doi.org/10.1162/neco.1997.9.8.1735.

19 Pfeiffer, M. and Pfeil, T. (2018). Deep learning with spiking neurons: opportunities and challenges. *Frontiers in Neuroscience* 12. https://doi.org/10.3389/fnins.2018.00774.

20 Roy, K., Panda, P., and Jaiswal, A. (2019). Towards spike-based machine intelligence with neuromorphic computing. *Nature* 575: 607–617.

2

Overview of Recent Advancements in Deep Learning and Artificial Intelligence

Vijaykrishnan Narayanan[1], Yu Cao[2], Priyadarshini Panda[3], Nagadastagiri Reddy Challapalle[1], Xiaocong Du[2], Youngeun Kim[3], Gokul Krishnan[2], Chonghan Lee[1], Yuhang Li[3], Jingbo Sun[2], Yeshwanth Venkatesha[3], Zhenyu Wang[2], and Yi Zheng[1]

[1] *School of Electrical Engineering and Computer Science, The Pennsylvania State University, University Park, PA, USA*
[2] *School of Electrical, Computer and Energy Engineering, Arizona State University, Tempe, AZ, USA*
[3] *School of Engineering & Applied Science, Yale University, New Haven, CT, USA*

Symbols and Acronyms

AE	autoencoders
BPTT	backpropagation through time
CTMC	continuous time Markov chain
CNN	convolutional neural network
DTMC	discrete-time Markov chain
DCSM	distinct class based splitting measure
GNN	graph neural network
GAE	graph autoencoders
HBPL	hierarchical Bayesian program learning
LSTM	long-short term memory
ML	machine learning
MCMC	Markov chain Monte Carlo
MLE	maximum likelihood estimation
MLP	multi-layer perceptron
NAS	network architecture search
OSL	one-shot learning
PCA	principal component analysis
RNN	recurrent neural networks
RL	reinforcement learning
RBL	restricted Boltzmann machine
STDP	spike-timing-dependent plasticity
SNN	spiking neural network
SGD	stochastic gradient descent
SVM	support vector machine
VPRSM	variable precision rough set model
ZSL	zero-shot learning

Advances in Electromagnetics Empowered by Artificial Intelligence and Deep Learning, First Edition.
Edited by Sawyer D. Campbell and Douglas H. Werner.

2.1 Deep Learning

Artificial intelligent (AI) systems have made profound impact on the entire society in the recent years. AI systems have achieved parity or even exceeding capabilities of humans in specialized tasks such as extracting visual information from images through object detection [1], classification [2, 3], and caption generation [4, 5]. The rapid adoption of machine learning (ML) approaches has been driven by the availability of large data sets available for training and the access to increased computational power provided by a new generation of machine learning hardware. Consequently, deep neural networks have become almost synonymous with AI systems in common parlance. The word cloud in Figures 2.2 and 2.3 emphasizes this trend where deep neural networks have emerged as the dominant. However, AI techniques and models are much more diverse than deep neural networks and were typically studied in the context of signal processing systems before the advent of deep neural network era. The word cloud in Figure 2.1 from work prior to 2016 shows this diversity.

This chapter will review the different styles of machine learning approaches. Figure 2.4 shows the taxonomy of machine learning showing different techniques. Intelligence cannot be measured by

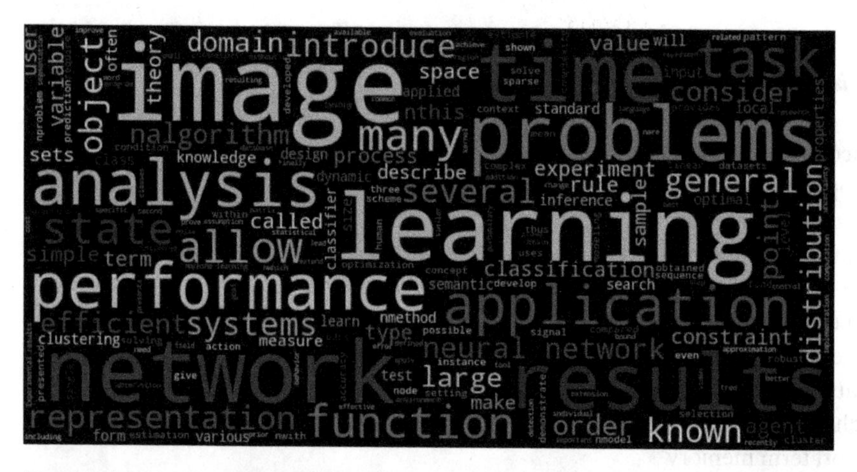

Figure 2.1 Wordcloud from 1990 to 2016.

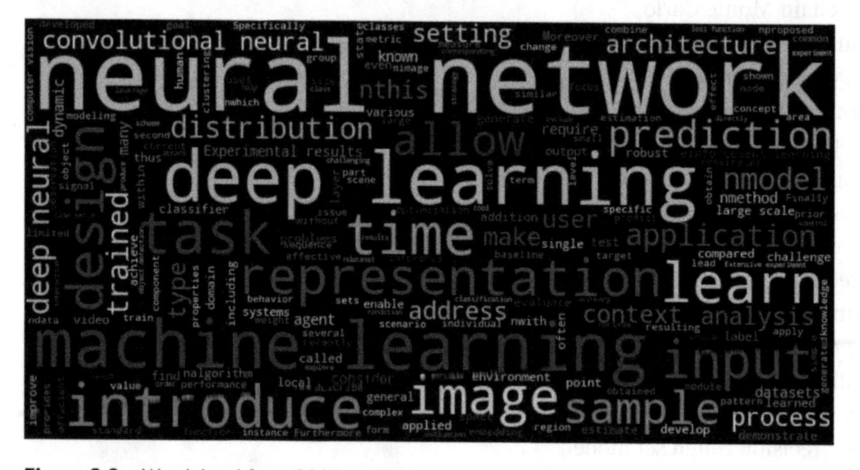

Figure 2.2 Wordcloud from 2017 to 2021.

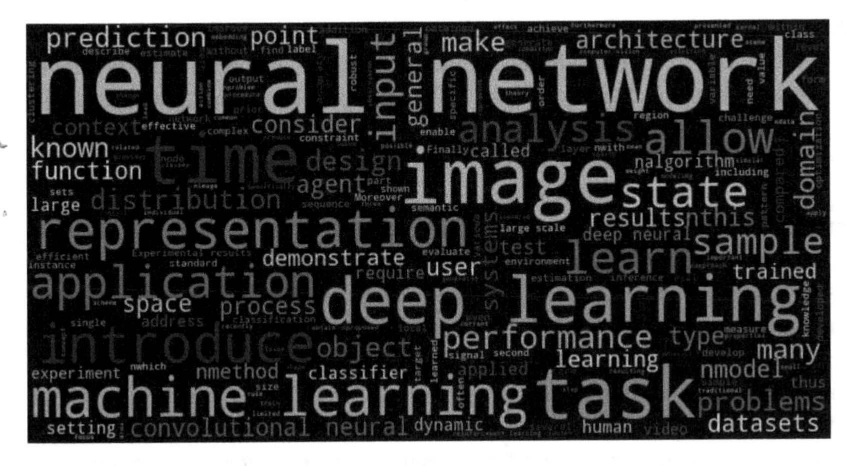

Figure 2.3 Wordcloud from 2020 to 2021.

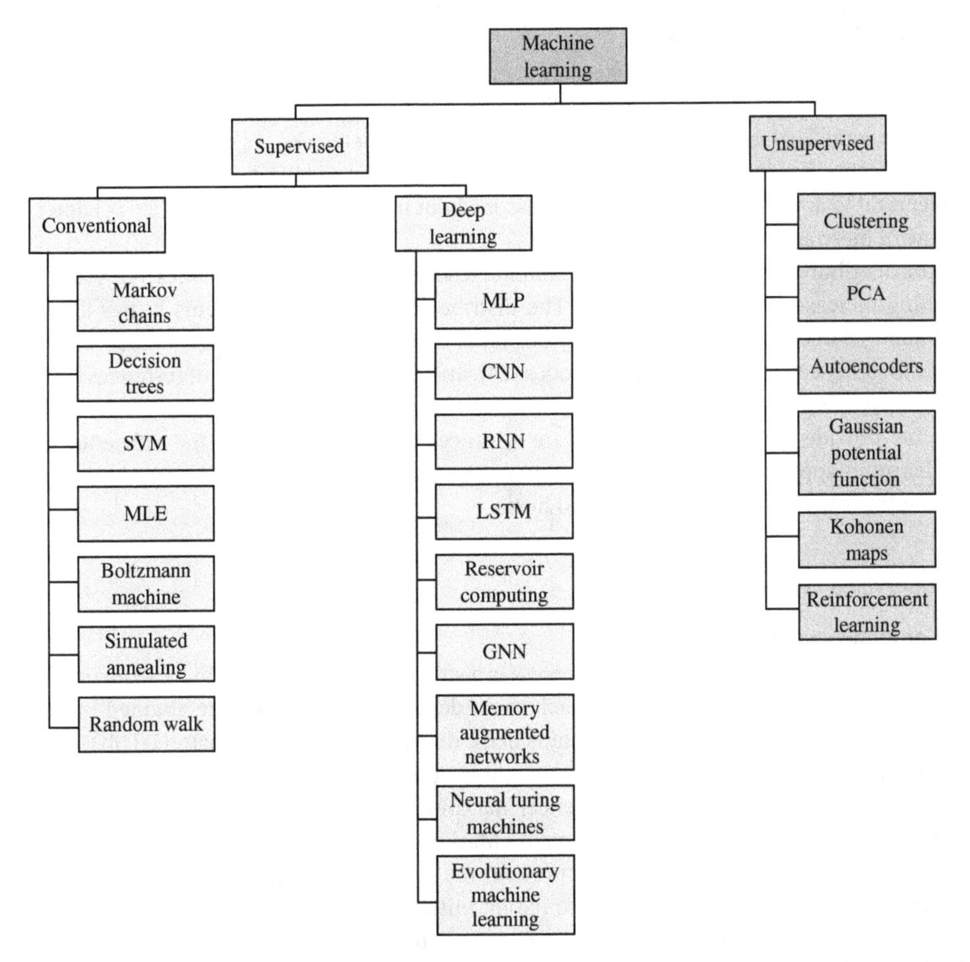

Figure 2.4 Taxonomy of machine learning showing different techniques. Broadly machine learning is classified into supervised and unsupervised learning.

how well a machine performs a single task or even several tasks. Instead, intelligence is determined by how a machine learns and stores knowledge about the world [6], enabling it to handle unanticipated tasks and new environments [7], learn rapidly without supervision [8], explain decisions [9], deduce the unobserved [10], and anticipate the likely outcomes [11]. Consequently, we will first review the category of supervised and unsupervised learning approaches. Unsupervised learning approaches are able to better adopt to novel situations without the need for large, annotated training sets. Supervised techniques include various statistically [2, 3] and biologically inspired models [12, 13]. Among biologically inspired models, neural network models have been a dominant approach. However, neural networks also lend themselves to unsupervised learning such as spike-timing-dependent plasticity inspired by spiking in human brain.

Recent advances in machine learning have involved the ability to learn continuously, rather than learn all possible cases. Consider a system that needs to learn to distinguish between 1000 possible classes to a system that is incrementally introduced to the 1000 classes. A key challenge in current machine learning approaches is catastrophic interference when learning a new class, making the system forget the ability to distinguish the earlier known classes [14]. Such incremental learning approaches clearly resonate with human behavior and is also key to deployment of AI systems in unsupervised and novel environments.

Neural machine translation approaches have enhanced the power of neural networks by evaluating an entire sequence rather than individual elements [15]. This quest drove interest in generative networks [16], autoencoders [17], and graph neural networks [18]. They also enabled rapid developments in sequence to sequence translation capabilities [19]. Graph networks have also been central to approaches that attempt to reason about machine inference [20]. This chapter covers some of these advances.

Due to the distribution of data sources and compute resources across different nodes, AI systems are becoming increasingly distributed [21]. The distributed nature of AI systems brings along unique challenges in concerns such as privacy of shared data [22], security of shared models [23], resource availability constraints at different nodes [24], and fairness in allocation of resources [25]. In this chapter, we introduce some of these distributed computing challenges.

Finally, we provide insights to the hardware advances that are enhancing the efficiency of machine learning approaches. The chapter contains resources to tools and data repositories to complement the learning of the topics covered here.

2.1.1 Supervised Learning

2.1.1.1 Conventional Approaches

Markov Chains A Markov chain or Markov process is a stochastic model describing a sequence of possible events in which the probability of each event depends only on the state attained in the previous event [26]. Two types of Markov chains exist, discrete-time Markov chain (DTMC) and continuous-time Markov chain (CTMC). Markov chains utilize a probability distribution that is determined by the current and past events. Hence, Markov chains possess the unique property of a memoryless system. The probability distribution in a Markov chain is represented as a $N \times N$ matrix with N events. Each entry (i, j) in the matrix represents the probability of the transition from the ith to jth event. Additionally, a Markov chain also has an initial state vector, represented as an $N \times 1$ matrix (a vector), that describes the probability distribution of starting at each of the N possible states. Entry i of the vector describes the probability of the chain beginning at state i. Markov chains have been used for multiple machine learning applications such as Markov chain Monte Carlo (MCMC) [27]. MCMC models are utilized when the model does not assign a zero probability to any state. Therefore, such models are employed as techniques for sampling from an energy-based model [28]. But, MCMC models require a theoretical guarantee on a case-by-case basis for accurate behavior. Other notable applications include anomaly detection [29] and time-series prediction [30] among others.

Decision Trees Decision trees are sequential models that combine a sequence of simple tests. Each test compares a numeric/nominal attribute against a threshold value/set of possible values [31]. Decision trees provide a more comprehensible model as compared to black-box models such as neural networks. For a given data point, a decision tree classifies it based on the proximity to the most frequently used class in the given partitioned region. The error rate is defined as the number of misclassified data points to the total number of data points. The problem of constructing optimal binary decision trees is an NP-complete problem and thus prior work has explored the efficient heuristics for constructing near-optimal decision trees.

There are two major phases in the induction of decision trees: (i) growth phase and (ii) pruning phase. The growth phase involves recursive partitioning of the training data resulting in a decision tree such that either each leaf node is associated with a single class or further partitioning of the given leaf would result in at least its child nodes being below some specified threshold. The pruning phase aims to generalize the decision tree. The tree generated in the growth phase is pruned to create a sub-tree that avoids over-fitting to the training data [32–34]. In each iteration, the algorithm considers the partition of the training set using the outcome from a discrete function. The choice of the function depends on the measure used to split the training set. After the selection of an appropriate split, each node further subdivides the training set into smaller subsets, until no split gains sufficient splitting measure or a stopping criterion is satisfied. Some examples of splitting measures include information gain, gain ratio, and gini value among others [31, 35, 36]. At the same time, Wang et al. [37] presented an approach for inducing decision trees by combining information entropy criteria with variable precision rough set model (VPRSM) and a node splitting measure termed as distinct class-based splitting measure (DCSM) for decision tree induction [38]. The complexity of the decision tree is controlled by the stopping criteria and the pruning method employed. Some of the common stopping criteria include all instances in the training set belonging to a single value of y, reaching the maximum tree depth, and number of cases in the terminal node being less than the minimum number of cases for parent nodes.

Support Vector Machine (SVM) Support vector machines (SVMs) utilize a function that separates observations belonging to one class from another based on the patterns extracted (features) from the training set [39]. The SVM generates a hyperplane that is used to determine the most probable class for the unseen data. Two main objectives of an SVM include low error rate for the classification and generalization across unseen data.

There are three stages in SVM analysis: (i) feature selection, (ii) training and testing the classifier, and (iii) performance evaluation. A pre-requisite for training an SVM classifier includes the transformation of the original raw training data into a set of features. The feature selection methods can be divided into three main types, embedded methods, filter methods, and wrapper methods. Embedding methods incorporate the feature selection into the classifier and the selection is performed automatically during the training phase of the SVM [40, 41]. Filter methods perform feature reduction before classification and compute the relevance measure on the training set to remove the least important elements. The feature reduction reduces redundancy in the raw data to increase the proportion of sample training data relative to the dimensionality of the features, aids the interpretation of the final classifier, and reduces computational load and accelerates the model.

Finally, within wrapper methods, the classifier is trained repeatedly using the feedback from every iteration to select a subset of features for the next iteration. The training of the SVM involves a labeled dataset wherein each training data point is associated with a label. The training process aims at optimizing w and b within the decision function $y = w \times x + b$. The final stage of the SVM analysis, performance evaluation, is done by evaluating the sensitivity, generalization, and the accuracy. To jointly evaluate accuracy and reproducibility, permutation testing is performed, where a hyperplane is estimated iteratively with randomly permuted class labels, across a window of hyperparameter values, for several resampled versions of the dataset. Applications of SVMs include neuroimaging [42, 43], cancer genomics [44], and forecasting [45] among others.

Maximum Likelihood Estimation (MLE) Maximum likelihood estimation (MLE) is the method of estimating the parameters of a probability distribution function for observed data. To achieve this, the likelihood function is maximized under the given probability distribution function such that the observed data is most probable. Consider a set of k examples $X = \{x^{(1)}, \dots, x^{(k)}\}$, drawn from an independent probability distribution $p_{data}(x)$. Let $p_{model}(x; \theta)$ be the family of probability distributions over the same space as that of θ. $p_{model}(x; \theta)$ maps the configuration of x to a real number estimating the true probability of $p_{data}(x)$ [28]. The maximum likelihood estimator for θ is defined as shown below:

$$\theta_{ML} = \underset{\theta}{\operatorname{argmax}}\, p_{model}(X; \theta) \tag{2.1}$$

Through this, the MLE algorithm minimizes the dissimilarity between the empirical distribution and p^*_{data} defined by the training set and the model distribution, by using KL divergence to measure the dissimilarity between the two. The minimization of the KL divergence is performed by minimizing the cross-entropy between the two distributions. As the number of samples increases, the MLE estimator becomes better in terms of the rate of convergence. Some of the properties of MLE include, the true distribution p_{data} must lie within the model p_{model} and the true distribution must correspond to one value of θ. Applications of MLE include linear regression and logistic regression.

Boltzmann Machine A Boltzmann machine is defined as a network of symmetrically connected, neuron-like units that make stochastic decisions about whether to be on or off [46]. Boltzmann machines utilize a simple learning algorithm [47] that allows them to discover interesting features that represent complex regularities in the training data. Boltzmann machines are used for two diverse computational problems. For a search problem, the weights on the connections are fixed and are used to represent a cost function. For a learning problem, the Boltzmann machine utilizes a set of binary data vectors and the machine learns to generate these vectors with high probability. To achieve this, the machine learns the weights on the connections by making small updates to the weights to reduce the cost function.

Learning within a Boltzmann machine can be classified into two, with hidden units and without hidden units. Consider the case without hidden units. Given a training set of state vectors or data, the learning within the Boltzmann machine aims at finding weights and biases to define a Boltzmann distribution in which the training vectors have high probability. To update the binary state for a given unit i, first, the Boltzmann machine computes the total input to the unit as shown below

$$Z_i = b_i + \sum_j s_j \times w_{i,j} \tag{2.2}$$

where $w_{i,j}$ is the weight on the connection between i and j, and s_j is 1 if unit j is on and 0 otherwise. Next, unit i turns on with a probability given by the logistic function as shown below:

$$P(s_i = 1) = \frac{1}{1 + e^{-z_i}} \tag{2.3}$$

A sequential update of the units in any order does not depend on their respective total inputs. Eventually, the network reaches an equilibrium state or a Boltzmann distribution in which the probability of the state vector is solely determined by the energy of the state vector relative to the energy of all possible binary state vectors. At the same time, learning in the presence of hidden units that act as latent variables (features). The features allow the Boltzmann machine to model distributions over visible state vectors that cannot be modeled by direct pairwise interactions between the visible units (input and output). The learning rule remains the same even in the presence of hidden units. Other types of Boltzmann machines include higher-order Boltzmann machine, conditional Boltzmann machine, and mean-field Boltzmann machines. Another variant of the Boltzmann machine is the restricted Boltzmann machine (RBL) [48]. RBL consists of a layer of visible units and a layer of hidden units with no visible–visible or hidden–hidden connections. Through this, the hidden units are conditionally independent given a visible vector. Hence, unbiased samples from $\langle s_i, s_j \rangle$ data can be obtained in a single parallel step. Sampling from the $\langle s_i, s_j \rangle$ model still requires multiple iterations that alternate between updating all the hidden units and the visible units in parallel.

2.1.1.2 Deep Learning Approaches

Convolutional Neural Networks In this section we will discuss the recent advancements in convolutional neural networks (CNNs) with focus on CNN structures, training methods, execution efficiency for both training and inference operations, and finally, open-source tools that help get started with implementation.

Background CNNs have been extensively used due to their ability to perform exceedingly well for a variety of machine learning tasks such as computer vision, speech recognition, and healthcare.

Model Structure Conventional CNNs consist of a set of layers connected in a sequential manner or with skip connections. In addition to convolutional layers, ReLU, pooling, batch-normalization are utilized for better performance. Figure 2.5 shows the typical structure of a convolution and fully

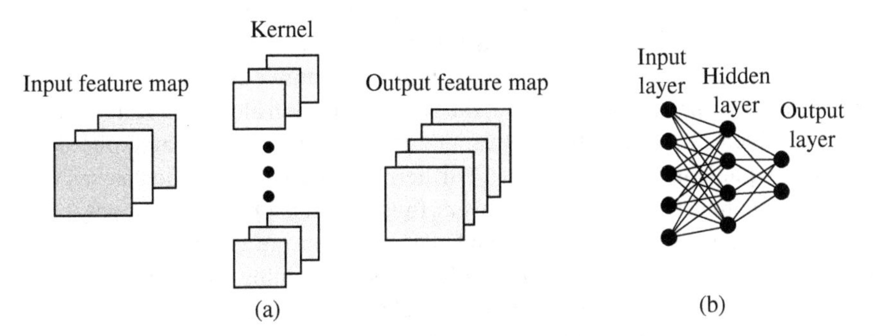

Figure 2.5 (a) Convolution operation within a CNN consisting of the IFM, kernel, and the OFM. The kernel window slides over the IFM to generate the OFM, (b) fully connected (FC) layer operation in a CNN. Each neuron within the FC layer is connected to a neuron in the subsequent layer. The edges represent weights of the FC layer.

connected layer. The sequential layers typically consist of a stack of convolutional (conv) layers that perform feature extraction from the input. Examples of conv layer kernels include 7×7, 5×5, 3×3, and 1×1. In addition, depth-wise convolutions proposed in MobileNet [49] break down a given $N \times N$ convolution into two parts. First, a $N \times 1$ is performed and the result is then run through a $1 \times N$ convolution. Depth-wise convolution results in better accuracy and lower hardware complexity. Pooling layers are utilized periodically to reduce the feature map size and in turn truncate noisy input. Finally, a set of classifier layers or fully connected (FC) layers are utilized to perform classification on the extracted features. The conv and FC layers consist of a set of weights that are trained to achieve best accuracy. Popular CNN structures include AlexNet [50], GoogleNet [51], ResNet [52], DenseNet [53], and SqueezeNet [54]. CNNs such as DenseNet and ResNet consist of skip connections from prior layers that result in a highly branched structure. Furthermore, the skip connections aim to improve the feature extraction process and are present within the conv layers only. But, conventional CNNs suffer from a wide range of drawbacks such as over-parameterization [55], higher hardware training and inference cost, difficulty in improving performance through wider and deeper networks, and vanishing gradient problem among others.

To address this, network architecture search (NAS) was introduced to automatically search for the most optimal neural network architecture based on the target design point. Design point is determined by the target application. For example, higher accuracy, better generalization, higher hardware efficiency, and lower data precision are some of the popular design points using NAS. The training methodology utilized by NAS is explained in the following section ("Training Methods" section). The training process removes the need for human intervention and the uncertainty associated with the choice of the hyperparameters utilized during the deep neural network (DNN) training process. The objective of NAS is to develop an optimized architecture by utilizing a set of building blocks. The building blocks include 1×1 conv, 3×3 conv, 5×5 conv, depth convolutions, skip connections with identity mapping, and maximum or average pooling. The blocks are chosen based on the NAS method to build the networks. Some of the popular techniques proposed include NasNet [56], FBNet [57], AmoebaNet [58], PNAS [59], ECONas [60], and MNasNet [61] among others.

Training Methods The process of training CNNs results in the optimal weight values that maximize the accuracy for the given task at hand. CNNs utilize a wide variety of training methods. The most popular training method is stochastic gradient descent (SGD). The training process utilizes backpropagation of the gradients of the loss function with respect to the trainable variables in the CNN. The backpropagation process utilizes chain rule within conventional calculus to perform the partial derivative evaluation [28]. The backpropagation methodology is utilized such that the loss function is approximated to be a convex function in a piecewise manner that can be optimized using the SGD process. With deeper and wider CNNs, the backpropagation algorithm suffers from the vanishing gradient problem. To address this, activation functions such as ReLU are utilized to remove the effect of small gradients within the CNN. Furthermore, architectures such as ResNet and DenseNet employ skip connections from earlier layers, thus allowing for gradient propagation through them. Other techniques such as dropout and regularization are employed to further improve the performance of CNN training. Finally, hardware-aware training methods have been introduced to further enhance the accuracy of the DNN model [62–66].

Other examples of CNN training methods include zero-shot learning (ZSL) [67], one-shot learning (OSL) [68], and evolutionary algorithms [58, 60]. ZSL is the ability to detect classes not seen during training. The condition is that the classes are not known during the supervised learning process. The attributes of an input image are predicted in the first stage, then its class label is inferred by searching the class that has the most similarity in terms of attributes. But most ZSL works assume that unseen data is used during the testing of the algorithm [69–72]. To address this, Scheirer et al. [73] proposed to generalize the ZSL task to the case where both seen and unseen classes are used at inference. Simultaneously, Frome et al. [74] propose to use label embeddings to operate on the generalized ZSL setting, while [75] utilize learning representations from images and classes through coupled linear regression of factorized joint embeddings. OSL aims at developing domain-specific features or inference procedures that contain highly discriminative properties for the given task. Hence, systems that utilize OSL provide high performance for similar instances but fail as a generalized or robust solution. Fe-Fei et al. [76, 77] developed a variational Bayesian framework for one-shot image classification. The authors utilized the premise that previously learned classes can be leveraged to help forecast future ones when very few examples are available from a given class. Lake et al. [78, 79] proposed to use hierarchical Bayesian program learning (HBPL) for OSL. HBPL is utilized to determine a structural explanation for the observed pixels. However, inference under HBPL leads to a large parameter space, resulting in an integration problem. Koch et al. [68] proposed to learn image representations using supervised metric-based approach with Siamese neural networks. The features from the network are then reused for OSL without any retraining.

Multi-layer Perceptron (MLP) Multi-layer perceptron (MLP) consists of a set of stacked layers that are similar to the FC layer within the CNN [80, 81]. Each MLP consists of an input layer with the number of neurons equal to the input size. A number of hidden layers are utilized within the MLP to perform the classification. The size and number of hidden layers can be varied depending on the task at hand. Finally, an output layer is used to represent the classes that are to be classified into. Each neuron within a layer is connected to a neuron within the next layer. The connections are represented as weights that can be learned during the training process. MLPs utilize backpropagation [82] to perform the weight updates and learn the task. In addition, the MLPs utilize activation functions such as ReLU and sigmoid [83, 84] across each layer. Compared to CNNs, MLPs have a faster training time due to the relatively simpler structure and smaller model size. At the same time, for inference, MLPs have a lower overhead on the hardware platform and have a faster inference time compared to CNNs. MLPs are utilized for a wide range of applications such as character recognition [85], solving differential equations [86], and recently for complex vision tasks [87].

Recurrent Neural Networks (RNN) and Variants Most DNNs provide high performance for independent data samples with no correlation. Recurrent neural networks (RNNs) provide a scalable solution to model data with temporal or sequential structure and varying length inputs and outputs, across different applications [15, 88, 89]. RNNs process sequential data one element at a time utilizing a connectionist model with the ability to selectively pass information. Through this, RNNs model input and/or output data consisting of a sequence of elements that are dependent. Further, RNNs can simultaneously model sequential and time dependencies at different scales. RNNs utilize a feedforward network that utilizes the edges that span adjacent time steps, introducing time to the model. RNNs do not have cycles among conventional edges, while edges that connect adjacent time steps, called recurrent edges, can form cycles. At time t, nodes with recurrent edges receive current input $x^{(t)}$ and also from hidden node values $h^{(t-1)}$ (previous state). The output $y^{(t)}$ at each time t depends on the hidden node values $h^{(t)}$ at time t. Furthermore, input $x^{(t-1)}$ at time $t-1$ influences the output $y^{(t)}$ at time t through recurrent connections. The following equations represent the recurrent operations within the RNN.

$$h^t = \sigma(W^{hx}x^t + W^{hh}x^{t-1} + b_h) \tag{2.4}$$

$$y^t = \text{softmax}(W^{yh}h^t + b_y) \tag{2.5}$$

The training of RNNs utilizes backpropagation through time (BPTT) [90]. But the long-range dependencies within the RNN result in significant training difficulties. Furthermore, they suffer from both vanishing gradient and gradient explosion problems. To solve the exploding gradient problem, truncated backpropagation through time (TBTT) can be employed [91]. At the same time, Sutskever et al. [92] showed training RNNs with a Hessian-free truncated Newton approach to generate text one character at a time.

Modern RNN architectures can be classified into two main categories. Long-short term memory (LSTM) introduces the memory cell, a unit of computation that replaces traditional nodes in the hidden layer of a network [93]. A memory cell is a composite unit, built from simpler nodes through a specific connection pattern and the novel inclusion of multiplicative nodes. LSTMs with the memory cell overcome the vanishing gradient problem, previously encountered in RNNs. Each memory cell contains a node with a self-connected recurrent edge of fixed weight one, ensuring that the gradient can pass across many time steps without vanishing or exploding. The main components within an LSTM are input node, input gate, internal state, forget gate, and output gate. The input node is the node that takes activation from the input layer $x^{(t)}$ at the current time step and from the hidden layer at the previous time step $h^{(t-1)}$. The input gate is a distinctive feature of the LSTM. A gate is implemented using the sigmoidal function that takes the activation from $x^{(t)}$ as well as from $h^{(t-1)}$. The value of the gate is used to multiply the value of another node. If the value of the gate is zero, then the node is cut-off, while if the value is one, the flow is passed through. Internal state refers to the state of the memory cell with the linear activation. The internal state consists of a self-connected recurrent edge with a fixed unit weight. Through the constant weight, across different time steps the error (backpropagation) flows without vanishing or exploding. The forget gate was introduced by Gers et al. [94]. Forget gates provide a mechanism to learn to flush the contents of the internal state. Finally, the output of the LSTM is generated by multiplying the internal state to the output gate. Different variations of the LSTM architecture have been proposed. Gers and Schmidhuber [95] proposed peephole connections that pass from the internal state directly to the input and output gates of that same node without having to be modulated by the output gate. Zaremba and Sutskever [96] proposed to use the *tanh* function instead of the sigmoid function.

Other variants of RNNs include bi-directional recurrent neural networks (BRNNs) proposed in [97]. BRNNs utilize two layers of hidden nodes. Both hidden layers are connected to the input and output. The two hidden layers are differentiated such that the first one has recurrent connections from the past time steps. At the same time, the direction of recurrent of connections within the second hidden layer is flipped, passing activation backward along the sequence. BRNNs are extended to BLSTMs to further improve the performance of LSTMs for applications such as handwriting recognition and phoneme classification [98, 99]. The neural Turing machine (NTM) extends RNNs with an addressable external memory [100]. NTMs improve the performance of RNNs on tasks such as sorting. The two main components of an NTM are the controller and memory matrix. The controller (recurrent or feedforward network) takes the input and returns output as well as passing instructions to and reading from the memory. The memory is represented by a large matrix of P memory locations, each of which is a vector of dimension Q. Additionally, a number of read and write heads facilitate the interaction between the controller and the memory matrix. NTMs are differentiable and can be trained using variants of stochastic gradient descent using BPTT.

Applications of RNNs, LSTMs, and NTMs include natural language representation [96, 101], natural language translation [15, 102], image captioning [103–105], and video encoding and captioning [106, 107].

Spiking Neural Networks (SNNs) Spiking neural networks (SNNs) [108, 109] have been studied as a prospective energy-efficient alternative over standard artificial neural networks (ANNs) including CNNs and RNNs. Notably, SNNs have different structures and functions compared to ANNs. SNNs process temporal information via a Leak-Integrate-and-Fire (LIF) neuron [110]. For discrete time t, the dynamic of LIF neuron with a membrane potential u_i^t can be formulated as follows [111, 112]:

$$u_i^t = \lambda u_i^{t-1} + \sum_j w_{ij} o_j^t \tag{2.6}$$

Here, λ is a leak factor, w_{ij} is the weight connection between pre-synaptic neuron j and post-synaptic neuron i. The neuron accumulates the given spike inputs and generates a spike output o_i^t, whenever the membrane potential exceeds the threshold θ:

$$o_i^t = \begin{cases} 1 & \text{if } u_i^t > \theta \\ 0 & \text{otherwise} \end{cases} \tag{2.7}$$

The membrane potential is reset to zero value after generating a spike. This integrate-and-fire behavior brings a non-differentiable point for a discrete transfer function, which is difficult to apply to conventional backpropagation used in ANN training [113].

To address this, various training algorithms have been proposed in the past decade. In early stage, a line of works [114, 115] used spike-timing-dependent plasticity (STDP) rule to train SNNs. Inspired by mammalian brain development, they increase or decrease the weight connection between two neurons based on their spike time. The STDP operation does not require a complicated backpropagation module, resulting in hardware-friendly implementation. However, STDP cannot be scaled up to the large-scale and complex tasks that have been widely studied in ANN. Another SNN training technique is an ANN–SNN conversion method which converts a pre-trained ANN model to a SNN model with weight or threshold scaling [116–118]. They approximate ReLU neuron and float activation with LIF activation and temporal spike signal, respectively. The ANN–SNN conversion method leverages a well-established ANN training technique, resulting in deep SNNs with high performance on complex tasks. Nonetheless, SNNs require thousands of time-steps to approximate the float value of ANNs. Recently, surrogate gradient learning has been proposed to address the non-differentiable nature of a LIF neuron by using an approximated backward gradient function [112, 113, 119]. Surrogate learning can learn the temporal dynamic of spikes, yielding a small number of time-steps.

Graph Neural Networks (GNNs) While DNNs effectively capture the hidden patterns within Euclidean data, increasing number of applications utilize graphs to represent data. For example, for e-commerce, a graph-based learning system can exploit the interactions between users and products to make highly accurate recommendations. Furthermore, the complexity of graphs and the underlying irregularity pose significant challenges to existing DNNs. Hence to address this, graph neural networks (GNNs) were introduced. GNNs can be categorized into three recurrent graph neural networks (RecGNNs) [18, 120, 121], convolutions graph neural networks (CGNNs) [122–124], and graph autoencoders (GAE) [125–127]. RecGNNs aim to learn node representations with recurrent neural architectures. RecGNNs assume a node within the graph constantly exchanges information with the neighboring nodes until a stable equilibrium is reached. Scarselli et al. [18] extend prior recurrent models to handle generic graphs spanning across cyclic, acyclic, directed, and undirected graphs. Based on an information diffusion mechanism, the state of the nodes is updated by exchanging neighborhood information recurrently until stable equilibrium. The update of a node's hidden state is as shown below:

$$h_n^t = \sum_{n \in N} f(X_n, X_{n,m}^e, X_n, h_n^{t-1}) \tag{2.8}$$

where $f(.)$ is the parametric function and $h_n^{(t)}$ is initialized randomly. To improve the efficiency of training of GNNs, Gallicchio and Micheli [121] proposed to use echo-state networks (GraphESN). GraphESN utilizes an encoder and an output layer. The encoder is randomly initialized and requires no training. GraphESN implements a state transition function to recurrently update node states until the graph converges. Finally, the output layer is trained by using the fixed node states as inputs. Simultaneously, Li et al. [128] proposed to use a gated recurrent unit within the GNN.

CGNNs were introduced to generalize the operation of convolution to graph data. CGNNs utilize an aggregation of a given node's features and the features from neighboring nodes. Furthermore, CGNNs stack multiple graph convolutional layers to extract high-level node representation. CGNNs then convolve the representation of the central node's neighbors' to derive the updated representation for the central node. Hamilton et al. [129] proposed to perform a sampling to obtain a fixed number of neighbors for each node. Through this, authors address the complexity due to the increased number of nodes. The graph convolutions are performed as shown below:

$$h_n^k = \sigma(W^{(k)} \cdot f_k(h_n^{k-1}, \{h_m^{(k-1)} \, \forall \, m \, \in S_{N(n)}\})) \tag{2.9}$$

where $h_m^{(t)} = x_n$, $f_k(.)$ is an aggregation function and $S_{N(n)}$ is a random sample of the neighbors of node n. Veličković et al. [130] utilize attention mechanisms to learn the relative weights between two connected nodes. The attention weights measure the connective strength between two nodes m and n. The attention mechanism employs the LeakyReLU function within it. Furthermore, graph attention network (GAT) utilizes a softmax function to ensure that the attention weights sum up to one over all the neighbors of node n. Other CGNN works have proposed variations of GAT [131] and LSTM-like gating mechanisms to further improve performance [122].

Training of CGNNs requires storing the whole graph data and intermediate states in the memory. Hence, the full training of CGNNs suffers from memory overflow. Hamilton et al. [129] proposed the use of batch-training for the CGNN. The root tree is sampled at each node by recursively sampling the node's neighborhood by K steps with a fixed sample size. For each sampled tree, the root node's hidden representation is computed by hierarchically aggregating the hidden node representations in a bottom to top approach. Chiang et al. [124] proposed to use a clustering algorithm to sample a subgraph. The operations are performed on the sampled subgraph. Due to the restricted neighborhood search within the graph, both larger and deeper graph architectures can be trained.

GAEs map nodes into a latent feature space and decode the graph information from latent representations. GAEs are used to learn network embeddings or generate new graphs. A low-dimensional vector is used to represent the node that preserve the node's topological information. GAEs learn network embeddings using an encoder to extract network embeddings. A decoder is used to enforce network embeddings to preserve the graph topological information (positive pointwise mutual information (PPMI) matrix and the adjacency matrix). Cao et al. [132] proposed to use a stacked denoising autoencoder (AE) to encode and decode PPMI matrix using an MLP. At the same time, Wang et al. [133] utilized a stacked AE to preserve the node first-order proximity and second-order proximity. Two loss functions on the outputs of the encoder and the outputs of the decoder are utilized. The first loss function enables the learned network embeddings to preserve the node first-order proximity by minimizing the distance between a node's network embedding and its neighbors' network embeddings. The second loss function enables the learned network embeddings to preserve the node second-order proximity by minimizing the distance between a node's inputs and its reconstructed inputs.

The applications of GNNs include computer vision (scene graph generation, point cloud classification, and action recognition) [134–137], natural language processing [138–140], and recommendation systems [141–143]. Other deep learning-based supervised learning methods include memory augmented neural networks [144, 145], evolutionary machine learning [146], and reservoir computing techniques [147].

2.1.2 Unsupervised Learning

2.1.2.1 Algorithm

Unsupervised learning refers to the process of extracting features from a distribution without any annotation for the data. Some applications of unsupervised learning include drawing samples from a distribution, learning to denoise data, and clustering data into different groups. The unsupervised learning algorithm aims to find the most optimal representation of the data. The optimal representation preserves maximum information about the input data x while utilizing constraints to ensure the representation is simpler than the data itself. The three main ways of defining the simpler representation are lower-dimensional representation, sparse representation, and independent representation. In this section, we detail some of the popular unsupervised learning methods that utilize the simpler representations.

Clustering Algorithms Clustering in the context of unsupervised learning utilizes non-labeled data. Clustering aims at separating finite unlabeled dataset into a finite and discrete set of hidden data structures [148, 149]. The input data is partitioned into certain number of clusters or groups. There is no universally agreed-upon definition for the same as detailed in [150]. The process of clustering can be broken down into four main steps, (i) feature selection and extraction, (ii) clustering algorithm design or selection, (iii) cluster validation, and (iv) results interpretation.

Clustering techniques can be classified into hierarchical clustering and partition clustering, based on the properties of clusters generated. Hierarchical clustering groups data into a sequence of partition from a single cluster to a cluster including all the data. Meanwhile, partition clustering divides data into predefined number of clusters without any hierarchical structure. The generation of clusters are based on the similarity measure or the distance of the data objects. A distance dissimilarity function on the data X is defined to satisfy symmetry, positivity, triangle inequality, and reflexivity as shown below

$$Symmetry: \; D(x_i, x_j) = D(x_j, x_i) \tag{2.10}$$

$$Positivity: \; D(x_i, x_j) \geq 0 \quad \forall \, x_i, x_j \tag{2.11}$$

$$Inequality: \; D(x_i, x_j) \leq D(x_i, x_k) + D(x_k, x_j) \quad \forall \, x_i, x_j, x_k \tag{2.12}$$

$$Reflexivity: \; D(x_i, x_j) = 0 \quad iff \; x_i = x_j \tag{2.13}$$

Similarly, a similarity function is utilized within the clustering algorithm. Some examples of similarity and dissimilarity functions within clustering algorithms include Minkowski distance [151], Euclidean distance [152], Mahalanobis distance [153], and cosine similarity [154]. Examples of clustering algorithms based on the above functions include K-means, DBSCAN, CURE, WaveCluster, and CLIQUE among others. Applications of clustering include bioinformatics, data pre-processing, DNA sequencing, and anomaly detection [155].

Principal Component Analysis (PCA) Principal component analysis (PCA), also known as Karhunen–Loeve expansion, is a mathematical algorithm that reduces the dimensionality of the data while retaining most of the variation in the data set. PCA performs the reduction by identifying the principal components along which the variation of data is the maximum. Each of the principal components is identified as a linear combination of the original variables. PCA utilizes linear factor models where latent factors h are extracted from a factorial distribution $p(h)$. The conditional variances for each of factors are equal. The conditional distribution utilized is shown below:

$$x = Wh + b + \sigma z \tag{2.14}$$

where z is a Gaussian noise. Iterative expectation-maximization algorithm is utilized to determine the parameters W and σ^2. The PCA model utilizes the observation that the variations within the data are captured using the latent variables h, within a reconstruction error σ^2. A lower σ leads to a more accurate PCA model across the high dimensional data. Applications of PCA include image compression, finding correlation within data, dimensionality reduction, and quantitative finance.

AutoEncoders AE are neural networks that can be trained to copy the input to the output. An internal hidden layer h is used to perform the mapping of the input to the output. There are two parts within the AE, encoder function $h = f(x)$ and a decoder function for reconstruction $r = g(h)$. The copy operation performed ensures that it is approximately done while not an exact copy of the input. Both the encoder and decoder functions f and g can be implemented using feedforward networks like MLP. Different types of AE have been proposed in the past like undercomplete AE, regularized AE, denoising AE, stochastic encoder and decoder, contractive AE, and predictive sparse decomposition AE. AE that have code dimensions lower than the input dimensions are referred to as undercomplete AE. Learning of an undercomplete AE allows the capture of the most salient features in the data. The learning of the undercomplete AE is performed by minimizing the loss function while penalizing the encoder function for being dissimilar from x. Another type of AE (regularized) is the sparse AE. In a sparse AE an additional penalty or sparse penalty is applied to the hidden state h during the training. Sparse AE are used for learning features for another task such as classification. The generated features from the sparse AE are sparse and are used for applications such as compression. Manifold learning utilizes unsupervised learning procedures based on methods such as nearest neighbor graphs. Each node within the graph is associated with a tangent plane that spans in the direction of variations in the vectors or data [156, 157]. Finally, contractive AE introduces an explicit regularizer on h encouraging the derivatives of f to be as small as possible [158]. Some of the applications of AE include information retrieval and semantic hashing [159].

Reinforcement Learning (RL) One of the primary goals of the field of AI is to produce fully autonomous agents that interact with their environments to learn optimal behaviors, improving over time through trial and error.

Reinforcement learning (RL) algorithms are autonomous agents that interact with the environment to learn optimal behavior and improve over time through trial and error. RL tries to maximize a reward signal instead of trying to find hidden structure as in the case of supervised learning. The four main elements within a RL system are the policy, reward signal, value function, and the model of the environment. A policy defines the learning agent's way of behaving at a given time. The policy maps the state of the environment to the actions to be taken at the given state. A reward signal defines the goal of the RL problem. For each step, the environment sends the RL agent a reward that is maximized by the algorithm. Through the reward signal, both the good and bad events are quantified. The value function specifies what is good for the RL algorithm in the long run. The value of the state is the total amount of the reward to the agent for the given state. Overall, the rewards determine the immediate intrinsic desirability of the state, while the value indicated the long-term desirability of the states by accounting for the rewards of the future state. Finally, the model of environment mimics the behavior of the environment. For a given state and action, the model predicts the resultant next state and next reward. RL can be modeled as a Markov decision process (MDP) that consists of a set of S states and the probability distribution of the starting states $p(s_0)$, set of actions A, transition dynamics $T(s_{t+1}|s_t, a_t)$ that maps the state-action pair at a given time t, an immediate instantaneous rewards function $R(s_t, a_t, s_{t+1})$, and a discount factor $\gamma \in [0,1]$. The policy P is to map the states to a probability distribution over actions. If the MDP is episodic, then the states are reset after the episode length T. The goal of the RL algorithm is to determine the optimal policy P^* that achieves maximum expected return from all states. Beyond value function-based RL, there exists a policy search-based RL. Policy search RL algorithms do not maintain a value function model, rather directly search for the optimal policy P^*. Neural network-based policy encoding has been employed in prior works to perform the policy search [160–162]. Conventional RL has been used for a wide variety of applications [163, 164], but limited to low-dimensional problems [165]. The advent of deep learning has resulted in deep reinforcement learning (DRL) that provides a more scalable solution. Some of the popular methods include deep Q-network (DQN) [166], GAE [167], model-based RL [168], hierarchical RL [169], multi-agent learning [170]. Applications of RL include games [171], robotics [172], natural language processing [173], and computer vision [174]. Other methods in unsupervised learning include Kohonen maps [175] and Gaussian potential function [176].

2.1.3 Toolbox

- **Conventional Techniques**
  https://github.com/TheAlgorithms/R
  https://scikit-learn.org/stable/
- **TensorFlow Models**
  https://github.com/tensorflow/tpu/tree/master/models
- **Pytorch Models**
  https://github.com/pytorch/examples
- **One-Shot Learning (OSL)**
  https://github.com/tensorfreitas/Siamese-Networks-for-One-Shot-Learning
- **Zero-Shot Learning (ZSL)**
  https://github.com/akshitac8/tfvaegan
- **Recurrent Neural Networks and Variants**
  https://github.com/pytorch/tutorials/blob/master/beginner_source/nlp/sequence_models_tutorial.py
- **Graph Neural Networks**
  https://github.com/pyg-team/pytorch_geometric
- **Repository of Modern and Conventional Algorithms**
  https://github.com/rasbt/deeplearning-models https://github.com/IntelLabs/distiller
- **Natural Language Processing**
  https://github.com/IntelLabs/nlp-architect

2.2 Continual Learning

2.2.1 Background and Motivation

The rapid advancement of computing and sensing technology has enabled many new applications, such as the self-driving vehicle, the surveillance drone, and the robotic system. Compared to conventional edge devices (e.g. cell phone or smart home devices), these emerging devices are required to deal with much more complicated, dynamic situations. One of the necessary attributes is the capability of continual learning: when encountering a sequence of tasks over time, the learning system should capture the new observation and update its knowledge in real time, without interfering or overwriting previously acquired knowledge. In order to learn a data stream continually, such a system should have the following features:

Online adaption: The system should be able to update its knowledge according to a continuum of data, without independent and identically distributed (i.i.d.) assumption on this data stream. For a dynamic system (e.g. a self-driving vehicle), it is preferred that such adaption is completed locally and in real time.

Preservation of prior knowledge: When new data arrives in a stream, previous data is very limited or even no longer exists. Yet the acquired knowledge from previous data should not be forgotten (i.e. overwritten or deteriorated due to the learning of new data). In other words, the prior distribution of the model parameters should be preserved.

Single-head evaluation: The network should be able to differentiate the tasks and achieve successful inter-task classification without the prior knowledge of the task identifier (i.e. which task current data belongs to). In the case of single-head, the network output should consist of all the classes seen so far. In contrast, multi-head evaluation only deals with intra-task classification where the network output only consists of a subset of all the classes.

Resource constraint: Resource usage such as the model size, the computation cost, and storage requirements should be bounded during continual learning from sequential tasks, rather than increasing proportionally or even exponentially over time.

2.2.2 Definitions

The continual learning problem can be formulated as follows: the machine learning system is continuously exposed to a stream of tasks $T_1, T_2, \dots,$ where each task consists of data from one or multiple classes. When the task $T_1 = \{X^1, \dots, X^{s-1}\}$ comes, our target is to minimize the loss function $\mathcal{L}(\mathcal{Y}; \mathcal{X}_{s-1}; \Theta)$ of this $(s-1)$-class classifier. Similarly, with the introduction of a new task T_2 with classes $\{X^s, \dots, X^t\}$, the target now is to minimize $\mathcal{L}(\mathcal{Y}; \mathcal{X}_t; \Theta)$ of this t-class classifier.

2.2.3 Algorithm

To learn the new task T_{new} while preserving the old knowledge from previous tasks T_{old}, current methods choose to either update the existing parameters in hope of finding a new balance point that gives the minimal loss over all tasks, or to dynamically expand the parameter space for the new knowledge while keeping the old parameters fixed. The former solution is achieved by introducing new regularization terms in the loss function to prevent big parameter drifting from T_{old} to T_{new}. The latter one aims to identify and freeze the most important parameters for T_{old} and using either the remaining parameters or the newly created ones to learn the new knowledge. In the following Sections 2.2.3.1–2.2.3.3, we will give a brief introduction to the most well-known solutions in each category.

2.2.3.1 Regularization

This category proposes to add a regularization term to the objective function while training each new task. Such regularization term penalizes big changes to the parameters that are of importance to the previous tasks so that the performance of T_{old} is protected.

- **Elastic weight consolidation (EWC)**
  The EWC [14] loss function is defined as:

$$\mathcal{L}(\theta) = \mathcal{L}_{new}(\theta) + \frac{\lambda}{2}\sum_i F_i(\theta_i - \theta^*_{old,i})^2 \tag{2.15}$$

  Along with the traditional classification loss $\mathcal{L}_{new}(\theta)$ to the new task, EWC proposed a new parameter-wise regularization constraint to force the updated solution θ^*_{new} stay close to the previous solution θ^*_{old}. The tightness of this constraint is evaluated by the empirical Fisher information matrix F_i, which characterizes the importance of each parameter θ_i toward the T_{old}. Such regularization term behaves like an elastic string to prevent the parameters from drifting too far away from its original solution.

  The Fisher F needs to be stored for each task independently, so that when the third tasks come, they can be used to evaluate the regularization loss separately between the new task and each of the old tasks.

- **Learning without Forgetting (LwF) and Incremental Classifier and Representation Learning (iCaRL)**
  The LwF [177] and iCaRL [178] share the similar loss function which is defined as,

$$\mathcal{L}(\theta) = \mathcal{L}_{new}(\theta) + \lambda\mathcal{L}_{old}(Y_o, \hat{Y}_o) \tag{2.16}$$

  where $\mathcal{L}_{new}(\theta)$ is the cross-entropy loss for the new task and $\mathcal{L}_{old}(Y_o, \hat{Y}_o)$ is the Knowledge Distillation loss. Different from EWC which forces the old and new parameters to be close, the LwF and iCaRL force the output Y_o of the new solution to be similar to its original output $\hat{Y}_o$. The difference between these two methods is the choice of using either the samples from T_{new} (LwF) or the stored examples of the T_{old} (iCaRL) as the inputs to calculate Y_o, $\hat{Y}_o$.

  For the stored samples management, iCaRL set a memory budget K as the total number of exemplars to be stored. When t classes has been learned, iCaRL will select $m = \frac{K}{t}$ samples per class to represent the best approximation of the corresponding data distribution. Specifically, they first collect the class-wise mean of the feature extracted from the trained model. Then iteratively add one example to the selected set such that the average of the features extracted from the current set is closest to that mean. This iteration will end once m samples have been selected.

- **Synaptic Intelligence (SI)**
  The SI [179] loss function is defined as:

$$\mathcal{L}(\theta) = \mathcal{L}_{new}(\theta) + \lambda\sum_i \Omega_i(\theta_i - \theta^*_{old,i})^2 \tag{2.17}$$

  This regularization term is very similar to the EWC where it also uses an important measure Ω_i to constrain the update on each individual parameter so that the new solution will be within the certain range of the old solution. This Ω_i is defined as follows:

$$\Omega_i = \sum_{t \in T_{old}} \frac{\omega_i^t}{\left(\Delta_i^t\right)^2 + \xi} \tag{2.18}$$

  ω_i^t characterizes the per-parameter contribution to changes in the total loss on each old task, which is approximated by the sum of each individual $\omega_{i,s}^t$ evaluated at each step s along the training trajectory. In each single step, $\omega_{i,s}^t$ is the product of the gradient $\frac{\partial \mathcal{L}}{\partial \theta_{i,s}}$ with the actual parameter update $\delta_{i,s}$, which can be considered as the approximation of the loss change caused by this parameter in this step. The sum of $\omega_{i,s}^t$ over every step gives the final importance measure of that parameter toward the current task.

Δ_i^v is defined as $\theta_i(t^v) - \theta_i(t^{v-1})$, which records the change of each parameter caused by learning the very last task v. The intuition interpretation is that if this Δ_i^v is small, which means very minor change has been made throughout the training on the old tasks, this parameter should be better and not change much during the training on the new tasks since it receives huge constraints in all the previous training.

Both ω_i^t and Δ_i^v help evaluate the overall per-parameter importance Ω_i to the previous tasks, setting constraints on the training on the new task to not drift away to its old solution.

2.2.3.2 Dynamic Network

When the new task is strongly associated with the previous task, some parameters can be shared and the optimal solution can be found across multiple tasks. All the regularization methods that have been introduced above are based on such assumption. However, if the relevance between tasks is not strong, a new parameter space is required. Therefore, some methods choose to dynamically expand the capacity of the neural network to continually absorb the new knowledge.

- **Progressive Neural Networks**
  The Progressive Neural Network [180] aims to completely avoid the catastrophic forgetting by keeping all the previous parameters fixed and using a brand-new neural network to learn the new task. Specifically, they propose to increase the amount of neurons in each layer and feed those new neurons with the outputs of both the old and newly created neurons from the previous layer. The number of new neurons per layer can be either constant or dynamically determined via pruning or compression techniques. All these new neurons compose to a "column" that is horizontally stacked to the original neural network and is only associated with that particular task. When another task comes, another "column" will be created. During the inference, a task label has to be provided in order to select which column of the neural network is to be activated for the specific task.
- **Dynamically Expandable Network (DEN)**
  This DEN [181] proposed three different techniques: (i) selective retraining, (ii) dynamic network expansion, and (iii) network split/duplication.
  Method (i) serves as the initial trial to learn the new task using the current model structure. If a desired loss is failed to reach, method (ii) will kick in to expand the network capacity. Specifically, they first add k new neurons W_l^k to each layer l, then train only W_l^k with a group sparsity regularization using the new task data. Such regularization will promote the sparsity of the network thus filtering out the redundant neurons in W_l^k. The remaining neurons will serve to capture the new features introduced by the new task.

  Method (iii) aims to reduce the semantic drift of each hidden unit after the training of the new task via the regularization term $\|\theta_{new} - \theta_{old}\|_2$. If the drift on the neuron is beyond a certain threshold, the algorithm will split this neuron into two copies so that both old and new tasks can find their optimal solution in this duplicated parameter space.

2.2.3.3 Parameter Isolation

Instead of searching for the optimal solution in the entire parameter space of a single neural network, some methods propose to identify and isolate certain number of the parameters so that each task-specific parameter group can be optimized for a particular task without changing other unrelated parameters.

- **PackNet**
  This method [182] proposes to use the pruning technique to promote the sparsity of the neural network so that any redundant parameters can be freed up for the future task. Specifically, they first use all available parameters θ_{all} to learn the new task, then prune certain percentage of the parameters whose absolute magnitudes are lower than a threshold. The remaining ones, denoted as θ_{task}, are identified as the most important parameters to that particular task and a second round of training is conducted only on these θ_{task}. Once the training is done, θ_{task} is fixed. The remaining parameters $\theta^*_{all} = \theta_{all} - \theta_{task}$ are available for the upcoming tasks to repeat these pruning, re-train, and fix procedure.
- **PathNet**
  Given a deep neural network, the PathNet [183] aims to search for the optimal path for each task using the binary tournament selection algorithm. When a new task come, they first randomly initialize multiple paths as the "population" for evolution. Then two paths are chosen randomly from the population and their fitness is evaluated by their training accuracy on the current task. The path with the lower fitness will be overwritten by a copy of the winner path plus some mutation. As this binary tournament continues, many paths start to overlap with each other since the path with higher fitness is more likely to survive. Eventually, the population converges to a single path which considered as the winner for that particular task. That path will be fixed. The remaining neurons will be used for the next round of evolution for the new task.

2.2.4 Performance Evaluation Metric

- *Single-head accuracy*: The network should be able to differentiate the tasks and achieve successful inter-task classification without the prior knowledge of the task identifier. To report the single-head overall accuracy if input data $\{X^1, \dots, X^t\}$ from all previous tasks have been observed so far, the model should be evaluated with testing data that was sampled uniformly and randomly from class 1 to class t and predict a label out of t classes $\{1, \dots, t\}$. For accuracy of each task, the model should be evaluated with testing data collected from each task T_i (supposing classes $\{1, \dots, g\}$) and predict a label out of t classes $\{1, \dots, t\}$ to report single-head accuracy of task T_i.
- *Multi-head accuracy*: Multi-head evaluation only deals with intratask classification where the network output only consists of a subset of all the classes. To report multi-head accuracy of each task T_i, the model should be evaluated with testing data collected from each task T_i (supposing classes $\{1, \dots, g\}$) and predict a label out of g classes $\{1, \dots, g\}$ to report multi-head accuracy of task T_i. Multi-head classification is more appropriate for multi-task learning than continual learning.
- *Memory budget*: The memory cost for storing the samples from previous tasks for memory rehearsal. For example, if input data are images, the number of stored images can be used as the metrics to evaluate memory budget.

2.2.5 Toolbox

- **Elastic Weight Consolidation (EWC)**
  https://github.com/ariseff/overcoming-catastrophic

- **Learning without Forgetting (LwF)**
  https://github.com/ngailapdi/LWF
- **Incremental Classifier and Representation Learning (iCaRL)**
  https://github.com/srebuffi/iCaRL
- **Progressive Neural Networks**
  https://github.com/arcosin/Doric
- **Dynamically Expandable Network (DEN)**
  https://github.com/jaehong31/DEN
- **PackNet**
  https://github.com/arunmallya/packnet
- **PathNet**
  https://github.com/kimhc6028/pathnet-pytorch

2.3 Knowledge Graph Reasoning

2.3.1 Background

Nowadays, knowledge graph is popular to be used for organizing, mining, and solving the knowledge which has been proved to be useful for analyzing the information of very large-scale database. The knowledge graph is a very powerful tool based on knowledge reasoning which is very important for the field of reasoning. The target of knowledge graph and knowledge reasoning is to extract the errors from data and infer more information from the existing data. For example, after detecting the different objects in a picture, we can use the knowledge reasoning to get the relations among these objects and understand these connections for better inferring the ideas in the picture. Also, the knowledge reasoning will give the feedback to enrich the graphs and help the researchers to realize the next step of applications.

2.3.2 Definitions

For the definition of the concepts of knowledge reasoning, the researchers have different understanding and give different concepts.

Zhang and Zhang [184] believed that knowledge reasoning is the process of collecting the information from a range of things, applying the different methods to analyze the data, calibrating the connections of things based on the knowledge graph, and making the decisions or inference on the existed facts. They also thought it is important to find the relationships among different objects or entities and then get the new ideas from the existed data. Tari [185] defined the knowledge reasoning as the process of inferring the new knowledge from the existing data and connections of things. Usually, we think that knowledge reasoning based on knowledge graph is the process of using existing knowledge to infer new knowledge. The target is to use different methods to infer the relations between the entities and create a more detailed knowledge graph.

For better understanding the knowledge reasoning and knowledge graph, we can use the following definition of reasoning over the knowledge graph: if we have a knowledge graph $KG = \langle E, R, T \rangle$ where E, T refer to the set of objects, R is the relations among the entities and the path of relation P. The edges in the connections of two nodes will build up the graph $(h, r, t) \in T$. Then, we want to infer the information which is not in the KG. The equation will be like this: $G' = \{(h, r, t) \mid h \in E, r \in R, t \in T, (h, r, t) \notin G\}$.

2.3.3 Database

- *WordNet*: WordNet is a lexical database of relations between words. Miller [186] introduced and explained this database. It contains the information for more than 200 languages. This database was first created in the Cognitive Science Laboratory of Princeton University by Professor George Armitage Miller in 1985. It only supported English at first. WordNet contains 155,327 words which are organized in 175,979 synsets. The total number of pairs of words is 207,016. It contains the lexical categories nouns, verbs, adjectives, and adverbs. The synsets are created from the connections of semantic and lexical relations. All the nouns and verbs are well organized with the hierarchies. The WordNet is very useful for the systems which need to solve information. The researchers use it for the word-sense disambiguation, information retrieval, automatic text classification, and summarization of text.
- *Wikidata*: Wikidata, which was introduced by Vrandecic and Krtoetzsch [187], is a collaborative, open, multilingual, editable, structured knowledge graph which was created by the Wikimedia foundation. The Wikipedia provides vast information to public who get the open data from the Wikidata. This database contains more than 200 different language versions of Wikipedia and people can edit it. Wikidata is a document-oriented based database and focuses on the items and statements which can provide the different topic and concept. In the main parts, the items consist of obligatorily which are related to the labels and descriptions and optionally which have multiple aliases and some statements. As of November 2018, this database was used by more than half of all English Wikipedia articles. It also was used by Apple's Siri and Amazon Alexa.
- *NELL*: Never-Ending Language Learning (NELL) system is a semantic machine learning system which was created by the research group of Carnegie Mellon University in 2010. Mitchell and Fredkin [188] showed that the database contains the basic semantic relationships between different categories of data. We can check items like cities, companies, and different sport teams. The target of NELL is to develop a system to answer the questions provided by natural language questions. By 2010, NELL had already contained 440,000 new items which also provide the knowledge base. It also accumulates more than 120 million beliefs created by reading the website, books, and articles.

2.3.4 Applications

- *Question answering system*: For answering the query question, we apply the Knowledge-based question and answering to find the answer. Besides, the knowledge graph needs to be completed because the question and answering needs the reasoning process. For example, we can use the knowledge reasoning and graph to understand the vision language problem of the computer vision tasks. Visual question answering is a popular question in which the computer will give the text-based question and answer about the image. In this process, the computer vision system can answer different natural language problems by using the knowledge reasoning and graph. The semantic and objects segmentation are all important to help the system to do the object detection and recognition. Applications like Apple's Siri, Microsoft's Cortana, and Amazon's Alexa all require the support of knowledge graph inference.
- *Recommendation systems*: The recommendation systems need the knowledge graph to help the creation of interlinks which will enrich the semantic information and find more connections. The connectivity between the users and data can be extracted as the nodes and paths which the knowledge reasoning systems need to complete. There are two methods for implementing the knowledge graph into the recommendation system: embedding-based and path-based methods.

Besides, Wang et al. [189] proposed the knowledge-aware recommendation algorithm which is named as knowledge-aware path recurrent network (KPRN). This idea will generate the connections for the entities and also find the relations. The connections between users and items are explored by the knowledge graph and the extra connections of inference can help get more information for the recommendation system. Researchers are also trying to mimic the propagation process to realize the knowledge reasoning via the GNNs.

2.3.5 Toolbox

- **WordNet**
  https://wordnet.princeton.edu/
- **Wikidata**
  https://www.wikidata.org/wiki/Wikidata:Main_Page
- **NELL**
  http://rtw.ml.cmu.edu/rtw/
- **Reasoning Over Knowledge Graph Paths for Recommendation**
  https://github.com/eBay/KPRN

2.4 Transfer Learning

2.4.1 Background and Motivation

Supervised learning has proven outstanding performance in computer vision based on the assumption that it can use a sufficient number of training samples. However, in real-world scenario, collecting a large number of training samples is difficult in many scenarios. To address this, transfer learning has been proposed, which aims to train neural networks on "source domain" where collecting data is easy and one can transfer the trained knowledge to "target domain" with a few (or zero) training samples. During the knowledge transfer process, it is natural that there is the performance degradation because of the differences observed from training (source domain) and test environments (target domain). Therefore, addressing these domain differences is the main problem in transfer learning. If the train set and the test set are from the similar distribution, the trained networks can capture useful features in the test dataset. On the other hand, if there is a large domain gap (e.g. train data from Amazon website and test data from real-world photo), there is huge performance degradation.

2.4.2 Definitions

Generally, transfer learning [190, 191] can be classified into three categories based on the label configuration of training (source domain) and test datasets (target domain). Here, domain can be simply defined as a marginal probability distribution (or feature space) of given data.

- *Inductive transfer learning*: Label information of both source domain and target domain is available (e.g. multi-task learning).
- *Transductive transfer learning*: Label information of target domain is not available. Only source domain label can be used for transfer learning (e.g. domain adaptation).
- *Unsupervised transfer learning*: Label information of both source domain and target domain is unknown (e.g. clustering).

In order to maximize the performance of neural networks on target domain, various algorithms have been proposed as introduced.

2.4.3 Algorithm

- **Distribution Difference Metric**

  Various metrics have been used to measure the distance between the feature space of two datasets. Let source dataset X_s and source dataset X_t represent distribution $P(X_s)$ and $P(X_t)$ in embedding space, respectively. Here, we introduce two representative metrics that are widely used in transfer learning.

  - **Maximum Mean Discrepancy (MMD)** [192] measured the Euclidean distance between empirical means of two probabilities, which can be formulated as:

$$MMD(P(X_s), P(X_t)) = ||\mathbb{E}[P(X_s)] - \mathbb{E}[P(X_t)]||^2_{\mathcal{H}} \tag{2.19}$$

  Due to its simplicity, several variants of MMD have been proposed [193, 194].

  - **Kullback–Leibler Divergence (KLD)** [195] calculate the expectation of the logarithmic difference between two distributions, which is not symmetrical (thus, $KLD(P(X_s)||P(X_t)) \neq KLD(P(X_t)||P(X_s))$).

$$KLD(P(X_s)||P(X_t)) = \sum_{x_s \in X_s, x_t \in X_t} P(X_s) \log \left(\frac{P(X_s)}{P(X_t)} \right) \tag{2.20}$$

  In addition, various metrics such as Jensen–Shannon Divergence ([196] and Bregman Divergence [197]) have been used for transfer learning.

- **Approaches to Transfer Learning**
  - *Instance transfer*: This approach assigns the target-domain labels (or score) to the source instance and leverages these instances to train networks on the target task. The prior approaches in this category calculate the score based on distance metrics such as Kullback–Leibler [198] or transform an instance to reproducing kernel Hilbert space (RKHS) [199]. Instance-transfer approach is suitable for the environment where a large number of labeled source domain samples and a small number of target domain samples are available.
  - *Feature representation transfer*: The objective of feature representation transfer is to reveal a universal feature representation that can represent well both source and target domains. For the inductive transfer learning setting, they generate a feature representation that is shared across related domains (or task) by learning multiple objective functions. If target-domain label information is not available, one popular approach designs the domain discriminator which is trained to predict the domain label from features, and at the same time, the feature extractor which is trained to deceive the domain discriminator. This approach is called domain adversarial training [200, 201], resulting in domain-invariant feature space.
- **Applications with Deep Learning**
  - *Using pre-trained model to downstream task*: By using a pre-trained model trained on a large-scale task, one can improve the performance of networks on a downstream task [202]. Firstly, they pre-train the model using a large-scale dataset such as ImageNet. After that, they replace the final classifier layers (or the part of networks) with a new classifier for downstream task. The networks are trained on a downstream dataset. Pre-trained networks provide improved training stability and performance.

- *Sequential input data*: This is a similar setting with continual learning where the input data is varying in sequence. The objective is to minimize catastrophic forgetting (i.e. networks forget previously learned information when the networks learn new information). Networks can use the prior knowledge to enhance learning capability on the current input. We refer to Section 2.2 for details.
- *Zero-shot translation*: The goal of zero-shot translation is to transfer pre-trained knowledge of cross-lingual translations to unseen language pairs. The representative example is Google's Neural Translation model (GNMT).

2.4.4 Toolbox

- **Domain Adaptation Toolbox**
  https://www.mathworks.com/matlabcentral/fileexchange/56704-a-domain-adaptation-toolbox
- **Awesome Transfer Learning**
  https://github.com/artix41/awesome-transfer-learning
- **NVIDIA TAO Toolkit**
  https://developer.nvidia.com/tao-toolkit

2.5 Physics-Inspired Machine Learning Models

2.5.1 Background and Motivation

In general, first-principle or physics-based numerical simulations are used to model and understand the physical processes in various application domains. These principles model the complex physical processes using the known understanding of the process using physical laws. However, there can be a case where the process is poorly understood or it is too complex to model based on the physics principles. In either case, it will lead to incorrect understanding and wrong predictions. These complex physics-based models also have parameters that need to be approximated or estimated from limited observed data. In addition, they consist of diverse components, and running these models at finer resolutions in space and time is computationally intensive and intractable at times.

Machine learning models have found tremendous successes in extracting knowledge from large volumes of data in various commercial applications such as computer vision, financial forecasting, and retail forecasting. In recent times, there has been a significant interest in applying ML models to understand the physical processes which are conventionally modeled using physics principles. However, black-box ML models have shown limited success owing to the inability to generalize to out-of-sample scenarios, inability to generate physically consistent results, and requirement of large data for better accuracy. Hence, rather than solely relying on a physics-only approach or the ML-only approach, researchers are now beginning to infuse the physics-based constraints or knowledge into the conventional ML models. The primary objective of physics-based ML is to improve the prediction performance by a better understanding of the underlying process, reducing the data or observations required to train the model, and improving the interpretability of the ML models. We briefly discuss various approaches to infuse physics into the ML models and also key application areas which are benefited from physics-based ML models in Section 2.5.2.

2.5.2 Algorithm

Figure 2.6 shows the overview of physics inspired machine learning approaches.

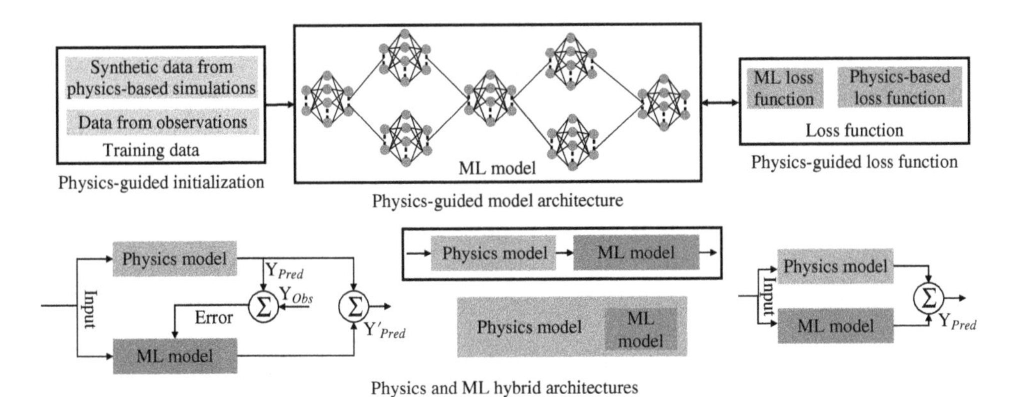

Figure 2.6 Physics-inspired machine learning models.

- **Physics-Guided Loss Function**
  Conventional ML models fail to generalize to out-of-sample scenarios as they cannot comprehend the complex relationships between many physical variables varying over time and space. In order to comprehend these complex relationships directly from limited observations or data and make the model physically consistent, researchers are incorporating physics-based constraints or the physical inconsistency term to the loss function of the ML models as shown below [203].

$$Loss = Loss_{TRN}(Y_{true}, Y_{pred}) + \gamma Loss_{PHY}(Y_{pred}) \tag{2.21}$$

  In addition to the conventional loss such as error between labels and predictions, loss function $Loss_{PHY}$ is added to make the model physically consistent. Whenever the prediction deviates from physical laws, the model will be penalized in the back-propagation. For example, in [204] researchers used the relationship between water density, water temperature, and lake depth parameters to incorporate the physics-based term in the loss function of the model used for predicting the lake temperatures. It is known that denser water sinks, and water temperature and density are inversely proportional. If the model predicts higher density for a point closer to the surface than a point further away from the surface, that is a physically inconsistent prediction and the model needs to be penalized. To incorporate this into the loss function, the averaged difference between the density of all the pairs of points is added to the loss function. In addition to this term, an additional hyper-parameter is added which determines how much weight should be given to this physics-based term. This hyper-parameter can be learned using grid search, etc. similar to other conventional hyper-parameters in ML models. Similarly, Jia et al. [205] used physics-based penalties for violating overall energy conservation, i.e. thermal energy gain should be consistent with overall thermodynamic fluxes in and out of the lake.

  Physics-based loss terms are incorporated in different ways such as PDE constraints in ML-based partial differential equation (PDE) solvers [206], constrained least-squares formulation to incorporate symmetries, and energy-preserving nonlinearities [207], incorporating various parameters such as atomic force, atomic energy, and kinetic and potential energy into the loss term of the ML models predicting molecular dynamics [208].
- **Physics-Guided Initialization** In most of the conventional ML models, the weight parameters are initialized according to a random distribution before the training. Weight initialization plays an important role in model convergence and poor initialization can lead to failure in training, etc. In recent times, transfer learning is used to initialize from a pre-trained model. In transfer learning, a pre-trained model on a related task using large data is fine-tuned to adapt to the task

at hand. Researchers are adopting a similar approach to incorporate physical knowledge into the model or weight parameter initialization to speed up the convergence of the training process and achieve better accuracy with limited observations or training data. Jia et al. [205] pre-train their model for predicting lake temperature dynamics using simulation or synthetic data generated by the physics-based model. This leads to a significant reduction in required training data (physical observations), and it also makes sure that the model is in a physically consistent state before it sees the observed data. Physics-guided initialization is also used in robotics [209], autonomous driving [210], chemical process modeling [211], etc. by pre-training the ML models on data/images generated using physics-based simulators and also transfer learning from ML modes used for physics-related tasks.

- **Physics-Guided Model Architecture** ML architectures incorporating specific physical characteristics of the problem they are solving have the potential to increase the interpretability of the model along with accuracy, etc. There are several techniques to incorporate physical characteristics or consistencies into the ML model architectures. Domain knowledge can be used to design connections between different layers or the nodes of the model, invariances, and symmetries in the physical systems can be encoded in the neural network using specific layers and operations, and infuse domain-specific layers in the networks for extracting specific information, etc.

  Muralidhar et al. [212] modeled the neural network architecture for calculating the drag force on particle suspension in moving fluids based on the physical equations. The pressure field and velocity field affect the pressure component and shear component of the drag force, respectively. Drag force can be modeled as a summation of the pressure component and shear component. The neural network is designed in a way to express the physically meaningful intermediate variables such as the pressure field, velocity field, pressure component, and shear component. Input layer processes the features and generates hidden representation; hidden representation is fed to two different branches, i.e. pressure field branch and velocity field branch. Outputs from these branches are combined using a convolution layer followed by a pooling layer; these features are further fed into two different branches, i.e. shear component branch and pressure component branch, which are further combined using a final fully connected layer. Because of the physics-inspired model architecture, the intermediate layer outputs from the pressure field layer/branch, velocity field layer/branch, etc. have the equivalent ground-truth variables from observations, and the error between predicted intermediate values and observed values can be included in the loss term along with the error in the final prediction. This further leads to richer training over the intermediate values along with inputs and outputs. Similarly, researchers in [213] fix some of the weight parameters in the model to physical constants in the known governing equations modeling the seismic wave propagation.

  Another approach in physics-inspired model architecture is to encode symmetries and variances in the physical systems into ML architectures. In [214], researchers embed rotational invariance in the neural network model to improve the prediction accuracy of turbulence modeling. Incorporating rotational invariance makes sure that identical flows with different axes directions will lead to similar predictions, thereby improving the overall model physical consistency and accuracy. They incorporate rotational invariance by using a higher-order multiplicative layer in the network which ensures that predictions lie on a rotationally invariant tensor basis.

- **Hybrid Architecture** Several techniques combine physics-based models and ML models which are running simultaneously to improve overall prediction accuracy, and for residual or error modeling. In [215, 216], researchers used a linear regression ML model to learn or predict the errors made by the physics-based model. ML model takes the input of the physics model and error in the prediction of the physics model, and predicts the bias that needs to be added to the output of the physics model to correct the error. In [204], researchers used the output of a physics-based model as one of the inputs to the ML model improved the overall accuracy of the ML model in lake temperature modeling. ML models are also used to replace one or more components of physics-based models which are not understood completely. Researchers in [217] observed that using a neural network model to estimate the variables in the turbulence closure model improved the overall accuracy of the physics-based fluid dynamics solver.

Predictions of the ML model and physics model can be combined adaptively based on the use case for better overall prediction. The relative weight of these models is determined by the use case, for example, in [218], researchers use ML predictions for short-range processes and physics-based predictions for long-range processes and adaptively combine them to obtain better accuracy in seismic activity prediction.

2.5.3 Applications

- **Improving the Accuracy of Physical Model** ML-based models are outperforming physics-based models in multiple domains due to availability of large amounts of data. However, for scientific applications, ML models showed limited success due to unavailability of large data, inability to generalize to out-of-sample scenarios, and inability to produce physically consistent results. Fusing physics-based knowledge with ML models can overcome the above challenges in using the ML models for scientific applications. In [205], researchers demonstrated that by infusing physics-based knowledge to ML model with physics-guided loss function and physics-guided initialization can provide much better prediction accuracy with significantly lesser training data, and also can generalize to out-of-sample scenarios. In some cases where the underlying processes are not completely understood, physics-based models might not have good accuracy. ML-based models integrated with process or physics-based knowledge can be leveraged in these cases when sufficient observations are available to yield better prediction accuracy.
- **Reducing Compute Complexity of Physical Model** Physics-based models simulating the physical world at fine-grained fidelity are extremely complex and require huge compute resources to run. Hence, many of the physics-based models run at a more coarse-grained fidelity to reduce the computational costs and model complexity by sacrificing the modeling accuracy. In [219], researchers used neural-network-based model to predict the finer-grained or finer-resolution variables from the coarser-resolution variables obtained by the physics-based model. This neural network-based approach demonstrated better accuracy as these models are extremely good at capturing non-linear relationships given enough observations. Conventional finite element method-based solvers for PDEs are extremely compute intensive. In [220], researchers used a neural network to learn and solve wide variety of PDEs demonstrating the benefits of ML models in solving scientific problems.
- **Obtain the Governing Equations of the Process Through ML Models** Several dynamic systems with large controlling parameters do not have formal analytical descriptions; abundant simulated/experimental data ML models are used to obtain the governing equations. In [221], researchers used techniques such as lasso, k-nearest neighbor (KNN), and neural networks to obtain the antenna design equations from several observed parameters, and demonstrated these equations can be directly used to design new antenna systems and reduce the design time and complexity significantly.

- **Synthetic Data Generation Using Physics-Based Constraints** ML models can learn the data distributions using supervised/unsupervised techniques and can generate the data much faster than physical simulations and has potential to generate new data. In [222], researchers used generative adversarial networks (GANs) to simulate heat conduction and fluid flow for data generating new data samples. In [223, 224], researchers incorporated physics-based loss terms in the GAN models to simulate the turbulent flows.

2.5.4 Toolbox

- **Physics guided loss function** for modeling lake temperatures https://github.com/arkadaw9/PGA_LSTM
- **Physics guided initialization** for autonomous driving https://github.com/Microsoft/AirSim
- **Physics Guided Model Architecture** https://github.com/nmuralid1/PhyNet
- **Hybrid Architecture** https://github.com/zhong1wan/data-assisted

2.6 Distributed Learning

2.6.1 Introduction

Machine learning over the past decade has been applied to increasingly complex problems with abundance of data. While graphics processing units (GPUs), with their high number of parallel hardware threads, have been successful in scaling up the computation necessary for training state-of-the-art machine learning models on large-scale data, it is often not enough to train such models on a single machine. It is only natural to scale out by employing multiple devices equipped with GPUs to accelerate the training process. Traditionally distributed learning refers to the process of sharing the workload of training to different nodes. This was mainly done using two different strategies, (i) Data Parallelism – where the data is divided among the workers and each worker trains a local copy of the model which is finally aggregated into a single model and (ii) Model Parallelism – when the model is too big, it is difficult to have the entire model in one GPU. Hence, subsets of layers are distributed across the worker nodes. Recent advances in compute technologies such as GPUs and TPUs [225] have enabled a multitude of machine learning algorithms and bolstered the demand for applications involving learning. This has also broadened the scope of distributed learning. With data-hungry algorithms such as large-scale deep neural networks, having more data is necessary to achieve high performance. However, the scale of data limits the training process in a centralized setting. There is an imperative need for distributing the machine learning workload across devices and turning the problem into a distributed system problem. This paradigm has opened up challenges in different fronts such as statistical, algorithmic, and systems.

In recent times, distributed learning is studied from different perspectives which can be broadly classified as:

- *Algorithms:* From the algorithm perspective the central question is – *How to remodel the standard centralized learning algorithms to a distributed setting?* Depending on the algorithm, there can be different ways of training and combining the solutions from different workers in the distributed system. The conventional way is to train the models separately on multiple workers and then take the average of the parameters. One of the major factors that needs to be considered here is communication between the nodes. There have been several innovations in reducing communication cost in distributed learning systems including different connection topologies such as ring, hierarchical, and peer-to-peer.

- *Optimization theory and statistics*: While traditionally the deep learning community relied on empirical results for validating the proposed algorithms, recently there has been a surge in theoretical understanding of the underlying optimization methods. In terms of statistical theory, the key challenges for distributed learning include proving the convergence and measuring the convergence rate. Additionally, theoretical understanding of the optimization gives useful insights toward estimating the amount of training data required to reach a solution. Recent works have explored theoretical bounds on the convergence rates in different cases of federated learning.
- *Systems*: On the systems side, large-scale Distributed and storage systems have been explored to enable robust distributed learning infrastructure. The key challenges include consistency and fault tolerance. Additionally, resource optimization in terms of compute and communication is also an open problem. With multiple devices/nodes involved in training, it is imperative to reduce the I/O between the nodes. Programming models also play a major role in developing robust systems. For example, TensorFlow uses a computation graph-based approach. We describe some of the popular frameworks in the Toolbox Section 2.6.4.

2.6.2 Definitions

Given a model f with parameters θ, the optimization problem in distributed learning system consisting of m clients is defined as,

$$\min_{\theta} f(\theta) = \frac{1}{m}\sum_{c=1}^{m} F_c(\theta) \tag{2.22}$$

where F_c is the objective function at client $c \in [m]$ with dataset D_c.

2.6.3 Methods

Here we describe some of the popular methods in achieving scalable and robust distributed learning. Figure 2.7 provides the overview of the distributed learning methods.

- *Federated learning (FL)*: As modern ML algorithms are being applied on sensitive data, it is necessary for developing algorithms that preserve data privacy. FL is a novel distributed training algorithm with data privacy guarantees that has received significant attention recently. With FL, instead of collecting the data into a central server, the model is trained at the edge devices and only the model is transferred between client and the server. This ensures the data privacy and is also scalable to large number of devices. The server periodically receives the gradients ∇_c from the clients and aggregates them to update the global model before broadcasting it back to all the clients. This process of communication of gradients is called as a *round*. Formally the updated model at round r can be expressed as:

$$\theta^{r+1} = \theta^r + \frac{1}{\sum_{c \in S^r} |D_c|} \sum_{c \in S^r} |D_c| \nabla_c^r \tag{2.23}$$

  Here, S^r is the set of participating clients that are communicating their gradients to the server. While standard federated learning has shown promising results, there are multiple challenges actively explored in the federated learning community. Major challenge is to handle varied distributions in the clients' data, commonly referred to as non-IID data. Toward this end,

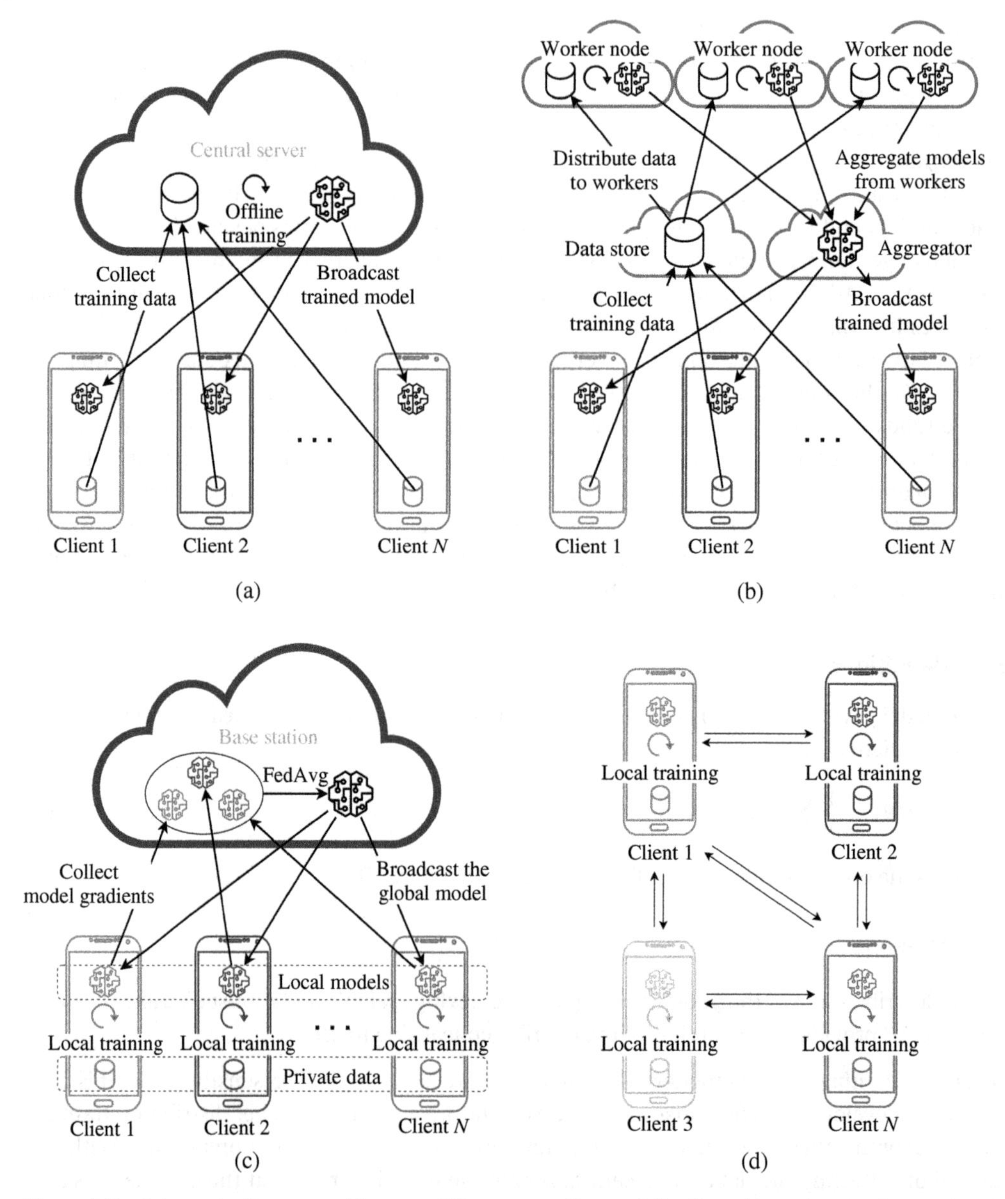

Figure 2.7 Schematic diagram to illustrate different paradigms of distributed learning. (a) Conventional Learning – the data is collected from the clients and the model is trained in an offline manner. (b) Distributed Learning – the dataset is collected at a single data store and then it is distributed across multiple worker nodes for training. (c) Federated Learning – In this case, the training is performed at the client, and model gradients are communicated to the base station which aggregates the gradients and updates the global model, and broadcasts it back to all the clients. Each of these processes is iterative and models at the clients are periodically updated. (d) Decentralized Learning – In this paradigm, there is no central server to coordinate the learning process. The individual clients collaborate in a peer-to-peer manner to share learned knowledge.

FedProx [226] introduced a proximal loss to regularize the client models from significantly deviating from the initial model received from the server. On the other hand, Karimireddy et al. [227] and Reddi et al. [228] tweaked the federated aggregation methods on the server to guide the model toward global optimum and avoid the adverse effects of non-IID data by using control variates and adaptive optimization, respectively. Additionally, since federated learning involves constant communication, if unchecked the total cost of communication can become prohibitive. For example, Nishio and Yonetani [229] reduced the communication cost by sampling the participating clients based on a set of predefined rules. On the other hand, the gradients can be compressed by pruning or quantizing before sharing with the server [230].

- *Optimization theory of distributed and federated learning*: Traditional Parallel Stochastic Gradient Descent (ParallelSGD) [231] extended the standard SGD method to a distributed setting by periodically taking the average of trained weights from all the worker computers and provided acceleration guarantees over standard SGD. Similar to this, there is a recent body of work providing convergence analysis of modern federated learning methods. For example, Reddi et al. [228] provided a generalized framework of adaptive algorithms for federated aggregation and provided their convergence analysis.
- *Privacy and security in distributed learning*: There is a significant attention in developing privacy preserving distributed learning algorithms. While federated learning ensures data privacy up to some extent given that the clients are only sharing the model gradients as opposed to raw data, it is shown that original data can be reconstructed from the gradients shared during the federated learning process [232]. Hence there are additional efforts toward maintaining privacy in federated learning systems. In addition to data privacy, several security threats have been identified which include data poisoning and backdoor attacks. Since the data is private to each client, the server has no control over the updates being sent from the clients. Hence, the clients can intentionally feed in updates from malicious data which in turn tampers with the global model [233]. A backdoor attack is a specific form of trojan attack where a malicious task is injected into the model without hampering the performance of the model on the original task [234].
- *Connection topology*: An important design decision to consider in developing a distributed machine learning system is how the constituent machines are organized. The popular topologies include Tree, Ring, parameter server, and peer-to-peer. The tree topology is strictly hierarchical where each node only communicates with a parent node and a set of child nodes. Nodes in this structure accumulate their gradient and communicate with their parent which in turn aggregates the gradients to update the model [235]. While this is easy to implement and scale to large number of clients, it is not communication efficient. In particular, the nodes with large number of children can become a bottleneck in terms of communication. To avoid this problem, ring topology limits the communication from each node to only its neighbors. This is commonly used strategy in multi-GPU implementations of model machine learning frameworks. In the parameter server approach, the common set of parameters are stored in a set of master nodes which is used as global shared memory and is read and written by all clients in parallel. Finally, fully decentralized peer-to-peer topology involves clients collaborating directly without the need for a controller. While this has the advantage of having no bottleneck as in case of centralized control, it poses new challenges such as communication synchronization.
- *Edge/cloud systems and On device AI*: With edge devices becoming more capable in terms of compute power and low-latency communication, modern machine learning algorithms are moving toward processing at edge or following a hybrid edge-cloud processing pattern. Fog computing is one such paradigm where the advantages of the cloud computing are brought closer to the edge where data is created and acted upon [236, 237].

2.6.4 Toolbox

- **Ray**
  https://github.com/ray-project/ray
- **Elephas**
  https://github.com/maxpumperla/elephas
- **FairScale**
  https://github.com/facebookresearch/fairscale
- **TensorFlowOnSpark**
  https://github.com/yahoo/TensorFlowOnSpark

2.7 Robustness

2.7.1 Background and Motivation

While neural networks have seen success in wide range of applications, several weaknesses in terms of robustness have been identified in deploying them in critical applications such as self-driving vehicles, anomaly detection, and speech recognition for voice commands. For example, in self-driving cars, the attacker can fool the model not to recognize stop signs [238]. It is also shown that adversarial attacks translate to physical world as well as the model running on the data from cameras can be fooled by tampering with the physical images. In speech recognition, adversarial samples can be generated that are not recognizable by humans to control voice-controlled devices [239]. Robustness in neural network systems is examined through different lenses ranging from simple noisy dat/labels to targeted adversarial attacks as shown in Figure 2.8. The performance of deep learning is heavily dependent on obtaining high-quality labeled data. This may not be always feasible given the scale of the applications currently being deployed. Hence, learning from noisy labels has become an important task in modern deep learning. In addition to natural noise in the datasets, there exists a branch of deep learning studying different ways of attacking deep learning models with specific adversarial inputs and their corresponding defense mechanisms. Adversarial inputs are formed by intentionally applying perturbations which are imperceptible by humans that makes the network misclassify the inputs with high confidence. The attack mechanism can be classified into black box, gray box, and white box attacks. In black box attacks, the attacker does not have access to the model information such as architecture and weights. In gray box attacks it is assumed that the attacker has the knowledge of the model structure but not the parameters, whereas in white box attacks the attacker has full information about the model including network

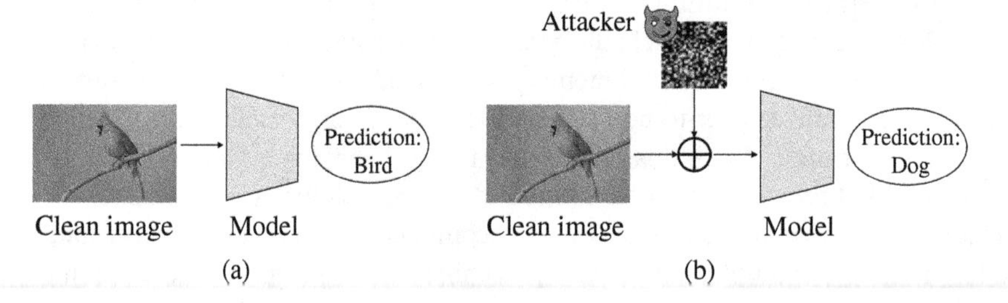

Figure 2.8 Illustration of the concept of robustness. (a) With a clean image, the network predicts correctly. (b) The attacker generates unrecognizable noise and adds it to the clean image, resulting in a wrong prediction. Source: Brian E Kushner/Adobe Stock.

structure and full set of weights. Additionally, there exist targeted attacks that aim to fool the model to output a specific class.

2.7.2 Definitions

We use the following notation in this section. The dataset is defined by set of features $\mathcal{X} = \{x_i\}_{i=1}^N$ and labels $\mathcal{Y} = \{y_i\}_{i=1}^N$. The set of model parameters is denoted by θ. Loss function is denoted by $\mathcal{L}$. We use the subscript x_{adv} to show the adversarial samples. The standard unperturbed inputs are often referred to as clean inputs.

2.7.3 Methods

2.7.3.1 Training with Noisy Data/Labels

This field of research examines the consequences of having noisy labels in the training dataset and several methods have been proposed to combat the effect of such noisy labels [240, 241]. In addition to deterioration in model performance, noisy labels have adverse effects in sample efficiency of training and require complexity of the model. Additionally, this noise in the labels can distort the observed class frequencies which can be of importance in training on datasets with class imbalance. At the same time, noisy labels can be advantageous in terms of privacy as a method to anonymize the dataset and also as a data augmentation method to increase the variability in the data. There are several proposed methods for suppressing the effect of noisy labels from different perspectives. For example, the model architecture can be made robust by introducing a noise adaptation layer after softmax to modify the output of the network based on estimated transition probability [242, 243]. Alternatively, using regularization techniques such as model pre-training [244], bilevel learning [245], robust early learning [246], adversarial training [247], label smoothing [248, 249], and mixup [250] are known to improve robustness in training on dataset with noisy labels.

2.7.3.2 Adversarial Attacks

Here we briefly describe some of the most popular methods of generating adversarial samples.

- *L-BFGS*: The sensitivity of neural networks to adversarial samples was identified in [251] who introduced a method called L-BFGS to compute adversarial perturbations with minimum L_p norm. Formally,

$$\min_x ||x - x_{adv}||_p \text{ subject to } f(x_{adv}) = y_{adv} \tag{2.24}$$

  Here, $||x - x_{adv}||_p$ is the L_p norm of the perturbations and y_{adv} is the target adversarial label ($y_{adv} \neq y$). While an approximate solution was proposed by Szegedy et al. [251] to minimize a hybrid loss, this problem is intractable.
- *FGSM*: The fast gradient sign method uses the gradients of the loss with respect to the input image to create an adversarial example. The objective of the attacker here is to perturb the input data in the direction opposite to that of gradient descent. This will increase the loss of the model on that input thereby reducing the performance of the model. Formally, given a model with parameters θ and input data x, the adversarial example x_{adv} is calculated as follows:

$$x_{adv} = x + \epsilon * \text{sign}(\nabla_x \mathcal{L}(\theta, x, y)) \tag{2.25}$$

  Here, $\mathcal{L}$ is the loss function. x and y are the data and labels of the dataset. ϵ is a small multiplier to keep the perturbations imperceptible.

- *BIM*: Basic iterative method, also referred to as iterative FGSM, extends the FGSM method by applying it iteratively with small step size and clipping the pixel values to constrain the adversarial sample to be in ϵ-neighborhood of the original sample.

$$x_{adv}^{(0)} = x,\ x_{adv}^{(N+1)} = Clip_{x,\epsilon}\{x_{adv}^{(N)} + \alpha * \text{sign}(\nabla_x \mathcal{L}(\theta, x_{adv}^{(N)}, y))\} \tag{2.26}$$

- *PGD*: A projected gradient descent (PGD) attack is a more powerful adversary which is also a multi-step variant of FGSM similar to BIM whereas instead PGD initializes to a random point in the region of interest defined by the L_∞ norm.

2.7.3.3 Defense Mechanisms

There is significant attention toward developing robust defense mechanisms to combat the adversarial attacks. Here we elaborate on some of the recent methods that have shown promising defense mechanism.

- *Adversarial training*: The current state-of-the-art defense against adversarial attacks is the adversarial training where adversarial samples are included in the training loop. This can be viewed as a data augmentation technique where the training data is expanded by including adversarial samples. This functions similar to standard data augmentation techniques such as adding randomly cropped and flipped samples to the training dataset.
- *Randomization*: This class of defense mechanisms aim at transforming the adversarial perturbations into random perturbations by using random input transformation [252, 253] or by adding random noise [254–256]. Additionally, activation pruning such as [257] provide a defense mechanism by randomly pruning out some of the activations in the network. Given that DNNs are fairly robust to random noise, this transformation will aid in nullifying the effects of adversarial perturbations.
- *Denoising*: Denoising methods aim to remove the adversarial effects from the inputs or from the extracted features. Adversarial inputs are detected by conventional denoising techniques such as bit reduction and image blurring [258, 259]. Further, GANs and AE-based feature extractors are also used to denoise adversarial inputs [260–262].

2.7.4 Toolbox

- **Torchattacks**
  https://github.com/Harry24k/adversarial-attacks-pytorch
- **Adversarial-robustness-toolbox**
  https://github.com/Trusted-AI/adversarial-robustness-toolbox
  adversarial attacks, robustness, defense mechanisms

2.8 Interpretability

2.8.1 Background and Motivation

While deep neural networks continue achieving state-of-the-art performance on real-world tasks, they also grow over a significantly large level of complexity. These deep neural networks contain millions of parameters and thus make it intricate to explore their underlying mechanism. We would like to highlight some importance of interpreting neural networks. First, people may question the real-world AI application if they are not sure deep learning can reach their reliability requirements. In the worst case, an error (such as adversarial attack) in neural network could cause catastrophic results (e.g. human lives and heavy financial loss). Another problem is the ethical and legal considerations. We should avoid any algorithmic discrimination. Certain applications like credit and insurance risk assessments should not have any fairness issues. Last, deep neural network is supposed to reveal some undiscovered knowledge when they get higher performance. Interpretability is a way to reveal it.

2.8.2 Definitions

Here, we use the definition by [263]: *Interpretability* is the ability to provide explanations in understandable terms to a human. Interpretability can be classified into 3 dimensions. In *Dimension 1*, interpretability methods are divided into passive approaches and active approaches. One can directly interpret the well-trained neural network with passive methods. Instead, active approaches require changing the network architecture or the training hyper-parameters to interpret the network. In *Dimension 2*, the methods are classified by their distinct types. The types of different interpretability often reveal different levels of explicitness. For example, the most explicit interpretability is generating logic rules. Given input $\boldsymbol{x}$ being a d-dimension vector, an example of logic rules of explaining output y can be given by

$$\begin{cases} \text{If}\,(\boldsymbol{x}_1 < a) \wedge (\boldsymbol{x}_2 > b) \wedge \dots & \text{then} \;\; y = 1 \\ \text{If}\,(\boldsymbol{x}_1 > c) \wedge (\boldsymbol{x}_2 \leq d) \wedge \dots & \text{then} \;\; y = 2 \\ \dots & \\ \text{If}\,(\boldsymbol{x}_d > u) \wedge (\boldsymbol{x}_d > v) \wedge \dots & \text{then} \;\; y = 3 \end{cases} \tag{2.27}$$

where a, b, c, d, u, v are some constants. Such kind of logic rules can give a straightforward explanation of neural networks. Other types of interpretability include explaining hidden semantics, network attribution, and input sample test. In *Dimension 3*, interpretability is divided into global and local methods. Global approaches treat the neural network and the dataset as a whole, while local approaches only explain the network on single or several input samples.

2.8.3 Algorithm

- *Activation maximization (AM)*: The goal of activation maximization is to optimize a synthetic input sample $\boldsymbol{x}$ and maximize the activation on a certain neuron. Mathematically, the objective can be described as

  $$\boldsymbol{x}^{\star} = \arg\max_{\boldsymbol{x}}(\text{Act}(\boldsymbol{x}, \theta - \lambda\Omega(\boldsymbol{x}))) \tag{2.28}$$

  where $\text{Act}(\boldsymbol{x}, \theta)$ is the activation value of the desired neuron given network input $\boldsymbol{x}$ and parameter θ. Ω is a regularizer and λ is the coefficient. In [264], the regularizer is defined as

  $$\Omega(\boldsymbol{x}) = \sum_{i,j} \left((\boldsymbol{x}_{i,j+1} - \boldsymbol{x}_{i,j}) + (\boldsymbol{x}_{i+1,j} - \boldsymbol{x}_{i,j})\right)^{\frac{\ell}{2}} \tag{2.29}$$

  which is called *total variance*. Normally, AM would maximize the score of the desired class to find the best representative images that the model can recognize. Total variance is applied to ensure the images are smooth.
- *Class activation mapping (CAM)*: CAM is a widely used visualization method in CNNs. Note that the convolutional network should have a global average pooling layer after the last convolutional layer. Let $f_k(x, y)$ be the kth channel activation at coordinates (x, y) in the last convolutional layer. The global average pooling of channel k can be represented by $F_k = \sum_{x,y} f_k(x, y)$. Then, to get the

network output score at cth class, a fully connected layer is required $S_c = \sum_k w_k^c F_k$. Overall, from activation $f_k(x,y)$ to network output at cth class S_c, we can rewrite it as

$$S_c = \sum_k w_k^c \sum_{x,y} f_k(x,y) = \sum x,y \sum_k w_k^c f_k(x,y) \tag{2.30}$$

Term $M_c(x,y) = \sum_k w_k^c f_k(x,y)$ is defined as the class activation map. M_c directly indicates the importance of activation at (x,y) leading to classification of an image to class c.

Apart from CAM, Grad-CAM utilizes the gradient information and does not require a global average pooling layer. To eliminate the dependence of w_k^c, Grad-CAM uses $\alpha_k^c = \frac{1}{Z}\sum_{x,y} \frac{\partial S_c}{\partial f_k(x,y)}$, which is the average gradient of network output score S_c with respect to kth channel activation (in the last convolutional layer). The Grad-CAM is calculated by $\text{ReLU}(\sum_k \alpha_k^c f_k(x,y))$.

2.8.4 ToolBox

- **Pytorch-grad-cam**
  https://github.com/jacobgil/pytorch-grad-cam
- **Awesome Machine Learning Interpretability**
  https://github.com/jphall663/awesome-machine-learning-interpretability
- **CVPR 2021 4th Tutorial on Interpretable Machine Learning for Computer Vision.**
  https://interpretablevision.github.io/

2.9 Transformers and Attention Mechanisms for Text and Vision Models

2.9.1 Background and Motivation

The Transformer model was proposed in Google's paper "Attention is All You Need" for neural machine translation [265]. Later the model was used to solve various problems including speech recognition and text-to-speech transformations. The model follows the existing seq2seq structure with encoders and decoders, but it solely uses attention mechanisms as mentioned in the name of the paper. The model's encoder–decoder structure without any recurrence and convolution operations resulted in better performance over the existing best results from RNNs.

In order to understand how the new architecture results in better performance both in accuracy and efficiency, we need to look at the drawbacks of the existing models that are based on RNNs [266]. For seq2seq tasks including language translation, it is crucial to figure out dependencies and connections of tokens in the sequential data. For example, the correlations of words in a sentence are important information for language translation. To extract such information, RNNs have loops in them, which can be thought of as multiple copies of the same network where a network is processing a word and passing a message to its successor network processing the next word. Such mechanism allows the past information of the previous words to persist throughout the propagation enabling processing sequences. However, such chain structure becomes ineffective when it comes to process long sentences. The past information from earlier words tends to get lost along the long chain, which we call vanishing gradient problem. This makes the model difficult to extract long-term dependencies between distant words in a sentence, which results in poor performance.

To mitigate the issue, LSTM networks are proposed to remember long-term information by keeping the intermediate outputs from each encoding step that is processing each word in a sentence and paying selective attention to these past hidden states and relating them to items in the output sequence [267]. However, the model was only able to partially solve the gradient vanishing problem since the limitation of the chain structure still remains where the sequence data have to move each cell in the loop processing each token at a time. Furthermore, the additional memory cells and selective attention require more compute resources and time to train the sequential model.

The Transformer model was proposed with scaled dot-product attention that is not based on the recurrent and sequential methods. This new attention mechanism processes a sentence as a whole and measures similarity scores between words in the given sentence rather than computing it word by word. Furthermore, the attention scores could be computed by assigning different representation subspaces to multiple heads, which empirically has shown the performance improvement.

2.9.2 Algorithm

- **Model Architecture**
  Figure 2.9 illustrates the full model architecture of the Transformer model. Within the model architecture, there are four main stages that process different tasks: Word Embedding, Encoder, Decoder, and Classification Head. We will focus on Word Embedding and Multi-head self-attention from Encoder and Decoder steps since these are newly introduced computation blocks in the Transformer paper.
- **Multi-head Self-Attention**
  Self-attention computes similarity score matrix is shown below:

$$\text{Attention}(Q, K, V) = \text{softmax}\left(\frac{QK^T}{\sqrt{d_k}}\right)V \tag{2.31}$$

$$a = \text{softmax}\left(\frac{QK^T}{\sqrt{d_k}}\right) \tag{2.32}$$

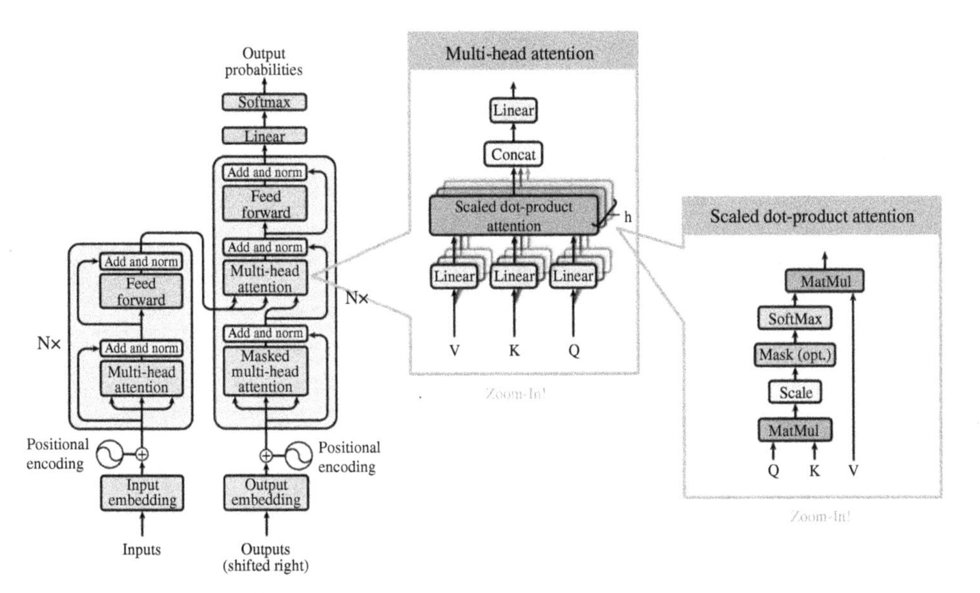

Figure 2.9 The full model architecture of the Transformer. The figure on the left shows the building blocks on the Transformer model. Both Encoder and the Decoder are composed of attention modules that can be stacked on top of each other. The attention module contains Multi-head attention and Feed Forward layers. The Multi-head attention computes scaled dot-product of the given hidden states. Source: [265]/Neural Information Processing Systems.

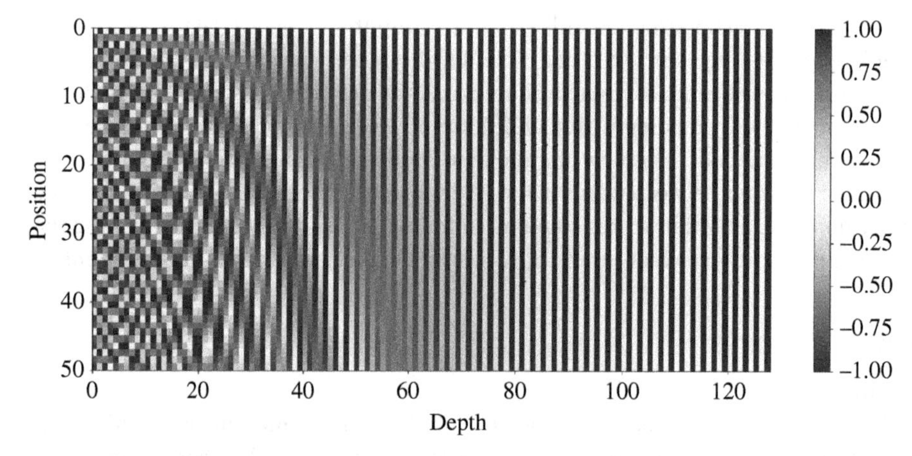

Figure 2.10 The visualization of the sinusoidal positional embedding.

where Q is a query matrix that contains vector representation of one word in a sentence, K is a key matrix that contains vector representation of all the words in the sentence, and V is a value matrix updated with the attention weights a that represent how each word of the sentence from Q is affected by all the other words in the sentence from K. Term $\frac{1}{\sqrt{d_k}}$ is the scaling factor to stabilize gradient propagation in the training stage. The *softmax* function is applied to have the attention matrix with a distribution between 0 and 1. The self-attention computation could apply multi-head mechanism where the Q, K, and V representations after the linear transformation could be split to multiple sub-representations and compute matrix multiplications in multiple heads in parallel.

- **Positional Encoding**
  Before starting to train the model, each word in the input sentences needs to be embedded into unique vector representations that the model could compute. While RNNs inherently take the order of word into account by sequentially processing word by word, the Transformer architecture needs its own way to embed the sequence order information. Thus, a sinusoid positional encoding is introduced to embed information about the order of the input sequence as shown below:

$$\begin{aligned} PE_{(pos,2i)} &= \sin\left(pos/10{,}000^{2i/d_{model}}\right) \\ PE_{(pos,2i+1)} &= \cos\left(pos/10{,}000^{2i/d_{model}}\right) \end{aligned} \tag{2.33}$$

The visualization of the sinusoidal positional embedding is shown in Figure 2.10

2.9.3 Application

- **Bidirectional Encoder Representations from Transformers (BERT)**
  BERT is a state-of-the-art model in a variety of natural language processing (NLP) tasks [268]. It is a much deeper and larger model that consists of Encoder blocks from the original Transformer architecture. The term "Bidirectional" denotes the same mechanism in the Transformer where the Transformer Encoder reads the entire sentence at once rather than word by word. The most important aspect of this massive model comes from its generalization, the ability to learn vast amount of contextual relations between words. This further helps in terms of quickly fine-tuning specific target NLP tasks and results in high-performance results.

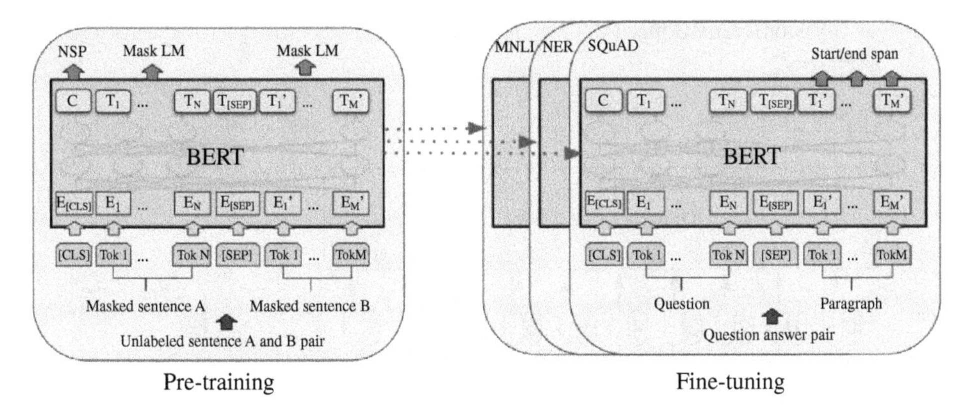

Figure 2.11 Block diagram for BERT pretraining and fine-tuning.

There are two main steps in training BERT for various NLP tasks: Pre-training and Fine-tuning as shown in Figure 2.11.

1. **Pre-training** is the step where the model learns general contextual relations between words in an unsupervised manner. The model is trained with massive dataset such as WikiText and Brown Corpus [269, 270]. There are two types of Pre-training methods: masked language modeling (MLM) and next sentence prediction (NSP). Based on the BERT paper [268], **MLM** replaces 15% of the words in each input sentence with a *MASK* token. Then the model is trained to predict the original words that are masked. In **NSP**, sentences are paired with each other and the model is trained to predict whether the second sentence in the pair is the subsequent sentence in the first sentence. 50% of the sentences are correctly paired in order and the other 50% of the sentences are randomly paired. Both methods are unsupervised, which avoids difficulties in preparing a large repository of specialized and labeled training data. They also enable training a general model that understands contextual relations between words in a large corpus of unlabeled data. This allows the model to be quickly fine-tuned to a target specific NLP tasks and perform better than other sequential models.
2. **Fine-tuning** allows to use BERT model for a specific task. There are various types of tasks for which BERT model could be fine-tuned: **Classifications** (such as sentiment analysis and next sentence classification), **Question Answering**, and **Named Entity Recognition**. Based on the task type, only a small head layer needs to be added to the pre-trained model.

- **Vision Transformer**

  After the success of Transformers in text-based tasks, the vision community applied the self-attention architecture in vision tasks. Google proposed Vision Transformer (ViT) that has better accuracy compared to CNNs which have been the state-of-the-art models in computer vision [271]. ViT is based on the self-attention mechanism introduced in the Transformer model. ViT represents an input image as a sequence of image patches, similar to the sequence of word embeddings in language models as shown in Figure 2.12.

2.9.4 Toolbox

- **Hugging Face**
  https://huggingface.co/
- **FAIRSEQ**
  https://github.com/pytorch/fairseq

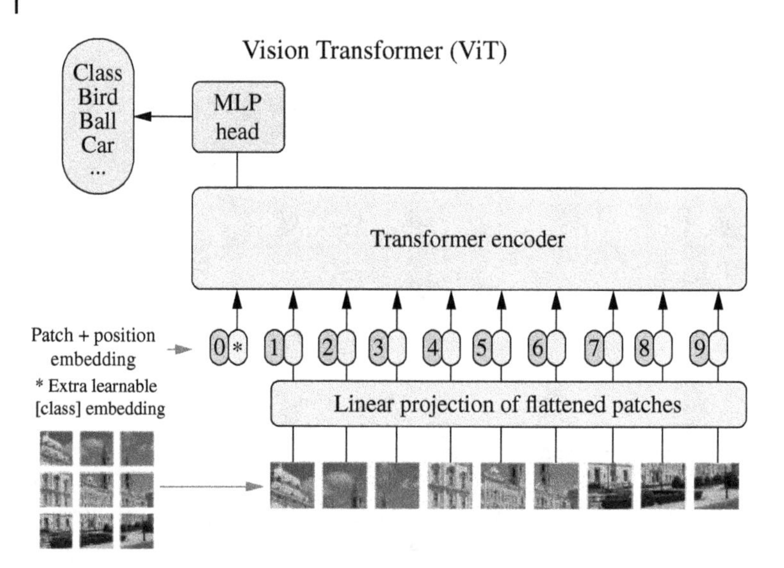

Figure 2.12 The full model architecture of Vision Transformer. Source: Alexey Dosovitskiy, Lucas Beyer, Alexander Kolesnikov, Dirk Weissenborn, Xiaohua Zhai, Thomas Unterthiner, Mostafa Dehghani, Matthias Minderer, Georg Heigold, Sylvain Gelly, et al. An image is worth 16x16 words: Trans97 formers for image recognition at scale. arXiv preprint arXiv:2010.11929, 2020.

- **AllenNLP**
  https://allennlp.org/
- **OpenNMT**
  https://opennmt.net/

2.10 Hardware for Machine Learning Applications

Hardware infrastructure, capable of processing large amounts of data with fast execution time, is one of the primary enablers for the current-generation machine learning revolution. The hardware used for machine learning applications comprise one or more of the below hardware architectures or processing systems.

- CPU – Central Processing Units
- GPU – Graphics Processing Units
- FPGA – Field Programmable Gate Arrays
- ASIC – Application-Specific Integrated Circuits

Figure 2.13 visualizes hardware specializations for deep learning workloads.

2.10.1 CPU

Central processing units (CPUs) are the most widely used computing platform owing to their general purpose computing, ease of programming, software support, and virtualization capabilities. However, they are inefficient in executing large number of matrix or vector operations in the machine learning networks due to lesser number of compute units. Hence, usage of CPUs is limited to the inference of neural networks. In recent years, there has been a significant effort in making CPUs more suitable for running machine learning or neural networks. Intel Incorporated dedicated vector instructions to support parallel computations in the machine learning networks in

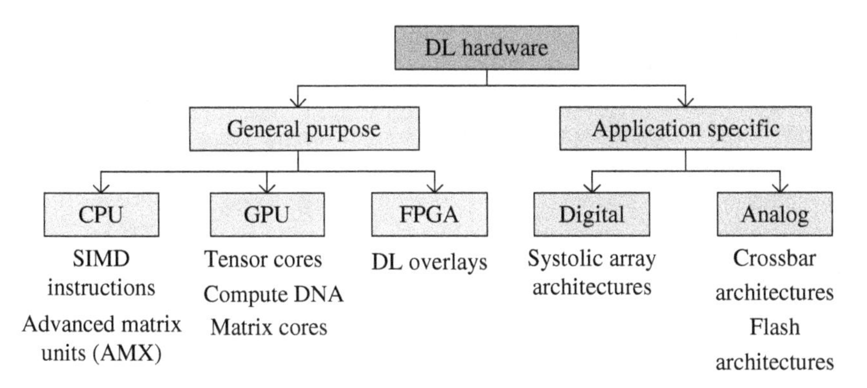

Figure 2.13 Hardware specializations for deep learning workloads. SIMD is short for single instruction/multiple data.

their Xeon architecture. Recently, Apple and Intel added customized advanced matrix processing units in their CPU architectures as a co-processor for efficient execution of parallel matrix/vector operations. In addition to the inference operation, CPUs are also widely used for training of sparse neural networks where computations are not structured and lead to inefficient execution in parallel architectures.

2.10.2 GPU

Graphics processing units (GPUs) are widely used for efficient execution of applications with huge parallelism such as graphics, weather predictions, and physics simulations. They consist of large number of compute cores capable of performing parallel floating point and integer arithmetic operations. As the GPUs are optimized for processing multiple computations simultaneously, they provide much faster execution times for machine learning network training and inference operations which consists of large arithmetic operations.

In addition to their compute capability, availability of parallel programming software such as CUDA accelerated the adaptation of GPUs for machine learning applications. GPU vendors such as Nvidia and AMD are optimizing their architecture for machine learning by introducing tensor cores and compute DNA cores, respectively, which incorporate custom dataflow for computations in the machine learning networks.

2.10.3 ASICs

Application-specific integrated circuit (ASIC) is an integrated circuit customized for a particular or custom use, rather than intended for general-purpose use like CPU. Even though GPUs have massive number of compute units, they can still run a variety of parallel algorithms (other than machine learning networks), and this supports general-purpose workloads, which makes them costlier and also power hungry, whereas ASICs are designed to execute a small set of computations more efficiently (low-latency and low power). This customization enables them to outperform GPUs and other hardware for specific machine learning applications.

However, ASICs incur significant capital and design time, and they are not programmable like FPGAs which make them less attractive to scale for emerging or newer machine learning architectures.

ASICs for ML can be grouped into digital and analog architectures. Digital architectures incorporate custom dataflow and memory hierarchy to efficiently support the parallel computations in ML networks along with requirement for large memory bandwidth. Google's tensor processing unit (TPU) introduced in [225] incorporates systolic array compute architecture along with high bandwidth off-chip memory to efficiently map the parallel computations and satisfy the large memory bandwidth requirements. In systolic array architectures, each processing element (PE) or node communicates only with its adjacent nodes or PEs, thereby reducing the data movement bottlenecks among the large number of parallel PEs. Analog architectures such as Mythic's analog matrix processor (AMP) illustrated in [272] and IBM's analog AI chip illustrated in [273] leverage the in situ compute capabilities of crossbar memory or storage organizations to perform the in-memory multiply-and-accumulate operation. In analog computing, multiply-and-accumulate operation is performed between weight parameters stored in the memory array and input features fed to the memory array by leveraging the in situ current summation operation across the bitlines. Analog computing achieves superior power efficiency due to in-memory computing (weight parameters need not be read and moved from the memory to PEs) and gaining lot of significance in edge computing applications [274–277]. Finally, chiplet-based in-memory architectures provide a scalable platform for large inference computing structures for edge applications [278, 279].

2.10.4 FPGA

Field programmable gate arrays (FPGAs) are integrated circuits with a programmable hardware fabric. Unlike other computing elements (CPUs, GPUs, ASICs), the circuitry inside an FPGA can be reprogrammed as needed. FPGAs are widely used as custom execution engines for machine learning applications due to computational efficiency and flexibility. In addition, FPGAs can be programmed for different existing and emerging data types and network structures efficiently. They offer low latencies and less power consumption for machine learning inference and are widely used for real-time applications such as computer vision [280].

FPGA vendors such as Intel and Xilinx are providing end-to-end software frameworks to map a machine learning model onto FPGAs efficiently with less design time.

Acknowledgment

This work was supported in part by SRC JUMP CBRIC Center and JUMP 2.0 PRISM center.

References

1 Redmon, J. and Farhadi, A. (2018). YOLOv3: An incremental improvement. *arXiv*.
2 Fix, E. and Hodges, J.L. (1989). Discriminatory analysis: nonparametric discrimination, consistency properties. *International Statistical Review / Revue Internationale de Statistique* 57 (3): 238–247. http://www.jstor.org/stable/1403797.
3 Hearst, M.A., Dumais, S.T., Osuna, E. et al. (1998). Support vector machines. *IEEE Intelligent Systems and their Applications* 13 (4): 18–28. https://doi.org/10.1109/5254.708428.
4 Liu, C., Mao, J., Sha, F., and Yuille, A. (2017). Attention correctness in neural image captioning. In *Proceedings of the AAAI Conference on Artificial Intelligence*, Vol. 31, 2017.
5 Farhadi, A., Hejrati, M., Sadeghi, A. et al. (2010). Every picture tells a story: generating sentences from images. In: *Computer Vision – ECCV 2010*. ECCV 2010. *Lecture Notes in*

Computer Science, vol. 6314 (ed. K. Daniilidis, P. Maragos, and N. Paragios), 15–29. Berlin, Heidelberg: Springer-Verlag. ISBN 978-3-642-15560-4. https://doi.org/10.1007/978-3-642-15561-1_2.

6 Joshi, A.V. (2020). *Essential Concepts in Artificial Intelligence and Machine Learning*, 9–20. Cham: Springer International Publishing. ISBN 978-3-030-26622-6. https://doi.org/10.1007/978-3-030-26622-6_2.

7 Mohseni, S., Wang, H., Yu, Z. et al. (2021). Practical machine learning safety: a survey and primer.

8 Gentleman, R. and Carey, V.J. (2008). *Unsupervised Machine Learning*. New York: Springer. ISBN 978-0-387-77239-4. https://doi.org/10.1007/978-0-387-77240-0_10.

9 Zerilli, J. (2020). Explaining machine learning decisions. http://philsci-archive.pitt.edu/19096/.

10 Bermúdez, L., Karlis, D., and Morillo, I. (2020). Modelling unobserved heterogeneity in claim counts using finite mixture models. *Risks* 8 (1). https://doi.org/10.3390/risks8010010.

11 Oermann, E.K., Rubinsteyn, A., Ding, D. et al. (2016). Using a machine learning approach to predict outcomes after radiosurgery for cerebral arteriovenous malformations. *Scientific Reports* 6: 21161. https://doi.org/10.1038/srep21161.

12 Fukushima, K. (2004). Neocognitron: a self-organizing neural network model for a mechanism of pattern recognition unaffected by shift in position. *Biological Cybernetics* 36: 193–202.

13 Maass, W. (1997). Networks of spiking neurons: the third generation of neural network models. *Neural Networks* 10 (9): 1659–1671. https://doi.org/10.1016/S0893-6080(97)00011-7.

14 Kirkpatrick, J., Pascanu, R., Rabinowitz, N. et al. (2017). Overcoming catastrophic forgetting in neural networks. *Proceedings of the National Academy of Sciences of the United States of America* 114 (13): 3521–3526. https://doi.org/10.1073/pnas.1611835114.

15 Sutskever, I., Vinyals, O., and Le, Q.V. (2014). Sequence to sequence learning with neural networks. *Advances in Neural Information Processing Systems (NIPS 2014)*, pp. 3104–3112.

16 Goodfellow, I.J., Pouget-Abadie, J., Mirza, M. et al. (2014). Generative adversarial networks. *Communications of the ACM* 63 (11): https://doi.org/10.1145/3422622.

17 Kramer, M.A. (1991). Nonlinear principal component analysis using autoassociative neural networks. *AIChE Journal* 37 (2): 233–243. https://doi.org/10.1002/aic.690370209.

18 Scarselli, F., Gori, M., Tsoi, A.C. et al. (2008). The graph neural network model. *IEEE Transactions on Neural Networks* 20 (1): 61–80.

19 Bahdanau, D., Cho, K., and Bengio, Y. (2016). Neural machine translation by jointly learning to align and translate.

20 Lecca, P. (2021). Machine learning for causal inference in biological networks: perspectives of this challenge. *Frontiers in Bioinformatics* 1. https://doi.org/10.3389/fbinf.2021.746712.

21 Shi, S., Wang, Q., and Chu, X. (2018). Performance modeling and evaluation of distributed deep learning frameworks on GPUs. *2018 IEEE 16th International Conference on Dependable, Autonomic and Secure Computing, 16th International Conference on Pervasive Intelligence and Computing, 4th International Conference on Big Data Intelligence and Computing and Cyber Science and Technology Congress (DASC/PiCom/DataCom/CyberSciTech).*

22 Shokri, R. and Shmatikov, V. (2015). Privacy-preserving deep learning. *Proceedings of the 22nd ACM SIGSAC Conference on Computer and Communications Security*, CCS '15, pp. 1310–1321. New York, NY, USA: Association for Computing Machinery. ISBN 9781450338325. https://doi.org/10.1145/2810103.2813687.

23 Hitaj, B., Ateniese, G., and Perez-Cruz, F. (2017). Deep models under the GAN: information leakage from collaborative deep learning. *CCS '17: Proceedings of the 2017 ACM SIGSAC Conference on Computer and Communications Security*, pp. 603–618.

24 Wang, S., Tuor, T., Salonidis, T. et al. (2019). Adaptive federated learning in resource constrained edge computing systems. *IEEE Journal on Selected Areas in Communications* 37 (6): 1205–1221. https://doi.org/10.1109/JSAC.2019.2904348.

25 Li, T., Sanjabi, M., Beirami, A., and Smith, V. (2020). Fair resource allocation in federated learning.

26 Gagniuc, P.A. (2017). *Markov Chains: From Theory to Implementation and Experimentation.* Wiley.

27 Andrieu, C., De Freitas, N., Doucet, A., and Jordan, M.I. (2003). An introduction to MCMC for machine learning. *Machine Learning* 50 (1): 5–43.

28 Goodfellow, I., Bengio, Y., and Courville, A. (2016). *Deep Learning*. MIT Press.

29 Ye, N. (2000). A Markov chain model of temporal behavior for anomaly detection. *Proceedings of the 2000 IEEE Systems, Man, and Cybernetics Information Assurance and Security Workshop*, Volume 166, p. 169. Citeseer.

30 Norberg, R. (1995). A time-continuous Markov chain interest model with applications to insurance. *Applied Stochastic Models and Data Analysis* 11 (3): 245–256.

31 Kotsiantis, S.B. (2013). Decision trees: a recent overview. *Artificial Intelligence Review* 39 (4): 261–283.

32 Esposito, F., Malerba, D., Semeraro, G., and Kay, J. (1997). A comparative analysis of methods for pruning decision trees. *IEEE Transactions on Pattern Analysis and Machine Intelligence* 19 (5): 476–491.

33 Fournier, D. and Crémilleux, B. (2002). A quality index for decision tree pruning. *Knowledge-Based Systems* 15 (1–2): 37–43.

34 Osei-Bryson, K.-M. (2007). Post-pruning in decision tree induction using multiple performance measures. *Computers & Operations Research* 34 (11): 3331–3345.

35 Safavian, S.R. and Landgrebe, D. (1991). A survey of decision tree classifier methodology. *IEEE Transactions on Systems, Man, and Cybernetics* 21 (3): 660–674.

36 Shih, Y.-S. (1999). Families of splitting criteria for classification trees. *Statistics and Computing* 9 (4): 309–315.

37 Wang, S., Wei, J., You, J., and Liu, D. (2006). ComEnVprs: A novel approach for inducing decision tree classifiers. In: *Advanced Data Mining and Applications. International Conference on Advanced Data Mining and Applications, Lecture Notes in Computer Science*, vol. 4093, 126–134. Berlin, Heidelberg: Springer-Verlag.

38 Chandra, B., Kothari, R., and Paul, P. (2010). A new node splitting measure for decision tree construction. *Pattern Recognition* 43 (8): 2725–2731.

39 Pisner, D.A. and Schnyer, D.M. (2020). Support vector machine. In: *Machine Learning*, 101–121. Elsevier.

40 Cuingnet, R., Chupin, M., Benali, H., and Colliot, O. (2010). Spatial and anatomical regularization of SVM for brain image analysis. *Neural Information Processing Systems (NIPS 2010).*

41 Mahmoudi, A., Takerkart, S., Regragui, F. et al. (2012). Multivoxel pattern analysis for fMRI data: a review. *Computational and Mathematical Methods in Medicine* 2012: 961257.

42 Orru, G., Pettersson-Yeo, W., Marquand, A.F. et al. (2012). Using support vector machine to identify imaging biomarkers of neurological and psychiatric disease: a critical review. *Neuroscience & Biobehavioral Reviews* 36 (4): 1140–1152.

43 Haxby, J.V., Connolly, A.C., and Guntupalli, J.S. (2014). Decoding neural representational spaces using multivariate pattern analysis. *Annual Review of Neuroscience* 37: 435–456.

44 Huang, S., Cai, N., Pacheco, P.P. et al. (2018). Applications of support vector machine (SVM) learning in cancer genomics. *Cancer Genomics & Proteomics* 15 (1): 41–51.

45 Ahmad, A.S., Hassan, M.Y., Abdullah, M.P. et al. (2014). A review on applications of ANN and SVM for building electrical energy consumption forecasting. *Renewable and Sustainable Energy Reviews* 33: 102–109.

46 Hinton, G.E. (2007). Boltzmann machine. *Scholarpedia* 2 (5): 1668.

47 Ackley, D.H., Hinton, G.E., and Sejnowski, T.J. (1985). A learning algorithm for Boltzmann machines. *Cognitive Science* 9 (1): 147–169.

48 Smolensky, P. (1986). Information Processing in Dynamical Systems: Foundations of Harmony Theory. *Technical Report*. March-September 1985 Colorado Univ at Boulder Dept of Computer Science, pp. 194–281.

49 Howard, A.G., Zhu, M., Chen, B. et al. (2017). MobileNets: Efficient convolutional neural networks for mobile vision applications. *arXiv preprint arXiv:1704.04861*.

50 Krizhevsky, A., Sutskever, I., and Hinton, G.E. (2012). ImageNet classification with deep convolutional neural networks. *Advances in Neural Information Processing Systems 25 (NIPS 2012)*, pp. 1097–1105.

51 Szegedy, C., Liu, W., Jia, Y. et al. (2015). Going deeper with convolutions. *Proceedings of the IEEE Conference on Computer Vision and Pattern Recognition*, pp. 1–9.

52 He, K., Zhang, X., Ren, S., and Sun, J. (2016). Deep residual learning for image recognition. *Proceedings of the IEEE Conference on Computer Vision and Pattern Recognition*, pp. 770–778.

53 Huang, G., Liu, Z., Van Der Maaten, L., and Weinberger, K.Q. (2017). Densely connected convolutional networks. *Proceedings of the IEEE Conference on Computer Vision and Pattern Recognition*, pp. 4700–4708.

54 Iandola, F.N., Han, S., Moskewicz, M.W. et al. (2016). SqueezeNet: AlexNet-level accuracy with 50x fewer parameters and <0.5 MB model size. *arXiv preprint arXiv:1602.07360*.

55 Krishnan, G., Ma, Y., and Cao, Y. (2020). Small-world-based structural pruning for efficient FPGA inference of deep neural networks. *2020 IEEE 15th International Conference on Solid-State & Integrated Circuit Technology (ICSICT)*, pp. 1–5. IEEE.

56 Zoph, B., Vasudevan, V., Shlens, J., and Le, Q.V. (2018). Learning transferable architectures for scalable image recognition. *Proceedings of the IEEE Conference on Computer Vision and Pattern Recognition*, pp. 8697–8710.

57 Wu, B., Dai, X., Zhang, P. et al. (2019). FBNet: Hardware-aware efficient convnet design via differentiable neural architecture search. *Proceedings of the IEEE/CVF Conference on Computer Vision and Pattern Recognition*, pp. 10734–10742.

58 Real, E., Aggarwal, A., Huang, Y., and Le, Q.V. (2019). Regularized evolution for image classifier architecture search. *Proceedings of the AAAI Conference on Artificial Intelligence*, Volume 33, pp. 4780–4789.

59 Liu, C., Zoph, B., Neumann, M. et al. (2018). Progressive neural architecture search. *Proceedings of the European Conference on Computer Vision (ECCV)*, pp. 19–34.

60 Zhou, D., Zhou, X., Zhang, W. et al. (2020). EcoNAS: Finding proxies for economical neural architecture search. *Proceedings of the IEEE/CVF Conference on Computer Vision and Pattern Recognition*, pp. 11396–11404.

61 Tan, M., Chen, B., Pang, R. et al. (2019). MnasNet: Platform-aware neural architecture search for mobile. *Proceedings of the IEEE/CVF Conference on Computer Vision and Pattern Recognition*, pp. 2820–2828.

62 Krishnan, G., Hazra, J., Liehr, M. et al. (2021). Design limits of in-memory computing: beyond the crossbar. *2021 5th IEEE Electron Devices Technology & Manufacturing Conference (EDTM)*, pp. 1–3. IEEE.

63 Krishnan, G., Sun, J., Hazra, J. et al. (2021). Robust RRAM-based in-memory computing in light of model stability. *2021 IEEE International Reliability Physics Symposium (IRPS)*, pp. 1–5. IEEE.

64 Charan, G., Mohanty, A., Du, X. et al. (2020). Accurate inference with inaccurate RRAM devices: a joint algorithm-design solution. *IEEE Journal on Exploratory Solid-State Computational Devices and Circuits* 6 (1): 27–35.

65 Charan, G., Hazra, J., Beckmann, K. et al. (2020). Accurate inference with inaccurate RRAM devices: statistical data, model transfer, and on-line adaptation. *2020 57th ACM/IEEE Design Automation Conference (DAC)*, pp. 1–6. IEEE.

66 Du, X., Krishnan, G., Mohanty, A. et al. (2019). Towards efficient neural networks on-a-chip: joint hardware-algorithm approaches. *2019 China Semiconductor Technology International Conference (CSTIC)*, pp. 1–5. IEEE.

67 Xian, Y., Lampert, C.H., Schiele, B., and Akata, Z. (2018). Zero-shot learning-a comprehensive evaluation of the good, the bad and the ugly. *IEEE Transactions on Pattern Analysis and Machine Intelligence* 41 (9): 2251–2265.

68 Koch, G., Zemel, R., and Salakhutdinov, R. (2015). Siamese neural networks for one-shot image recognition. *ICML Deep Learning Workshop*, Volume 2. Lille.

69 Kodirov, E., Xiang, T., Fu, Z., and Gong, S. (2015). Unsupervised domain adaptation for zero-shot learning. *Proceedings of the IEEE International Conference on Computer Vision*, pp. 2452–2460.

70 Li, X., Guo, Y., and Schuurmans, D. (2015). Semi-supervised zero-shot classification with label representation learning. *Proceedings of the IEEE International Conference on Computer Vision*, pp. 4211–4219.

71 Li, X. and Guo, Y. (2015). Max-margin zero-shot learning for multi-class classification. *Proceedings of the 18th International Conference on Artificial Intelligence and Statistics*, pp. 626–634. PMLR.

72 Zhang, Z. and Saligrama, V. (2016). Zero-shot recognition via structured prediction. In: *Computer Vision – European Conference on Computer Vision, Lecture Notes in Computer Science*, vol. 9911 (ed. B. Leibe, J. Matas, N. Sebe, and M. Welling), 533–548. Cham: Springer.

73 Scheirer, W.J., de Rezende Rocha, A., Sapkota, A., and Boult, T.E. (2013). Toward open set recognition. *IEEE Transactions on Pattern Analysis and Machine Intelligence* 35 (7): 1757–1772. https://doi.org/10.1109/TPAMI.2012.256.

74 Frome, A., Corrado, G.S., Shlens, J. et al. (2013). DeViSE: A deep visual-semantic embedding model. *Advances in neural information processing systems 26 (NIPS 2013).*

75 Zhang, H., Shang, X., Yang, W. et al. (2016). Online collaborative learning for open-vocabulary visual classifiers. *Proceedings of the IEEE Conference on Computer Vision and Pattern Recognition*, pp. 2809–2817.

76 Fe-Fei, L., Fergus, R., and Perona, P. (2003). A Bayesian approach to unsupervised one-shot learning of object categories. *Proceedings 9th IEEE International Conference on Computer Vision*, pp. 1134–1141. IEEE.

77 Fei-Fei, L., Fergus, R., and Perona, P. (2006). One-shot learning of object categories. *IEEE Transactions on Pattern Analysis and Machine Intelligence* 28 (4): 594–611.

78 Lake, B., Salakhutdinov, R., Gross, J., and Tenenbaum, J. (2011). One shot learning of simple visual concepts. *Proceedings of the Annual Meeting of the Cognitive Science Society*, Volume 33.

79 Lake, B., Salakhutdinov, R., and Tenenbaum, J. (2012). Concept learning as motor program induction: a large-scale empirical study. *Proceedings of the Annual Meeting of the Cognitive Science Society*, Volume 34.

80 Rosenblatt, F. (1961). Principles of Neurodynamics. Perceptrons and the Theory of Brain Mechanisms. *Technical Report*. Buffalo, NY: Cornell Aeronautical Lab Inc, pp. 245–248.
81 Marvin, M. and Seymour, P. (1969). *Perceptrons*. Cambridge, MA: MIT Press.
82 Rumelhart, D.E., Hinton, G.E., and Williams, R.J. (1986). Learning representations by back-propagating errors. *Nature* 323 (6088): 533–536.
83 Agarap, A.F. (2018). Deep learning using rectified linear units (RELU). *arXiv preprint arXiv:1803.08375*.
84 Nwankpa, C., Ijomah, W., Gachagan, A., and Marshall, S. (2018). Activation functions: comparison of trends in practice and research for deep learning. *arXiv preprint arXiv:1811.03378*.
85 Cho, S.-B. (1997). Neural-network classifiers for recognizing totally unconstrained handwritten numerals. *IEEE Transactions on Neural Networks* 8 (1): 43–53.
86 Kumar, M. and Yadav, N. (2011). Multilayer perceptrons and radial basis function neural network methods for the solution of differential equations: a survey. *Computers & Mathematics with Applications* 62 (10): 3796–3811.
87 Tolstikhin, I., Houlsby, N., Kolesnikov, A. et al. (2021). MLP-Mixer: An all-MLP architecture for vision. *35th Conference on Neural Information Processing Systems*.
88 Jordan, M.I. (1997). Serial order: a parallel distributed processing approach. *Advances in Psychology* 121: 471–495.
89 Lipton, Z.C., Berkowitz, J., and Elkan, C. (2015). A critical review of recurrent neural networks for sequence learning. *arXiv preprint arXiv:1506.00019*.
90 Werbos, P.J. (1990). Backpropagation through time: what it does and how to do it. *Proceedings of the IEEE* 78 (10): 1550–1560.
91 Williams, R.J. and Zipser, D. (1989). A learning algorithm for continually running fully recurrent neural networks. *Neural Computation* 1 (2): 270–280.
92 Sutskever, I., Martens, J., and Hinton, G.E. (2011). Generating text with recurrent neural networks. *ICML*.
93 Hochreiter, S. and Schmidhuber, J. (1997). Long short-term memory. *Neural Computation* 9 (8): 1735–1780.
94 Gers, F.A., Schmidhuber, J., and Cummins, F. (2000). Learning to forget: continual prediction with LSTM. *Neural Computation* 12 (10): 2451–2471.
95 Gers, F.A. and Schmidhuber, J. (2000). Recurrent nets that time and count. *Proceedings of the IEEE-INNS-ENNS International Joint Conference on Neural Networks. IJCNN 2000. Neural Computing: New Challenges and Perspectives for the New Millennium*, Volume 3, pp. 189–194. IEEE.
96 Zaremba, W. and Sutskever, I. (2014). Learning to execute. *arXiv preprint arXiv:1410.4615*.
97 Schuster, M. and Paliwal, K.K. (1997). Bidirectional recurrent neural networks. *IEEE Transactions on Signal Processing* 45 (11): 2673–2681.
98 Graves, A. and Schmidhuber, J. (2005). Framewise phoneme classification with bidirectional LSTM and other neural network architectures. *Neural Networks* 18 (5–6): 602–610.
99 Graves, A., Liwicki, M., Fernández, S. et al. (2008). A novel connectionist system for unconstrained handwriting recognition. *IEEE Transactions on Pattern Analysis and Machine Intelligence* 31 (5): 855–868.
100 Graves, A., Wayne, G., and Danihelka, I. (2014). Neural turing machines. *arXiv preprint arXiv:1410.5401*.
101 Socher, R., Manning, C.D., and Ng, A.Y. (2010). Learning continuous phrase representations and syntactic parsing with recursive neural networks. *Proceedings of the NIPS-2010 Deep Learning and Unsupervised Feature Learning Workshop*, Volume 2010, pp. 1–9.

102 Auli, M., Galley, M., Quirk, C., and Zweig, G. (2013). Joint language and translation modeling with recurrent neural networks. *Proceedings of EMNLP.*

103 Vinyals, O., Toshev, A., Bengio, S., and Erhan, D. (2015). Show and tell: a neural image caption generator. *Proceedings of the IEEE Conference on Computer Vision and Pattern Recognition*, pp. 3156–3164.

104 Karpathy, A. and Fei-Fei, L. (2015). Deep visual-semantic alignments for generating image descriptions. *Proceedings of the IEEE Conference on Computer Vision and Pattern Recognition*, pp. 3128–3137.

105 Mao, J., Xu, W., Yang, Y. et al. (2014). Deep captioning with multimodal recurrent neural networks (M-RNN). *arXiv preprint arXiv:1412.6632.*

106 Srivastava, N., Mansimov, E., and Salakhudinov, R. (2015). Unsupervised learning of video representations using LSTMS. *International Conference on Machine Learning*, pp. 843–852. PMLR.

107 Venugopalan, S., Rohrbach, M., Donahue, J. et al. (2015). Sequence to sequence-video to text. *Proceedings of the IEEE International Conference on Computer Vision*, pp. 4534–4542.

108 Roy, K., Jaiswal, A., and Panda, P. (2019). Towards spike-based machine intelligence with neuromorphic computing. *Nature* 575 (7784): 607–617.

109 Mostafa, H. (2017). Supervised learning based on temporal coding in spiking neural networks. *IEEE Transactions on Neural Networks and Learning Systems* 29 (7): 3227–3235.

110 Izhikevich, E.M. (2003). Simple model of spiking neurons. *IEEE Transactions on Neural Networks* 14 (6): 1569–1572.

111 Wu, Y., Deng, L., Li, G. et al. (2019). Direct training for spiking neural networks: faster, larger, better. *Proceedings of the AAAI Conference on Artificial Intelligence*, Volume 33, pp. 1311–1318.

112 Kim, Y. and Panda, P. (2021). Revisiting batch normalization for training low-latency deep spiking neural networks from scratch. *Frontiers in Neuroscience* 15: 773–954.

113 Neftci, E.O., Mostafa, H., and Zenke, F. (2019). Surrogate gradient learning in spiking neural networks. *IEEE Signal Processing Magazine* 36: 61–63.

114 Bi, G.-q. and Poo, M.-m. (1998). Synaptic modifications in cultured hippocampal neurons: dependence on spike timing, synaptic strength, and postsynaptic cell type. *Journal of Neuroscience* 18 (24): 10464–10472.

115 Hebb, D.O. (2005). *The Organization of Behavior: A Neuropsychological Theory*. Psychology Press.

116 Sengupta, A., Ye, Y., Wang, R. et al. (2019). Going deeper in spiking neural networks: VGG and residual architectures. *Frontiers in Neuroscience* 13: 95.

117 Diehl, P.U., Neil, D., Binas, J. et al. (2015). Fast-classifying, high-accuracy spiking deep networks through weight and threshold balancing. *2015 International Joint Conference on Neural Networks (IJCNN)*, pp. 1–8. IEEE.

118 Rueckauer, B., Lungu, I.-A., Hu, Y. et al. (2017). Conversion of continuous-valued deep networks to efficient event-driven networks for image classification. *Frontiers in Neuroscience* 11: 682.

119 Lee, J.H., Delbruck, T., and Pfeiffer, M. (2016). Training deep spiking neural networks using backpropagation. *Frontiers in Neuroscience* 10: 508.

120 Gori, M., Monfardini, G., and Scarselli, F. (2005). A new model for learning in graph domains. *Proceedings. 2005 IEEE International Joint Conference on Neural Networks, 2005*, Volume 2, pp. 729–734. IEEE.

121 Gallicchio, C. and Micheli, A. (2010). Graph echo state networks. *The 2010 International Joint Conference on Neural Networks (IJCNN)*, pp. 1–8. IEEE.

122 Liu, Z., Chen, C., Li, L. et al. (2019). GeniePath: Graph neural networks with adaptive receptive paths. *Proceedings of the AAAI Conference on Artificial Intelligence*, Volume 33, pp. 4424–4431.

123 Xu, K., Hu, W., Leskovec, J., and Jegelka, S. (2018). How powerful are graph neural networks? *arXiv preprint arXiv:1810.00826.*

124 Chiang, W.-L., Liu, X., Si, S. et al. (2019). Cluster-GCN: An efficient algorithm for training deep and large graph convolutional networks. *Proceedings of the 25th ACM SIGKDD International Conference on Knowledge Discovery & Data Mining*, pp. 257–266.

125 Simonovsky, M. and Komodakis, N. (2018). GraphVAE: Towards generation of small graphs using variational autoencoders. In: *Artificial Neural Networks and Machine Learning. International Conference on Artificial Neural Networks, Lecture Notes in Computer Science*, vol. 11139 (ed. V. Køurková, Y. Manolopoulos, B. Hammer et al.), 412–422. Cham: Springer.

126 Ma, T., Chen, J., and Xiao, C. (2018). Constrained generation of semantically valid graphs via regularizing variational autoencoders. *arXiv preprint arXiv:1809.02630.*

127 De Cao, N. and Kipf, T. (2018). MolGAN: An implicit generative model for small molecular graphs. *arXiv preprint arXiv:1805.11973.*

128 Li, Y., Tarlow, D., Brockschmidt, M., and Zemel, R. (2015). Gated graph sequence neural networks. *arXiv preprint arXiv:1511.05493.*

129 Hamilton, W., Ying, Z., and Leskovec, J. (2017). Inductive representation learning on large graphs. *Advances in Neural Information Processing Systems 30 (NIPS 2017).*

130 Veličković, P., Cucurull, G., Casanova, A. et al. (2017). Graph attention networks. *arXiv preprint arXiv:1710.10903.*

131 Zhang, J., Shi, X., Xie, J. et al. (2018). GAAN: Gated attention networks for learning on large and spatiotemporal graphs. *arXiv preprint arXiv:1803.07294.*

132 Cao, S., Lu, W., and Xu, Q. (2016). Deep neural networks for learning graph representations. *Proceedings of the AAAI Conference on Artificial Intelligence*, Volume 30.

133 Wang, D., Cui, P., and Zhu, W. (2016). Structural deep network embedding. *Proceedings of the 22nd ACM SIGKDD International Conference on Knowledge Discovery and Data Mining*, pp. 1225–1234.

134 Xu, D., Zhu, Y., Choy, C.B., and Fei-Fei, L. (2017). Scene graph generation by iterative message passing. *Proceedings of the IEEE Conference on Computer Vision and Pattern Recognition*, pp. 5410–5419.

135 Wang, Y., Sun, Y., Liu, Z. et al. (2019). Dynamic graph cnn for learning on point clouds. *ACM Transactions On Graphics (TOG)* 38 (5): 1–12.

136 Qi, S., Wang, W., Jia, B. et al. (2018). Learning human-object interactions by graph parsing neural networks. *Proceedings of the European Conference on Computer Vision (ECCV)*, pp. 401–417.

137 Li, Z., Du, X., and Cao, Y. (2020). GAR: Graph assisted reasoning for object detection. *Proceedings of the IEEE/CVF Winter Conference on Applications of Computer Vision*, pp. 1295–1304.

138 Marcheggiani, D. and Titov, I. (2017). Encoding sentences with graph convolutional networks for semantic role labeling. *arXiv preprint arXiv:1703.04826.*

139 Bastings, J., Titov, I., Aziz, W. et al. (2017). Graph convolutional encoders for syntax-aware neural machine translation. *arXiv preprint arXiv:1704.04675.*

140 Marcheggiani, D., Bastings, J., and Titov, I. (2018). Exploiting semantics in neural machine translation with graph convolutional networks. *arXiv preprint arXiv:1804.08313.*

141 Berg, R., Kipf, T.N., and Welling, M. (2017). Graph convolutional matrix completion. *arXiv preprint arXiv:1706.02263.*

142 Ying, R., He, R., Chen, K. et al. (2018). Graph convolutional neural networks for web-scale recommender systems. *Proceedings of the 24th ACM SIGKDD International Conference on Knowledge Discovery & Data Mining*, pp. 974–983.

143 Monti, F., Bronstein, M.M., and Bresson, X. (2017). Geometric matrix completion with recurrent multi-graph neural networks. *arXiv preprint arXiv:1704.06803*.

144 Santoro, A., Bartunov, S., Botvinick, M. et al. (2016). Meta-learning with memory-augmented neural networks. *International Conference on Machine Learning*, pp. 1842–1850. PMLR.

145 Santoro, A., Bartunov, S., Botvinick, M. et al. (2016). One-shot learning with memory-augmented neural networks. *arXiv preprint arXiv:1605.06065*.

146 Al-Sahaf, H., Bi, Y., Chen, Q. et al. (2019). A survey on evolutionary machine learning. *Journal of the Royal Society of New Zealand* 49 (2): 205–228.

147 Lukoševičius, M. and Jaeger, H. (2009). Reservoir computing approaches to recurrent neural network training. *Computer Science Review* 3 (3): 127–149.

148 Baraldi, A. and Alpaydin, E. (2002). Constructive feedforward ART clustering networks. II. *IEEE Transactions on Neural Networks* 13 (3): 662–677.

149 Cherkassky, V. and Mulier, F.M. (2007). *Learning from Data: Concepts, Theory, and Methods*. Wiley.

150 Everitt, B., Landau, S., and Leese, M. (2001). *Cluster Analysis*, 4e. London: Arnold.

151 Hathaway, R.J., Bezdek, J.C., and Hu, Y. (2000). Generalized fuzzy c-means clustering strategies using L/sub p/norm distances. *IEEE Transactions on Fuzzy Systems* 8 (5): 576–582.

152 MacQueen, J. (1967). Some methods for classification and analysis of multivariate observations. *Proceedings of the 5th Berkeley Symposium on Mathematical Statistics and Probability*, Volume 1, pp. 281–297. Oakland, CA, USA.

153 Mao, J. and Jain, A.K. (1996). A self-organizing network for hyperellipsoidal clustering (HEC). *IEEE Transactions on Neural Networks* 7 (1): 16–29.

154 Li, B. and Han, L. (2013). Distance weighted cosine similarity measure for text classification. *International Conference on Intelligent Data Engineering and Automated Learning*, pp. 611–618. Springer.

155 Xu, R. and Wunsch, D. (2005). Survey of clustering algorithms. *IEEE Transactions on Neural Networks* 16 (3): 645–678.

156 Roweis, S.T. and Saul, L.K. (2000). Nonlinear dimensionality reduction by locally linear embedding. *Science* 290 (5500): 2323–2326.

157 Weinberger, K.Q., Sha, F., and Saul, L.K. (2004). Learning a kernel matrix for nonlinear dimensionality reduction. *Proceedings of the 21st International Conference on Machine Learning*, p. 106.

158 Rifai, S., Mesnil, G., Vincent, P. et al. (2011). Higher order contractive auto-encoder. In: *Machine Learning and Knowledge Discovery in Databases. Joint European Conference on Machine Learning and Knowledge Discovery in Databases, Lecture Notes in Computer Science*, vol. 6912 (ed. D. Gunopulos, T. Hofmann, D. Malerba, and M. Vazirgiannis), 645–660. Berlin, Heidelberg: Springer-Verlag.

159 Salakhutdinov, R. and Hinton, G. (2009). Semantic hashing. *International Journal of Approximate Reasoning* 50 (7): 969–978.

160 Heess, N., Hunt, J.J., Lillicrap, T.P., and Silver, D. (2015). Memory-based control with recurrent neural networks. *arXiv preprint arXiv:1512.04455*.

161 Levine, S., Finn, C., Darrell, T., and Abbeel, P. (2016). End-to-end training of deep visuomotor policies. *The Journal of Machine Learning Research* 17 (1): 1334–1373.

162 Lillicrap, T.P., Hunt, J.J., Pritzel, A. et al. (2015). Continuous control with deep reinforcement learning. *arXiv preprint arXiv:1509.02971.*

163 Kohl, N. and Stone, P. (2004). Policy gradient reinforcement learning for fast quadrupedal locomotion. *IEEE International Conference on Robotics and Automation, 2004. Proceedings. ICRA'04. 2004*, Volume 3, pp. 2619–2624. IEEE.

164 Ng, A.Y., Coates, A., Diel, M. et al. (2006). Autonomous inverted helicopter flight via reinforcement learning. In: *Experimental Robotics IX*, 363–372. Springer.

165 Arulkumaran, K., Deisenroth, M.P., Brundage, M., and Bharath, A.A. (2017). Deep reinforcement learning: a brief survey. *IEEE Signal Processing Magazine* 34 (6): 26–38.

166 Mnih, V., Kavukcuoglu, K., Silver, D. et al. (2015). Human-level control through deep reinforcement learning. *Nature* 518 (7540): 529–533.

167 Schulman, J., Moritz, P., Levine, S. et al. (2015). High-dimensional continuous control using generalized advantage estimation. *arXiv preprint arXiv:1506.02438.*

168 Oh, J., Guo, X., Lee, H. et al. (2015). Action-conditional video prediction using deep networks in atari games. *arXiv preprint arXiv:1507.08750.*

169 Vezhnevets, A.S., Osindero, S., Schaul, T. et al. (2017). Feudal networks for hierarchical reinforcement learning. *International Conference on Machine Learning*, pp. 3540–3549. PMLR.

170 Sukhbaatar, Szlam, A., and Fergus, R. (2016). Learning multiagent communication with backpropagation. *Advances in Neural Information Processing Systems 29 (NIPS 2016)*, pp. 2244–2252.

171 Yannakakis, G.N. and Togelius, J. (2018). *Artificial Intelligence and Games*. Springer.

172 Zhu, Y., Mottaghi, R., Kolve, E. et al. (2017). Target-driven visual navigation in indoor scenes using deep reinforcement learning. *2017 IEEE International Conference on Robotics and Automation (ICRA)*, pp. 3357–3364. IEEE.

173 Guu, K., Pasupat, P., Liu, E.Z., and Liang, P. (2017). From language to programs: bridging reinforcement learning and maximum marginal likelihood. *arXiv preprint arXiv:1704.07926.*

174 Liu, F., Li, S., Zhang, L. et al. (2017). 3DCNN-DQN-RNN: A deep reinforcement learning framework for semantic parsing of large-scale 3D point clouds. *Proceedings of the IEEE International Conference on Computer Vision*, pp. 5678–5687.

175 Oja, E. and Kaski, S. (1999). *Kohonen Maps*. Elsevier.

176 Chen, C.-L., Chen, W.-C., and Chang, F.-Y. (1993). Hybrid learning algorithm for Gaussian potential function networks. *IEE Proceedings D-Control Theory and Applications*, Volume 140, pp. 442–448. IET.

177 Li, Z. and Hoiem, D. (2018). Learning without forgetting. *IEEE Transactions on Pattern Analysis and Machine Intelligence* 40 (12): 2935–2947. https://doi.org/10.1109/TPAMI.2017.2773081.

178 Rebuffi, S.-A., Kolesnikov, A., Sperl, G., and Lampert, C.H. (2016). iCaRL: incremental classifier and representation learning. *arXiv:1611.07725 [cs, stat].* http://arxiv.org/abs/1611.07725. arXiv: 1611.07725.

179 Zenke, F., Poole, B., and Ganguli, S. (2017). Continual learning through synaptic intelligence. *arXiv:1703.04200 [cs, q-bio, stat].* http://arxiv.org/abs/1703.04200. arXiv: 1703.04200.

180 Rusu, A.A., Rabinowitz, N.C., Desjardins, G. et al. (2016). Progressive neural networks. *arXiv:1606.04671 [cs].* http://arxiv.org/abs/1606.04671. arXiv: 1606.04671.

181 Yoon, J., Yang, E., Lee, J., and Hwang, S.J. (2017). Lifelong learning with dynamically expandable networks. *arXiv:1708.01547 [cs]*, http://arxiv.org/abs/1708.01547. arXiv: 1708.01547.

182 Mallya, A. and Lazebnik, S. (2018). PackNet: Adding multiple tasks to a single network by iterative pruning. *2018 IEEE/CVF Conference on Computer Vision and Pattern Recognition*, pp. 7765–7773. Salt Lake City, UT: IEEE. ISBN 978-1-5386-6420-9. https://doi.org/10.1109/CVPR.2018.00810. https://ieeexplore.ieee.org/document/8578908/.

183 Fernando, C., Banarse, D., Blundell, C. et al. (2017). PathNet: Evolution channels gradient descent in super neural networks. *CoRR*, abs/1701.08734. http://arxiv.org/abs/1701.08734.

184 Zhang, B. and Zhang, L. (1992). *Theory and Applications of Problem Solving*. Elsevier Science Inc.

185 Tari, L. (2013). *Knowledge Inference*. New York: Springer. ISBN 978-1-4419-9863-7. https://doi.org/10.1007/978-1-4419-9863-7_166.

186 Miller, G.A. (1995). WordNnet: A lexical database for English. *Communications of the ACM* 38 (11): 39–41.

187 Vrandecic, D. and Krtoetzsch, M. (2014). Wikidata: a free collaborative knowledgebase. *Communications of the ACM* 57 (10): 78–85.

188 Mitchell, T. and Fredkin, E. (2012). Never-ending language learning. *IEEE International Conference on Big Data.*

189 Wang, X., Wang, D., Xu, C. et al. (2018). Explainable reasoning over knowledge graphs for recommendation. *Proceedings of the AAAI Conference on Artificial Intelligence* 33 (01): AAAI-19, IAAI-19, EAAI-20.

190 Zhuang, F., Qi, Z., Duan, K. et al. (2020). A comprehensive survey on transfer learning. *Proceedings of the IEEE* 109 (1): 43–76.

191 Pan, S.J. and Yang, Q. (2009). A survey on transfer learning. *IEEE Transactions on Knowledge and Data Engineering* 22 (10): 1345–1359.

192 Borgwardt, K.M., Gretton, A., Rasch, M.J. et al. (2006). Integrating structured biological data by kernel maximum mean discrepancy. *Bioinformatics* 22 (14): e49–e57.

193 Pan, S.J., Tsang, I.W., Kwok, J.T., and Yang, Q. (2010). Domain adaptation via transfer component analysis. *IEEE Transactions on Neural Networks* 22 (2): 199–210.

194 Ghifary, M., Kleijn, W.B., and Zhang, M. (2014). Domain adaptive neural networks for object recognition. In: *Pacific Rim International Conference on Artificial Intelligence: Trends in Artificial Intelligence. PRICAI 2014, Lecture Notes in Computer Science*, vol. 8862 (ed. D.N. Pham and S.B. Park), 898–904. Cham: Springer.

195 Kullback, S. and Leibler, R.A. (1951). On information and sufficiency. *The Annals of Mathematical Statistics* 22 (1): 79–86.

196 Dagan, I., Lee, L., and Pereira, F. (1997). Similarity-based methods for word sense disambiguation. *arXiv preprint cmp-lg/9708010.*

197 Bregman, L.M. (1967). The relaxation method of finding the common point of convex sets and its application to the solution of problems in convex programming. *USSR Computational Mathematics and Mathematical Physics* 7 (3): 200–217.

198 Sugiyama, M., Suzuki, T., Nakajima, S. et al. (2008). Direct importance estimation for covariate shift adaptation. *Annals of the Institute of Statistical Mathematics* 60 (4): 699–746.

199 Huang, J., Gretton, A., Borgwardt, K. et al. (2006). Correcting sample selection bias by unlabeled data. *Advances in Neural Information Processing Systems 19 (NIPS 2006)*, pp. 601–608.

200 Bousmalis, K., Trigeorgis, G., Silberman, N. et al. (2016). Domain separation networks. *NIPS*, pp. 343–351.

201 Tzeng, E., Hoffman, J., Saenko, K., and Darrell, T. (2017). Adversarial discriminative domain adaptation. *CVPR*, pp. 7167–7176.

202 Marcelino, P. (2018). Transfer learning from pre-trained models. *Towards Data Science.*

203 Karpatne, A., Atluri, G., Faghmous, J.H. et al. (2017). Theory-guided data science: a new paradigm for scientific discovery from data. *IEEE Transactions on Knowledge and Data Engineering* 29 (10): 2318–2331. https://doi.org/10.1109/TKDE.2017.2720168.

204 Karpatne, A., Watkins, W., Read, J.S., and Kumar, V. (2017). Physics-guided neural networks (PGNN): an application in lake temperature modeling. *CoRR*, abs/1710.11431. http://arxiv.org/abs/1710.11431.

205 Jia, X., Willard, J., Karpatne, A. et al. (2019). Physics guided RNNs for modeling dynamical systems: a case study in simulating lake temperature profiles. *ArXiv*, abs/1810.13075.

206 Zhu, Y., Zabaras, N., Koutsourelakis, P.-S., and Perdikaris, P. (2019). Physics-constrained deep learning for high-dimensional surrogate modeling and uncertainty quantification without labeled data. *Journal of Computational Physics* 394: 56–81. https://doi.org/10.1016/j.jcp.2019.05.024.

207 Loiseau, J.-C. and Brunton, S.L. (2018). Constrained sparse Galerkin regression. *Journal of Fluid Mechanics* 838: 42–67. https://doi.org/10.1017/jfm.2017.823.

208 Zhang, L., Han, J., Wang, H. et al. (2018). Deep potential molecular dynamics: a scalable model with the accuracy of quantum mechanics. *Physical Review Letters* 120: 143001. https://doi.org/10.1103/PhysRevLett.120.143001.

209 Bousmalis, K., Irpan, A., Wohlhart, P. et al. (2018). Using simulation and domain adaptation to improve efficiency of deep robotic grasping. *2018 IEEE International Conference on Robotics and Automation (ICRA)*, pp. 4243–4250.

210 Shah, S., Dey, D., Lovett, C., and Kapoor, A. (2017). AirSim: High-fidelity visual and physical simulation for autonomous vehicles. *Field and Service Robotics*. https://arxiv.org/abs/1705.05065.

211 Yan, W., Hu, S., Yang, Y. et al. (2011). Bayesian migration of Gaussian process regression for rapid process modeling and optimization. *Chemical Engineering Journal* 166: 1095–1103.

212 Muralidhar, N., Bu, J., Cao, Z. et al. (2020). PhyNet: Physics guided neural networks for particle drag force prediction in assembly. *SDM*.

213 Sun, J., Niu, Z.W., Innanen, K.A. et al. (2020). A theory-guided deep-learning formulation and optimization of seismic waveform inversion. *Geophysics* 85 (2): R87–R99.

214 Ling, J., Kurzawski, A., and Templeton, J. (2016). Reynolds averaged turbulence modelling using deep neural networks with embedded invariance. *Journal of Fluid Mechanics* 807: 155–166. https://doi.org/10.1017/jfm.2016.615.

215 Forssell, U. and Lindskog, P. (1997). Combining semi-physical and neural network modeling: an example of its usefulness. *IFAC Proceedings Volumes* 30 (11): 767–770.

216 Thompson, M.L. and Kramer, M.A. (1994). Modeling chemical processes using prior knowledge and neural networks. *Aiche Journal* 40: 1328–1340.

217 Parish, E.J. and Duraisamy, K. (2016). A paradigm for data-driven predictive modeling using field inversion and machine learning. *Journal of Computational Physics* 305: 758–774. https://doi.org/10.1016/j.jcp.2015.11.012.

218 Paolucci, R., Gatti, F., Infantino, M. et al. (2018). Broadband ground motions from 3D physics-based numerical simulations using artificial neural networks. *Bulletin of the Seismological Society of America* 108: 1272–1286.

219 Sharifi, E., Saghafian, B., and Steinacker, R. (2019). Downscaling satellite precipitation estimates with multiple linear regression, artificial neural networks, and spline interpolation techniques. *Journal of Geophysical Research Atmospheres* 124. https://doi.org/10.1029/2018JD028795.

220 Li, Z., Kovachki, N., Azizzadenesheli, K. et al. (2021). Fourier neural operator for parametric partial differential equations.

221 Sharma, Y., Zhang, H.H., and Xin, H. (2020). Machine learning techniques for optimizing design of double T-shaped monopole antenna. *IEEE Transactions on Antennas and Propagation* 68 (7): 5658–5663. https://doi.org/10.1109/TAP.2020.2966051.

222 Farimani, A.B., Gomes, J., and Pande, V.S. (2017). Deep learning the physics of transport phenomena. *ArXiv*, abs/1709.02432.

223 Yang, Z., Wu, J.-L., and Xiao, H. (2020). Enforcing deterministic constraints on generative adversarial networks for emulating physical systems.

224 Wu, J.-L., Kashinath, K., Albert, A. et al. (2020). Enforcing statistical constraints in generative adversarial networks for modeling chaotic dynamical systems. *Journal of Computational Physics* 406: 109–209. https://doi.org/10.1016/j.jcp.2019.109209.

225 Jouppi, N.P., Young, C., Patil, N. et al. In-datacenter performance analysis of a tensor processing unit. *Proceedings of the 44th Annual International Symposium on Computer Architecture*, pp. 1–12.

226 Li, T., Sahu, A.K., Zaheer, M. et al. (2020). Federated optimization in heterogeneous networks. *Proceedings of Machine Learning and Systems* 2: 429–450.

227 Karimireddy, S.P., Kale, S., Mohri, M. et al. (2020). SCAFFOLD: stochastic controlled averaging for federated learning. *International Conference on Machine Learning*, pp. 5132–5143. PMLR.

228 Reddi, S., Charles, Z., Zaheer, M. et al. (2020). Adaptive federated optimization. *arXiv preprint arXiv:2003.00295.*

229 Nishio, T. and Yonetani, R. (2019). Client selection for federated learning with heterogeneous resources in mobile edge. *ICC 2019-2019 IEEE International Conference on Communications (ICC)*, pp. 1–7. IEEE.

230 Albasyoni, A., Safaryan, M., Condat, L., and Richtárik, P. (2020). Optimal gradient compression for distributed and federated learning. *arXiv preprint arXiv:2010.03246.*

231 Zinkevich, M., Weimer, M., Smola, A.J., and Li, L. (2010). Parallelized stochastic gradient descent. *NIPS*, Volume 4, p. 4. Citeseer.

232 Geiping, J., Bauermeister, H., Dröge, H., and Moeller, M. (2020). Inverting gradients–how easy is it to break privacy in federated learning? *arXiv preprint arXiv:2003.14053.*

233 Mu noz-González, L., Biggio, B., Demontis, A. et al. (2017). Towards poisoning of deep learning algorithms with back-gradient optimization. *Proceedings of the 10th ACM Workshop on Artificial Intelligence and Security*, pp. 27–38.

234 Sun, Z., Kairouz, P., Suresh, A.T., and McMahan, H.B. (2019). Can you really backdoor federated learning? *arXiv preprint arXiv:1911.07963.*

235 Agarwal, A., Chapelle, O., Dudík, M., and Langford, J. (2014). A reliable effective terascale linear learning system. *The Journal of Machine Learning Research* 15 (1): 1111–1133.

236 Yi, S., Li, C., and Li, Q. (2015). A survey of fog computing: concepts, applications and issues. *Proceedings of the 2015 Workshop on Mobile Big Data*, pp. 37–42.

237 Bonomi, F., Milito, R., Zhu, J., and Addepalli, S. (2012). Fog computing and its role in the Internet of Things. *Proceedings of the First Edition of the MCC Workshop on Mobile Cloud Computing*, pp. 13–16.

238 Eykholt, K., Evtimov, I., Fernandes, E. et al. (2018). Robust physical world attacks on deep learning visual classification. *Proceedings of the IEEE Conference on Computer Vision and Pattern Recognition*, pp. 1625–1634.

239 Carlini, N., Mishra, P., Vaidya, T. et al. (2016). Hidden voice commands. *25th USENIX Security Symposium (USENIX Security 16)*, pp. 513–530, Austin, TX: USENIX Association. ISBN 978-1-931971-32-4. https://www.usenix.org/conference/usenixsecurity16/technical-sessions/presentation/carlini (accessed 29 March 2023).
240 Frenay, B. and Verleysen, M. (2014). Classification in the presence of label noise: a survey. *IEEE Transactions on Neural Networks and Learning Systems* 25 (5): 845–869. https://doi.org/10.1109/TNNLS.2013.2292894.
241 Song, H., Kim, M., Park, D. et al. (2021). Learning from noisy labels with deep neural networks: a survey. *IEEE Transactions on Neural Networks and Learning Systems*, pp. 1–19.
242 Chen, X. and Gupta, A. (2015). Webly supervised learning of convolutional networks. *Proceedings of the IEEE International Conference on Computer Vision (ICCV).*
243 Xiao, T., Xia, T., Yang, Y. et al. (2015). Learning from massive noisy labeled data for image classification. *Proceedings of the IEEE Conference on Computer Vision and Pattern Recognition*, pp. 2691–2699.
244 Hendrycks, D., Lee, K., and Mazeika, M. (2019). Using pre-training can improve model robustness and uncertainty. *International Conference on Machine Learning*, pp. 2712–2721. PMLR.
245 Jenni, S. and Favaro, P. (2018). Deep bilevel learning. *Proceedings of the European Conference on Computer Vision (ECCV)*, pp. 618–633.
246 Xia, X., Liu, T., Gong, C. et al. (2020). Robust early-learning: hindering the memorization of noisy labels. *International Conference on Learning Representations.*
247 Goodfellow, I.J., Shlens, J., and Szegedy, C. (2014). Explaining and harnessing adversarial examples. *arXiv preprint arXiv:1412.6572.*
248 Pereyra, G., Tucker, G., Chorowski, J. et al. (2017). Regularizing neural networks by penalizing confident output distributions. *arXiv preprint arXiv:1701.06548.*
249 Lukasik, M., Bhojanapalli, S., Menon, A., and Kumar, S. (2020). Does label smoothing mitigate label noise? *International Conference on Machine Learning*, pp. 6448–6458. PMLR.
250 Zhang, H., Cisse, M., Dauphin, Y.N., and Lopez-Paz, D. (2017). mixup: Beyond empirical risk minimization. *arXiv preprint arXiv:1710.09412.*
251 Szegedy, C., Zaremba, W., Sutskever, I. et al. (2013). Intriguing properties of neural networks. *arXiv preprint arXiv:1312.6199.*
252 Xie, C., Wang, J., Zhang, Z. et al. (2017). Mitigating adversarial effects through randomization. *arXiv preprint arXiv:1711.01991.*
253 Guo, C., Rana, M., Cisse, M., and Van Der Maaten, L. (2017). Countering adversarial images using input transformations. *arXiv preprint arXiv:1711.00117.*
254 Liu, X., Cheng, M., Zhang, H., and Hsieh, C.-J. (2018). Towards robust neural networks via random self-ensemble. *Proceedings of the European Conference on Computer Vision (ECCV)*, pp. 369–385.
255 Lecuyer, M., Atlidakis, V., Geambasu, R. et al. (2019). Certified robustness to adversarial examples with differential privacy. *2019 IEEE Symposium on Security and Privacy (SP)*, pp. 656–672. IEEE.
256 Li, B., Chen, C., Wang, W., and Carin, L. (2019). Certified adversarial robustness with additive noise. *Advances in Neural Information Processing Systems 32 (NIPS 2019).*
257 Dhillon, G.S., Azizzadenesheli, K., Lipton, Z.C. et al. (2018). Stochastic activation pruning for robust adversarial defense. *arXiv preprint arXiv:1803.01442.*
258 Xu, W., Evans, D., and Qi, Y. (2017). Feature squeezing: detecting adversarial examples in deep neural networks. *arXiv preprint arXiv:1704.01155.*

259 Xu, W., Evans, D., and Qi, Y. (2017). Feature squeezing mitigates and detects carlini/wagner adversarial examples. *arXiv preprint arXiv:1705.10686.*

260 Samangouei, P., Kabkab, M., and Chellappa, R. (2018). Defense-GAN: Protecting classifiers against adversarial attacks using generative models. *arXiv preprint arXiv:1805.06605.*

261 Shen, S., Jin, G., Gao, K., and Zhang, Y. (2017). APE-GAN: Adversarial perturbation elimination with GAN. *arXiv preprint arXiv:1707.05474.*

262 Meng, D. and Chen, H.. MagNet: A two-pronged defense against adversarial examples. *Proceedings of the 2017 ACM SIGSAC Conference on Computer and Communications Security*, pp. 135–147.

263 Zhang, Y., Tiňo, P., Leonardis, A., and Tang, K. (2021). A survey on neural network interpretability. *IEEE Transactions on Emerging Topics in Computational Intelligence* 5 (5): 726–742.

264 Mahendran, A. and Vedaldi, A. (2015). Understanding deep image representations by inverting them. *Proceedings of the IEEE Conference on Computer Vision and Pattern Recognition*, pp. 5188–5196.

265 Vaswani, A., Shazeer, N., Parmar, N. et al. (2017). Attention is all you need. *Advances in Neural Information Processing Systems 30 (NIPS 2017)*, pp. 5998–6008.

266 Rumelhart, D.E., Hinton, G.E., and Williams, R.J. (1985). Learning Internal Representations by Error Propagation. *Technical Report March-September 1985.* California University San Diego La Jolla Inst for Cognitive Science.

267 Sak, H., Senior, A.W., and Beaufays, F. (2014). Long short-term memory recurrent neural network architectures for large scale acoustic modeling.

268 Devlin, J., Chang, M.-W., Lee, K., and Toutanova, K. (2018). BERT: Pre-training of deep bidirectional transformers for language understanding. *arXiv preprint arXiv:1810.04805.*

269 Merity, S., Xiong, C., Bradbury, J., and Socher, R. (2016). Pointer sentinel mixture models.

270 Bird, S., Klein, E., and Loper, E. (2009). *Natural Language Processing with Python: Analyzing Text with the Natural Language Toolkit*. O'Reilly Media, Inc.

271 Dosovitskiy, A., Beyer, L., Kolesnikov, A. et al. (2020). An image is worth 16x16 words: transformers for image recognition at scale. *arXiv preprint arXiv:2010.11929.*

272 Mythic (2021). M1076 analog matrix processor. https://www.mythic-ai.com/product/m1076-analog-matrix-processor/.

273 IBM (2021). Analog AI chip. https://analog-ai.mybluemix.net/.

274 Shafiee, A., Nag, A., Muralimanohar, N. et al. (2016). ISAAC: A convolutional neural network accelerator with in-situ analog arithmetic in crossbars. *2016 ACM/IEEE 43rd Annual International Symposium on Computer Architecture (ISCA)*, pp. 14–26.

275 Chi, P., Li, S., Xu, C. et al. (2016). PRIME: A novel processing-in-memory architecture for neural network computation in ReRAM-based main memory. *2016 ACM/IEEE 43rd Annual International Symposium on Computer Architecture (ISCA)*, pp. 27–39.

276 Krishnan, G., Mandal, S.K., Chakrabarti, C. et al. (2020). Interconnect-aware area and energy optimization for in-memory acceleration of DNNs. *IEEE Design & Test* 37 (6): 79–87.

277 Mandal, S.K., Krishnan, G., Chakrabarti, C. et al. (2020). A latency-optimized reconfigurable NoC for in-memory acceleration of DNNs. *IEEE Journal on Emerging and Selected Topics in Circuits and Systems* 10 (3): 362–375.

278 Krishnan, G., Mandal, S.K., Chakrabarti, C. et al. (2021). System-level benchmarking of chiplet-based IMC architectures for deep neural network acceleration. *2021 IEEE 14th International Conference on ASIC (ASICON)*, pp. 1–4. IEEE.

279 Krishnan, G., Mandal, S.K., Pannala, M. et al. (2021). SIAM: Chiplet-based scalable in-memory acceleration with mesh for deep neural networks. *ACM Transactions on Embedded Computing Systems (TECS)* 20 (5s): 1–24.

280 Ma, Y., Krishnan, G., Cao, Y. et al. (2021). SWIFT: Small-world-based structural pruning to accelerate DNN inference on FPGA. *The 2021 ACM/SIGDA International Symposium on Field-Programmable Gate Arrays*, p. 148.

Section II

Advancing Electromagnetic Inverse Design with Machine Learning

3

Breaking the Curse of Dimensionality in Electromagnetics Design Through Optimization Empowered by Machine Learning

N. Anselmi[1], *G. Oliveri*[1], *L. Poli*[1], *A. Polo*[1], *P. Rocca*[1,2], *M. Salucci*[1], *and A. Massa*[1,3,4,5,6]

[1] *ELEDIA Research Center (ELEDIA@UniTN – University of Trento), DICAM – Department of Civil, Environmental, and Mechanical Engineering, Trento, Italy*
[2] *ELEDIA Research Center (ELEDIA@XIDIAN – Xidian University), Xi'an, Shaanxi Province, China*
[3] *ELEDIA Research Center (ELEDIA@TSINGHUA – Tsinghua University), Haidian, Beijing, China*
[4] *ELEDIA Research Center (ELEDIA@UESTC – UESTC), School of Electronic Science and Engineering University of Electronic Science and Technology of China, Chengdu, China*
[5] *School of Electrical Engineering Tel Aviv University, Tel Aviv, Israel*
[6] *ELEDIA Research Center (ELEDIA@UIC – University of Illinois Chicago), Chicago, IL, USA*

3.1 Introduction

Traditionally, full wave solvers (*FWS*) have been considered as very accurate and general-purpose numerical tools for solving Maxwell's equations in time and/or frequency domain [1–4]. For this reason, they have been often employed for the reliable modeling and analysis of complex electromagnetic (*EM*) devices and systems. However, they typically require high computational resources and their simulation time exponentially grows with the dimension of the discretization mesh [4]. Therefore, their exploitation in *EM* design problems is often regarded as the bottleneck of the overall synthesis procedure, especially when addressing the iterative minimization of a properly defined cost function quantifying the mismatch between simulated and target performance [5, 6]. A more efficient alternative is represented by analytic/semi-analytic solvers, which exhibit a lower computational cost. Nevertheless, they often involve approximations to simplify the involved equations and generally they neglect complex *EM* phenomena and interactions (e.g. mutual coupling) which could cause unacceptable deviations of the real performance of the designed devices. Furthermore, they can be only defined for some classes of devices/systems with rather canonical/simple layouts, their exploitation being limited to the solution of specific synthesis problems [7].

Owing to such considerations, robust and reliable *EM* design methods should jointly leverage on (*i*) the convergence toward the global minimum of the cost function without being trapped into false solutions/local minima and (*ii*) the accurate (*FWS*-based) modeling of the *EM* behavior of each trial solution, so that the final design reliably fits all user-defined requirements. Concerning the goal (*i*), "trial-and-error" or "parametric analysis" procedures [8] where each component of the system undergoes an independent design process are generally sub-optimal and they cannot be employed when there are several interdependent descriptors because of the extremely large number of necessary parametric evaluations. Alternatively, Evolutionary Algorithms (*EA*s) have been widely employed in *EM* design problems because of their "hill-climbing" features overcoming local minima issues without requiring, at the same time, the analytic knowledge of the cost function and of its derivative [9–11]. However, the "price to pay" for an exhaustive exploration of the solution

Advances in Electromagnetics Empowered by Artificial Intelligence and Deep Learning, First Edition.
Edited by Sawyer D. Campbell and Douglas H. Werner.

space with respect to deterministic/local search strategies is the exploitation of nature-inspired collaboration/competition mechanisms among a finite set of $P > 1$ agents, rapidly determining the unaffordable computational cost of the overall synthesis procedure [9]. On the one hand, an effective recipe to counteract such a limitation relies on the improvement of the *EA*s convergence rate either by identifying a minimum-cardinality set of K design variables (allowing to keep as low as possible P, since $P \propto K$), or by exploiting *a priori* information to form an initial population of trial solutions already belonging to the so-called "attraction basin" of the global minimum to reduce the number of iterations, I. On the other hand, the computational cost of *EA*s can be significantly alleviated by reducing the time required to compute the cost function associated to each trial guess [5].

Within this context, the System-by-Design (*SbD*) is an emerging paradigm aimed at enabling an effective and computationally affordable exploitation of *EA*s for solving complex *EM* design problems [5, 6, 12]. In other words, the *SbD* is defined as "*how to deal with complexity*" and its goal is the "*task-oriented design, definition, and integration of system components to yield EM devices with user-desired performance having the minimum costs, the maximum scalability, and suitable reconfigurability properties*" [5]. Toward this end, the *SbD* performs a task-oriented design without focusing on a local matching objective, treating instead the whole system and its complexity during the synthesis process. Moreover, thanks to its high modularity it can be potentially customized to every design problem by suitably choosing, implementing, and interconnecting its functional blocks. Thanks to such features, several *SbD* implementations have been proposed in the recent scientific literature including, for instance, the design of metamaterial lenses [13–17], nanostructures [18], wide-angle impedance layers (*WAIM*s) [19–21], reflectarrays [22], fractal radiators [23], airborne radomes [6], *5G* arrays [5], automotive radar antennas [5], time-modulated arrays (*TMA*s) [5], holographic *EM* skins [24], and polarizers [25] (Table 3.1).

Table 3.1 Recent state-of-the-art works on *SbD* as applied to several *EM* design problems.

References	Year	Designed devices
[15]	2014	Metamaterial Lenses
[19]	2015	*WAIM*s
[20]	2017	*WAIM*s
[18]	2017	Nanostructures
[16]	2018	Metamaterial Lenses
[17]	2018	Metamaterial Lenses
[23]	2019	Fractal Radiators
[14]	2019	Metamaterial Lenses
[22]	2020	Reflectarrays
[21]	2021	*WAIM*s
[25]	2021	Polarizers
[6]	2021	Airborne Radomes
[5]	2021	Microstrip Arrays
[5]	2021	*TMA*s
[5]	2021	Automotive Radar Antennas
[24]	2021	*EM* Skins

Besides such design-oriented applications, it is worth mentioning that the *SbD* paradigm proved to be very flexible and suitable also for addressing other complex *EM* problems involving global optimization tasks such as, for instance, inverse scattering ones [26].

This chapter is organized as follows. After reviewing the fundamental concepts and driving ideas of the *SbD* (Section 3.2), its customization to representative state-of-the-art *EM* design problems is detailed (Section 3.3). Finally, some conclusions and final remarks along with a discussion on the envisaged trends are drawn (Section 3.4).

3.2 The *SbD* Pillars and Fundamental Concepts

In the following, let us indicate as $\underline{\xi} = \{\xi_k;\ k = 1, \dots, K\}$ the set of K degrees-of-freedom (*DoF*s) describing the solution of the design problem at hand. The goal of the *SbD* is two-fold [5]:

1. To find the best solution $\underline{\xi}^{(SbD)}$ complying with

$$\left|\Phi\left\{\underline{\xi}^{(SbD)}\right\} - \Phi\left\{\underline{\xi}^{(GO)}\right\}\right| \leq \zeta \tag{3.1}$$

   where $\Phi\{\ .\ \}$ is the cost function, $\underline{\xi}^{(GO)}$ is the global optimum, and ζ is the allowed maximum deviation;
2. To find $\underline{\xi}^{(SbD)}$ within a *reasonable* time interval $\Delta t^{(SbD)}$, i.e. yielding a remarkable time saving with respect to a standard *EA*-based solution strategy

$$\Delta t^{(SbD)} \ll \Delta t^{(EA)} \tag{3.2}$$

   being $\Delta t^{(EA)} = (P \times I) \times \Delta t^{(FWS)}$, where $\Delta t^{(FWS)}$ is the *CPU* time of a single *FWS*-based evaluation of $\Phi\{\ .\ \}$.

According to the *SbD* paradigm, the ambitious goals (3.1) and (3.2) are addressed by suitably defining, implementing, and interconnecting problem-driven functional blocks, each performing a specific sub-task [5, 6, 12] (Figure 3.1). Generally speaking, although several customizations and implementations of the *SbD* synthesis loop have been proposed depending on the solved problem (see Section 3.3), it is possible to define four main functional blocks [5] (Figure 3.1):

1. *Requirements and Constraints Definition* (*RCD*) – This block is aimed at mathematically expressing the requirements and a set of physical-admissibility constraints;
2. *Problem Formulation* (*PF*) – This block formulates the design problem as a global optimization one. Toward this end, it defines the "smartest" model of the solution described by a minimum-cardinality set of *DoF*s $\underline{\xi}$ as well as it mathematically expresses the cost function $\Phi\{\underline{\xi}\}$ encoding the mismatch between the project requirements/constraints (*RCD* Block) and the evaluated performance of the guess design described by $\underline{\xi}$;
3. *Cost Function Computation* (*CFC*) – The purpose of this block is to provide the most efficient yet accurate assessment of the cost function associated to each p-th ($p = 1, \dots, P$) trial solution $\underline{\xi}_i^{(p)}$ explored at the i-th ($i = 1, \dots, I$) iteration of the optimization process;
4. *Solution Space Exploration* (*SSE*) – This block performs the effective sampling of the K-dimensional search space (*PF* Block) to find $\underline{\xi}^{(SbD)}$. Toward this end, it generates a population of trial solutions $\mathcal{P}_{i=0} = \left\{\underline{\xi}_{i=0}^{(p)};\ p = 1, \dots, P\right\}$ and evolves it through the iterations by applying a set of properly defined operators $\mathcal{F}\{\ .\ \}$ (i.e. $\mathcal{P}_i = \mathcal{F}\left\{\mathcal{P}_{i-1}\right\}, i = 1, \dots, I$) [9].

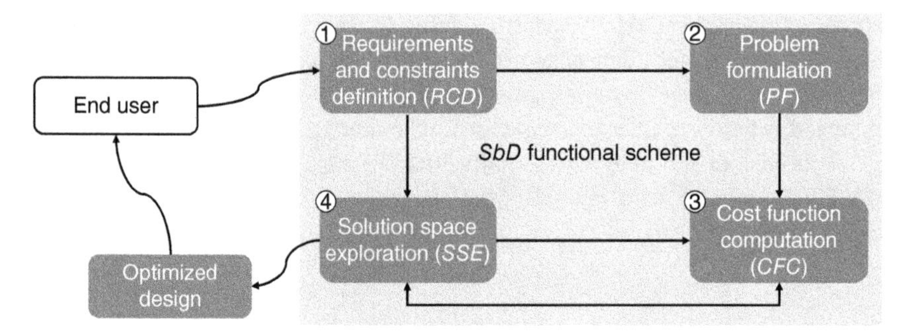

Figure 3.1 *SbD paradigm* – General functional scheme.

In the following, let us briefly describe more in detail the fundamental concepts behind the implementation of the *PF*, *SSE*, and *CFC* blocks, moving to Section 3.3 the discussion of the *RCD* block with some illustrative examples. Concerning the *PF* block (Figure 3.1), the definition of the problem *DoF*s $\underline{\xi}$ should comply with the following guidelines: (*i*) their number, *K*, should be kept as low as possible to limit the necessary agents/trial solutions used in the *SSE* block, while (*ii*) guaranteeing a high flexibility of the model and the existence of a solution fitting the *RCD* block requirements [5]. In order to limit the cardinality of $\underline{\xi}$, sensitivity analyses can be performed to determine which descriptors have the largest influence on the optimized key performance indicators (*KPI*s). Otherwise, "smart" models could be exploited to derive a minimum set of *DoF*s. This is the case, for instance, of the design problem in [6], where a spline-based representation was adopted by Salucci et al. to model the internal profile of electrically large airborne radomes (see Section 3.3.4). Accordingly, the *DoF*s became the location of a limited set of spline control points describing the radome shape with high flexibility (Figure 3.2a), resulting in a remarkable shrinking of the solution space with respect to a pixel-based representation requiring a much larger number of descriptors (i.e. defining the presence/absence of dielectric material in each pixel – Figure 3.2b) [6].

As for the second task of the *PF* block, it is worth noticing that the cost function must be carefully defined since (*i*) it is the unique link between the synthesis strategy and the underlying *EM* physics and (*ii*) it determines the overall complexity of the *K*-dimensional landscape that must be explored by means of the *SSE* block [5].

Dealing with the *CFC* block (Figure 3.1), as anticipated the evaluation of $\Phi\{\underline{\xi}\}$ exclusively based on *FWS*s is the main bottleneck of standard *EA*-based design procedures, preventing their use in high-complexity problems where $\Delta t^{(EA)}$ becomes practically unfeasible. To significantly alleviate such an issue, the *SbD* relies on the exploitation of computationally expedite but accurate forward analysis tools and/or digital twins (*DT*s) formulated within the machine learning (*ML*) framework [27, 28]. Concerning these latter, several approaches have been proposed to substitute, totally or partially, the repeated time-consuming use of *FWS*s with *DT*s based on, for instance, Support Vector Regression (*SVR*) [23] or Gaussian Process Regression (*GPR*) [22]. Such techniques are based on the generation of a training database $\mathcal{D}$ of T properly selected input/output (*I/O*) pairs $\left(\mathcal{D} = \left[\left(\underline{\xi}^{(t)};\ \Phi\left\{\underline{\xi}^{(t)}\right\}\right);\ t = 1, \dots, T\right]\right)$ to build a fast surrogate model $\tilde{\Phi}\{\ .\ \}$ of $\Phi\{\ .\ \}$ to be exploited by the *SSE* block [27]. In this context, it is worth noticing that the selection of the minimum-cardinality set of *DoF*s by means of the *PF* block perfectly fits the need for counteracting the negative effects of the so-called "curse-of-dimensionality" [29], as done in the first step of "three-steps" learning-by-examples (*LBE*) strategies, which is devoted at

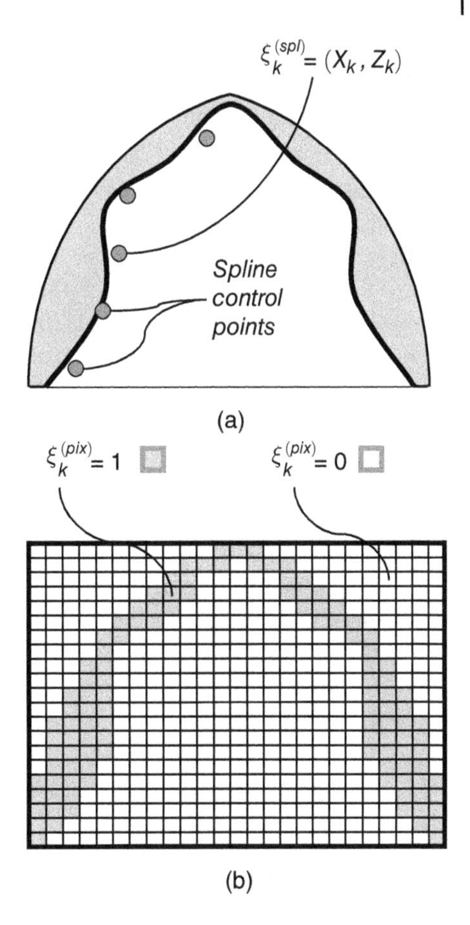

Figure 3.2 *SbD paradigm (PF block)* – Example sketch of the (a) pixel-based and (b) spline-based modeling of the internal profile of an airborne radome. Source: [6] ©[2022] IEEE.

minimizing K and therefore limiting the number of necessary training samples [5]. Owing to the previous considerations, when using a *DT* model the overall *SbD* time can be approximated to $\Delta t^{(SbD)} \approx T \times \Delta t^{(FWS)}$, yielding a time saving of [5]

$$\Delta t_{sav} = \frac{\Delta t^{(EA)} - \Delta t^{(SbD)}}{\Delta t^{(EA)}} \approx \frac{(P \times I) - T}{(P \times I)} \tag{3.3}$$

with respect to a standard approach, and accordingly a "rule-of-thumb" for the successful application of the *SbD* is to guarantee that $T \ll (P \times I)$ [5].

Finally, concerning the *SSE* block (Figure 3.1), the choice of the most suitable optimization engine is of paramount importance to enable a *SbD* implementation with competitive global search capabilities. As a matter of fact, according to the "*no free-lunch theorem*" (*NFL*) of optimization [30], there is no optimal tool for any optimization problem. In other words, since "*it is not possible to a-priori identify the best combination of the functional blocks of the SbD able to perform well on every possible problem*" [5], great care must be put in selecting the most suitable optimization tool for the synthesis problem at hand. In this framework, advanced *SSE* implementations have been proposed based on the "collaboration" between the optimizer and the *DT* model to trigger "reinforcement learning" procedures and overcome the limitations of *BARE-SbD* approaches based on the straightforward replacement of the *FWS* with its *ML*-based surrogate [5].

In conclusion, it is worth observing that a complex *EM* design problem generally involves multiple logical layers, each sub-part of the overall synthesis procedure being addressed in a different hierarchical level. Some examples of *SbD* layers are [6]:

1. *Material-by-Design* and *Metamaterial-by-Design* – Address the task-oriented design of advanced systems comprising artificial materials whose constituent properties are driven by the device functional requirements [13];
2. *Architecture-by-Design* – This layer is concerned with the definition of the optimal overall system layout and architecture (e.g. the feeding network);
3. *Algorithm-by-Design* – Implements control strategies including, for instance, those related with the real-time calibration and tuning of synthesized devices/systems;
4. *Measurements-by-Design* – Deals with the integration of the design stage with the successive testing procedures for antenna qualification [31].

3.3 *SbD* at Work in *EMs* Design

This section is aimed at presenting some illustrative examples of state-of-the-art implementations of the *SbD* for addressing complex *EM* design problems (Table 3.1). More precisely, the following applicative contexts are discussed in the following: elementary radiators (Section 3.3.1), reflectarrays (Section 3.3.2), and metamaterial lenses (Section 3.3.3). Other *SbD* customizations for the synthesis of *EM* skins, radomes, and polarizers, are resumed in Section 3.3.4 as well.

3.3.1 Design of Elementary Radiators

The efficient design of compact radiators operating over multiple bands suitable for the Internet of Things (*IoT*) has been addressed within the *SbD* paradigm by Salucci et al. [23]. In this framework, microstrip antennas with perturbed fractal shapes proved to be effective in realizing low-profile and low-cost multi-band antennas. As a matter of fact, they are more flexible with respect to standard/unperturbed layouts since they can break the fixed relationship among the working resonances [32]. In this framework, the *SbD* has been applied by Salucci et al. [23] to define a scalable synthesis tool which can be efficiently re-used for several different objectives/designs and fractal layouts (e.g. Sierpinski, Hilbert, Koch fractal patches). The design procedure, whose functional scheme is shown in Figure 3.3, consists in the following blocks [23]:

- *Solution Space Exploration* (*SSE*): iteratively generates, according to a global optimization strategy based on the Particle Swarm Optimization (*PSO*), a sequence of trial solutions (i.e. perturbed fractal layouts) converging to the design targets;
- *Physical Response Emulator* (*PRE*): performs the fast computation of the *EM* behavior of each guess radiator by predicting the Q resonant frequencies $\{\tilde{f}_q;\ q = 1, \ldots, Q\}$ of each *SSE*-Block-generated guess fractal geometry;
- *Physical Objective Assessment* (*POA*): evaluates the "quality" of each trial solution/design outputted by the *SSE* block by computing the matching between its resonant frequencies, $\{\tilde{f}_q;\ q = 1, \ldots, Q\}$, and the desired ones, $\{f_q;\ q = 1, \ldots, Q\}$.

More in detail, the design has been formulated as the minimization of the following cost function [23]

$$\Phi\left\{\underline{\xi}\right\} = \sum_{q=1}^{Q}\left[f_q - \tilde{f}_q\left(\underline{\xi}\right)\right]^2 \tag{3.4}$$

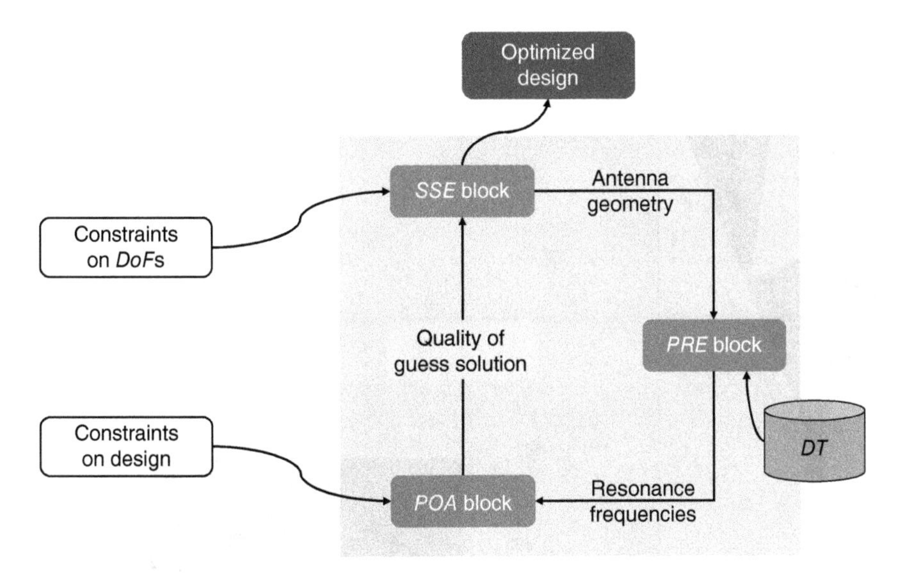

Figure 3.3 *SbD customization to fractal radiators design* – Functional flowchart of the *SbD*-based method for the synthesis of multi-band perturbed fractal microstrip radiators. Source: Adapted from [23].

where $\underline{\xi}$ are the fractal shape descriptors (e.g. $\underline{\xi} = \{w_a, h_a, w_1, w_2, w_3, w_4, w_5, h_f, h_g\}$ for dual-band Sierpinski Gasket antennas – Figure 3.4a and b). As for the *PRE* block (Figure 3.3), a *ML* strategy based on *SVR* has been adopted to build a fast *DT* of the method of moments (*MoM*) *FWS* [27]. In order to enhance and speed-up the *SVR* training phase, the orthogonal array (*OA*) sampling technique [33] has been adopted for the generation of low-cardinality/highly informative training sets, since it enforces that the training locations span the search space uniformly while minimizing the number of necessary *I/O* samples. It is worth noticing that the proposed *SbD* loop can be straightforwardly applied to several applicative scenarios differing for the operative bands since the training phase is performed once and off-line for a given perturbed fractal shape [27].

To give a proof of the effectiveness of the resulting *SbD* synthesis process, the design of a dual-band ($Q = 2$) antenna working in the LTE-2100 ($f_1 = 2.045$ (GHz)) and the LTE-3500 ($f_2 = 3.5$ (GHz)) bands as obtained in Salucci et al. [23] is discussed hereinafter. The plot of the cost function versus the iteration index is shown in Figure 3.4c, showing that the *SbD* converged in approximately $I_{conv} \simeq 100$ iterations, resulting in a time saving of about five orders of magnitude with respect to a standard optimization. As for the frequency response of the designed radiator, the plot of the input reflection coefficient in Figure 3.4d confirms the fitting of the synthesis objectives in terms of impedance matching in the two target bands (i.e. $\left|S_{11}\left(f_q\middle|\underline{\xi}^{(SbD)}\right)\right| \leq S_{11}^{th}$ for $q = \{1;\ 2\}$, being $S_{11}^{th} = -10$ (dB)) [23].

Still concerning the application of the *SbD* paradigm to the synthesis of elementary radiators, Massa and Salucci presented in [5] the design of a slotted substrate integrated waveguide (*SIW*) antenna for 77 (GHz) automotive radar applications (Figure 3.5a). The design has been efficiently addressed by means of a novel "confidence-enhanced" *SbD* strategy leveraging on both the global search features of the *PSO* and the capability of *GPR*-based *DT*s to provide a "reliability index" of the predictions enabling a "reinforced learning" strategy that updates the *DT* during the optimization

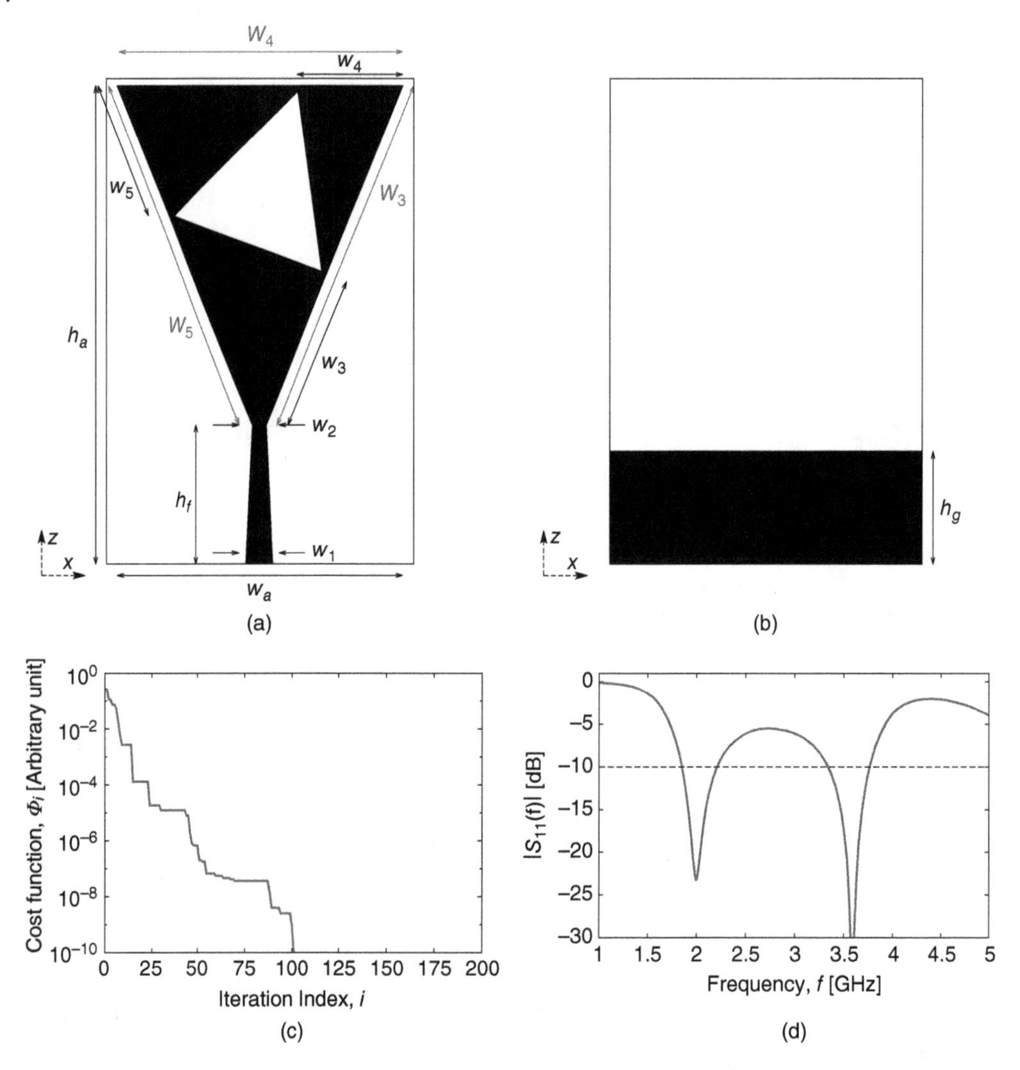

Figure 3.4 *SbD customization to fractal radiators design* - (a) Front and (b) back views of the planar Sierpinski Gasket fractal antenna with corresponding *DoF*s [23]; (c) Evolution of the cost function and (d) optimized reflection coefficient behavior of an SbD-optimized dual-band radiator working in the LTE-2100 ($f_1 = 2.045$ (GHz)) and the LTE-3500 ($f_2 = 3.5$ (GHz)) bands. Source: [23] / Springer Nature / Public Domain CC BY 4.0.

by adaptively selecting additional training samples [27]. The minimization of the following cost function has been considered [5]

$$\Phi\left\{\underline{\xi}\right\} = \sum_{\Psi\in\Theta}\Phi_\Psi\left\{\underline{\xi}\right\} \tag{3.5}$$

where $\Theta = \left\{S_{11};\ SLL;\ HPBW;\ BDD\right\}$ are the optimized *KPI*s and

$$\Phi_\Psi\left\{\underline{\xi}\right\} = \frac{1}{Q}\sum_{q=1}^{Q}\frac{\mathcal{H}\left\{\Psi\left(f_q\middle|\underline{\xi}\right) - \Psi^{th}\right\}}{\left|\Psi^{th}\right|}; \qquad \Psi\in\Theta \tag{3.6}$$

are the cost terms related to the reflection coefficient ($\Psi = S_{11}$), the sidelobe level ($\Psi = SLL$), the half-power beamwidth ($\Psi = HPBW$), and the beam direction deviation ($\Psi = BDD$), $\mathcal{H}\{\ .\ \}$ being

Heaviside's function and superscript "th" denoting the user-defined threshold value. The synthesis method was applied to a *SIW* comprising $N = 18$ slots etched on a Rogers *RO3003* substrate ($\varepsilon_r = 3.0$, $\tan\delta = 0.0013$) of thickness $h = 1.27 \times 10^{-1}$ (mm), yielding a search space of dimension $K = 17$ (Figure 3.5a). In order to enable a robust design taking into account mutual coupling, the design was performed simulating the radiation features of the central embedded element within a linear arrangement of $M = 7$ equally spaced identical replicas, with the $(M - 1) = 6$ neighboring

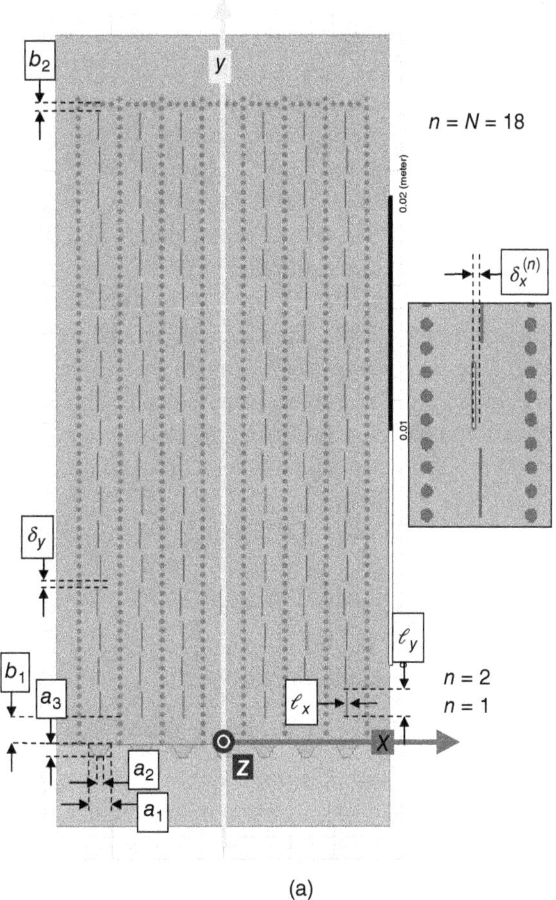

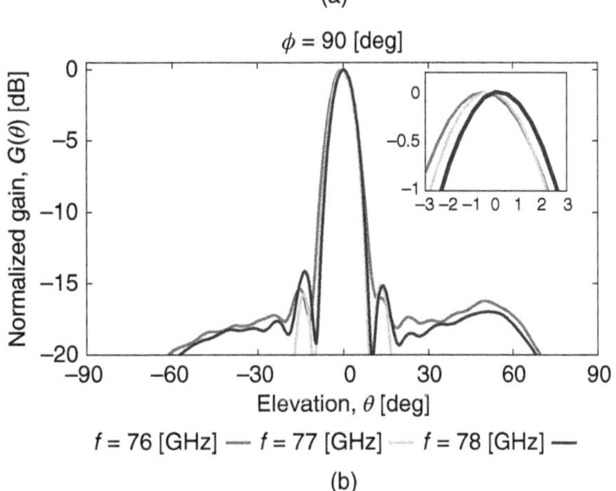

Figure 3.5 *SbD customization to automotive antennas design* – (a) Geometry of the 77 (GHz) slotted *SIW* antenna [5] and (b) normalized gain pattern of the *SbD*-optimized solution Source: [5] ©[2021] IEEE.

radiators terminated on matched loads. Exploiting $P = 20$ agents and $I = 200$ iterations, a time saving of $\Delta t_{sav} = 92.87\%$ has been yielded by the *SbD* over a standard optimization (this latter being unfeasible since $\Delta t^{(EA)} \approx 200$ (days)) [5]. An almost-perfect matching of all requirements has been obtained with a very good stability of the radiation pattern over the frequency band of interest ($f \in [76,\ 78]$ (GHz)), as denoted in the plot of Figure 3.5b [5].

3.3.2 Design of Reflectarrays

The effective and computationally efficient task-oriented synthesis of wide reflectarray antennas (*RA*s) comprising arbitrarily complex unit cells (*UC*s) has been addressed within the *SbD* paradigm by Oliveri et al. [22]. In this framework, it is worth pointing out that complex *UC*s comprising a large number of *DoF*s can be exploited in order to enhance the beam control capabilities of *RA*s. However, this comes at the cost of a more complex design process (since the search space dimensionality exponentially grows with the number of *DoF*s of the *UC*) and of a more difficult mapping of the relation between the *UC* descriptors and the *RA* radiation features. Such problems can be only partially tackled exploiting offline *FWS*-computed databases, since their generation is computationally unfeasible when complex geometries are at hand [34]. To enable the computationally efficient design of *RA*s, the synthesis problem has been formulated in [22] as a multi-scale one where the *DoF*s are defined at the micro-scale as the geometric descriptors of each *UC*, while the objectives are defined at the macro-scale level in terms of requirements on the radiation characteristics of the complete *RA* layout. More precisely, the arising design problem has been formulated in terms of the global minimization of the following cost function

$$\Phi\left\{\underline{\xi}\right\} = \sum_{\beta\in\{\psi,\chi\}} \alpha_\beta \Phi_\beta\left(\underline{\xi}\right) + \alpha_\Gamma \Phi_\Gamma\left(\underline{\xi}\right) \tag{3.7}$$

where $\underline{\xi} \triangleq \left\{\underline{g}_{su};\ s = 1, \dots, S,\ u = 1, \dots, U\right\}$ is the complete set of *RA* descriptors, $(S \times U)$ being the total number of *UC*s, and $\underline{g}_{su} = \left\{g_{su}^d;\ d = 1, \dots, D\right\}$ being the set of micro-scale geometric descriptors of each $(s,\ u)$-th *UC*. Moreover,

$$\Phi_\beta\left(\underline{\xi}\right) \triangleq \int_0^{2\pi}\int_0^{\pi} \Delta E_\beta\left(\theta, \varphi|\,\underline{\xi}\right) \sin(\theta)\, d\theta d\varphi \tag{3.8}$$

quantifies the mask mismatch for the β-component of the radiated far-field, defined according to Ludwig's theory [35] as the component along the directions $\hat{\psi} \triangleq \sin(\varphi)\hat{\theta} + \cos(\varphi)\hat{\varphi}$ ($[\beta = \psi \rightarrow E_\psi\left(\theta, \varphi|\,\underline{\xi}\right)]$ or $\hat{\chi} = \cos(\varphi)\hat{\theta} - \sin(\varphi)\hat{\varphi}$ $[\beta = \chi \rightarrow E_\chi\left(\theta, \varphi|\,\underline{\xi}\right)]$, respectively. Moreover, $\Delta E_\beta\left(\theta, \varphi|\,\underline{\xi}\right) \triangleq \left[\left|E_\beta\left(\theta, \varphi|\,\underline{\xi}\right)\right| - \mathcal{M}_\beta(\theta, \varphi)\right] \times \mathcal{H}\left[\left|E_\beta\left(\theta, \varphi|\,\underline{\xi}\right)\right| - \mathcal{M}_\beta(\theta, \varphi)\right]$, $\mathcal{M}_\beta(\theta, \varphi)$ being the macro-scale requirement in terms of radiation mask for the β-component of the field. As for the term Φ_Γ, it has been defined to control/reduce the discrepancies between adjacent *UC*s as needed in many realistic *RA*s implementations [22]

$$\Phi_\Gamma\left(\underline{\xi}\right) \triangleq \frac{1}{(S-1)\times(U-1)} \sum_{s=1}^{S-1}\sum_{u=1}^{U-1} \sqrt{\frac{1}{D}\sum_{d=1}^{D}\left(\left|g_{su}^d - g_{s,u+1}^d\right|^2 + \left|g_{su}^d - g_{s+1,u}^d\right|^2\right)}. \tag{3.9}$$

The arising synthesis problem has been addressed by means of a modular *SbD* iterative strategy consisting in the following functional blocks (Figure 3.6):

- *Cost Function Computation* – this block mathematically defines and computes the macro-scale objectives subject to the user-defined guidelines;

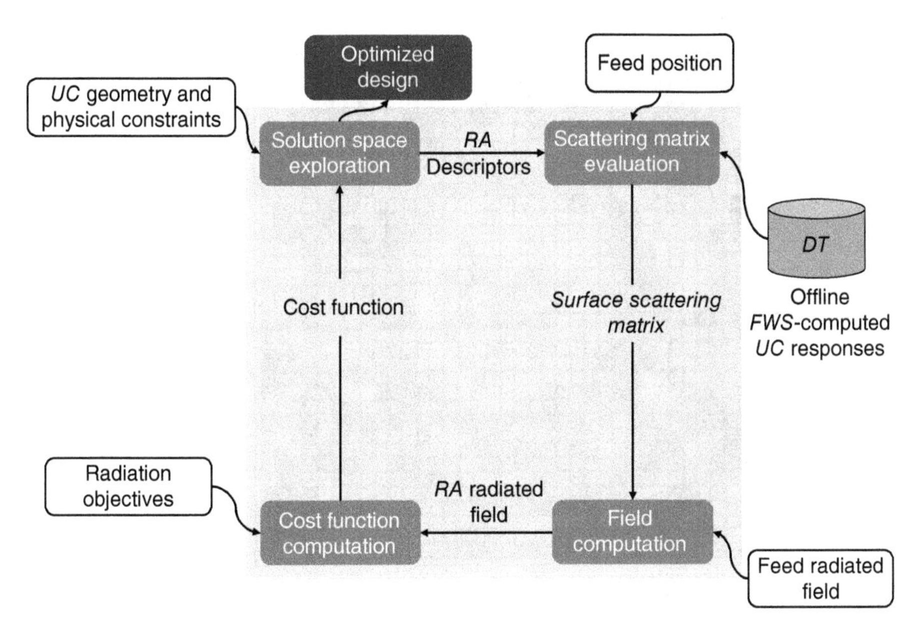

Figure 3.6 *SbD customization to reflectarrays design* – Functional flowchart of the *SbD*-based method for the multi-scale task-oriented synthesis of reflectarray antennas Source: Adapted from [22].

- *Field Computation* – the aim of this block is to link the *EM* behavior of a *UC* and the overall *RA* radiating features exploiting local periodicity assumptions;
- *Scattering Matrix Evaluation* – this block efficiently predicts the physical relationship between the micro-scale *UC* descriptors and the resulting *EM* response;
- *Solution Space Exploration* – this block is devoted to minimize the cost function (3.7).

The goal was to handle the synthesis of wide *RA*s with a very high computational efficiency, by shifting the most cumbersome calculations to a preliminary offline phase, aimed at training a *DT* based on *GPR*. Toward this end, the solution of the task-oriented multi-scale problem, comprising hundreds/thousands of continuous variables (i.e. $K = U \times S \times D$), has been effectively found by means of an iterative optimization scheme not requiring the simultaneous handling of the whole set of *DoF*s that would imply convergence issues due to the huge dimensionality of the problem at hand. Therefore, the adopted nested strategy in [22] considered the iterative minimization of (3.7) only with respect to the (s, u)-th entry of $\underline{\xi}$ while keeping the remaining $(U \times S) - 1$ *UC*s unaltered.

To provide an example of the capabilities of such a *SbD* strategy, let us refer to the test case in Oliveri et al. [22] concerned with the synthesis of a large *RA* comprising $(U \times S) = (35 \times 35)$ Phoenix *UC*s. Regardless of the very high dimensionality of the solution space ($K \simeq 10^3$), the designed layout in Figure 3.7a turned out to provide a very good matching of the *SLL* requirement ($SLL_{\chi}^{th} = -25$ (dB)) on the co-polar field component, E_{χ}, as also verified by means of the *FWS*-computed pattern in Figure 3.7b and the corresponding mismatch with the pattern mask in Figure 3.7c.

3.3.3 Design of Metamaterial Lenses

Oliveri et al. proposed in [15] an innovative *SbD* methodology for the synthesis of isotropic and non-magnetic metamaterial lenses within the transformation electromagnetics (*TE*) framework [36]. *TE* techniques enforce the equivalence between the radiated field distribution of a virtual

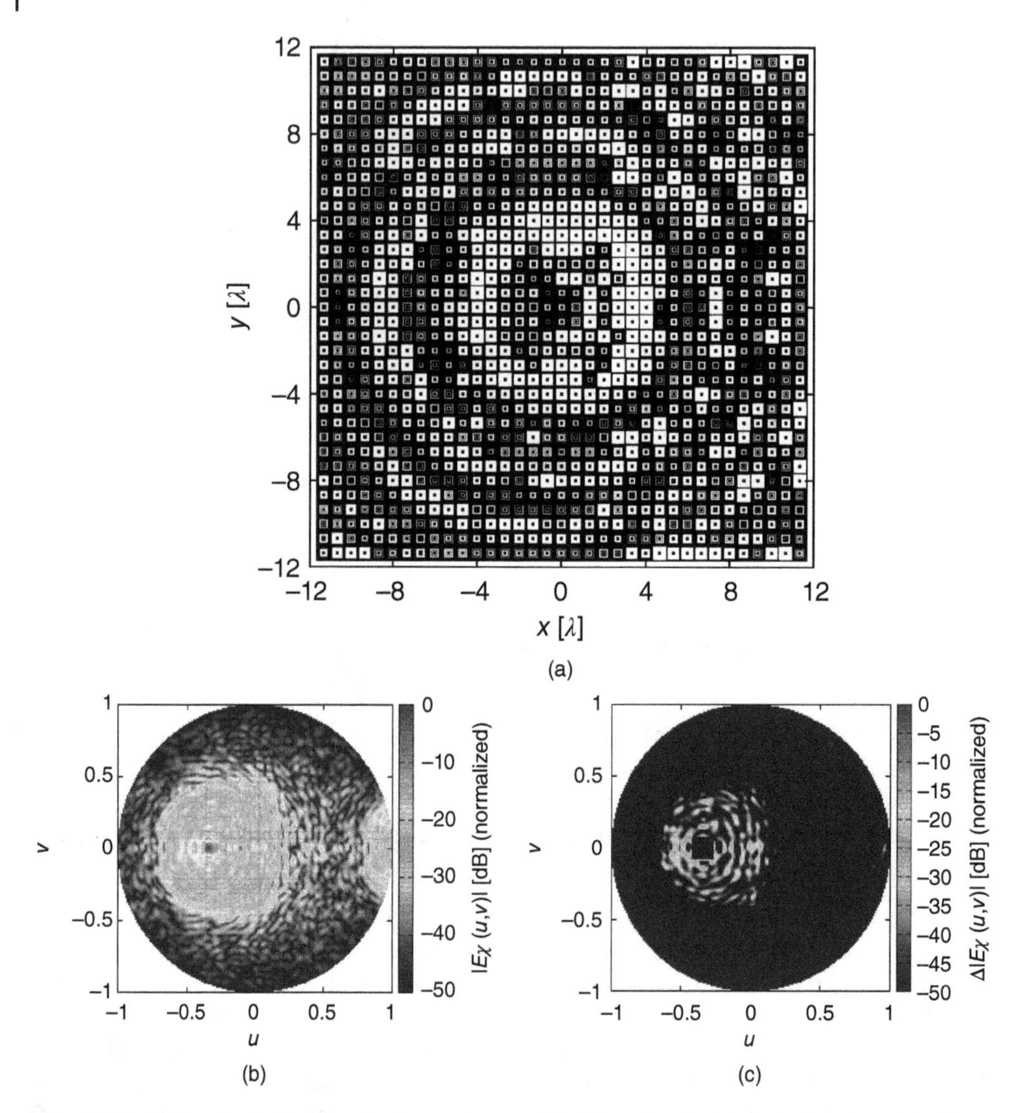

Figure 3.7 *SbD customization to reflectarrays design* – (a) *SbD*-synthesized layout of a microstrip *RA* comprising $(U \times S) = (35 \times 35)$ Phoenix *UC*s and corresponding *FWS*-computed distribution of (b) the co-polar pattern, and (c) co-polar mask mismatch Source: [22] ©[2020] IEEE.

geometry with known *EM* properties and a physical layout with user-defined geometry and properly synthesized materials. The solution of the field manipulating problem at hand is therefore obtained by exploiting the invariance of Maxwell's equations to coordinate transforms in order to derive the mapping function between the two domains [36]. However, when arbitrary contours are at hand, *TE* techniques cannot *a priori* guarantee the generation of isotropic materials, leading to several implementation issues from a practical/manufacturing point of view. Such a problem can be mitigated by properly defining the descriptors of the virtual/physical contours [14, 15]. However, the arising synthesis problem is rather complex since it requires suitable strategies to cope with the non-linearity of the underlying relationships and time-expensive simulations of the physical layout embedding the *TE*-generated lens.

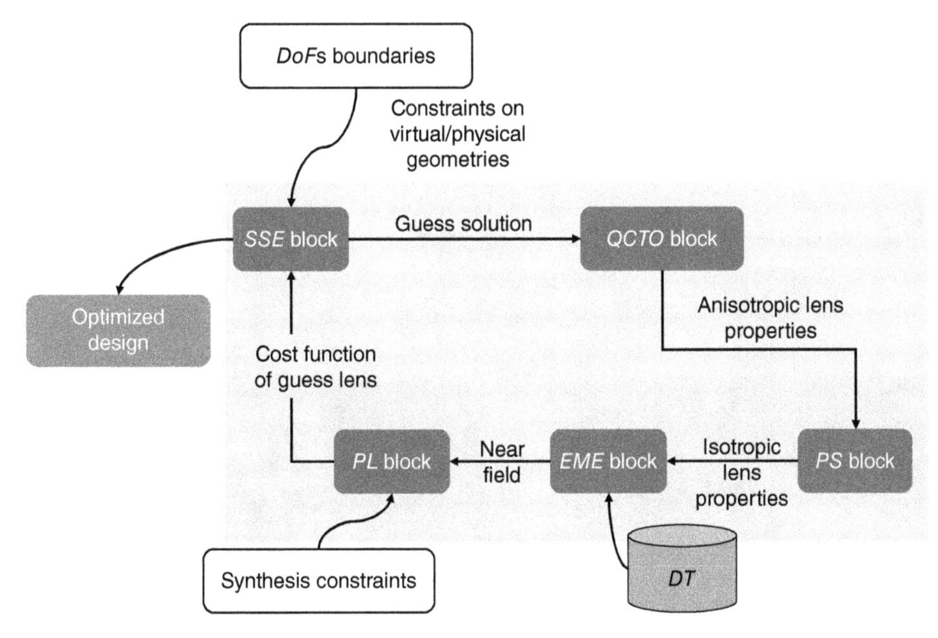

Figure 3.8 *SbD customization to metamaterial lenses design* – Functional flowchart of the *SbD*-based method for the synthesis of isotropic metamaterial lenses for conformal array transformation Source: Adapted from[15].

The proposed *SbD* solution scheme in [15] (Figure 3.8) is based on the following functional blocks:

- *Solution Space Exploration* (*SSE*) – Generates a set of guess geometries exploring the solution space at hand to reach the design objectives;
- *Quasi Conformal Transformation Optics* (*QCTO*) – Computes the dielectric properties of the physical lens starting from the mapping between virtual and physical domains;
- *Physical Simplification* (*PS*) – Derives an isotropic approximation of the lens material synthesized by the *QCTO* block;
- *EM Emulator* (*EME*) – Predicts the field distribution radiated by a given trial guess solution;
- *Physical Linkage* (*PL*) – Determines the physical admissibility of each guess solution (lens distribution) and computes the cost function.

As for the cost function, it has been defined as follows [15]

$$\Phi\left(\underline{\xi}\right) = \Phi_{NF}\left(\underline{\xi}\right) + \Phi_{SLL}\left(\underline{\xi}\right) + \Phi_{HPBW}\left(\underline{\xi}\right) \tag{3.10}$$

where $\underline{\xi}$ contains the geometric descriptors of the virtual (Ω) and physical (Ω') contours (Figure 3.9a), respectively, and

$$\Phi_{NF}\left(\underline{\xi}\right) = \frac{\int \left| E'\left(\mathbf{r}' | \underline{\xi}\right) - E\left(\mathbf{r} | \underline{\xi}\right)\Big]_{\mathbf{r}=\mathbf{r}'} \right| d\mathbf{r}'}{\int \left| E\left(\mathbf{r} | \underline{\xi}\right)\Big]_{\mathbf{r}=\mathbf{r}'} \right| d\mathbf{r}'} \tag{3.11}$$

quantifies the mismatch between the field distribution $E(\mathbf{r})$ radiated by the virtual arrangement and that generated by the physical layout, $E'\left(\mathbf{r}'\right)$, as predicted by means of a *GPR*-based *DT* (built in the *EME* block (Figure 3.8)) starting from the dielectric distribution within the lens. Moreover,

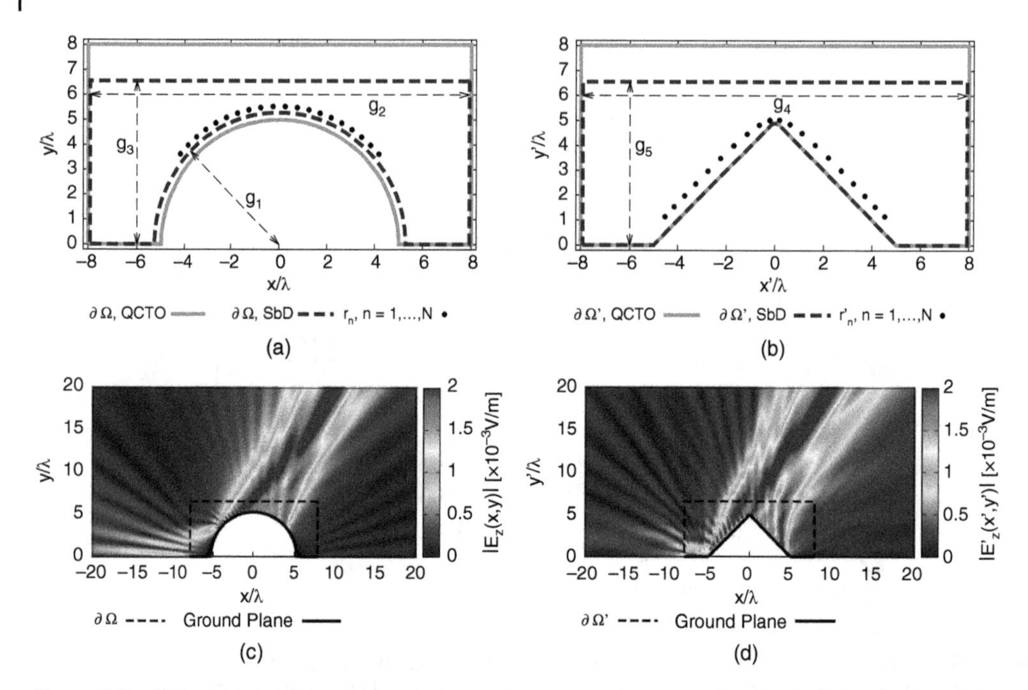

Figure 3.9 *SbD customization to metamaterial lenses design* – Geometry of (a) virtual and (b) physical layouts [15]; Plot of the electric field radiated by the (c) virtual and (d) isotropic *SbD* layouts when steering the main beam towards the direction $\phi_s = 60$ (deg) Source: [15]. ©[2014] IEEE.

$\Phi_{SLL}\left(\underline{\xi}\right) = \left|\frac{SLL'\left(\underline{\xi}\right)-SLL\left(\underline{\xi}\right)}{SLL\left(\underline{\xi}\right)}\right|$ and $\Phi_{HPBW}\left(\underline{\xi}\right) = \left|\frac{HPBW'\left(\underline{\xi}\right)-HPBW\left(\underline{\xi}\right)}{HPBW\left(\underline{\xi}\right)}\right|$ are the normalized deviations of the physical array *SLL* and *HPBW* from the corresponding virtual quantities, respectively.

The effectiveness of the *SbD* loop of Figure 3.8 has been assessed by Oliveri et al. [15] when dealing with the conformal transformation of a virtual scenario comprising $N = 20$ sources located in free-space within a circularly eroded Ω (Figure 3.9a) into a physical geometry represented by a corner-eroded lens covering a corner-conformal array (Figure 3.9b), resulting into the set of $K = 5$ *DoF*s $\underline{\xi} = \{g_k;\ k = 1, \dots, K\}$ (Figure 3.9a,b). The outcome of the *SbD* synthesis is shown in Figure 3.9 where a comparison between the near field plots of the virtual radiated field (Figure 3.9c) and that generated by the physical layout (Figure 3.9d) is given when considering a steering of the main beam toward the direction $\phi_s = 60$ (deg). The result clearly indicates that a good matching between the two distributions outside the lens support has been obtained, clearly indicating the effectiveness of the proposed strategy.

Within the same *TE* framework, a *SbD* approach has been proposed by Salucci et al. [14] for the efficient synthesis of miniaturized linear arrays exploiting reduced-complexity isotropic lenses and a spline-based approach to model arbitrary transformation contours with a limited number of *DoF*s.

3.3.4 Other *SbD* Customizations

Oliveri et al. proposed in [24] an innovative *SbD*-based strategy for the synthesis of holographic static passive smart *EM* skins (*SPSS*s) enabling advanced beam-forming capabilities within the emerging "Smart *EM* Environment" framework [37] (Figure 3.10a). Differently from more sophisticated but expensive technologies such as Reconfigurable Intelligent Surfaces (*RIS*s) [38, 39], *SPSS*s

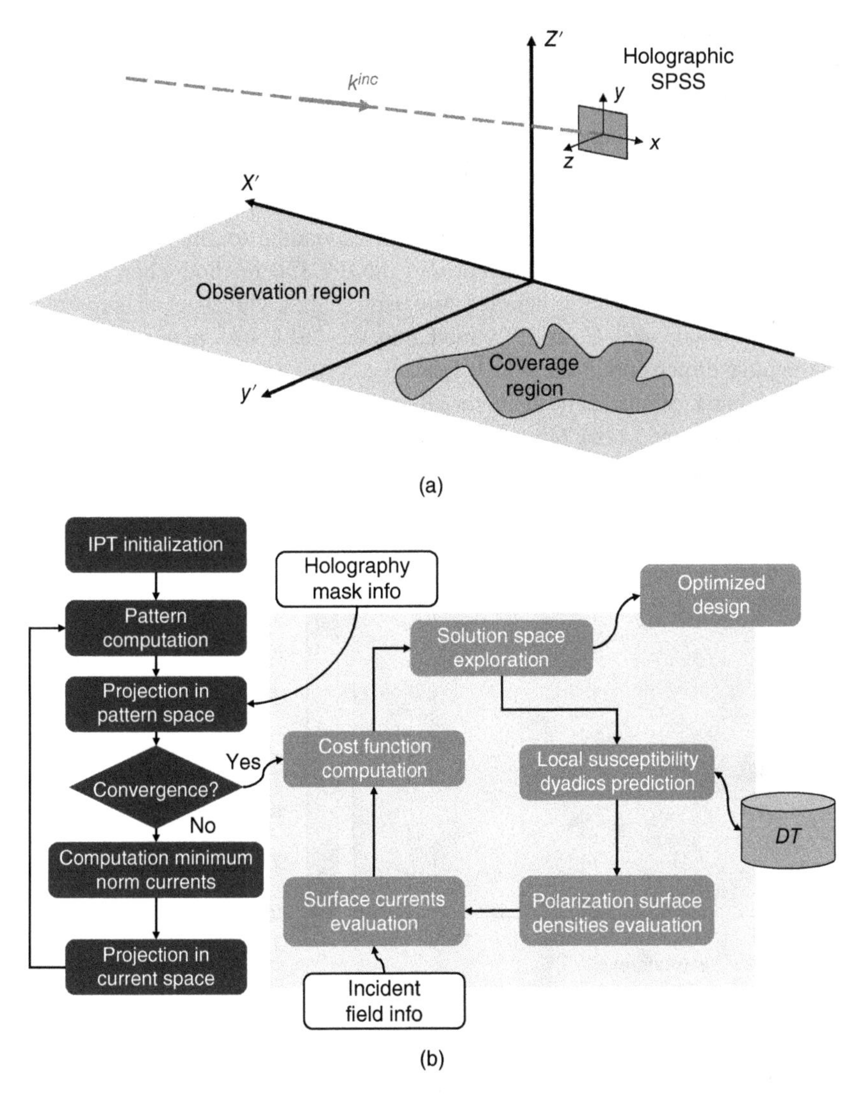

Figure 3.10 *SbD Customization to Holographic SPSSs Design* - (*a*) Problem geometry and (*b*) functional flowchart of the *IPT-SbD* method for the synthesis of holographic *SPSS*s Source: [24]. ©[2021] IEEE.

can afford complex pattern footprints with simple, light, and inexpensive layouts, improving the coverage in selected areas and reducing "blind spots" in urban scenarios through engineered metasurfaces exhibiting unconventional wave manipulation capabilities [24]. However, their design involves several challenges including the need for (*i*) a robust beam shaping not requiring real-time calibration/tuning operations, and (*ii*) a sufficiently large extension to guarantee an adequate level of reflected power toward the coverage region (Figure 3.10a).

Therefore, a huge dimensionality of the search space is generally involved since a very large number of *SPSS* micro-scale *DoF*s have to be optimized. Furthermore, unlike reflectarrays (Section 3.3.2), the synthesis of *SPSS*s is characterized by a limited control/knowledge of the direction of the incident beam coming from the basestation antenna. Owing to such considerations, the arising multi-scale *EM* design problem has been formulated in [24] within the Generalized Sheet Transition Condition (*GSTC*) theoretical framework. Then, it has been solved exploiting the combination of a local search technique based on the Iterative Projection Technique (*IPT*) [40] and a customized version of the *SbD* paradigm, resulting in the functional scheme of Figure 3.10b. More in detail, a two-step strategy has been implemented to (a) deduce through the *IPT* the ideal surface currents radiating the desired footprint pattern and (b) optimize by means of the *SbD* the descriptors of the *SPSS* matching those reference distributions. As for this latter, it comprises a "Local Susceptibility Dyadics Prediction" block (Figure 3.10b) aimed at estimating, by means of a *GPR*-based surrogate model, the susceptibility tensors corresponding to any trial guess of the *UC* descriptors. To provide a proof of the effectiveness of the *IPT-SbD* method when dealing with advanced beam-forming tasks involving detailed footprint shapes, Oliveri et al. reported in [24] the design of a large (400×400 *UC*s) *SPSS* matching the "ELEDIA" pattern mask of Figure 3.11a. Despite the huge dimension of the search space (i.e. $K = 1.6 \times 10^5$), the synthesized layout faithfully fulfills the mask requirement as pictorially confirmed by the footprint pattern within the observation region (Figure 3.11b). Moreover, it is worth pointing out that such a result has been yielded with an overall execution time four orders of magnitude lower with respect to a standard optimization [24].

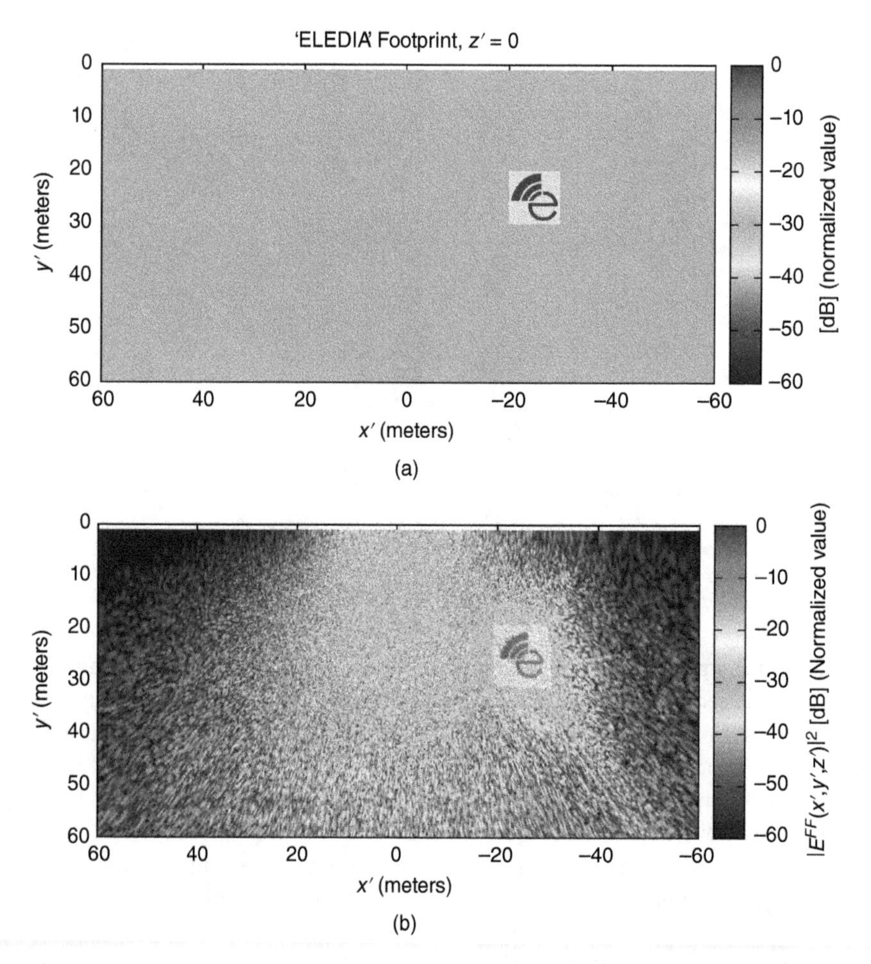

Figure 3.11 *SbD customization to holographic SPSSs design* – Plot of the (a) "ELEDIA" footprint pattern mask and (b) footprint pattern radiated in the observation region by a *SbD*-designed *SPSS* comprising 400×400 *UC*s Source: [24]. ©[2021] IEEE.

The *SbD* paradigm has been also applied by Salucci et al. [6] to the synthesis of electrically large airborne radomes. Within this context, the material and the internal structure of a radome play a fundamental role in minimizing the overall attenuation, scattering, and depolarization of the *EM* waves radiated by the enclosed antenna [41]. However, because of the large scale of the *EM* problem to solve, *FWS*-based analyses are typically affordable only within local refinement phases, being otherwise substituted with simplified models (e.g. based on transmission line theory and/or ray tracing tools) to enable global optimization. To overcome such limitation, a *SbD* approach has been introduced in [6] to synthesize with high computational efficiency airborne radomes minimizing

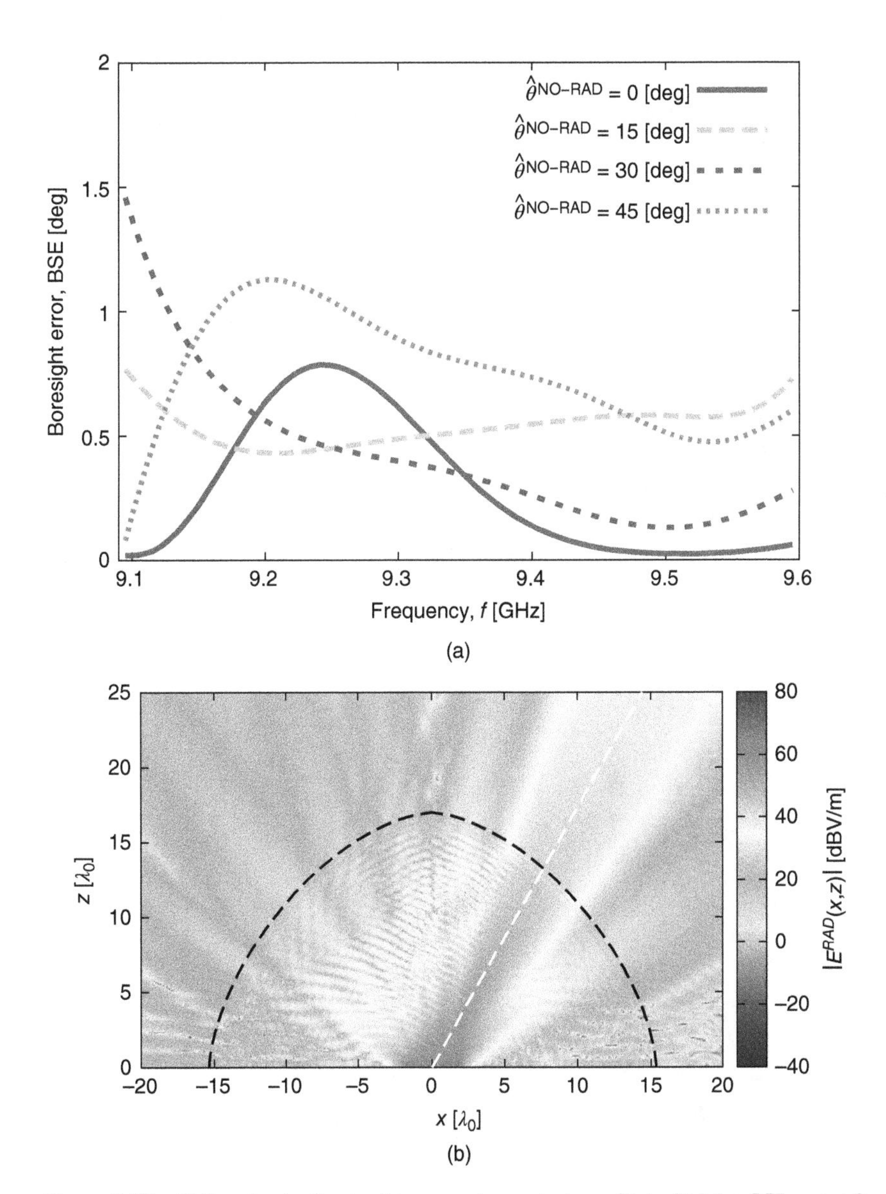

Figure 3.12 *SbD customization to airborne radomes design* – Plot of (a) the *BSE* versus frequency for a linear array enclosed within the *SbD*-optimized spline-shaped radome and (b) electric field magnitude of the radome-enclosed array when $\hat{\theta}^{NO-RAD} = 30$ (deg) Source: [6]. ©[2022] IEEE.

the bore-sight error (*BSE*) of the enclosed phased antenna array. Toward this end, the problem has been recast as a global optimization one aimed at minimizing the mismatch between the array beam pointing direction when the radome is absent, $\hat{\theta}^{NO-RAD}(f_q, \theta_v)$, or present, $\hat{\theta}^{RAD}\left(f_q, \theta_v \middle| \underline{\xi}\right)$, respectively, when setting linear phase shift excitations to steer the main lobe toward the v-th ($v = 1, \ldots, V$) direction θ_v at the q-th ($q = 1, \ldots, Q$) frequency sample f_q. As for the problem *DoF*s $\underline{\xi}$, as previously mentioned (see Section 3.2) they correspond to the position of K control points of a spline function modeling the inner profile of the radome (Figure 3.2a). It is worth remarking that such a definition follows the *SbD* guidelines to yield (*i*) a flexible representation guaranteeing the existence of a solution as well as (*ii*) an effective recipe against the curse-of-dimensionality to generate an accurate *DT* from a limited set of training samples. The arising *SbD* methodology proved to be effective in synthesizing an ogive-shaped radome with circular base diameter $A = 30.7$ (λ_0) and length $L = 17$ (λ_0), λ_0 being the free-space wavelength at the central frequency $f_0 = 9.595$ (GHz) [6]. As a matter of fact, optimizing $K = 5$ spline control points it was possible to minimize the *BSE* of a linear array ($BSE < 1.5$ (deg) – Figure 3.12a) within an operative band of $\Delta f = 500$ (MHz) centered at f_0, with highly directive and well-shaped beam patterns pointed towards the desired directions (e.g. $\hat{\theta}^{NO-RAD} = 30$ (deg) – Figure 3.12b). Thanks to the adopted *SbD* strategy, a non-negligible computational efficiency with a time saving of $\Delta t_{sav} = 87.5\%$ was yielded [6].

Finally, it is worth mentioning the work done by Arnieri et al. [25] dealing with the design of dual-band reflection-mode wide-bandwidth and wide-angle scanning circular polarizers. In such a context, a suitably customized *SbD* strategy has been proposed to improve the performance of analytically synthesized devices with very high computational efficiency. Accordingly, a two-step design strategy has been implemented where the first step consists in deriving a fast yet reliable guess of the polarizer *UC* descriptors starting from their equivalent lumped element values. Such a preliminary guess has been exploited to perform a *FWS*-based fine-tuning of the final layout, by optimizing the resulting axial ratio (*AR*) values within the operative bandwidth with a *SbD*-based procedure. To assess the performance of the arising hybrid synthesis approach, a prototype of a dual-band polarizer was realized and shown in [25] (Figure 3.13a). It turned out that the measured

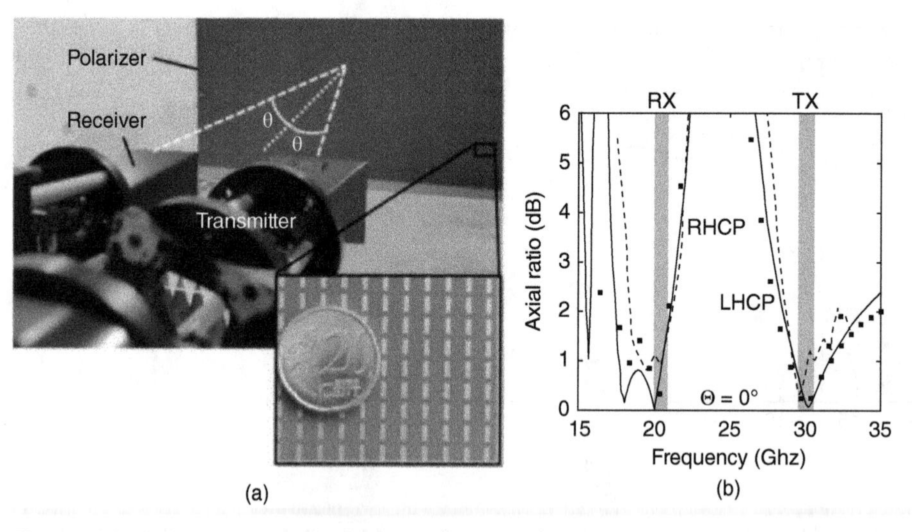

Figure 3.13 *SbD customization to polarizers design* – (a) Picture of the manufactured *SbD*-designed dual-band linear-to-circular polarizer and (b) plot of the measured (dashed line), *FWS*-simulated (continuous line), and analytic (square dots) *AR* versus frequency Source: [25]. ©[2022] IEEE.

AR was not only complying with the design objectives, but also close to the predicted performance as seen from the *SbD* synthesis loop (Figure 3.13b).

3.4 Final Remarks and Envisaged Trends

The *SbD* has recently emerged as "*a functional ecosystem to handle complexity in the design of complex EM systems*" [5, 6]. It enables the computationally efficient solution of complex *EM* synthesis problems by means of a suitable problem-driven selection and interconnection of functional blocks sharing the goal of fitting the required system performance and functionality. A review of the most recent advances in the application of the *SbD* paradigm was illustrated in this chapter, pointing out the following key aspects:

- thanks to its high modularity and depending on the available resources and problem objectives, the *SbD* allows the designer to choose/customize the best functional scheme and the most suitable (i.e. problem-oriented) techniques for the implementation of each sub-task;
- The definition of "smart" solution models plays a fundamental role to derive a limited set of highly representative/flexible *DoF*s enabling at the same time an enhanced global exploration of the functional landscape and a mitigation of the curse-of-dimensionality;
- The *NFL* theorem and concepts originally formulated for global optimization problems [30] straightforwardly apply to the *SbD* as well, since efficient procedures and robust designs can be yielded only if a proper choice/integration of each functional block is carefully undertaken.

Although successful customizations have been documented in different *EM* design contexts (Section 3.3), many open challenges still exist to develop highly efficient *SbD* strategies enabling, for instance, the synthesis of unconventional antenna architectures with realistic radiators [42], the real-time control and tuning of reconfigurable metasurfaces and *EM* skins [39], the integration with multi-objective optimization strategies [43], and the exploitation of multi-physics forward modeling tools. Furthermore, the extension of current methodologies to embrace high-fidelity prediction methods formulated within the deep learning framework [44] must be carefully studied. Finally, innovative strategies for reducing the "hunger" of training data must be explored exploiting, for instance, Compressive Sampling (*CS*)-based strategies [45] capable of overcoming Nyquist's theoretical bound to build highly informative and reduced-size training sets.

Acknowledgments

This work benefited from the networking activities carried out within the Project "SPEED" (Grant No. 6721001) funded by National Science Foundation of China under the Chang-Jiang Visiting Professorship Program and the Project "National Centre for HPC, Big Data and Quantum Computing (CN HPC)" funded by the European Union-NextGenerationEU within the PNRR Program (CUP: E63C22000970007). Views and opinions expressed are however those of the author(s) only and do not necessarily reflect those of the European Union or the European Research Council. Neither the European Union nor the granting authority can be held responsible for them. The authors acknowledge the Italian Ministry of Education, Universities and Research (MUR) in the framework of the project DICAM-EXC (Departments of Excellence 2023-2027, grant L232/2016). A. Massa wishes to thank E. Vico for her never-ending inspiration, support, guidance, and help.

References

1 Davidson, D. (2011). *Computational Electromagnetics for RF and Microwave Engineering*. Cambridge: Cambridge University Press.
2 Jin, J.-M. (2014). *The Finite Element Method in Electromagnetics*. Wiley-IEEE Press.
3 Harrington, R.F. (1993). *Field Computation by Moment Methods*. Wiley-IEEE Press.
4 Taflove, A. and Hagness, S.C. (2005). *Computational Electrodynamics: The Finite-Difference Time-Domain Method*. Artech House.
5 Massa, A. and Salucci, M. (2022). On the design of complex *EM devices and systems through the system-by-design paradigm - a framework for dealing with the computational complexity. IEEE Transactions on Antennas and Propagation* 70 (2): 1328–1343. https://doi.org/10.1109/TAP.2021.3111417.
6 Salucci, M., Oliveri, G., Hannan, M.A., and Massa, A. (2022). System-by-design paradigm-based synthesis of complex systems: the case of spline-contoured 3D radomes. *IEEE Antennas and Propagation Magazine* 64 (1): 72–83. https://doi.org/10.1109/MAP.2021.3099719.
7 Kim, J.H., Chun, H.J., Hong, I.C. et al. (2014). Analysis of FSS radomes based on physical optics method and ray tracing technique. *IEEE Antennas and Wireless Propagation Letters* 13: 868–871.
8 Vegni, C. and Bilotti, F. (2002). Parametric analysis of slot-loaded trapezoidal patch antennas. *IEEE Transactions on Antennas and Propagation* 50 (9): 1291–1298.
9 Rocca, P., Benedetti, M., Donelli, M. et al. (2009). Evolutionary optimization as applied to inverse problems. *Inverse Problem* 25: 123003, pp. 1–41.
10 Rocca, P., Oliveri, G., and Massa, A. (2011). Differential evolution as applied to electromagnetics. *IEEE Antennas and Propagation Magazine* 53 (1): 38–49.
11 Goudos, S. (2021). *Emerging Evolutionary Algorithms for Antennas and Wireless Communications*. Stevenage: SciTech Publishing Inc.
12 Massa, A., Oliveri, G., Rocca, P., and Viani, F. (2014). System-by-design: a new paradigm for handling design complexity. *8th European Conference on Antennas and Propagation (EuCAP 2014)*, The Hague, The Netherlands (6–11 April 2014), pp. 1180–1183.
13 Massa, A. and Oliveri, G. (2016). Metamaterial-by-design: theory, methods, and applications to communications and sensing - editorial. *EPJ Applied Metamaterials* 3 (E1): 1–3.
14 Salucci, M., Tenuti, L., Gottardi, G. et al. (2019). A system-by-design method for efficient linear array miniaturization through low-complexity isotropic lenses. *Electronics Letters* 55 (8): 433–434.
15 Oliveri, G., Tenuti, L., Bekele, E. et al. (2014). An SbD-QCTO approach to the synthesis of isotropic metamaterial lenses. *IEEE Antennas and Wireless Propagation Letters* 13: 1783–1786.
16 Salucci, M., Oliveri, G., Anselmi, N., and Massa, A. (2018). Material-by-design synthesis of conformal miniaturized linear phased arrays. *IEEE Access* 6: 26367–26382.
17 Salucci, M., Oliveri, G., Anselmi, N. et al. (2018). Performance enhancement of linear active electronically-scanned arrays by means of MbD-synthesized metalenses. *Journal of Electromagnetic Waves and Applications* 32 (8): 927–955.
18 Nagar, J., Campbell, S.D., Ren, Q. et al. (2017). Multiobjective optimization-aided metamaterials-by-design with application to highly directive nanodevices. *IEEE Journal on Multiscale and Multiphysics Computational Techniques* 2: 147–158.
19 Oliveri, G., Viani, F., Anselmi, N., and Massa, A. (2015). Synthesis of multilayer WAIM coatings for planar phased arrays within the system-by-design framework. *IEEE Transactions on Antennas and Propagation* 63 (6): 2482–2496.

20 Oliveri, G., Salucci, M., Anselmi, N., and Massa, A. (2017). Multiscale system-by-design synthesis of printed WAIMs for waveguide array enhancement. *IEEE Journal on Multiscale and Multiphysics Computational Techniques* 2: 84–96.

21 Oliveri, G., Polo, A., Salucci, M. et al. (2021). SbD-Based synthesis of low-profile WAIM superstrates for printed patch arrays. *IEEE Transactions on Antennas and Propagation* 69 (7): 3849–3862.

22 Oliveri, G., Gelmini, A., Polo, A. et al. (2020). System-by-design multi-scale synthesis of task-oriented reflectarrays. *IEEE Transactions on Antennas and Propagation* 68 (4): 2867–2882.

23 Salucci, M., Anselmi, N., Goudos, S., and Massa, A. (2019). Fast design of multiband fractal antennas through a system-by-design approach for NB-IoT applications. *EURASIP Journal on Wireless Communications and Networking* 2019 (1): 68–83.

24 Oliveri, G., Rocca, P., Salucci, M., and Massa, A. (2021). Holographic smart EM skins for advanced beam power shaping in next generation wireless environments. *IEEE Journal on Multiscale and Multiphysics Computational Techniques* 6: 171–182.

25 Arnieri, E., Salucci, M., Greco, F. et al. (2022). An equivalent circuit/system-by-design approach to the design of reflection-type dual-band circular polarizers. *IEEE Transactions on Antennas and Propagation* 70 (3): 2364–2369. https://doi.org/10.1109/TAP.2021.3111511.

26 Salucci, M., Poli, L., Rocca, P., and Massa, A. (2022). Learned global optimization for inverse scattering problems: matching global search with computational efficiency. *EEE Transactions on Antennas and Propagation* 70 (8): 6240–6255.

27 Massa, A., Oliveri, G., Salucci, M. et al. (2018). Learning-by-examples techniques as applied to electromagnetics. *Journal of Electromagnetic Waves and Applications* 32 (4): 516–541.

28 Forrester, A.I.J., Sobester, A., and Keane, A.J. (2008). *Engineering Design via Surrogate Modelling: A Practical Guide*. Hoboken, NJ: Wiley.

29 Jimenez-Rodriguez, L.O., Arzuaga-Cruz, E., and Velez-Reyes, M. (2007). Unsupervised linear feature-extraction methods and their effects in the classification of high-dimensional data. *IEEE Transactions on Geoscience and Remote Sensing* 45 (2): 469–483.

30 Wolper, D.H. and Mcready, W.G. (1997). No free lunch theorem for optimization. *IEEE Transactions on Evolutionary Computation* 1 (1): 67–82.

31 Salucci, M., Migliore, M.D., Oliveri, G., and Massa, A. (2018). Antenna measurements-by-design for antenna qualification. *IEEE Transactions on Antennas and Propagation* 66 (11): 6300–6312.

32 Werner, D.H. and Ganguly, S. (2003). An overview of fractal antenna engineering research. *IEEE Antennas and Propagation Magazine* 45 (1): 38–57.

33 Hedayat, A., Sloane, N., and Stufken, J. (2013). *Orthogonal Arrays*. New York: Springer.

34 Salucci, M., Tenuti, L., Oliveri, G., and Massa, A. (2018). Efficient prediction of the EM response of reflectarray antenna elements by an advanced statistical learning method. *IEEE Transactions on Antennas and Propagation* 66 (8): 3995–4007.

35 Ludwig, A. (1973). The definition of cross polarization. *IEEE Transactions on Antennas and Propagation* 21 (1): 116–119.

36 Kwon, D.-H. and Werner, D.H. (2010). Transformation electromagnetics: an overview of the theory and applications. *IEEE Antennas and Propagation Magazine* 52 (1): 24–46.

37 Massa, A., Benoni, A., Da Rú, P. et al. (2021). Designing smart electromagnetic environments for next-generation wireless communications. *Telecom* 2 (2): 213–221.

38 Basar, E., Di Renzo, M., De Rosny, J. et al. (2019). Wireless communications through reconfigurable intelligent surfaces. *IEEE Access* 7: 116753–116773.

39 Di Renzo, M., Zappone, A., Debbah, M. et al. (2020). Smart radio environments empowered by reconfigurable intelligent surfaces: how it works, state of research, and the road ahead. *IEEE Journal on Selected Areas in Communications* 38 (11): 2450–2525.

40 Rocca, P., Haupt, R.L., and Massa, A. (2009). Sidelobe reduction through element phase control in subarrayed array antennas. *IEEE Antennas and Wireless Propagation Letters* 8: 437–440.

41 Kozakoff, D.J. (2010). *Analysis of Radome-Enclosed Antennas*, 2e. Norwood, MA: Artech House.

42 Rocca, P., Oliveri, G., Mailloux, R.J., and Massa, A. (2016). Unconventional phased array architectures and design methodologies - a review. *Proceedings of the IEEE* 104 (3): 544–560.

43 Nagar, J. and Werner, D.H. (2017). A comparison of three uniquely different state of the art and two classical multiobjective optimization algorithms as applied to electromagnetics. *IEEE Transactions on Antennas and Propagation* 65 (3): 1267–1280.

44 Massa, A., Marcantonio, D., Chen, X. et al. (2019). DNNs as applied to electromagnetics, antennas, and propagation - a review. *IEEE Antennas and Wireless Propagation Letters* 18 (11): 2225–2229.

45 Massa, A., Rocca, P., and Oliveri, G. (2015). Compressive sensing in electromagnetics – a review. *IEEE Antennas and Propagation Magazine* 57 (1): 224–238.

4

Artificial Neural Networks for Parametric Electromagnetic Modeling and Optimization

Feng Feng[1], *Weicong Na*[2], *Jing Jin*[3], *and Qi-Jun Zhang*[4]

[1] *School of Microelectronics, Tianjin University, Tianjin, China*
[2] *Faculty of Information Technology, Beijing University of Technology, Beijing, China*
[3] *College of Physical Science and Technology, Central China Normal University, Wuhan, China*
[4] *Department of Electronics, Carleton University, Ottawa, ON, Canada*

4.1 Introduction

Parametric electromagnetic (EM) modeling and optimization techniques are integral parts of the field of microwave design. Parametric models can be constructed based on the relationship between EM response and geometric variables. The constructed parametric model enables rapid simulation and optimization using different geometric variables, which can then be used in high-standard designs.

Artificial neural network (ANN) is a common technique for parameterized modeling and optimization of the design [1–9]. With proper training, ANN can characterize the relationship between EM responses and geometric variables. After the training process, ANN can provide rapid and precise predictions of the EM performances in passive elements; furthermore, it can be exploited in succeeding high-standard design of microwave circuits. Over the past few decades, ANN has been applied to parametric EM modeling and optimization of various microwave circuits [10–32]. A deep neural network (DNN) refers to a neural network with multiple hidden layers [33]. While modeling very complicated daedal relationships, such as modeling with high-dimensional filters with many input variables, DNNs have better performance than shallow neural networks (a neural network with a small number of hidden layers) [33].

To enhance the accuracy and reliability of ANN modeling for microwave devices, the knowledge-based neural network (KBNN) model has been exploited [11, 34–37]. The knowledge-based model combines existing knowledge in the form of an empirical or equivalent circuit model with neural networks to open up a more efficient and more precise model. The knowledge-based method combines the neural network with future knowledge such as the equivalent circuit [11, 34, 35] and analysis formula [25, 36]. Leveraging this knowledge can help improve the learning and generalization capabilities of the overall model and speed up the model development process [37]. Space mapping (SM) technique is recognized in engineering optimization and modeling in the microwave area. Various SM approaches have been developed to map the future knowledge (i.e. coarse models) to EM performances of microwave devices [38–63]. The neural network model based on SM utilizes a large number of existing empirical models, reduces the number of EM evaluations required for training, improves model precision and universality, and reduces the complexity of neural network topology compared to pure ANN methods.

Advances in Electromagnetics Empowered by Artificial Intelligence and Deep Learning, First Edition.
Edited by Sawyer D. Campbell and Douglas H. Werner.

Another advanced approach for parametric EM modeling and microwave device optimization, namely a combined neural network and transfer function modeling method, has been presented [64–71]. This method combines a neural network and transfer function and can be used even without an accurate equivalent circuit model. Moreover, using this method we can represent the change of EM characteristics of passive devices with frequency in the form of a transfer function.

In the subsequent part of this chapter, we first describe the general structure and training of ANN in Section 4.2. In Section 4.3, the structure and deep learning algorithm of the hybrid DNN technology for microwave models are presented. A unified KBNN method and its training algorithm are discussed in Section 4.4. The parametric modeling methods using combined ANN and transfer function are described in Section 4.5. The ANN-based surrogate optimization methods are introduced in Section 4.6. Section 4.7 concludes the chapter and discusses the future directions of ANN for parametric EM modeling and optimization.

4.2 ANN Structure and Training for Parametric EM Modeling

There are two kinds of basis elements in an ANN, namely, the neurons and links between them [1]. Every link has a weighting parameter. Every neuron receives and processes the information from other neurons, and produces a corresponding output [2]. Neurons in ANN are normally divided into multiple layers, including the input layer, the output layer, and hidden layers [2]. The input and the output layer are the first and the last layers in the ANN, respectively. The hidden layers are the other layers located between the first and the last layers.

Motivated by different microwave applications, many ANN-based parametric modeling techniques have been investigated based on various ANN structures, such as multilayer perception (MLP) neural networks [32], dynamic neural networks [72, 73], radial basis function (RBF) neural networks [74, 75], recurrent neural networks (RNNs) [76–78], time-delay neural networks (TDNNs) [79], state-space dynamic neural networks (SSDNNs) [80, 81], and deep neural networks (DNNs) [33].

One of the most popular ANN structures is the MLP, whose structure is shown in Figure 4.1. There are usually an input layer, an output layer, and at least a hidden layer in an MLP [2]. In an MLP, the information is processed in a feedforward way. First, the input layer receives external information. Then, the subsequent hidden layers process the information. Finally, the output layer generates the final output responses.

An ANN needs to be well trained using corresponding data so as to learn the nonlinear EM behavior of microwave device/circuit [1]. The first step for developing an ANN model is to determine the model inputs and outputs according to the original EM problems. In general, geometrical/physical parameters are taken as the inputs for ANN-based parametric EM modeling, while EM responses are taken as the outputs. Data generating, data scaling, training, validating, and testing are the typical steps forming the systematic process to develop an ANN model. To train the ANN model, data samples representing the relationships between inputs and outputs of the original EM problem need to be generated. Typically, the generated data samples should be grouped into three sets. Depending on the purposes, the three sets of data are named as the training data, the validation data, and the test data, respectively. During training, the weighting parameters in ANN model are adjusted to make the relationship between the model inputs and outputs match the training data as well as possible. To monitor the quality of training, the validation data are used. It is also used to verify whether the training process should be stopped. After the training process has been finished,

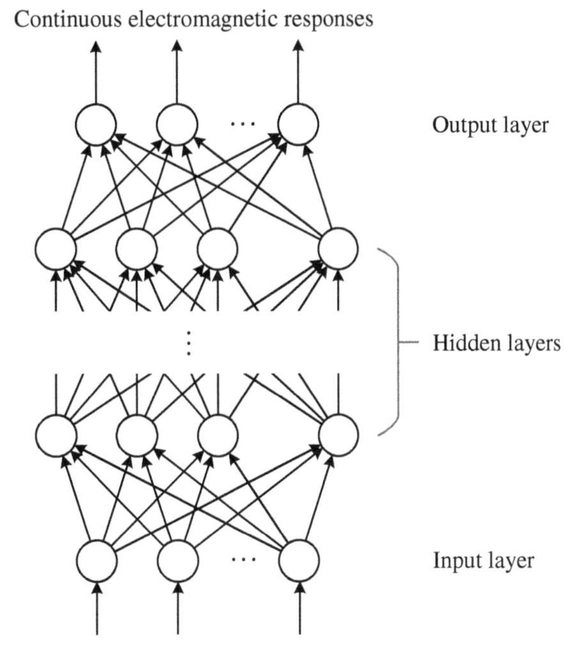

Figure 4.1 Illustration of the MLP structure for parametric EM modeling. Source: [1] Zhang and Gupta 2000/ Association for Computing Machinery.

the test data, which have never been used during training, are used to independently examine the model's accuracy and its ability to generalize.

4.3 Deep Neural Network for Microwave Modeling

DNN has been presented for modeling microwave devices/circuits with high dimensional inputs. This section introduces a hybrid DNN structure combining smooth ReLUs and sigmoid functions for microwave modeling. Compared to the shallow ANN, the hybrid DNN is able to handle microwave modeling problems with higher dimensions. In addition, compared to the classical DNN with only ReLUs, the hybrid DNN is able to achieve higher model accuracy, while using even fewer hidden neurons.

4.3.1 Structure of the Hybrid DNN

The hybrid DNN is a fully connected neural network whose number of hidden layers is three or more [33]. In the hybrid DNN, there are two types of activation functions, which are the smooth ReLUs and sigmoid functions. In this way, the vanishing gradient problem would be avoided during the DNN training, and the number of hidden neurons would be reduced in the same time [33]. Suppose that only sigmoid functions are used, there would be the vanishing gradient problem during the DNN training process. If only smooth ReLUs are used, the DNN would require more hidden neurons.

Let $\boldsymbol{x} = [x_1\ x_2\ x_3\ \dots\ x_n]^T$ and $\boldsymbol{y} = [y_1\ y_2\ y_3\ \dots\ y_m]^T$ be defined as the inputs and outputs of the hybrid DNN model, respectively, where n and m are the number of model inputs and outputs, respectively. Suppose there are p hidden layers in the hybrid DNN using sigmoid functions and q hidden layers using smooth ReLUs. The total number of layers in the hybrid DNN will be

$L = p + q + 2$. After constructing the structure of the hybrid DNN, the next step is to train the hybrid DNN model by adjusting values of weighting parameters.

4.3.2 Training of the Hybrid DNN

Suppose for the inputs $\boldsymbol{x}_k$, the desired outputs of the hybrid DNN are represented by $\boldsymbol{d}_k$. Therefore, the training data samples for training the hybrid DNN can be represented by $(\boldsymbol{x}_k, \boldsymbol{d}_k)$, $k = 1, 2, \ldots, T_r$, where T_r represents the number of training samples. To train the hybrid DNN, the standard error function [1] is used, which is defined as

$$E(\boldsymbol{w}) = \sum_{k=1}^{T_r} \left(\frac{1}{2} \sum_{j=1}^{m} (y_j(\boldsymbol{x}_k, \boldsymbol{w}) - d_{jk})^2 \right) \tag{4.1}$$

in which $y_j(\boldsymbol{x}_k, \boldsymbol{w})$ represents the jth output of the hybrid DNN model for $\boldsymbol{x}_k$, and d_{jk} represents the jth element in $\boldsymbol{d}_k$.

Following is the process of developing the hybrid DNN for high dimensional microwave applications [33].

Step 1: Generate training and test data randomly within a pre-defined range. Decide the number of hidden neurons per hidden layer and the model error threshold, E_r. Initialize p and q by setting $p = 3$ and $q = 0$.
Step 2: Train the DNN that has p sigmoid hidden layers. Calculate the training error, E_{train}. If $p = 3$, go to Step 4; otherwise, go to Step 3.
Step 3: Compare E_{train} and E_b. If $E_{train} \geq E_b$, which means that the vanishing gradient problem exists, go to Step 7; otherwise, go to Step 4.
Step 4: Compare E_{train} and E_r. If $E_{train} > E_r$, which means that the training is in a state of underlearning, set $E_b = E_{train}$, add a new sigmoid layer, i.e. $p = p + 1$, and return to Step 2; otherwise, go to Step 5.
Step 5: Test the trained DNN. Define the test error as E_{test}. If $E_{test} > E_r$, i.e. the training is in a state of overlearning, go to Step 6; otherwise, stop the training process.
Step 6: Train the DNN with more training data, calculate E_{train}, and return to Step 4.
Step 7: Remove the last added sigmoid hidden layer, i.e. $p = p - 1$, and train the DNN to minimize the training error. After training, add a layer with classical ReLUs to the trained model, i.e. $q = 1$.
Step 8: Train the hybrid DNN and calculate E_{train}.
Step 9: Compare E_{train} and E_r. If $E_{train} > E_r$, which means that the training is in a state of underlearning, add a new hidden layer with classical ReLUs, i.e. $q = q + 1$, and return to Step 8; otherwise, go to Step 10.
Step 10: Test the trained hybrid DNN model. If $E_{test} > E_r$, add more training samples, and return to Step 8; otherwise, go to Step 11.
Step 11: Replace all classical ReLUs in the trained hybrid DNN using smooth ReLUs.
Step 12: Refine the model accuracy by training the hybrid DNN. Stop the training process.

4.3.3 Parameter-Extraction Modeling of a Filter Using the Hybrid DNN

The hybrid DNN technique is illustrated by the development of the parameter-extraction model for a sixth-order filter example [33]. The parameter-extraction model is used to extract the coupling matrix from given S-parameters.

In this application example, the center frequency is 11,785.5 MHz, and the bandwidth is 56.2 MHz [83]. The ideal coupling matrix is

$$\boldsymbol{M}_{ideal} = \begin{bmatrix} -0.0473 & 0.8489 & 0 & 0 & 0 & 0 \\ 0.8489 & -0.0204 & 0.6064 & 0 & 0 & 0 \\ 0 & 0.6064 & -0.0305 & 0.5106 & 0 & -0.2783 \\ 0 & 0 & 0.5106 & 0.0005 & 0.7709 & 0 \\ 0 & 0 & 0 & 0.7709 & -0.0026 & 0.7898 \\ 0 & 0 & -0.2783 & 0 & 0.7898 & 0.0177 \end{bmatrix} \tag{4.2}$$

According to (4.2), the $\boldsymbol{M}_{ideal}$ matrix consists of 12 nonzero coupling parameters. Therefore, the model outputs of the hybrid DNN are $\boldsymbol{y} = [M_{11}\ M_{22}\ M_{33}\ M_{44}\ M_{55}\ M_{66}\ M_{12}\ M_{23}\ M_{34}\ M_{45}\ M_{56}\ M_{36}]^T$. The model inputs are S_{11} in dB at 41 frequency samples in the frequency range of 11,720–11,850 MHz, i.e. $\boldsymbol{x} = [dB(S_{11}(f_1))\ dB(S_{11}(f_2)) \ldots dB(S_{11}(f_{40}))\ dB(S_{11}(f_{41}))]^T$, where $f_1, f_2, \ldots, f_{40}, f_{41}$ are forty-one frequency samples. The dimension of the model inputs, i.e. 41 dimensions, is high. For every nonzero coupling parameter, a tolerance of ± 0.3 is used as the range for the training and test data. In this pre-defined range, random samples are generated for coupling parameters. Then, the corresponding filter is simulated to get the S-parameters. The S-parameters and coupling parameters are swapped to obtain the training and test data. After data generation, the hybrid DNN technique is used for the parameter-extraction modeling of this example. 200 hidden neurons are used for each hidden layer. After training, the final hybrid DNN model is composed of one input layer, eight sigmoid hidden layers, four smooth ReLU hidden layers, and one output layer. The training and test error of the trained DNN model are 1.31% and 1.79%, respectively. Besides, the model outputs are smooth and the derivatives are continuous [33]. Figure 4.2 shows the structure of the obtained hybrid DNN model for this example.

Once developed, the trained DNN parameter-extraction model can be used to extract the coupling matrix for different detuned filters. Here, two detuned filters are used to test the developed model. The desired coupling matrix and the extracted coupling matrix from the DNN model for the slightly detuned case are

$$\boldsymbol{M}_{desired} = \begin{bmatrix} 0.0127 & 0.9489 & 0 & 0 & 0 & 0 \\ 0.9489 & 0.0296 & 0.7064 & 0 & 0 & 0 \\ 0 & 0.7064 & -0.0805 & 0.4106 & 0 & -0.1783 \\ 0 & 0 & 0.4106 & 0.0505 & 0.6709 & 0 \\ 0 & 0 & 0 & 0.6709 & -0.0526 & 0.6898 \\ 0 & 0 & -0.1783 & 0 & 0.6898 & -0.0323 \end{bmatrix} \tag{4.3}$$

and

$$\boldsymbol{M}_{extracted} = \begin{bmatrix} 0.0173 & 0.9483 & 0 & 0 & 0 & 0 \\ 0.9483 & 0.0215 & 0.6993 & 0 & 0 & 0 \\ 0 & 0.6993 & -0.0753 & 0.4080 & 0 & -0.1828 \\ 0 & 0 & 0.4080 & 0.0500 & 0.6715 & 0 \\ 0 & 0 & 0 & 0.6715 & -0.0520 & 0.6918 \\ 0 & 0 & -0.1828 & 0 & 0.6918 & -0.0412 \end{bmatrix} \tag{4.4}$$

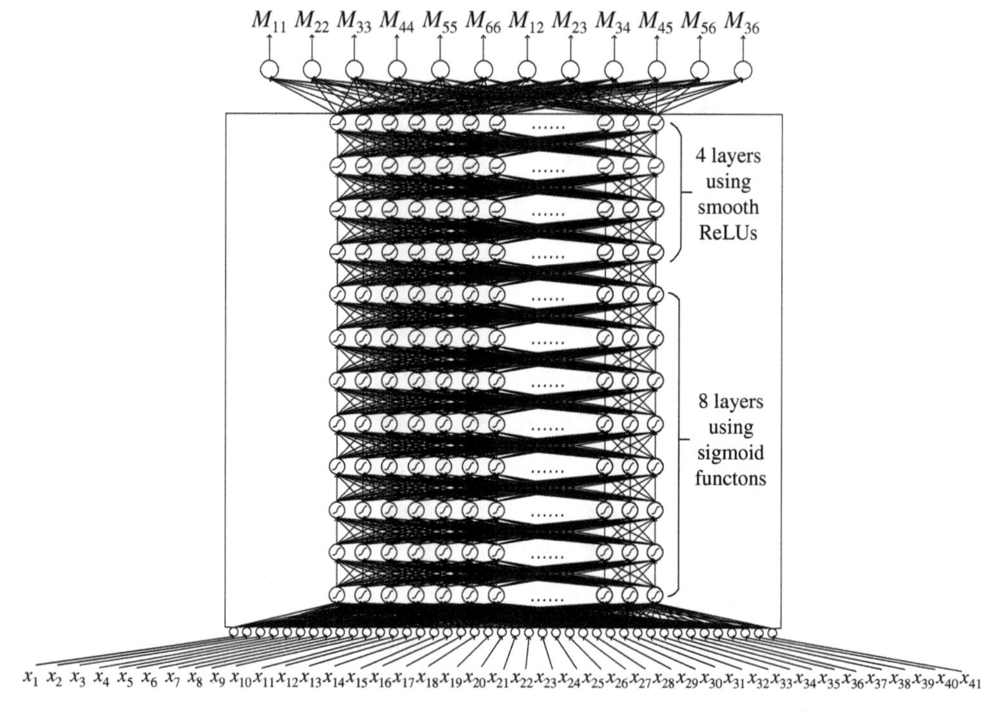

Figure 4.2 The structure of the hybrid DNN parameter-extraction model with continuous inputs and continuous outputs for the sixth-order filter. The input neurons represent S-parameters (continuous variables) at 41 frequency points, while the output neurons represent filter coupling parameters (continuous variables). It is a 14-layer hybrid DNN. The hidden layers include eight sigmoid layers and four smooth ReLU layers. Source: [33] Zhang et al. 2019/IEEE.

The desired coupling matrix and the extracted coupling matrix from the DNN model for the highly detuned case are

$$\boldsymbol{M}_{desired} = \begin{bmatrix} 0.2027 & 0.5989 & 0 & 0 & 0 & 0 \\ 0.5989 & 0.2296 & 0.8564 & 0 & 0 & 0 \\ 0 & 0.8564 & -0.2805 & 0.7606 & 0 & -0.0283 \\ 0 & 0 & 0.7606 & -0.2495 & 0.5209 & 0 \\ 0 & 0 & 0 & 0.5209 & -0.2526 & 0.5398 \\ 0 & 0 & -0.0283 & 0 & 0.5398 & 0.2677 \end{bmatrix} \tag{4.5}$$

and

$$\boldsymbol{M}_{extracted} = \begin{bmatrix} 0.1940 & 0.6153 & 0 & 0 & 0 & 0 \\ 0.6153 & 0.1811 & 0.8747 & 0 & 0 & 0 \\ 0 & 0.8747 & -0.2501 & 0.7481 & 0 & -0.0183 \\ 0 & 0 & 0.7481 & -0.2589 & 0.4957 & 0 \\ 0 & 0 & 0 & 0.4957 & -0.2312 & 0.5299 \\ 0 & 0 & -0.0183 & 0 & 0.5299 & 0.2659 \end{bmatrix} \tag{4.6}$$

Table 4.1 Comparison of shallow ANN and DNNs that have similar number of weighting parameters.

Structure of neural network	Hidden neurons per layer	Number of training parameters	Training error	Test error
shallow neural network (two sigmoid hidden layers)	514	292k	4.22%	4.89%
deep neural network (eight sigmoid hidden layers)	200	292k	1.95%	2.40%

Source: [33] Zhang and coworkers 2019/IEEE.

Table 4.2 Comparison between the Hybrid DNN and Pure ReLU DNNs.

Structure of deep neural networks	Training error	Test error
14-layer hybrid deep neural network[a)]	1.31%	1.79%
14-layer pure ReLU network	2.68%	3.16%
16-layer pure ReLU network	2.43%	3.00%
20-layer pure ReLU network	2.20%	2.73%

a) The 14-layer hybrid DNN has eight sigmoid hidden layers and four smooth ReLU hidden layers.
Source: [33] Zhang and coworkers 2019/IEEE.

According to (4.3)–(4.6), the extracted coupling parameters match well with the desired coupling parameters for both two test cases, which shows that the developed DNN is accurate and reliable for coupling parameter extraction for different cases within the training range.

For comparison purposes, a shallow ANN is used to develop the parameter-extraction model for this example. The shallow ANN has two hidden layers, where each hidden layer has 514 neurons. The number of training parameters in the shallow ANN is similar to that of the DNN having eight hidden layers and 200 neurons per hidden layer.

Table 4.1 shows the comparison between the shallow ANN and the DNN [33]. According to Table 4.1, when the shallow ANN and the DNN have the same activation functions and the similar number of training parameters, the DNN can improve the model accuracy more effectively as compared to the shallow ANN.

The hybrid DNN is also compared with the pure ReLU DNNs to demonstrate the benefit of combining smooth ReLUs and sigmoid functions. The comparison results are shown in Table 4.2 [33]. All hybrid DNN and the pure ReLU DNNs as shown in Table 4.2 have 200 neurons per hidden layer. According to Table 4.2, the hybrid DNN can achieve higher modeling accuracy using even fewer hidden neurons as compared to the pure ReLU DNNs.

4.4 Knowledge-Based Parametric Modeling for Microwave Components

In this chapter, automated knowledge-based parametric modeling with unified and elegant model structure adaptation algorithm [83] is described. A new formulation using l_1 optimization is presented to automatically determine the mapping neural network (NN) structures in the knowledge-based parametric model. An extended and unified structure for knowledge-based parametric model is presented to include different kinds of mapping structures. To train this unified model, a training algorithm with l_1 optimization is also presented and derived in this chapter. The most critical part of this training algorithm is the utilization of the l_1 optimization, which has the distinctive property for feature selection within the training process [82, 84]. After l_1 optimization, the weights in the mapping NNs include zero weights and non-zero weights. The mapping structure corresponding to the zero weights can be ignored and deleted in the final model. Therefore, taking use of this property, the type of the mappings in a knowledge-based parametric model and the nonlinearity of each mapping NN can be directly obtained from the l_1 optimization solutions. In this way, the final mapping structure is automatically adjusted by the presented method to achieve an accurate knowledge-based parametric model with a suitable and compact structure. It takes shorter time than conventional knowledge-based parametric modeling in which the knowledge-based model is a fixed structure.

4.4.1 Unified Knowledge-Based Parametric Model Structure

We define $\boldsymbol{x} = [x_1, x_2, \ldots, x_n]^T$ to denote a vector containing n inputs of the knowledge-based model. $\boldsymbol{x}$ usually involves the physical geometrical parameters of a microwave component. We define $\boldsymbol{y} = [y_1, y_2, \ldots, y_m]^T$ to denote a vector containing m outputs of the knowledge-based parametric model. $\boldsymbol{y}$ usually involves the EM response of the microwave component, such as S-parameters. There have been many empirical models to represent the $\boldsymbol{x} - \boldsymbol{y}$ relationship. However, the existing empirical model usually has limited accuracy and cannot always satisfy the accuracy requirement with the change of the values of the physical geometrical parameters. In this situation, it is necessary to use knowledge-based modeling with mapping structure to map the difference between the existing empirical model and the new EM data, therefore retaining the computation efficiency of the empirical model while preserving the accuracy of these new EM data. However, the mapping structure in the knowledge-based parametric model cannot be determined in a straightforward way because it is affected by many aspects. In this chapter, a unified knowledge-based parametric model structure is presented. Direct connections between the input neurons and output neurons are added for each mapping NN, so that the mapping NN becomes a combination of two-layer and three-layer perceptions. This presented mapping NN has the capability to represent either a linear mapping or a nonlinear mapping. The presented entire knowledge-based parametric model contains three mixed linear/nonlinear mapping NNs, i.e. the input mapping, the output mapping, and the frequency mapping, respectively, as shown in Figure 4.3.

To represent the model structure and training of the model, various symbols are defined. We define $\boldsymbol{f}_{map1}$ as the input mapping, f_{map2} as the frequency mapping and $\boldsymbol{f}_{map3}$ as the output mapping, respectively. Each of these mappings is a NN. We define H_1, H_2, and H_3 to denote the number of hidden neurons in these three NNs, respectively. We also define $\boldsymbol{u}_{map1}$, $\boldsymbol{u}_{map2}$, and $\boldsymbol{u}_{map3}$ to denote the vectors containing the weights for the direct input–output neuron connections in three NNs, respectively, and define $\boldsymbol{u}$ as a vector containing $\boldsymbol{u}_{map1}$, $\boldsymbol{u}_{map2}$, and $\boldsymbol{u}_{map3}$, i.e. $\boldsymbol{u} = [\boldsymbol{u}_{map1}^T\ \boldsymbol{u}_{map2}^T\ \boldsymbol{u}_{map3}^T]^T$. We define $\boldsymbol{t}_{map1}$, $\boldsymbol{t}_{map2}$, and $\boldsymbol{t}_{map3}$ to denote the vectors containing the weights for the input-hidden neuron connections in three NNs, respectively. We also define $\boldsymbol{v}_{map1}$, $\boldsymbol{v}_{map2}$, and $\boldsymbol{v}_{map3}$ to denote the vectors containing the weights between the hidden-output neuron connections in three NNs, respectively, and define $\boldsymbol{v}$ as a vector containing $\boldsymbol{v}_{map1}$, $\boldsymbol{v}_{map2}$, and $\boldsymbol{v}_{map3}$, i.e. $\boldsymbol{v} = [\boldsymbol{v}_{map1}^T\ \boldsymbol{v}_{map2}^T\ \boldsymbol{v}_{map3}^T]^T$. Therefore, the weights in $\boldsymbol{u}$ can be regarded as the linear mapping weights, while the weights in $\boldsymbol{v}$ can be regarded as the nonlinear mapping weights.

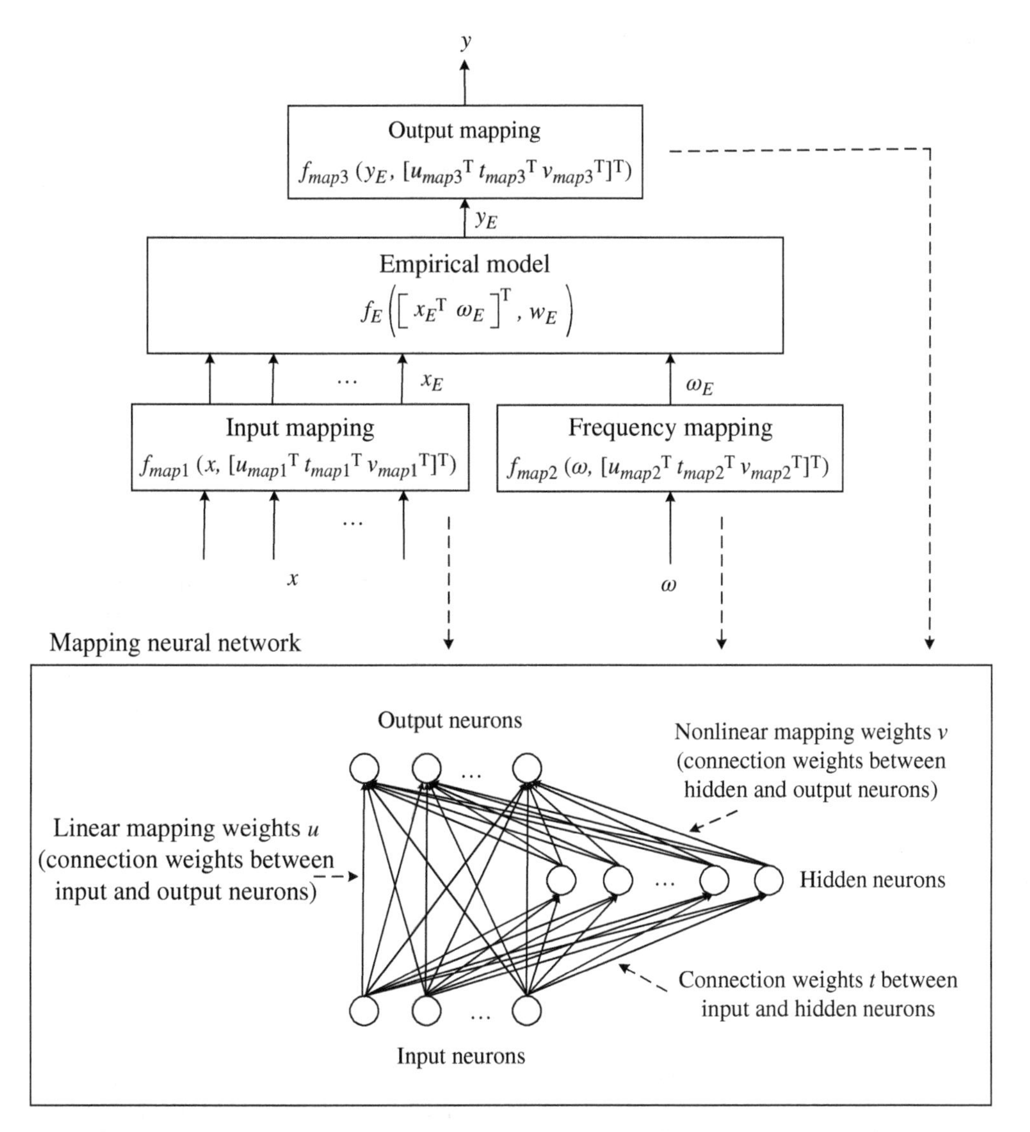

Figure 4.3 Unified knowledge-based parametric model structure. Source: Adapted from [83].

Let $\boldsymbol{x}_E = [x_{E1}, x_{E2}, \dots, x_{En}]^T$ represent the geometrical inputs of the empirical model, ω_E represent the frequency input of the empirical model, and $\boldsymbol{y}_E = [y_{E1}, y_{E2}, \dots, y_{En}]^T$ represent the outputs of the empirical model, respectively. Therefore,

$$\boldsymbol{y}_E = \boldsymbol{f}_E\left([\boldsymbol{x}_E^T \ \omega_E]^T, \boldsymbol{w}_E\right) \tag{4.7}$$

is the formulation of the empirical model [83] where $\boldsymbol{w}_E$ contains the parameters in the existing empirical model. When the empirical model is not accurate enough to match the new EM data, mappings are used to adjust the model. The input mapping aims to map the values of physical geometrical parameters to make $\boldsymbol{y}_E$ at the mapped new set of values fit the new EM data better. $\boldsymbol{x}$ and $\boldsymbol{x}_E$ represent the original and the modified values of the geometrical parameters which is given to the empirical model, respectively. A NN is suitable to represent the input mapping between $\boldsymbol{x}$

and $\boldsymbol{x}_E$. In the presented method, we define the input mapping NN as [83]

$$\boldsymbol{x}_E = \boldsymbol{f}_{map1}(\boldsymbol{x}, \boldsymbol{w}_{map1}) \tag{4.8}$$

where $\boldsymbol{w}_{map1} = [\boldsymbol{u}_{map1}^T \ \boldsymbol{t}_{map1}^T \ \boldsymbol{v}_{map1}^T]^T$ are the weights in $\boldsymbol{f}_{map1}$. The input mapping NN structure is formulated as follows to include both linear and nonlinear mappings, where the ith output of $\boldsymbol{f}_{map1}$ is [83]

$$f_{map1,i} = \sum_{j=0}^{n}(u_{map1,ij} \cdot x_j) + \sum_{k=0}^{H_1}(v_{map1,ik} \cdot z_{map1,k}), \quad i = 1, 2, \ldots, n \tag{4.9a}$$

$$z_{map1,k} = \begin{cases} \sigma\left(\sum_{j=0}^{n} t_{map1,kj} \cdot x_j\right), & k \neq 0 \\ 1, & k = 0 \end{cases} \tag{4.9b}$$

$$x_0 = 1 \tag{4.9c}$$

where $z_{map1,k}$ is the kth hidden neuron output in $\boldsymbol{f}_{map1}$. x_0 and $z_{map1,0}$ are regarded as the bias parameters in $\boldsymbol{f}_{map1}$. $\sigma(\cdot)$ is the sigmoid activation function. $u_{map1,ij}$, $v_{map1,ik}$, and $t_{map1,kj}$ are the connection weights of $\boldsymbol{f}_{map1}$, where i, j, and k are the indices of the neurons in the output layer, hidden layer, and input layer, respectively. Therefore, the input mapping includes linear/nonlinear or no mapping by only one NN.

The frequency mapping is useful when there is a frequency shift between the response of the EM data and the empirical model. The frequency mapping NN structure is formulated as [83]

$$\begin{aligned} \omega_E &= f_{map2}(\omega, \ \boldsymbol{w}_{map2}) \\ &= \sum_{b=0}^{1}(u_{map2,1b} \cdot \omega_b) + \sum_{c=0}^{H_2}(v_{map2,1c} \cdot z_{map2,c}) \end{aligned} \tag{4.10a}$$

$$z_{map2,c} = \begin{cases} \sigma\left(\sum_{b=0}^{1} t_{map2,cb} \cdot \omega_b\right), & c \neq 0 \\ 1, & c = 0 \end{cases} \tag{4.10b}$$

$$\omega_0 = 1 \tag{4.10c}$$

where $\boldsymbol{w}_{map2}$ is the vector containing the weights in f_{map2}, i.e. $\boldsymbol{w}_{map2} = [\boldsymbol{u}_{map2}^T \ \boldsymbol{t}_{map2}^T \ \boldsymbol{v}_{map2}^T]^T$. We define $z_{map2,c}$ as the cth hidden neuron output in f_{map2}. ω_0 and $z_{map2,0}$ are regarded as the bia parameters in f_{map2}. $u_{map2,1b}$, $v_{map2,1c}$, and $t_{map2,cb}$ represent the connection weights in f_{map2}, where 1, b, and c represent the indices of the neurons in the input layer, output layer, and hidden layer, respectively. Therefore, the frequency mapping includes linear/nonlinear or no mapping by only one NN.

The output mapping aims to improve the robustness and reliability of the model. The output mapping NN structure is formulated as [82]

$$\boldsymbol{y} = \boldsymbol{f}_{map3}(\boldsymbol{y}_E^{\prime} \boldsymbol{w}_{map3}) \tag{4.11}$$

where $\boldsymbol{w}_{map3}$ is the vector containing the weights in $\boldsymbol{f}_{map3}$, i.e. $\boldsymbol{w}_{map3} = [\boldsymbol{u}_{map3}^T \ \boldsymbol{t}_{map3}^T \ \boldsymbol{v}_{map3}^T]^T$. Similarly as (4.9a) and (4.10a), the internal structure of the output mapping neural network is formulated to include linear and nonlinear mappings as follows, where the rth output of the output mapping is [83]

$$f_{map3,r} = \sum_{p=0}^{m}(u_{map3,rp} \cdot y_{E,p}) + \sum_{q=0}^{H_3}(v_{map3,rq} \cdot z_{map3,q}), \quad r = 1, 2, \ldots, m \tag{4.12a}$$

$$z_{map3,q} = \begin{cases} \sigma\left(\sum_{p=0}^{m} t_{map3,qp} \cdot y_{E,p}\right), & q \neq 0 \\ 1, & q = 0 \end{cases} \tag{4.12b}$$

$$y_{E,0} = 1 \tag{4.12c}$$

where $z_{map3,q}$ is the qth hidden neuron output of $\boldsymbol{f}_{map3}$. $y_{E,0}$ and $z_{map3,0}$ are regarded as the bias parameters in $\boldsymbol{f}_{map3}$. $u_{map3,rp}$, $v_{map3,rq}$, and $t_{map3,qp}$ represent the connection weights in $\boldsymbol{f}_{map3}$, where p, q, and r represent the indices of the neurons in the input layer, hidden layer, and output layer, respectively. Therefore, the presented output mapping includes linear/nonlinear or no mapping by only one NN.

Let $\boldsymbol{w} = [\boldsymbol{w}_{map1}^T\ \boldsymbol{w}_{map2}^T\ \boldsymbol{w}_{map3}^T]^T$ be defined as a vector containing all the neural network weights. Combining all the mappings together with the empirical model, the overall $\boldsymbol{x}$-$\boldsymbol{y}$ relationship can be represented as [83]

$$\begin{aligned} \boldsymbol{y} &= \boldsymbol{f}_{map3}(\boldsymbol{y}_E, \boldsymbol{w}_{map3}) \\ &= \boldsymbol{f}_{map3}\left(\boldsymbol{f}_E\left(\left[\boldsymbol{x}_E^T\ \omega_E\right]^T, \boldsymbol{w}_E\right), \boldsymbol{w}_{map3}\right) \\ &= \boldsymbol{f}_{map3}\left(\boldsymbol{f}_E\left(\left[\boldsymbol{f}_{map1}^T(\boldsymbol{x}, \boldsymbol{w}_{map1})\ f_{map2}(\omega, \boldsymbol{w}_{map2})\right]^T, \boldsymbol{w}_E\right), \boldsymbol{w}_{map3}\right). \end{aligned} \tag{4.13}$$

This input and output relationship is determined by both the empirical model and various mappings. In this way, we are allowed to change the model by changing the mappings, which includes changing the topology of the mapping structure and changing the values of weights in the mapping NNs.

All cases of mappings are included in this presented unified and elegant knowledge-based model, i.e. the combination of no/linear/nonlinear mapping and input/frequency/output mapping. Generally speaking, a simple modeling problem with a narrow modeling range requires linear mappings while a complicated modeling problem with a wider modeling range requires nonlinear mappings. The quantitative decision of use which kind of mapping depends on the specific modeling problem and cannot be known before the modeling process. In addition, because the input/frequency/output mapping play different roles on the modeling results, each of them can be linear or nonlinear [82].

Existing automated knowledge-based model structure adaptation methods usually try different mapping structures in a brute force sequential manner. Then these different modeling results are compared to choose the most suitable one. To automatically select the appropriate final mapping structure, a new training method with l_1 optimization is presented to train the unified structure.

4.4.2 Training with l_1 Optimization of the Unified Knowledge-Based Parametric Model

Training data are generated before the model training. The training data are defined as $(\boldsymbol{x}^{(l)}, \boldsymbol{d}^{(l)})$, $l = 1, 2, \ldots, N_L$, where $\boldsymbol{d}^{(l)}$ represents the desired model output for the lth training sample $\boldsymbol{x}^{(l)}$, and N_L represents the number of training samples.

To automatically select the mapping structure, two-stage model training is performed with l_1 optimization. In the first stage, the presented training determines whether each of the three mappings in the knowledge-based model should be linear or nonlinear. In the second stage, if a linear mapping exists, the algorithm determines whether it can be further deleted or not. After the two-stage training, the presented algorithm automatically produces the knowledge-based model with appropriate mapping for a given modeling problem.

In the first-stage training, the l_1 norms of the weights $\boldsymbol{v}$ are added to the training error function [1], formulated as [83]

$$\begin{aligned} E^1_{train}(\boldsymbol{w}) = & \sum_{l=1}^{N_L} \left(\frac{1}{2} \left\| \boldsymbol{y}(\boldsymbol{x}^{(l)}, \boldsymbol{w}) - \boldsymbol{d}^{(l)} \right\|^2 \right) + \sum_{i=1}^{n} \sum_{k=0}^{H_1} \lambda^1_{map1,ik} \left| v_{map1,ik} \right| \\ & + \sum_{c=0}^{H_2} \lambda^1_{map2,1c} \left| v_{map2,1c} \right| + \sum_{r=1}^{m} \sum_{q=0}^{H_3} \lambda^1_{map3,rq} \left| v_{map3,rq} \right| \end{aligned} \tag{4.14}$$

where $\lambda^1_{map1,ik}$, $\lambda^1_{map2,1n}$, and $\lambda^1_{map3,rq}$ are non-negative constants. Since the weights $\boldsymbol{v}$ are nonlinear mapping weights, the purpose of the first-stage training is to adjust $\boldsymbol{w}$ to make the training error as small as possible and make as many $\boldsymbol{v}$ to zeros as possible.

In the presented first-stage training, we take advantage of the distinctive property of l_1 norms that some large values in $\boldsymbol{v}$ can be kept [82, 84, 85]. The solution of the optimization may well be a few non-zeros v's, while others are zeros. In this way, the l_1 norms automatically select the important components of $\boldsymbol{v}$. It is these nonzero v's that relate to the selected nonlinear mapping structure. If all v's are equal to zeros, it is linear mapping; otherwise, it is nonlinear mapping. In this way, the decision of using whether linear or nonlinear mapping is automatically made by the presented algorithm after the first-stage training.

As we described before, there exist three mappings in the presented unified knowledge-based parametric model and each of them can be linear or nonlinear. After the first-stage training, if these three mappings are all nonlinear, the modeling process stops and the final knowledge-based model has three nonlinear mappings. Otherwise, the algorithm performs to the second-stage training. In the second-stage training, the weights $\boldsymbol{u}$ are retained, while the weights in the nonlinear mapping NNs are no longer trained by setting as constants. The second-stage training decides whether the linear mapping NNs of the first-stage training solution can be further deleted or not. We formulate the second-stage training error function with l_1 norms as [83]

$$\begin{aligned} E^2_{train}(\boldsymbol{u}) = & \sum_{l=1}^{N_L} \left(\frac{1}{2} \left\| \boldsymbol{y}(\boldsymbol{x}^{(l)}, \boldsymbol{w}) - \boldsymbol{d}^{(l)} \right\|^2 \right) + \sum_{i=1}^{n} \sum_{j=0, j \neq i}^{n} \lambda^2_{map1,ij} |u_{map1,ij}| \\ & + \sum_{i=1}^{n} \lambda^2_{map1,ii} |u_{map1,ii} - 1| + \lambda^2_{map2,10} |u_{map2,10}| + \lambda^2_{map2,11} |u_{map2,11} - 1| \\ & + \sum_{r=1}^{m} \sum_{p=0, p \neq r}^{m} \lambda^2_{map3,rp} |u_{map3,rp}| + \sum_{r=1}^{m} \lambda^2_{map3,rr} |u_{map3,rr} - 1| \end{aligned} \tag{4.15}$$

where $\lambda^2_{map1,ij}$, $\lambda^2_{map2,10}$, $\lambda^2_{map2,11}$, and $\lambda^2_{map3,rp}$ are non-negative constants. $\lambda^2_{map1,ij}$, or $\lambda^2_{map2,10}$ and $\lambda^2_{map2,11}$, or $\lambda^2_{map3,rp}$ will be equal to zero if the input mapping $\boldsymbol{f}_{map1}$, the frequency mapping f_{map2}, or the output mapping $\boldsymbol{f}_{map3}$ is nonlinear. After the second-stage training, if

$$\begin{cases} u_{map1,ij} = 0, \ i \neq j \\ u_{map1,ij} = 1, \ i = j \end{cases} \tag{4.16}$$

i.e.

$$\boldsymbol{x}_E = \boldsymbol{x} \tag{4.17}$$

$\boldsymbol{f}_{map1}$ is unnecessary. Similarly, if

$$\begin{cases} u_{map2,10} = 0 \\ u_{map2,11} = 1 \end{cases} \tag{4.18}$$

f_{map2} is unnecessary. If

$$\begin{cases} u_{map3,rp} = 0, \; r \neq p \\ u_{map3,rp} = 1, \; r = p \end{cases} \tag{4.19}$$

$\boldsymbol{f}_{map3}$ is unnecessary. Otherwise, the final model needs linear mapping. After above procedures, the presented training decides whether a mapping is necessary or not automatically [83].

4.4.3 Automated Knowledge-Based Model Generation

The presented training algorithm with l_1 optimization is used to train the presented unified knowledge-based parametric model. The topology and type of the mapping structure is automatically determined. We define the testing error $E_{test}(\boldsymbol{w}^*)$ as [83]

$$E_{test}(\boldsymbol{w}^*) = \sum_{v=1}^{N_V} \left(\frac{1}{2} \left\| \boldsymbol{y}(\boldsymbol{x}^{(v)}, \boldsymbol{w}^*) - \boldsymbol{d}^{(v)} \right\|^2 \right) \tag{4.20}$$

where $\boldsymbol{w}^*$ is a vector of all optimal NN weights after the training process, $\boldsymbol{d}^{(v)}$ is the desired response of the vth testing sample $\boldsymbol{x}^{(v)}$, and N_V is the total number of testing data.

Let E_d be defined as the user-required error threshold. Once $E_{test} \leq E_d$ is achieved in the first-stage training, we automatically check the values of $\boldsymbol{v}$ in the input/frequency/output mapping NNs. If three mappings are all nonlinear, we stop and the final model has three nonlinear mappings. Otherwise, the second-stage training is performed to make the decision whether the linear mapping is further simplified/deleted or not. In the second-stage training, the weights in the linear mapping are retained to achieve $E_{test} \leq E_d$ again.

Figure 4.4 shows the flowchart of automated model generation of the knowledge-based parametric model.

4.4.4 Knowledge-Based Parametric Modeling of a Two-Section Low-Pass Elliptic Microstrip Filter

We consider the knowledge-based parametric modeling of a low-pass microstrip filter [63, 86, 87]. EM data are generated using simulator *CST Microwave Studio* and design of experiment DOE sampling method to train and test the parametric model. The geometrical input parameters of the model are $\boldsymbol{x} = \left[L_1 \; L_2 \; L_{c1} \; L_{c2} \; W_c \; G_c\right]^T$ and the output of the model is the magnitude of S_{21}. The existing equivalent circuit model [70, 87] is used as the empirical model in this example. Because the empirical model is still not accurate enough after optimization, mappings are added using the presented method. For the model training, the values of $\lambda^1_{map1,ik}$, $\lambda^1_{map2,1c}$ and $\lambda^1_{map3,rq}$ in (4.14) are equal to 10.

Knowledge-based models are developed for two different cases for comparison purpose. The modeling range for Case 1 is narrow, while for Case 2 is wider, as shown in Table 4.3. In this modeling example, the testing error threshold is 2%.

Table 4.4 and Figure 4.5 show the modeling results. For Case 1, after the two-stage training with l_1 optimization, the final mapping structure only contains a linear input mapping. We set the values of $\lambda^2_{map1,ij}$, $\lambda^2_{map2,10}$, $\lambda^2_{map2,11}$, and $\lambda^2_{map3,rp}$ in (4.15) to be 10. All nonlinear mapping weights $\boldsymbol{v}_{map1}$, $\boldsymbol{v}_{map2}$,

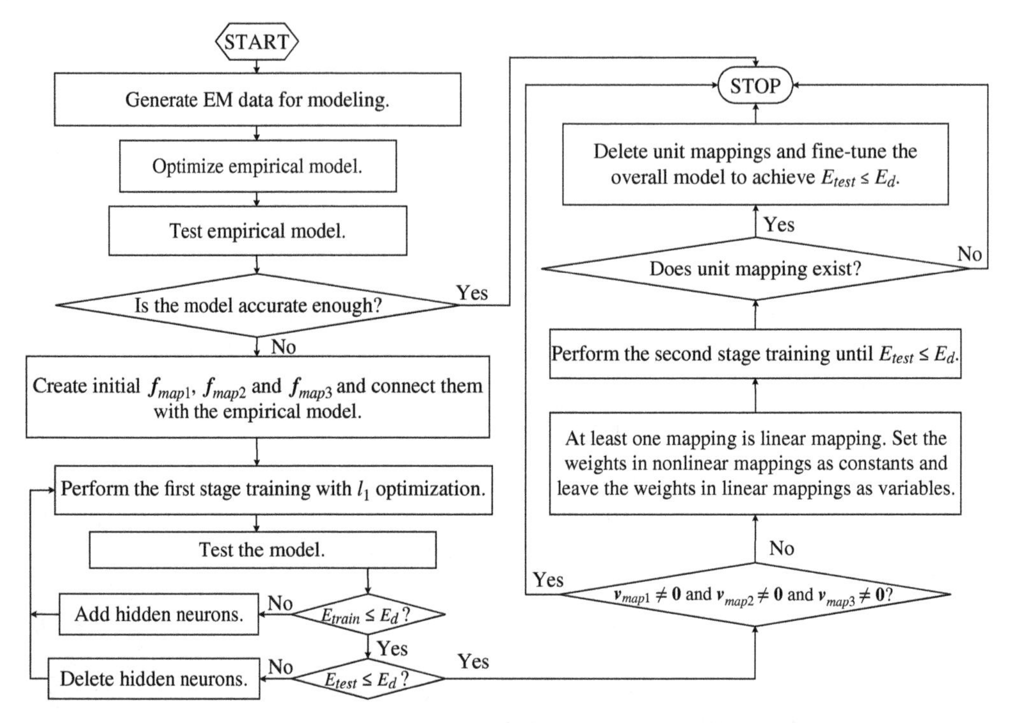

Figure 4.4 Flowchart of the automated model generation for knowledge-based parametric model. Source: Adapted from [82].

Table 4.3 Modeling range for the low-pass filter example.

	Training data		Testing data	
	Min	**Max**	**Min**	**Max**
Input		**Case 1**		
L_1 (mil)	44.88	48.82	45.28	48.42
L_2 (mil)	129.92	137.8	130.32	137.4
L_{c1} (mil)	159.84	167.72	160.24	167.32
L_{c2} (mil)	44.88	48.82	45.28	48.42
W_c (mil)	6.3	7.85	6.5	7.68
G_c (mil)	1.96	2.76	2.05	2.72
Freq (GHz)	1	4	1	4
		Case 2		
L_1 (mil)	40.15	50	40.55	49.6
L_2 (mil)	125.2	144.88	126	143.7
L_{c1} (mil)	155	174.8	156	174
L_{c2} (mil)	39.4	50	40.16	49.2
W_c (mil)	5.9	0.84	6.3	9.45
G_c (mil)	1.96	3.54	2.08	3.35
Freq (GHz)	1	4	1	4

Source: Adapted from [83].

Table 4.4 Modeling results for low-pass filter example.

	Case 1	Case 2
Final mapping structure	Linear input mapping	Nonlinear input mapping + Nonlinear frequency mapping
No. of Training data	81*101	81*101
No. of Testing data	64*101	64*101
Training error	1.89%	1.63%
Testing error	1.92%	1.97%

Source: Adapted from [83].

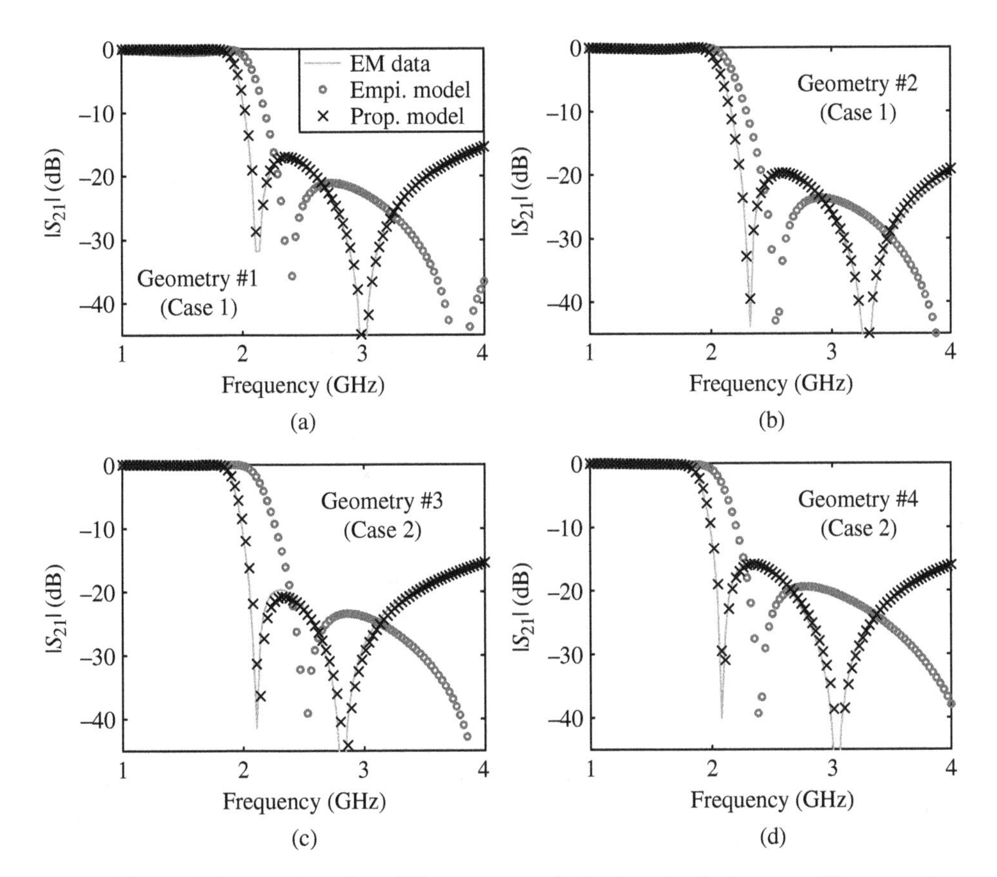

Figure 4.5 Modeling results at four different geometrical values for the low-pass filter example: (a) $\boldsymbol{x} = [45.28, 3137.4, 167.32, 45.67, 7.09, 2.36]^T$ (mil), (b) $\boldsymbol{x} = [48.42, 136.61, 160.63, 46.06, 7.48, 2.72]^T$ (mil), (c) $\boldsymbol{x} = [40.55, 138.58, 168.9, 48.03, 9.06, 2.28]^T$ (mil) and (d) $\boldsymbol{x} = [44.49, 136.22, 171.26, 43.31, 8.27, 2.36]^T$ (mil). Source: [82] Na et al. 2017/IEEE.

and $\boldsymbol{v}_{map3}$ are almost zeros, i.e. all mappings are linear. In addition, both the frequency mapping and the output mapping are unit mapping, so that they are unnecessary in the final model structure. This modeling result is consistent with our cognition that because the modeling range is narrow, a linear mapping is sufficient to achieve an accurate model.

Table 4.5 Comparison between the model from the unified KBNN modeling method and the models from Brute Force Modeling method with different combinations of mappings for the low-pass filter modeling problem.

Mapping structure	Testing error (%)	
	Case 1	Case 2
Optimal mapping structure by Unified KBNN modeling: **Linear input mapping**	**1.92**	/
Optimal mapping structure by Unified KBNN modeling: **(Nonlinear (Input + Frequency) mapping**	/	**1.97**
No mapping	16.43	14.33
Linear input mapping	1.92	2.38
Nonlinear input mapping	1.88	2.04
Linear output mapping	24.95	13.80
Nonlinear output mapping	7.73	9.10
Linear (Input + Frequency + Output) mapping	1.89	2.17
Nonlinear (Input + Frequency + Output) mapping	1.85	1.96

Source: [83] Na et al. 2017/IEEE.

For Case 2, l_1 optimization is also performed for the model development. After l_1 optimization, v_{map1} and v_{map2} are non-zeros and v_{map3} are zeros. We set $\lambda^2_{map1,ij}$, $\lambda^2_{map2,10}$ and $\lambda^2_{map2,11}$ in (4.15) to be 0, and set the values of $\lambda^2_{map3,rp}$ in (4.15) to be 10. In addition, because the output mapping is unit mapping and not needed, a nonlinear input mapping and a nonlinear frequency mapping are automatically chosen as the final mapping structure. From the modeling results in Cases 1 and 2, we can see that the presented unified automated knowledge-based parametric modeling algorithm uses l_1 optimization to select the topology and type of mapping structure for the given modeling problem automatically.

In addition, the modeling results of the presented algorithm with existing brute force modeling method are also compared, as listed in Table 4.5. In brute force modeling, different mapping structures have to be tried one-by-one to select an accurate and compact knowledge-based model. Only some examples of these trials are shown in Table 4.5. In reality, there are more possibilities of the mappings in a knowledge-based parametric model.

Furthermore, we compare the modeling results using the presented knowledge-based parametric model with adaptive mapping structure and the fixed structured knowledge-based model. We consider a fixed model structure with three nonlinear mappings and fewer training data are used to develop models in Cases 1 and 2. Table 4.6 shows the modeling results. We can see that the model with three nonlinear mappings has larger testing error, which cannot meet the user-desired accuracy, than the presented knowledge-based model. This means the model with fixed three nonlinear mappings is over-learning. In this situation, we need to add more training data and continue the training process to finally achieve a model that meets the user-required accuracy. Because the presented unified knowledge-based modeling algorithm with l_1 optimization preserves the model

Table 4.6 Comparison of the modeling results using fixed mapping structure and the presented adaptive mapping structure.

Final mapping structure	Case 1		Case 2	
	3 nonlinear mappings (fixed)	Linear input mapping (presented)	3 nonlinear mappings (fixed)	Nonlinear input mapping (presented)
No. of Training data	49*101		49*101	
No. of Testing data	64*101		64*101	
Training error	1.21%	1.45%	1.87%	1.74%
Testing error	2.82%	1.92%	2.43%	1.98%

Source: [83] Na et al. 2017/IEEE.

accuracy and makes the model to be as compact as possible, fewer data are needed to develop the model. Therefore, when the number of training data is limited, it is more efficient to use the presented modeling method than simply choosing a fixed mapping structure, e.g. three nonlinear mappings. On the other hand, when the number of training data is sufficient, the model developed by the presented method is more compact and simpler than the model with fixed three nonlinear mappings, making it much easier to be used for high-level circuit design [83].

4.5 Parametric Modeling Using Combined ANN and Transfer Function

An improved knowledge-based modeling method that combines neural networks and transfer functions (neuro-transfer function or neuro-TF) is utilized to implement parametric modeling of EM performances effectively [64–71]. The knowledge-based method is applicative in cases where equivalent circuits or empirical models are not available. When using the neuro-TF approach, transfer functions are used as prior knowledge, expressing the highly nonlinear relationship between electromagnetic response and frequency. With the prior knowledge, the residual relationship between the coefficients of the transfer function and the geometric parameters has considerable linearity. It is much simpler for a neural network to learn this relationship than to aimlessly learn the entirely nonlinear frequency response. Hence, the efficiency is higher when using neuro-TF for training. The overall composition of the neural-TF model is displayed in Figure 4.6.

4.5.1 Neuro-TF Modeling in Rational Form

Rational form of a neural-TF model includes a rational transport function and a neural network [65]. Let $\boldsymbol{x}$ denote the vector representation of the input of the neural network, the outputs are the coefficients in the transport function. The transfer function coefficients are extracted from the EM response using the commonly used vector-fitting method [88]. The neural network must be trained in order to learn the non-linear relation between the input and output of the neural

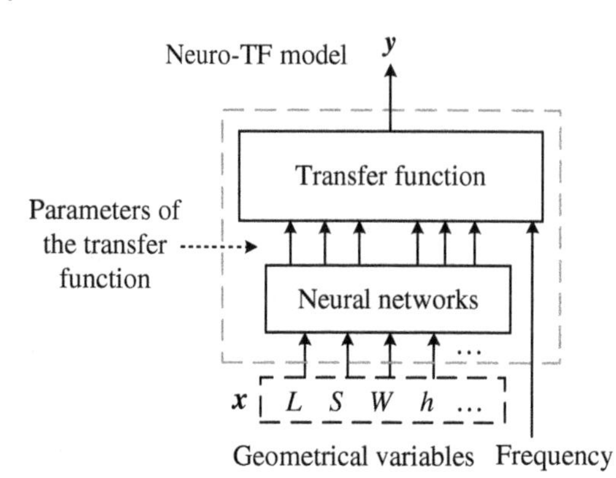

Figure 4.6 Overall composition of the neural-TF model. Source: (Adapted from [65]).

network. Let H be the transport function output. The transport function in rational form is

$$H(\boldsymbol{x}, \boldsymbol{w}, s) = \frac{\sum_{i=0}^{N} a_i(\boldsymbol{x}, \boldsymbol{w}) s^i}{1 + \sum_{i=1}^{N} b_i(\boldsymbol{x}, \boldsymbol{w}) s^i} \tag{4.21}$$

where s is the complex frequency in Laplace transform; N is the order of the transport function; a_i and b_i are coefficients of the rational transport function; $\boldsymbol{w}$ represents the internal neural network weights. The neural network learns the non-linear relation between a_i, b_i, and the design parameters.

Set $\boldsymbol{y}$ be the vector representation of the neuro-TF model output, containing the real and imaginary components of H. $\boldsymbol{d}$ is the vector form of the train data exported from the EM emulators. Through the training process, the weight $\boldsymbol{w}$ of the neural network is continuously updated, so that the error between $\boldsymbol{y}$ and $\boldsymbol{d}$ [65] can be reduced to the minimum.

4.5.2 Neuro-TF Modeling in Zero/Pole Form

Model order reduction (MOR) technique is another approach to build a neuro-TF model, more specifically, the matrix Padé via Lanczos (MPVL) algorithm, to calculate the poles and zeros in the transport function to address the order-changing problem. Two main parts are identified in the model: zero-pole-based transfer functions in complicated propagation spaces and neural networks for learning the association between poles/zeros/gain and geometric parameters. The transport function in zero/pole form can be expressed as follows [71]:

$$H(\omega) = \gamma(\omega) K \frac{\prod_{i=1}^{q-1} (\gamma(\omega) - \gamma_0 - z_i)}{\prod_{i=1}^{q} (\gamma(\omega) - \gamma_0 - p_i)} + c \tag{4.22}$$

where $H(\omega)$ is the frequency response of the microwave passive component at frequency ω; $\gamma(\omega)$ is a frequency variable (i.e. the propagation constant); p_i, z_i, and K represent the poles, the zeros, and the gain coefficient of the transport function, severally; q is the order of the transfer function which is determined by the Pade via Lanczos (PVL) algorithm; γ_0 is dependent on the expansion point in frequency, and c is a constant. The zeros and poles in $H(\omega)$ are referred to as the zeros and poles in the "γ-domain" (the complex propagation domain) [71].

Each time the geometrical parameters are changed, p_i, z_i, and K need to be recomputed by MPVL and eigendecomposition. However, there may not exist clear correspondences between the poles/zeros before and after recalculation. As a result, if these raw poles/zeros are used to forecast the newly changed poles/zeros of the geometric parameters, the solution may be wrong. This is the so-called pole and zero mismatch problem [71].

4.5.3 Neuro-TF Modeling in Pole/Residue Form

Using a pole/residual form of the transport function is yet another modeling approach in neuro-TF. The neuro-TF model in the form of pole/residue format is made up of a pole-residue-based transport function and an ANN. The design parameters and frequencies of the model are used as inputs, and the electromagnetic response is used as output. The transport function based on pole/residue format can be expressed as [68]

$$H(s) = \sum_{i=1}^{N} \frac{r_i}{s - p_i} \tag{4.23}$$

where r_i and p_i denote the residues and poles in the transport function, severally. N is the order of the transport function.

4.5.4 Vector Fitting Technique for Parameter Extraction

When using neuro-TF modeling, for every geometric sampling, the vector fitting approach is firstly used [88] to find a set of poles and residues with the least order transport function. Use $\boldsymbol{p}$ and $\boldsymbol{r}$ to denote the poles and residues, respectively. What is known about the vector fitting procedure is the EM behavior $\boldsymbol{d}$, i.e. S parameter, for all the geometric samplings. By setting up frequency offset and frequency scaling techniques during vector fitting process, all the poles and residues in complex-valued form can be obtained [68].

Set $\boldsymbol{c}_k$ be the vector representation of the poles and residues in the transport function of the kth geometric sampling after the process of vector fitting, which can be expressed as

$$\boldsymbol{c}_k = \begin{bmatrix} \boldsymbol{p}^{(k)} \\ \boldsymbol{r}^{(k)} \end{bmatrix} = [p_1^{(k)} p_2^{(k)} \cdots p_{n_k}^{(k)} r_1^{(k)} r_2^{(k)} \cdots r_{n_k}^{(k)}]^T \tag{4.24}$$

where n_k is the order of the transport belonging to the kth geometric sampling. $\boldsymbol{p}^{(k)}$ is the vector representation of all the poles of the kth geometric sampling, i.e. $\boldsymbol{p}^{(k)} = [p_1^{(k)} p_2^{(k)} \cdots p_{n_k}^{(k)}]^T$. In a similar way, $\boldsymbol{r}^{(k)}$ is the vector representation of all the poles of the kth geometric sampling, i.e. $\boldsymbol{r}^{(k)} = [r_1^{(k)} r_2^{(k)} \cdots r_{n_k}^{(k)}]^T$.

Due to the nonlinearity and uncertainty of the relationship between the pole/residues and geometrical parameters, neural networks are used to establish the mapping relationship between the pole/residues and the geometric parameters. Because different values of geometric parameters have various minimum orders, the number of poles and residues in $\boldsymbol{c}_k$ may vary with the changing geometric samplings. The training of neural networks will become more difficult as a result of this variation. An improved pole-residue tracing approach is employed to deal with the variations in the transport functions order.

4.5.5 Two-Phase Training for Neuro-TF Models

The training of the neuro-TF model adopts a two-phase training algorithm [65]. In the first phase, an initial training procedure is carried out. In this process, the weights in the neural network are changed in order to study the relationship between the parameters in the transport function and

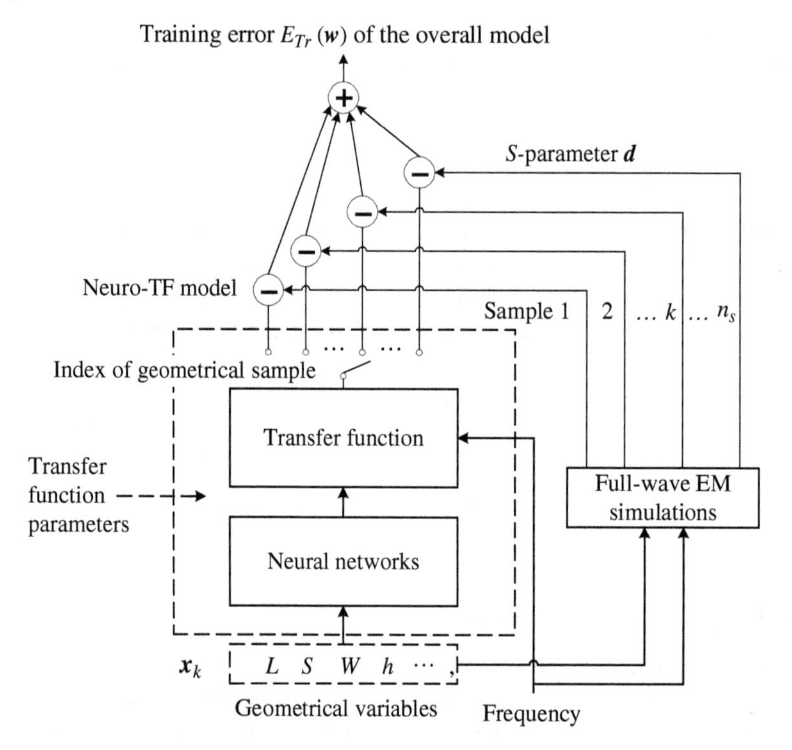

Figure 4.7 Refinement training procedure for the general form of the neuro-TF. Source: (Adapted from [65]).

the geometric variables. Set k be the label of the training examples, $k \in T_r$, where T_r denotes a collection of labels containing training examples under varying geometric variables. The train data in this one phase is $(\boldsymbol{x}_k, \boldsymbol{C}_k)$, i.e. examples of geometric variables are used as inputs to the model and the transfer function parameters as model outputs, where $\boldsymbol{C}_k$ denotes the set including every coefficient of the kth training example. After the initial training of the neural networks, training in phase two is carried out to further improve the ultimate model.

A model refining to improve the final neuro-TF model is performed during the second phase. Train data in this phase is denoted as $(\boldsymbol{x}_k, \boldsymbol{d}_k)$, $k \in T_r$, i.e. examples of geometric variables as inputs, the corresponding EM responses as outputs to the model. Figure 4.7 has illustrated the realization process of refining training for the entire model. The neuro-TF in process of refining training is constituted by a transport function of (4.21) of rational form and neural networks, the original starting points of which are the optimum solutions obtained from the preliminary training. Training and testing process of the neuro-TF are simultaneously included in the model refining procedure.

The training error E_{Tr} is the difference between the training data and the EM response. When the overall model training error falls below an already-defined threshold E_t, training in phase two (i.e. the refining process) terminates. After finishing the training course, the performance of the neural-TF model after training is checked using a group of independent test data that is never utilized in previous training. The test error E_{Te} represents the difference between the testing data and the EM response. Suppose that both training and testing errors are less than the threshold, denoted as E_t, the refining procedure stops, and the neuro-TF model after training can be further utilized in advanced designs. If not, the quantity of neurons in the hidden layer is regulated, and the training of the entire model is performed iteratively until the termination condition is satisfied.

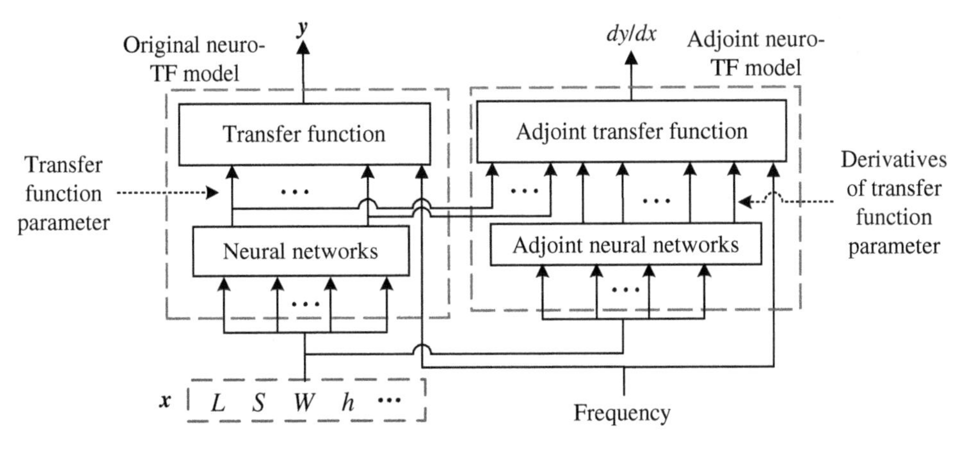

Figure 4.8 The structure of the sensitivity-analysis-based neuro-TF model. Source: [70] Feng et al. 2017/IEEE.

4.5.6 Neuro-TF Model Based on Sensitivity Analysis

Sensitivity analysis is a progressive technique for neuro-TF modeling with less train data. Figure 4.8 illustrates a neuro-TF model based on sensitivity analysis. It is composed of two submodels, namely the original and the sensitivity-analysis-based adjoint neural-TF models [70]. $\boldsymbol{x}$ is the input common to both submodels. Adjoint neuro-TF model's output is the derivative of the initial neuro-TF model's output $\boldsymbol{y}$ with respect to the input $\boldsymbol{x}$, indicated as $\boldsymbol{dy}/\boldsymbol{dx}$.

Give the meaning of the initial neural networks is the neural networks in the initial neuro-TF model. Similarly, neural networks in the adjoint neuro-TF model are represented by adjoint neural networks [9]. A mapping between the derivatives of parameters in the transport function and the geometric variables is established by an adjoint neural network.

Both EM simulation and susceptibility data are utilized when training the neuro-TF model which is based on sensitivity analysis. In the case of the same training dataset, the model trained with EM susceptibility information is more precise than the model trained without EM susceptibility information. That is, by utilizing sensitive information, high modeling precision can be achieved with small amount of train data. The model based on sensitivity analysis adopts a two-step training method [70], which is the initial and refinement training process.

When the training process is in the first phase, the initial and adjoint neural networks should be in the training procedure at the same time. Set $\hat{\boldsymbol{c}}_k$ and $\hat{\boldsymbol{A}}_k$ be the coefficients in the transport function and the derivatives of these coefficients with respect to $\boldsymbol{x}$, separately. The train data of the initial neural networks can be expressed as $(\boldsymbol{x}_k, \hat{\boldsymbol{c}}_k)$, and the training data of the adjoint neural networks are $(\boldsymbol{x}_k, \hat{\boldsymbol{A}}_k)$. Let $\boldsymbol{c}(\boldsymbol{x}, \boldsymbol{w})$ be the vector representation of the neural networks outputs, i.e. coefficients of the transport function. The error of the training process in the first step can be expressed as

$$E_{Pre}(\boldsymbol{w}) = \frac{1}{2n_s}\sum_{k=1}^{n_s}\|\boldsymbol{c}(\boldsymbol{x}_k, \boldsymbol{w}) - \hat{\boldsymbol{c}}_k\|^2 + \frac{1}{2n_s}\sum_{k=1}^{n_s}\left\|\frac{\partial \boldsymbol{c}(\boldsymbol{x}_k, \boldsymbol{w})}{\partial \boldsymbol{x}^T} - \hat{\boldsymbol{A}}_k\right\|_F^2 \tag{4.25}$$

where n_s denotes the overall number of training dataset; $\|\cdot\|$ and $\|\cdot\|_F$ represent the L_2 and Frobenius norms, separately.

After the first step of training, the second step of training, that is, the refining training is used to improve the entire model [70]. In the second step, the train data for the initial and adjoint neural-TF models are $(\boldsymbol{x}_k, \boldsymbol{d}_k)$, and $(\boldsymbol{x}_k, \boldsymbol{d}'_k)$, separately. During this time, the initial and the adjoint neural-TF models will be trained simultaneously. The error of the training process in the second step can be expressed as

$$\begin{aligned} E_{Tr}(\boldsymbol{w}) &= E_{orig}(\boldsymbol{w}) + E_{adj}(\boldsymbol{w}) \\ &= \frac{1}{2n_s}\sum_{k=1}^{n_s}\sum_{j=1}^{n_y} a_j \left\| y_j(\boldsymbol{x}_k, \boldsymbol{w}) - d_{k,j} \right\|^2 + \frac{1}{2n_s}\sum_{k=1}^{n_s}\sum_{j=1}^{n_y}\sum_{i=1}^{n_x} b_{j,i} \left\| \frac{\partial y_j(\boldsymbol{x}_k, \boldsymbol{w})}{\partial x_i} - d'_{k,j,i} \right\|^2 \end{aligned} \tag{4.26}$$

where E_{orig} denotes the training error between the EM simulation response and the initial model. E_{adj} represents the training error between the susceptibility information and the adjoint model. n_s is the total quantity of sets of the training data. n_x and n_y is the quantity of components in $\boldsymbol{x}$ and $\boldsymbol{y}$, separately. $d_{k,j}$ denotes the simulation response of the kth data for the jth output. $d'_{k,j,i}$ is the kth data's susceptibility information of the jth output concerning to the ith input. a_j and $b_{j,i}$ are the weights of the initial and the adjoint models, separately. When the total training error is below a pre-defined threshold, the overall process stops. Despite the complexity of training for the neuro-TF model which is based on susceptibility analysis, the use of the model after training is quite effortless. The adjoint model can be merely utilized in the course of training. The initial model is available to precise susceptibility information after training. As a result, the final model (i.e. the original model itself) is uncomplicated enough to be utilized later for advanced level of designs.

4.5.7 A Diplexer Example Using Neuro-TF Model Based on Sensitivity Analysis

A diplexer is utilized to demonstrate the sensitivity-analysis-based neuro-TF modeling approach [70]. Diplexer integral structure is displayed in Figure 4.9. D_1, D_2, D_3, D_4, D_5, and D_6 are the displacement from the terminus of every microstripe to its equivalents edge.

The input variables of the model include six geometric variables and frequency, i.e. $[D_1\ D_2\ D_3\ D_4\ D_5\ D_6\ \omega]^T$. There are 42 output variables, including the real/imaginary part of S_{11}, S_{21}, and S_{31}, and their differential coefficients concerning six input geometrical parameters. The EM emulator is employed to produce EM susceptibility information for the real and imaginary parts of S_{11}, S_{21}, and S_{31} with respect to six input variables D_1, D_2, D_3, D_4, D_5, and D_6.

Two diverse situations are utilized to elaborate the neuro-TF modeling based on susceptibility analysis. For the purpose of comparability, the sensitivity analysis ANN approach [9], and an already-existing neural-TF model that does not use sensitivity information [68] are used for

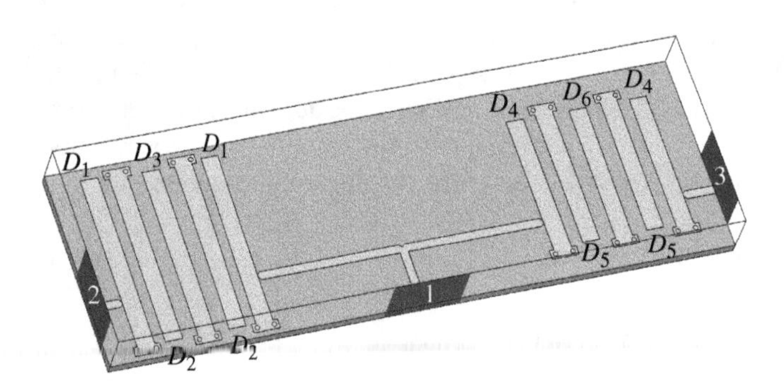

Figure 4.9 Example of a 3-D framework of the diplexer for EM emulation. Source: [70] Feng et al. 2017/IEEE.

modeling. Table 4.7 shows the comparison of training error and test error in the two methods, including the sensitivity analysis ANN Modeling approach with small amounts of data (25 sets of train data), already-existing modeling method without using sensitivity information with small pieces of data (25 packs of train data), already-existing modeling method without using sensitivity information with considerable amounts of data (81 sets of train data), and the neuro-TF modeling based on susceptibility analysis with small pieces of data (25 packs of train data) [70]. All approaches' training and test errors are comparatively low due to the geometric dimensions varying in a small range in Situation 1. In Situation 2, the sensitivity analysis ANN and neural-TF models that do not have sensitivity information using small amounts of data could not obtain high modeling precision due to a wide geometric range.

The comparisons of differential coefficients of the real component of S_{11} concerning sensitivity variables D_1 and D_2 for one training sample by the neuro-TF model that does not use sensitivity information with small amounts of data and a mass of data, the neuro-TF model based on sensibility analysis utilizing small amounts of data and CST sensibility analysis are displayed in Figure 4.10. As

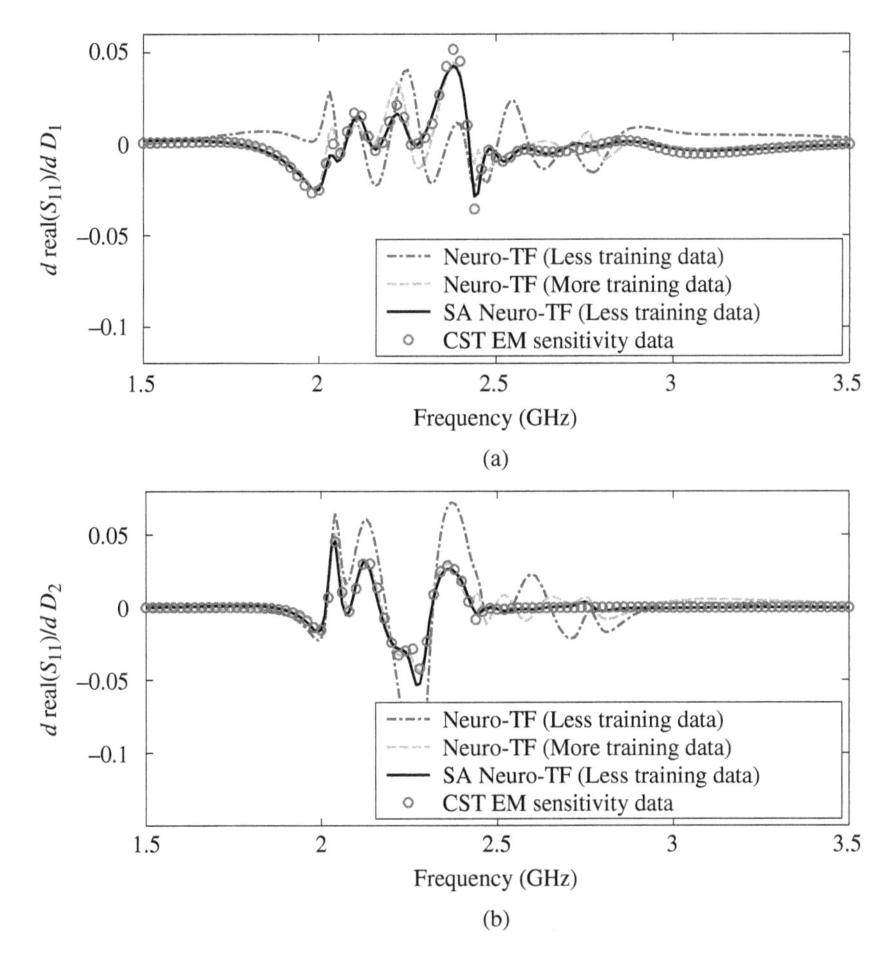

Figure 4.10 Sensitivity comparison of neuro-TF models in place without susceptibility information utilizing less and more data, the neuro-TF model based on sensibility analysis utilizing less data and CST EM sensitive data of the diplexer example: (a) differential coefficient of the real part of S_{11} concerning the sensitive variable D_1 and (b) differential coefficient of the real part of S_{11} concerning the sensitive variable D_2. Source: Adapted from [70].

Table 4.7 Comparison of EM optimization times for two microwave examples with different methods.

	Model type	Number of hidden neurons	Average training error	Average testing error
Case1(Small range)	SAANN model (25 training samples)	30	0.906%	0.924%
	Neuro-TF model (25 training samples)	10	0.867%	0.932%
	SA neuro-TF model (25 training samples)	12	0.892%	0.915%
Case2(Wider range)	SAANN model (25 training samples)	36	1.165%	11.564
	Neuro-TF model (25 training samples)	12	1.095%	9.264%
	SAANN model (81 training samples)	18	1.135%	1.368%
	Neuro-TF model (81 training samples)	12	1.065%	1.254%

Source: [70] Feng et al. 2017/IEEE.

shown in Figure 4.10, sensitivity information from the trained neuro-TF model based on sensibility analysis is more precise than that from the other methods in the comparison.

For two different test patterns #1 and #2, the output comparisons for the neural-TF model based on sensibility analysis and the neural-TF model that do not use information of sensitivity are shown in Figure 4.11. As illustrated in Figures 4.7 and 4.11, the neural-TF modeling based on sensitivity analysis [70] requires less training data than the non-sensitivity neural-TF model [68] and can

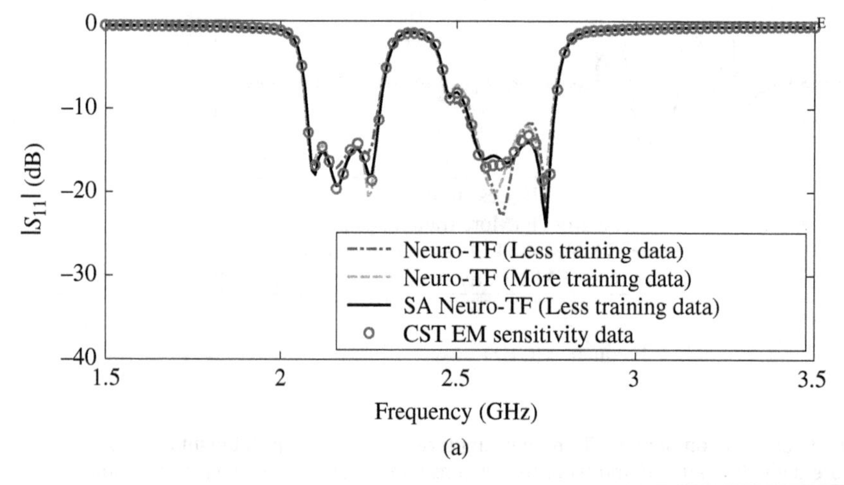

(a)

Figure 4.11 Comparisons of S_{11} of the neuro-TF model that does not use sensibility information with less and more data, model based on sensibility analysis with less data and CST EM data: (a) test pattern #1 and (b) test pattern #2 of the diplexer example. Source: [70] Feng et al. 2017/IEEE.

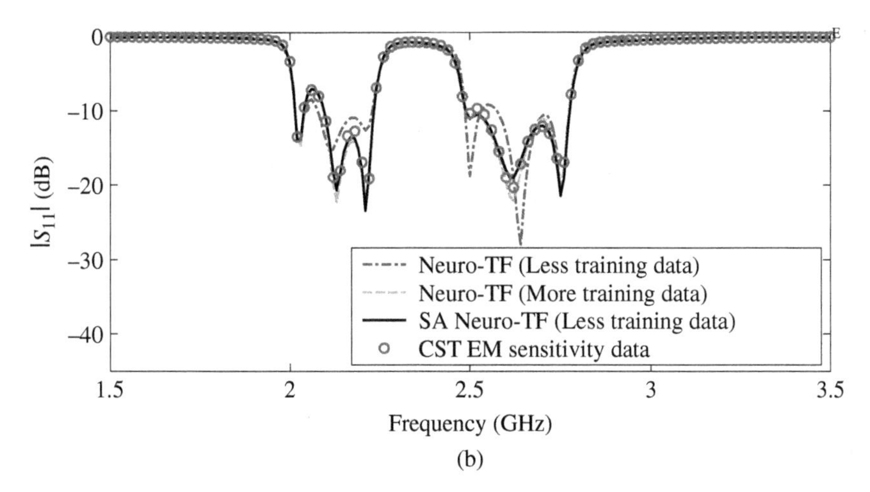

Figure 4.11 *(Continued)*

achieve better modeling accuracy. As a consequence, by lessening the quantity of train data, the modeling method based on susceptibility analysis can greatly speed up the model building process.

4.6 Surrogate Optimization of EM Design Based on ANN

4.6.1 Surrogate Optimization and Trust Region Update

Traditional EM optimization typically performs fine model simulation only once at each iteration. The trust region algorithm, on the other hand, utilizes several fine model simulations within a considerably large range near a pre-selected centerpoint, providing a wealth of information for calculating the orientation of the next iterative update. To anticipate the EM performance of an unknown, selected region, simulations of fine models are performed at the center point and its large neighborhood. After each iteration, the position of the center point is updated using the trust region formulation. It can be seen from the equations in [89] that the coefficients of TF are determined merely by geometric variables. $\boldsymbol{w}_r$ denotes the weight vector for each factor in the numerator or denominator. The value of $\boldsymbol{w}_r$ is updated after each optimization iteration, aiming to minimize the difference $\boldsymbol{w}_*$ of EM responses between the surrogate model and the fine model at the ith optimization iteration, which can be formulated as [89]

$$\boldsymbol{w}^* = \arg\min_{\boldsymbol{w}} \sum_{k=1}^{N_s} \sum_{\omega\in\Omega} \left\| R_s - R_f \right\| = \arg\min_{\boldsymbol{w}} \sum_{k=1}^{N_s} \sum_{\omega\in\Omega} \left\| H - R_f \right\| \tag{4.27}$$

where ω denotes all frequencies needed to calculate the EM response of the fine model and N_s represents the number of samplings in the confidence interval. Refined surrogate model matching will be acquired for DOE samplings within all the confidence intervals.

Surrogate model optimization method combined with the transfer function and trust region technique is expressed as using a sampling approach to realize the distribution of several data samples, using parallel processors for calculation, and using a trust region frame for optimization update. Orthogonal sampling is utilized to produce several sampling points where the subspace is sampled at an equal density and the sampled points are orthorhombic. Compared with the star distributing sampling strategy, the surrogate model can be available within a larger domain due to the orthotropic taking of samples surrounding the centerpoint. Moreover, fewer sampling points are used after orthogonal sampling than in the full-grid distributing method [65].

After determining the trust region, several samples are produced around the centerpoint in each iteration using a DOE [65] sampling technique. An implementation of the DOE sampling technique that generates sampling points near the centerpoint is illustrated in Figure 4.1. The centerpoint will move to the next region as optimization progresses; accordingly, all the sampling points move to the new region as the center point moves. As a result, variation of the sample values will change as the optimization proceeds. The trust region for each optimization iteration varies according to the trust radius δ^{new}. r_a is an index [89] which decides if the trust radius expands or keeps the same value as the preceding iteration, as shown in the following equation [89]

$$\delta^{new} = \begin{cases} c_1\delta^{(i)}, & r_a < 0.1 \\ \min\left(c_2\delta^{(i)}, \boldsymbol{\Delta}^{\max}\right), & r_a > 0.75 \\ \delta^{(i)}, & \text{Otherwise} \end{cases} \tag{4.28}$$

where $\Delta^{\max}$ is the maximum size of each design variable. $c_1 = 0.69$ and $c_2 = 1.3$. The parallel computational method is used to perform fine model calculations on several samples. Utilize parallel computing method to speed up the generation of data.

4.6.2 Neural TF Optimization Method Based on Adjoint Sensitivity Analysis

For the purpose of minimizing the sum of squares of the residual between the fine and the surrogate models, optimization formulas for the surrogate model after training are expressed as [90]

$$\begin{aligned} E^k(w) &= \alpha E_0^k(w) + \beta E_a^k(w) \\ &= \alpha \sum_{i=1}^{n_s}\sum_{j=1}^{n_f} \left\| y_s\left(\boldsymbol{x}_i^k, \boldsymbol{w}, s_j\right) - y_f\left(\boldsymbol{x}_i^k, s_j\right) \right\|^2 \\ &\quad + \beta \sum_{i=1}^{n_s}\sum_{j=1}^{n_f} \left\| \tilde{y}_s\left(\boldsymbol{x}_i^k, \boldsymbol{w}, s_j\right) - \tilde{y}_f\left(\boldsymbol{x}_i^k, s_j\right) \right\|^2 \end{aligned} \tag{4.29}$$

$$\boldsymbol{w}^k = \arg \min_{w} E^k(\boldsymbol{w}) \tag{4.30}$$

where y_f and $\tilde{y}_f$ denote the EM response of the fine model and the adjoint EM sensitivity information, separately. n_s represents the number of fine models selected. E^k represents the training error between the initial and adjoint models at iteration k. E_0^k and E_α^k represent the training error of the neuro-TF model and the model with adjoint sensitivity, separately. n_f represents the number of selected frequency points, and α and β are weight coefficients. $\boldsymbol{w}_k$ is a vector representation of the optimum coefficients in iteration k. Precise sensibility information is able to accelerate the optimization of the surrogate model and reduce iteration times.

4.6.3 Surrogate Model Optimization Based on Feature-Assisted of Neuro-TF

The utilization of characteristic parameters accelerates the optimization of the surrogate model. Characteristic parameters are usually taken as characteristic frequencies and their corresponding frequency responses located in particular regions, e.g. zeroes of transmission or reflection of filters [91]. The purpose of introducing feature parameters is to avoid the appearance of local minima and to obtain the optimum solution as soon as possible. The framework of the neuro-TF based

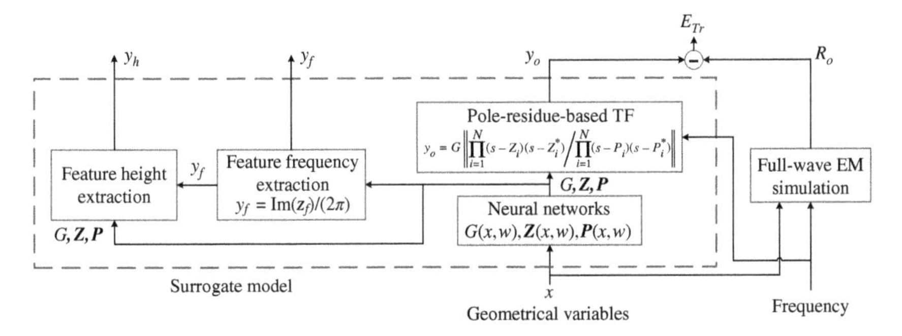

Figure 4.12 Schematic diagram of the overall structure of the neuro-TF model based on feature-assisted. Poles, residues, and gain as coefficients of the transport function. Source: Adapted from [91].

on feature-assisted is illustrated in Figure 4.12. The pole-residue form of the neuro-TF model is adopted, and the characteristic parameters are used as the neural networks' outputs.

After obtaining a well-trained neural network, to proceed with the surrogate optimization, multiple features should be extracted, including characteristic frequencies y_l^f and heights y_h. The feature frequencies relate to the imaginary parts of the TF residues. The extraction formula can be expressed as [91]

$$y_l^f(\boldsymbol{x}, \boldsymbol{w}) = \frac{\text{Im}\left(Z_{Q_l}(\boldsymbol{x}, \boldsymbol{w})\right)}{2\pi}, \quad l = 1, 2, \ldots, N_f \tag{4.31}$$

where l=1,

$$Q_l = \underset{q\in\{1,\ldots,N_k\}}{\arg\min} \left\{ \sum_{j=1}^{n_s} \left\| \text{Re}\left(z_q^j\right) \right\| \right\} \tag{4.32}$$

$l \geq 2$,

$$Q_l = \underset{\substack{q\in[1,\ldots,N_k\} \\ q\notin\{Q_1,\ldots,Q_{l-1}\}}}{\arg\min} \left\{ \sum_{j=1}^{n_s} \left\| \text{Re}\left(z_q^j\right) \right\| \right\} \tag{4.33}$$

where z_q^j denotes the data of the qth TF residue for sampling j. n_s is the total number of training examples. y_h represents the amplitude of the S parameter at the intermediate frequency of the two eigenfrequencies. Once the characteristic parameters have been extracted, EM optimization using surrogate models can be realized utilizing neural-TF models based on feature-assisted [91].

4.6.4 EM Optimization of a Microwave Filter Utilizing Feature-Assisted Neuro-TF

The feature-assisted neuro-TF approach can be demonstrated by designing a fifth-order microwave filter with cylindrical metal posts located in the middle of all the apertures and cavities [91], as shown in Figure 4.13, where H_{c1}, H_{c2}, and H_{c3} are altitudes of cylinders with the larger radius; W_1, W_2, W_3, and W_4 are breadths of the apertures for each section. Seven input design parameters are included in this filter, denoted by $\boldsymbol{x} = [H_{c1}\ H_{c2}\ H_{c3}\ W_1\ W_2\ W_3\ W_4]^T$. Outputs of the filter contain the amplitude of S_{11}, and the characteristic frequencies as well as characteristic heights corresponding to the EM responses of the given structure. Imaginary parts of residues are used as the characteristic frequencies, and amplitudes of the EM response at the middle frequency of

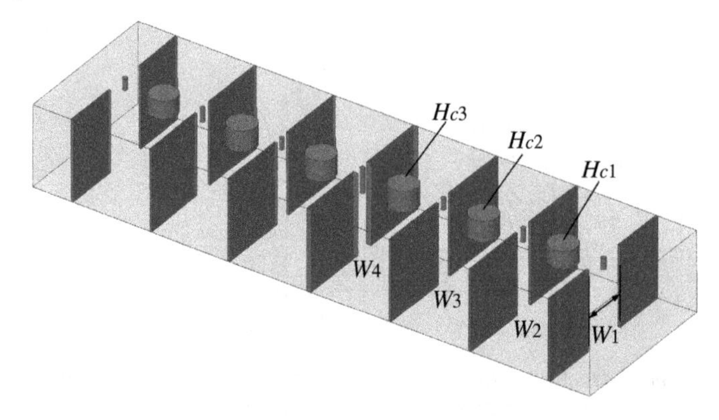

Figure 4.13 Structure of the inter-digital bandpass filter with five geometric variables for EM optimization. Source: Adapted from [91].

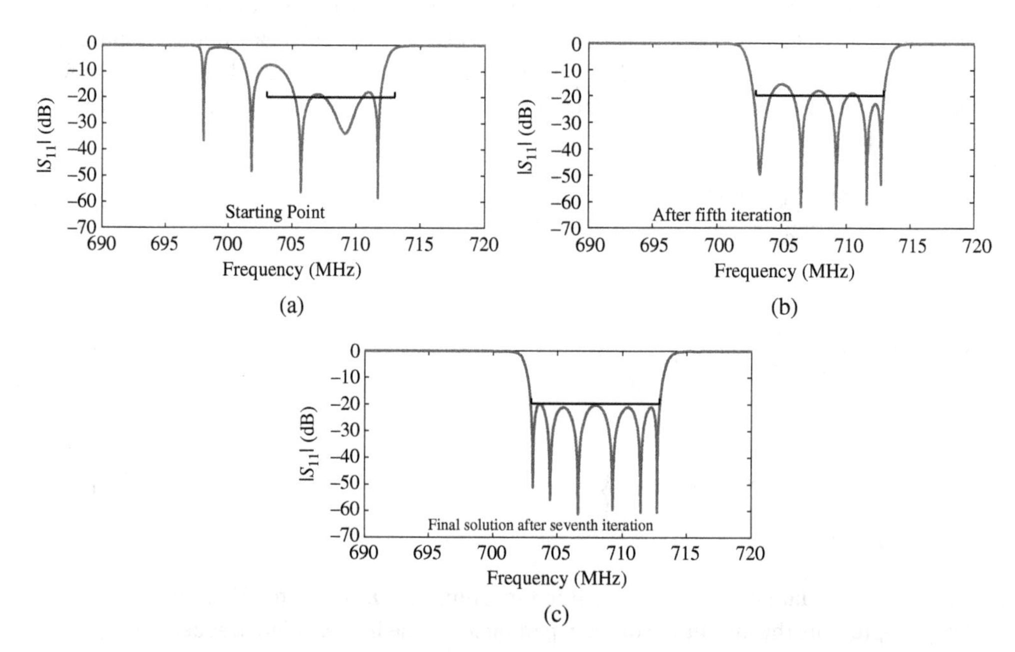

Figure 4.14 EM responses of the multi-feature-assisted surrogate optimization. (a) EM response at the starting point. (b) EM responses after five iterations. (c) Desired response obtained after iteration 7. Source: Adapted from [91].

two characteristic frequencies are used as the characteristic heights. *NeuroModelerPlus* software is employed to perform all simulation operations of the entire model.

Design specification is considered to be $|S_{11}| \leq -20$ dB in the frequency range of 703–713 MHz. Starting point is chosen as $\boldsymbol{x}^0 = [43\ 50.5\ 50.5\ 115\ 51\ 50\ 55]^T$ (mm). Feature-assisted neuro-TF approach is utilized to accomplish the model optimization. *HFSS* softwave is employed to generate EM information and complete the validation process for optimization results. After seven optimization iterations, the optimized geometrical solution $\boldsymbol{x}^7 = [42.072\ 50.343\ 50.473\ 115.78\ 50.123\ 44.179\ 47.612]^T$ (mm) is obtained. Figure 4.14 indicates the EM responses of the given structure at the optimized starting point, after five iterations, and the desired EM response after seven iterations [91]. When the optimization starting point does not quite match the desired response, the feature-assisted method is more likely to go away from local minimization than the no-feature-assisted approach.

4.7 Conclusion

This chapter has introduced both the fundamentals and advanced formulations of ANN techniques for parameterized modeling and optimization problems. General ANN structure and training have been introduced along with an advanced DNN approach. In order to further enhance the precision and robust performance of modeling, a KBNN model combined with priori knowledge is proposed. Advanced ANN techniques that combine the ANN with transfer function have also been discussed. Finally, surrogate modeling optimization approaches based on ANN are presented.

Future directions of ANN for parametric EM modeling and optimization can include exploiting more DNN methods in EM modeling and optimization, discovering new ways to combine various prior knowledge with neural networks, applying ANN models for EM yield analysis and optimization, expanding ANN techniques from EM design to multi-physics design, and so on. The research of ANN techniques for parametric EM modeling and optimization is still a direction of continuous development and updating.

References

1 Zhang, Q.J. and Gupta, K.C. (2000). *Neural Networks for RF and Microwave Design*. Norwood, MA: Artech House.

2 Zhang, Q.J., Gupta, K.C., and Devabhaktuni, V.K. (2003). Artificial neural networks for RF and microwave design - from theory to practice. *IEEE Transactions on Microwave Theory and Techniques* 51 (4): 1339–1350.

3 Rayas-Sánchez, J.E. (2004). EM-based optimization of microwave circuits using artificial neural networks: the state-of-the-art. *IEEE Transactions on Microwave Theory and Techniques* 52 (1): 420–435.

4 Steer, M.B., Bandler, J.W., and Snowden, C.M. (2002). Computer-aided design of RF and microwave circuits and systems. *IEEE Transactions on Microwave Theory and Techniques* 50 (3): 996–1005.

5 Mkadem, F. and Boumaiza, S. (2011). Physically inspired neural network model for RF power amplifier behavioral modeling and digital predistortion. *IEEE Transactions on Microwave Theory and Techniques* 59 (4): 913–923.

6 Root, D.E. (2012). Future device modeling trends. *IEEE Microwave Magazine* 13 (7): 45–59.

7 Yu, H., Chalamalasetty, H., and Swaminathan, M. (2019). Modeling of voltage-controlled oscillators including I/O behavior using augmented neural networks. *IEEE Access* 7: 38973–38982.

8 Liao, S., Kabir, H., Cao, Y. et al. (2011). Neural-network modeling for 3-D substructures based on spatial EM-field coupling in finite-element method. *IEEE Transactions on Microwave Theory and Techniques* 59 (1): 21–38.

9 Sadrossadat, S.A., Cao, Y., and Zhang, Q.J. (2013). Parametric modeling of microwave passive components using sensitivity-analysis-based adjoint neural-network technique. *IEEE Transactions on Microwave Theory and Techniques* 61 (5): 1733–1747.

10 Na, W.C. and Zhang, Q.J. (2013). Automated parametric modeling of microwave components using combined neural network and interpolation techniques. *2013 IEEE MTT-S International Microwave Symposium Digest (MTT)*, Seattle, WA, USA (02–07 June 2013).

11 Na, W. and Zhang, Q.J. (2014). Automated knowledge-based neural network modeling for microwave applications. *IEEE Microwave and Wireless Components Letters* 24 (7): 499–501.

12 Gosal, G., Almajali, E., McNamara, D., and Yagoub, M. (2015). Transmitarray antenna design using forward and inverse neural network modeling. *IEEE Antennas and Wireless Propagation Letters* 15 (1): 1483–1486.

13 Watson, P.M. and Gupta, K.C. (1996). EM-ANN models for microstrip vias and interconnects in multilayer circuits. *IEEE Transactions on Microwave Theory and Techniques* 44 (12): 2495–2503.

14 Zhang, Q.J. and Nakhla, M.S. (1994). Signal integrity analysis and optimization of VLSI interconnects using neural network models. *Proceedings of IEEE International Symposium on Circuits and Systems - ISCAS '94* London, England (30 May 1994 - 02 June 1994), pp. 459–462.

15 Veluswami, A., Nakhla, M.S., and Zhang, Q.J. (1997). The application of neural networks to EM-based simulation and optimization of interconnects in highspeed VLSI circuits. *IEEE Transactions on Microwave Theory and Techniques* 45 (5): 712–723.

16 Huang, A.D., Zhong, Z., Wu, W., and Guo, Y.X. (2016). An artificial neural network-based electrothermal model for GaN HEMTs with dynamic trapping effects consideration. *IEEE Transactions on Microwave Theory and Techniques* 64 (8): 2519–2528.

17 Wang, F. and Zhang, Q.J. (1997). Knowledge based neural models for microwave design. *IEEE Transactions on Microwave Theory and Techniques* 45 (12): 2333–2343.

18 Devabhaktuni, V.K., Xi, C., Wang, F., and Zhang, Q.J. (2002). Robust training of microwave neural models. *International Journal of RF and Microwave Computer-Aided Engineering* 12 (1): 109–124.

19 Watson, P.M. and Gupta, K.C. (1997). Design and optimization of CPW circuits using EM-ANN models for CPW components. *IEEE Transactions on Microwave Theory and Techniques* 45 (12): 2515–2523.

20 Zhang, J., Ma, K., Feng, F., and Zhang, Q.J. (2015). Parallel gradient-based local search accelerating particle swarm optimization for training microwave neural network models. *2015 IEEE MTT-S International Microwave Symposium*, Phoenix, AZ (17–22 May 2015), pp. 1–3.

21 Fedi, G., Manetti, S., Pelosi, G., and Selleri, S. (2000). Design of cylindrical posts in rectangular waveguide by neural network approach. *IEEE AP-S International Symposium Digest*, Salt Lake City, UT (16–21 July 2000), pp. 1054–1057.

22 Ding, X., Devabhaktuni, V.K., Chattaraj, B. et al. (2004). Neural-network approaches to electromagnetic-based modeling of passive components and their applications to high-frequency and high-speed nonlinear circuit optimization. *IEEE Transactions on Microwave Theory and Techniques* 52 (1): 436–449.

23 Devabhaktruni, V.K., Yagoub, M., and Zhang, Q.J. (2001). A robust algorithm for automatic development of neural-network models for microwave applications. *IEEE Transactions on Microwave Theory and Techniques* 49 (12): 2282–2291.

24 Fang, Y., Yagoub, M.C.E., Wang, F., and Zhang, Q.J. (2000). A new macromodeling approach for nonlinear microwave circuits based on recurrent neural networks. *IEEE Transactions on Microwave Theory and Techniques* 48 (12): 2335–2344.

25 Bandler, J.W., Ismail, M.A., Rayas-Sanchez, J.E., and Zhang, Q.J. (1999). Neuromodeling of microwave circuits exploiting space-mapping technology. *IEEE Transactions on Microwave Theory and Techniques* 47 (12): 2417–2427.

26 Creech, G.L., Paul, B.J., Lesniak, C.D. et al. (1997). Artificial neural networks for fast and accurate EM-CAD of microwave circuits. *IEEE Transactions on Microwave Theory and Techniques* 45 (5): 794–802.
27 Na, W., Zhang, W., Yan, S., and Liu, G. (2019). Automated neural-based modeling of microwave devices using parallel computation and interpolation approaches. *IEEE Access* 7: 73929–73937.
28 Yan, S., Zhang, Y., Jin, X. et al. (2018). Multi-physics parametric modeling of microwave passive components using artificial neural networks. *Progress In Electromagnetics Research M* 72: 79–88.
29 Zhang, W., Feng, F., Gongal-Reddy, V.M.R. et al. (2018). Space mapping approach to electromagnetic centric multiphysics parametric modeling of microwave components. *IEEE Transactions on Microwave Theory and Techniques* 66 (7): 3169–3185.
30 Xu, J., Yagoub, M.C.E., Ding, R., and Zhang, Q.J. (2003). Exact adjoint sensitivity analysis for neural-based microwave modeling and design. *IEEE Transactions on Microwave Theory and Techniques* 51 (1): 226–237.
31 Nikolova, N.K., Bandler, J.W., and Bakr, M.H. (2004). Adjoint techniques for sensitivity analysis in high-frequency structure CAD. *IEEE Transactions on Microwave Theory and Techniques* 52 (1): 403–419.
32 Rizzoli, V., Neri, A., Masotti, D., and Lipparini, A. (2002). A new family of neural network-based bidirectional and dispersive behavioral models for nonlinear RF/microwave subsystems. *International Journal of RF and Microwave Computer-Aided Engineering* 12 (1): 51–70.
33 Jin, J., Zhang, C., Feng, F. et al. (2019). Deep neural network technique for high-dimensional microwave modeling and applications to parameter extraction of microwave filters. *IEEE Transactions on Microwave Theory and Techniques* 67 (10): 4140–4155.
34 Cao, Y. and Wang, G. (2007). A wideband and scalable model of spiral inductors using space-mapping neural network. *IEEE Transactions on Microwave Theory and Techniques* 55 (12): 2473–2480.
35 Rayas-Sánchez, J.E. and Gutierrez-Ayala, V. (2006). EM-based Monte Carlo analysis and yield prediction of microwave circuits using linear-input neural-output space mapping. *IEEE Transactions on Microwave Theory and Techniques* 54 (12): 4528–4537.
36 Devabhaktuni, V.K., Chattaraj, B., Yagoub, M.C.E., and Zhang, Q.J. (2003). Advanced microwave modeling framework exploiting automatic model generation, knowledge neural networks, and space mapping. *IEEE Transactions on Microwave Theory and Techniques* 51 (7): 1822–1833.
37 Kabir, H., Zhang, L., Yu, M. et al. (2010). Smart modeling of microwave device. *IEEE Microwave Magazine* 11 (3): 105–118.
38 Bandler, J.W., Cheng, Q.S., Dakroury, S.A. et al. (2004). Space mapping: the state of the art. *IEEE Transactions on Microwave Theory and Techniques* 52 (1): 337–361.
39 Koziel, S., Cheng, Q.S., and Bandler, J.W. (2008). Space mapping. *IEEE Microwave Magazine* 9 (6): 105–122.
40 Bandler, J.W., Cheng, Q.S., Nikolova, N.K., and Ismail, M.A. (2004). Implicit space mapping optimization exploiting preassigned parameters. *IEEE Transactions on Microwave Theory and Techniques* 52 (1): 378–385.
41 Bandler, J.W., Ismail, M.A., and Rayas-Sanchez, J.E. (2002). Expanded space-mapping EM-based design framework exploiting preassigned parameters. *IEEE Transactions on Circuits and Systems–I: Fundamental Theory and Applications* 49 (12): 1833–1838.
42 Koziel, S., Bandler, J.W., and Madsen, K. (2009). Space mapping with adaptive response correction for microwave design optimization. *IEEE Transactions on Microwave Theory and Techniques* 57 (2): 478–486.

43 Bandler, J.W., Hailu, D.M., Madsen, K., and Pedersen, F. (2004). A space mapping interpolating surrogate algorithm for highly optimized EM-based design of microwave devices. *IEEE Transactions on Microwave Theory and Techniques* 52 (11): 2593–2600.

44 Ayed, R.B., Gong, J., Brisset, S. et al. (2012). Three-level output space mapping strategy for electromagnetic design optimization. *IEEE Transactions on Magnetics* 48 (2): 671–674.

45 Zhang, L., Xu, J., Yagoub, M.C.E. et al. (2005). Efficient analytical formulation and sensitivity analysis of neuro-space mapping for nonlinear microwave device modeling. *IEEE Transactions on Microwave Theory and Techniques* 53 (9): 2752–2767.

46 Bakr, M.H., Bandler, J.W., Ismail, M.A. et al. (2000). Neural space-mapping optimization for EM-based design. *IEEE Transactions on Microwave Theory and Techniques* 48 (12): 2307–2315.

47 Gutierrez-Ayala, V. and Rayas-Sanchez, J.E. (2010). Neural input space mapping optimization based on nonlinear two-layer perceptrons with optimized nonlinearity. *International Journal of RF and Microwave Computer-Aided Engineering* 20 (5): 512–526.

48 Feng, F. and Zhang, Q.J. (2015). Neural space mapping optimization for EM design. *2015 Asia-Pacific Microwave Conference (APMC)*, Nanjing, China (06–09 December 2015), pp. 1–3.

49 Gorissen, D., Zhang, L., Zhang, Q.J., and Dhaene, T. (2011). Evolutionary neuro-space mapping technique for modeling of nonlinear microwave devices. *IEEE Transactions on Microwave Theory and Techniques* 59 (2): 213–229.

50 Koziel, S., Bandler, J.W., and Madsen, K. (2006). A space mapping framework for engineering optimization: theory and implementation. *IEEE Transactions on Microwave Theory and Techniques* 54 (10): 3721–3730.

51 Koziel, S., Bandler, J.W., and Cheng, Q.S. (2011). Tuning space mapping design framework exploiting reduced electromagnetic models. *IET Microwaves, Antennas and Propagation* 5 (10): 1219–1226.

52 Meng, J., Koziel, S., Bandler, J.W. et al. (2008). Tuning space mapping: a novel technique for engineering design optimization. *2008 IEEE MTT-S International Microwave Symposium Digest*, Atlanta, Georgia (15–20 June 2008), pp. 991–994.

53 Zhang, L., Aaen, P.H., and Wood, J. (2012). Portable space mapping for efficient statistical modeling of passive components. *IEEE Transactions on Microwave Theory and Techniques* 60 (3): 441–450.

54 Zhang, C., Feng, F., and Zhang, Q.J. (2013). EM optimization using coarse and fine mesh space mapping. *2013 Asia-Pacific Microwave Conference Proceedings (APMC)*, Seoul, Korea, December 2013, pp. 824–826.

55 Feng, F., Zhang, C., Gongal-Reddy, V.M.R., and Zhang, Q.J. (2014). Knowledge-based coarse and fine mesh space mapping approach to EM optimization. *International Conference on Numerical Electromagnetic Modeling and Optimization*, Pavia, Italy (14–16 May 2014), pp. 1–4.

56 Koziel, S., Ogurtsov, S., Bandler, J.W., and Cheng, Q.S. (2013). Reliable space-mapping optimization integrated with EM-based adjoint sensitivities. *IEEE Transactions on Microwave Theory and Techniques* 61 (10): 3493–3502.

57 Koziel, S., Cheng, Q.S., and Bandler, J.W. (2014). Fast EM modeling exploiting shape-preserving response prediction and space mapping. *IEEE Transactions on Microwave Theory and Techniques* 62 (3): 399–407.

58 Fong, F., Gongal-Reddy, V.M.R., Zhang, C. et al. (2016). Recent advances in parallel EM optimization approaches. *IEEE MTT-S International Conference on Microwave Millimeter Wave Technology*, Beijing, China (05–08 June 2016), pp. 1–3.

59 Feng, F., Gongal-Reddy, V.M.R., Zhang, S., and Zhang, Q.J. (2015). Recent advances in space mapping approach to EM optimization. *Proceedings of Asia-Pacific Microwave Conference*, Nanjing, China (06–09 December 2015), pp. 1–3.

60 Garcia-Lamperez, A., Llorente-Romano, S., Salazar-Palma, M., and Sarkar, T.K. (2004). Efficient electromagnetic optimization of microwave filters and multiplexers using rational models. *IEEE Transactions on Microwave Theory and Techniques* 52 (2): 508–521.

61 Garcia-Lamperez, A. and Salazar-Palma, M. (2016). Multilevel aggressive space mapping applied to coupled-resonator filters. *IEEE MTT-S International Microwave Symposium Digest*, San Francisco, CA (22–27 May 2016), pp. 1–4.

62 Sans, M., Selga, J., Velez, P. et al. (2015). Automated design of common-mode suppressed balanced wideband bandpass filters by means of aggressive space mapping. *IEEE Transactions on Microwave Theory and Techniques* 63 (12): 3896–3908.

63 Feng, F., Zhang, C., Gongal-Reddy, V.M.R. et al. (2014). Parallel space-mapping approach to EM optimization. *IEEE Transactions on Microwave Theory and Techniques* 62 (5): 1135–1148.

64 Gongal-Reddy, V.M.R., Feng, F., and Zhang, Q.J. (2015). Parametric modeling of millimeter-wave passive components using combined neural networks and transfer functions. *Global Symposium on Millimeter Waves (GSMM)*, Montreal, QC, Canada (25–27 May 2015), pp. 1–3.

65 Cao, Y., Wang, G., and Zhang, Q.J. (2009). A new training approach for parametric modeling of microwave passive components using combined neural networks and transfer functions. *IEEE Transactions on Microwave Theory and Techniques* 57 (11): 2727–2742.

66 Guo, Z., Gao, J., Cao, Y., and Zhang, Q.J. (2012). Passivity enforcement for passive component modeling subject to variations of geometrical parameters using neural networks. *2012 IEEE/MTT-S International Microwave Symposium Digest*, Montreal, QC, Canada (17–22 June 2012), pp. 1–3.

67 Cao, Y., Wang, G., Gunupudi, P., and Zhang, Q.J. (2013). Parametric modeling of microwave passive components using combined neural networks and transfer functions in the time and frequency domains. *International Journal of RF and Microwave Computer-Aided Engineering* 23 (1): 20–33.

68 Feng, F., Zhang, C., Ma, J., and Zhang, Q.J. (2016). Parametric modeling of EM behavior of microwave components using combined neural networks and pole-residue-based transfer functions. *IEEE Transactions on Microwave Theory and Techniques* 64 (1): 60–77.

69 Feng, F. and Zhang, Q.J. (2015). Parametric modeling using sensitivity-based adjoint neuro-transfer functions for microwave passive components. *IEEE MTT-S International Conference on Numerical Electromagnetic and Multiphysics Modeling and Optimization*, Ottawa, Canada (11–14 August 2015), pp. 1–3.

70 Feng, F., Gongal-Reddy, V.M.R., Zhang, C. et al. (2017). Parametric modeling of microwave components using adjoint neural networks and pole-residue transfer functions with EM sensitivity analysis. *IEEE Transactions on Microwave Theory and Techniques* 65 (6): 1955–1975.

71 Zhang, J., Feng, F., Zhang, W. et al. (2020). A novel training approach for parametric modeling of microwave passive components using Padé via Lanczos and EM sensitivities. *IEEE Transactions on Microwave Theory and Techniques* 68 (6): 2215–2233.

72 Cao, Y., Xu, J.J., Devabhaktuni, V.K. et al. (2003). An adjoint dynamic neural network technique for exact sensitivities in nonlinear transient modeling and high-speed interconnect design. *IEEE MTT-S International Microwave Symposium Digest*, Philadelphia, PA, USA (08–13 June 2003), pp. 165–168.

73 Xu, J., Yagoub, M.C.E., Ding, R., and Zhang, Q.J. (2002). Neural based dynamic modeling of nonlinear microwave circuits. *IEEE Transactions on Microwave Theory and Techniques* 50 (12): 2769–2780.

74 Stievano, I.S., Maio, I.A., and Canavero, F.G. (2002). Parametric macromodels of digital I/O ports. *IEEE Transactions on Advanced Packaging* 25 (5): 255–264.

75 Isaksson, M., Wisell, D., and Ronnow, D. (2005). Wide-band dynamic modeling of power amplifiers using radial-basis function neural networks. *IEEE Transactions on Microwave Theory and Techniques* 53 (11): 3422–3428.

76 O'Brien, B., Dooley, J., and Brazil, T.J. (2006). RF power amplifier behavioral modeling using a globally recurrent neural network. *IEEE MTT-S International Microwave Symposium Digest*, San Francisco, CA (11–16 June 2006), pp. 1089–1092.

77 Fang, Y.H., Yagoub, M.C.E., Wang, F., and Zhang, Q.J. (2000). A new macromodeling approach for nonlinear microwave circuits based on recurrent neural networks. *IEEE Transactions on Microwave Theory and Techniques* 48 (12): 2335–2344.

78 Yan, S., Zhang, C., and Zhang, Q.J. (2014). Recurrrent neural network technique for behavioral modeling of power amplifier with memory effects. *International Journal of RF and Microwave Computer-Aided Engineering* 25 (4): 289–298.

79 Liu, T., Boumaiza, S., and Ghannouchi, F.M. (2004). Dynamic behavioral modeling of 3G power amplifiers using real-valued time-delay neural networks. *IEEE Transactions on Microwave Theory and Techniques* 52 (3): 1025–1033.

80 Cao, Y., Ding, R.T., and Zhang, Q.J. (2004). A new nonlinear transient modelling technique for high-speed integrated circuit applications based on state-space dynamic neural network. *IEEE MTT-S International Microwave Symposium Digest*, Fort Worth, TX (06–11 June 2004), pp. 1553–1556.

81 Cao, Y., Ding, R.T., and Zhang, Q.J. (2006). State-space dynamic neural network technique for high-speed IC applications: modeling and stability analysis. *IEEE Transactions on Microwave Theory and Techniques* 54 (6): 2398–2409.

82 Bandler, J.W., Chen, S.H., and Daijavad, S. (1986). Microwave device modeling using efficient l_1 optimization: a novel approach. *IEEE Transactions on Microwave Theory and Techniques* 34 (12): 1282–1293.

83 Na, W., Feng, F., Zhang, C., and Zhang, Q.J. (2017). A unified automated parametric modeling algorithm using knowledge-based neural network and l1 optimization. *IEEE Transactions on Microwave Theory and Techniques* 65 (3): 729–745.

84 Bandler, J.W., Kellermann, W., and Madsen, K. (1987). A nonlinear l_1 optimization algorithm for design, modeling, and diagnosis of networks. *IEEE Transactions on Circuits and Systems* 34 (2): 174–181.

85 Hald, J. and Madsen, K. (1985). Combined LP and quasi-Newton methods for nonlinear l_1 optimization. *SIAM Journal on Numerical Analysis* 22 (1): 68–80.

86 Tu, W.H. and Chang, K. (2006). Microstrip eeliptic-function low-pass filters using distributed elements or slotted ground structure. *IEEE Transactions on Microwave Theory and Techniques* 54 (10): 3786–3792.

87 Esfandiari, R., Maku, D., and Siracusa, M. (1983). Design of interdigitated capacitors and their application to gallium arsenide monolithic filters. *IEEE Transactions on Microwave Theory and Techniques* 31 (1): 57–61.

88 Gustavsen, B. and Semlyen, A. (1999). Rational approximation of frequency domain responses by vector fitting. *IEEE Transactions on Power Delivery* 14 (3): 1052–1061.

89 Gongal-Reddy, V.-M.-R., Zhang, S., Zhang, C., and Zhang, Q.-J. (2016). Parallel computational approach to gradient based EM optimization of passive microwave circuits. *IEEE Transactions on Microwave Theory and Techniques* 64 (1): 44–59.

90 Feng, F., Na, W., Liu, W. et al. (2020). Parallel gradient-based em optimization for microwave components using adjoint- sensitivity-based neuro-transfer function surrogate. *IEEE Transactions on Microwave Theory and Techniques* 68 (9): 3606–3620.

91 Feng, F., Na, W., Liu, W. et al. (2020). Multifeature-assisted neuro-transfer function surrogate-based EM optimization exploiting trust-region algorithms for microwave filter design. *IEEE Transactions on Microwave Theory and Techniques* 68 (2): 531–542.

5

Advanced Neural Networks for Electromagnetic Modeling and Design

Bing-Zhong Wang[1], *Li-Ye Xiao*[2], *and Wei Shao*[1]

[1] *School of Physics, University of Electronic Science and Technology of China, Institute of Applied Physics, Chengdu, China*
[2] *Department of Electronic Science, Xiamen University, Institute of Electromagnetics and Acoustics, Xiamen, China*

5.1 Introduction

An artificial neural network (ANN) derives its computing power through its parallel distributed structure and its ability to learn and therefore generalize. Generalization refers to the neural network's reasonable outputs for inputs that are not encountered during training. These two information-processing capabilities make it possible for ANNs to obtain good approximate solutions to complex electromagnetic problems. For the design of microwave components, circuits, or antennas, full-wave electromagnetic simulations are usually called by an optimization algorithm hundreds of times to converge to an optimal result. The time-consuming simulations result in a heavy computational burden to complete the design. Due to the accurate and fast simulation ability, the neural network has been introduced to learn the relationship between geometrical variables and electromagnetic responses with a training process. Once the training process is completed, the trained ANN can be a good alternative to electromagnetic simulations to significantly speed up the design.

This chapter discusses some improved techniques suitable for efficient ANN modeling in electromagnetics. First, two semi-supervised learning schemes are introduced to model microwave passive components, based on the dynamic adjustment kernel extreme learning machine (DA-KELM) and radial basis function (RBF) neural network. Second, some effective ANN techniques are proposed for antennas and arrays, including multi-parameter modeling in antenna modeling, inverse ANN for multi-objective design, and prior knowledge for periodic array modeling. Third, an autoencoder neural network, which is used to reduce the dimensionality of data from time-domain full-wave simulations, calculates wave propagation in uncertain media.

5.2 Semi-Supervised Neural Networks for Microwave Passive Component Modeling

5.2.1 Semi-Supervised Learning Based on Dynamic Adjustment Kernel Extreme Learning Machine

If a neural model doesn't achieve satisfactory accuracy, a simple and direct way is to retrain it with an updated training dataset. However, current models are mostly time-consuming because

Advances in Electromagnetics Empowered by Artificial Intelligence and Deep Learning, First Edition.
Edited by Sawyer D. Campbell and Douglas H. Werner.

re-collecting a new training dataset and re-executing the training process are required. Generally, there is a great overlap between the old training dataset and the new one. If the overlapped information is used effectively, the retraining cost will be reduced greatly. A DA-KELM, which contains increased learning, reduced learning, and hybrid learning in the interior structure, is proposed to improve accuracy and speed up retraining with the ability to capture the overlap between the old training dataset and the new one. In DA-KELM, the increased learning is employed for solving the under-fitting problem when training samples are insufficient to obtain accurate results. When the quantity of training samples is too large to obtain accurate results, the reduced learning is utilized to solve this over-fitting problem. If the range of the training dataset involves shifting or expansion, the hybrid learning, which includes the increased learning and reduced learning, is used to obtain the final results conveniently and directly.

On the other hand, most of the reported models for parametric modeling of EM behaviors are based on supervised learning, in which the labeled sampling data from full-wave EM simulations ought to be sufficient for ANN training. Thus, the number of full-wave simulations is the main factor that influences the effectiveness of collecting training and testing samples. Here, from the concept of semi-supervised learning, a model tries to obtain the satisfactory accuracy with much less time consumption. In the model, only a few training samples come from full-wave simulations, while a considerable number of unlabeled samples are produced by the model itself, which leads to a fast ANN modeling process. The semi-supervised learning model contains the initial training and self-training. In the initial training process, a DA-KELM is trained with the labeled training samples from full-wave simulations to ensure its good convergence. In the self-training process, another DA-KELM is duplicated from the existing one and the unlabeled samples of different geometrical variables are inputted to each DA-KELM. After the test with the labeled testing samples, the inner parameters from the more accurate DA-KELM are selected and assigned to the other one. The self-training iteration is repeated till the model accuracy is satisfied.

5.2.1.1 Dynamic Adjustment Kernel Extreme Learning Machine

The extreme learning machine (ELM), a single-hidden layer feed-forward neural network (SLFNN), is suitable for nondifferential activation functions, and, to a large extent, it prevents some troubling issues, such as stopping criteria, learning rate, training epochs, and local minima [1]. By fixing input weights and hidden layer bias, ELM transforms its learning into a matrix calculation, which is performed at a much faster learning speed than the back propagation (BP) algorithm in the traditional ANNs [2]. Due to the efficient calculation, ELM has been successfully employed for modeling microwave components [3, 4].

The original training dataset $\{(\boldsymbol{x}_j, \boldsymbol{t}_j) | \boldsymbol{x}_j \in \mathbb{R}^n, \boldsymbol{t}_j \in \mathbb{R}^m\}$, where $\boldsymbol{x}_j = [x_{j1}, x_{j2}, \ldots, x_{jn}]^T$ is the input, $\boldsymbol{t}_j = [t_{j1}, t_{j2}, \ldots, t_{jm}]^T$ is the corresponding target, $j = 1, 2, \ldots, N$, and N is the total number of distinct samples, is denoted as $G = (\boldsymbol{X}, \boldsymbol{T})$. A standard SLFNN with L hidden nodes and activation function $g(\boldsymbol{x})$ could be mathematically modeled as

$$\sum_{i=1}^{L} \beta_i g_i(\boldsymbol{x}_j) = \sum_{i=1}^{L} \beta_i g(\boldsymbol{w}_i \cdot \boldsymbol{x}_j + b_i) = \boldsymbol{o}_j \tag{5.1}$$

where $\boldsymbol{w}_i = [w_{i1}, w_{i2}, \ldots, w_{in}]^T$ is the weight vector connecting the ith hidden node to the input ones, $\boldsymbol{\beta}_i = [\beta_{i1}, \beta_{i2}, \ldots, \beta_{im}]^T$ is the weight vector connecting the ith hidden node to the output ones, b_i is the threshold of the ith hidden node, and $\boldsymbol{o}_j = [o_{j1}, o_{j2}, \ldots, o_{jm}]^T$ is the jth output vector of SLFNN [5]. With zero error, i.e., $\sum_{j=1}^{N} \|\boldsymbol{o}_j - \boldsymbol{t}_j\| = 0$, (5.1) can be compactly expressed as

$$\boldsymbol{H\beta} = \boldsymbol{T} \tag{5.2}$$

where $\boldsymbol{\beta} = [\beta_{ij}]_{L\times m}$, $\boldsymbol{T} = [t_{ij}]_{N\times m}$ and $\boldsymbol{H}$ is the hidden layer output matrix that could be presented as

$$\boldsymbol{H}(\boldsymbol{w}_1, \boldsymbol{w}_2, \dots, \boldsymbol{w}_L, b_1, b_2, \dots, b_L, \boldsymbol{x}_1, \boldsymbol{x}_2, \dots, \boldsymbol{x}_L) = \begin{bmatrix} g(\boldsymbol{w}_1 \cdot \boldsymbol{x}_1 + b_1) & g(\boldsymbol{w}_2 \cdot \boldsymbol{x}_1 + b_2) & \cdots & g(\boldsymbol{w}_L \cdot \boldsymbol{x}_1 + b_L) \\ g(\boldsymbol{w}_1 \cdot \boldsymbol{x}_2 + b_1) & g(\boldsymbol{w}_2 \cdot \boldsymbol{x}_2 + b_2) & \cdots & g(\boldsymbol{w}_L \cdot \boldsymbol{x}_2 + b_L) \\ \vdots & \vdots & \vdots & \vdots \\ g(\boldsymbol{w}_1 \cdot \boldsymbol{x}_N + b_1) & g(\boldsymbol{w}_2 \cdot \boldsymbol{x}_N + b_2) & \cdots & g(\boldsymbol{w}_L \cdot \boldsymbol{x}_N + b_L) \end{bmatrix}_{N\times L} \tag{5.3}$$

With the kernel function, the kernel extreme learning machine (KELM) is an extended version of ELM [6]. Compared with ELM, the number of hidden nodes (the dimensionality of the hidden layer feature space) from KELM can be infinite, which guarantees good generalization performance and low computational cost via the kernel trick. To date, KELM has attracted increasing interests of different applications from researchers and engineers [5, 7–12].

In ELM, $g(\boldsymbol{x})$ can be referred as an explicit nonlinear mapping, i.e. the so-called ELM random feature mapping, which maps the data from the input space to the ELM feature space. If $g(\boldsymbol{x})$ in (5.1) is replaced with $\phi(\boldsymbol{x})$, the KELM model could be presented as

$$\min_{\alpha} \left\{ L_{\text{KELM}} = \frac{1}{2}\|\boldsymbol{\alpha}\|_2^2 + \frac{C}{2}\sum_{i=1}^{N} \xi_i^2 \right\} \tag{5.4a}$$

$$\text{s.t. } \phi(\boldsymbol{x}_i)^T \boldsymbol{\alpha} + \xi_i = t_i, i = 1, \dots, N \tag{5.4b}$$

Here L_{KELM} is the weighted sum of the empirical risk (i.e. sum error square $\sum_{i=1}^{N} \xi_i^2$) and structural risk (i.e. $\|\boldsymbol{\alpha}\|_2^2$ which is derived from maximizing the distance of the margin separating classes [13–15]). A model with good generalization ability should have the tradeoff between the two risks [16]. $\boldsymbol{\alpha}$ is equivalent to $\boldsymbol{\beta}$ in (5.1), C is a user-specified parameter which provides a tradeoff between the training error and the model structural complexity, ξ_i is the training error with respect to the training datum $\boldsymbol{x}_i$, and $\boldsymbol{\phi}(\boldsymbol{x})$ is induced by a kernel function $k(\boldsymbol{x},\cdot)$ satisfying Mercer's conditions where $k(\boldsymbol{x},\cdot)$ is a symmetric positive semi-definite function [17]. These conditions are commonly used in the support vector machine (SVM), which is an implicit and infinite dimensional system. Thus, the nonlinear problem in the input space becomes a solvable linear one in the feature space. With the representer theorem [18–22], the solution $\boldsymbol{\alpha}$ of (1.2.4) can be written as

$$\boldsymbol{\alpha} = \sum_{i=1}^{\text{N}} \beta_{\text{i}} \phi(\text{x}_{\text{i}}) \tag{5.5}$$

Plugging (5.5) and (5.4b) into (5.4a), we get

$$L_{\text{KELM}} = \frac{1}{2}\boldsymbol{\beta}^T \boldsymbol{K} \boldsymbol{\beta} + \frac{C}{2}(\boldsymbol{T} - \boldsymbol{K}\boldsymbol{\beta})^T(\boldsymbol{T} - \boldsymbol{K}\boldsymbol{\beta}) \tag{5.6}$$

where $\boldsymbol{K}$ is the hidden layer output matrix in KELM and it could be presented as

$$\boldsymbol{K}(\boldsymbol{x}_1, \boldsymbol{x}_2, \dots, \boldsymbol{x}_N) = \begin{bmatrix} k(\boldsymbol{x}_1, \boldsymbol{x}_1) & k(\boldsymbol{x}_1, \boldsymbol{x}_2) & \cdots & k(\boldsymbol{x}_1, \boldsymbol{x}_N) \\ k(\boldsymbol{x}_2, \boldsymbol{x}_1) & k(\boldsymbol{x}_2, \boldsymbol{x}_2) & \cdots & k(\boldsymbol{x}_2, \boldsymbol{x}_N) \\ \vdots & \vdots & \vdots & \vdots \\ k(\boldsymbol{x}_N, \boldsymbol{x}_1) & k(\boldsymbol{x}_N, \boldsymbol{x}_2) & \cdots & k(\boldsymbol{x}_N, \boldsymbol{x}_N) \end{bmatrix}_{N\times N} \tag{5.7}$$

Letting $\frac{dL_{\text{KELM}}}{d\boldsymbol{\beta}} = 0$ gives birth to

$$(\boldsymbol{K}/C + \boldsymbol{K}^T K)\widehat{\boldsymbol{\beta}}_{\text{KELM}} = \boldsymbol{K}^T T \tag{5.8}$$

If the matrix $\boldsymbol{K}/C + \boldsymbol{K}^T\boldsymbol{K}$ is singular, a damp factor μ, which is usually a small positive number, can be added along its ridge to get the unique solution. Thus, (5.8) becomes

$$(\boldsymbol{K}/C + \boldsymbol{K}^T K + \mu \boldsymbol{I})\widehat{\boldsymbol{\beta}}_{\text{KELM}} = \boldsymbol{K}^T\boldsymbol{T} \tag{5.9}$$

Plugging $\boldsymbol{U} = \boldsymbol{K}^T\boldsymbol{K}$ and $\boldsymbol{V} = \boldsymbol{K}^T\boldsymbol{T}$ into (5.9), we get

$$\widehat{\boldsymbol{\beta}}_{\text{KELM}} = (\boldsymbol{K}/C + \boldsymbol{U} + \mu \boldsymbol{I})^{-1}\boldsymbol{V} \tag{5.10}$$

Additionally, $\widehat{\boldsymbol{\beta}}_{\text{KELM}}$ can be obtained from (5.10) as well even if $\boldsymbol{K}/C + \boldsymbol{U}$ is singular, for the purpose of improving the matrix condition number to get a stable solution. u_{ij} and v_{ij} are the elements of $\boldsymbol{U}$ and $\boldsymbol{V}$, respectively,

$$u_{ij} = \sum_{n=1}^{N} k_{in}^T k_{nj} = \sum_{n=1}^{N} k_{ni} k_{nj} = \sum_{n=1}^{N} k(\boldsymbol{x}_n, \boldsymbol{x}_i) k(\boldsymbol{x}_n, \boldsymbol{x}_j) \tag{5.11}$$

$$v_{ij} = \sum_{n=1}^{N} k_{in}^T t_{nj} = \sum_{n=1}^{N} k_{ni} t_{nj} = \sum_{n=1}^{N} k(\boldsymbol{x}_n, \boldsymbol{x}_i) t_{nj} \tag{5.12}$$

Based on (5.10), the output of KELM model could be obtained as

$$f_{\text{KELM}}(\boldsymbol{x}) = \sum_{i=1}^{N} \widehat{\beta}_i k(\boldsymbol{x}, \boldsymbol{x}_i) \tag{5.13}$$

where $\widehat{\beta}_i$ is the ith element of $\widehat{\boldsymbol{\beta}}_{\text{KELM}}$.

According to (5.7), we can get the hidden layer output matrix $\boldsymbol{K}$ from $\boldsymbol{X}$. Thus, we have $(\boldsymbol{K}, \boldsymbol{T})$. The newly arrived training dataset $\{(x_j, t_j) | \boldsymbol{x}_j \in \mathbb{R}^n, t_j \in \mathbb{R}^m, j = N+1, N+2, \ldots, N+\Delta N\}$ is denoted as $\Delta G = (\Delta \boldsymbol{X}, \Delta \boldsymbol{T})$. Then, according to (5.7), we can get the hidden layer output matrix $\Delta \boldsymbol{K}$ from $\Delta \boldsymbol{X}$. Thus, we also have $(\Delta \boldsymbol{K}, \Delta \boldsymbol{T})$.

Now, the new training dataset with the arrived ΔG is $\text{G}^+ = (\boldsymbol{X}^+, \boldsymbol{T}^+) = \{(\boldsymbol{x}_j, \boldsymbol{t}_j) | \boldsymbol{x}_j \in \mathbb{R}^n, \boldsymbol{t}_j \in \mathbb{R}^m, j = 1, 2, \ldots, N+\Delta N\}$, where $\boldsymbol{X}^+ = [\boldsymbol{X}\ \Delta \boldsymbol{X}]^T$ and $\boldsymbol{T}^+ = [\boldsymbol{T}\ \Delta \boldsymbol{T}]^T$. The new hidden layer output matrix $\boldsymbol{K}^+$ can be easily derived from $\boldsymbol{K}$, $\Delta \boldsymbol{K}_1$ and $\Delta \boldsymbol{K}$, i.e. $\boldsymbol{K}^+ = \begin{bmatrix} \boldsymbol{K} & \Delta \boldsymbol{K}_1 \\ \Delta \boldsymbol{K}_1^T & \Delta \boldsymbol{K} \end{bmatrix}$, where

$$\Delta \boldsymbol{K}_1 = \begin{bmatrix} k(\boldsymbol{x}_1, \boldsymbol{x}_{N+1}) & \cdots & k(\boldsymbol{x}_1, \boldsymbol{x}_{N+\Delta N}) \\ \vdots & \vdots & \vdots \\ k(\boldsymbol{x}_N, \boldsymbol{x}_{N+1}) & \cdots & k(\boldsymbol{x}_N, \boldsymbol{x}_{N+\Delta N}) \end{bmatrix}_{N \times \Delta N} \tag{5.14}$$

$$\Delta \boldsymbol{K} = \begin{bmatrix} k(\boldsymbol{x}_{N+1}, \boldsymbol{x}_{N+1}) & \cdots & k(\boldsymbol{x}_{N+1}, \boldsymbol{x}_{N+\Delta N}) \\ \vdots & \vdots & \vdots \\ k(\boldsymbol{x}_{N+\Delta N}, \boldsymbol{x}_{N+1}) & \cdots & k(\boldsymbol{x}_{N+\Delta N}, \boldsymbol{x}_{N+\Delta N}) \end{bmatrix}_{\Delta N \times \Delta N} \tag{5.15}$$

Thus, $(\boldsymbol{K}^+, \boldsymbol{T}^+)$ could be obtained.

According to the matrix multiplication operator, we have

$$\boldsymbol{K}^{+T}\boldsymbol{K}^+ = \begin{bmatrix} \boldsymbol{U} + \Delta \boldsymbol{K}_1 \Delta \boldsymbol{K}_1^T & \boldsymbol{K}^T \Delta \boldsymbol{K}_1 + \Delta \boldsymbol{K}_1 \Delta \boldsymbol{K} \\ \Delta \boldsymbol{K}_1^T \boldsymbol{K} + \Delta \boldsymbol{K}^T \Delta \boldsymbol{K}_1^T & \Delta \boldsymbol{K}_1^T \Delta \boldsymbol{K}_1 + \Delta \boldsymbol{K}^T \Delta \boldsymbol{K} \end{bmatrix} \tag{5.16}$$

$$\boldsymbol{K}^{+T}\boldsymbol{T}^+ = \begin{bmatrix} \boldsymbol{V} + \Delta \boldsymbol{K}_1 \Delta \boldsymbol{T} & \Delta \boldsymbol{K}_1^T \boldsymbol{T} + \Delta \boldsymbol{K}^T \Delta \boldsymbol{T} \end{bmatrix}^T \tag{5.17}$$

Letting $\boldsymbol{U}^+ = \boldsymbol{K}^{+T}\boldsymbol{K}^+$, $\Delta \boldsymbol{U}_{11} = \boldsymbol{U} + \Delta \boldsymbol{K}_1 \Delta \boldsymbol{K}_1^T$, $\Delta \boldsymbol{U}_{12} = \Delta \boldsymbol{U}_{21}^T = \boldsymbol{K}^T \Delta \boldsymbol{K}_1 + \Delta \boldsymbol{K}_1 \Delta \boldsymbol{K}$, and $\Delta \boldsymbol{U}_{22} = \Delta \boldsymbol{K}_1^T \Delta \boldsymbol{K}_1 + \Delta \boldsymbol{K}^T \Delta \boldsymbol{K}$, similar to (5.11), the elements δu_{ij}^{11}, δu_{ij}^{12}, and δu_{ij}^{22} of matrices $\Delta \boldsymbol{U}_{11}$, $\Delta \boldsymbol{U}_{12}$, and $\Delta \boldsymbol{U}_{22}$ can be, respectively, expressed as

$$\delta u_{ij}^{11} = u_{ij} + \sum_{k=N+1}^{N+\Delta N} k(\boldsymbol{x}_i, \boldsymbol{x}_k) k(\boldsymbol{x}_j, \boldsymbol{x}_k) \tag{5.18}$$

$$\delta u_{ij}^{12} = \sum_{k=1}^{N+\Delta N} k(\boldsymbol{x}_k, \boldsymbol{x}_i) k(\boldsymbol{x}_k, \boldsymbol{x}_{N+j}) \tag{5.19}$$

$$\delta u_{ij}^{22} = \sum_{k=1}^{N+\Delta N} k(\boldsymbol{x}_k, \boldsymbol{x}_{N+i}) k(\boldsymbol{x}_k, \boldsymbol{x}_{N+j}) \tag{5.20}$$

Letting $\boldsymbol{V}^+ = \boldsymbol{K}^{+T}\boldsymbol{T}^+$, $\Delta \boldsymbol{V}_1 = \boldsymbol{V} + \Delta \boldsymbol{K}_1 \Delta \boldsymbol{T}$, and $\Delta \boldsymbol{V}_2 = \Delta \boldsymbol{K}_1^T \boldsymbol{T} + \Delta \boldsymbol{K}^T \Delta \boldsymbol{T}$, the elements δv_{ij}^1 and δv_{ij}^2 of matrices $\Delta \boldsymbol{V}_1$ and $\Delta \boldsymbol{V}_2$ can be expressed as

$$\delta v_{ij}^1 = v_{ij} + \sum_{k=N+1}^{N+\Delta N} k(\boldsymbol{x}_k, \boldsymbol{x}_i) t_{kj} \tag{5.21}$$

$$\delta v_{ij}^2 = \sum_{k=1}^{N+\Delta N} k(\boldsymbol{x}_k, \boldsymbol{x}_{N+i}) t_{kj} \tag{5.22}$$

Finally, according to (5.10), the new output weight vector $\hat{\boldsymbol{\beta}}_{\text{KELM}}$ could be obtained as follows,

$$\hat{\boldsymbol{\beta}}_{\text{KELM}} = (\boldsymbol{K}^+/C + \boldsymbol{U}^+ + \mu \boldsymbol{I})^{-1} \boldsymbol{V}^+ \tag{5.23}$$

Compared with the conventional KELM, DA-KELM doesn't need to re-calculate $\boldsymbol{K}$ in $\boldsymbol{K}^+$ when the number of training data is increased. Because of the increased learning process, the calculation result of $\boldsymbol{K}$ which has been obtained in the previous learning process can be employed directly. So do $\boldsymbol{U}$ in $\boldsymbol{U}^+$ and $\boldsymbol{V}$ in $\boldsymbol{V}^+$. For the overlapped elements in DA-KELM, the amount of $2N^2 + N$ element calculations could be saved.

Similar to the increased learning, with the reduced training dataset $\overline{\Delta} G = (\overline{\Delta} \boldsymbol{X}, \overline{\Delta} \boldsymbol{T})$, the new output weight vector $\hat{\boldsymbol{\beta}}_{KELM}$ is obtained as follows,

$$\hat{\boldsymbol{\beta}}_{\text{KELM}} = (\boldsymbol{K}^-/C + \boldsymbol{U}^- + \mu \boldsymbol{I})^{-1} \boldsymbol{V}^- \tag{5.24}$$

where $\boldsymbol{K}^-$, $\boldsymbol{U}^-$, and $\boldsymbol{V}^-$, which are subsets of original $\boldsymbol{K}$, $\boldsymbol{U}$, and $\boldsymbol{V}$, respectively, are the new matrices based on the reduced learning. So, the reduced learning does not need to calculate $\boldsymbol{K}^-$, $\boldsymbol{U}^-$, and $\boldsymbol{V}^-$ or to execute EM simulations any more.

The hybrid learning includes two parts, the reduced learning and increased learning. The decreased training dataset $\left\{ \left(\boldsymbol{x}_j^-, \boldsymbol{t}_j^- \right) \middle| \boldsymbol{x}_j \in \mathbb{R}^n, \boldsymbol{t}_j \in \mathbb{R}^m, j = N - \Delta N^- + 1, N - \Delta N^- + 2, \ldots, N \right\}$ is denoted as $\Delta G^- = (\Delta \boldsymbol{X}^-, \Delta \boldsymbol{T}^-)$ and the increased training dataset $\left\{ \left(\boldsymbol{x}_j^+, \boldsymbol{t}_j^+ \right) \middle| \boldsymbol{x}_j^+ \in \mathbb{R}^n, \ \boldsymbol{t}_j^+ \in \mathbb{R}^m, \right.$ $\left. j = N - \Delta N^- + 1, N - \Delta N^- + 2, \ldots, N - \Delta N^- + \Delta N^+ \right\}$ is denoted as $\Delta G^+ = (\Delta X^+, \Delta T^+)$. According to (5.7), we get the hidden layer output matrices $\Delta \boldsymbol{K}^-$ from $\Delta \boldsymbol{X}^-$ and $\Delta \boldsymbol{K}^+$ from $\Delta \boldsymbol{X}^+$, respectively. Thus, $(\Delta \boldsymbol{K}^-, \Delta \boldsymbol{X}^-)$ and $(\Delta \boldsymbol{K}^+, \Delta \boldsymbol{X}^+)$ could be obtained.

Now, the new training dataset after the operation of ΔG$^-$ and ΔG$^+$ is $G' = (\boldsymbol{X} - \Delta \boldsymbol{X}^- + \Delta \boldsymbol{X}^+$, $\boldsymbol{T} - \Delta \boldsymbol{T}^- + \Delta \boldsymbol{T}^+)$. The new hidden layer output matrix $\boldsymbol{K}$' can be easily derived from $\boldsymbol{K}$, ΔK^-, and ΔK^+. Thus, based on $\boldsymbol{K} = \begin{bmatrix} \boldsymbol{K}^- & \Delta \boldsymbol{K}_1^- \\ \left(\Delta \boldsymbol{K}_1^-\right)^T & \Delta \boldsymbol{K}^- \end{bmatrix}$, $\text{T} = [\boldsymbol{T}^- \ \Delta \boldsymbol{T}^-]^T$, $\boldsymbol{K}' = \begin{bmatrix} \boldsymbol{K}^- & \Delta \boldsymbol{K}_1^+ \\ \left(\Delta \boldsymbol{K}_1^+\right)^T & \Delta \boldsymbol{K}^+ \end{bmatrix}$, and $\text{T}' = [\text{T}^- \Delta T^+]^T$, we can get $(\boldsymbol{K}', \boldsymbol{T}')$.

According to the matrix multiplication operator, we have

$$\boldsymbol{K}'^T \boldsymbol{K}' = \begin{bmatrix} \boldsymbol{K}^{-T}\boldsymbol{K}^- + \Delta \boldsymbol{K}_1^+ \left(\Delta \boldsymbol{K}_1^+\right)^T & \boldsymbol{K}^{-T} \Delta \boldsymbol{K}_1^+ + \Delta \boldsymbol{K}_1^+ \Delta \boldsymbol{K}^+ \\ \left(\Delta \boldsymbol{K}_1^+\right)^T \boldsymbol{K}^- + (\Delta \boldsymbol{K}^+)^T \left(\Delta \boldsymbol{K}_1^+\right)^T & \left(\Delta \boldsymbol{K}_1^+\right)^T \Delta \boldsymbol{K}_1^+ + (\Delta \boldsymbol{K}^+)^T \Delta \boldsymbol{K}^+ \end{bmatrix} \tag{5.25}$$

Letting $\boldsymbol{U}' = \boldsymbol{K}'^T \boldsymbol{K}'$, $\Delta \boldsymbol{U}'_{11} = \boldsymbol{K}^{-T}\boldsymbol{K}^- + \Delta \boldsymbol{K}_1^+ \left(\Delta \boldsymbol{K}_1^+\right)^T$, $\Delta \boldsymbol{U}'_{12} = \Delta \boldsymbol{U}'^T_{21} = \boldsymbol{K}^{-T} \Delta \boldsymbol{K}_1^+ + \Delta \boldsymbol{K}_1^+ \Delta \boldsymbol{K}^+$, and $\Delta \boldsymbol{U}'_{22} = \left(\Delta \boldsymbol{K}_1^+\right)^T \Delta \boldsymbol{K}_1^+ + (\Delta \boldsymbol{K}^+)^T \Delta \boldsymbol{K}^+$, the elements $\delta u_{ij}'^{11}$, $\delta u_{ij}'^{12}$, and $\delta u_{ij}'^{22}$ of matrices $\Delta \boldsymbol{U}'_{11}$, $\Delta \boldsymbol{U}'_{12}$, and $\Delta \boldsymbol{U}'_{22}$ can be, respectively, expressed as

$$\delta u_{ij}^{\prime 11} = \sum_{k=1}^{N-\Delta N^{-}+\Delta N^{+}} k(\boldsymbol{x}_i, \boldsymbol{x}_k)\, k(\boldsymbol{x}_j, \boldsymbol{x}_k) \tag{5.26}$$

$$\delta u_{ij}^{\prime 12} = \sum_{k=1}^{N-\Delta N^{-}+\Delta N^{+}} k(\boldsymbol{x}_k, \boldsymbol{x}_i)\, k\left(\boldsymbol{x}_k, \boldsymbol{x}_{N-\Delta N^{-}+j}\right) \tag{5.27}$$

$$\delta u_{ij}^{\prime 22} = \sum_{k=1}^{N-\Delta N^{-}+\Delta N^{+}} k\left(\boldsymbol{x}_k, \boldsymbol{x}_{N-\Delta N^{-}+i}\right)\, k\left(\boldsymbol{x}_k, \boldsymbol{x}_{N-\Delta N^{-}+j}\right) \tag{5.28}$$

Similarly, according to the matrix multiplication operator, we have

$$\boldsymbol{K}'^{T}\boldsymbol{T}' = \begin{bmatrix} (\boldsymbol{K}^{-})^{T}\boldsymbol{T}^{-} + \Delta\boldsymbol{K}_1^{+}\Delta\boldsymbol{T}^{+} \\ \left(\Delta\boldsymbol{K}_1^{+}\right)^{T}\boldsymbol{T}^{-} + (\Delta\boldsymbol{K}^{+})^{T}\Delta\boldsymbol{T}^{+} \end{bmatrix} \tag{5.29}$$

Letting $\boldsymbol{V}' = \boldsymbol{K}'^{T}\boldsymbol{T}'$, $\Delta\boldsymbol{V}_1' = (\boldsymbol{K}^{-})^{T}\boldsymbol{T}^{-} + \Delta\boldsymbol{K}_1^{+}\Delta\boldsymbol{T}^{+}$, and $\Delta\boldsymbol{V}_2' = \left(\Delta\boldsymbol{K}_1^{+}\right)^{T}\boldsymbol{T}^{-} + (\Delta\boldsymbol{K}^{+})^{T}\Delta\boldsymbol{T}^{+}$, the elements $\delta v_{ij}^{\prime 1}$ and $\delta v_{ij}^{\prime 2}$ of matrices $\Delta\boldsymbol{V}'_1$ and $\Delta\boldsymbol{V}'_2$ can be expressed as

$$\delta v_{ij}^{\prime 1} = \sum_{k=1}^{N-\Delta N^{-}+\Delta N^{+}} k(\boldsymbol{x}_k, \boldsymbol{x}_i) t_{kj} \tag{5.30}$$

$$\delta v_{ij}^{\prime 2} = \sum_{k=1}^{N-\Delta N^{-}+\Delta N^{+}} k\left(\boldsymbol{x}_k, \boldsymbol{x}_{N-\Delta N^{-}+i}\right) t_{kj} \tag{5.31}$$

Based on (5.10), the new output weight vector $\widehat{\boldsymbol{\beta}}_{\mathrm{KELM}}$ could be obtained

$$\widehat{\boldsymbol{\beta}}_{\mathrm{KELM}} = (\boldsymbol{K}'/C + \boldsymbol{U}' + \mu\boldsymbol{I})^{-1}\boldsymbol{V}' \tag{5.32}$$

Compared with the conventional KELM, DA-KELM could utilize the overlapped elements to speed up re-training due to the direct utilization of previous calculation results. Meanwhile, to construct a new training dataset with a suitable size for efficient calculation in the hybrid learning, the reduced learning should be processed firstly.

To evaluate the performance of the increased learning, the reduced learning and hybrid learning from DA-KELM, the conventional KELM is employed as a comparison model to approximate the SinC function, which is a commonly used instance [23, 24].

$$y(x) = \mathrm{sinc}(x) = \begin{cases} \frac{\sin(\pi x)}{\pi x} & x \neq 0 \\ 1 & x = 0 \end{cases} \tag{5.33}$$

The evaluation setting is similar to [23]. A training set (x_i, y_i) and a testing set (x'_i, y'_i) contain N data, respectively. x_i is distributed in the range of $(-\pi, \pi)$ uniformly. Large uniform noise distributed in $[-0.2, 0.2]$ has been added to all the training samples, while testing data remain noise free, as shown in Figure 5.1. Here, a performance index, root mean squared error (RMSE), is defined by

$$\mathrm{RMSE} = \sqrt{\frac{\sum_{i=1}^{N}(y_i - \widehat{y}_i)^2}{N}} \tag{5.34}$$

where y_i is the calculation result from (5.33) and $\widehat{y}_i$ is the output value of DA-KELM.

Here the popular Gaussian kernel function is used [24]

$$k(\boldsymbol{x}_i, \boldsymbol{x}_j) = e^{-\frac{\|x_i - x_j\|_2^2}{2\gamma^2}} \tag{5.35}$$

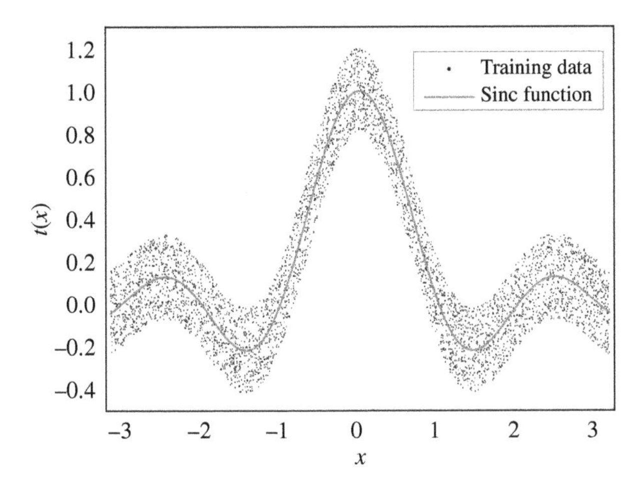

Figure 5.1 SinC function with noise distribution. Source: Xiao et al. [25], ©2018 IEEE.

where γ is the kernel parameter. The regularization parameter C in (5.10) and γ have been determined by leave-one-out cross validation with the values of 10^{-1} and 2^{-2} from 144 different pairs, respectively [26]. The value of μ in (5.10) is set as 2^{-8}.

DA-KELM was firstly trained with 4000 samples. Secondly, with the increased learning, 1000 samples have been directly added to continue training instead of rebuilding a training dataset. Thirdly, with the reduced learning, DA-KELM is trained by subtracting 2000 samples from the 5000 training samples, and the results from 3000 training samples could be obtained conveniently. Finally, with the hybrid learning, 2000 and 1000 samples were added and subtracted, respectively, and the results of a new training dataset with 4000 samples could be obtained directly instead of rebuilding a training dataset to resume training. On the contrary, KELM was separately trained with corresponding training samples during the process.

There are 20 significant vectors assigned for KELM and DA-KELM, and 50 trials have been conducted for the calculations. The average results of RMSE and standard deviation (Dev) are shown in Table 5.1.

The process of DA-KELM model could be illustrated as follows. Firstly, the original training dataset is built with N samples. Then, DA-KELM is trained with the obtained training dataset, and the trained DA-KELM is tested to estimate whether current accuracy satisfies the requirement. If both the training error and testing error are large, DA-KELM should be re-trained with the increased learning to overcome the under-fitting problem. If there exists a big gap between the training error and testing error, DA-KELM should be re-trained with the reduced learning until a good solution is achieved. If the range of the training dataset is shifted or expanded, DA-KELM should be trained with the hybrid learning to obtain the final results conveniently and directly. The flowchart of the training process is shown in Figure 5.2.

5.2.1.2 Semi-Supervised Learning Based on DA-KELM

There are two training processes in the semi-supervised learning model, initial training and self-training. The former consists of one DA-KELM and the latter consists of two DA-KELMs.

There are N_0 labeled training datasets obtained from full-wave simulations for initial training. The jth training dataset is presented as $\{\boldsymbol{Ini.G_j}=(\boldsymbol{Ini.x_j}, \boldsymbol{Ini.t_j}) \mid \boldsymbol{Ini.x_j} \in \mathbb{R}^n, \boldsymbol{Ini.t_j} \in \mathbb{R}^m\}$, where $\boldsymbol{Ini.x_j} = [Ini.x_{j1}, Ini.x_{j2}, \dots, Ini.x_{jn}]^T$ is the input of a set of geometrical variables, n is the number

Table 5.1 Performance Comparison for Learning Noise-Free Function: SinC.

			KELM	DA-KELM
Original training	**Training samples**		**4000**	**4000**
	CPU time (s)	Training	0.0645	0.0642
		Testing	0.0248	0.0247
	Training	RMSE	0.1112	0.1108
		Dev	0.0030	0.0030
	Testing	RMSE	0.0074	0.0074
		Dev	0.0024	0.0024
Increased learning	**Training Samples**		**5000**	**+1000**
	CPU time (s)	Training	0.0725	0.0231
		Testing	0.0248	0.0247
	Training	RMSE	0.0947	0.0942
		Dev	0.0029	0.0028
	Testing	RMSE	0.0062	0.0063
		Dev	0.0023	0.0022
Reduced learning	**Training Samples**		**3000**	**−2000**
	CPU time (s)	Training	0.0454	0.0304
		Testing	0.0248	0.0248
	Training	RMSE	0.1001	0.1005
		Dev	0.0032	0.0033
	Testing	RMSE	0.0081	0.0081
		Dev	0.0026	0.0027
Hybrid learning	**Training Samples**		**4000**	**+2000-1000**
	CPU time (s)	Training	0.0644	0.0365
		Testing	0.0244	0.0238
	Training	RMSE	0.1109	0.1110
		Dev	0.0029	0.0030
	Testing	RMSE	0.0074	0.0072
		Dev	0.0023	0.0023

Source: Xiao et al. [25], ©2018 IEEE.)

of geometrical variables, $\boldsymbol{Ini.t}_j = [Ini.t_{j1}, Ini.t_{j2}, \ldots, Ini.t_{jm}]^T$ is the output of the corresponding transfer function (TF) coefficients from the simulated result (i.e. real and imaginary parts of S-parameters), and m is related to the order of TF.

The pole-residue-based TF, an effective TF form in EM simulations [27], is chosen in this study. It is presented as

$$H(s) = \sum_{i=1}^{Q} \frac{r_i}{s - p_i} \tag{5.36}$$

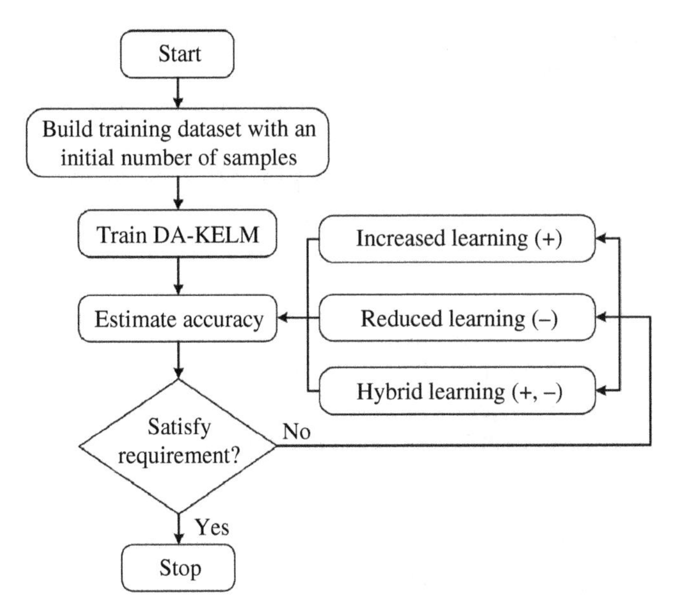

Figure 5.2 Flowchart of the training process of DA-KELM. Source: Xiao et al. [25], ©2018 IEEE.

where p_i and r_i are the pole and residue coefficients of TF, respectively, and $Q = m/2$ is the TF order. The vector fitting technique [25] is employed to obtain the poles and residues of the TF corresponding to the simulated result. The pole-residue tracking technique is used to solve the order-changing problem here [27].

After DA-KELM is trained with the labeled training datasets of ***Ini.G***, the initial error, *Ini.e*, of the model could be obtained. The process of the initial training is shown in Figure 5.3.

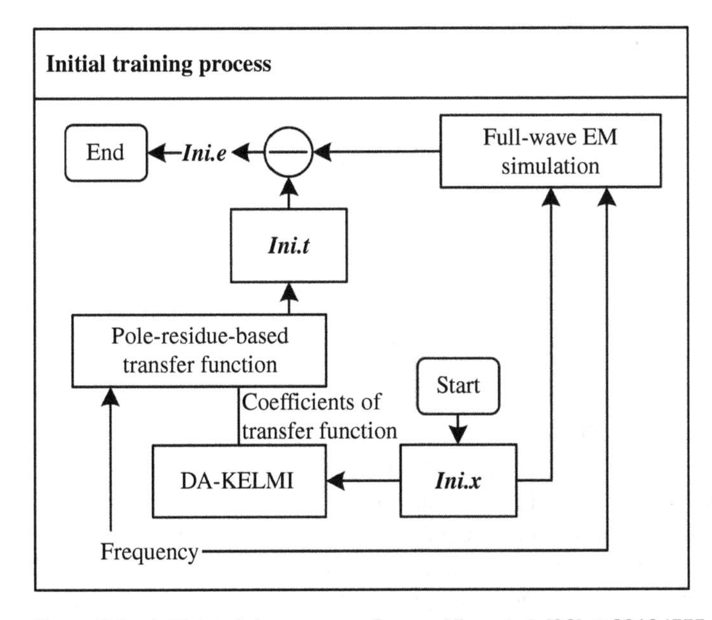

Figure 5.3 Initial training process. Source: Xiao et al. [28], ©2019 IEEE.

Before the self-training process, another DA-KELM is duplicated from the existing one obtained in the first training process. For the ith iteration step of the self-training process, as shown in Figure 5.4, two unlabeled input datasets of geometrical variables, $\boldsymbol{Self_1.x_j^i} = \left[Self_1.x_{j1}^i, Self_1.x_{j2}^i, \ldots, Self_1.x_{jn}^i\right]^T$ and $\boldsymbol{Self_2.x_j^i} = \left[Self_2.x_{j1}^i, Self_2.x_{j2}^i, \ldots, Self_2.x_{jn}^i\right]^T$, are input to the two DA-KELMs, respectively, where $j = 1, 2, \ldots, N_i$, and N_i is the total number of input datasets in the ith iteration step. Here, N_i is much smaller than N_0, and $\boldsymbol{Self_1.x_j^i}$ $\boldsymbol{Self_2.x_j^i}$ satisfy the following relationship

$$\boldsymbol{Self_1.x_j^i} \cap \boldsymbol{Self_2.x_j^i} \neq \boldsymbol{Self_1.x_j^i} \cup \boldsymbol{Self_2.x_j^i} \tag{5.37}$$

Thus, $\boldsymbol{Self_1.t_j^i} = \left[Self_1.t_{j1}^i, Self_1.t_{j2}^i, \ldots, Self_1.t_{jm}^i\right]^T$ and $\boldsymbol{Self_2.t_j^i} = \left[Self_2.t_{j1}^i, Self_2.t_{j2}^i, \ldots, Self_2.t_{jm}^i\right]^T$ could be obtained with the increased learning technique from DA-KELM$_1$ and DA-KELM$_2$, respectively. Two sets of self-training datasets obtained from DA-KELM$_1$ and DA-KELM$_2$ are presented as $\boldsymbol{Self_1.G^i} = (\boldsymbol{Self_1.x^i}, \boldsymbol{Self_1.t^i})$ and $\boldsymbol{Self_2.G^i} = (\boldsymbol{Self_2}.x^i, \boldsymbol{Self_2.t^i})$, respectively. Then $\boldsymbol{Self_2.G^i}$ and $\boldsymbol{Self_1.G^i}$ are employed to train DA-KELM$_1$ and DA-KELM$_2$, respectively, and the errors of the two DA-KELMs, e_1^i and e_2^i, could be obtained by using a set of labeled testing samples. At the same time, the output weight vectors of $\hat{\boldsymbol{\beta}}_1^i$ in DA-KELM$_1$ and $\hat{\boldsymbol{\beta}}_2^i$ in DA-KELM$_2$ could be obtained. If $e_1^i > e_2^i$, the value of $\hat{\boldsymbol{\beta}}_1^i$ will be substituted with that of $\hat{\boldsymbol{\beta}}_2^i$, and vice versa.

For the $(i+1)$th iteration step of the self-training process, $\boldsymbol{Self_1.x^{i+1}}$ and $\boldsymbol{Self_2.x^{i+1}}$ are input to obtain $\boldsymbol{Self_1.t^{i+1}}$ and $\boldsymbol{Self_2.t^{i+1}}$ from DA-KELM$_1$ and DA-KELM$_2$, respectively. In the $(i+1)$th step, the two DA-KELMs with more accurate inner parameters result in the better $\boldsymbol{Self_1.t^{i+1}}$ and $\boldsymbol{Self_2.t^{i+1}}$ than $\boldsymbol{Self_1.t^i}$ and $\boldsymbol{Self_2.t^i}$. Thus, with the training of $\boldsymbol{Self_1.G^{i+1}}$ and $\boldsymbol{Self_2.G^{i+1}}$, the accuracy of DA-KELM can be enhanced in the $(i+1)$th step. With the testing of the labeled data, the less accurate inner parameters of DA-KELM are updated, and then the more accurate self-training dataset can be produced to further train DA-KELM. The self-learning iteration is repeated until one of the testing errors is less than a specified error of e_{spec}.

Compared with the traditional training method in which the training datasets are all collected from full-wave EM simulations, the proposed model only involves a small number of labeled samples.

5.2.1.3 Numerical Examples

In this section, the mean absolute percentage error (MAPE) is used for calculating training and testing errors here:

$$\text{MAPE} = \frac{1}{M}\sum_{m=1}^{M}\left|\frac{y_m - \hat{y}_m}{y_m}\right| \times 100\% \tag{5.38}$$

where y_m is the mth value in the testing dataset, $\hat{y}_m$ is the corresponding output from the two models, and M represents the length of the dataset.

Example 5.1 ***Hybrid Learning in DA-KELM for a Quadruple-Mode Filter***

In the optimization process, it is common for shifting the value range of the optimized variables to obtain good results. For this case, the hybrid learning is a better choice than the increased learning and reduced learning. In the hybrid learning scheme, the reduced learning and increased learning simultaneously deal with the shifted dataset to obtain the final result conveniently and directly, without step-by-step calculation. To evaluate the effectiveness of the hybrid learning, the modeling process of a quadruple-mode filter in Figure 5.5 is provided, where the height and diameter of the

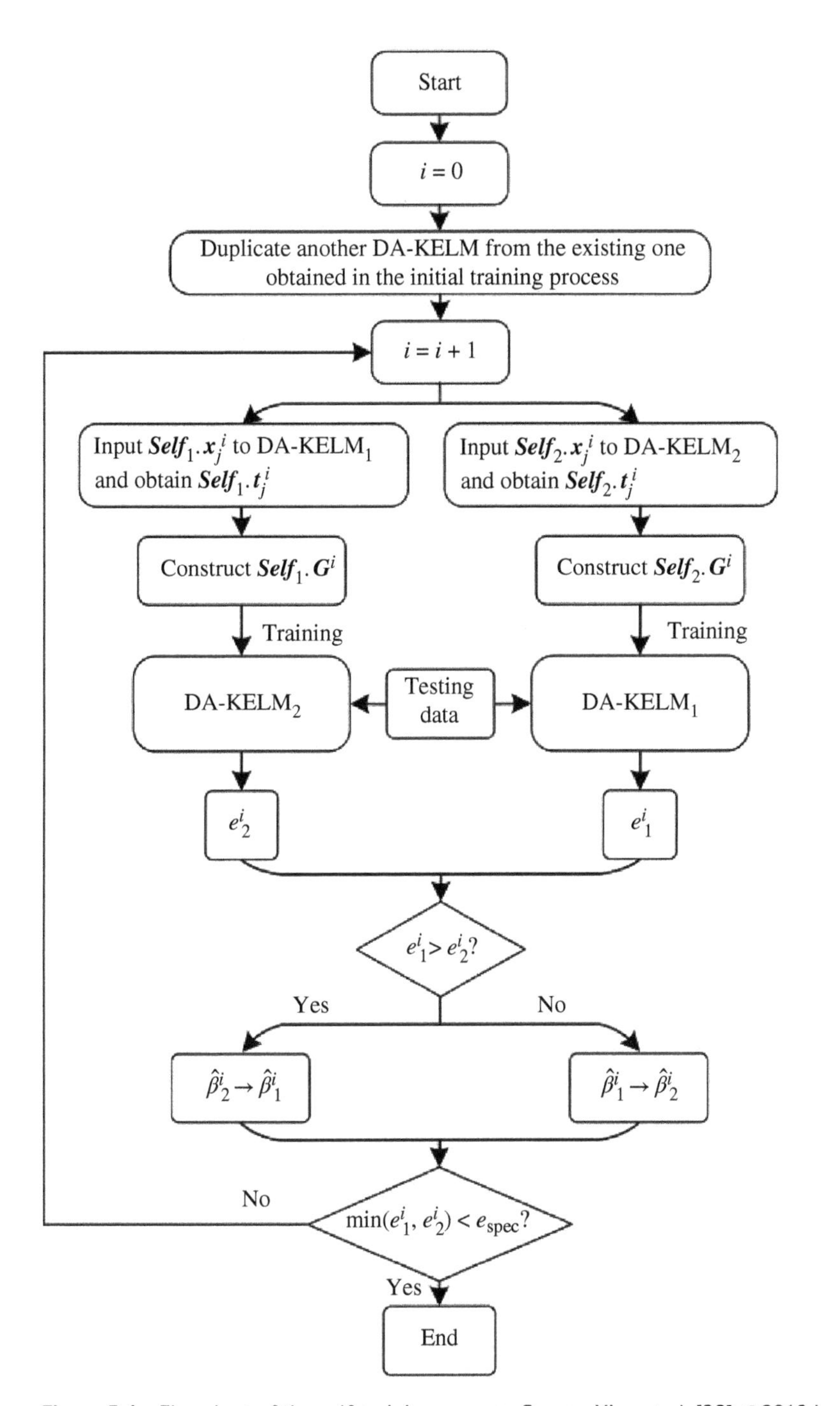

Figure 5.4 Flowchart of the self-training process. Source: Xiao et al. [28], ©2019 IEEE.)

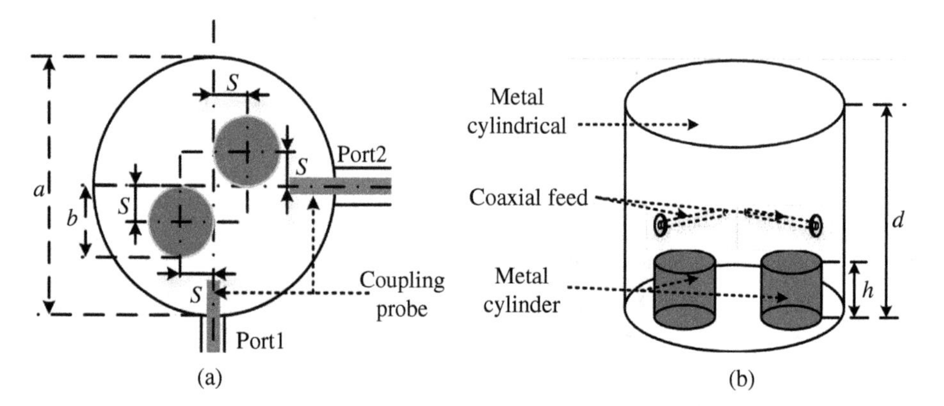

Figure 5.5 Structure of the quadruple-mode filter. (a) Top view and (b) Three-dimensional view. Source: Xiao et al. [25], ©2018 IEEE.)

Table 5.2 Definition of seven-level training and seven-level testing data for the quadruple-mode filter.

Geometrical variables		Training Data (49 samples) Min	Max	Step	Testing Data (49 samples) Min	Max	Step
Original training dataset	a (mm)	50.1	51.3	0.2	50.2	51.4	0.2
	b (mm)	15.5	16.1	0.1	15.55	16.15	0.1
	S (mm)	7.4	8	0.1	7.45	8.05	0.1
	h (mm)	15.5	16.1	0.1	15.55	16.15	0.1
	d (mm)	50.1	51.3	0.2	50.2	51.4	0.2
Shifting training dataset	a (mm)	50.5	51.7	0.2	50.6	51.8	0.2
	b (mm)	15.7	16.3	0.1	15.75	16.35	0.1
	S (mm)	7.6	8.2	0.1	7.65	8.25	0.1
	h (mm)	15.7	16.3	0.1	15.75	16.35	0.1
	d (mm)	50.5	51.7	0.2	50.6	51.8	0.2

Source: Xiao et al. [25], ©2018 IEEE.)

cavity are d and a, and the height and diameter of the two perturbation metal cylinders in the cavity are h and b [29]. The model has five input geometrical variables, i.e. $\boldsymbol{X} = [a\ b\ S\ h\ d]^T$. Frequency is an additional input parameter with an original range of 1–5 GHz. The model has two outputs, the real part R_{S11} and imaginary part I_{S11} of S_{11}, i.e. $\boldsymbol{y} = [R_{S11}\ I_{S11}]^T$. In this example, the number N of the original samples is set as 49, and n and m are 5 and 16, respectively.

The DA-KELM model is originally trained with 49 training samples defined with the seven-level design of experiment (DOE) method [30], and it is tested with other 49 testing samples from the seven-level DOE method, as shown in Table 5.2. The training and testing errors in Table 5.9 are 0.4654% and 0.7486%, respectively. When the selection range of the original dataset is changed, as shown in Table 5.2, 15 original training samples which are not in the shifting range are subtracted and 15 new training samples which appear in the shifting range are added to the training dataset. After the hybrid learning process, the obtained training and testing errors in Table 5.3 are 0.4653% and 0.7398%, respectively.

Table 5.3 Results of hybrid learning for the quadruple-mode filter.

		KELM	DA-KELM
Original training	Training samples	49	49
	Calculation time (s)	5074.45	5074.53
	Training error	0.4716%	0.4654%
	Testing error	0.7394%	0.7486%
Hybrid learning	Training samples	49	−15(original)+15(new)
	Calculation time (s)	5074.45	1415.59
	Training error	0.4703%	0.4653%
	Testing error	0.7314%	0.7398%

Source: Xiao et al. [25]/IEEE.

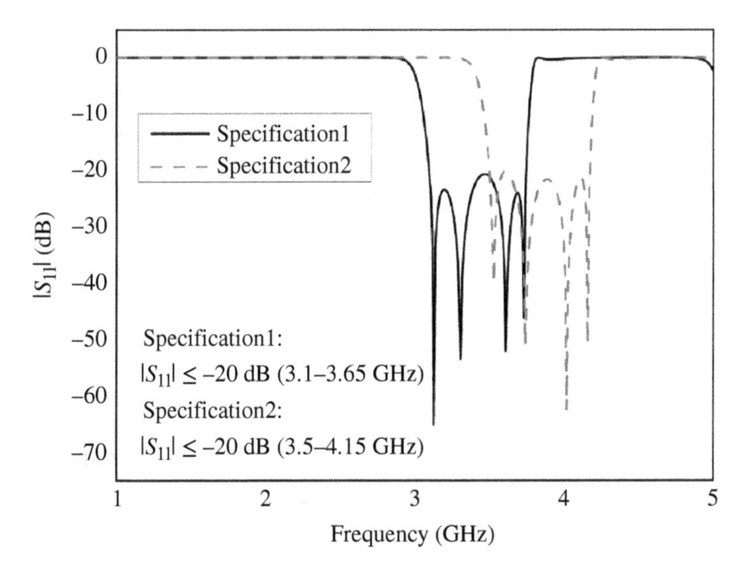

Figure 5.6 Optimization results of S_{11} from DA-KELM. Source: Xiao et al. [25], ©2018 IEEE.

Similarly, the conventional KELM is selected as a comparison and it is trained with 49 samples. When the selection range of the original dataset is changed, it is trained with new updated 49 samples. From Table 5.3, the calculation time of DA-KELM is much shorter than that of KELM.

The trained model as a substitute for the time-consuming EM simulation is also applied to the optimization design. As shown in Figure 5.6, two separate filters are optimized to reach two different design specifications. The initial values are $\boldsymbol{x} = [51\ 16\ 7\ 16\ 51]^T$. The optimization with the flower pollination algorithm (FPA) [33] of the quadruple-mode filter is performed by calling DA-KELM repeatedly. The optimized geometrical values for the two separate filters are: $\boldsymbol{x}_{opt1} = [50.0014\ 14.4064\ 7.4097\ 14.8101\ 44.8997]^T$, $\boldsymbol{x}_{opt2} = [49.9915\ 13.9814\ 7.1998\ 14.7911\ 45.3092]^T$. The optimization costs only about 30 seconds to achieve the optimal solution for each filter, shown in Table 5.4. Compared with the EM simulation, DA-KELM can save considerable time in filter design.

Table 5.4 Running time of optimization process based on direct EM optimization and the DA-KELM.

	CPU time of model development	
	Direct EM optimization	**DA-KELM model**
Specification 1	10 h	30 s
Specification 2	11 h	30 s
Total	21 h	1.68 h (training) + 60 s

Source: Xiao et al. [25], ©2018 IEEE.

Example 5.2 ***Semi-Supervised DA-KELM for Microstrip-to-Microstrip Vertical Transition***
For comparison, a Microstrip-to-Microstrip (MS-to-MS) vertical transition in [34] is employed to evaluate the semi-supervised learning model. In Figure 5.7, there are six input geometrical variables, i.e. $\boldsymbol{x} = [L_m\ W_m\ W_s\ L_{s1}\ L_{s2}\ L_{s3}]^T$. The frequency range is from 1 to 15 GHz. The real part R_{S11} and imaginary part I_{S11} of S_{11} are the two outputs of the model, i.e. $\boldsymbol{y} = [R_{S11}\ I_{S11}]^T$. The *HFSS* software with the full-wave EM simulation produces the labeled training and testing data for modeling.

As same as the selection of training samples in [25], the DOE method is used to sample both training and testing data [30]. 36 labeled samples with the six-level DOE method defined in Table 5.5 are used as training samples for DA-KELM. Meanwhile, 36 training samples with six-level DOE method defined in a wide parameter range as another case are also added for a general evaluation. The collection of 49 testing samples with seven-level DOE method in Table 5.5 costs about 3698.11s.

First, 15 samples from the 36 labeled training samples are randomly selected to train the semi-supervised learning model. 15 training samples are not sufficient, and the testing errors are 4.8418% for Case 1 and 5.2891% for Case 2. Then, the semi-supervised learning model carries out its self-training operation to train itself. Each self-training iteration step involves five unlabeled training samples, which are randomly selected from 100 unlabeled training samples with the ten-level DOE method, and the 49 labeled testing samples. The labeled and unlabeled training samples are defined with different levels of the DOE method because the number of the labeled samples is always less than that of the unlabeled ones. Meanwhile, the minimum and maximum

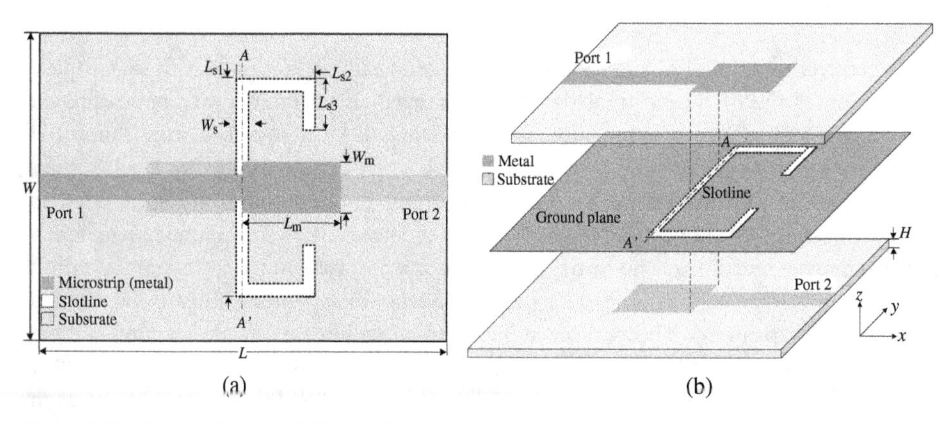

Figure 5.7 Schematic of the MS-to-MS vertical transition where $W = 20$ mm, $L = 30$ mm, and $H = 0.813$ mm: (a) top view and (b) Layout. Source: Xiao et al. [28], ©2019 IEEE.

Table 5.5 Definition of six-level training and seven-level testing data for the MS-to-MS vertical transition.

		Training data (36 samples)			Testing data (49 samples)		
Geometrical variables		**Min**	**Max**	**Step**	**Min**	**Max**	**Step**
Case 1	L_m (mm)	4.8	6.3	0.3	4.85	6.65	0.3
	W_m (mm)	1.3	2.8	0.3	1.45	2.95	0.25
	W_s (mm)	0.15	0.65	0.1	0.21	0.69	0.08
	L_{s1} (mm)	9.6	11.6	0.4	9.9	11.7	0.3
	L_{s2} (mm)	1.2	1.7	0.1	1.26	1.74	0.08
	L_{s3} (mm)	2.05	2.55	0.1	2.11	2.59	0.08
Case 2	L_m (mm)	4.425	6.675	0.45	4.4	7.1	0.45
	W_m (mm)	0.925	3.175	0.45	1.075	3.325	0.375
	W_s (mm)	0.025	0.775	0.15	0.09	0.81	0.12
	L_{s1} (mm)	9.1	12.1	0.6	9.45	12.15	0.45
	L_{s2} (mm)	1.075	1.825	0.15	1.14	1.86	0.12
	L_{s3} (mm)	1.925	2.675	0.15	1.99	2.71	0.12

Source: Xiao et al. [28]/IEEE.

Table 5.6 Modeling results for the MS-to-MS vertical transition.

		Model in [25]	**Semi-supervised learning model**
Case 1	Labeled training samples	36	15
	Training time (s)	2716.98	1132.09 (initial training) + 10.98 (self-training)
	Training error	0.6718%	0.6702%
	Testing error	0.8726%	0.8733%
Case 2	Labeled training samples	36	15
	Training time (s)	2718.15	1132.12 (initial training) + 10.67 (self-training)
	Training error	0.6981%	0.6978%
	Testing error	0.8947%	0.8946%

Source: Xiao et al. [28]/IEEE.

values of each variable defined in the DOE method for labeled and unlabeled training sampling should be the same.

After 17 iteration steps of the self-training process, the results from the semi-supervised learning model are shown in Table 5.6. Meanwhile, the results from DA-KELM are also shown in Table 5.6. Figure 5.8 plots the curves of testing errors with the iteration steps. The testing errors in Figure 5.8 gradually decrease in the self-training process for both Cases 1 and 2.

Compared with the DA-KELM model in [25], the semi-supervised learning model only utilizes 15 labeled training samples, leading to a reduction of 58% for collecting training samples from *HFSS*. After 17 iteration steps of the self-training process, the training and testing errors of the semi-supervised learning model are at the same level as the DA-KELM model for both Cases 1 and 2.

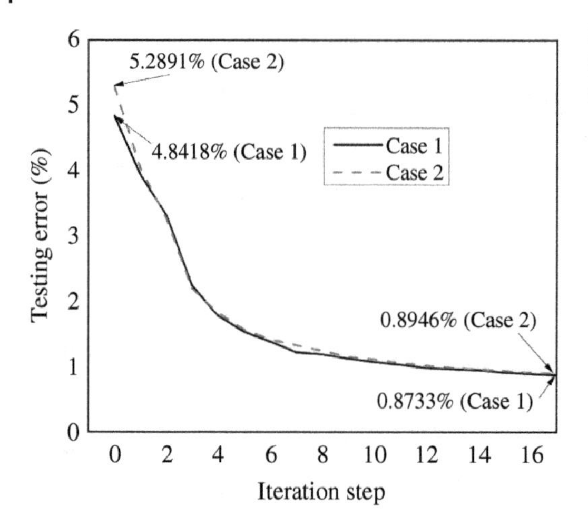

Figure 5.8 Testing errors with the iteration steps for the MS-to-MS vertical transition. Source: Xiao et al. [28]/IEEE.

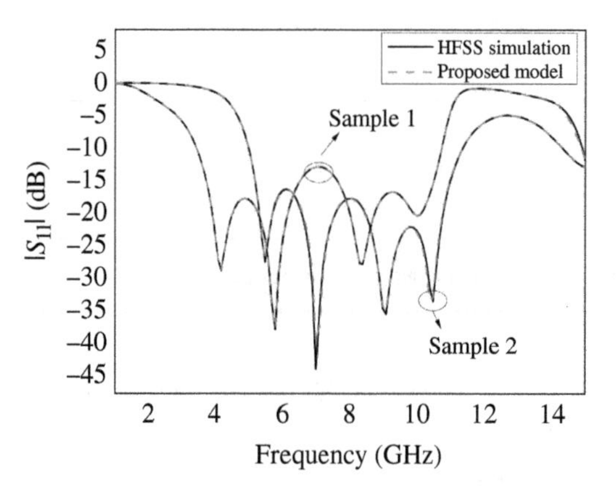

Figure 5.9 Comparison of $|S_{11}|$ for Samples 1 and 2, where the samples are out of the range of training dataset. Source: Xiao et al. [28]/IEEE.

Similar to [25], two other examples, $\boldsymbol{x}'_1 = [4.423\ 0.919\ 0.779\ 9.05\ 1.071\ 1.923]^T$ and $\boldsymbol{x}'_2 = [6.676\ 3.179\ 0.778\ 12.12\ 1.827\ 2.68]^T$, which are out of the range of the training dataset, are employed to evaluate the semi-supervised learning model. Taking the semi-supervised learning model of Case 2 for example, from Figure 5.9, the model can still maintain good simulation performance with the average MAPE value of 0.8817% even the geometrical variables are out of the range of training dataset.

After the training process, the trained semi-supervised learning model of Case 2 could be the alternative to the time-consuming EM simulation for the optimization design of MS-to-MS vertical transitions. Two different design specifications are realized in Figure 5.10 by repeatedly calling the model by FPA. The initial values of geometrical variables are set as $\boldsymbol{x}_{initial} = [5.5\ 2\ 0.5\ 10\ 1.5\ 2.3]^T$. The optimized geometrical values for the two separate specifications are: $\boldsymbol{x}_{opt1} = [6.4\ 2.8\ 0.3\ 10.0\ 1.3\ 2.4]^T$ and $\boldsymbol{x}_{opt2} = [6.4\ 2.0\ 0.6\ 11.5\ 1.7\ 2.5]^T$.

From Table 5.7, one can see that the semi-supervised learning model and the DA-KELM model cost much less optimization time than the optimization with full wave EM simulations even though the training time is involved. Furthermore, the semi-supervised learning model is more efficient than the DA-KELM model due to its fewer labeled training samples.

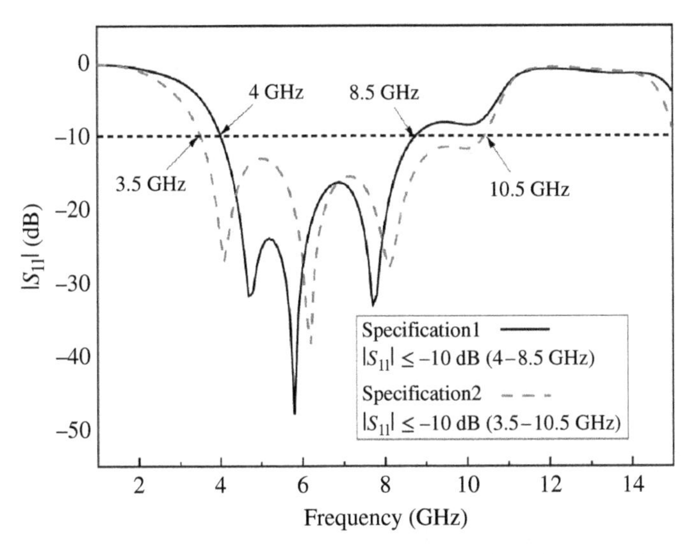

Figure 5.10 Optimization results of the MS-to-MS vertical transitions. Source: Xiao et al. [28]/IEEE.

Table 5.7 Optimization time of the MS-to-MS vertical transitions.

	Direct EM optimization	Model in [25]	Proposed model
Specification 1	3.0 h	30 s	30 s
Specification 2	3.0 h	30 s	30 s
Total	6.0 h	0.76 h (training) + 60 s	0.32 h (training) + 60 s

Source: Xiao et al. [28]/IEEE.

5.2.2 Semi-Supervised Radial Basis Function Neural Network

In this section, a semi-supervised radial basis function neural network (SS-RBFNN) model with an effective sampling strategy is proposed to enhance the modeling accuracy and convergence rate in the training process. After evaluating the current training, the sampling strategy selects suitable training samples to ensure each sub-region of the whole sampling region with the same level of training and testing accuracy. Compared with the conventional semi-supervised learning model, SS-RBFNN simplifies the modeling process. It does not need to be conducted between two separated ANNs, and only one RBFNN implements both supervised and unsupervised learning. SS-RBFNN also includes increased learning function to largely enhance the calculation efficiency when the model needs to utilize more training samples.

5.2.2.1 Semi-Supervised Radial Basis Function Neural Network

RBFNN uses RBFs as the activation function. The network output is a linear combination of RBF of the inputs and neuron parameters. With the training dataset $\{(\boldsymbol{x}_j, \boldsymbol{t}_j)|\boldsymbol{x}_j \in \mathbb{R}^n, \boldsymbol{t}_j \in \mathbb{R}^m\}$, where $\boldsymbol{x}_j = [x_{j1}, x_{j2}, \ldots, x_{jn}]^T$ is the input, $t_j = [t_{j1}, t_{j2}, \ldots, t_{jm}]^T$ is the corresponding target, $j = 1, 2, \ldots, N$, and N is the total number of distinct samples, the output $\boldsymbol{o}_j$ of RBFNN, which has L hidden nodes, can be described as follows:

$$\sum_{l=1}^{L} \boldsymbol{w}_l \varphi(\|\boldsymbol{x}_j - c_l\|) = \boldsymbol{o}_j \tag{5.39}$$

where $\boldsymbol{w}_l = [w_{l1}, w_{l2}, \ldots, w_{lm}]^T$ is the output weight, c_l denotes the lth center, $||\cdot||$is Euclidean distance, and $\varphi(\cdot)$ represents RBF.

In the training process, the training error or cost function is written as:

$$L = \frac{1}{2}||\boldsymbol{W}||^2 + \frac{\sigma}{2}||\boldsymbol{T} - \boldsymbol{O}||^2 = \frac{1}{2}||\boldsymbol{W}||^2 + \frac{\sigma}{2}||\boldsymbol{T} - \boldsymbol{HW}||^2 \tag{5.40}$$

where $||\cdot||$denotes the Frobenius norm, $\boldsymbol{T} = [\boldsymbol{t}_1, \boldsymbol{t}_2, \ldots, \boldsymbol{t}_N]^\mathrm{T}$, $\boldsymbol{O} = [\boldsymbol{o}_1, \boldsymbol{o}_2, \ldots, \boldsymbol{o}_N]^\mathrm{T}$, $\boldsymbol{W} = [\boldsymbol{w}_1, \boldsymbol{w}_2, \ldots, \boldsymbol{w}_L]^T$, $||\boldsymbol{W}||$ is an added regular term to avoid over-fitting, $\sigma > 0$ is a tunable parameter, which provides a tradeoff between the training errors and the regularized item, $\boldsymbol{H}$ is the hidden-layer output matrix, and $\boldsymbol{W}$ is the output weight matrix. $\boldsymbol{H}$ can be presented as:

$$\boldsymbol{H} = \begin{bmatrix} \varphi(||\boldsymbol{x}_1 - c_1||) & \cdots & \varphi(||\boldsymbol{x}_1 - c_L||) \\ \vdots & \ddots & \vdots \\ \varphi(||\boldsymbol{x}_N - c_L||) & \cdots & \varphi(||\boldsymbol{x}_N - c_L||) \end{bmatrix}_{N \times L} \tag{5.41}$$

and the optimal $\boldsymbol{W}$ is

$$\boldsymbol{W}^* = \arg\min_{W \in \mathbb{R}^L} \left(\frac{1}{2}||\boldsymbol{W}||^2 + \frac{\sigma}{2}||\boldsymbol{T} - \boldsymbol{HW}||^2\right) = \left(\boldsymbol{H}^T\boldsymbol{H} + \frac{\boldsymbol{I}_L}{\sigma}\right)^{-1} \boldsymbol{H}^T\boldsymbol{T}. \tag{5.42}$$

To satisfy two assumptions which are the basis construction rules in semi-supervised learning [35], based on the manifold regularization framework, the related function between two patterns of x_i and $\boldsymbol{x}_j$ can be described as:

$$L_m = \frac{1}{2}\sum_{i,j} r_{ij}||P(\boldsymbol{t}|\boldsymbol{x}_i) - P(\boldsymbol{t}|\boldsymbol{x}_j)||^2 \tag{5.43}$$

where $P(\boldsymbol{t}|\boldsymbol{x}_i)$ is the conditional probability and r_{ij} is the weight between two patterns of $\boldsymbol{x}_i$ and $\boldsymbol{x}_j$. Only if $\boldsymbol{x}_i$ and $\boldsymbol{x}_j$ are close, e.g. $\boldsymbol{x}_i$ is among the k nearest neighbors of $\boldsymbol{x}_i$ or $\boldsymbol{x}_j$ is among the k nearest neighbors of $\boldsymbol{x}_i$, the weight between $\boldsymbol{x}_i$ and $\boldsymbol{x}_j$ can be nonzero. r_{ij} is computed with the Gaussian function of $e^{-||x_i - x_j||^2/2\sigma^2}$, or simply fixed to 1. The similarity matrix is sparse and it is described as $\boldsymbol{R} = [r_{ij}]$.

According to [23], (5.43) can be approximated by the following expression:

$$\hat{L}_m = \frac{1}{2}\sum_{i,j} r_{ij}||\hat{\boldsymbol{t}}_i - \hat{\boldsymbol{t}}_j||^2 \tag{5.44}$$

where $\hat{\boldsymbol{t}}_i$ and $\hat{\boldsymbol{t}}_j$ are the corresponding prediction of $\boldsymbol{x}_i$ and $\boldsymbol{x}_j$, respectively. The matrix form of (5.44) can be simply expressed as:

$$\hat{L}_m = \mathrm{Tr}(\hat{\boldsymbol{T}}^T\boldsymbol{L}\hat{\boldsymbol{T}}) \tag{5.45}$$

where $\mathrm{Tr}(\cdot)$ denotes the trace of a matrix, $\boldsymbol{L} = \boldsymbol{D} - \boldsymbol{R}$ is known as graph Laplacian, and $\boldsymbol{D}$ is a diagonal matrix with its diagonal elements of $D_{ii} = \sum_j r_{ij}$.

In semi-supervised learning, the labeled and unlabeled data in the training dataset are denoted as $\{\boldsymbol{X}^l, \boldsymbol{T}^l\} = \left\{\boldsymbol{x}_i^l, \boldsymbol{t}_i^l\right\}_{i=1}^{N_l}$ and $\boldsymbol{X}^u = \left\{\boldsymbol{x}_i^u\right\}_{i=1}^{N_u}$, respectively, where N_1 and N_u are the numbers of labeled and unlabeled data. With the manifold regularization, (5.42) is modified as SS-RBFNN:

$$L_{\text{SS-RBFNN}} = \frac{1}{2}||\boldsymbol{W}||^2 + \frac{1}{2}\sum_{i=1}^{N}\sigma_i||\boldsymbol{T}_i - \boldsymbol{H}_i\boldsymbol{W}||^2 + \frac{\lambda}{2}\mathrm{Tr}(\boldsymbol{O}^T\boldsymbol{LO}) \tag{5.46}$$

and it can be further presented as:

$$L_{\text{SS-RBFNN}} = \frac{1}{2}||\boldsymbol{W}||^2 + \frac{1}{2}\left\|\sigma^{\frac{1}{2}}(\tilde{\boldsymbol{T}} - \boldsymbol{HW})\right\|^2 + \frac{\lambda}{2}\mathrm{Tr}(\boldsymbol{W}^T\boldsymbol{H}^T\boldsymbol{LHW}) \tag{5.47}$$

where $\widetilde{\boldsymbol{T}} \in \mathbb{R}^{(N_l+N_u)\times m}$ is the augmented training target with its first l rows equal to $\boldsymbol{T}_l$ and the rest equal to 0, σ is a $(N_l+N_u)\times(N_l+N_u)$ diagonal matrix with its first N_l diagonal elements of $[\boldsymbol{\sigma}]_{ii} = \sigma_i$ and the rest equal to 0. The gradient of the objective function with respect to $\boldsymbol{W}$ is

$$\nabla L_{\text{SS-RBFNN}} = \boldsymbol{W} + \boldsymbol{H}^T\sigma(\widetilde{\boldsymbol{T}} - \boldsymbol{HW}) + \lambda \boldsymbol{H}^T\boldsymbol{LHW} \tag{5.48}$$

and then the optimal $\boldsymbol{W}$ is

$$\begin{aligned}\boldsymbol{W}^* &= (\boldsymbol{I}_L + \boldsymbol{H}^T\boldsymbol{\sigma}\boldsymbol{H} + \lambda\boldsymbol{H}^T\boldsymbol{LH})^{-1}\boldsymbol{H}^T\boldsymbol{\sigma}\widetilde{\boldsymbol{T}}\\ &= (\boldsymbol{I}_L + \boldsymbol{H}^T\boldsymbol{\sigma}\boldsymbol{H} + \lambda\boldsymbol{H}^T(\boldsymbol{D}-\boldsymbol{R})\boldsymbol{H})^{-1}\boldsymbol{H}^T\boldsymbol{\sigma}\widetilde{\boldsymbol{T}}\\ &= (\boldsymbol{I}_L + \boldsymbol{H}^T\boldsymbol{\sigma}\boldsymbol{H} + \lambda(\boldsymbol{H}^T\boldsymbol{DH} - \boldsymbol{H}^T\boldsymbol{RH}))^{-1}\boldsymbol{H}^T\boldsymbol{\sigma}\widetilde{\boldsymbol{T}}\end{aligned} \tag{5.49}$$

Plugging $\boldsymbol{U} = \boldsymbol{H}^T\boldsymbol{\sigma}\boldsymbol{H}, \boldsymbol{V} = \boldsymbol{H}^T\boldsymbol{DH}, \boldsymbol{P} = \boldsymbol{H}^T\boldsymbol{RH}$, and $\boldsymbol{Q} = \boldsymbol{H}^T\boldsymbol{\sigma}\widetilde{\boldsymbol{T}}$ into (5.49), we obtain

$$\boldsymbol{W}^* = (\boldsymbol{I}_L + \boldsymbol{U} + \lambda(\boldsymbol{V} - \boldsymbol{P}))^{-1}\boldsymbol{Q} \tag{5.50}$$

Additionally, even if $\boldsymbol{U} + \lambda(\boldsymbol{V} - \boldsymbol{P})$ is singular, (5.50) is also valid to improve the condition number of the matrix for a stable solution. The elements of $\boldsymbol{U}$, $\boldsymbol{V}$, $\boldsymbol{P}$, and $\boldsymbol{Q}$, u_{ij}, v_{ij}, p_{ij}, and q_{ij}, are, respectively, expressed as follows:

$$u_{ij} = \sum_{k=1}^{N} h_{ik}^{\mathrm{T}} h_{kj}\sigma_{kk} = \sum_{k=1}^{N} h_{ki}h_{kj}\sigma_{kk} = \sum_{k=1}^{N} \varphi(\|x_k - c_i\|)\varphi(\|x_k - c_j\|)\sigma_{kk} \tag{5.51}$$

$$v_{ij} = \sum_{k=1}^{N} h_{ik}^{\mathrm{T}} h_{kj}D_{kk} = \sum_{k=1}^{N} h_{ki}h_{kj}D_{kk} = \sum_{k=1}^{N} \varphi(\|x_k - c_i\|)\varphi(\|x_k - c_j\|)D_{kk} \tag{5.52}$$

$$p_{ij} = \sum_{k=1}^{N}\left(\sum_{l=1}^{N} h_{il}^{\mathrm{T}} r_{lk}\right)h_{kj} = \sum_{k=1}^{N}\left(\sum_{l=1}^{N} h_{li}r_{lk}\right)h_{kj} \tag{5.53}$$

$$q_{ij} = \sum_{k=1}^{N} h_{ik}^{\mathrm{T}} t_{kj}\sigma_{kk} = \sum_{k=1}^{N} h_{ki}t_{kj}\sigma_{kk} = \sum_{k=1}^{N} \varphi(\|x_k - c_i\|)t_{kj}\sigma_{kk} \tag{5.54}$$

When more training samples are needed, the newly arrived training dataset $\{(\boldsymbol{x}_j, \boldsymbol{t}_j)|\boldsymbol{x}_j \in \mathbb{R}^n, \boldsymbol{t}_j \in \mathbb{R}^m, j = N+1, N+2, \ldots, N+\Delta N\}$ is denoted as $\Delta\boldsymbol{G} = (\Delta\boldsymbol{X}, \Delta\boldsymbol{T})$ and the hidden-layer output matrix $\Delta\boldsymbol{H}$ can be obtained from $\Delta\boldsymbol{X}$ according to (5.41).

The new training dataset with $\Delta\boldsymbol{G}$ is $\boldsymbol{G}^+ = (\boldsymbol{X}^+, \boldsymbol{T}^+) = \{(\boldsymbol{x}_j, \boldsymbol{t}_j)|\boldsymbol{x}_j \in \mathbb{R}^n, \boldsymbol{t}_j \in \mathbb{R}^m, j = 1, 2, \ldots, N+\Delta N\}$, where $\boldsymbol{X}^+ = [\boldsymbol{X}\ \ \Delta\boldsymbol{X}]^T$ and $\boldsymbol{T}^+ = [\widetilde{\boldsymbol{T}}\ \ \Delta\boldsymbol{T}]^T$. The new hidden-layer output matrix $\boldsymbol{H}^+ = [\boldsymbol{H}\ \ \Delta\boldsymbol{H}]^T$ can be easily derived from $\boldsymbol{H}$ and $\Delta\boldsymbol{H}$, where

$$\Delta\boldsymbol{H} = \begin{bmatrix} \varphi(\|\boldsymbol{x}_{N+1} - c_1\|) & \cdots & \varphi(\|\boldsymbol{x}_{N+1} - c_L\|)\\ \vdots & \vdots & \vdots\\ \varphi(\|\boldsymbol{x}_{N+\Delta N} - c_1\|) & \cdots & \varphi(\|\boldsymbol{x}_{N+\Delta N} - c_L\|)\end{bmatrix}_{\Delta N\times L} \tag{5.55}$$

Meanwhile, $\boldsymbol{\sigma}$ is updated to $\boldsymbol{\sigma}^+$ as the following formula:

$$\boldsymbol{\sigma}^+ = \begin{bmatrix} \boldsymbol{\sigma} & \boldsymbol{\sigma}_1\\ \boldsymbol{\sigma}_1^T & \Delta\boldsymbol{\sigma}\end{bmatrix} \tag{5.56}$$

where $\boldsymbol{\sigma}_1 = 0$ and $[\Delta\boldsymbol{\sigma}]_{ii} = \boldsymbol{\sigma}_i, i = N+1, N+2, \ldots, N+\Delta N$.

$\boldsymbol{R}$ is updated to $\boldsymbol{R}^+$ as:

$$\boldsymbol{R}^+ = \begin{bmatrix} \boldsymbol{R} & \boldsymbol{R}_1\\ \boldsymbol{R}_1^T & \Delta\boldsymbol{R}\end{bmatrix} \tag{5.57}$$

where

$$\boldsymbol{R}_1 = \begin{bmatrix} r_{1,N+1} & \cdots & r_{1,N+\Delta N} \\ \vdots & \vdots & \vdots \\ r_{N,N+1} & \cdots & r_{N,N+\Delta N} \end{bmatrix} \tag{5.58}$$

and

$$\Delta\boldsymbol{R} = \begin{bmatrix} r_{N+1,N+1} & \cdots & r_{N+1,N+\Delta N} \\ \vdots & \vdots & \vdots \\ r_{N+\Delta N,N+1} & \cdots & r_{N+\Delta N,N+\Delta N} \end{bmatrix} \tag{5.59}$$

Similarly, $\boldsymbol{D}$ is updated to $\boldsymbol{D}^+$ as:

$$\boldsymbol{D}^+ = \begin{bmatrix} \boldsymbol{D} & \boldsymbol{D}_1 \\ \boldsymbol{D}_1^T & \Delta\boldsymbol{D} \end{bmatrix} \tag{5.60}$$

where $\boldsymbol{D}_1 = 0$ and $[\Delta\boldsymbol{D}]_{ii} = \sum_{j=N+1}^{N+\Delta N} \boldsymbol{w}_{ij}$, $i = N+1, N+2, \ldots, N+\Delta N$. Thus, $\boldsymbol{H}^{+T}\boldsymbol{\sigma}^+\boldsymbol{H}^+$, $\boldsymbol{H}^{+T}\boldsymbol{D}^+\boldsymbol{H}^+$, $\boldsymbol{H}^{+T}\boldsymbol{\sigma}^+\widetilde{\boldsymbol{T}}^+$, and $\boldsymbol{H}^{+T}\boldsymbol{R}^+\boldsymbol{H}^+$ can be obtained with the matrix multiplication operator as follows:

$$\boldsymbol{H}^{+T}\boldsymbol{\sigma}^+\boldsymbol{H}^+ = \boldsymbol{H}^T\boldsymbol{\sigma}\boldsymbol{H} + \Delta\boldsymbol{H}^T\Delta\boldsymbol{\sigma}\Delta\boldsymbol{H} = \boldsymbol{U} + \Delta\boldsymbol{H}^T\Delta\boldsymbol{\sigma}\Delta\boldsymbol{H} \tag{5.61}$$

$$\boldsymbol{H}^{+T}\boldsymbol{D}^+\boldsymbol{H}^+ = \boldsymbol{H}^T\boldsymbol{D}\boldsymbol{H} + \Delta\boldsymbol{H}^T\Delta\boldsymbol{D}\Delta\boldsymbol{H} = \boldsymbol{V} + \Delta\boldsymbol{H}^T\Delta\boldsymbol{D}\Delta\boldsymbol{H} \tag{5.62}$$

$$\boldsymbol{H}^{+T}\boldsymbol{\sigma}^+\widetilde{\boldsymbol{T}}^+ = \boldsymbol{H}^T\boldsymbol{\sigma}\widetilde{\boldsymbol{T}} + \Delta\boldsymbol{H}^T\Delta\boldsymbol{\sigma}\Delta\widetilde{\boldsymbol{T}} = \boldsymbol{Q} + \Delta\boldsymbol{H}^T\Delta\boldsymbol{\sigma}\Delta\widetilde{\boldsymbol{T}} \tag{5.63}$$

and

$$\begin{aligned} \boldsymbol{H}^{+T}\boldsymbol{R}^+\boldsymbol{H}^+ &= \boldsymbol{H}^T\boldsymbol{R}\boldsymbol{H} + \Delta\boldsymbol{H}^T\boldsymbol{R}_1^T\boldsymbol{H} + \boldsymbol{H}^T\boldsymbol{R}_1\Delta\boldsymbol{H} + \Delta\boldsymbol{H}^T\Delta\boldsymbol{R}\Delta\boldsymbol{H} \\ &= \boldsymbol{P} + \Delta\boldsymbol{H}^T\boldsymbol{R}_1^T\boldsymbol{H} + \boldsymbol{H}^T\boldsymbol{R}_1\Delta\boldsymbol{H} + \Delta\boldsymbol{H}^T\Delta\boldsymbol{R}\Delta\boldsymbol{H} \end{aligned} \tag{5.64}$$

Letting $\boldsymbol{U}^+ = \boldsymbol{H}^{+T}\boldsymbol{\sigma}^+\boldsymbol{H}^+ = \boldsymbol{U} + \Delta\boldsymbol{H}^T\Delta\boldsymbol{\sigma}\Delta\boldsymbol{H}$, $\boldsymbol{V}^+ = \boldsymbol{H}^{+T}\boldsymbol{D}^+\boldsymbol{H}^+ = \boldsymbol{V} + \Delta\boldsymbol{H}^T\Delta\boldsymbol{D}\Delta\boldsymbol{H}$, $\boldsymbol{Q}^+ = \boldsymbol{H}^{+T}\boldsymbol{\sigma}^+\widetilde{\boldsymbol{T}}^+ = \boldsymbol{Q} + \Delta\boldsymbol{H}^T\Delta\boldsymbol{\sigma}\Delta\widetilde{\boldsymbol{T}}$, and $\boldsymbol{P}^+ = \boldsymbol{H}^{+T}\boldsymbol{R}^+\boldsymbol{H}^+ = \boldsymbol{P} + \Delta\boldsymbol{H}^T\boldsymbol{R}_1^T\boldsymbol{H} + \boldsymbol{H}^T\boldsymbol{R}_1\Delta\boldsymbol{H} + \Delta\boldsymbol{H}^T\Delta\boldsymbol{R}\Delta\boldsymbol{H}$, the elements of $\boldsymbol{U}^+$, $\boldsymbol{V}^+$, $\boldsymbol{Q}^+$, and $\boldsymbol{P}^+$, u_{ij}^+, v_{ij}^+, q_{ij}^+, and p_{ij}^+, can be, respectively, presented as:

$$u_{ij}^+ = u_{ij} + \sum_{k=N+1}^{N+\Delta N} \varphi(\|\boldsymbol{x}_k - c_i\|)\varphi(\|\boldsymbol{x}_k - c_j\|)\sigma_{kk} \tag{5.65}$$

$$v_{ij}^+ = v_{ij} + \sum_{k=N+1}^{N+\Delta N} \varphi(\|\boldsymbol{x}_k - c_i\|)\varphi(\|\boldsymbol{x}_k - c_j\|)D_{kk} \tag{5.66}$$

$$q_{ij}^+ = q_{ij} + \sum_{k=N+1}^{N+\Delta N} \varphi(\|\boldsymbol{x}_k - c_i\|)t_{kj}\sigma_{kk} \tag{5.67}$$

and

$$p_{ij}^+ = p_{ij} + \sum_{k=1}^{N}\left(\sum_{l=N+1}^{N+\Delta N} h_{li}r_{kl}\right)h_{kj} + \sum_{k=1}^{\Delta N}\left(\sum_{l=1}^{N} h_{li}r_{l,N+k}\right)h_{N+k,j} + \sum_{k=1}^{\Delta N}\left(\sum_{l=N+1}^{N+\Delta N} h_{li}r_{l,N+k}\right)h_{N+k,j} \tag{5.68}$$

Finally, the updated output weight vector can be deduced as:

$$\boldsymbol{W}^* = (\boldsymbol{I}_L + \boldsymbol{U}^+ + \lambda(\boldsymbol{V}^+ - \boldsymbol{P}^+))^{-1}\boldsymbol{Q}^+ \tag{5.69}$$

With this method, the recalculation of $\boldsymbol{U}$, $\boldsymbol{V}$, $\boldsymbol{P}$, and $\boldsymbol{Q}$ in $\boldsymbol{U}^+$, $\boldsymbol{V}^+$, $\boldsymbol{Q}^+$, and $\boldsymbol{P}^+$ is not necessary and the number of element calculations of $4L^2$ can be saved.

5.2.2.2 Sampling Strategy

To overcome the non-uniform error distribution and slow convergence and ensure each sub-region of the whole sampling region with the same level of training and testing accuracy, a sampling strategy is presented in this section. A two-dimensional input space (x_1, x_2) in Figure 5.11 is taken as an example. The DOE method is employed to determine the original training and testing datasets.

When the training process starts, N_s labeled training samples are randomly selected from the training dataset as the first subset to train SS-RBFNN. Then all the testing samples are employed to evaluate the training result of each training sample, which directly reflects the current training performance. The training result of training sample (x_i, x_j) in the first training iteration can be presented as the following formula:

$$V^1_{Train}(x_i, x_j) = \frac{V^1_{Test}(x'_i, x'_j) + V^1_{Test}(x'_{i+1}, x'_j) + V^1_{Test}(x'_i, x'_{j+1}) + V^1_{Test}(x'_{i+1}, x'_{j+1})}{4} \tag{5.70}$$

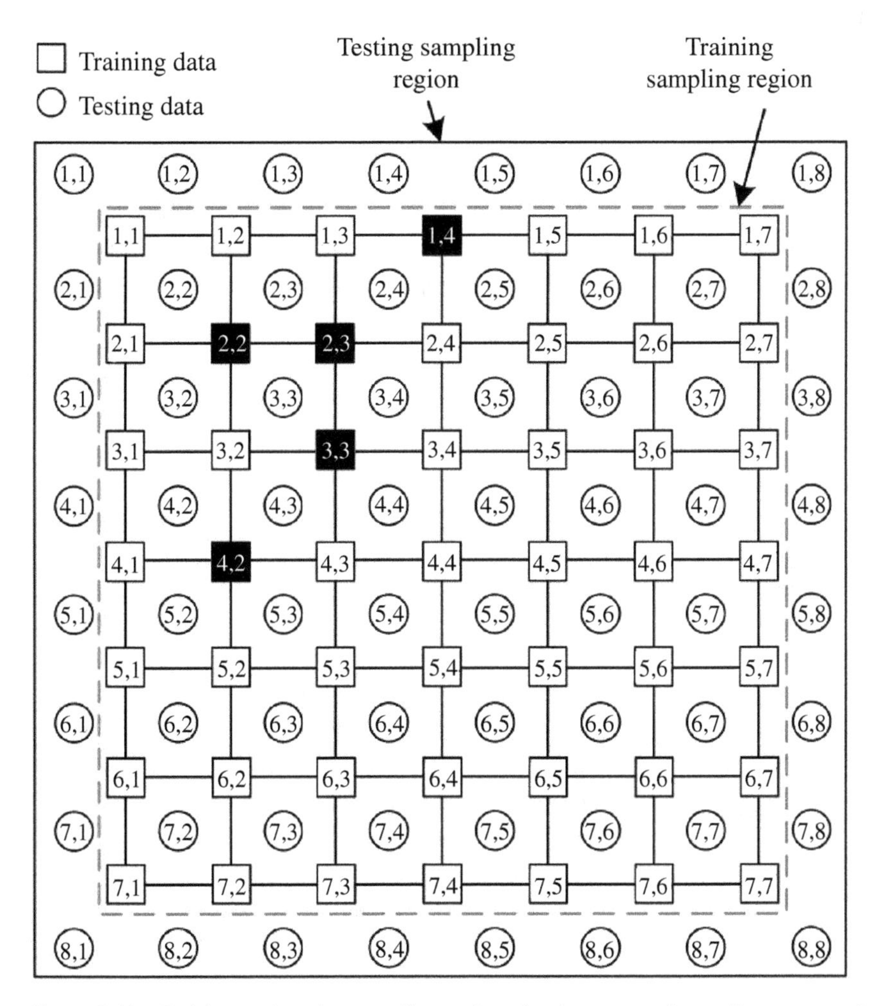

Figure 5.11 Training and testing sampling regions for the proposed sampling strategy. Source: Xiao et al. [32]/IEEE.

where $V^1_{Test}(x\prime_i, x\prime_j) = 1 - \left|\hat{t\prime}^1_{ij} - t\prime_{ij}\right| / t\prime_{ij}$ is the result from testing sample (x'_i, x'_j), t'_{ij} is the corresponding target of (x'_i, x'_j), and $\hat{t\prime}^1_{ij}$ is the corresponding output of SS-RBFNN from (x'_i, x'_j). As shown in Figure 5.11, the result of training sample (x_1, x_4) in the first training iteration, $V^1_{Train}(x_1, x_4)$, is determined with $V^1_{Test}(x\prime_1, x\prime_4)$, $V^1_{Test}(x\prime_1, x\prime_5)$, $V^1_{Test}(x\prime_2, x\prime_4)$, and $V^1_{Test}(x\prime_2, x\prime_5)$. Thirdly, we rank the training samples in ascending order of training results and select the first N_s samples into the previous subset to construct the current subset of training samples. The obtained subset is used to train the model with increased learning. The training result of each sample in the current iteration can also be obtained with the evaluation from all the testing samples, and then other N_s training samples are selected to continue the training process.

From selecting training samples to evaluating training performance, the process is repeated till the whole sampling region has the same level of training and testing accuracy. To satisfy a given accuracy, the unlabeled training data can be added to the training subset to continue training the model with increased learning after labeled data run out.

5.2.2.3 SS-RBFNN With Sampling Strategy

As shown in Figure 5.12, the process of SS-RBFNN with sampling strategy based on the pole-residue-based TF can be illustrated as follows. First, N_s labeled training samples are randomly selected from the training dataset as the first subset to train SS-RBFNN. Then all the testing samples are employed to evaluate current training performance and obtain the training result of each training sample. Thirdly, according to the training result, N_s other samples are selected into the previous subset to construct the current subset, and the obtained subset is employed to continue training SS-RBFNN with increased learning. The process, which does not involve the unlabeled training data until the labeled ones run out, is repeated till the whole sampling region reaches the satisfying modeling performance.

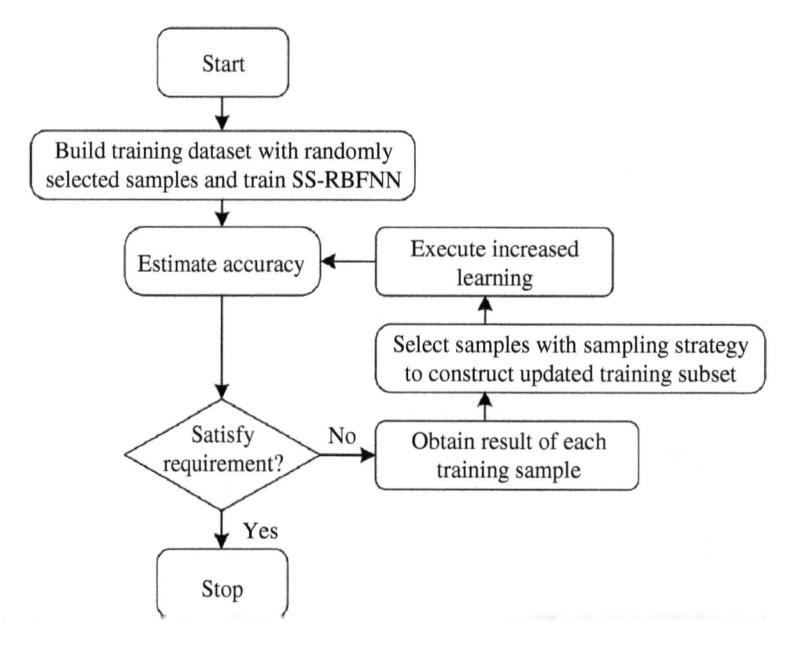

Figure 5.12 Flowchart of the training process of SS-RBFNN with sampling strategy. Source: Xiao et al. [32]/IEEE.

Example 5.3 ***SS-RBFNN With Sampling Strategy for Microstrip-to-Microstrip Vertical Transition***

An MS-to-MS vertical transition [36] is taken as an example to evaluate the proposed model. As shown in Figure 5.13, eight geometrical variables, i.e. $\boldsymbol{x} = [W_m\ L_m\ W_s\ W_{s1}\ L_s\ L_{s1}\ L_{s2}\ L_{s3}]^T$, are set as the inputs of the model. The real part R_{S11} and imaginary part I_{S11} of S_{11} with the frequency range from 0 to 6 GHz are set as the two outputs of the model, i.e. $\boldsymbol{y} = [R_{S11}\ I_{S11}]^T$.

With seven levels and eight levels in the DOE method, 49 labeled training samples and 64 testing samples are defined in Table 5.8, respectively. Meanwhile, with a wider parameter range, two other sets of 49 training samples and 64 testing samples are also defined to evaluate the model. The model costs 20089.16s and 20093.41s to collect of 64 testing samples for the two cases in Table 5.8, respectively. To evaluate the calculation accuracy and efficiency of the proposed model, the semi-supervised model in [28] and the semi-supervised RBFNN in [31] are selected as the comparison models.

The comparison models and the proposed model are firstly trained with 28 labeled training samples for Case 1 and 32 ones for Case 2, respectively. Due to different sampling strategies, the labeled training subsets for the three models are different. The comparison models are trained with unlabeled data which are randomly selected from 256 unlabeled training samples defined with the 16-level DOE method, while the proposed model employs the new sampling strategy to select unlabeled samples. Each iteration step of unlabeled-data training involves four unlabeled samples for the comparison model in [28] and the proposed model. The comparison model in [31] does not contain increased learning function, so it is directly trained with 100 unlabeled training samples which are randomly selected from the 256 unlabeled training samples.

The results of the labeled-data training and unlabeled-data training are shown in Table 5.9. The comparison model in [28] and the proposed model both contain the iterative process of increased learning. The comparison model in [28] produces average testing errors of 2.7248% for Case 1 and 3.5399% for Case 2 in the whole sampling region after 23 and 30 iteration steps of unlabeled-data training, respectively. With 16 and 21 iteration steps, the proposed model achieves less testing errors of 0.6502% for Case 1 and 0.6739% for Case 2, respectively. Because the comparison model in [31] does not involve the iterative process of increased learning, it costs the shortest time in unlabeled-data training among the three semi-supervised models but leads to the worst modeling performance.

Figure 5.14 plots the curves of the testing error versus iteration step in unlabeled-data training for the comparison model in [28] and the proposed model. The results from the comparison model

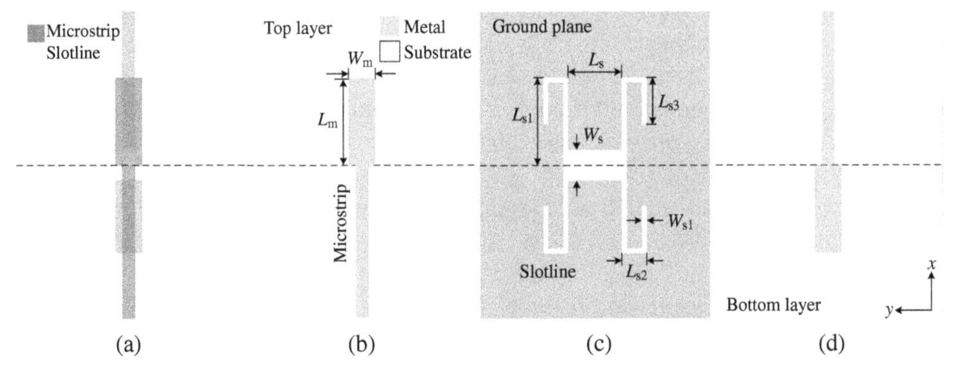

Figure 5.13 Schematic of the MS-to-MS vertical transition. (a) Top view. (b) Top layer. (c) Ground plane. (d) Bottom layer. Source: Xiao et al. [32]/IEEE.

Table 5.8 Definition of seven-level training and eight-level testing data for the MS-to-MS vertical transition.

Geometrical variables		Training data (49 samples)			Testing data (64 samples)		
		Min	Max	Step	Min	Max	Step
Case 1	W_m (mm)	3.82	3.94	0.02	3.81	3.95	0.02
	L_m (mm)	20.18	20.78	0.1	20.13	20.83	0.1
	W_s (mm)	1.35	1.65	0.05	1.325	1.675	0.05
	W_{s1} (mm)	0.35	0.65	0.05	0.325	0.675	0.05
	L_s (mm)	9.58	10.18	0.1	9.53	10.23	0.1
	L_{s1} (mm)	10.47	13.47	0.5	10.22	13.72	0.5
	L_{s2} (mm)	2.15	2.75	0.1	2.1	2.8	0.1
	L_{s3} (mm)	9.63	10.23	0.1	9.58	10.28	0.1
Case 2	W_m (mm)	3.79	3.97	0.03	3.775	3.985	0.03
	L_m (mm)	20.03	20.93	0.15	19.955	21.005	0.15
	W_s (mm)	1.275	1.725	0.075	1.2375	1.7625	0.075
	W_{s1} (mm)	0.275	0.725	0.075	0.2375	0.7625	0.075
	L_s (mm)	9.43	10.33	0.15	9.355	10.405	0.15
	L_{s1} (mm)	9.72	14.22	0.75	9.345	14.595	0.75
	L_{s2} (mm)	2	2.9	0.15	1.925	2.975	0.15
	L_{s3} (mm)	9.48	10.38	0.15	9.405	10.455	0.15

Source: Xiao et al. [32]/IEEE.

in [31] are not provided in Figure 5.14 due to its non-iterative process in unlabeled-data training. In the stage of labeled-data training, the two models cost almost the same calculation time because the time is mainly spent in full-wave simulations for collecting labeled samples. In the stage of unlabeled-data training, different from the comparison model in which the updated training samples are randomly selected, the sampling strategy can evaluate the current error distribution to help the proposed model update the training samples with a faster convergence speed, as shown in Figure 5.14.

To further evaluate the performance of the proposed model, the testing error distributions from the comparison model in [28] and the proposed model are provided in Figure 5.15. One can see that the comparison model cannot achieve satisfying performance out of the sampling region. However, the proposed model can achieve stable and accurate performance in the whole testing region, which indicates that the proposed sampling strategy makes the model more efficient. Meanwhile, in the training-data region, the comparison model in [28] also shows the non-uniform testing error distribution.

Once the training with the labeled and unlabeled data is accomplished, the trained model can also be used to optimize MS-to-MS vertical transitions by calling FPA repeatedly. The initial value of each geometrical variable is set as the mid-value in its training range. A given specification and the obtained optimal solution are provided in Figure 5.16, and its corresponding geometrical values are $\boldsymbol{x}_{opt}$ = [3.84 20.780 1.35 0.5 10.08 11.47 2.55 10.23]. The optimization time to reach the specification is recorded in Table 5.10. The proposed model costs much less CPU time than the direct EM simulation for optimization. Next, the optimized transition was fabricated for measurement. Its measured results and fabricated prototype are also given in Figure 5.16.

Table 5.9 Results of model learning for the MS-to-MS vertical transition.

			Model in [28]	**Model in [31]**	**Proposed model**
Case 1	Labeled-data training	Number of training samples	28	28	28
		Calculation time (s)	8789.93 s	8784.18 s	8790.76 s
		Training error	6.1803%	8.6545%	3.9318%
		Testing error	7.8945%	9.1158%	4.5902%
	Unlabeled-data training	Calculation time (s)	64.87 s	16.39 s	19.39 s
		Training error	1.1084%	4.5128%	0.5847%
		Testing error (in the sampling region)	1.2103%	4.9892%	0.6502%
		Testing error (out of the sampling region)	5.5969%	6.8687%	0.6519%
		Testing error (whole sampling region)	2.7248%	6.3716%	0.6502%
Case 2	Labeled-data training	Number of training samples	32	32	32
		Calculation time (s)	10741.28 s	10750.76 s	10742.61 s
		Training error	7.7938%	9.1539%	4.3846%
		Testing error	9.4594%	9.4257%	5.2817%
	Unlabeled- data training	Calculation time (s)	143.61 s	18.17 s	23.54 s
		Training error	1.5403%	4.8634%	0.6381%
		Testing error (in the sampling region)	1.5772%	5.6319%	0.6739%
		Testing error (out of the sampling region)	7.3497%	7.1641%	0.6757%
		Testing error (whole sampling region)	3.5399%	6.6855%	0.6739%

Source: Xiao et al. [32]/IEEE.

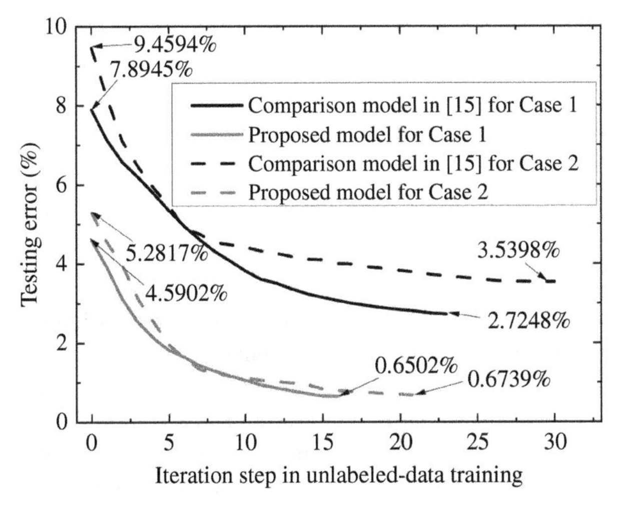

Figure 5.14 Testing errors versus the iteration steps for the MS-to-MS vertical transition. Source: Xiao et al. [32]/IEEE.

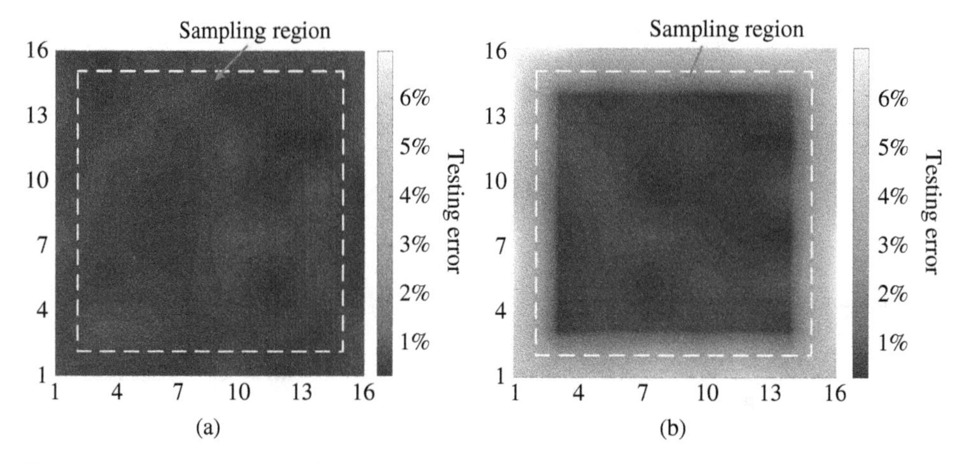

Figure 5.15 Testing error distributions in the testing region. (a) Proposed model. (b) Comparison model in [28]. Source: Xiao et al. [32]/IEEE.

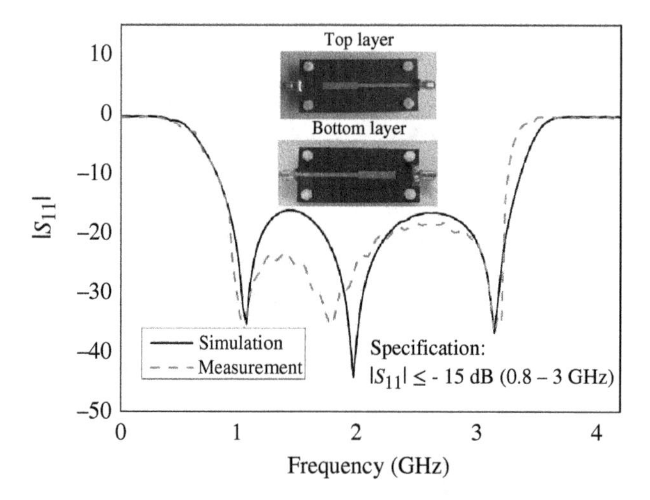

Figure 5.16 Simulated and measured $|S_{11}|$ of the optimized MS-to-MS vertical transition. Source: Xiao et al. [32]/IEEE.

5.3 Neural Networks for Antenna and Array Modeling

5.3.1 Modeling of Multiple Performance Parameters for Antennas

Due to the limit of the traditional neuro-TF model structure, only one performance parameter could be obtained from the output when geometrical variables are input. For antenna design, the operating frequency, bandwidth, gain, radiation pattern, and so on should be considered simultaneously. The result of a single S-parameter from neuro-TF models is insufficient to describe antenna performance. To solve the issue of parametric modeling of antennas with multi-parameters, an ANN model with a new structure is presented to analyze a Fabry–Perot (FP) resonator antenna in this section. The model consists of three parallel branches which could simultaneously output the accurate results of S-parameter, gain, and radiation pattern. Each branch works independently in the training and testing processes.

Table 5.10 Optimization time based on the direct EM simulation and the proposed model for the MS-MS vertical transition.

	Direct EM simulation	Proposed model
Optimization	15 h	30 s
Total	15 h	2.45 h (training) + 30 s

Source: Xiao et al. [32]/IEEE.

The whole process of the ANN model, as shown in Figure 5.17, consists of the training process and the testing process. In the training process, the training data are firstly obtained with the vector fitting technique which extracts the coefficients (or poles/residues) of TF from EM responses. Then in each branch which represents a certain antenna performance parameter, the original training data are classified into several categories for ANN and SVM training based on the TF orders to reduce the internal interference. The training data in each category are used to train the corresponding ANN to learn the nonlinear mapping between the geometrical variables and the coefficients of TF. Meanwhile, SVM [38] is trained to learn the relationship between geometrical variables and the order of TF in each branch. In the testing process, the geometrical variables are firstly input into the trained SVM and a matched category can be obtained. Then the geometrical variables are input into the corresponding ANN to output the pole and residue coefficients of TF.

The structure of Branch 1 for *S*-parameter is shown in Figure 5.18. To accurately map the relationship between geometrical variables and *S*-parameter with different TF orders, several ANNs are contained in this branch. $\boldsymbol{x}$ is a vector of the geometrical variables, representing the inputs of Branch 1, and the TF coefficients of *S*-parameter are set as outputs of ANNs. The pole-residue-based TF is chosen in this study. It is presented as

$$H(s) = \sum_{i=1}^{Q} \frac{r_i}{s - p_i} \tag{5.71}$$

where p_i and r_i are the pole and residue coefficients of TF, respectively, and Q is the order of TF.

The initial training data of neural networks are obtained with the vector fitting technique [33]. With vector fitting, we obtain the poles and residues of the TF corresponding to a given set of *S*-parameters. However, different *S*-parameter curves may lead to different TF orders. It is hard for the chaotic TF orders to train ANN accurately. Thus, according to the orders of TF, the original training samples are classified into different categories of C_k, ($k = 1, 2, \ldots, K_1$), where K_1 is the total number of categories in Branch 1. The samples with the same TF order are classified into a category, and the order of each category could be presented as Q_k ($k = 1, 2, \ldots, K_1$).

Since the relationship between the poles/residues and $\boldsymbol{x}$ is nonlinear and unknown, ANNs are employed to learn this nonlinear relationship through the training process. $\boldsymbol{O} = \{O_1, \ldots, O_W\}$ is a vector representing the outputs of the EM simulations (i.e. real and imaginary parts of *S*-parameters), where W is the number of the sample points of frequency. $\boldsymbol{O'} = \{O'_1, \ldots, O'_W\}$ is a vector representing real and imaginary parts of the outputs of the pole-residue-based TF. The objective here is to minimize the error between $\boldsymbol{O}$ and $\boldsymbol{O'}$ for different $\boldsymbol{x}$, by adjusting the internal weights and thresholds of ANN. It is worth noting that one category is only used to train one ANN model named as ANN_k, ($k = 1, 2, \ldots, K_1$).

At the same time, the training samples are also used to train an SVM model, which determines the TF orders of $\boldsymbol{x}$ for classification during the testing process, as shown in Figure 5.19. $\boldsymbol{x}$ is the input of SVM. $\boldsymbol{Q'} = \{Q'_1, \ldots, Q'_{K1}\}$ is a vector representing the output of SVM and

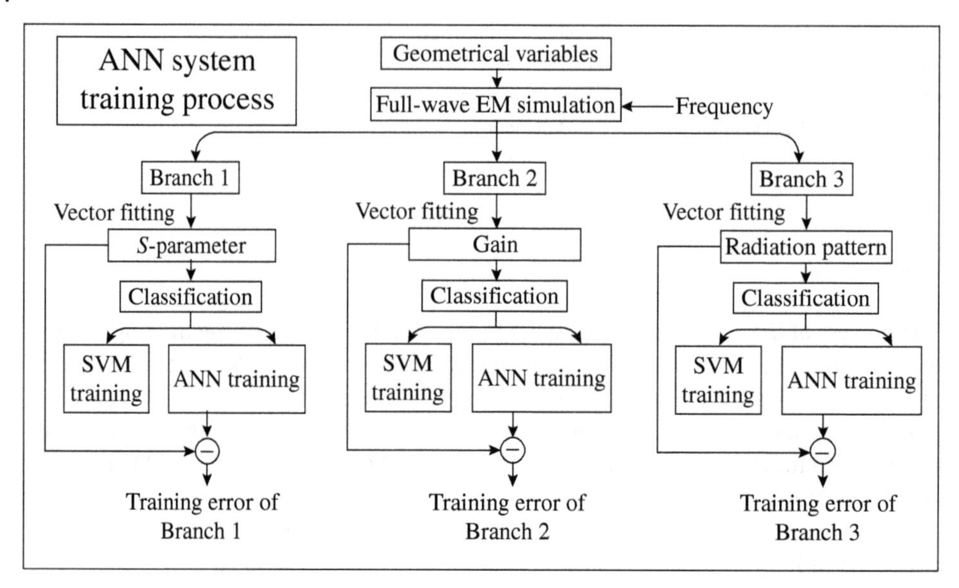

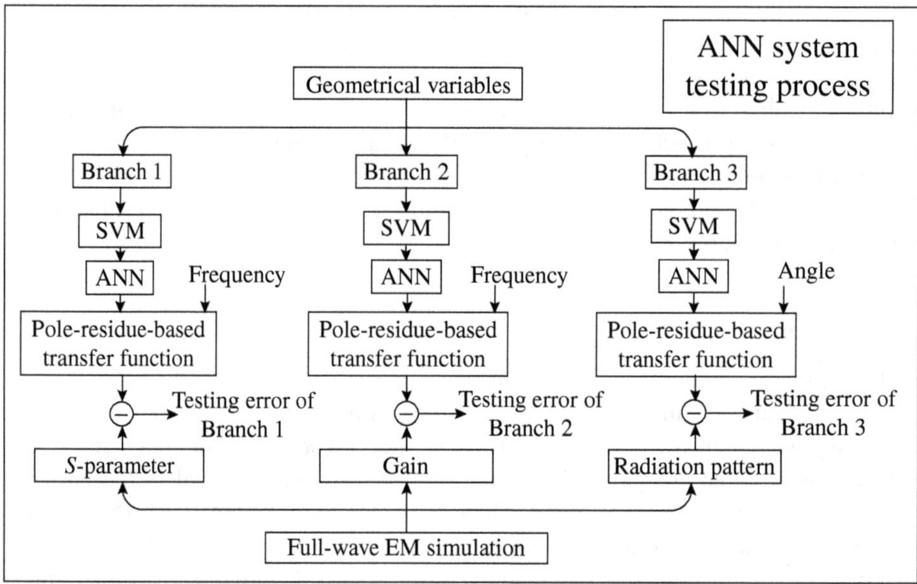

Figure 5.17 Whole process of the ANN model. Source: Xiao et al. [37]/IEEE.

$\boldsymbol{Q} = \{Q_1, \ldots, Q_{K1}\}$ is a vector representing the actual order of TF. The training objective is to minimize the error between $\boldsymbol{Q}'$ and $\boldsymbol{Q}$ for different $\boldsymbol{x}$ by adjusting the internal weights and thresholds of SVM.

The structure of Branch 2 for gain is similar to Branch 1. In Branch 3, the frequency in TF is replaced by the angle for radiation pattern.

Example 5.4 ***Modeling of Multiple Performance Parameters for FP Resonator Antenna***
The FP resonator antenna is a kind of highly directive antenna [39], which is formed by placing an electromagnetic band gap (EBG) structure as a partially reflective surface (PRS) in front of a simple primary radiator with a ground plane. Its main advantages include high radiation efficiency and low complexity compared with the conventional antennas [40]. An FP resonator antenna proposed

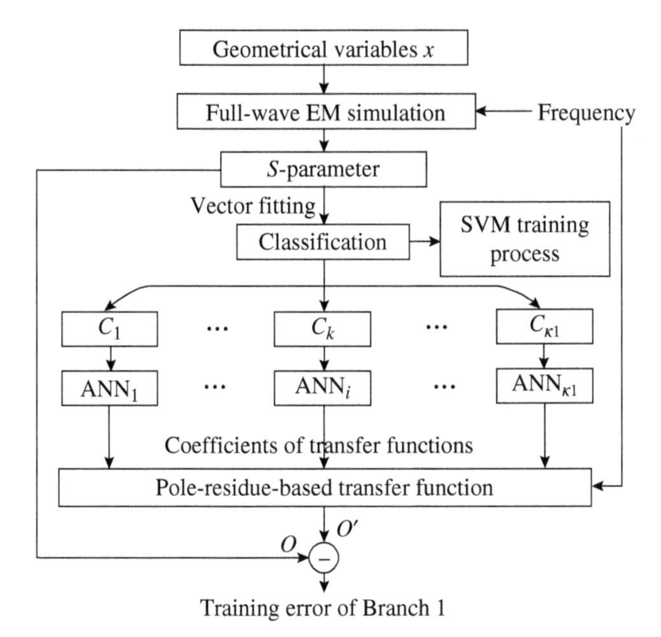

Figure 5.18 Training process of Branch 1. Source: Xiao et al. [37]/IEEE.

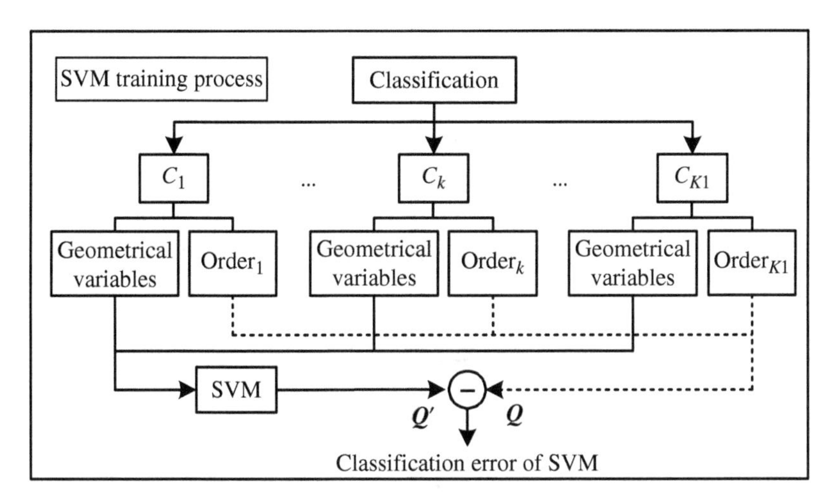

Figure 5.19 Training process of SVM. Source: Xiao et al. [37]/IEEE.

in [41] is considered as an example to evaluate the proposed ANN model. The EBG structure of this antenna, as shown in Figure 5.20, is a combination of two complementary frequency selective surface (FSS) structures (square patches and square apertures). The EBG structure with a dimension of 72 mm × 72 mm ($2.4\lambda \times 2.4\lambda$ at 10 GHz) and 77 unit cells is arranged and applied as the PRS to the design of the FP resonator antenna. At each corner of the 9×9 rectangular array, one unit cell is eliminated. The metal patches and the square apertures are fabricated on the top and at the bottom of the substrate with a thickness of T and a relative dielectric constant of ε_r, respectively. The dimension ($l_b \times l_b$) of the unit cell is 8 mm × 8 mm (less than $\lambda/3$ at 10 GHz). l_p and l_a are the dimensions of the patch and aperture, respectively. The lateral dimension of the antenna is of the same size as the EBG structure.

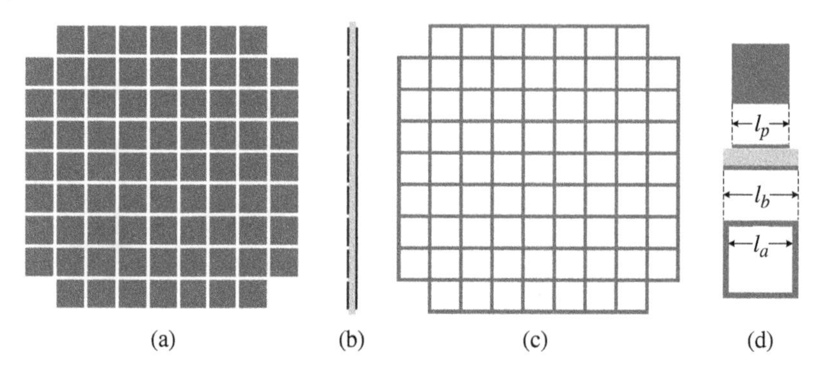

Figure 5.20 EBG structure of the FP resonator antenna: (a) top view, (b) side view, (c) bottom view, and (d) unit cell. Source: Xiao et al. [37]/IEEE.

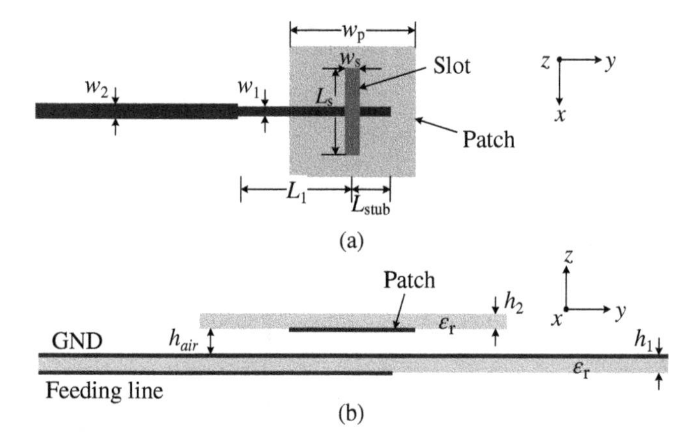

Figure 5.21 Structure of the feeding antenna: (a) top view and (b) side view. Source: Xiao et al. [37]/IEEE.

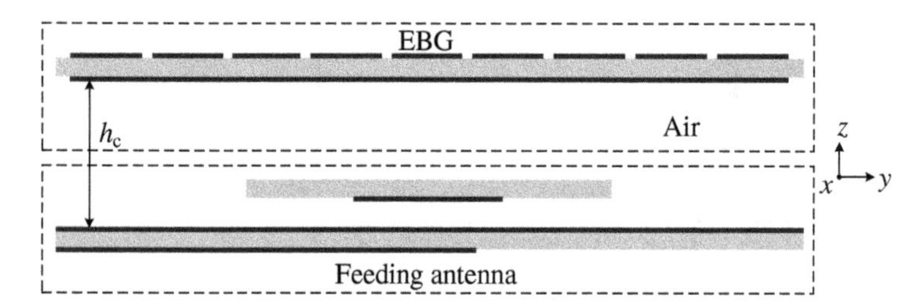

Figure 5.22 Structure of the FP resonator antenna. Source: Xiao et al. [37]/IEEE.

The feeding structure which is an integral part of the FP resonator antenna is shown in Figure 5.21. The parasitic patch is designed on the Rogers RT/duroid 5880 substrate ($\varepsilon_r = 2.2$ and $\tan\delta = 0.0009$). The feeding antenna is spaced from the ground plane by an air gap and it is coupled with the feed line through a slot in the ground plane. The parameters of the antenna are as follows: $w_p = 9.3$ mm, $w_1 = 1.2$ mm, $w_2 = 2.3$ mm, $w_s = 2.3$ mm, $L_1 = 9.5$ mm, $L_s = 8.2$ mm, $L_{stub} = 3$ mm, $h_{air} = 2.5$ mm, and $h_1 = h_2 = 0.787$ mm.

Figure 5.22 shows the whole structure of the FP resonator antenna. The feeding antenna is placed at the center of the cavity. h_c is the cavity height between the EBG layer and the ground plane.

Table 5.11 Definition of training and testing data for the FP resonator antenna.

Geometrical variables		Training data (64 samples)			Testing data (36 samples)		
		Min	Max	Step	Min	Max	Step
Case 1	l_p (mm)	5.6	6.3	0.1	5.65	6.15	0.1
	l_a (mm)	5.2	5.9	0.1	5.25	5.75	0.1
	h_c (mm)	14.5	15.55	0.15	14.575	15.325	0.15
Case 2	l_p (mm)	5.4	6.45	0.15	5.475	6.225	0.15
	l_a (mm)	5	6.05	0.15	5.075	5.825	0.15
	h_c (mm)	14.25	16	0.25	14.375	15.625	0.25

Source: Xiao et al. [37]/IEEE.

Table 5.12 Number of samples in each category for S-parameter.

	Category	Number of samples
Case 1	Category 1 (eight order)	18
	Category 2 (nine order)	24
	Category 3 (ten order)	22
Case 2	Category 1 (eight order)	10
	Category 2 (nine order)	12
	Category 3 (ten order)	14
	Category 4 (eleven order)	17
	Category 5 (twelve order)	11

Source: Xiao et al. [37]/IEEE.

Three geometrical variables of the EBG structure, i.e. $\boldsymbol{x} = [l_p\ l_a\ h_c]^T$, which play important roles in the performance of the antenna, are set as the input variables of the whole model.

The model is applied to two different cases, i.e. Case 1 with a narrow parameter range and Case 2 with a wide one. In the both cases, the training and testing data from the DOE method with eight levels (64 training samples) and six levels (36 testing samples) are shown in Table 5.11, where the values of h_c are set around $\lambda/2$ and the values of l_p and l_a are set around $\lambda/5$ at 10 GHz. The total time for training-data generation from EM simulations is 21.33 hours, and the total time for testing-data generation is 12 hours. Meanwhile, the Hecht–Nelson method [42] is used to determine the node number of the hidden layer: the node number of the hidden layer is $2n + 1$ when the node number of the input layer is n. *HFSS* software performs the full-wave EM simulation and generates the training and testing data for modeling.

In Branch 1, the TF order of S-parameter varies slightly from eight to ten among the samples of different geometrical variables in Case 1. In Case 2, the TF order varies from eight to twelve. The training samples are divided into proper categories according to their TF orders, as shown in Table 5.12, for ANN training.

Meanwhile, the geometrical variables and corresponding TF orders from training samples are set as the input and output of SVM for training, respectively. For 36 testing samples, the classifying results are shown in Figure 5.23. The classification precision of the trained SVM is 97.22%. After the

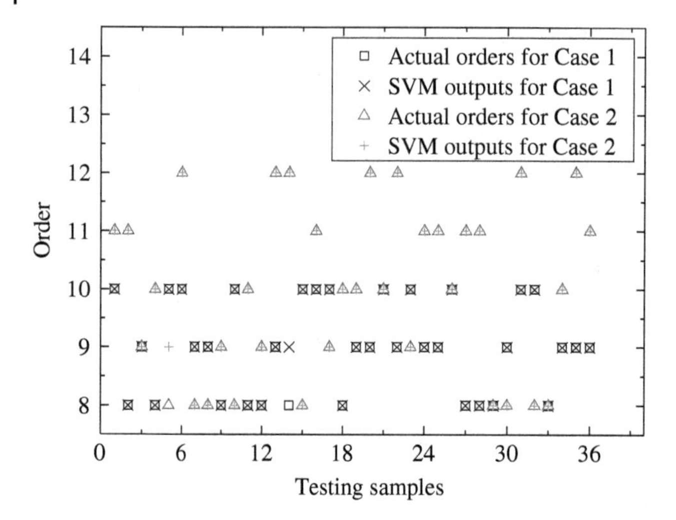

Figure 5.23 Classifying results of *S*-parameter from SVM for Cases 1 and 2. Source: Xiao et al. [37]/IEEE.

Table 5.13 Modeling results of the proposed model for the FP resonator antenna.

S-parameter	Average training MAPE	Average testing MAPE
Case 1	0.399%	0.615%
Case 2	0.424%	0.672%
Gain	Average training MAPE	Average testing MAPE
Case 1	0.873%	0.954%
Case 2	0.857%	0.971%
Radiation pattern (E-plane)	Average training MAPE	Average testing MAPE
Case 1	2.645%	2.964%
Case 2	2.448%	2.912%
Radiation pattern (H-plane)	Average training MAPE	Average testing MAPE
Case 1	2.662%	2.973%
Case 2	2.441%	2.933%

Source: Xiao et al. [37]/IEEE.

modeling process, the average training MAPEs of Branch 1 are 0.399% for Case 1 and 0.424% for Case 2, while the average testing MAPEs are 0.615% and 0.672%.

In Branch 2, the TF coefficients from a given set of broadside radiation gain (z-direction in Figure 5.21) of the FP resonator antenna are set as the outputs of ANNs. After the modeling process, the average training MAPEs of gain are 0.873% for Case 1 and 0.857% for Case 2, while the average testing MAPEs of gain are 0.954% and 0.971%.

In Branch 3, ANNs are used for mapping the geometrical variables onto the TF coefficients which are extracted from two given sets of radiation patterns of E-plane and H-plane at 10 GHz, respectively. Different from the above two branches, the angle in TF replaces the frequency, with an original range of $[-2\pi, 2\pi]$ for radiation pattern in this branch.

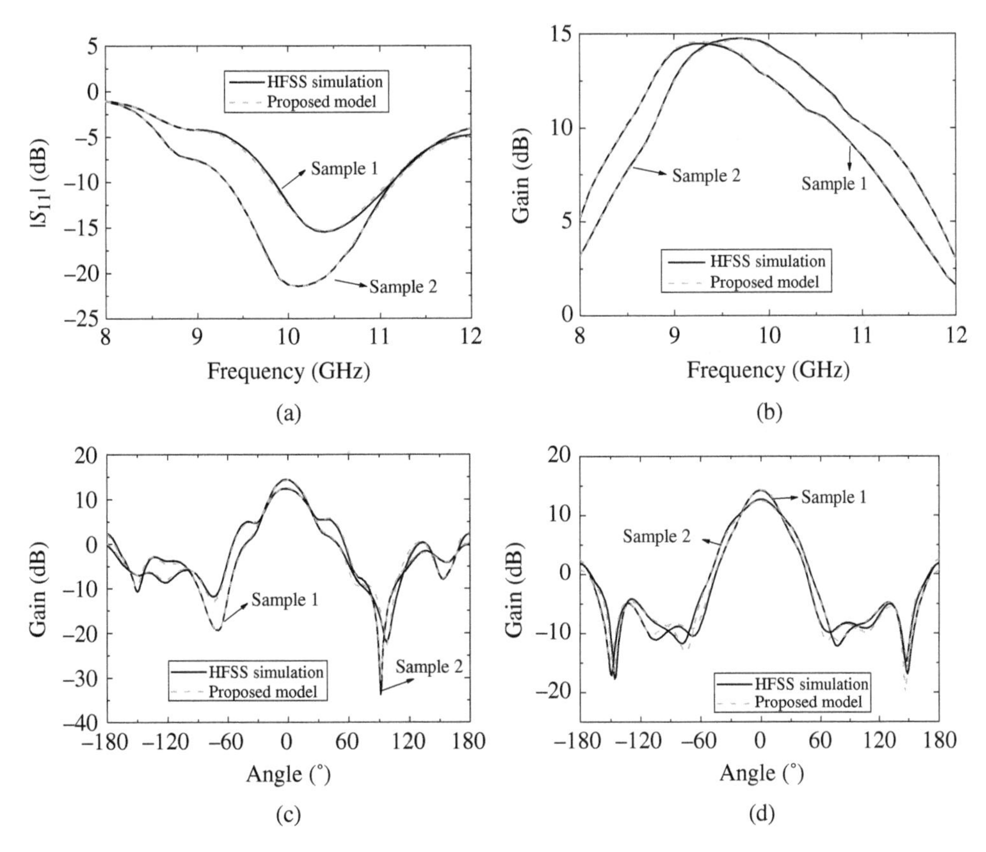

Figure 5.24 Comparison between the model and HFSS for Samples 1 and 2. (a) S_{11}, (b) gain, (c) E-plane pattern @10GHz, and (d) H-plane pattern @10GHz, where the samples are in the range of training data. Source: Xiao et al. [37]/IEEE.

After the modeling process, the average training MAPEs are 2.645% for Case 1 and 2.448% for Case 2, while the average testing MAPEs are 2.964% and 2.912%. Due to the uneven curve of radiation pattern, the training and testing MAPEs are bigger than those in Branches 1 and 2. The whole modeling results of the model are shown in Table 5.13.

Figure 5.24 shows the outputs of two different test geometrical samples of the FP resonator antenna with the proposed model and HFSS simulation. The variables for the two samples in the range of the training data are $\boldsymbol{x}_1 = [5.62\ 5.44\ 14.63]^T$ and $\boldsymbol{x}_2 = [5.87\ 5.51\ 15.36]^T$. The average MAPEs of S-parameter for the two samples are 0.605% and 0.569%. The average MAPEs of gain for the two samples are 0.964% and 0.969%. The average MAPEs of the E-plane pattern for the two samples are 2.936% and 2.954%, and they are 2.934% and 2.951% for the H-plane pattern. It is observed that the model can achieve good accuracy when the geometrical samples are never used in training.

Two other test geometrical samples, which are selected out of the range of the training data, are chosen to evaluate the model. The variables for the two samples are $\boldsymbol{x}'_1 = [5.3\ 4.9\ 14.2]^T$ and $\boldsymbol{x}'_2 = [6.5\ 4.9\ 16.1]^T$. As shown in Figure 5.25, the average MAPEs of S-parameter for the two samples are 0.553% and 0.515%. The average MAPEs of gain for the two samples are 0.959% and 0.961%. The average MAPEs of the E-plane pattern for the two samples are 2.913% and 2.947%, and they are 2.946% and 2.933% for the H-plane pattern. It is observed that the model can achieve good accuracy even though the geometrical samples are out of the range of the training data.

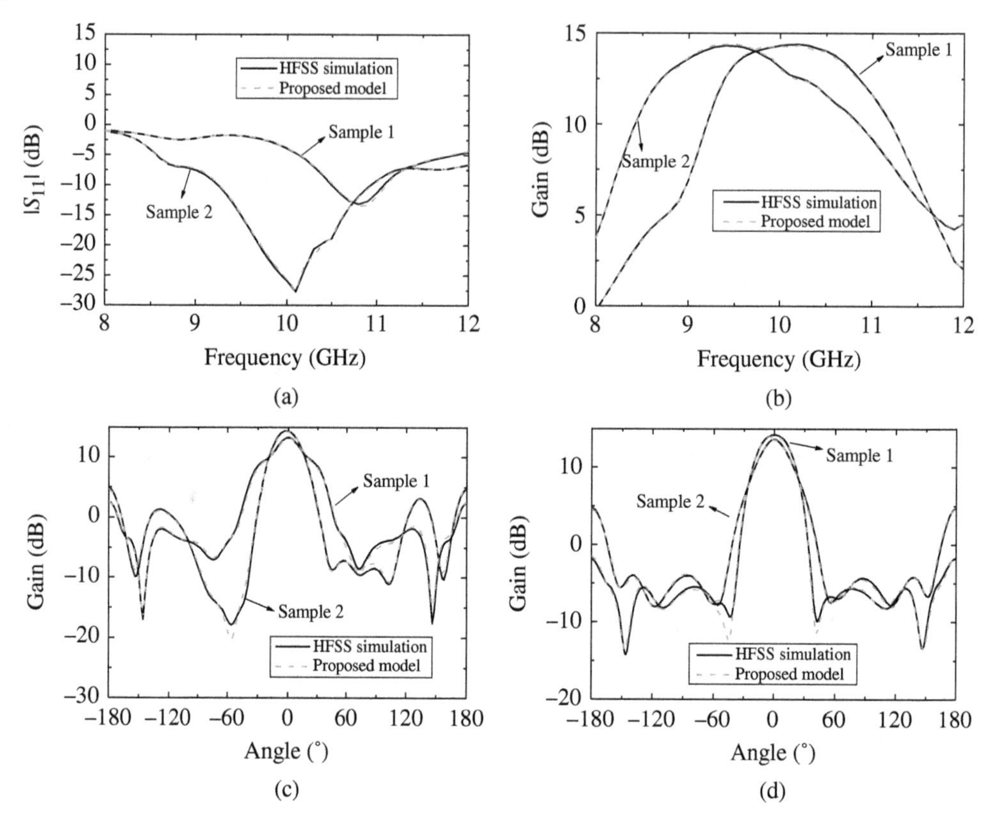

Figure 5.25 Comparison between the model and HFSS for Samples 1 and 2. (a) S_{11}, (b) gain, (c) E-plane pattern @10GHz, and (d) H-plane pattern @10GHz, where the samples are out of the range of training data. Source: Xiao et al. [37]/IEEE.

Once the training is completed, the trained model which is a substitute for the time-consuming EM simulation can be applied to the design optimization. As an example of using the trained model for antenna design, two separate FP resonator antennas are optimized to reach two different design specifications.

Optimization Objectives of Specification 1

(1) $|S_{11}| \leq -10$ dB at the frequency range of 8.75–11.25 GHz.
(2) Relative 3 dB gain bandwidth reaches 32%.
(3) Gain of the main lobe $G_{max} \geq 12.5$ dB.

Optimization objectives of Specification 2

(1) $|S_{11}| \leq -10$ dB at the frequency range of 10–11 GHz.
(2) Relative 3 dB gain bandwidth reaches 21%.
(3) Gain of the main lobe $G_{max} \geq 14$ dB.

With the non-dominated sorting genetic algorithm II (NSGA-II) [43], the optimization of the FP resonator antennas is performed by calling the trained model repeatedly. The initial variables are chosen as $\boldsymbol{x}_{initial} = [15\ 5.5\ 6]^T$. The optimization spends only about 60 seconds to achieve the optimal solution for each antenna, shown in Figure 5.26. The optimized geometrical values for the two separate antennas are $\boldsymbol{x}_{opt1} = [14.748\ 5.189\ 5.904]^T$ and $\boldsymbol{x}_{opt2} = [14.733\ 6.011\ 6.401]^T$ which are selected from the Pareto fronts.

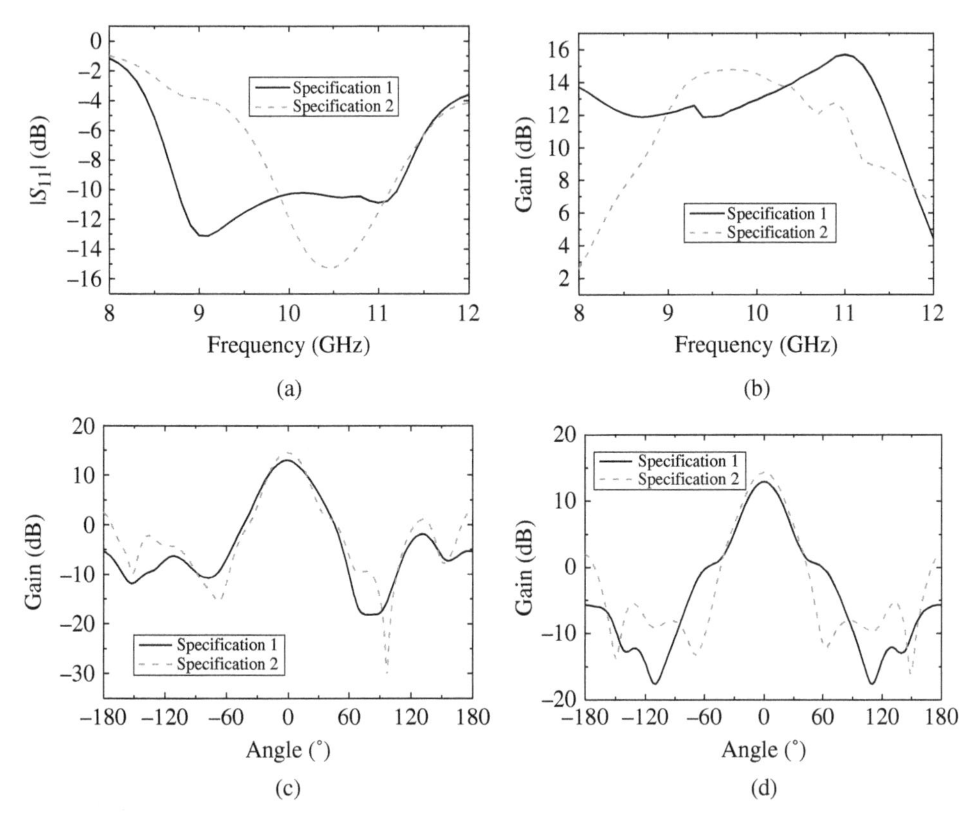

Figure 5.26 Optimization results of Specifications 1 and 2 from the trained model: (a) S_{11} parameter, (b) gain, (c) E-plane pattern @10GHz, and (d) H-plane pattern @10GHz. Source: Xiao et al. [37]/IEEE.

Table 5.14 Running time of direct EM optimization and the proposed model.

	CPU time of model development	
	Direct EM optimization	**Proposed model**
Antenna 1	45 h	60 s
Antenna 2	49 h	60 s
Total	94 h	33.33 h (modeling) + 120 s

Source: Xiao et al. [37]/IEEE.

Compared with the directive EM optimization in which the EM simulations are repeatedly called by NSGA-II, the design using the trained model could save considerable time as shown in Table 5.14.

5.3.2 Inverse Artificial Neural Network for Multi-objective Antenna Design

Most of the ANN models are forward ones for modeling, where geometrical variables are set as the input while EM responses are the output. However, a design process is an inverse problem that the geometrical variables should be found according to a given EM objective. In this process, the trained ANN model needs to be repetitively called by an optimization algorithm until a satisfying solution is achieved. With the increase of input dimension, more training and optimization costs are required. Meanwhile, in practice, TF can bring extra errors from vector fitting [27]. Thus, it is labored for the simulation and optimization of a component with large dimension of geometrical variables as the input. For multi-objective antenna design, some important performance indexes, such as the operating frequency, gain, and radiation pattern, should be considered simultaneously. In [37], a forward ANN was introduced to model antennas with three performance indexes. However, the enhancement of input dimension of the geometrical variables often increases the training cost. In addition, an extra multi-objective optimization operation should be involved to execute the whole design program.

In this section, an inverse ANN model is proposed to carry out the multi-objective antenna design, where antenna performance indexes, i.e. S-parameter (or voltage standing wave ratio [VSWR]), gain, and radiation pattern, are set as the input of the proposed model and corresponding geometrical variables are set as the output. Meanwhile, to solve the multi-value problem, ELM is introduced to obtain the final value effectively.

There are five layers, including the input layer, the first hidden layer, the first output layer, the second hidden layer, and the second output layer, in the customized inverse model. As shown in Figure 5.27, the first three layers construct the first part which consists of three independent branches for the S-parameter (or VSWR), gain, and radiation pattern, respectively. The first part is employed to map the relationship from the three performance indexes to geometrical variables. However, due to lack of the constraint condition, the outputs of these three branches are usually different. To obtain a satisfying unique solution, the second part, which includes the fourth and fifth layers, integrates the outputs from the three branches in the first part and maps them to the actual geometrical variables.

The input layer nodes are filled with the column vector $\overline{\mathbf{x}}_j = \left[\overline{\mathbf{x}}_j^1, \overline{\mathbf{x}}_j^2, \overline{\mathbf{x}}_j^3\right]^T$, where $\overline{\mathbf{x}}_j^1 = \left[x_{1j}^1, x_{2j}^1, \ldots, x_{M_1j}^1\right]^T \in \mathbf{C}^{M_1}$, $\overline{\mathbf{x}}_j^2 = \left[x_{1j}^2, x_{2j}^2, \ldots, x_{M_2j}^2\right]^T \in \mathbf{R}^{M_2}$, and $\overline{\mathbf{x}}_j^3 = \left[x_{1j}^3, x_{2j}^3, \ldots, x_{M_3j}^3\right]^T \in \mathbf{R}^{M_3}$ are the complex-valued S-parameter, real-valued gain, and real-valued radiation pattern, respectively. The subscript j denotes the jth training dataset, and M_1, M_2, and M_3 are the subset dimensions in the training dataset. The output column $\overline{\mathbf{o}}_j = [o_{1j}, o_{2j}, \ldots, o_{Nj}]^T \in \mathbf{R}^N$ is the geometrical variables and its dimension N is their total number. It is evaluated by

$$\left|\overline{\mathbf{o}}_j = \overline{\overline{\boldsymbol{a}}} g_{real}\left(\overline{\overline{\omega}}_{real} \bigcup_{k=1}^{3} \left(\left|\overline{\overline{\beta}}_k g_k\left(\overline{\overline{\omega}}_k \overline{\mathbf{x}}_j^k + \overline{\mathbf{b}}_k\right)\right|\right) + \overline{\mathbf{p}}\right)\right. \tag{5.72}$$

where $j = 1, 2, \ldots, P$ when there are totally P sets of training data. If $k = 1$ indicates the input of the S-parameter branch, $\overline{\overline{\omega}}_1$ is an $L \times M$ complex-valued random weight matrix, $\overline{\overline{\beta}}_1$ is a $K \times L$ complex-valued weight matrix connecting the neurons of the first hidden layer to those of the first output layer, and $\overline{\mathbf{b}}_1$ is a column threshold vector with the complex-valued random number of L. Meanwhile, the activation function g_1 is a complex activation function and the output of the first hidden layer is the L complex values. If k = 2 or 3 indicates the input of the gain or radiationpattern branch, $\overline{\overline{\omega}}_{2,3}$, $\overline{\overline{\beta}}_{2,3}$, and $\overline{\mathbf{b}}_{2,3}$ are the real-valued weight matrices. Similarly, $g_{2,3}$ are real-valued activation functions.

Then, the outputs of the three branches are integrated as a union set. $\overline{\overline{\boldsymbol{a}}}$, $\boldsymbol{g}_{real}$, $\overline{\overline{\omega}}_{real}$, and $\overline{\mathbf{p}}$, which take the similar effects as $\overline{\overline{\beta}}_k$, g_k, $\overline{\overline{\omega}}_k$, and $\overline{\mathbf{b}}_k$, respectively, work for the next layer of the model with real values. $\overline{\overline{\boldsymbol{a}}}$ connects the neurons of the second hidden layer and the output layer. The dimensions of $\overline{\overline{\boldsymbol{a}}}$, $\overline{\overline{\omega}}_{real}$, and $\overline{\mathbf{p}}$ are $N \times S$, $S \times K$, and $S \times 1$, respectively.

With many times of experiments in this section, an inverse hyperbolic function expressed in (5.73) is selected as the complex activation function of g_1 and the sigmoid function in (5.74) is

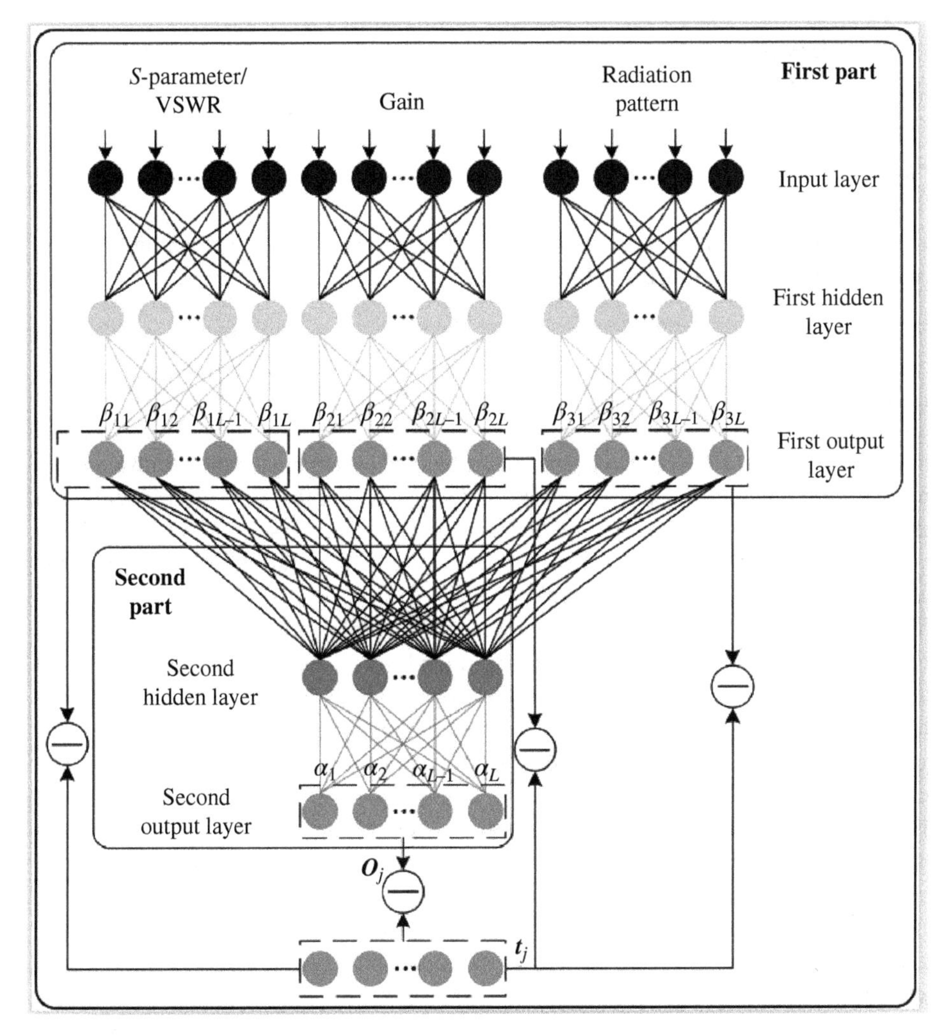

Figure 5.27 Structure of the proposed model, where the first part contains three parallel branches and the second part is ELM. Source: Xiao et al. [37]/IEEE.

selected as the real activation function of $g_{2,3}$ and g_{real}.

$$g_1 = \text{arcsin}\,h(x) = \int_0^x \frac{dt}{(1+t^2)^{1/2}} \tag{5.73}$$

$$g_{2,3} = g_{real} = \text{sigmoid}(x) = \frac{1}{1+e^{-x}} \tag{5.74}$$

The actual values of geometrical variables of jth training sample are denoted as $\bar{\mathbf{t}}_j = [t_{1j}, t_{2j}, \dots, t_{Nj}]^T \in \mathbf{R}^N$. The mismatch between the predicted values of the first output layer and the actual values of geometrical variables should be minimized, i.e. $\sum_{j=1}^{P} \|\bar{\mathbf{o}}_j - \bar{\mathbf{t}}_j\| = 0$, to obtain the optimized weight matrices $\bar{\bar{\boldsymbol{\alpha}}}$ and $\bar{\bar{\boldsymbol{\beta}}}_k$ of the both hidden layers. Thus, the relationship between the weight matrix $\bar{\bar{\boldsymbol{\beta}}}_k$ of the first hidden layer and the true values of geometrical variables $\bar{\mathbf{t}}_j$ for all the training datasets can be compactly expressed as

$$\bar{\bar{\boldsymbol{\beta}}}\,\bar{\bar{\mathbf{G}}} = \bar{\bar{\mathbf{T}}} \tag{5.75}$$

where $\overline{\overline{\mathbf{T}}} = [\bar{\mathbf{t}}_1; \bar{\mathbf{t}}_2; \ldots \bar{\mathbf{t}}_P]_{N\times P}$ and $\overline{\overline{\mathbf{G}}}$, with the dimension of $L \times P$, is the combination of all the column vectors of the output of the first hidden layer. Thus, the optimized complex weight matrix $\overline{\overline{\beta}}$ could be computed by

$$\overline{\overline{\beta}} = \overline{\overline{\mathbf{T}}}\,\overline{\overline{\mathbf{G}}}^{\dagger} \tag{5.76}$$

where the complex matrix $\overline{\overline{\mathbf{G}}}^{\dagger}$ is the Moor–Penrose generalized inverse of $\overline{\overline{\mathbf{G}}}$. Following the same procedure, we can obtain the optimized real matrix

$$\overline{\overline{\alpha}} = \overline{\overline{\mathbf{T}}}\,\overline{\overline{\mathbf{G}}}_{real}^{\dagger} \tag{5.77}$$

where $\overline{\overline{G}}_{real}$, with the dimension of $S \times P$, is the combination of all the column vectors of the output of the second hidden layer. We can see that the proposed model can be trained with a very low cost since the unknown weights of each hidden layer can be obtained by solving the matrix inverse only once. The number of the hidden layer is set to two, and the node numbers of both hidden layers are determined with the Hecht–Nelson method. Each hidden layer has $2n + 1$ nodes where n is the node number of the input layer, i.e. $2M_1 + 1$, $2M_2 + 1$, and $2M_3 + 1$ for the three branches in the first hidden layer and $2N + 1$ for the second hidden layer.

The above description shows that the output weight matrices of the first and second hidden layers are independently solved in the model. The first hidden layer is classified into three different branches to output the values of geometrical variables separately, which helps the model to recognize each index better. The second hidden layer is used to integrate the outputs from the three branches and map them to the actual geometrical variables, as well as to perform error correction to further enhance the modeling accuracy.

The training and testing processes of the model are shown in Figure 5.28. The training process aims to make the error between the predicted geometrical variables and actual ones minimum by adjusting the inner weights of the model. In the testing process, the S-parameter, gain, and radiation pattern could be simultaneously input into the trained model to obtain the testing results.

For forward models, because of the restriction of the model structure and the strong sensitivity of the geometrical variable change, the higher dimension of the input of the forward models means the more complex nonlinear mapping. Therefore, the required number of training samples in the forward models grows greatly to meet the high accuracy request. For the inverse model, the sensitivity from VSWR, gain, and radiation pattern to the geometrical variables is not strong. Furthermore, TF, which is extracted with vector fitting, may bring extra errors in the training and testing processes of the forward models. Thus, compared with the forward models, only a small number of training samples are involved to handle the large dimension of geometrical variables in the inverse model.

Example 5.5 ***Inverse Model for Multi-mode Resonant Antenna***

A multi-mode resonant antenna [44] is chosen as the example for the evaluation of the model. The antenna structure which involves an FR4 (relative permittivity $\varepsilon_r = 4.3$, relative permeability $\mu_r = 1$, and loss tangent $\tan\delta = 0.025$) substrate is shown in Figure 5.29. The S-parameter, gain, and radiation pattern are set as the input of the model. Twelve geometrical variables, i.e. $\bar{\mathbf{t}} = [P_x\, P_y\, W_x\, W_1\, L_0\, L_1\, L_2\, R_{in1}\, R_{in2}\, D_1\, D_2\, D_x]^T$, are the output. The frequency range is from 0 to 10 GHz.

Forty-nine training and 49 testing samples defined with seven levels in the DOE method are, respectively, shown in Table 5.15, including a narrow value range in Case 1 and a wide range in Case 2. The collection of the 49 training or testing samples costs about 4.9 hours. To evaluate the

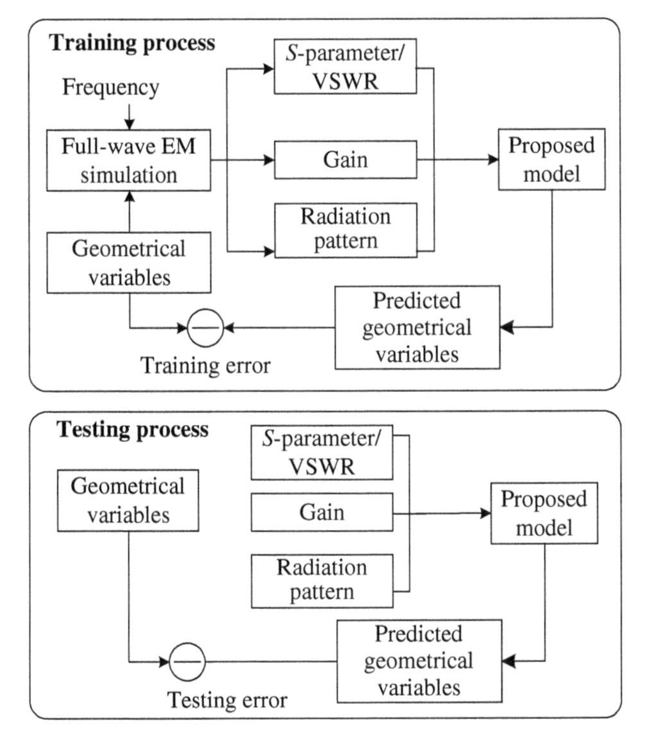

Figure 5.28 Training and testing processes of the inverse model. Source: Xiao et al. [37]/IEEE.

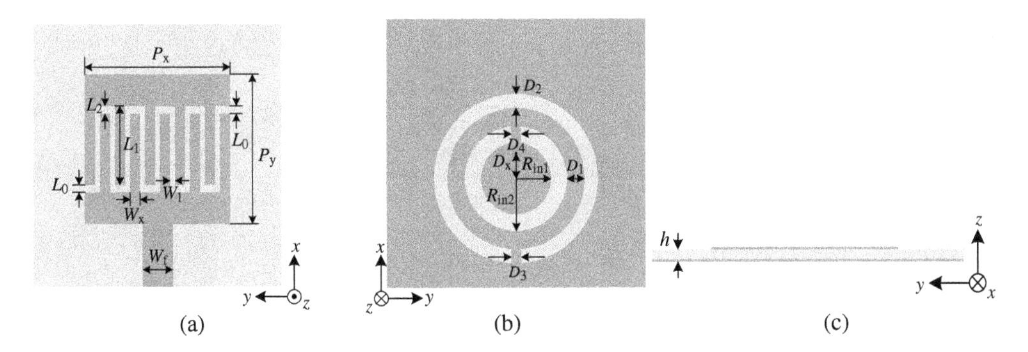

Figure 5.29 Geometry of the multi-mode resonant antenna, where $h = 1.57$ mm and $W_f = 5$ mm. (a) Top view. (b) Bottom view. (c) Front view. Source: Xiao et al. [37]/IEEE.

modeling performance of the model, a traditional inverse ELM with three layers, i.e. the input layer, hidden layer, and output layer, is employed as a comparison model with the same input and output parameters. The node number of the hidden layer from the comparison model is also determined with the Hecht–Nelson method. The hidden layer has $2n + 1$ nodes where n is the node number of the input layer.

After the modeling process, Table 5.16 shows the output results with the average training and testing MAPEs from the two models. It can be seen that more accurate results are obtained from the proposed model than the traditional inverse ELM in [32]. Because the performance indexes are indistinguishably intermixed in the traditional inverse ELM, the chaotic environment for the model training leads to an unsatisfying result.

Table 5.15 Definition of seven-level training and seven-level testing data for the multi-mode resonant antenna.

Geometrical variables	Case 1						Case 2					
	Training data (49 samples)			Testing data (49 samples)			Training data (49 samples)			Testing data (49 samples)		
	Min	Max	Step	Min	Max	Step	Min	Max	Step	Min	Max	Step
P_x (mm)	17.5	20.5	0.5	17.75	20.75	0.5	16.9	21.1	0.7	17.25	21.45	0.7
P_y (mm)	17.5	20.5	0.5	17.75	20.75	0.5	16.9	21.1	0.7	17.25	21.45	0.7
W_x (mm)	1.3	2.5	0.2	1.4	2.6	0.2	1	2.8	0.3	1.15	2.95	0.3
W_1 (mm)	0.85	1.15	0.05	0.9	1.2	0.05	0.7	1.3	0.1	0.75	1.35	0.1
L_0 (mm)	0.35	0.65	0.05	0.4	0.7	0.05	0.2	0.8	0.1	0.25	0.85	0.1
L_1 (mm)	7.25	7.85	0.1	7.3	7.9	0.1	7.1	8	0.15	0.6	1.8	0.15
L_2 (mm)	0.35	0.65	0.05	0.4	0.7	0.05	0.2	0.8	0.1	0.25	0.85	0.1
R_{in1} (mm)	2.4	3.6	0.2	2.5	3.7	0.2	2.25	3.75	0.25	2.5	4	0.25
R_{in2} (mm)	4.2	4.8	0.1	4.25	4.85	0.1	3.9	5.1	0.2	4	5.2	0.2
D_1 (mm)	1.35	1.65	0.05	1.4	1.7	0.05	1.2	1.8	0.1	1.25	1.85	0.1
D_2 (mm)	1.85	2.15	0.05	1.9	2.2	0.05	1.7	2.3	0.1	1.75	2.35	0.1
D_x (mm)	0	1.2	0.2	0.1	1.3	0.2	0	2.4	0.4	0.2	2.6	0.4

Source: Xiao et al. [37]/IEEE.

Table 5.16 Results of the proposed model and the inverse ELM model for the multi-mode resonant antenna.

	Average training MAPE		Average testing MAPE	
	Inverse ELM model [32]	Proposed model	Inverse ELM model [32]	Proposed model
Case 1	6.2519%	0.5814%	6.8215%	0.6198%
Case 2	6.5641%	0.6125%	7.1365%	0.6314%

Source: Xiao et al. [37]/IEEE.

To further evaluate the proposed model, a pre-defined structure with the geometrical variables of $\bar{\mathbf{t}}_{\text{Target}} = [18.0\ 19.0\ 1.0\ 1.0\ 0.6\ 7.6\ 0.6\ 2.5\ 4.8\ 1.5\ 2.1\ 0.5]^T$, which does not appear in the training dataset, is introduced to generate EM responses. When the EM responses provided by the pre-defined structure are input into the trained proposed model and traditional inverse ELM, the outputs of the antenna structures of $\bar{\mathbf{t}}_{\text{Proposed}} = [17.9874\ 18.9816\ 1.0049\ 0.9942\ 0.6030\ 7.6347\ 0.6031\ 2.5143\ 4.7770\ 1.5083\ 2.1104\ 0.4977]^T$ and $\bar{\mathbf{t}}_{\text{ELM}} = [16.7630\ 17.5163\ 1.0635\ 0.9349\ 0.6412\ 8.1829\ 0.6432\ 2.6941\ 5.0968\ 1.6013\ 2.2270\ 0.4625]^T$ are obtained. The MAPEs of the outputs are 0.5151% for the proposed model and 6.9629% for the traditional inverse ELM. Furthermore, the two predicted structures are simulated with the *CST* software and their performance indexes are also shown in Figure 5.30. From Figure 5.30, one can see that the simulated results from the proposed model are more accurate than those from the traditional inverse ELM according to the simulated results from the pre-defined structure.

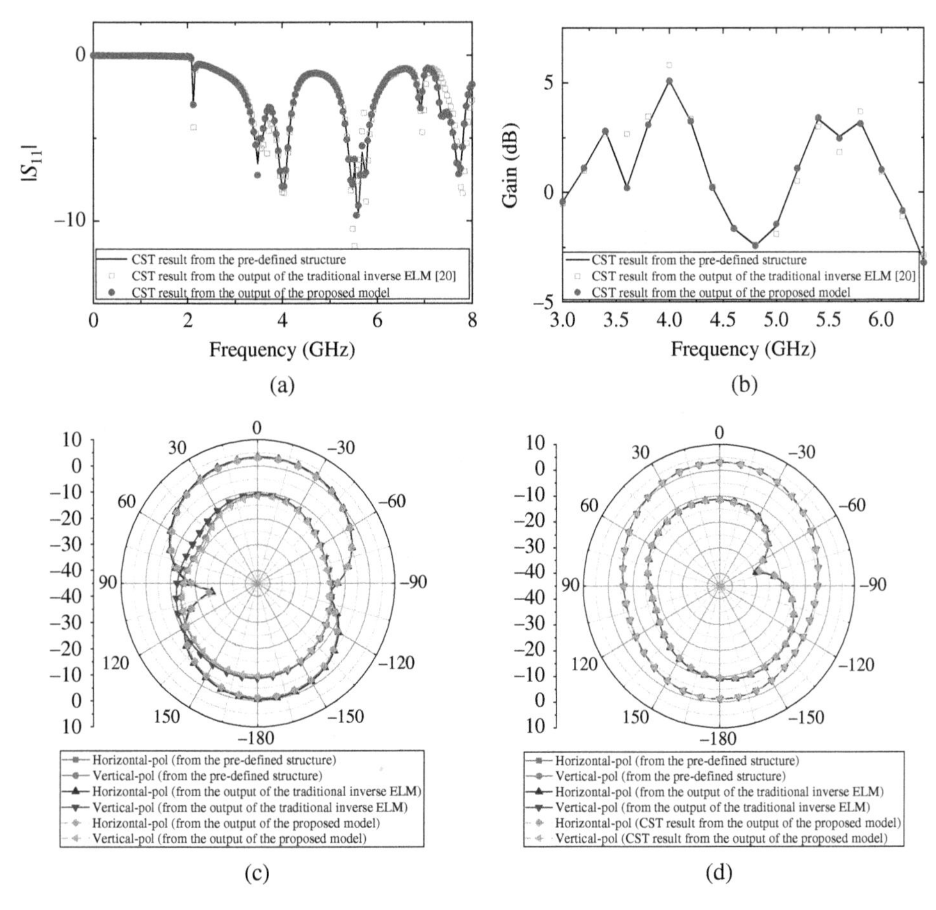

Figure 5.30 Simulated results of *CST*. (a) $|S_{11}|$. (b) Gain. (c) Radiation pattern at 4.4 GHz in the *xoz* plane. (d) Radiation pattern at 4.4 GHz in the *yoz* plane. Source: Xiao et al. [37]/IEEE.

Next, a multi-mode resonant antenna is designed and the design objectives of the antenna are as follows:

(1) $|S_{11}| \leq -10$ dB in the frequency ranges of 4.15–4.5 GHz and 5.9–6.3 GHz;
(2) Relative 3 dB gain bandwidth reaches 30%;
(3) Gain of the main lobe $G_{max} \geq 4.5$ dB at 3.8 and 5.8 GHz.

In the design process, the above performance indexes only need to be input into the trained model, and then the geometrical variables of the antenna can be directly obtained. The obtained results are $\bar{\mathbf{t}}_{\text{Optimal}} = [19\ 19\ 1.24\ 0.68\ 0.23\ 6.9\ 0.23\ 1.72\ 4.62\ 1.26\ 2.1\ 0]^T$, and the corresponding performance indexes from the *CST* software and measurement are shown in Figure 5.31. The simulated and measured results are in good agreement from Figure 5.31. When the multi-objective flower pollination algorithm (MOFPA) [45] and EM full-wave simulation of *CST* are employed to implement the design, it costs about 29 hours to reach the performance indexes. However, the proposed model only costs 4.9 hours, which contain sampling time and training time, to complete the design. In the meantime, compared with the antenna that has a single band in [44], the optimized antenna has an extra high frequency band.

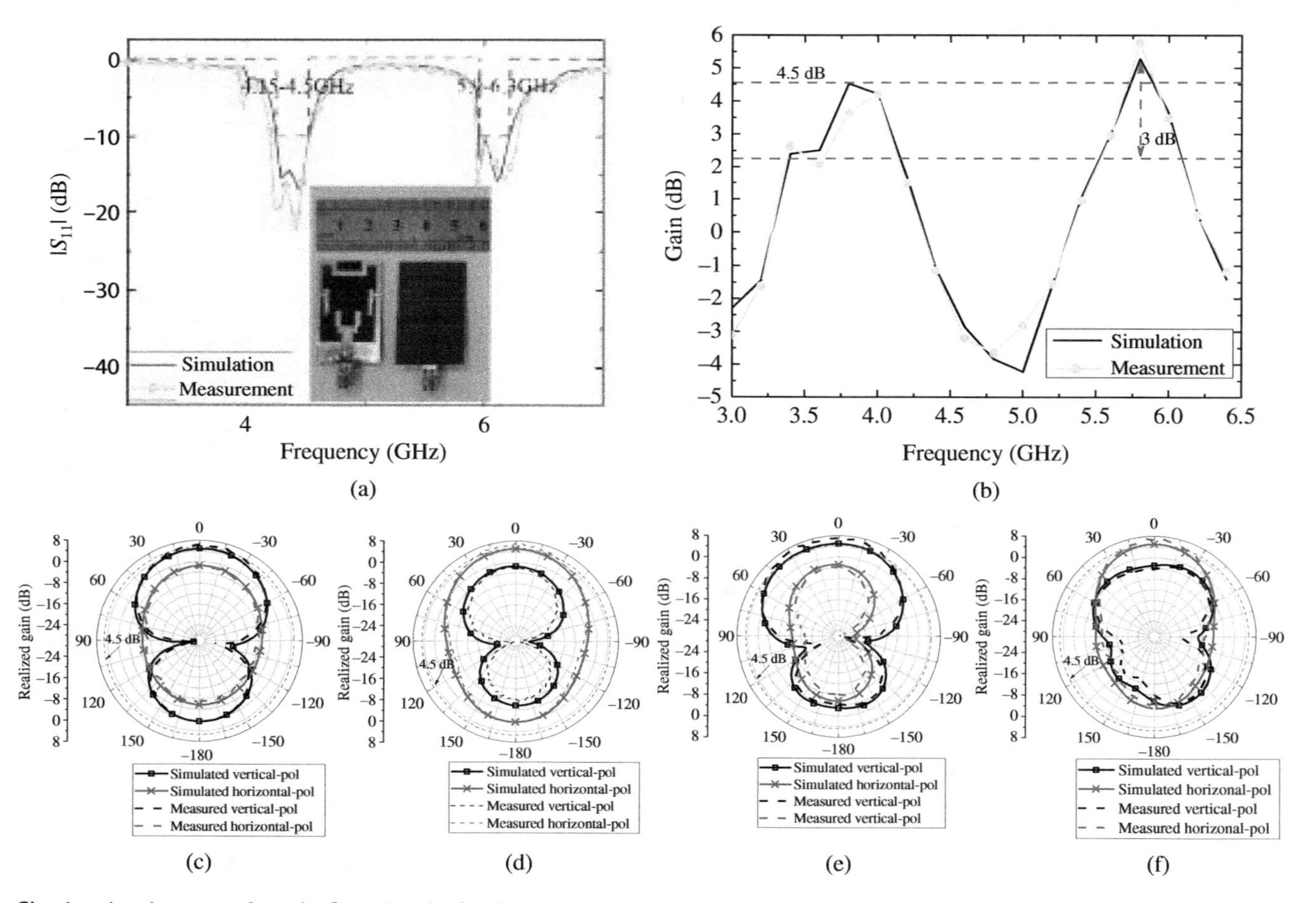

Figure 5.31 Simulated and measured results from the obtained structure of the multi-mode resonant antenna. (a) $|S_{11}|$. (b) Gain. (c) Radiation pattern at 3.8 GHz in the *xoz* plane. (d) Radiation pattern at 3.8 GHz in the *yoz* plane. (e) Radiation pattern at 5.8 GHz in the *xoz* plane. (f) Radiation pattern at 5.8 GHz in the *yoz* plane. Source: Xiao et al. [37]/IEEE.

5.3.2.1 Knowledge-Based Neural Network for Periodic Array Modeling

An effective multi-grade ANN model is introduced for the efficient modeling of finite periodic arrays in this section. Considering mutual coupling and array environment, the proposed model is designed with two sub-ANNs. The first is called element-ANN. It is constructed according to the array elements, where the geometrical variables are set as input and the corresponding coefficients of TF are the output. Due to the same structure of each element, the EM response and TF coefficients of each element are the same without considering mutual coupling. When the training data in element-ANN are collected, only a single element with the absorbing boundary condition is simulated in a full-wave solver. Each element is represented with an element-ANN, which provides prior knowledge for the modeling of the next stage. Then the output of element-ANNs, a set of the TF coefficients corresponding to each element, is collected to construct a matrix. If mutual coupling among all elements and other non-linear influences from array environment are involved, the second-grade ANN, called array-ANN, is trained. Thus the connection between element-ANN and array-ANN could effectively map the relationship between the element geometrical variables and the array EM responses. To a certain extent, the model separates the element information from the array information, which alleviates the chaotic problem in ANN processing.

There are two grades in the whole training process of the proposed ANN model, i.e. element-ANN training for the element and array-ANN training for the whole array.

In the first-grade training process, element-ANN is employed to learn the relationship between the geometrical variables of an element and its EM response without involving mutual coupling and array environment. Considering a two-dimensional array in which the number of elements is $K \times M$ (K is the element number in each row and M is the number in each column), there are $K \times M$ element-ANNs in the model.

Then the output of element-ANNs, a set of the coefficients of TF corresponding to all elements, is collected to construct a matrix with the dimension of $K \times M \times Q$, where Q is the order of TF. The matrix is as the input of the second-grade training process. When the coupling effect of all elements and array environment has to be considered, array-ANN is trained to obtain the relationship between the matrix and the EM response of the whole array. Thus, the proposed multi-grade ANN model decomposes the modeling of finite periodic arrays into two stages that involve the information of the element response and the whole array response, separately. Meanwhile, the proposed model could also realize multi-parameter modeling [37] to conveniently describe the performance of arrays.

Without considering mutual coupling and array environment, the diagram of the training process of element-ANN is plotted in Figure 5.32. The training dataset of element-ANN is presented as $\{(\boldsymbol{x}_j, \boldsymbol{t}_j) | \boldsymbol{x}_j \in \mathbb{R}^n, \boldsymbol{t}_j \in \mathbb{R}^m\}$, where $\boldsymbol{x}_j = [x_{j1}, x_{j2}, \ldots, x_{jn}]^T$ is the input corresponding to n geometrical variables, $\boldsymbol{t}_j = [t_{j1}, t_{j2}, \ldots, t_{jm}]^T$ is the output corresponding to m TF coefficients from EM simulations (i.e. real and imaginary parts of S-parameters), $j = 1, 2, \ldots, N$, and N is the total number of training samples. The actual output of element-ANN is presented as $\boldsymbol{t}'_j = \left[t'_{j1}, t'_{j2}, \ldots, t'_{jm}\right]^T$.

The pole-residue-based TF is used here and the poles and residues of TF corresponding to the EM simulation results of the element with the absorbing boundary condition are obtained with the vector fitting technique. The pole-residue tracking technique is used to solve the order-changing problem.

Element-ANN is employed to learn the nonlinear and unknown relationship between the poles/residues and $\boldsymbol{x}$ through the training process. The objective of this process is to make the error between $\boldsymbol{t}_j$ and $\boldsymbol{t}'_j$ achieve the minimum value with different $\boldsymbol{x}$, by continually revising the internal weights and thresholds of element-ANN. It is noted that only a single element-ANN needs to be

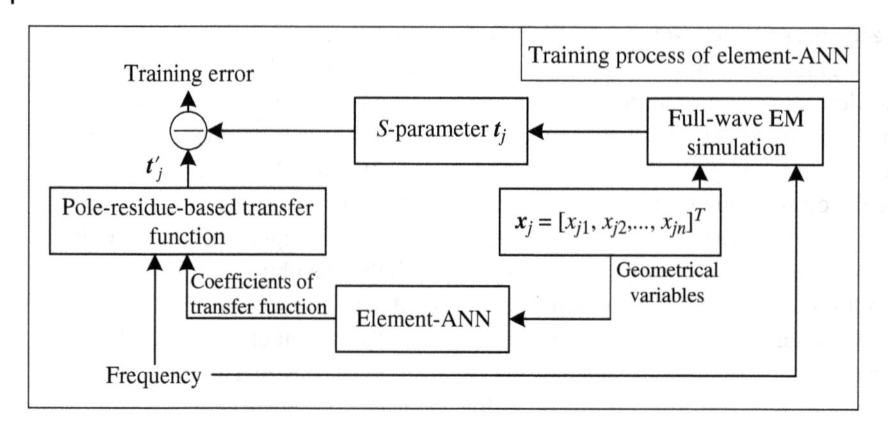

Figure 5.32 Training process of element-ANN. Source: Xiao et al. [28]/IEEE.

trained for a periodic array because the output of each element is the same without considering mutual coupling and array environment.

The diagram of the training process of array-ANN for the whole array is plotted in Figure 5.33. The output from element-ANNs is set as the input of array-ANN. The jth input sample of array-ANN is denoted as $\boldsymbol{T}_j = \left[\boldsymbol{t}'_j\right]^T_{K\times M}$ which is constructed with the output of element-ANNs based on $\boldsymbol{x}_j$. The target of array-ANN is denoted as $\boldsymbol{O}_j$ which is the corresponding TF coefficients from EM simulation of the whole array based on $\boldsymbol{x}_j$ and the corresponding output of array-ANN is denoted as $\boldsymbol{O}'_j$. Here, array-ANN is employed to learn the nonlinear relationship between $\boldsymbol{T}_j$ and $\boldsymbol{O}_j$ through the training process, and the objective is to make the error between $\boldsymbol{O}_j$ and $\boldsymbol{O}_j$ minimum, by updating the internal weights and thresholds of array-ANN.

Example 5.6 ***Multi-Grade ANN for U-Slot Microstrip Array***

A single-layer wideband U-slot array introduced in [46] is selected as the numerical example to evaluate the multi-grade model. Figure 5.34 shows the structure of the array and the array consists of 2 × 2 U-slot elements fed by microstrip lines. The model has seven input geometrical variables, i.e. $\boldsymbol{x} = [x_1\, x_2\, x_3\, x_4\, x_5\, x_6\, x_7]^T$. The node number of the hidden layer is also determined based on the Hecht–Nelson method [42]. Frequency is related to the input parameter, with an original range of 5–10 GHz. S_{11} and H-plane radiation pattern with the multi-parameter modeling method proposed in [37] are the two outputs of the model.

Based on the definition of seven and six levels in the DOE method, 49 training samples and 36 testing samples from *HFSS* simulations are illustrated in Table 5.17. Element-ANN and array-ANN cost 1.83 and 3.27 hours for the training-data generation from EM simulations, respectively. Element-ANN and array-ANN cost 1.35 and 2.4 hours for the testing-data generation, respectively.

To evaluate the validity of the proposed multi-grade ANN model, the model proposed in [37] is selected as a comparison model. The comparison model is directly built on the basis of the EM response of the whole array. The input and output of the comparison model are $\boldsymbol{x}$ and $\boldsymbol{O}'_j$, the same as the proposed multi-grade ANN model. Compared with the multi-grade ANN model, however, the comparison model doesn't consider the element response specially. The training- and testing-data of the comparison model are the same as those of the array-ANN model. In Table 5.18,

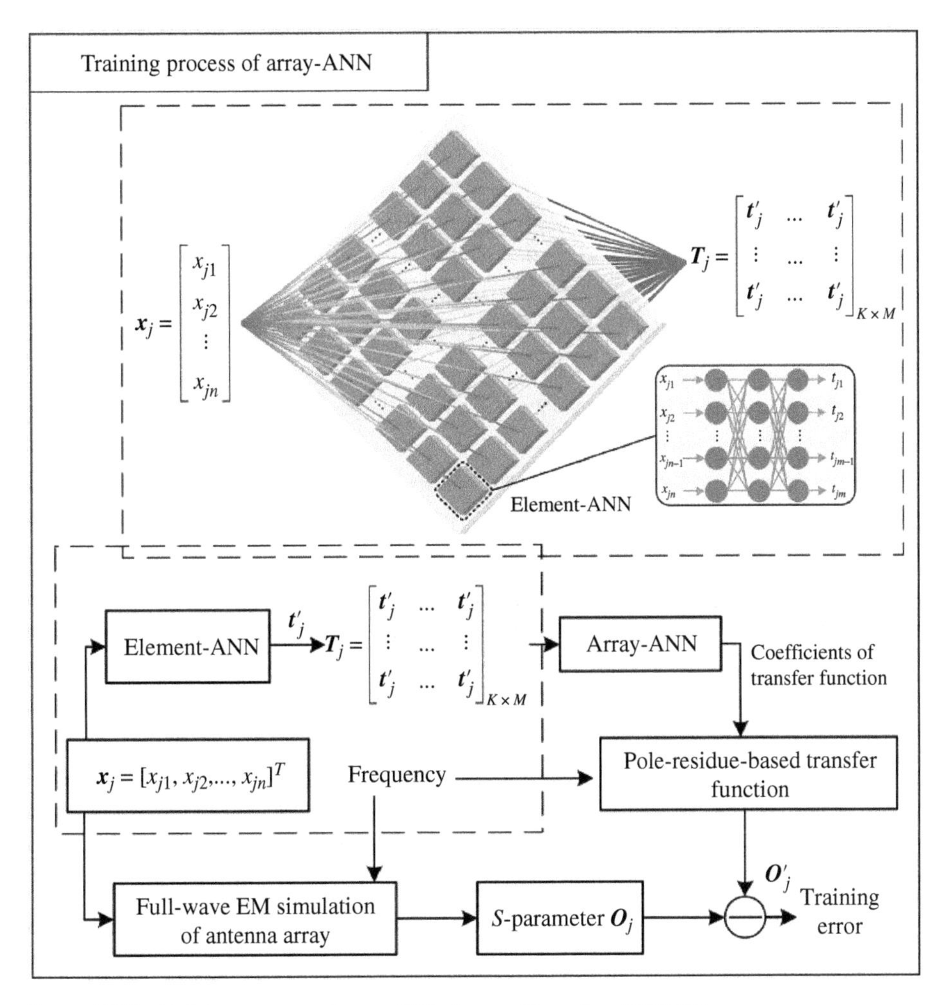

Figure 5.33 Training process of array-ANN. Source: Xiao et al. [28]/IEEE.

the results from the proposed model are more accurate than those from the comparison model. Element-ANN provides prior knowledge for array-ANN in the training process at the expense of a few more sampling data.

Two sets of geometrical variables in the range of training data, i.e. $\boldsymbol{x}_1 = [7.98\ 5.63\ 23.57\ 8.46\ 0.49\ 12.16\ 1.27]^T$ and $\boldsymbol{x}_2 = [9.48\ 5.61\ 23.61\ 9.16\ 0.47\ 13.63\ 1.31]^T$, are utilized to evaluate the proposed model. The proposed model can achieve high accuracy, with the average MAPE values of 0.5819% in $|S_{11}|$ and 1.6279% in H-plane radiation pattern. The two sets of geometrical variables were never used in the training process. Figure 5.35 plots the curves of $|S_{11}|$ of the two different arrays with the proposed model and *HFSS* simulation.

For two other sets of geometrical variables out of the range of training data, i.e. $\boldsymbol{x'}_1 = [6.47\ 3.45\ 23.48\ 7.42\ 0.17\ 11.4\ 0.18]^T$ and $\boldsymbol{x'}_2 = [9.67\ 6.72\ 26.61\ 10.53\ 1.47\ 18.33\ 1.53]^T$, the proposed model can obtain the average MAPE value of 0.6381% in $|S_{11}|$ and 1.6379% in H-plane pattern. Figure 5.36 plots the curves of $|S_{11}|$ of the two different arrays with the proposed model and *HFSS* simulation. The curves from the ANN model and EM simulation are in good agreement.

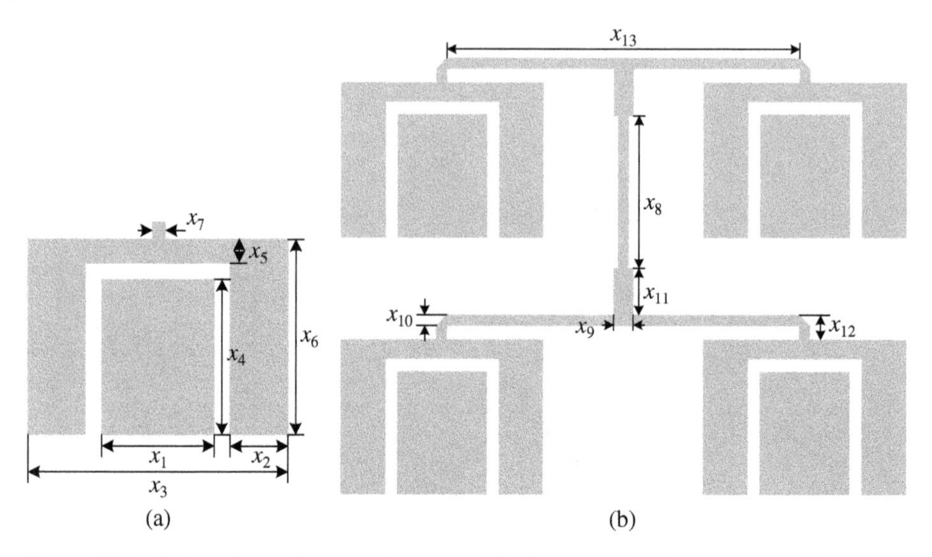

Figure 5.34 U-slot microstrip array, where $x_8 = 19.63$ mm, $x_9 = 6.26$ mm, $x_{10} = 1.95$ mm, $x_{11} = 5.00$ mm, $x_{12} = 2.95$ mm, $x_{13} = 31.11$ mm, and the thickness and relative permittivity of the substrate are 2.7 and 3.5 mm, respectively. (a) Structure of the U-Slot microstrip element. (b) Structure of the array. Source: Xiao et al. [28]/IEEE.

Table 5.17 Definition of seven-level training and six-level testing data for the U-slot microstrip element and array.

Geometrical variables	Training data (49 samples)			Testing data (36 samples)		
	Min	Max	Step	Min	Max	Step
x_1 (mm)	6.5	9.5	0.5	6.75	9.25	0.5
x_2 (mm)	3.5	6.5	0.5	3.75	6.25	0.5
x_3 (mm)	23.5	26.5	0.5	23.75	26.25	0.5
x_4 (mm)	7.5	10.5	0.5	7.75	10.25	0.5
x_5 (mm)	0.2	1.4	0.2	0.3	1.3	0.2
x_6 (mm)	12	18	1	12.5	17.5	1
x_7 (mm)	0.2	1.4	0.2	0.3	1.3	0.2

Source: Xiao et al. [28]/IEEE.

After the training process, the trained model could be the alternative to the time-consuming EM solver for the optimization design. As an example of using the trained model for array design, two separate U-slot microstrip arrays are optimized to reach two different design specifications.

The following are the optimization objectives of Array 1:

(1) $|S_{11}| \leq -10$ dB in the frequency range of 11.5–13.5 GHz;
(2) Gain of the main lobe $G_{max} \geq 10$ dB.

The optimization objectives of Array 2 are:

(1) $|S_{11}| \leq -10$ dB in the frequency range of 9.5–13.5 GHz;
(2) Gain of the main lobe $G_{max} \geq 12$ dB.

Table 5.18 Results of the proposed model and comparison model for the U-slot microstrip array.

	Average training MAPE		Average testing MAPE	
	Comparison model	Proposed model	Comparison model	Proposed model
S-parameter	1.4187%	0.5748%	1.4701%	0.6003%
H-plane pattern	2.7619%	1.6208%	2.8901%	1.6419%

Source: Xiao et al. [28]/IEEE.

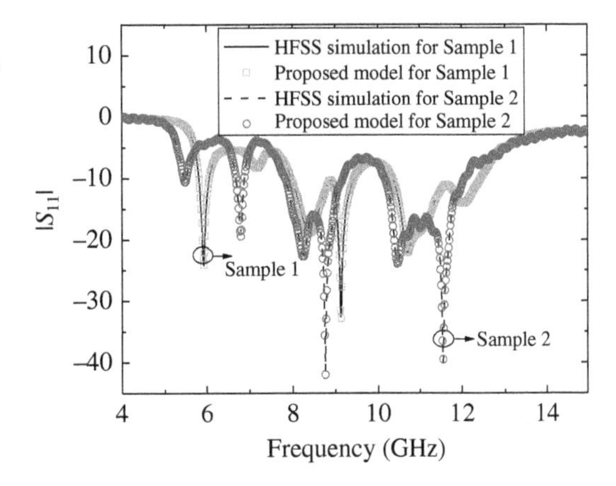

Figure 5.35 $|S_{11}|$ from the ANN model and HFSS for Samples 1 and 2, where the samples are in the range of training data. Source: Xiao et al. [28]/IEEE.

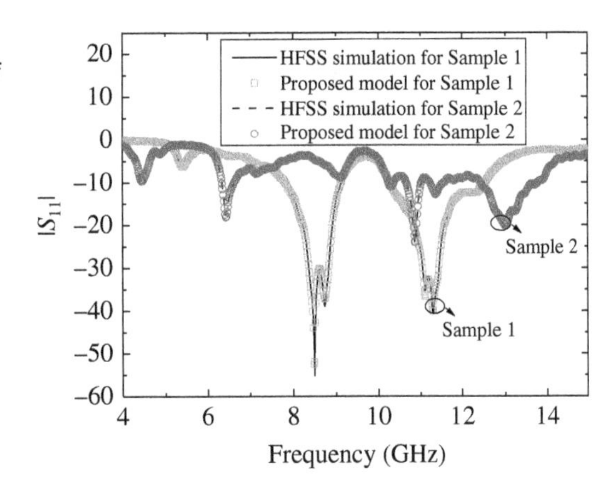

Figure 5.36 $|S_{11}|$ from the ANN model and HFSS for Samples 1 and 2, where the samples are out of the range of training data. Source: Xiao et al. [28]/IEEE.

The two design specifications are realized in Figure 5.37 by repeatedly calling the multi-grade ANN model with the MOFPA [45]. The initial values of geometrical variables are set as $\boldsymbol{x} = [9\ 6\ 25\ 9\ 0.5\ 13\ 0.7]^T$. The optimized geometrical variables for the two separate arrays are $\boldsymbol{x}_{Array1} = [7.614\ 4.587\ 26.614\ 12.157\ 1.548\ 13.143\ 0.895]^T$ and $\boldsymbol{x}_{Array2} = [9.468\ 8.947\ 24.794\ 9.546\ 0.987\ 12.986\ 1.158]^T$. In Table 5.19, the optimization based on the ANN model only consumes about 45 seconds to obtain the optimal solution for each array. Thus, compared with the EM simulation, the multi-grade ANN model can save considerable time in array designs even if the training time is involved.

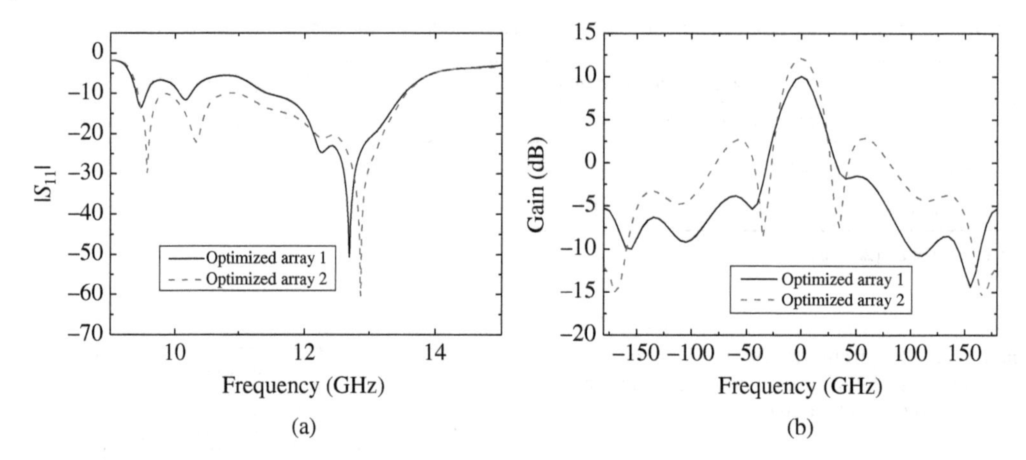

Figure 5.37 Optimization results of Arrays 1 and 2 from the trained model. (a) $|S_{11}|$. (b) H-plane pattern @10GHz. Source: Xiao et al. [28]/IEEE.

Table 5.19 Running time of the optimization process based on the EM simulation and the multi-grade model.

	CPU time	
	EM simulation	**Multi-grade ANN model**
Array 1	16 h	45 s
Array 2	15 h	45 s
Total	31 h	5.1 h (training) + 90 s

Source: Xiao et al. [28]/IEEE.

5.4 Autoencoder Neural Network for Wave Propagation in Uncertain Media

5.4.1 Two-Dimensional GPR System with the Dispersive and Lossy Soil

The parameter values of complex media can be uncertain in reality, and uncertainty quantification (UQ) in the simulation of wave propagation is essential. Here a two-dimensional (2-D) ground penetrating radar (GPR) system is simulated with the auxiliary differential equation (ADE) finite-difference time-domain (FDTD) [47], and the soil is considered as a non-magnetic medium with the frequency-dependent dielectric permittivity represented by a two-term Debye model. The numerical simulation of GPRs relies on a set of input parameters which can affect the electromagnetic pulses and then the survey of an object. In practice, the exact values of the inputs are always unknown, leading to uncertainties in the output of the simulation [48].

The traditional UQ method is Monte Carlo simulation (MCS) [49] which requires running the full-wave FDTD simulation several thousand times to converge, resulting in a high computational cost. Furthermore, the sampling of propagating waves in FDTD usually produces high-dimensional data. To overcome the problem, in this section, an efficient surrogate model based on ANN is introduced to imitate the GPR calculation. An autoencoder neural network is pre-trained and introduced into the surrogate model to map the high-dimensional outputs to a suitable low-dimensional space. Since high-dimensional data which lead to the well-known curse of dimensionality usually

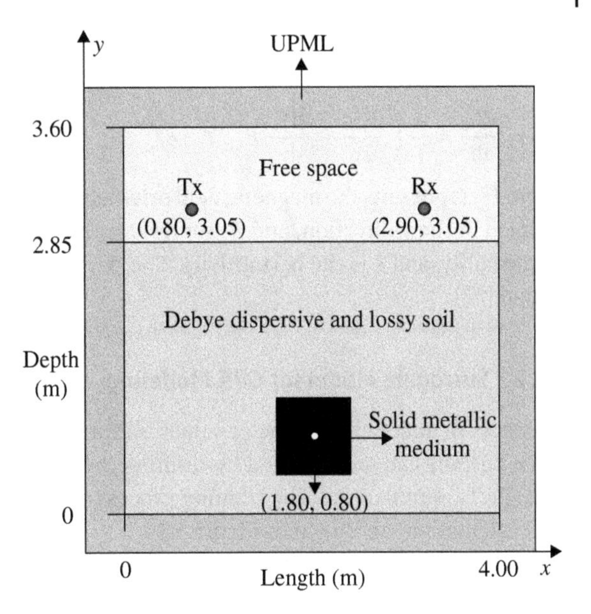

Figure 5.38 2-D GPR system with the dispersive and lossy soil. Source: Cheng et al. [51]/IEEE.

lies in a lower-dimensional manifold, so-called "intrinsic dimensionality space" [50]. The autoencoder neural network, a multi-layer perceptron (MLP) with symmetric structure, aims to learn a compressed representation for an input through minimizing its reconstruction error. The autoencoder neural network is composed of two networks, an "encoder" network and a symmetric "decoder" network. The encoder network transforms the input from a high-dimensional space into features or codes in a low-dimensional space and the symmetric decoder network is used reconstruct the input from corresponding codes [50]. The two networks are trained jointly by tuning the weights of the decoder network first and the encoder network next.

Figure 5.38 shows a two-dimensional (2-D) GPR system. The solid metallic target which is a square is buried in the dispersive and lossy soil. Tx and Rx are the transmitter and the receiver, respectively. Here, the soil material strongly exhibits dispersive properties in the operating frequency range of the GPR system, and the uncertain parameter values of the dispersive materials introduce uncertainties in the simulation result of propagating waves. The soil is considered as a nonmagnetic medium with the frequency-dependent dielectric permittivity, and it is represented by a two-term Debye model with a static conductivity σ_s. The parameters of Debye model are obtained by measurement. They have uncertainties due to measuring tools, manual operation, or other reasons. Assuming there are seven input parameters with uncertain values of $\varepsilon_\infty(\theta)$, $\varepsilon_s(\theta)$, $A_p(\theta)$, $\tau_p(\theta)$ $(p = 1, 2)$ and $\sigma_s(\theta)$ in the complex relative permittivity $\varepsilon_r(\omega, \theta)$, where $\varepsilon_\infty(\theta)$ is the electric permittivity at infinite frequency, $\varepsilon_s(\theta)$ is the static electric permittivity, $A_p(\theta)$ is the pole amplitude, and $\tau_p(\theta)$ is the relaxation time. The form of $\varepsilon_r(\omega, \theta)$ is

$$\varepsilon_r(\omega, \theta) = \varepsilon_\infty(\theta) + \sum_{p=1}^{2} \frac{(\varepsilon_s(\theta) - \varepsilon_\infty(\theta))A_p(\theta)}{1 + j\omega\tau_p(\theta)} + \frac{\sigma_s(\theta)}{j\omega\varepsilon_0} \tag{5.78}$$

where ω is the angular frequency and θ is a random variable.

Maxwell's equations for wave propagating in a 2-D domain are

$$\frac{\partial H_x}{\partial t} = -\frac{1}{\mu}\frac{\partial E_z}{\partial y} \tag{5.79}$$

$$\frac{\partial H_y}{\partial t} = \frac{1}{\mu}\frac{\partial E_z}{\partial x} \tag{5.80}$$

$$\frac{\partial E_z}{\partial t} = \frac{1}{\varepsilon}\left(\frac{\partial H_y}{\partial x} - \frac{\partial H_x}{\partial y}\right) \tag{5.81}$$

where H_x represents the magnetic field oriented in the x-direction, H_y represents the magnetic field oriented in the y-direction, and E_z represents the electric field oriented in the z-direction. μ is the permeability, and ε is the permittivity. The ADE-FDTD method is employed to simulate the GPR system [52].

5.4.2 Surrogate Model for GPR Modeling

After a set of uncertain parameter values is input into the surrogate model, the statistical quantities of the outputs can be evaluated by running the surrogate model instead of running thousands of ADE-FDTD simulations. The training process and testing process of the new surrogate model for GPR calculation are shown in Figure 5.39.

The input dataset for the training of the surrogate model consists of two parts: the uncertain parameters of soil and the electric field data from ADE-FDTD. In the training process of the proposed surrogate model, the uncertain parameters of soil $\mathbf{I} = \{\mathbf{I}_1, \mathbf{I}_2, \mathbf{I}_3, \ldots, \mathbf{I}_M\}$ ($\mathbf{I}_m \in R^S$ $(1 \leq m \leq M)$ represents an S-dimensional vector) and the encoder outputs $\mathbf{C} = \{\mathbf{C}_1, \mathbf{C}_2, \mathbf{C}_3, \ldots, \mathbf{C}_M\}$ ($\mathbf{C}_m \in R^d$ represents a d-dimensional vector) construct the datasets of training samples, where M is the number of training samples. In the testing process, for a new set of uncertain inputs $\mathbf{I} = \{\mathbf{I}_1, \mathbf{I}_2, \mathbf{I}_3, \ldots, \mathbf{I}_N\}$, the compressed outputs $\mathbf{C}' = \{\mathbf{C}'_1, \mathbf{C}'_2, \mathbf{C}'_3, \ldots, \mathbf{C}'_N\}$ ($\mathbf{C}'_n \in R^d$ $(1 \leq n \leq N)$ represents a d-dimensional vector, and N is the number of testing samples) can be obtained from the trained ANN, and then with the trained decoder, the predicted outputs $\mathbf{U}' = \{\mathbf{U}'_1, \mathbf{U}'_2, \mathbf{U}'_3, \ldots,$

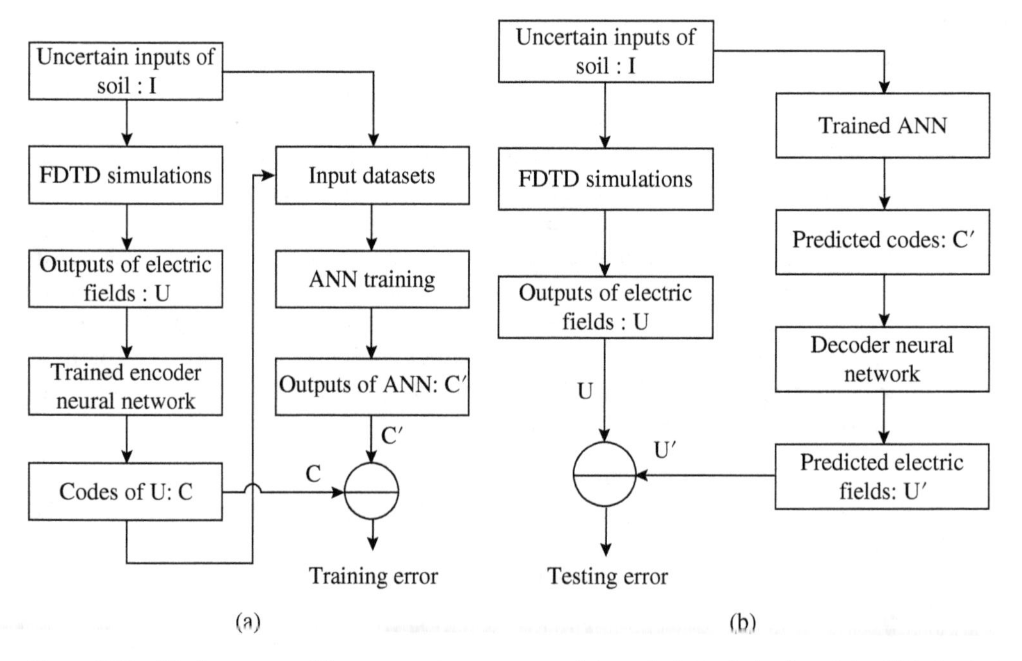

Figure 5.39 Whole process of the proposed surrogate model to mimic the behavior of the GPR simulation. (a) Training process. (b) Testing process. Source: Cheng et al. [51]/IEEE.

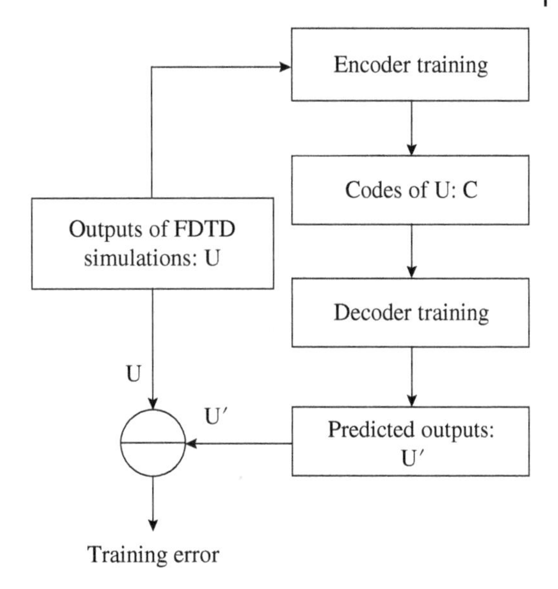

Figure 5.40 Training process of an autoencoder neural network. Source: Cheng et al. [51]/IEEE.

$\mathbf{U}'_N\}$ ($\mathbf{U}'_n \in R^D$ is a D-dimensional vector) corresponding to the new set of inputs **I** are obtained. The trained surrogate model predicts the outputs in GPR systems efficiently.

The encoder and decoder used in the surrogate model are from an autoencoder neural network which needs to be trained before constructing the surrogate model. The training process of an autoencoder neural network is presented in Figure 5.40 [50]. The encoder maps the input data from a high-dimensional space into codes in a low-dimensional space, and the decoder reconstructs the input data from the corresponding codes. Given the training data $\mathbf{U} = \{\mathbf{U}_1, \mathbf{U}_2, \mathbf{U}_3, \ldots, \mathbf{U}_L\}$ ($\mathbf{U}_l \in R^D$ $(1 \leq l \leq L)$ represents a D-dimensional vector), the encoder transforms the input matrix of **U** into a hidden representation of $\mathbf{C} = \{\mathbf{C}_1, \mathbf{C}_2, \mathbf{C}_3, \ldots, \mathbf{C}_L\}$ ($\mathbf{C}_l \in R^d$) through activation functions, where $d = D$, and L is the number of training samples of the autoencoder neural network. Then, the matrix of **C** is transformed back to a reconstruction matrix of $\mathbf{U}' = \{\mathbf{U}'_1, \mathbf{U}'_2, \mathbf{U}'_3, \ldots, \mathbf{U}'_L\}$ by the decoder. The input data **U** are the electric fields or the magnetic fields calculated from ADE-FDTD, and D is the number of time steps.

5.4.3 Modeling Results

As described above, the 2-D GPR system used for modeling is presented in Figure 5.38. The computational domain is 4.00 m × 3.60 m along the x and y directions, respectively. The center of the solid metallic medium is (1.80 m, 0.80 m) and its size is 0.30 m × 0.30 m. The metallic medium is buried 2.05 m below ground level. The position of Tx is (0.80 m, 3.05 m), and the position of Rx is (2.90 m, 3.05 m). The excitation is the Blackmann–Harris pulse source [53]. The parameters of Debye model are presented in Table 5.20 which are obtained by measurement. There are uncertainties in these parameter values. Here the moisture of soil is 2.5%, and the variation in each parameter is 10%. Figure 5.41 shows that the ADE-FDTD output E_z of the 2-D GPR system observed at the position of the receiver changes with seven uncertain input parameters of $\varepsilon_\infty(\theta)$, $\varepsilon_s(\theta)$, $A_p(\theta)$, $\tau_p(\theta)$ $(p = 1, 2)$, and $\sigma_s(\theta)$ in the complex relative permittivity $\varepsilon_r(\omega, \theta)$. It is shown that the output uncertainty with a variation of 10% in each uncertain input is much larger than that with a variation of 5% in each uncertain input.

There are seven uncertain parameters in ADE-FDTD simulations and the variation in each uncertain parameter is 10%. The inputs and outputs in the FDTD simulation are both normalized before

Table 5.20 Model parameters for the dispersive and lossy soil.

Moisture	ε_∞	σ_s (mS/m)	A_1	A_2	τ_1(ns)	τ_2(ns)
2.5%	3.20	0.397	0.75	0.30	2.71	0.108

Source: Cheng et al. [51]/IEEE.

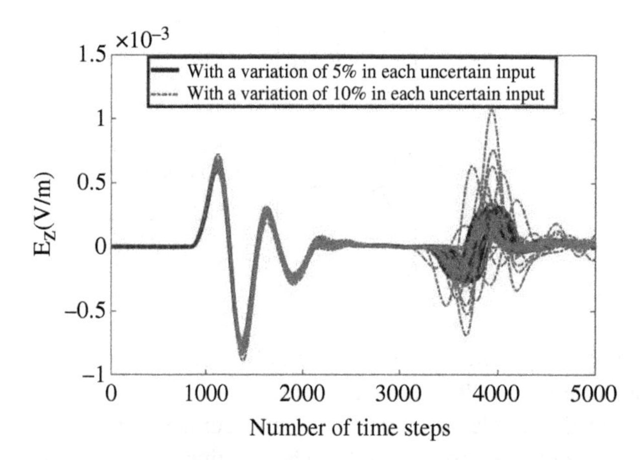

Figure 5.41 Output E_z for 60 sets of inputs from FDTD simulations. Source: Cheng et al. [51]/IEEE.

Table 5.21 Hyperparameters of the ANNs.

	Batch size	Number of epochs	Number of neurons in each hidden layer
Surrogate model	50	1000	1000, 1000
Autoencoder	50	5000	1000, 1000, 200, 1000, 1000

Source: Cheng et al. [51]/IEEE.

the surrogate model is trained. The dimensions of FDTD output and its code are 5000 and 200, respectively. It is required to run 100 FDTD simulations to collect the training samples of the autoencoder neural network, and the proposed surrogate model requires 200 training samples. The details of the hyperparameters of the proposed surrogate model and the autoencoder neural network are presented in Table 5.21.

Once the surrogate model is trained, it can be used for various analyses such as quantifying the uncertainty in the simulation results. When there are seven uncertain parameters in ADE-FDTD simulations and the variation in each uncertain parameter is 10%, the mean and standard deviation evaluated with the surrogate model are presented in Figure 5.42. The results of the proposed surrogate model are compared with those of MCS, and they agree well with each other in Figure 5.42. The CPU time required for the new model includes three parts: (i) the time to run FDTD simulations to collect the training data, (ii) the time to train ANNs, and (iii) the time to predict the outputs corresponding to a new set of inputs. The details of the CPU time of the two methods are presented in Table 5.22. Compared with running a thousand FDTD simulations in MCS, the proposed surrogate model largely reduces the number of FDTD simulations and improves the efficiency.

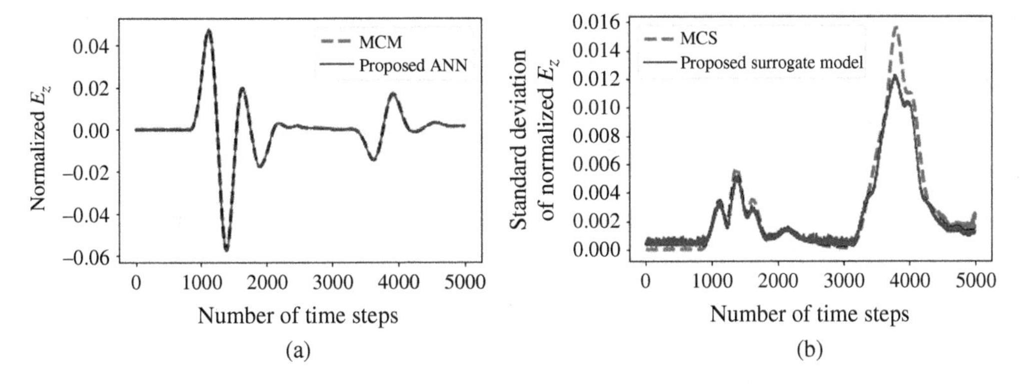

Figure 5.42 Seven uncertain input parameters of dispersive and lossy soil in the GPR simulation with the variation of 10% in each parameter. (a) mean of normalized E_z. (b) standard variance of normalized E_z. Source: Cheng et al. [51]/IEEE.

Table 5.22 CPU Time of the surrogate model and MCS.

Methods	Number of simulations	CPU Time (s)
MCS	1000	1046375.60
Surrogate model	200	262274.56+2204.33+2.10

Source: Cheng et al. [51]/IEEE.

References

1 Huang, G.B., Zhu, Q.Y., and Siew, C.K. (2006). Extreme learning machine: theory and applications. *Neurocomputing* 70 (1–3): 489–501.

2 Shi, X., Wang, J., Liu, G. et al. (2016). Application of extreme learning machine and neural networks in total organic carbon content prediction in organic shale with wire line logs. *Journal of Natural Gas Science and Engineering* 33: 687–702.

3 Zhang, C., Zhu, Y., Cheng, Q. et al. (2017). Extreme learning machine for the behavioral modeling of RF power amplifiers. In: *IEEE MTT-S International Microwave Symposium*, Honolulu, HI, USA, 558–561.

4 Xiao, L., Shao, W., Liang, T., and Wang, B. (2017). Efficient extreme learning machine with transfer functions for filter design. In: *IEEE MTT-S International Microwave Symposium*, Honolulu, HI, USA, 2017, 555–557.

5 Wei, Y., Xiao, G., Deng, H. et al. (2015). Hyperspectral image classification using FPCA-based kernel extreme learning machine. *Optik* 126 (23): 3942–3948.

6 Huang, G.-B., Wang, D.H., and Lan, Y. (2011). Extreme learning machines: a survey. *International Journal of Machine Learning and Cybernetics* 2 (2): 107–122.

7 Bi, X., Zhao, X., Wang, G. et al. (2015). Distributed extreme learning machine with kernels based on mapreduce. *Neurocomputing* 149 (Part A): 456–463.

8 Deng, W.-Y., Zheng, Q.-H., and Wang, Z.-M. (2014). Cross-person activity recognition using reduced kernel extreme learning machine. *Neural Networks* 53: 1–7.

9 Fu, H., Vong, C.-M., Wong, P.-K., and Yang, Z. (2016). Fast detection of impact location using kernel extreme learning machine. *Neural Computing and Applications* 27 (1): 121–130.

10 Liu, X., Wang, L., Huang, G.-B. et al. (2015). Multiple kernel extreme learning machine. *Neurocomputing* 149 (Part A): 253–264.

11 Shamshirband, S., Mohammadi, K., Chen, H.-L. et al. (2015). Daily global solar radiation prediction from air temperatures using kernel extreme learning machine: a case study for Iran. *Journal of Atmospheric and Solar - Terrestrial Physics* 134: 109–117.

12 Wong, P.K., Wong, K.I., Vong, C.M., and Cheung, C.S. (2015). Modeling and optimization of biodiesel engine performance using kernel-based extreme learning machine and cuckoo search. *Renewable Energy* 74: 640–647.

13 Vapnik, V.N. (1995). *The Nature of Statistical Learning Theory*. New York: Springer.

14 Cristianini, N. and Shawe-Taylor, J. (2000). *An Introduction to Support Vector Machines*. Cambridge: Cambridge University Press.

15 Fung, G. and Mangasarian, O.L. (2001). Proximal support vector machine classifiers. In: *Knowledge Discovery and Data Mining (KDD)*, 77–86. San Francisco, CA: Springer.

16 Deng, W., Zheng, Q., and Chen, L. (2009, 2009). Regularized extreme learning machine. In: *IEEE Symposium on Computational Intelligence and Data Mining*, Nashville, TN, USA, 389–395.

17 Mercer, J. (1909). Functions of positive and negative type and their connection with the theory of integral equations. *Philosophical Transactions of the Royal Society A* 209 (441–458): 415–446.

18 Huang, G.-B., Zhou, H., Ding, X., and Zhang, R. (2012). Extreme learning machine for regression and multiclass classification. *IEEE Transactions on Systems, Man, and Cybernetics. Part B, Cybernetics* 42 (2): 513–529.

19 Bottou, L., Chapelle, O., Decoste, D., and Weston, J. (2007). Training a support vector machine in the primal. *Neural Computation* 19 (5): 1155–1178.

20 Scholköpf, B., Herbrich, R., and Smola, A.J. (2001). A generalized representer theorem. In: *Lecture Notes in Computer Science*, 416–426.

21 Zhao, Y. and Sun, J. (2008). Robust support vector regression in the primal. *Neural Networks* 21 (10): 1548–1555.

22 Zhao, Y. and Sun, J. (2009). Recursive reduced least squares support vector regression. *Pattern Recognition* 42 (5): 837–842.

23 Zhao, Y.P. (2016). Parsimonious kernel extreme learning machine in primal via Cholesky factorization. *Neural Networks* 80: 95–109.

24 An, S., Liu, W., and Venkatesh, S. (2007). Fast cross-validation algorithms for least squares support vector machine and kernel ridge regression. *Pattern Recognition* 40 (8): 2154–2162.

25 Xiao, L., Shao, W., Ding, X., and Wang, B. (2018). Dynamic adjustment kernel extreme learning machine for microwave component design. *IEEE Transactions on Microwave Theory and Techniques* 66 (10): 4452–4461.

26 Feng, F., Zhang, C., Ma, J., and Zhang, Q.J. (2016). Parametric modeling of EM behavior of microwave components using combined neural networks and pole-residue-based transfer functions. *IEEE Transactions on Microwave Theory and Techniques* 64 (1): 60–77.

27 Gustavsen, B. and Semlyen, A. (1999). Rational approximation of frequency domain responses by vector fitting. *IEEE Transactions on Power Delivery* 14 (3): 1052–1061.

28 Xiao, L.-Y., Shao, W., Ding, X. et al. (2019). Parametric modeling of microwave components based on semi-supervised learning. *IEEE Access* 7: 35890–35897.

29 Wong, S.W., Feng, S.F., Zhu, L., and Chu, Q.X. (2015). Triple- and quadruple-mode wideband bandpass filter using simple perturbation in single metal cavity. *IEEE Transactions on Microwave Theory and Techniques* 63 (10): 1–9.

30 Schmidt, R. and Launsby, R.G. (1992). *Understanding Industrial Designed Experiments*. Colorado Springs, CO, USA: Air Force Academy.

31 Mohamed, F.A.H., Friedhelm, S., and Günther, P. (2010). Semi-supervised learning for tree-structured ensembles of RBF networks with co-training. *Neural Networks* 23: 497–509.

32 Xiao, L., Shao, W., Jin, F.-L. et al. (2020). Efficient Inverse "Extreme learning machine for parametric design of metasurfaces". *IEEE Antennas and Wireless Propagation Letters* 19: 992–996.

33 Yang, X.S. (2012). Flower pollination algorithm for global optimization. In: *Proceedings of the 11th international Conference on Unconventional Computation and Natural Computation (UCNC)*, Orléans, France, September 2012, 242–243.

34 Yang, L., Zhu, L., Choi, W.-W., and Tam, K.-W. (2017). Analysis and design of wideband microstripto-microstrip equal ripple vertical transitions and their application to bandpass filters. *IEEE Transactions on Microwave Theory and Techniques* 65 (8): 2866–2877.

35 Huang, G., Song, S., Gupta, J.N.D., and Wu, C. (2014). Semi-supervised and unsupervised extreme learning machines. *IEEE Trans. Cybernetics* 44 (12): 2405–2417.

36 Yang, L., Zhu, L., Zhang, R. et al. (2019). Novel multilayered ultra-broadband bandpass filters on high-impedance slotline resonators. *IEEE Transactions on Microwave Theory and Techniques* 67 (1): 129–139.

37 Xiao, L., Shao, W., Jin, F.-L., and Wang, B.-Z. (2018). Multiparameter modeling with ANN for antenna design. *IEEE Transactions on Antennas and Propagation* 66 (7): 3718–3723.

38 Cortes, C. and Vapnik, V. (1995). Support-vector networks. *Machine Learning* 20 (3): 273–297.

39 Feresidis, A.P. and Vardaxoglou, J.C. (2001). High gain planar antenna using optimized partially reflective surfaces. *IEE Proceedings - Microwaves, Antennas and Propagation* 148 (6): 345–350.

40 Weily, A.R., Esselle, K.P., Bird, T.S., and Sanders, B.C. (2006). High gain circularly polarised 1-D EBG resonator antenna. *Electronics Letters* 42 (18): 1012–1013.

41 Wang, N., Liu, Q., Wu, C. et al. (2014). Wideband Fabry-Perot resonator antenna with two complementary FSS layers. *IEEE Transactions on Antennas and Propagation* 62 (5): 2463–2471.

42 Hecht-Nielsen, S.R. (1987). Kolmogorov's mapping neural network existence theorem. In: *IEEE Joint Conf. on Neural Networks*, New York, *USA*, vol. 3, 11–14.

43 Deb, K., Pratap, A., Agarwal, S., and Meyarivan, T. (2002). A fast and elitist multiobjective genetic algorithm: NSGA-II. *IEEE Transactions on Evolutionary Computation* 6 (2): 182–197.

44 Ha, J., Kwon, K., Lee, Y., and Choi, J. (2012). Hybrid mode wideband patch antenna loaded with a planar metamaterial unit cell. *IEEE Transactions on Antennas and Propagation* 60 (2): 1143–1147.

45 Yang, X., Karamanoglua, M., and He, X. (2013). Multi-objective flower algorithm for optimization. *Procedia Computer Science* 18: 861–868.

46 Wang, H., Huang, X., and Fang, D.G. (2008). A single layer wideband U-slot microstrip patch antenna array. *IEEE Antennas and Wireless Propagation Letters* 7: 9–12.

47 A. Taflove and S. Hagness, *Computational Electrodynamics: The Finite-Difference Time- Domain Method*, 2nd ed. Boston, MA, USA: Artech House, 2000.

48 Sudret, B. (2007). Uncertainty propagation and sensitivity analysis in mechanical models contributions to structural reliability and stochastic spectral methods. In: *Habilitation ` a Diriger des Recherches*, vol. 147. Clermont-Ferrand, France: Universite Blaise Pascal.

49 McKay, M.D., Beckman, R.J., and Conover, W.J. (1979). A comparison of three methods for selecting values of input variables in the analysis of output from a computer code. *Technometrics* 21 (2): 239–245.

50 Liu, W., Wang, Z., and Liu, X. (2017). A survey of deep neural network architectures and their applications. *Neurocomputing* 234: 11–26.

51 Cheng, X., Zhang, Z.Y., and Shao, W. (2020). *IEEE Access* 8: 218323–218330.

52 Cheng, X., Shao, W., and Wang, K. (2019). Uncertainty analysis in dispersive and lossy media for ground-penetrating radar modeling. *IEEE Antennas and Wireless Propagation Letters* 18 (9): 1931–1935.

53 Teixeira, F.L., Cho Chew, W., Straka, M. et al. (1998). Finite-difference time-domain simulation of ground penetrating radar on dispersive, inhomogeneous, and conductive soils. *IEEE Transactions on Geoscience and Remote Sensing* 36 (6): 1928–1937.

Section III

Deep Learning for Metasurface Design

6

Generative Machine Learning for Photonic Design

Dayu Zhu, Zhaocheng Liu, and Wenshan Cai

School of Electrical and Computer Engineering, Georgia Institute of Technology, Atlanta, GA, USA

6.1 Brief Introduction to Generative Models

6.1.1 Probabilistic Generative Model

In machine-learning-based photonic design methodologies, two major classes of machine learning models are used: discriminative models and generative models [1]. A discriminative model predicts the probability of label y conditioned on the input data x, i.e. $p(y|x)$, while the generative model learns a probability distribution of x, $p(x)$, or a joint probability of x and y, $p(x,y)$. Intuitively, a discriminative model is a function $f(x)$, transforming the input data into a label or value $y = f(x)$. Thus, discriminative models are always related to classification and regression tasks in supervised learning schemes. A generative model, on the other hand, can map the data x into a compact representation, and with some sampling algorithms, we can retrieve more data that is similar to the input dataset. In photonic design tasks, a generative model is often used to learn the distribution of a predefined dataset. Searching for an optimal design inside the distribution can be much faster than the original optimization problem without the prior information. Unlike deterministic optimization schemes, due to the stochastic property of a generative model, identifying a global optimal solution using a generative design strategy is more probable. We will derive a rigorous analysis on why generative models in photonic design could work better than traditional optimization schemes in Section 6.1.2.

6.1.2 Parametrization and Optimization with Generative Models

Before we dive into design examples with generative models, it is useful to develop some theory background to help us understand why generative models could be preferred in photonic inverse design problems. In traditional design methods, a device can be represented using a few geometric parameters, a pixelated image, or some advanced curves such as non-uniform rational B-spline (NURBS) [2]. One strategy of identifying the optimal design is through gradient-based optimization methods such as the adjoint method [3, 4]. There could be some problems when carrying out complex optimizations through this strategy. The two most infamous ones are the local optimum problem and high dimensionality problem. In the following discussion, we will model the design procedure as a probabilistic model and see how these difficulties can be mitigated with generative design strategies.

Advances in Electromagnetics Empowered by Artificial Intelligence and Deep Learning, First Edition.
Edited by Sawyer D. Campbell and Douglas H. Werner.

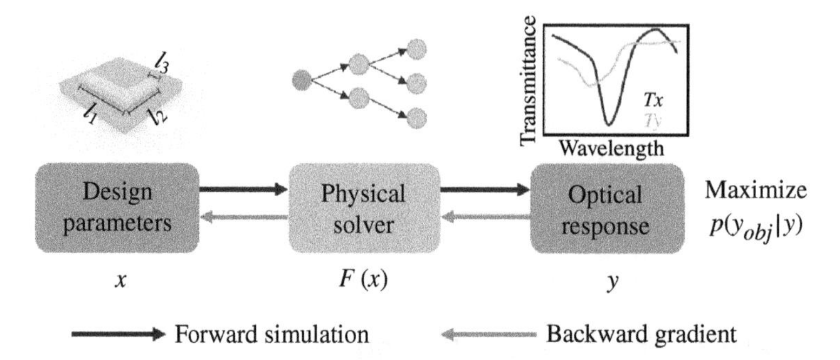

Figure 6.1 The gradient-based optimization, which consists of the forward simulation and the backward gradient process.

6.1.2.1 Probabilistic Model for Gradient-Based Optimization

The gradient-based optimization process is illustrated in the flowchart Figure 6.1. We define the input and output of a physical system as x and y, respectively. In photonic design, x can be the design parameters and y may represent the optical responses. The output can be transformed from the input by $y = F(x)$, where F is a deterministic function representing physical simulation. We denote the design objective as y_{obj}. We can formulate the gradient-based optimization process as:

$$p(y_{obj}|x) = \frac{1}{\sqrt{2\pi}\sigma} \exp\left(-\frac{1}{2\sigma^2}(y - y_{obj})^2\right) \tag{6.1}$$

The above function is a Gaussian distribution and can be interpreted as the probability of achieving the y_{obj} given x. Here σ is a hyperparameter of the optimization, the meaning of which will be explained later. The logarithm of the probability is more convenient to work with:

$$L = \log p(y_{obj}|x) = -\log\sqrt{2\pi}\sigma - \frac{1}{2\sigma^2}||y(x) - y_{obj}||^2 \tag{6.2}$$

In gradient-based optimization, we aim to maximize the log likelihood in Eq. (6.2) by gradient descent method. The gradient of the log probability in the probabilistic model is

$$\nabla_x L = -\frac{1}{\sigma^2}(y - y_{obj})\frac{\partial y}{\partial x} \tag{6.3}$$

In a deterministic formulation of the photonic optimization, we usually define a mean squared error (MSE) loss function $L_{MSE} = 1/2||y(x) - y_{obj}||^2$, and try to minimize the loss. However, we can find that the log likelihood gradient and MSE loss gradient is only off by a constant proportional to $1/\sigma^2$. Thus, we can conclude the probabilistic model of the gradient-based optimization is almost identical to its deterministic MSE counterpart. The only difference is the hyperparameter σ, which affects the step size of the optimization.

6.1.2.2 Sampling-Based Optimization

Now, we consider sampling multiple designs x^i and optimize the overall performance of all the designs. We denote the optical responses and the probability of achieving the design objectives as y^i and y^i_{obj}, respectively. Assuming each design is independent, we thus can calculate the probability of all samples $p(y_{all})$ as

$$p(y_{all}) = \prod_i p\left(y^i_{obj}\right) \tag{6.4}$$

The log likelihood of $p(y_{all})$ has the form

$$L_{smpl} = \log p(y_{all}) = -N \log \sqrt{2\pi}\sigma - \frac{1}{2\sigma^2} \sum_i \left\| y^i - y^i_{obj} \right\|^2 \tag{6.5}$$

where N is the sample size. To maximize the log probability, we compute the gradient:

$$\nabla_x L_{smpl} = \sum_i \nabla_x L_i \tag{6.6}$$

where $L_i = \left\| y^i - y^i_{obj} \right\|^2$. Equation (6.6) indicates that the sampling algorithm takes into account all the design errors. Hence, the designed structure can be more robust as compared to the designs through deterministic optimization schemes. Indeed, such a sampling approach is able to identify the global solution – the optimal solution which maximizes $p(y_{obj})$ instead of $p(y_{obj} \mid x)$. To see that, we expand the probability of achieving the objective $p(y_{obj})$ with respect to the design parameters x:

$$p(y_{obj}) = \int p(y_{obj}, x)dx = \int p(y_{obj}|x)p(x)dx = \mathbb{E}_x[p(y_{obj}|x)] \tag{6.7}$$

The log probability of $p(y_{obj})$ can be expanded as

$$\begin{aligned} \log p(y_{obj}) &= \log \mathbb{E}_x[p(y_{obj}|x)] \geq \mathbb{E}_z[\log p(y_{obj}|x)] \\ &= \mathbb{E}_z\left[-\log \sqrt{2\pi}\sigma - \frac{1}{2\sigma^2}\|y - y_{obj}\|^2\right] \qquad \propto L_{smpl} \end{aligned} \tag{6.8}$$

We used Jensen's inequality in the above equation. From Eq. (6.8), we know that maximizing L_{smpl} is equivalent to maximizing the lower bound of $\log p(y_{obj})$. Thus, we can conclude that optimizing through the sampling algorithm is more likely to converge to a global solution than deterministic optimization methods.

6.1.2.3 Generative Design Strategy

Although the sampling algorithm can identify global optima, it barely yields a satisfiable solution in practice. This is because the geometric parameter x is not a good representation due to the roughness of the design space and rareness of the local/global optima, and thus, it is challenging to locate the only optima in the original x representation. Therefore, we seek to reparametrize x through another smooth function with more parameters w and optimize the newly introduced parameters. In doing do, we automatically create multiple sets of w that could possibly represent good candidates of x. Identifying any set of w is more feasible than locating the only few optima in the original x representation. We take a step further and consider the generative model

$$x = G_w(z) \tag{6.9}$$

where G_w is a function parametrized by w and maps a random vector z to design space x. We can assume z is a sample from a distribution $f_Z(z)$. In a generative strategy, instead of directly optimizing x, we choose to optimize w such that the generated x has the desired optical responses. Usually, we use a deterministic generator, but we can always formulate the generator as a probabilistic process $p(x \mid z)$. The distribution $p(x)$ can be derived from $p(x \mid z)$ through:

$$p(x) = \int p(x|z)p(z)\, dz \tag{6.10}$$

In practice, the $p(x)$ may have no closed form, but we can optimize the L_{smpl} without knowing $p(x)$. For a deterministic generator, the gradient of the L_{smpl} with respect to w can be derived through the chain rule:

$$\nabla_w L_{smpl} = \sum_i \nabla_x L_i \frac{\partial x}{\partial w} \tag{6.11}$$

For a stochastic generator, reparameterization [5] is required to produce gradients flow in optimization. L_{smpl} can be improved by iteratively subtracting the gradient from it. After the optimization, a generative model G_w is able to produce a population of samples that satisfy the design objectives. The best individual is selected as the final design. Additionally, we can also maximize the distribution $p(y_{obj})$ through a similar manner. After some derivation, the gradient of $p(y_{obj})$ is identical to the form proposed in Ref. [6].

6.1.2.4 Generative Adversarial Networks in Photonic Design

In the above discussion, we did not consider the prior distribution of the design parameters $p(x)$. In fact, we can identify x by looking for the maximizing $p(x \mid y_{obj})$. According to Bayes' theorem, for each sample, we can rewrite $p\left(x^i | y^i_{obj}\right)$ as

$$p\left(x^i | y^i_{obj}\right) = \frac{p\left(y^i_{obj} | x^i\right) p(x^i)}{p\left(y^i_{obj}\right)} \propto p\left(y^i_{obj} | x^i\right) p(x^i) \tag{6.12}$$

Assuming each sample is independent, we can have the log likelihood of all samples as:

$$L' = \log \prod_i p\left(x^i | y^i_{obj}\right) \tag{6.13}$$

Expanding L' we have:

$$L' = \sum_i \log p\left(y^i_{obj} | x^i\right) + \sum_i \log p(x^i) = NL_{MSE} + \sum_i \log p(x^i) \tag{6.14}$$

Equation (6.14) contains an additional term and essentially tries to maximize the $p(x)$ through sampling. Unfortunately, calculating $p(x)$ requires the inversion of the generator function G. The inversion does not always exist because the generator is mostly like a surjective function. Besides, this method requires massive sampling to accurately estimate $p(x)$, which suffers from large variance for problems with a high dimensional design space. Hence, such a method does not provide a practical algorithm. However, if we have prior data guiding us to where in the optimization space we should look at, we can approximate the likelihood in a faster manner.

Suppose the prior data has a distribution $q(x)$, we can enforce the distribution of $p(x)$ to be similar to the prior distribution $q(x)$. To achieve that, we replace the distribution $p(x)$ in Eq. (6.14) with the Jensen–Shannon (JS) divergence of the two distributions. The revised loss function can be written as:

$$L_G = NL_{MSE} + \lambda D_{JS}(p(x) \| q(x)) \tag{6.15}$$

Expanding $D_{JS}(p\|q) = D_{KL}\left(p \| \frac{p+q}{2}\right) + D_{KL}\left(q \| \frac{p+q}{2}\right)$, we have

$$L_G = NL_{MSE} + \lambda \int \left(p \log \frac{p}{p+q} + q \log \frac{q}{p+q}\right) dx + const. \tag{6.16}$$

Define function $D(x) = \frac{q(x)}{p(x)+q(x)}$, the above function changes to:

$$\begin{aligned} L_G &= L_{MSE} + \lambda \int (p \log(1 - D(x)) + q \log D(x))\, dx + const. \\ &= L_{MSE} + \lambda \mathbb{E}_{x \sim p(x)}[1 - D(x)] + \lambda \mathbb{E}_{x \sim q(x)}[D(x)] + const. \end{aligned} \tag{6.17}$$

The gradient of L is:

$$\nabla_w L_G = \nabla_w L_{MSE} - \lambda \mathbb{E}_{x \sim p(x)}[\nabla_w D(x)] = \nabla_w L_{MSE} - \lambda \mathbb{E}_{z \sim p(z)} \left[\frac{\partial D(x)}{\partial x} \frac{\partial x}{\partial w} \right] \tag{6.18}$$

It looks like we have solved the problem. However, we still need to derive the explicit form of $p(x)$ in Eq. (6.18). Although it is possible to construct an invertible neural network [7], we can avoid building the special network by parametrizing the function $D_\theta(x)$ with some weights θ, and learning the optimal θ such that the distance between $p(x)$ and $q(x)$ can be measured. This is the basic idea of a generative adversarial network (GAN) [8]. In the training of a GAN, we define the loss function for $D_\theta(x)$:

$$L_D = -\mathbb{E}_{x \sim p(x)}[1 - D_\theta(x)] - \mathbb{E}_{x \sim q(x)}[D_\theta(x)] \tag{6.19}$$

Combined with the loss of the generator (Eq. (6.19)), we can approximate the prior distribution $q(x)$ while optimizing the parameters x in the training process of the GAN. Detailed training methods of a GAN are introduced in Ref. [9, 10].

6.1.2.5 Discussion

From the above derivation, we can see that the generative design strategy has two benefits over gradient-based optimization: (i) it optimizes the performances of a population of designs, and thus the global optimum is more likely to be found, (ii) it considers the prior distribution of design parameters, which can help us speed up the design. In some cases, if the design space is highly dimensional, generative models also compress the design space into a lower dimensional latent space. Optimization on the low dimensional space can be much faster. Here, we only take GANs as an example to illustrate why generative designs should be considered. It is the same with other generative models. In photonic optimization, in addition to the generative design strategy illustrated above, we can utilize generative models to capture the distribution of datasets to provide insights on design, or to perform dimensionality reduction to simplify the optimization. Regardless of what design strategy is utilized, a generative model usually benefits the optimization because it can lead to global optimum and speed up the optimization. These benefits stand out prominently for the design with parameters in a very high dimensional space.

6.2 Generative Model for Inverse Design of Metasurfaces

6.2.1 Generative Design Strategy for Metasurfaces

In this section, we present metasurfaces as the target design with generative strategy [11–13]. Directly using the strategy illustrated in Section 6.1, we can explore random topologies of metasurfaces for desired optical responses [9]. In this section, our aim is to identify the topology of a metasurface unit structure given the desired optical responses. As a representative and generalizable case study, we applied a generative design strategy to the design of metasurfaces with prescribed spectral behavior under linearly polarized illumination. The general unit cell

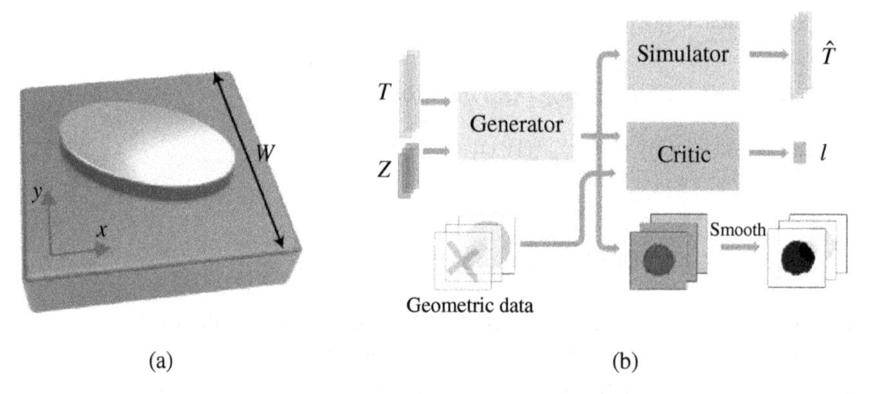

Figure 6.2 Generative model for the design of metasurfaces. (a) The unit cell of the metasurface in the inverse design problem. (b) Network architecture of the generative model for photonic design.

of the metasurface used in this model is shown in Figure 6.2a, which has a single layered gold pattern in a square lattice, situated on a glass substrate. Other common parameters include a lattice constant of $w = 340$ nm and a thickness of the gold structure set to $d = 50$ nm. Throughout the training, the unit cell structure is represented as a binary image of 64×64 pixels, in which 1 stands for gold and 0 for void (air). Each transmission spectrum T_{ij} is represented as a 44-entry vector with equal frequency intervals.

The network architecture of the generative design strategy is illustrated Figure 6.2b. We divide the network into three parts: a simulator (S), a generator (G), and a critic (D). The primary goal is to train an overfitted generator, which produces metasurface patterns in response to given input spectra T such that the Euclidean distance between the spectra of the generated pattern T' and the input spectra T is minimized. The simulator is a pretrained model with fixed weights, serving as a surrogate model. The training data for the generator is made up of several geometric classes such as circles, arcs, crosses, ellipses, rectangles, and so forth. These geometric data specify a region where the topology of a meta structure is likely to appear. The generator learns the distribution of these data and tries to propose valid candidates for the optimization.

The network architecture is capable of generating metasurface patterns in response to an arbitrarily input, on-demand set of optical spectra, whose T_{ij} components and frequency range of interest are defined by the user. At the input of the generator, we specify a set of transmittance spectra T_{ij} from 170 to 600 THz, which corresponds to a wavelength range of 500 nm to 1.8 μm. In the following discussion, we define a set of patterns to be used as a test set and denote this test set as s, the spectra of each s as T. Once these spectra are passed through the network architecture, a pattern is retrieved which we denote as s' in correspondence to s. In verification, the generated set s' is FEM-simulated with the spectra defined as T'. Unless mentioned otherwise, in the following experiments we set the number of patterns being parallelly searched at each run to be 40 and the total iterations of the training to be 50,000. For each target spectrum, valid patterns may occur at different stages of the training process whenever the losses of the simulator and critic are both reasonably small.

6.2.2 Model Validation

To illustrate the overall competence of the generative framework, we use the spectra, T, of randomly selected test samples from each geometric class as the input and allow the network to

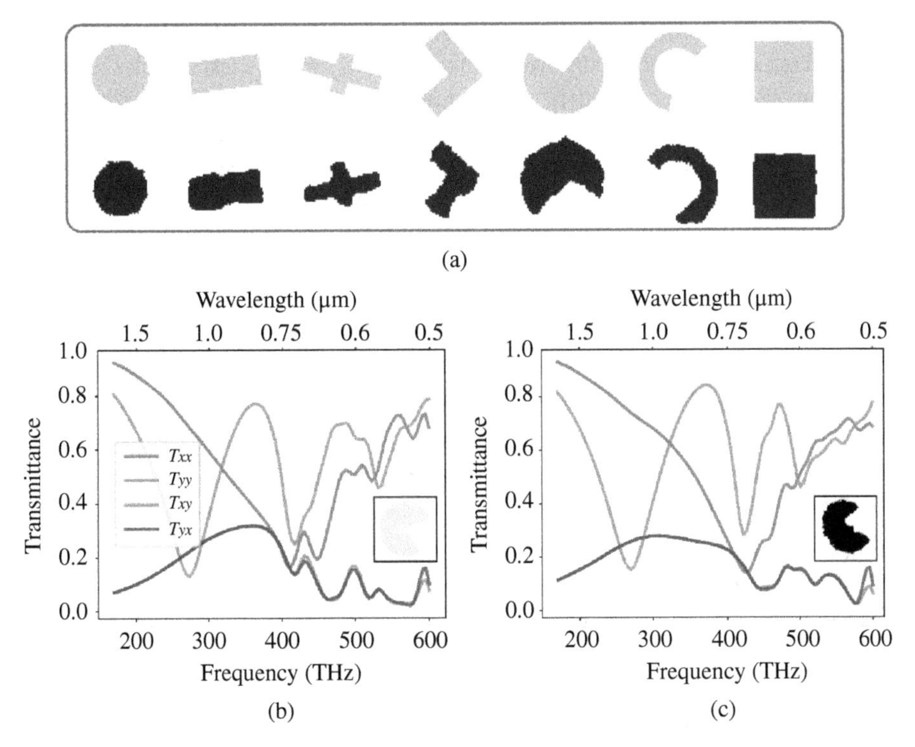

Figure 6.3 Generating patterns with mixed classes of geometric data. (a) Test patterns and the corresponding generated patterns. (b and c) Transmittance spectra of a test pattern (b) and its corresponding generated pattern (c), with the pattern depicted as the inset in each panel.

seek proper patterns based on these spectra. In this situation, we ensure the existence of solutions by using actual spectra of real patterns as the "target" or input. In each test, the critic is fed with 1000 data points (i.e. geometrical shapes) that are randomly generated from the same geometric class. Figure 6.3a shows representative samples from such experiments for each class of the geometry. The first row depicts the test samples s, and the second row shows the corresponding samples s' generated by the network. The geometric patterns in each pair of the two rows agree very well, partially because the full spectra input T_{ij} to the generator substantially narrows down the possible solutions. Because the generator does not receive any direct information on the geometry for the input spectra, it may uncover equivalent patterns s' that are different from the test structures s while yielding the same spectral behavior. Such examples can be found in the cross, sector, and arc cases in Figure 6.3a, where the discovered patterns s' are mirror-flipped counterparts of s with the same optical responses under linearly polarized illuminations. Unit cells of the metasurfaces are placed as insets. Comparing the two sets allows us to conclude that the network has successfully identified the correct structure to replicate the spectra with only minor deviations. We also note that the geometric data fed into the critic network does not necessarily contain the right shape of the resultant solutions. If the type of the right pattern s for the required spectra is contained in the geometric data at the input, the nature of the GAN will lead to a decent chance of identifying the structure s as a proper candidate; otherwise, the critic will guide the generator to produce patterns with geometric features similar to those of the right geometry.

To exemplify the generality and versatility of the constructed network, we feed into the critic mixed data from all classes of geometry with over 8000 data points, and the test samples are also randomly selected from all possible geometries. An example of a discovered structure with mixed geometric input is presented in Figure 6.3b,c. The Figure 6.3 plot the FEM-simulated spectra of the test pattern s and the generated pattern s' in response to the spectral demand, respectively. Although no specific class of geometry is indicated during the training, the generator is able to reach a pattern, s', that not only geometrically resembles s but more importantly possesses transmittance spectra, T', nearly identical to that of T. In general, if the input geometric data contains more than one topology that satisfies the spectral demand, the network may generate some or all of them in a probabilistic manner. Moreover, by changing the distribution of the geometric data, the user may achieve diverse solutions in response to the same spectral request, thereby mitigating the potential degeneracy.

6.2.3 On-demand Design Results

In practice, the desired spectra at the input are user defined, and the existence of solutions is not guaranteed. This is particularly true when certain constraints are applied to the metasurface design. For instance, in the present study parameters such as the materials used, unit cell size, and thickness of the patterned layer are all predefined in the training data. Nevertheless, when the simulator is sufficiently robust, the network is still able to unearth the best possible pattern that yields spectra T' with minimized deviation from the input spectra T. To demonstrate this feature, here we design a metasurface with the desired, user-defined spectra behavior shown in Figure 6.4a: (i) T_{xx} and T_{yy} are two Gaussian-like resonances with randomly chosen mean μ, variance σ, and amplitude a, and (ii) T_{xy} and T_{yx} are zero. The generated pattern along with its spectra T' is shown in Figure 6.4b. Although there exists no exact solution to spectral demand described above, the network eventually generates patterns whose spectra share common features with the input spectra including the resonance frequency, the spectral bandwidth, and the transmission magnitude.

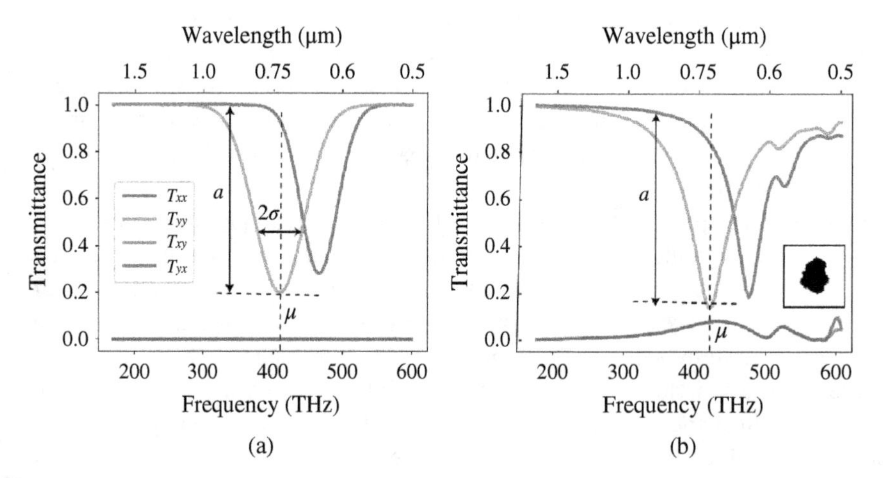

Figure 6.4 Inverse design of metasurface with on-demand spectra. (a) Desired transmittance spectra. (b) The designed unit cell and its spectra.

6.3 Gradient-Free Optimization with Generative Model

In the previous metasurface design, we have leveraged gradient-based optimization methods and generative models to optimize the topology of a metasurface structure. In fact, we can also use gradient-free optimization algorithms in a generative design strategy. In such a case, generative models are used to learn a prior distribution of design examples and exclude designs that are unlikely to appear in actual scenarios. Gradient-free optimizers iteratively refine the latent vectors of the generative model based on the evaluated performance of the designs. If the dimensionality of the design space is high, generative models also serve as a dimensionality reduction tool.

6.3.1 Gradient-Free Optimization Algorithms

Gradient-free optimization is a family of optimization algorithms that do not require derivative information. They are often used in design problems when the simulation time is relatively short or gradients are difficult to calculate. Some gradient-free examples include Bayesian optimization (BO) [14], evolution strategy (ES) [15], particle swarm optimization (PSO) [16], genetic algorithms [17, 18], random search [19], and many more. Due to the lack of gradient information, these algorithms are expected to require many more evaluation iterations than gradient-based algorithms. However, the intrinsically stochastic properties of gradient-free methods lead to a higher likelihood of converging to global optimal solutions. The major disadvantage of gradient-free methods in photonic optimization is its dimensionality problem – a high dimensionality can require an exponential increase of simulations. Generative models, due to the aforementioned characteristics, are a remedy to this problem. Coupling generative models with gradient-free method can both speed up the optimization process and increase the likelihood of identifying the global optimal solutions.

6.3.2 Evolution Strategy with Generative Parametrization

In this section, we present a generative design strategy utilizing an ES, a gradient-free optimization algorithm [20]. The design goal is the same as the one presented in Section 6.2. We will see that coupling gradient-free optimization and generative models can yield equivalently good results as the gradient-based generative design strategy. It is noteworthy that the choice of the gradient-free optimization algorithm is very flexible. BO and PSO can also produce similar design candidates.

6.3.2.1 Generator from VAE

In this design strategy, there is no fundamental difference between the type of generative model. In a gradient-free generative design strategy, a generator in the generative model is trained to produce topologies similar to a prior dataset. Here we try to use another generative model, variational autoencoder (VAE), to assist the fast, global optimization.

A VAE is composed of an encoder and a decoder (G). The decoder is the generator we need in optimization. The encoder transforms the input geometric data s represented by binary images into mean vectors μ and standard deviation vectors σ. Latent vectors v are sampled from a Gaussian distribution $v \sim N(\mu, \sigma)$. The decoder G then reconstructs v back to the geometric information. In doing so, pattern topologies with similar features are mapped to the same region in the latent space, so that the decoded patterns with similar features can be continuously varied by perturbing the latent vectors. Informally, the objective of the training is to optimize the weights of the decoder

G so that the randomly sampled v can be mapped to topologies that are similar to the training dataset. To enforce global continuity of the latent space, additional regularization is needed during the training of the VAE. A more detailed training process can be found in Ref. [5, 21].

6.3.2.2 Evolution Strategy

The gradient-free optimization algorithm we chose for the design problem is ES. Figure 6.5 illustrates the flowchart of the ES framework consolidating a deep generative model. In the initialization stage, a population of all individuals is constructed. Each individual contains two random vectors – a latent vector v and a mutation strength m. The mutation strength m is a vector with the same dimension as v and will be used to randomly vary the value of v in a later step. After initialization, all individuals are reconstructed to their corresponding patterns s through the generator G. The recovered structures are then simulated through either a physical simulator or a neural network approximator. In order to avoid repeated simulations and expedite the optimization process, we adopt a neural network (S) trained with simulated data to predict the physical responses $\hat{q} = S(s)$ of the structures. With the design objective q and the simulated result $\hat{q}$, we calculate a fitness score r, representing the agreement of q and $\hat{q}$, for each individual. The value of the fitness score is calculated through a fitness function $r = F(\hat{q},q)$, where the form of F is designed by the user based on the goal of the optimization. For example, if q is a vector representing the spectral responses, the fitness function can be defined as $F(\hat{q},q) = \|q-\hat{q}\|$.

After the evaluation of fitness scores r, the algorithm enters the optimization stage that emulates natural selection. In detail, μ best individuals with the highest scores are selected as parents (P), and their reproduction leads to the next generation of children (Q) with a total number of individuals λ. A latent vector v_{child} of a child is created from two parent vectors v_1 and v_2 via two manners, (i) crossover, which randomly exchanges segments of vectors of the two parents, and (ii) interpolation, which linearly interpolates the vectors of the two parents as $v_{child} = \alpha v_1+(1-\alpha)v_2$ with random weight $\alpha\in[0, 1]$. The mutation strength of a child m_{child} is derived in the same manner as v_{child}.

The λ reproduced individuals together with the μ parents constructs the next generation, which is known as the $(\mu + \lambda)$ strategy. In the mutation stage, noises sampled from the normal distributions $\delta_v = N(0, m)$ are added to the latent vector as $v + \delta_v$ to avoid stagnancy of the evolution. The mutated population forms the next generation of the population in the optimization process. The whole algorithm then performs the iterative process of reconstruction, simulation, evaluation, selection, reproduction, and mutation until a satisfactory individual with an optimized score is identified. During the optimization, we occasionally remove individuals v that stay unchanged for

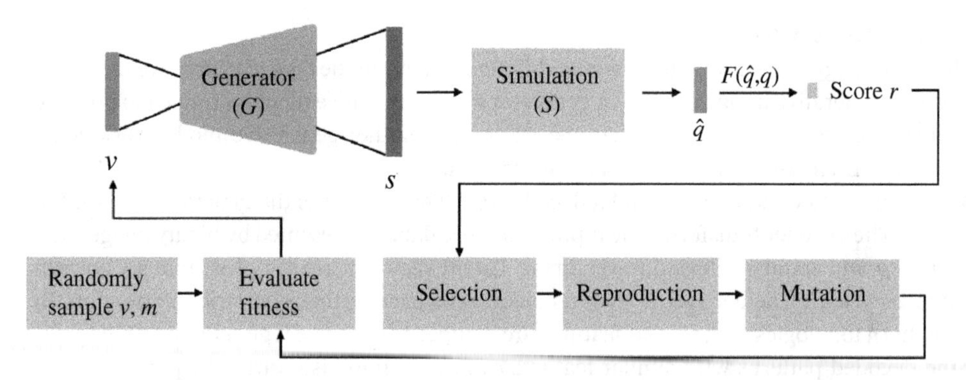

Figure 6.5 Flowchart of the evolutionary strategy (ES) framework. The developed approach employs an ES algorithm with a loop of evaluation of fitness, selection, reproduction, and mutation.

a few iterations to reduce the unnecessarily repeated simulation and speed up the convergence. The algorithm stops once the design criteria are achieved or the maximum iteration is reached.

6.3.2.3 Model Validation

We evaluate the performance of the ES-based generative design strategy with design examples of meta structures with the same configuration in Section 6.2. Figure 6.6 presents six examples of the input spectra T (solid lines) and FEM simulated spectra T' (dashed lines) in conjunction with the corresponding patterns s and s' in the unit cell. The generative approach captures the prominent features of the input transmittance and generates structures with minimal discrepancy between T and T'. As the ES algorithm only considers the distance between T and $\widehat{T}$, any shapes, similar to or distinct from s, can be identified as an optimal solution as long as the score $F(T, \widehat{T})$ is maximized. This feature also allows us to unearth different possible topologies in response to a given T in multiple runs.

6.3.2.4 On-demand Design Results

We further test the capacity of the proposed framework for the inverse design of photonic structures, based on arbitrary, user-defined input spectra. Four sets of experiments for such on-demand design have been performed, and the results are shown in Figure 6.7. In Figure 6.7, the input spectra T_{xx} and T_{yy} are set to be two randomly chosen, Gaussian-shaped curves, and T_{xy} and T_{yx} are set to zero throughout the frequency range of interest. The results of the FEM-simulated spectra of the identified patterns (dashed lines) faithfully match the input spectra in terms of both the spectral location and the bandwidth. It is worth noting that the independent manipulation of T_{xx}, T_{yy}, T_{xy}, T_{yx} is not readily achievable by conventional human design approaches, while our framework is able to accomplish simultaneous control of T_{xx}, T_{yy} and suppression of T_{xy} and T_{yx}. In the on-demand design cases, we note the accuracies cannot achieve 95% as in the previous

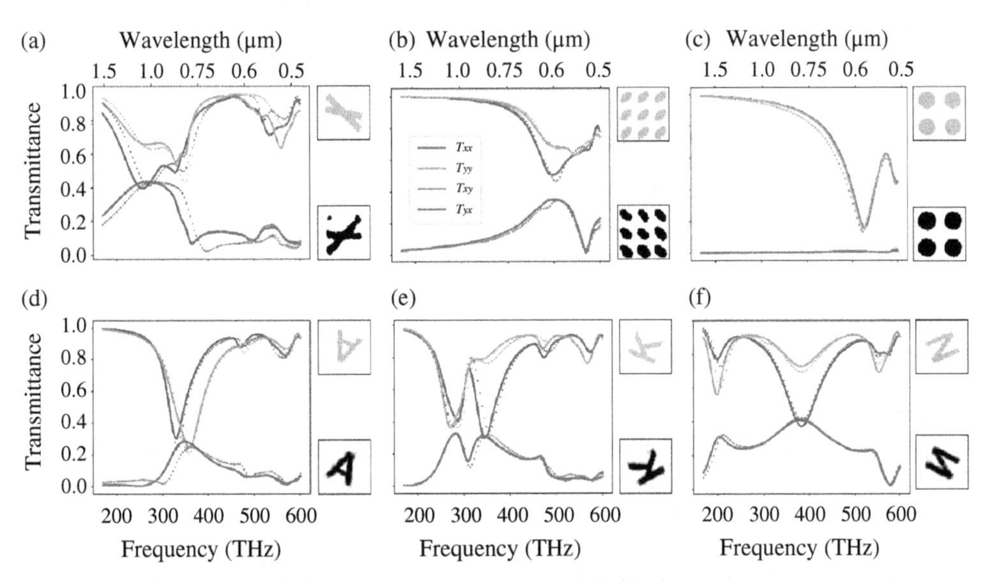

Figure 6.6 Test results of designed photonic structures through ES algorithm and the generative model. (a–c) Examples of test structures with geometric patterns. (d–f) Examples of test structures with letter-shaped patterns. The test patterns and the designed patterns are presented on the upper right and lower right of each panel, respectively. The desired spectra are shown in solid lines, while the simulated spectra of designed patterns are in dashed lines.

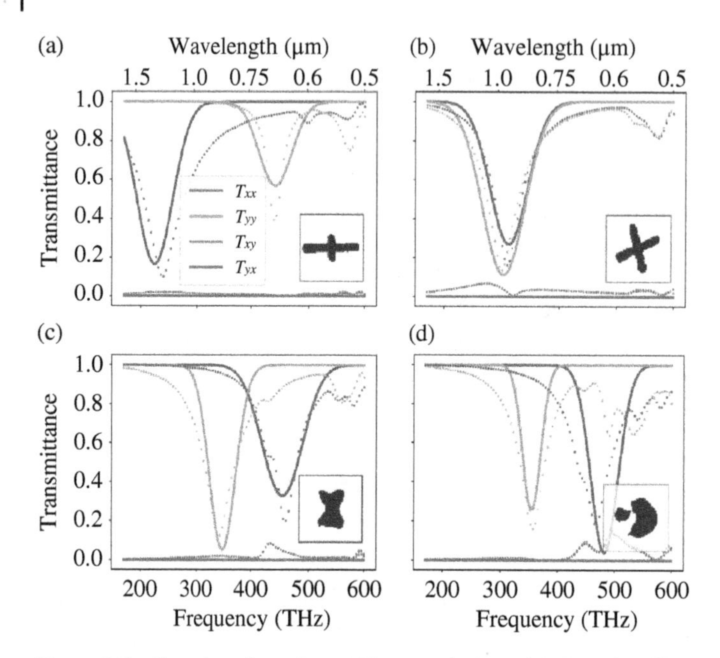

Figure 6.7 Samples of on-demand inverse design with Gaussian-like target spectra. (a and b) Cross-shaped designs with target and simulated spectra. (c and d) Freeform-shaped designs with desired and simulated spectra. The target spectra are shown in solid lines, while the simulated spectra of designed patterns are in dashed lines.

test/validation cases. This is because the input spectra may not have exact corresponding photonic structures. However, as a global optimization algorithm, ES is able to avoid local minima unlike local optimizations and identify a globally optimal structure with significantly higher likelihood. On the other hand, relying on the stochasticity of ES, a global optimum can be confirmed through multiple runs of the optimization with random initialization.

6.3.3 Cooperative Coevolution and Generative Parametrization

The above design strategy only identifies one meta-structure. In more complicated applications, we often need to design a unit cell composed with multiple meta-structures. We name the larger unit cell as meta-molecule and the meta-structures that make up the larger cell as meta-atoms. Due to the increase of design parameters, optimizing the whole meta-molecule at once is not realistic. Here we present a greedy evolutionary optimization strategy call cooperative coevolution (CC) as the gradient-free optimization algorithm to resolve this problem [22].

6.3.3.1 Cooperative Coevolution

To identify a meta-molecule with multiple meta-atoms while avoiding exponentially increased simulation time for the entire molecule, we introduce a CC into the generative design framework. The CC algorithm looks like an ES algorithm with minor modification to allow the joint optimization of multiple classes of populations. Specifically, we divide the design task of the whole meta-molecule into the design of independent meta-atoms, and iteratively optimize the latent vector of each meta-atom. The algorithm treats the population of latent vectors v of meta-atoms as a species. In each iteration of the evolution, the CC algorithm picks one species for update while assuming all other species are optimized. The algorithm then decodes the latent vectors in the species through the trained generator, evaluates the fitness scores of the decoded meta-atoms, and optimizes the species by performing loops of bio-inspired operations including selection, reproduction, and mutation. These operation steps are also adopted in conventional evolution strategies. The CC algorithm performs the above steps and iterates in a round-robin fashion for all species (latent vectors of meta-atoms) until all their fitness scores reach the desired criteria.

6.3.3.2 Diatomic Polarizer

We first leverage the generative design framework to design metasurfaces composed of "diatomic" nanostructures for polarization conversion. A diatomic meta-molecule by our definition is a meta-molecule consisting of two distinct meta-atoms. In such meta-molecules, the coupling between adjacent meta-atoms is sufficiently weak, in which case we can approximate the scattered far-field light from the meta-molecule as the superposition of the waves separately scattered by the two atoms. Based on this assumption, we inversely design a series of diatomic meta-molecules that are able to convert linearly polarized incident light into prescribed polarization states at the transmitted side. Figure 6.8a presents the designed meta-molecules that rotate the polarization angle of x-polarized incident light to 15°, 30°, 45°, and 60°, as well as the design that converts the incident light to a circularly polarized (CP) light, operated at a wavelength of 800 nm. The plots of the converted polarization, computed by the network simulator and FEM full-wave simulation, are illustrated in Figure 6.8b,c, respectively. As our designed devices are ultrathin metallic metasurfaces composed of discrete meta-atoms, they inevitably suffer from compromised performance. Circumventing this problem requires the metasurfaces to be multi-layered or dielectric.

6.3.3.3 Gradient Metasurface

We further utilize our framework to inversely design metasurfaces with a gradient phase distribution. Such metasurfaces are conventional examples of metasurfaces for diverse functionalities such as the generalized Snell's law, beam steering, and meta-holography. The meta-molecules in our gradient metasurfaces for wavefront control and polarization manipulation are composed of eight meta-atoms. In such a meta-molecule, every two adjacent atoms should scatter light with an equal amplitude and a constant phase difference $\Delta\phi$. This requirement cannot be formulated into a single objective function to be minimized by traditional optimization techniques. To circumvent this problem, we design a fitness function of a meta-atom only associated with itself and its adjacent neighbor. For example, to optimize the ith meta-atom S_i in the meta-molecule, we define its fitness function $F(S_i, S_{i-1})$ which is only locally associated with the $(i-1)^{\text{th}}$ meta-atom. As a greedy algorithm, the CC does not guarantee optimal solutions, but empirically, it converges to a solution with the desired amplitude requirement and phase distribution in our experiments.

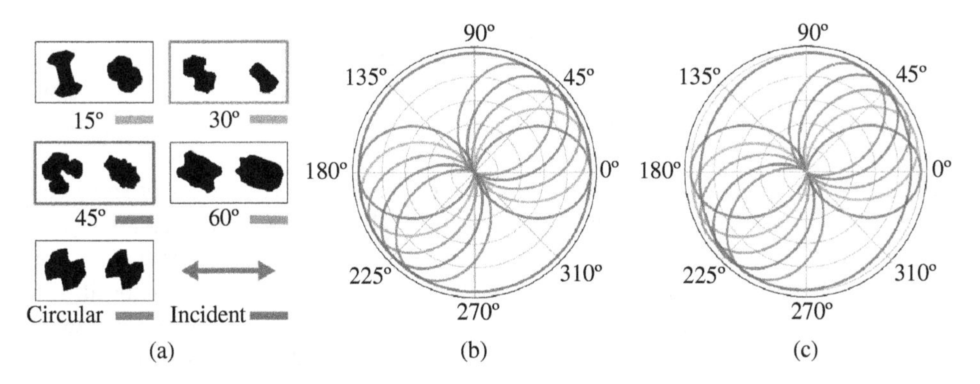

Figure 6.8 Inversely designed diatomic meta-molecules for polarization manipulation. (a) Designed meta-molecules. (b) Converted polarization states predicted by the network simulator. (c) FEM simulations of the converted polarization states.

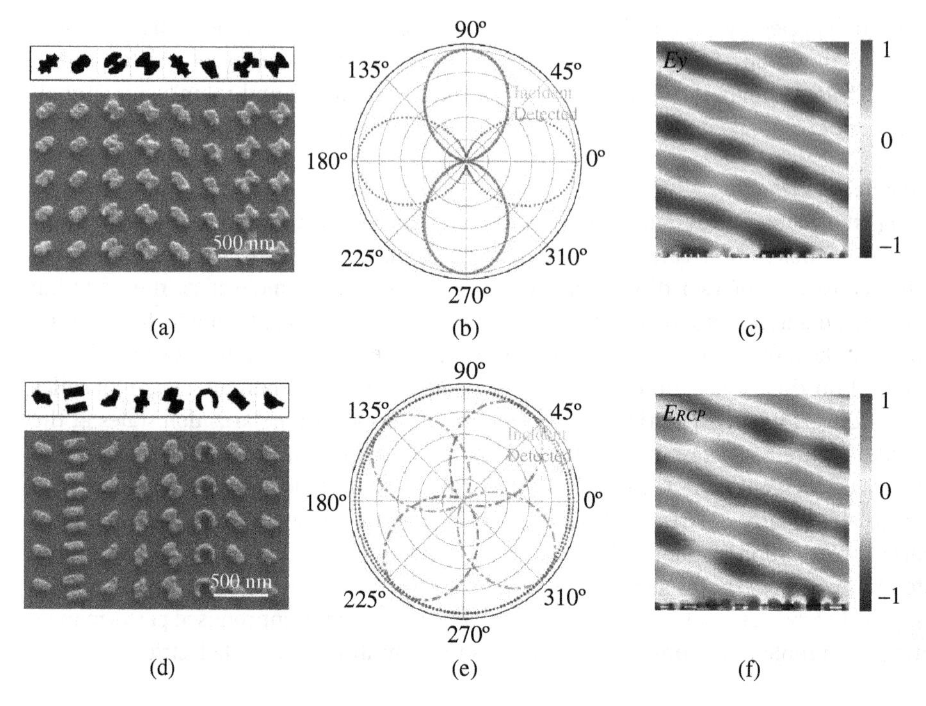

Figure 6.9 Inversely designed meta-molecules for gradient metasurfaces. (a-c) Gradient metasurface for linear polarized light, with (a) meta-molecule and SEM image (b) measured polarization states of the incident and deflected light (c) simulated electric field E_y. (d-f) Gradient metasurface for circular polarized light, with (a) meta-molecule and SEM image (b) measured polarization states of the incident and deflected light (c) simulated electric field E_{RCP}. Source: (a and d) Zhaocheng Liu et al. (2020)/John Wiley & Sons.

We apply the generative framework to the design of an eight-atom meta-molecule, which can convert x-polarized incident light to its y-polarized counterpart and deflect the cross-polarized light to a specific angle. We chose the central operating wavelength of the metasurface as 800 nm to facilitate further experimental verifications. Figure 6.9a shows the unit cell of the identified metasurface, together with the SEM image of the corresponding fabricated sample. We conducted polarization analysis of the diffracted light on the transmitted side, as illustrated in Figure 6.9b where the dashed line indicates the measured polarization state. As compared to the incident polarization (dash-dotted line), the converted polarization is perfectly orthogonal to the incident one as demanded by the design objective. The measured transmittance of the deflected light is 10.9%. Figure 6.9c presents the FEM simulated distribution of the electric field E_y emerging from the metasurface, which further illustrates the accurate phase gradient and amplitude distribution induced by the eight meta-atoms.

Another example we present here is the inverse design and experimental verification of a gradient metasurface for CP light. The envisioned metasurface should be able to convert left circularly polarized (LCP) light to its cross-polarization (RCP) and bring about an additional phase gradient to the converted portion of light for beam steering. At the design wavelength of 800 nm, one of the identified solutions and its fabricated sample are shown in Figure 6.9d. We characterized the polarization states of the incident and converted light in the polar graph as in Figure 6.9e, where the dash-dotted and dashed lines indicate the measured intensity with and without a quarter waveplate, respectively. The measured transmittance of the RCP light is 9.89%. The orthogonality of the polarizations measured with a quarter waveplate unambiguously proves the device flips the circular polarization from LCP to RCP. The simulated electric field E_{RCP} (Figure 6.9f) under the LCP incidence confirms the correct polarization, phase gradient, and amplitude distribution of the design. Unlike traditional circular gradient metasurfaces, where only geometric phase contributes to the phase distribution, the phase gradient of the new design is jointly induced by the geometric phase φ_{geo} and the material-induced phase delay φ_{mat}. When the incidence is LCP, φ_{geo} and φ_{mat} constructively contribute to the desired phase gradient; however, under the RCP incidence, φ_{geo} and φ_{mat} destructively contribute to the phase profile, causing asymmetrical behavior of the metasurface Since the generative framework does not consider the particular physical mechanism during the inverse design process, nor does it require predefined constraints or human intervention, it tends to discover metaphotonic devices and novel photonic phenomena with complex light-matter interactions.

6.4 Design Large-Scale, Weakly Coupled System

In Section 6.3 we have presented a generative inverse design methodology for single layered photonics. In this section, we will show that generative models can assist the design of much more complicated meta-optic systems. Although metasurfaces can present unconventional light controllability, they are limited by various downsides such as low efficiency and limited functionalities. Such single-layered devices can hardly be multifunctional, since the optical responses of a unit cell at different polarizations, frequencies, and spatial positions are correlated with each other in an entangled manner. Achieving complicated and multiple functions calls for designing meta-optic systems with multiple layers of metasurfaces. A meta-optic system can be multilayered, multifunctional, large-scale, and nonperiodic, which introduces exponential number of degrees of freedom. We will show that, under appropriate approximation, generative design strategies are able to assist the design of such a large-scale optimization problem. As shown in Figure 6.10,

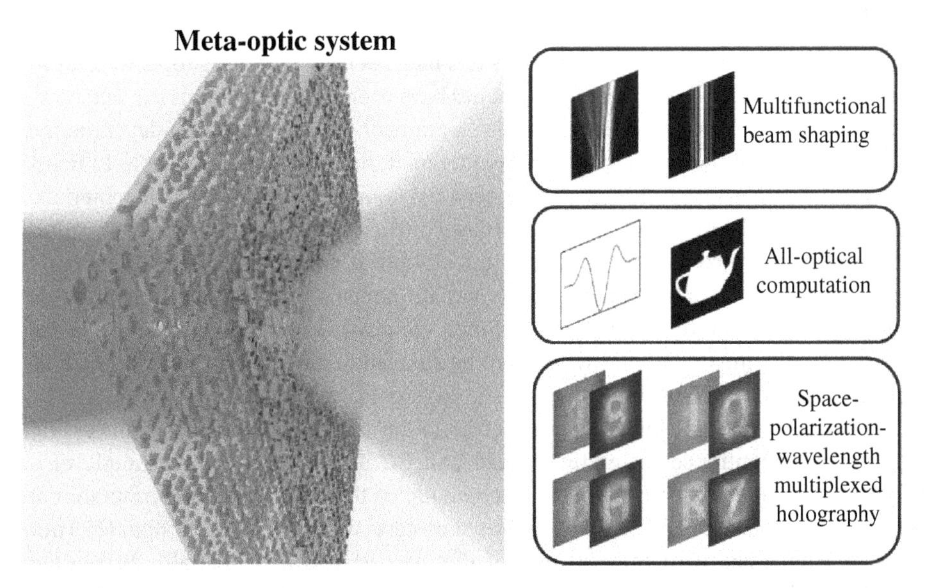

Figure 6.10 Generative strategy for the design of large-scale meta-optic system. A meta-optic system can be highly complicated, and be used in multifunctional beam shaping, all-optical computation, as well as multiplexed holography, etc.

A designed meta-optic system can achieve novel functionality, such as multifunctional beam shaping, all-optical computation, as well as highly multiplexed holography [23].

6.4.1 Weak Coupling Approximation

The generative optimization strategy we used is similar to the one presented in Section 3.2. In the design problem of single-layered metasurface, we have assumed that the adjacent meta-atoms have weak coupling effects. This allows us to fast approximate the optical responses during optimization. It should be noted that, if the coupling effect is a concern in a design problem, we can choose a larger unit cell with multiple meta-atoms as the building block in the optimization. Similar approaches have been applied in large-scale metalens inverse design [24, 25].

Since we aim at designing a multilayer structure, we need to find a fast method to approximate the stacked structure. The approach we adopted is (i) retrieving the amplitude/phase responses of a single-layered structure and (ii) inferring the performance of the layered structure from the single-layered elements using the wave matrix method [26]. The input–output relation of the layered structure can be modeled as:

$$\begin{pmatrix} E_{out}^{+} \\ E_{out}^{-} \end{pmatrix} = M_{inter}^{(1)} M_{delay}^{(2)} M_{inter}^{(2)} \dots M_{delay}^{(n-1)} M_{inter}^{(n-1)} \begin{pmatrix} E_{in}^{+} \\ E_{in}^{-} \end{pmatrix} \tag{6.20}$$

where $M_{\text{inter}}^{(i)}$ represents the wave matrix of each layer of the structure, and $M_{\text{delay}}^{(i)}$ is the wave matrix for the spacer between adjacent layers. This approximation requires no coupling between layers. With the fast evaluation method using various approximations, we can utilize the generative design strategies with CC optimization to design multilayered meta-optic systems. Due to the linear time complexity of the CC algorithm, we can design very large-scale meta-systems with ultra-high degree of freedom, such as up to 10^6.

6.4.2 Analog Differentiator

Calculation and signal-processing with metasurfaces has long been an intriguing topic to achieve all-optical computation [27–29]. A number of approaches have been proposed on this subject; however, some require design of a metasurface and two bulky gradient-index structures independently, and others are based on multilayered homogeneous metamaterials but only applicable to math operations with even symmetry and the design may need too many layers or unrealistic materials. Inspired by the theoretical background of the multilayer metamaterial method, we can exploit the meta-optic system to realize certain functions for all-optical computation. The meta-optic system is an ideal candidate to achieve the spatially variant Green's function, which represents wavenumber-dependent transmittance in k-space. Since the spatially distributed supercells are not necessarily symmetric, the meta-optic system can be adapted to any linear operation, such as derivation, integration, and convolution.

We present a meta-optic system for all-optical computation and all-optical signal processing designed with the aforementioned generative model and CC algorithm. In our example of a meta-system for second-order differentiation, the schematic of the three-layered computational meta-optic system is illustrated in Figure 6.11a. In this showcase, when the incident light (659 nm in the polymer) is y-polarized with the real part of the electric field as a spatially distributed function along the x-axis, the real part of the y-polarized output wave will be proportional to the second-order derivative of the input function. Figure 6.11b depicts the structure of the designed meta-optic system. The system has three layers, each of which has 25 spatially variant unit cells

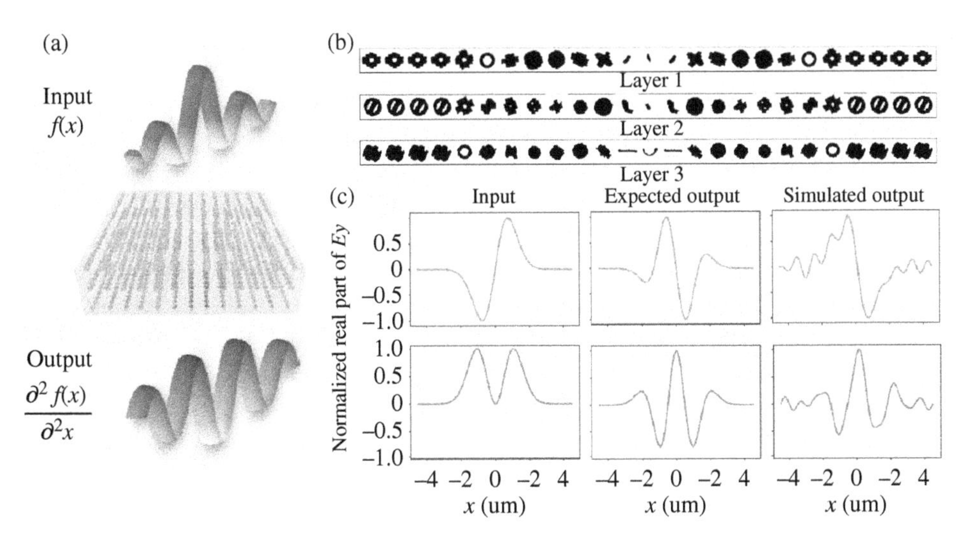

Figure 6.11 A designed second-order differentiator. (a) The functioning schematic of the analog differentiator. (b) The structure of the differentiator. (c) The simulated performance of the differentiator.

along the x direction and is periodic along the y direction, with a spacer of 200 nm between adjacent layers. We define Layer 1 on the input side, and Layer 3 for output. To evaluate the performance of the differentiator, we tested it with two different inputs. The two images on the left column of Figure 6.11c present the input functions, while the middle and the right columns represent the corresponding target outputs and the simulated outputs of the meta-optic system, respectively. The great similarities between the targets and the simulated results of the device validate the performance of the differentiator. Since the meta-optic system consists of only 25 unit cells along the x direction for the representation and operation of a continuous function, the computational resolution of the differentiation is limited due to the discretization in the spatial domain. Increasing the number of unit cells of the metasurface is expected to substantially suppress the ripples in the simulated output. To take one step forward, if a two-dimensional image is the input, the differentiator will detect the second-order edges of the image along the x direction. While if the differentiator is designed to be spatially variant in two dimensions, it will be able to distinguish all the second-order edges of an input image.

6.4.3 Multiplexed Hologram

Next, we present a multilayer multifunctional meta-optic system that can achieve space-polarization-wavelength multiplexed hologram. This meta-hologram functions at wavelengths (in the polymer) of both 562 and 659 nm, in both x and y polarizations, with nine operational positions along the propagation, and projects a total of 36 holographic images: numerical digits 0–9 and capital letters A–Z. Metasurface has been exploited as one of the most promising media to achieve holography, with high imaging quality and ultrathin thickness [30]. Most of such optical meta-holograms are static and monofunctional, which leads to limited information-storing capacity of the holography. Multiplexed metasurfaces may open up the vistas to encode an enormous amount of information into a single hologram. Some of the recent works have shown the possibility to realize a multiplexed hologram in spatial, polarization, or wavelength channel, and some have attempted two of the channels [31–33]. Here, we present a meta-hologram that utilizes all three channels, with no upper limit on the total amount of displayed images [34].

The hologram is neither phase-only nor amplitude-only, but with mixed phase and amplitude information carried in one integrated system. It consists of three layers, with an overall size of 2000 by 2000 supercells. We first virtually align all the objective images on axis and parallel to the hologram plane at different distances (100–740 μm from the meta-hologram, with 80 μm interval). Although we only care about the magnitude of displayed images, we still need to introduce random phase to the images in order to suppress the crosstalk between different image planes [35]. Next, we apply a Sommerfeld–Fresnel transformation (Huygens propagation) to obtain the amplitude and phase distribution on the hologram plane as derived from the image planes. The final amplitude and phase distribution is the weighted sum of the projections, where the weight is proportional to the square root of the distance from the meta-hologram to a particular image. The weight is introduced to ensure that each image can be recovered by the hologram with both high fidelity and sufficient brightness. To reduce the computational complexity, the objective field amplitude after the hologram is binarized into the values of 0.75 or 0 with a threshold of 0.3. The phase is also discretized into three values within the 2π period: $-\pi$, -1/3 π, and 1/3 π. Therefore, for each polarization at a specific wavelength, there will be four possible amplitude and phase combinations: 0, −0.75, 0.75∠−1/3 π, and 0.75∠1/3 π. Since the hologram is designed to work at two different wavelengths (562 and 659 nm) and two polarization states (x and y polarizations), there are totally $(4^2)^2 = 256$ amplitude and phase combinations, which means the same number of unique supercells should be used to constitute the meta-hologram. Once the 256 supercells are designed by the algorithm, any hologram that operates for both x and y polarizations at the two wavelengths can be designed in a matter of seconds, since different holograms are simply different maps of the

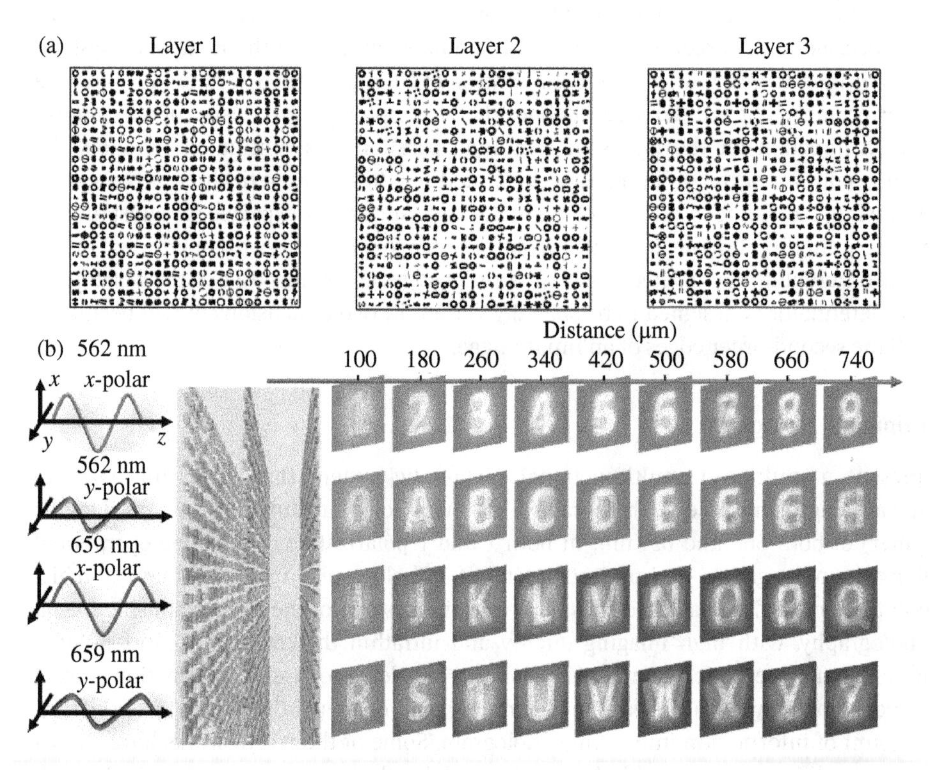

Figure 6.12 A designed space-polarization-wavelength multiplexed hologram. (a) A part of the structure of the meta-hologram. (b) Simulated performance of the hologram. The projected 36 images are the numerical digits 0 to 9, and capital letters A to Z.

same collection of supercells. In this example, the hologram consists of 2000×2000 supercells, and Figure 6.12a displays a small portion of the overall structure. We assume the incidence light illuminates from Layer 1, and the output light beyond Layer 3 automatically forms the projected images at the expected locations. Figure 6.12b shows the 36 simulated holographic images, and the color tones of the images are applied to distinguish the shorter (562 nm) and longer (659 nm) operating wavelengths. Most of the holographic images are formed with clear outlines and fine features. The image quality does not degrade at large distances, and the fidelity of images is satisfactory at each polarization and each wavelength. As for the scalability, the 256 unique supercells can be readily adapted to design any space-polarization-wavelength multiplexed holograms for the polarization states and operating wavelengths specified before. It is worth noting that our method can be extended to the design of holograms at more operating wavelengths and image positions. Adding one wavelength with both polarizations will increase the number of unique supercells by 16 times.

6.5 Auxiliary Methods for Generative Photonic Parametrization

The parametrization method using generative models can be further enhanced using a couple of methods that utilize physical information. In this section, we will include a few methods that can assist the photonic representations using generative models.

6.5.1 Level Set Method

The level set method is an approach to represent a topology of a photonic structure. In n-dimensional structure, a level set function φ is a $n + 1$ dimensional surface, and the level set can be defined as:

$$\{(x, y, z) | \varphi(x, y, z) = c\} \tag{6.21}$$

In topology optimization, the constant c is often chosen as 0. The benefit of using a level set method, as compared to image-based topology optimization, in photonic optimization is that the structure is always binarized and the number of parameters can be greatly reduced. Various level set methods, including bicubic interpolation [36], non-uniform rational basis [37], radial basis function interpolation [38], have been investigated in the design of photonic and mechanical structures. When proper constraints, the level set methods also allow the accurate control of fabrication constraints such as minimum line width, maximum curvature, and minimum feature size [39].

In the context of generative representation, we can always use a probabilistic model to generate the parameters of a level set function. Although it could lead to increased optimization iterations due to the vast trainable parameters, it helps the convergence toward global optimal solutions, as discussed in Section 6.1. The gradient of the parameters p of a level set can be calculated through boundary integral.

$$\frac{\partial}{\partial p} f(p) = \int_{\Omega} \frac{\partial f}{\partial \varphi} \frac{\partial \varphi}{\partial p} \frac{1}{|\nabla \varphi|} dx \tag{6.22}$$

where $f(p)$ is the loss function and Ω is the level set. The term $\partial f / \partial \varphi$ is the gradient from either an adjoint method or a surrogate model. $\partial \varphi / \partial p$ and $| \nabla \varphi|$ can be analytically derived or approximated depending on the parametrization method.

6.5.2 Fourier Level Set

In the design of diffractive components, especially low-index structures, the far field intensity distribution can be approximated by the Fourier transform of the output intensity/phase profile from the structure. Indeed, optimization of phase profiles in hologram displays is achieved through iterative Fourier transforms [40, 41]. However, for large-angle, non-paraxial diffractive components, simply using Fourier transforms cannot yield accurate far field intensity distributions. We need a physical optimization method that utilizes a simulation algorithm such as RCWA [42, 43]. One design approach is through optimizing a gray-scaled image that represents the structure using adjoint method. For diffractive problems, we could also explore level set methods.

Due to the relation between far-field intensity distributions and near-field phase profiles, we can parametrize the device with Fourier coefficients [44]. To ensure the device is fully binarized, we construct the 3D surface using Fourier series and take the zero-level set as its topology. Since Fourier series can produce complex values, we need to enforce some symmetries to the coefficients. For example, in a 2D design problem, the 3 by 3 coefficient matrix can have 9 DOFs with the following form:

$$v = \begin{bmatrix} v_1 + iv_2 & v_3 + iv_4 & v_5 + iv_6 \\ v_7 + iv_8 & v_9 & v_7 - iv_8 \\ v_5 - iv_6 & v_3 - iv_4 & v_1 - iv_2 \end{bmatrix} \tag{6.23}$$

where all v_i's are real numbers. To retrieve the structure in image representation, we first pad the matrix to the desired pixel resolutions, and then perform FFT to find the level set function, and finally take out the zero level set to represent the device topology. The gradient of v with respect to the loss function can be calculated like other level set methods.

When designing few-dot diffractive optical elements (DOE), we can directly utilize the Fourier level set method without the generative design pipeline. Suppose we need to design a non-paraxial DOE with 3 by 3 diffractive orders, we only need to optimize nine coefficients as in Eq. (6.23). With such small design parameters, gradient-free optimization methods such as BO and evolutionary optimization can efficiently identify the solution. Figure 6.13 shows four designs with desired far field intensity distributions with the Fourier level set and evolutionary algorithms. For complex design problems, gradients from the modeling method such as adjoint method can help reduce the optimization time.

6.5.3 Implicit Neural Representation

In the level set method, the level set function is a continuous function that, if represented in images, requires infinite resolution. Converting between continuous and image-based representations could be problematic in some situations. Usually, an expensive up-sampling is unavoidable when retrieving the 3D level set function from a pixelated image. Implicit neural representation (INR) is able to resolve this problem.

Unlike networks with image-based representation that transform the data in Cartesian coordinates, INR receives image coordinates and produces the values at the specified coordinate. This process is shown in Figure 6.14. Because the input is continuous coordinates, the output can have arbitrarily infinite resolution. Such a method is widely used in graphics, rendering, and other computer vision tasks [45]. In a level set representation, the output values, which compose the level set function, can be between −1 and 1. The topology of the device with infinite resolution is the 0-level set of the output.

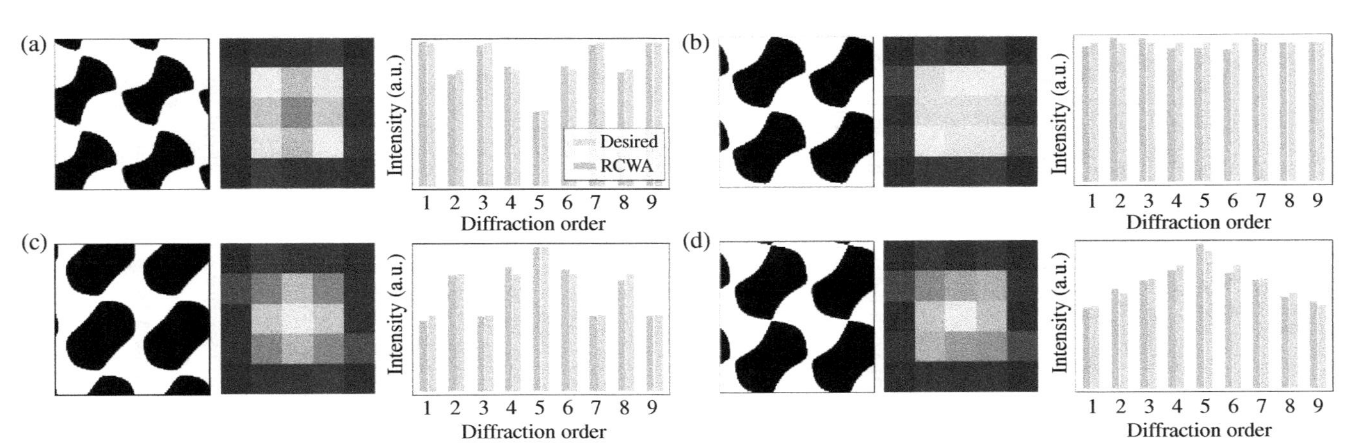

Figure 6.13 Inverse design with the level set method. (a and c) Examples of designed DOEs with on-demand diffraction intensity distributions. (b and d) Examples of designed DOEs with uniform and gradient diffraction intensity distributions, respectively. Each panel presents the unit cell of the DOE, the simulated efficiency of all the diffraction orders, and the comparison of the desired and designed intensity distributions.

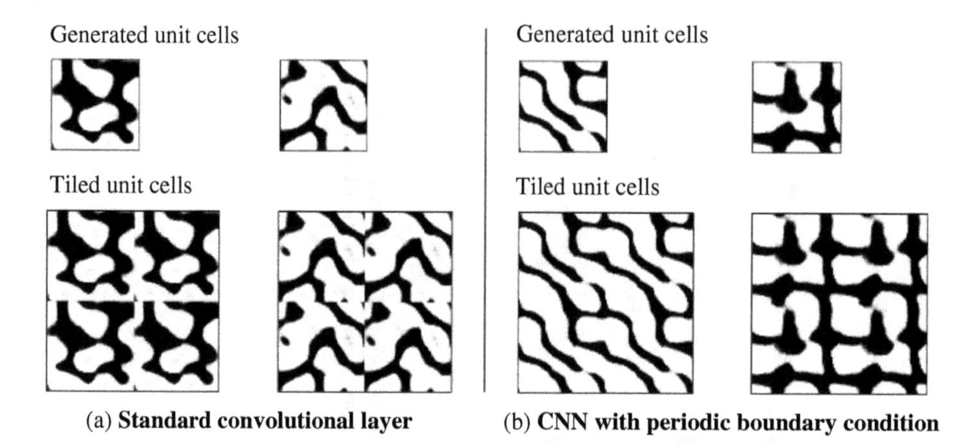

Figure 6.14 Randomly generated patterns from (a) a standard CNN and (b) a CNN with enforcing periodic boundary conditions.

In level set methods represented by INR, the parameters are the weights w of the network. In generative models, an additional latent vector can be included to produce random patterns. When representing photonic topology, networks with INR have increased controllability over image-based networks [22, 46].

Another benefit of using INR is that the network allows the accurate calculation of derivatives of a level set function with respect to its coordinates and weights. In modern deep learning frameworks such as PyTorch [47] and TensorFlow [48], a single-line code can do this. This feature enables fast, accurate calculation of curvature and feature size of the topology of a device. Fabrication constraints that are related to maximum curvature and minimum feature size can be formulated as an analytical cost function with automatic differentiation, which can be readily incorporated in the optimization process.

In addition to the generative model, INR networks also perfectly fit physical modeling because in actual physical system, space and time is not discretized. Some works related to physical inspired neural networks and INRs [49, 50] could be a future direction to data-driven physical modeling and design.

6.5.4 Periodic Boundary Conditions

Periodic boundary conditions are commonly seen in physical systems such as gratings, metasurfaces, and metamaterials. Generating a structure with periodic boundary conditions is essential to achieve the desired performance in these device designs. Level set methods can achieve periodic boundary conditions by enforcing the left/upper parameters and right/bottom ones to be identical. Some level set methods, such as the Fourier level set, naturally have periodic boundary conditions. However, in some cases, we need to produce a periodic structure from a network. We can also enforce periodicity using some tricks.

Take convolutional neural networks (CNNs) [51, 52] as an example. To generate a periodic structure, we need to enforce the periodicity in every layer of the transformation. A simple remedy to a traditional CNN layer is to pad the input with sufficient zeros and crop the output with correct shape. In machine learning frameworks, padding/cropping operations can maintain the gradients and does not incur additional computational burden. In Figure 6.14, we present two set of randomly generated structures with and without enforcing the periodicity. In this experiment, the training dataset only contains periodic images. The traditional CNN is not able to learn the periodicity in all stages of training. Periodicity can also be enforced in INR in a similar way [53].

Periodic networks could be more useful in predictive models. For example, in a metasurface, each unit cell can contain one meta-atom. Although every two adjacent meta-atoms are not in physical contact, the fields can have periodic coupling effects. If a traditional CNN or implicit network is used to learn the field behavior, it ignores the coupling effect. The coupling can be strong enough to affect the device performance. A periodic network can naturally bring about the field information in inference, and thus can improve the prediction accuracy while reducing the size of the training dataset.

6.6 Summary

In this chapter, we have presented generative design strategies for the design and optimization of photonic structures with prescribed design objectives. Through the modeling of a generative strategy and practical design examples using generative models, we have shown that utilizing a generative strategy can avoid certain disadvantages of traditional optimization methods and help the optimization converge to a global optimum in an expedite way. A generative design strategy can be constructed with gradient-based or gradient-free optimization algorithms, allowing flexible optimization schemes for various design requirements. With appropriate approximation, a generative design strategy is able to explore a tremendous design space. In our examples demonstrated here, the number of design parameters can range from 10^3, such as in meta-structures, to 10^6, as seen in a multiplexed hologram. We realize there are remaining challenges in generative design strategies. For example, a generative model requires certain prior datasets of the candidate structures. It is not viable, in most cases, to assure whether the optimal design lies within the distribution defined in the prior dataset. In addition, generative design methods typically require vast sampling in order to guide to a globally optimized solution. It may not be the most effective approach for devices with long simulation time. That said, we optimistically anticipate rapid advances of machine learning technique to bring about new design strategies, overcome current optimization difficulties, and substantially boost the discovery and development of advanced photonic devices in essential and unconventional applications, including optical communications, high-resolution displays, virtual/augmented reality, sensing technologies, and so many more.

References

1 LeCun, Y., Bengio, Y., and Hinton, G. (2015). Deep learning. *Nature* 521: 436.
2 Gálvez, A. and Iglesias, A. (2012). Particle swarm optimization for non-uniform rational B-spline surface reconstruction from clouds of 3D data points. *Information Sciences* 192: 174.
3 Lalau-Keraly, C.M., Bhargava, S., Miller, O.D., and Yablonovitch, E. (2013). Adjoint shape optimization applied to electromagnetic design. *Optics Express* 21: 21693.
4 Hughes, T.W., Minkov, M., Williamson, I.A., and Fan, S. (2018). Adjoint method and inverse design for nonlinear nanophotonic devices. *ACS Photonics* 5: 4781.
5 Doersch, C. (2016). Tutorial on variational autoencoders, arXiv preprint arXiv:1606.05908.
6 Jiang, J. and Fan, J.A. (2019). Global optimization of dielectric metasurfaces using a physics-driven neural network. *Nano Letters* 19: 5366.

7 Jacobsen, J.-H., Smeulders, A., and Oyallon, E. (2018). i-Revnet: deep invertible networks, arXiv preprint arXiv:1802.07088.

8 Goodfellow, I., Pouget-Abadie, J., Mirza, M. et al. (2014). Generative adversarial nets, presented at *Advances in Neural Information Processing Systems*, Montreal, Canada.

9 Liu, Z., Zhu, D., Rodrigues, S.P. et al. (2018). Generative model for the inverse design of metasurfaces. *Nano Letters* 18: 6570.

10 Gulrajani, I., Ahmed, F., Arjovsky, M. et al. (2017). Improved training of wasserstein gans, arXiv preprint arXiv:1704.00028.

11 Smith, D., Pendry, J., and Wiltshire, M. (2004). Metamaterials and negative refractive index. *Science* 305: 788.

12 Bomzon, Z.E., Biener, G., Kleiner, V., and Hasman, E. (2002). Space-variant Pancharatnam–Berry phase optical elements with computer-generated subwavelength gratings. *Optics Letters* 27: 1141.

13 Yu, N. and Capasso, F. (2014). Flat optics with designer metasurfaces. *Nature Materials* 13: 139.

14 Pelikan, M., Goldberg, D.E., and Cantú-Paz, E. (1999). BOA: the Bayesian optimization algorithm, presented at. In: *Proceedings of the Genetic and Evolutionary Computation Conference GECCO-99, Orlando, FL, USA*, 525–532.

15 Hansen, N., Müller, S.D., and Koumoutsakos, P. (2003). Reducing the time complexity of the derandomized evolution strategy with covariance matrix adaptation (CMA-ES). *Evolutionary Computation* 11: 1.

16 Trelea, I.C. (2003). The particle swarm optimization algorithm: convergence analysis and parameter selection. *Information Processing Letters* 85: 317.

17 Sivanandam, S. and Deepa, S. (2008). Genetic algorithms. In: *Introduction to Genetic Algorithms*, 15. Springer.

18 Srinivas, M. and Patnaik, L.M. (1994). Genetic algorithms: a survey. *Computer* 27: 17.

19 Solis, F.J. and Wets, R.J.-B. (1981). Minimization by random search techniques. *Mathematics of Operations Research* 6: 19.

20 Liu, Z., Raju, L., Zhu, D., and Cai, W. (2020). A hybrid strategy for the discovery and design of photonic structures. *IEEE Journal on Emerging and Selected Topics in Circuits and Systems* 10: 126.

21 Kingma, D.P. and Welling, M. (2013). Auto-encoding variational bayes, arXiv preprint arXiv:1312.6114.

22 Liu, Z., Zhu, D., Lee, K.T. et al. (2020). Compounding meta-atoms into metamolecules with hybrid artificial intelligence techniques. *Advanced Materials* 32: 1904790.

23 Zhu, D., Liu, Z., Raju, L. et al. (2021). Building multifunctional metasystems via algorithmic construction. *ACS Nano* 15: 2318.

24 Lin, Z. and Johnson, S.G. (2019). Overlapping domains for topology optimization of large-area metasurfaces. *Optics Express* 27: 32445.

25 An, S., Zheng, B., and Shalaginov, M.Y. (2021). Deep Convolutional Neural Networks to Predict Mutual Coupling Effects in Metasurfaces, arXiv preprint arXiv:2102.01761.

26 Raeker, B.O. and Grbic, A. (2019). Compound metaoptics for amplitude and phase control of wave fronts. *Physical Review Letters* 122: 113901.

27 Silva, A., Monticone, F., Castaldi, G. et al. (2014). Performing mathematical operations with metamaterials. *Science* 343: 160.

28 Wu, W., Jiang, W., Yang, J. et al. (2017). Multilayered analog optical differentiating device: performance analysis on structural parameters. *Optics Letters* 42: 5270.

29 Zhou, Y., Zheng, H., Kravchenko, I.I., and Valentine, J. (2020). Flat optics for image differentiation. *Nature Photonics* 14: 316.
30 Ni, X., Kildishev, A.V., and Shalaev, V.M. (2013). Metasurface holograms for visible light. *Nature Communications* 4: 2807.
31 Huang, L., Mühlenbernd, H., Li, X. et al. (2015). Broadband hybrid holographic multiplexing with geometric metasurfaces. *Advanced Materials* 27: 6444.
32 Wen, D., Yue, F., Li, G. et al. (2015). Helicity multiplexed broadband metasurface holograms. *Nature Communications* 6: 8241.
33 Li, L., Cui, T.J., Ji, W. et al. (2017). Electromagnetic reprogrammable coding-metasurface holograms. *Nature Communications* 8: 1.
34 Zhu, D., Liu, Z., Raju, L. et al. (2021). Building multi-functional meta-optic systems through deep learning, presented at. In: *2021 Conference on Lasers and Electro-Optics (CLEO), San Jose, CA, USA*, FTu4H-3.
35 Makey, G., Yavuz, Ö., Kesim, D.K. et al. (2019). Breaking crosstalk limits to dynamic holography using orthogonality of high-dimensional random vectors. *Nature Photonics* 13: 251.
36 Vercruysse, D., Sapra, N.V., Su, L. et al. (2019). Analytical level set fabrication constraints for inverse design. *Scientific Reports* 9: 1.
37 Khoram, E., Qian, X., Yuan, M., and Yu, Z. (2020). Controlling the minimal feature sizes in adjoint optimization of nanophotonic devices using b-spline surfaces. *Optics Express* 28: 7060.
38 Luo, Z., Tong, L., and Kang, Z. (2009). A level set method for structural shape and topology optimization using radial basis functions. *Computers & Structures* 87: 425.
39 Piggott, A.Y., Petykiewicz, J., Su, L., and Vučković, J. (2017). Fabrication-constrained nanophotonic inverse design. *Scientific Reports* 7: 1.
40 Wyrowski, F. and Bryngdahl, O. (1988). Iterative Fourier-transform algorithm applied to computer holography. *Journal of the Optical Society of America A* 5: 1058.
41 Wang, B., Dong, F., Li, Q.-T. et al. (2016). Visible-frequency dielectric metasurfaces for multiwavelength achromatic and highly dispersive holograms. *Nano Letters* 16: 5235.
42 Moharam, M., Grann, E.B., Pommet, D.A., and Gaylord, T. (1995). Formulation for stable and efficient implementation of the rigorous coupled-wave analysis of binary gratings. *Journal of the Optical Society of America A* 12: 1068.
43 Moharam, M., Pommet, D.A., Grann, E.B., and Gaylord, T. (1995). Stable implementation of the rigorous coupled-wave analysis for surface-relief gratings: enhanced transmittance matrix approach. *Journal of the Optical Society of America A* 12: 1077.
44 Liu, Z., Zhu, Z., and Cai, W. (2020). Topological encoding method for data-driven photonics inverse design. *Optics Express* 28: 4825.
45 Park, J.J., Florence, P., Straub, J. et al. (2019). Deepsdf: learning continuous signed distance functions for shape representation, presented at. In: *Proceedings of the IEEE/CVF Conference on Computer Vision and Pattern Recognition, Long Beach, CA, USA*, 165–174.
46 Stanley, K.O. (2007). Compositional pattern producing networks: a novel abstraction of development. *Genetic Programming and Evolvable Machines* 8: 131.
47 Paszke, A., Gross, S., Massa, F. et al. (2019). Pytorch: an imperative style, high-performance deep learning library. *Advances in Neural Information Processing Systems* 32: 8026.
48 Abadi, M., Barham, P., Chen, J. et al. (2016). Tensorflow: a system for large-scale machine learning, presented at. In: *12th USENIX symposium on operating systems design and implementation (OSDI 16)*.
49 Xie, Y., Takikawa, T., Saito, S. et al. (2021). Neural Fields in Visual Computing and Beyond, arXiv preprint arXiv:2111.11426.

50 Bountouridis, D., Harambam, J., Makhortykh, M. et al. (2019). SIREN: a simulation framework for understanding the effects of recommender systems in online news environments, presented at. In: *Proceedings of the Conference on Fairness, Accountability, and Transparency, Atlanta, GA, USA*, 150–159.

51 LeCun, Y., Bottou, L., Bengio, Y., and Haffner, P. (1998). Gradient-based learning applied to document recognition. *Proceedings of the IEEE* 86: 2278.

52 Krizhevsky, A., Sutskever, I., and Hinton, G.E. (2012). Imagenet classification with deep convolutional neural networks, presented at. In: *Advances in Neural Information Processing Systems*, 84–90. Lake Tahoe, USA.

53 Liu, Z. (2021). Algorithmic design of photonic structures with deep learning, Doctoral dissertation, Georgia Institute of Technology.

7

Machine Learning Advances in Computational Electromagnetics

Robert Lupoiu and Jonathan A. Fan

Department of Electrical Engineering, Stanford University, Stanford, CA, USA

7.1 Introduction

The field of computational electromagnetics (CEM) involves the development of algorithms that can numerically model Maxwell's equations with high accuracy. It initiated during the first computing revolution in the 1960s, during which a wide range of partial differential equation solvers were developed across many fields of science and engineering, including structural analysis, fluid flow, heat transfer, and seismology [1–3]. Conventionally, CEM algorithms have been developed to be general, with the capability to solve electromagnetics problems with few to no prior assumptions about the problem details. Accurate solutions to Maxwell's equations are achieved with quantitative convergence criteria, and rigorous frameworks have been developed to quantify the trade-off between numerical approximation and solution accuracy. To date, these methods have been developed to a high level of maturity [4–7].

The machine learning revolution of the last decade has presented entirely new opportunities for CEM, with the potential to dramatically accelerate the simulation and design of electromagnetic systems. Electromagnetic systems posed for CEM modeling are amenable to being framed in the context of machine learning because they are not arbitrary, but can instead be described in terms of application domains that each involve highly limited geometric layouts, material libraries, and electromagnetic objectives. For example, chip-based silicon photonic circuits typically involve the patterning of thin films consisting of a handful of materials and limited operating wavelengths. Structures within a technological domain therefore use strongly related physics that can be "learned" and generalized using machine learning models. CEM is particularly well suited for deep learning because conventional solvers can be readily used to curate large batches of training data, and they can also be used to quantify accuracy and convergence of electromagnetic field solutions.

In this chapter, we will discuss emergent concepts at the interface of the data sciences and conventional CEM algorithms. We will summarize the mechanics of conventional CEM algorithms, including those based on finite differences, finite elements, and the method of moments. These algorithms provide a foundation for understanding how to numerically discretize a CEM problem and find accurate solutions with rigorous convergence criteria. We will then discuss how machine learning algorithms can augment CEM solvers. Some of these concepts involve deep networks that are trained end-to-end to serve as surrogate CEM solvers, while others are embedded within the framework of conventional algorithms to leverage the numerical stability and accuracy of these methods.

Advances in Electromagnetics Empowered by Artificial Intelligence and Deep Learning, First Edition.
Edited by Sawyer D. Campbell and Douglas H. Werner.

7.2 Conventional Electromagnetic Simulation Techniques

This section covers an introduction to three conventional CEM techniques that are the basis for most academically and commercially used CEM software: the Finite Difference (FD) Method, Finite Element Method (FEM), and the Method of Moments (MoM). The machine learning methods surveyed in Sections 7.3-7.5 either directly integrate with these methods or exploit insights provided by their formulations. The finite difference methods involve approximating the differential form of Maxwell's equations with the finite difference formalism, and they include time domain (FDTD) and frequency domain (FDFD) formulations [4, 6]. FEM is a frequency domain solver in which the volumetric domain is subdivided into a generally irregular grid of voxels that describe electromagnetic fields as a combination of primitive basis functions [4, 7]. MoM is a frequency domain solver based on the integral form of Maxwell's equations and is used to evaluate currents and fields at the surfaces of homogeneous media [4–6]. Each technique has its own set of strengths and weaknesses regarding computational complexity, accuracy, and scaling. The proper choice of a solver for a given problem strongly depends on problem geometry, source configuration, and boundary conditions.

7.2.1 Finite Difference Frequency (FDFD) and Time (FDTD) Domain Solvers

With the finite difference methods, electromagnetic fields are discretized on a regular grid of voxels, and Maxwell's equations in the differential form are formulated to relate these discrete electric and magnetic field components. The discrete electric and magnetic field positions are defined on a "Yee" grid (Figure 7.1), which staggers the field component positions and ensures that the differential relationships between electric and magnetic fields as defined by Maxwell's equations are numerically stable [4]. The Yee grid was introduced by Kane Yee in the 1960s in his original formulation of the FDTD method [8], where he showed that field staggering mitigates the significant discretization errors inherent to collocated grid representations. The Yee-based representations of Ampere's law and Faraday's law in the frequency and time domains, with derivatives approximated using first-order central difference expressions, are summarized in Table 7.1. These expressions for FDTD and FDFD are very similar, with the key difference being the assumption of harmonic wave phenomena with frequency ω in the frequency domain formulation.

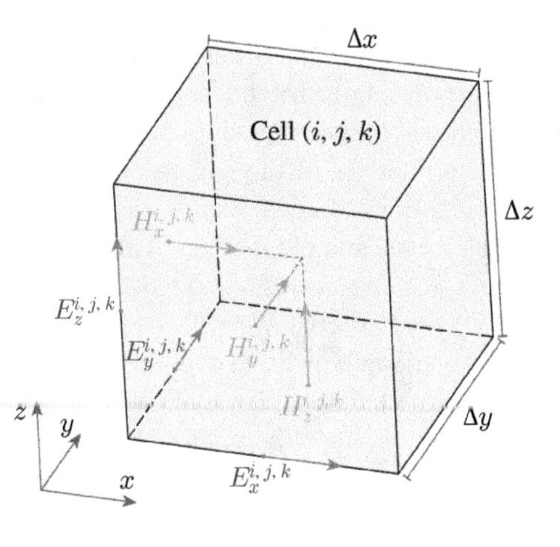

Figure 7.1 The structuring of the six field components within a single cell of the Yee grid formalism. The electric and magnetic field components are not collocated, but rather staggered and non-overlapping, which enforces computational stability.

Table 7.1 Faraday's Law and Ampere's Law expressed in the differential continuous time domain, FDFD, and FDTD formulations.

	Faraday's Law
Differential continuous Ttme domain formulation	$\nabla \times \mathbf{E} = -\frac{\partial \mathbf{B}}{\partial t} - \mathbf{M} \Longrightarrow \begin{cases} \frac{\partial E_z(t)}{\partial y} - \frac{\partial E_y(t)}{\partial z} = -\mu \frac{\partial H_x(t)}{\partial t} - M_x(t) \\ \frac{\partial E_x(t)}{\partial z} - \frac{\partial E_z(t)}{\partial x} = -\mu \frac{\partial H_y(t)}{\partial t} - M_y(t) \\ \frac{\partial E_y(t)}{\partial x} - \frac{\partial E_x(t)}{\partial y} = -\mu \frac{\partial H_z(t)}{\partial t} - M_z(t) \end{cases}$
FDFD formulation	$\begin{cases} \frac{E_z^{i,j+1,k} - E_z^{i,j,k}}{\Delta y} - \frac{E_y^{i,j,k+1} - E_y^{i,j,k}}{\Delta z} = -i\omega\mu_x^{i,j,k} H_x^{i,j,k} - M_x^{i,j,k} \\ \frac{E_x^{i,j,k+1} - E_x^{i,j,k}}{\Delta z} - \frac{E_z^{i+1,j,k} - E_z^{i,j,k}}{\Delta x} = -i\omega\mu_y^{i,j,k} H_y^{i,j,k} - M_y^{i,j,k} \\ \frac{E_y^{i+1,j,k} - E_y^{i,j,k}}{\Delta x} - \frac{E_x^{i,j+1,k} - E_x^{i,j,k}}{\Delta y} = -i\omega\mu_z^{i,j,k} H_z^{i,j,k} - M_z^{i,j,k} \end{cases}$
FDTD formulation	$\begin{cases} \frac{E_z^{i,j+1,k}(t) - E_z^{i,j,k}(t)}{\Delta y} - \frac{E_y^{i,j,k+1}(t) - E_y^{i,j,k}(t)}{\Delta z} = -\mu_x^{i,j,k} \frac{H_x^{i,j,k}(t+\frac{\Delta t}{2}) - H_x^{i,j,k}(t-\frac{\Delta t}{2})}{\Delta t} - M_x^{i,j,k}(t) \\ \frac{E_x^{i,j,k+1}(t) - E_x^{i,j,k}(t)}{\Delta z} - \frac{E_z^{i+1,j,k}(t) - E_z^{i,j,k}(t)}{\Delta x} = -\mu_y^{i,j,k} \frac{H_y^{i,j,k}(t+\frac{\Delta t}{2}) - H_y^{i,j,k}(t-\frac{\Delta t}{2})}{\Delta t} - M_y^{i,j,k}(t) \\ \frac{E_y^{i+1,j,k}(t) - E_y^{i,j,k}(t)}{\Delta x} - \frac{E_x^{i,j+1,k}(t) - E_x^{i,j,k}(t)}{\Delta y} = -\mu_z^{i,j,k} \frac{H_z^{i,j,k}(t+\frac{\Delta t}{2}) - H_z^{i,j,k}(t-\frac{\Delta t}{2})}{\Delta t} - M_z^{i,j,k}(t) \end{cases}$
	Ampere's Law
Differential continuous time domain formulation	$\nabla \times \mathbf{H} = \frac{\partial \mathbf{D}}{\partial t} + \mathbf{J} \Longrightarrow \begin{cases} \frac{\partial H_z(t)}{\partial y} - \frac{\partial H_y(t)}{\partial z} = \varepsilon \frac{\partial E_x(t)}{\partial t} + J_x(t) \\ \frac{\partial H_x(t)}{\partial z} - \frac{\partial H_z(t)}{\partial x} = \varepsilon \frac{\partial E_y(t)}{\partial t} + J_y(t) \\ \frac{\partial H_y(t)}{\partial x} - \frac{\partial H_x(t)}{\partial y} = \varepsilon \frac{\partial E_z(t)}{\partial t} + J_z(t) \end{cases}$
FDFD formulation	$\begin{cases} \frac{H_z^{i,j+1,k} - H_z^{i,j,k}}{\Delta y} - \frac{H_y^{i,j,k+1} - H_y^{i,j,k}}{\Delta z} = i\omega\varepsilon_x^{i,j,k} E_x^{i,j,k} + J_x^{i,j,k} \\ \frac{H_x^{i,j,k+1} - H_x^{i,j,k}}{\Delta z} - \frac{H_z^{i+1,j,k} - H_z^{i,j,k}}{\Delta x} = i\omega\varepsilon_y^{i,j,k} E_y^{i,j,k} + J_y^{i,j,k} \\ \frac{H_y^{i+1,j,k} - H_y^{i,j,k}}{\Delta x} - \frac{H_x^{i,j+1,k} - H_x^{i,j,k}}{\Delta y} = i\omega\varepsilon_z^{i,j,k} E_z^{i,j,k} + J_z^{i,j,k} \end{cases}$
FDTD formulation	$\begin{cases} \frac{H_z^{i,j+1,k}(t+\frac{\Delta t}{2}) - H_z^{i,j,k}(t+\frac{\Delta t}{2})}{\Delta y} - \frac{H_y^{i,j,k+1}(t+\frac{\Delta t}{2}) - H_y^{i,j,k}(t+\frac{\Delta t}{2})}{\Delta z} = \varepsilon_x^{i,j,k} \frac{E_x^{i,j,k}(t+\Delta t) - E_x^{i,j,k}(t)}{\Delta t} + J_x^{i,j,k}(t+\frac{\Delta t}{2}) \\ \frac{H_x^{i,j,k+1}(t+\frac{\Delta t}{2}) - H_x^{i,j,k}(t+\frac{\Delta t}{2})}{\Delta z} - \frac{H_z^{i+1,j,k}(t+\frac{\Delta t}{2}) - H_z^{i,j,k}(t+\frac{\Delta t}{2})}{\Delta x} = \varepsilon_y^{i,j,k} \frac{E_y^{i,j,k}(t+\Delta t) - E_y^{i,j,k}(t)}{\Delta t} + J_y^{i,j,k}(t+\frac{\Delta t}{2}) \\ \frac{H_y^{i+1,j,k}(t+\frac{\Delta t}{2}) - H_y^{i,j,k}(t+\frac{\Delta t}{2})}{\Delta x} - \frac{H_x^{i,j+1,k}(t+\frac{\Delta t}{2}) - H_x^{i,j,k}(t+\frac{\Delta t}{2})}{\Delta y} = \varepsilon_z^{i,j,k} \frac{E_z^{i,j,k}(t+\Delta t) - E_z^{i,j,k}(t)}{\Delta t} + J_z^{i,j,k}(t+\frac{\Delta t}{2}) \end{cases}$

The FD formulations use central finite difference derivative approximations.

In FDTD, Ampere's law and Faraday's law are formulated as update equations and are iteratively evaluated in a time-marching process with time step Δt to compute the time evolution of electromagnetic fields across the entire domain. Consider the update of field components at cell position (i, j, k) at time step t, in the absence of magnetic or current sources. First, using the FDTD formulation of Faraday's Law (Table 7.1), the magnetic field components are calculated at time step $t + \frac{\Delta t}{2}$ using the previously determined magnetic field components at time step $t - \frac{\Delta t}{2}$ and the electric field components from the current time step, t. This update equation for the H_x field component, for example, is readily evaluated as:

$$H_x^{i,j,k}\left(t+\frac{\Delta t}{2}\right)=\frac{\Delta t}{\mu_x^{i,j,k}}\left(\frac{E_y^{i,j,k+1}(t)-E_y^{i,j,k}(t)}{\Delta z}-\frac{E_z^{i,j+1,k}(t)-E_z^{i,j,k}(t)}{\Delta y}\right)+H_x^{i,j,k}\left(t-\frac{\Delta t}{2}\right) \tag{7.1}$$

Update equations for the other field components have similar forms. Next, the electric field components are determined at time step $t+\Delta t$ using the FDTD formulation of Ampere's Law, using the newly calculated $H(t+\frac{\Delta t}{2})$ and previously calculated $E(t)$ field components.

While the FDTD approach to solving Maxwell's equations can apply in principle to arbitrary domains, there exist practical limitations in computational memory that limit the domain size and time step granularity. In order for the algorithm to be stable, the time step must be smaller than the stability bound, which is a function of wave velocity and voxel size. Domains featuring high refractive index media or small feature sizes therefore necessitate slow time stepping procedures. In addition, the requirement of voxels with regular shapes, such as cuboidal, means that curvilinear surfaces can be difficult to model unless the Yee grid is specified to be very fine, adding to further increased computer memory requirements and reduced Δt. Note that it is necessary to reduce the time step size for finer meshes in order to maintain the stability of the algorithm, as a consequence of the *Courant–Friedrichs–Levy (CFL)* stability condition [9].

In FDFD, steady-state solutions to Maxwell's equations are directly evaluated within the full domain. In this formalism, the relationships between electric and magnetic fields at all voxels are expressed through Faraday's law and Ampere's law, which can be written in the form of vector and matrix relations. Faraday's law is:

$$\boldsymbol{C_e}\vec{\boldsymbol{e}}=-i\omega\boldsymbol{T_\mu}\vec{\boldsymbol{h}}-\vec{\boldsymbol{m}} \tag{7.2}$$

$\vec{\boldsymbol{e}}$ and $\vec{\boldsymbol{h}}$ are vectors that include field values at every voxel in the domain, the matrix $\boldsymbol{C_e}$ includes derivative expressions that relate electric field values between neighboring voxels based on finite difference relations, $\boldsymbol{T_\mu}$ contains permeability values in the simulation domain as a function of position, and $\vec{\boldsymbol{m}}$ represents magnetic currents. Ampere's law is:

$$\boldsymbol{C_h}\vec{\boldsymbol{h}}=i\omega\boldsymbol{T_\varepsilon}\vec{\boldsymbol{e}}+\vec{\boldsymbol{j}} \tag{7.3}$$

The matrix $\boldsymbol{C_h}$ includes derivative expressions that relate magnetic field values between neighboring voxels based on finite difference relations, $\boldsymbol{T_\varepsilon}$ contains permittivity values in the simulation domain as a function of position, and $\vec{\boldsymbol{j}}$ represents electric currents. These equations can be combined together to produce a wave equation in terms of electric or magnetic fields. For electric fields, this expression is:

$$(\boldsymbol{C_h}\boldsymbol{T_\mu}^{-1}\boldsymbol{C_e}-\omega^2\boldsymbol{T_\varepsilon})\vec{\boldsymbol{e}}=-i\omega\vec{\boldsymbol{j}}-\boldsymbol{C_h}\boldsymbol{T_\mu}^{-1}\vec{\boldsymbol{m}} \tag{7.4}$$

This expression has the form of $\boldsymbol{A}\vec{\boldsymbol{x}}=\vec{\boldsymbol{b}}$, where $\boldsymbol{A}$ and $\vec{\boldsymbol{b}}$ contain known permittivity, permeability, and source information about the domain, and $\vec{\boldsymbol{x}}$ represents the unknown fields. The fields are then readily evaluated as $\vec{\boldsymbol{x}}=\boldsymbol{A}^{-1}\vec{\boldsymbol{b}}$.

From this formalism, we see that while FDTD and FDFD use similar book keeping tools in the form of the Yee grid, the algorithms use very different computational flows. While FDTD can be readily parallelized, FDFD involves the inversion of a large sparse matrix and thus cannot be directly parallelized. In terms of computational evaluation, there are no straightforward strategies to speed up FDTD simulations, which utilize update equations with basic algebraic relations. However, various mathematical tricks can be applied to FDFD to accelerate the inversion of sparse matrices, improving its computational scaling. In addition, advanced matrix inversion methods based on generalized minimal residual method (GMRES) [10] or related Krylov subspace methods [11] can be preconditioned with prior knowledge of the system to further accelerate solution convergence.

An important consideration for both FD methods and CEM simulators more generally is the specification of proper boundary conditions along the simulation domain boundary. For bounded problems in which the fields properly terminate within the simulation domain, boundary conditions such as Perfect Electric Conductors (PECs) can be used. For unbounded problems involving field propagation to regions outside of the domain, Absorbing Boundary Conditions (ABCs) are required that absorb incoming waves without back-reflections, effectively simulating unbounded wave propagation. One highly effective ABC model is the perfectly matched layer (PML), which utilizes an artificial anisotropic absorbing material to convert propagating waves to exponentially decaying waves without backreflection [12]. PMLs are the *de facto* modern ABC because they can robustly suppress reflections from absorbed waves over a wide range of frequencies and incident angles [5, 13]. However, while PMLs are highly effective, they require thicknesses on the order of half a free space wavelength, thereby adding significant computational overhead to the wave simulation procedure.

7.2.2 The Finite Element Method (FEM)

The FEM is the most widely used scientific computing technique for solving differential equations. It was initially developed by mechanical engineers to solve solid mechanics problems in the 1940s, and it became an integral tool for CEM in the 1980s with the advent of reliable ABCs [5]. In this section, we will focus on FEM concepts based on the weighted residual method, which is a standard formulation of the technique [7]. The steps for solving Maxwell's equations with this version of the FEM algorithm are delineated in Sections 7.2.2.1–7.2.2.3:

7.2.2.1 Meshing

The solution domain, Ω, is first subdivided into generally irregularly shaped voxels termed *elements*. For a 2D domain, these are commonly triangles or quadrilaterals, and for 3D domains, they are tetrahedra, pyramids, prisms, or hexahedra. Triangles in 2D and tetrahedra in 3D are most often used, as they can accurately represent arbitrarily shaped objects, including those with curvilinear layouts. A unique feature of FEM elements is that they can be adaptively scaled to different sizes within the domain, leading to computationally efficient representations of electromagnetic fields without loss of accuracy. For example, small element sizes are typically specified in domain regions featuring rapidly varying spatial field profiles while larger element sizes are specified in regions featuring slowly varying spatial field profiles. The meshing procedure is typically performed with specialized programs, which use heuristic criteria such as geometric feature size and refractive index distributions in the domain to achieve accurate adaptive meshing without knowledge of the ground-truth fields.

7.2.2.2 Basis Function Expansion

The fields within each finite element are approximated as $f(\mathbf{r}) \approx \sum_{i=1}^{n}, f_i\varphi_i(\mathbf{r})$, where $\varphi_i(\mathbf{r})$ are primitive basis functions serving as analytic descriptions of the electromagnetic fields and f_i are unknown primitive basis function coefficients that are to be determined. The primitive basis functions are typically chosen to be generic low-order polynomials, and they are specified in a manner such that field continuity between neighboring elements is consistent with Maxwell's equations. The total number of primitive basis functions within an element scales with the number of nodes within the element, n. The use of more nodes leads to the use of higher-order basis functions and more accurate descriptions of the fields, at the expense of computational memory and overhead.

7.2.2.3 Residual Formulation

To solve our differential equation problem, our goal is to specify the unknown primitive basis function coefficients such that error between $f(\mathbf{r})$ and the ground-truth field solutions is minimized. This error to be minimized can be framed in terms of the residual $r = \mathcal{L}(f) - s$, where $\mathcal{L}$ is an integro-differential operator defined by Maxwell's equations and s is the source. In the weighted residual method, the problem is cast in the weak form: we solve the system of equations $\langle w_i, r \rangle = \int_\Omega w_i r d\Omega$, where w_i is a set of N weighting functions, and we set these weighted residual expressions to zero. In the case where the weighting functions are chosen to be the same as the basis functions, this formulation is termed the Galerkin method. The problem can now be expressed in the form $\mathbf{A}\vec{x} = \vec{b}$, where $\mathbf{A}$ is a matrix that corresponds to Maxwell-based terms, $\vec{x}$ is the vector containing the unknown coefficients, and $\vec{b}$ is a vector of source terms, and it is solved by inverting $\mathbf{A}$.

7.2.3 Method of Moments (MoM)

Whereas FD and FEM solve for electric and magnetic fields using a volume meshing approach, MoM is a variational method that solves for J and Q (i.e., ρ) in Maxwell's equations using a surface meshing approach. More specifically, the field solution is expressed as a superposition of integrals composed of the problem's sources and a *Green's function*. With sufficient information about the known field in the problem setup, we task MoM with determining the unknown sources [4]. With surface meshing, MoM is memory efficient and fast, and it is particularly well suited for modeling unbounded systems comprising homogeneous media, such as scattering from a perfectly conducting metal structure [5]. Most generally, the MoM solves linear equations of the form:

$$\mathcal{L}(\varphi) = f \tag{7.5}$$

where $\mathcal{L}$ is a linear operator, φ is the unknown quantity, and f is the excitation function enforced by the set of equations that dictate the physics of the problem [5].

Similar to FEM, Eq. (7.5) is numerically solved by determining the unknown coefficients of a basis function expansion of the solution,

$$\varphi = \sum_{n=1}^{N} a_n v_n \tag{7.6}$$

where v_n are pre-selected basis functions and a_n are the scaling coefficients for which a numerical solution is determined [5].

Plugging the expansion of Eq. (7.6) back into Eq. (7.5), a system of equations can be formulated to solve for the unknown expansion coefficients by applying weighting functions, similarly to the method presented for finding the FEM solution in Section 7.2.2. That is, we again encounter a problem in the form $\mathbf{A}\vec{x} = \vec{b}$, where $\mathbf{A}$ is a matrix containing weighted Maxwell-based terms, $\vec{x}$ is the vector containing the unknown expansion coefficients, and $\vec{b}$ is a vector of weighted forcing function terms, and it is solved by inverting $\mathbf{A}$. After determining $\vec{x}$, the problem solution is given by inputting the basis function coefficients into Eq. (7.6).

7.3 Deep Learning Methods for Augmenting Electromagnetic Solvers

A judicious approach to overcoming the limitations inherent to classical CEM techniques is augmenting existing algorithms with deep learning methods. Such approaches benefit from the combined computational stability advantages of classical CEM and the inference abilities of deep learning. As part of a burgeoning effort to improve conventional CEM using deep learning, techniques have been developed that augment all of the primary methods introduced in Section 7.2. We introduce here several deep learning augmentation methods that offer impressive potential for computational acceleration over conventional CEM algorithms.

7.3.1 Time Domain Simulators

7.3.1.1 Hardware Acceleration

In FDTD, the update equations for an individual voxel require field data only from nearest neighbor voxels. As such, the evaluation of field updates across the full domain can be readily subdivided and evaluated in a parallelizable manner. This parallelization follows many of the mathematical operations naturally performed by machine learning algorithms, which are executed on graphics processing unit (GPU) or tensor processing unit (TPU) computing hardware. With the *RCNN-FDTD (i.e., recurrent convolutional neural network-finite difference time domain)* algorithm [14], FDTD is cast in the formalism of machine learning, providing a natural software-hardware interface for parallelizing and accelerating FDTD simulations. The FD operator is expressed as a convolutional neural network (CNN) kernel and the time marching procedure is formulated in the recurrent neural network (RNN) framework [14], leading to speedups of 4.5 times over comparable CPU-parallelized code with no decrease in accuracy.

To illustrate the mapping of conventional FDTD to *RCNN-FDTD*, consider the procedure for solving for the H_x field component at an arbitrary time step for 2D TM modes (E_z, H_x, H_y). The FDTD algorithm employs Faraday's law to determine H_x at the next time step, requiring the most updated values of H_x and E_z. The update equation for H_x can be written generally as:

$$H_{x_{i,j}}^{t+1} = W_1 \cdot H_{x_{i,j}}^{t} + W_2 \cdot \left(E_{z_{i,j+1}}^{t+\frac{1}{2}} - E_{z_{i,j}}^{t+\frac{1}{2}} \right) \quad (7.7)$$

W_1 is termed the *spatial coefficient matrix*, W_2 is termed the *temporal coefficient matrix*, and both matrices can be directly read off from the conventional update equations. E_z and H_y are determined in a similar fashion from their respective FD update equations. In the formalism of machine learning, W_1 and W_2 can be directly described as CNN and RNN kernels, respectively. This setup is illustrated in Figure 7.2.

With this method, no modification is made to the FDTD algorithm itself; rather its casting as cascaded neural networks allows straightforward hardware acceleration with GPUs and TPUs. As such, W_1 and W_2 are deterministically formulated based on the setup of the FDTD algorithm and no network training is performed. Taking advantage of the highly efficient parallelization made possible by the machine learning community to accelerate FDTD, the *RCNN* scheme illustrates a subtle consequence of the machine learning revolution. Although FD algorithms were previously parallelizable before the advent of accessible machine learning software and hardware, implementing and employing such algorithms required specialized knowledge of how to parallelize algorithms in code. With the wealth of machine learning resources currently available, code parallelization is a trivial task if it can be expressed within the formalism of machine learning algorithms.

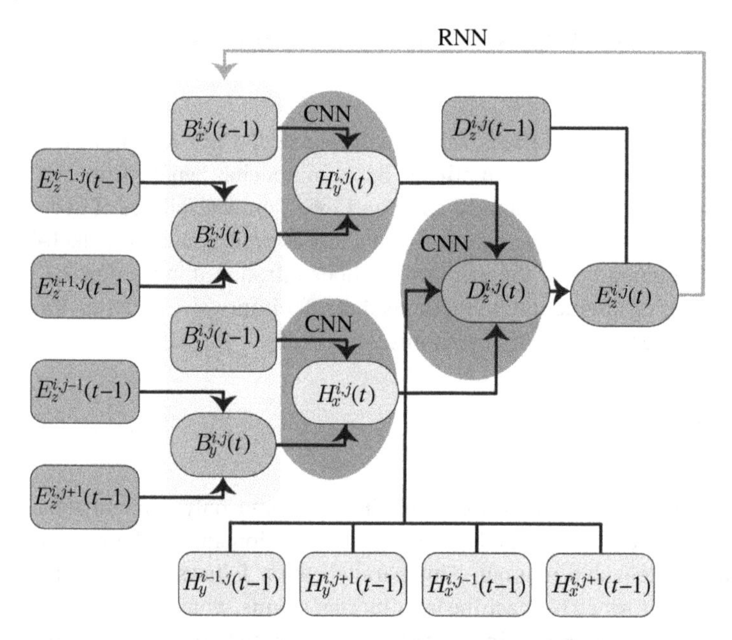

Figure 7.2 The *RCNN* update scheme, which expresses the FDTD formalism within the framework of CNN and RNN operations to take advantage of the accelerating software and hardware developed by the machine learning community. CNN operations collect and process field values from neighboring cells to perform the finite spatial differencing, after which the time-stepping is performed using an RNN operation. No learnable parameters are used in this scheme.

7.3.1.2 Learning Finite Difference Kernels

The conventional FDTD algorithm involves a fundamental trade-off between accuracy and time step size: greater simulation accuracy requires finer spatial grid resolution, which is accommodated by the requirement of smaller time steps to maintain stability [9] and more computational memory. While the parallelization concepts from Section 7.3.1.1 can help mitigate some of this computational loading, the trade-off still remains. A conceptually new way to address this trade-off is to consider FDTD simulations with spatially coarse grids and large time steps, and to specify W_1 and W_2 as learnable kernels trained from data. These kernels can be treated as super-resolution operators and have the potential to utilize learned nonlinear interpolation to produce electromagnetic field updates with the accuracy of spatially fine simulations.

The first attempt to develop learned FD kernels formulated the FDTD stepping process using either RNN or CNN operations, with learned coefficients that approximated the finite differencing approximations [15]. The RNN-only formulation of Yee grid field prediction is similar to the *RCNN-FDTD* method, as both employ a simple RNN architecture to obtain the field values at succeeding time steps. The main difference in the structure of the algorithms is that instead of collecting the scaled field values of interest from neighboring voxels through convolutional operations, the RNN-only algorithm obtains the raw voxel values of all immediate neighboring voxels as a vector:

$$x_t = [u_{n+1,m}(t), u_{n-1,m}(t), u_{n,m}(t), u_{n,m+1}(t), u_{n,m-1}(t)] \tag{7.8}$$

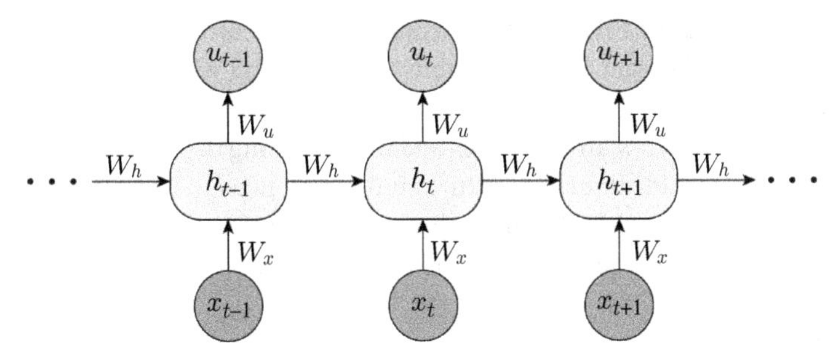

Figure 7.3 A basic recurrent neural network (RNN) architecture. The time sequential inputs of the RNN are processed one step at a time, with a preceding operation's processed result serving as part of the next time step's input. Note that the same set of learned weights are used at each time step of the process depicted in the figure. This "unfolded" representation is thus unfolded in time and does not depict multiple layers of a neural network.

As illustrated in Figure 7.3, this is then processed using a basic recurrent neural network (RNN). Mathematically, the computational graph of these basic RNNs may be formulated as two governing equations,

$$\begin{cases} h_t = \sigma_h(W_x x_t + W_h h_{t-1}) \\ u_t = \sigma_u(W_u h_t) \end{cases} \tag{7.9}$$

where x and y are the inputs and outputs of the RNN at any one time, respectively, h are the values of the hidden layers, W are the learned weight matrices of either the input, hidden, or output layers, and σ is the neural network activation function. RNNs are uniquely positioned to efficiently process sequential data because the same architecture is capable of processing arbitrarily sized inputs without increasing the number of parameters. For FDTD-like field predictions, one pass of a voxel's state vector, x_t, through the RNN yields its value at the following time step, $u_{n,m}(t+1)$.

The efficacy of the RNN kernel learning method is analyzed by training it to solve the 2D wave equation in a square domain without structures and Dirichlet boundary conditions upon a Dirac delta excitation in a fixed location at time step 0. The training data is composed of the field values over 100 time steps in the simulation domain, which is discretized into 11x11 voxels. The trained model is evaluated at a fixed time step for an excitation location different than the one used in the training data. The trained RNN's predictions maintain the ground truth's shapes and features on the relatively restricted dataset it was tested on in [15], but the average relative error obtained was a substantially high 15%.

Although the RNN is superior at processing sequential data, its architecture is not designed to inherently process the differential spatial relationships between neighboring voxels, which is a key aspect of the FDTD algorithm. Thus, it is more natural to express FDTD steps in the formalism of the CNN, where voxel values are updated as combinations of themselves and neighboring voxel values. In this particular introductory study, the input to the CNN is the entire 2D field at time step t, and the output is the entire field at the next time step, $t + \Delta t$. The chosen neural network architecture is composed of a single convolutional layer, followed by a pooling layer and two fully connected layers. The CNN was trained using the same dataset as the RNN, but tested instead using the same excitation source as used for training at a time step beyond that present in the training dataset. This network architecture and testing procedure led to much higher accuracy compared to the RNN approach, with an average relative error of less than 3%.

While the first attempt to learn generalizable FDTD kernels using machine learning falls short of demonstrating an accurate, generalizable technique, it establishes promising results and the potential for related concepts to transcend the limitations of the conventional FDTD method. It is anticipated that more advanced neural network architectures and more training data can lead to a more scalable and generalizable time domain simulator. Future studies will need to be quantitatively benchmarked with equivalent conventional FDTD algorithms to demonstrate the efficacy of machine learning to efficiently establish super-resolution methods with learned FDTD kernels.

7.3.1.3 Learning Absorbing Boundary Conditions

Machine learning models can also reduce the computational burden of FDTD simulations by producing a one-voxel-thick ABC model that performs similarly to much thicker conventional PMLs [16]. The proposed model utilizes a fully connected neural network with a single hidden layer, and its performance is demonstrated for a 2-D TE_z problem. The training data is accrued by collecting the PML field values for randomly selected points, P, on the simulation domain-PML boundary. For each point, the H_z, E_x, and E_y field values are each collected for the four closest positions to P on the Yee grid at time step t of the FDTD simulation to form the input data set. The corresponding H_z, E_x, and E_y field components are collected at the single closest positions to P for time step $t + \Delta t$, forming the output data set. The training data is collected for several randomized incident angles of illumination. Throughout the entirety of the time-stepped FDTD test simulation, it is demonstrated that the machine learning FDTD method is capable of maintaining an error rate below 1%, which is comparable to that of a five-voxel-thick conventional PML. This study demonstrates that machine learning is a viable solution for decreasing the compute requirements of unbounded CEM problems.

7.3.2 Augmenting Variational CEM Techniques Via Deep Learning

FEM and MoM are prime candidates for acceleration via machine learning due to the steep computational scaling for matrix inversion, resulting in long simulation times for problems with large sets of basis functions or large simulation domains. The approaches explored to date involve either attempting to directly predict the final solution within the framework of a specific method or to use machine learning to reduce the dimensionality of the problem [17, 18]. Our focus in this section will be on the latter, which takes advantage of the full rigor of the variational method, including deterministic solution error bounding and estimation [17].

Dimensionality reduction of variational techniques can be achieved by employing a neural network to learn an ultra-reduced, but highly expressive, problem-dependent set of basis functions that are in turn used to find rigorous solutions [17, 18]. The neural network is trained to predict the coefficients, a, of the complete set of primitive basis functions, F, given the rigorously solved coefficients of an analog of the problem formed by a significantly reduced basis. The rigorous solution to the reduced problem is computationally inexpensive compared to the full problem, as significantly smaller matrices must be inverted. Neural networks are well-suited to learn the relationship between the reduced basis and full basis solutions because of their proven strength of empirically establishing accurate relationships between high-dimensional datasets and their low-dimensional representations [17].

To mitigate the impact of deviations between the neural network-predicted coefficients and ground-truth values, the network-predicted coefficients are used to construct *macro* basis functions, F_{macro}, which comprise a linear combination of primitive basis functions weighted by the predicted coefficients. The primitive basis functions selected for these macro basis functions are chosen to be distinct from those imposing boundary conditions in the problem, $F_{boundary}$. Each of the macro basis functions is therefore defined as:

$$f_{macro} = \sum_{I_{macro}} a_i f_i \quad (7.10)$$

where $I_{macro} = \{i \in [1...N] \mid f_i \notin F_{boundary}\}$. The set of basis functions composed of the union between F_{macro} and $F_{boundary}$ is denoted by $\overline{F}$ and is composed of much fewer elements than the total number of primitive basis functions N. The CEM problem is once again solved using the conventional variational method with the $\overline{F}$ basis. The careful distinction of the macro and boundary basis functions as separate entities in $\overline{F}$ allows for an accurate, convergent solution throughout the entirety of the domain.

This technique of utilizing macro basis functions in conjunction with the variational method performs significantly better than more naive techniques. For a simple 1-D scattered field prediction problem, the macro basis technique was benchmarked against conventionally solving the problem using the same number of primitive basis functions as in $\overline{F}$ and also against the direct neural network predicted solution. For the real component of the calculated field, the macro basis function formulation method achieved an average root-mean-square error (RMSE) of about 0.15 on the test dataset, whereas the conventional approach attained an average RMSE of about 0.3 and the direct neural network prediction realized an average RMSE of about 0.6.

The macro basis function formulation technique results in significant reductions in computational complexity for conventional CEM solvers. Conventional iterative methods for performing the variational calculation face computational complexity of $O(N^2)$ and direct methods suffer from $O(N^3)$ scaling. By reducing the number of basis functions used by the variational algorithm using the macro function technique (i.e., a reduction from $|F|$ to $|\overline{F}|$) by a factor of γ, iterative and direct solvers experience speedups of a factor of $1/\gamma^2$ and $1/\gamma^3$, respectively [17]. The computational cost for solving the initial dimensionally reduced problem is typically negligible and thus ignored in this analysis. These computational savings are significant as problem domains scale in size and complexity. It is also noted that this method can also extend to FDFD techniques by expressing the field as a weighted sum of Dirac-delta basis functions that are centered on the Yee grid sample points. Here, the "reduced basis" is a coarser grid than that of the original problem, thus framing the task of this acceleration technique as super-resolution algorithm.

7.4 Deep Electromagnetic Surrogate Solvers Trained Purely with Data

In this section, we will discuss how direct solution predictions of electromagnetic problems can be cast as data-driven computer vision problems solved via neural networks. These deep networks are tasked with learning a one-to-one mapping between a set of structures and their respective boundary conditions to an output consisting of the set of electromagnetic field distributions. In this manner, the network is tasked with learning the operator that maps device structure to electric fields, where an operator by definition maps a function to another function. More specifically, it learns Green's function operators that map spatial dielectric structure to field for a given source. This task of predicting the electromagnetic field value at each pixel position of the inputted "image" is referred to as *pixel-wise regression* and is the most demanding class of computer vision problems (Figure 7.4) [19, 20]. To perform this learning task, training data is generated using

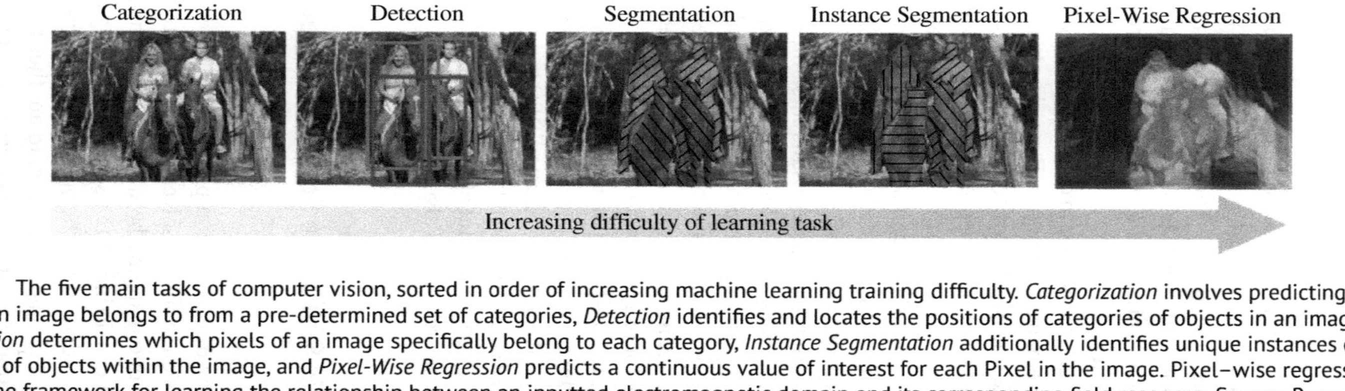

Figure 7.4 The five main tasks of computer vision, sorted in order of increasing machine learning training difficulty. *Categorization* involves predicting which category an image belongs to from a pre-determined set of categories, *Detection* identifies and locates the positions of categories of objects in an image, *Segmentation* determines which pixels of an image specifically belong to each category, *Instance Segmentation* additionally identifies unique instances of known categories of objects within the image, and *Pixel-Wise Regression* predicts a continuous value of interest for each Pixel in the image. Pixel–wise regression provides the framework for learning the relationship between an inputted electromagnetic domain and its corresponding field response. Source: Reproduced with permission from Leonid Sigal, delivered at the University of British Columbia, 2021.

conventional CEM solvers, which provides a straightforward pathway to producing up to millions of highly accurate input–output data pairs needed for training machine learning models.

In an initial demonstration of a surrogate solver, a deep CNN [21] was trained to predict the electrostatic response of arbitrarily shaped dielectric ellipsoids in 2-D 64×64 pixel and 3-D $64 \times 64 \times 64$ pixel domains excited by a point source located in one of eleven positions on a line in the periphery of the simulation domain [22]. The CNN's first input channel is composed of the dielectric distribution within the full domain, and the second is the Euclidean distance from each voxel to the excitation point source. The single-channel output of the CNN is the electromagnetic response of the dielectric distribution on a restricted output domain of 32×32 pixels for the 2-D case or $32 \times 32 \times 32$ pixels for the 3-D case, located in the center of the full-size input domain. The input–output training data pairs were obtained using a finite difference solver that calculated Poisson's equation constrained by Dirichlet boundary conditions. The model was trained using an Adam optimizer [23] for a loss function that took the squared $L^2 - norm$ between the base-10 logarithms of the CNN-predicted and ground-truth fields. The network architecture is relatively generic and not explicitly configured for the task of pixel-wise regression. As depicted in Figure 7.5, six consecutive 3D convolution encoding operations encode the inputted source and dielectric permittivity information, which is then decoded into corresponding field values throughout the computational domain by two consecutive 3D convolution operations.

The most generalized problems undertaken by the CNN in Figure 7.5 involved the modeling of the electrostatic response of four ellipsoids for the 2-D case and two arbitrarily shaped ellipsoids for

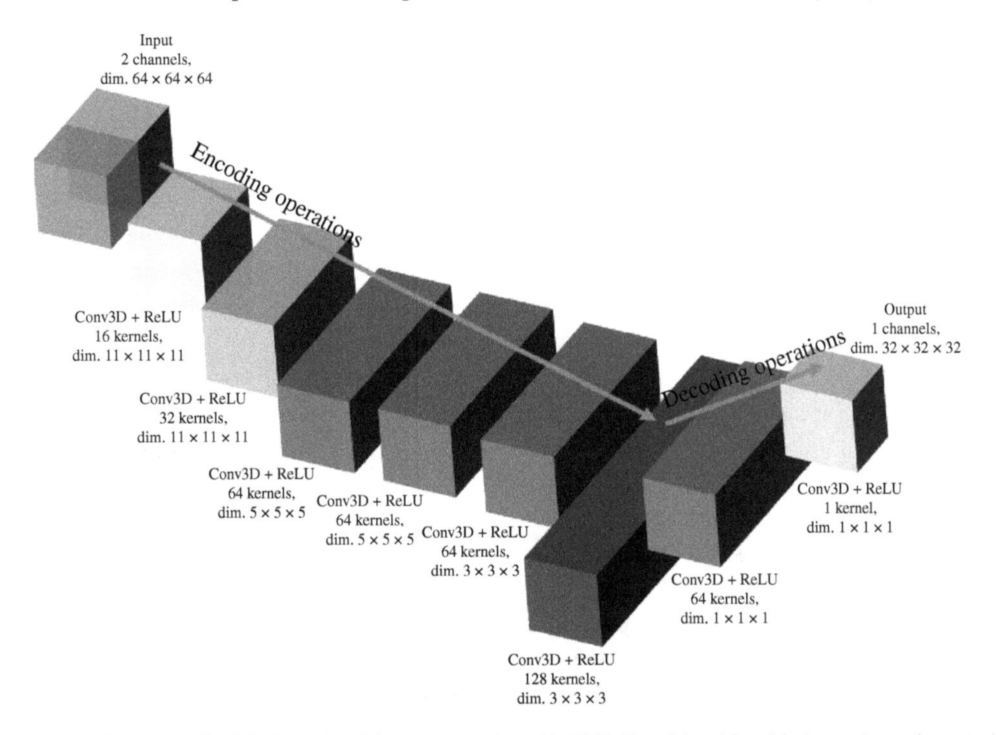

Figure 7.5 The 3D CNN-based architecture employed in [22]. Two $64 \times 64 \times 64$ channels are inputted, one containing the dielectric information from the input domain and the other the distance between each pixel and the point source. The neural network input then undergoes an encoding and decoding process executed by 3D convolutional operations to output a one-channel $32 \times 32 \times 32$ map of the electromagnetic response in a simulation domain centered in the input domain. The number of kernels and their respective dimensions that yield each processed block are labeled. Source: "Adapted from [22]."

the 3-D case. In the former case, the ellipsoids are always centered in each quadrant of the domain, with a fixed major axis length and variable minor axis length, rotation angle, and dielectric permittivity. The 3-D case contains one ellipsoid centered in each half of the simulation domain, each with variable semiaxes and permittivity. The 2-D CNN was trained with 40,000 training samples and tested with 10,000 samples. Both the 2-D and 3-D variations of the simulator achieved relative errors below 3%. While this preliminary demonstration was applied to a relatively narrow range of variable input parameters, it introduced the potential for deep networks to serve as surrogate electromagnetic solvers.

To model broader classes of electromagnetics systems with high accuracy, more careful consideration of the neural network architecture is required. In this aim, convolutional-based encoder–decoder neural networks that first encode the input image and then decode its latent representation have been developed by the machine learning community for pixel-wise regression tasks, making them well suited for modeling entire classes of electromagnetic systems. One of the highest performing segmentation architectures is the U-Net [24], depicted in Figure 7.6. The network is a symmetric, fully convolutional encoder–decoder neural network composed of a series of "convolutional blocks," which consist of a series of convolution, batch normalization, and ReLU nonlinear activation operations. The first part of the network performs *encoding* and contracts in spatial dimension size from one convolutional block to the next using max pooling operations at the end of each block. The second part of the network performs *decoding* and is composed of blocks that are symmetrically sized compared to the encoding part of the network, with the

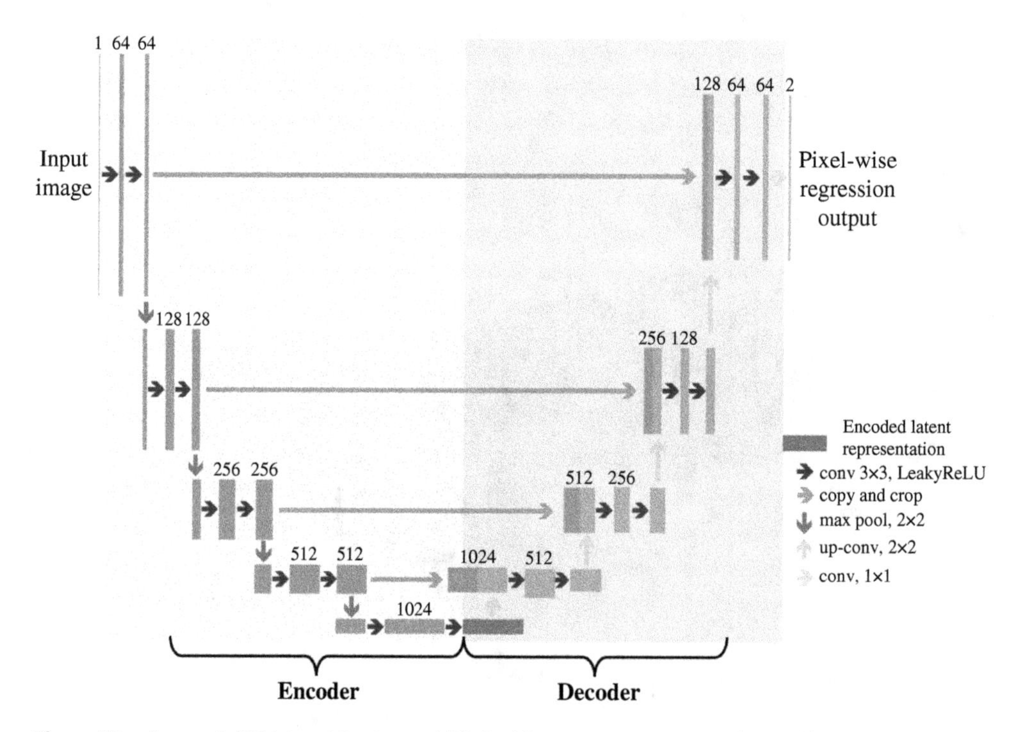

Figure 7.6 A generic U-Net architecture, which in this case accepts a one-channel input and produces a two-channel output. The first half of the U-Net consists of a series of *encoding* "convolutional blocks," which are followed by a symmetrical series of *decoding* "convolutional blocks," which are presented against an orange background. Spatial dimension reduction of the input is achieved in the encoding portion of the network using max pool operations, and the input dimensions are restored in the decoding portion by up-sampling using the transposed convolution operation.

expansion of the spatial dimensions being realized through upsampling operations executed at the end of each convolutional block. Residual connections are employed within each residual block, adding the result of the first convolution operation to the batch-normalized result at the end of the block. Shortcut connections are also used between the result of each encoding convolutional block before the max pooling operation to the input of each decoding block, with the connections made between symmetric blocks of each half of the network.

The direct connections between the two halves of the network between the same resolution levels are key to preserving physically relevant information at each pixel between the input and output images of the network. Besides the residual and shortcut connections facilitating the flow of gradients during the training process to prevent the *vanishing gradient problem* [25], they help to capture a strong relationship between the inputted dielectric geometry and the characteristics of the outputted wave profile. They also preserve a direct relationship between dielectric discontinuities and learned boundary conditions that are localized in the pixels of the output that are proximal to those containing the discontinuity region in the input. Owing to these features of its architecture, the U-Net scales well to the processing of large amounts of training data and has been successfully applied to both 2-D and 3-D domains [26, 27].

Preliminary U-Net-based networks accurately predicted the magnetic field response of arbitrarily superimposed ellipsoid and polygonal-shaped dielectric scatterers with a permittivity ranging from 2 to 10, under the illumination of arbitrarily angled plane waves [28]. With this approach, the dielectric scatterer image input of the neural network is augmented with a second channel containing the excitation source, demonstrating that a single trained network is capable of incorporating general simulation configurations. The U-Net can also be modified to function with 3D convolutions, as shown in Figure 7.7 [26]. Over a domain of size $45 \times 45 \times 10$ pixels, the U-Net was trained to predict the full electromagnetic response of one or more randomly arranged cuboidal dielectric blocks with fixed height upon normally incident plane wave illumination. The neural network, trained with 28,000 training samples, achieved an average cross-correlation of 0.953 with the ground-truth fields, with only 2.3% of tested devices exhibiting markedly poor outlier performance, under 0.80. These field solutions could be used in conjunction with other physics-based calculations, such as near-to-far-field transformations, to accurately compute other physical quantities, such as far-field scattering profiles. Upon calculation of such secondary physical quantities using the same neural network trained using 28,000 devices, the near-field was manifestly incorrect (with a ground-truth cross-correlation under 0.80) for only 5.3% outlier devices, and the far-field was similarly significantly incorrect for 10.1% of devices. The occurrence of failed predictions is the result of the U-Net providing a non-physically rigorous solution, which has a risk of producing

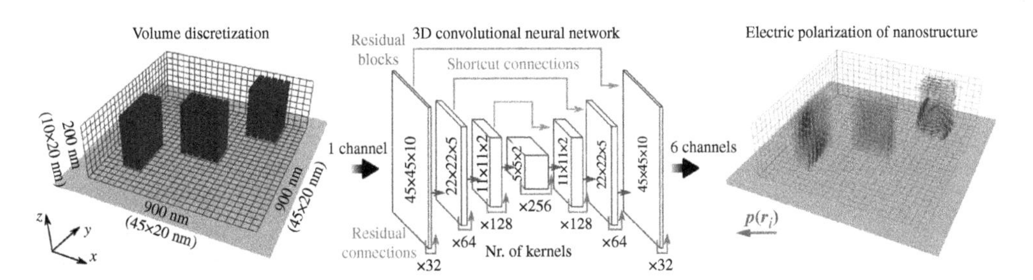

Figure 7.7 The U-Net fashioned in [26] utilizes 3-D convolutions to process an arbitrary configuration of 3-D dielectric pillars (as shown) or 2-D metallic polygons into the corresponding real and imaginary parts of the electromagnetic field components. Source: Reproduced with permission from [26]. ©2022 American Chemical Society.

significantly erroneous results. For example, the neural network performs poorly in regimes of resonance, which are acutely underrepresented in the randomly generated training dataset. The same study also demonstrated that a U-Net can be generalized to non-dielectric devices, by training one to predict the field response of arbitrarily shaped planar metallic polygons [26].

7.5 Deep Surrogate Solvers Trained with Physical Regularization

Although state-of-the-art segmentation neural networks, including the U-Net, are capable of learning the relationship between complex structures and the corresponding electromagnetic field responses, the approach of training solely based on input-output ground-truth data pairs has its limits. As electromagnetic fields are constrained by Maxwell's equations, incorporating this knowledge explicitly into the loss functions can lead to more robust field solutions. This section focuses on methods that exploit the governing physics of a solution space to train more rigorous general machine learning solvers: physics-informed neural networks (PINNs), physics-informed neural networks with hard constraints (hPINNs), and WaveY-Net.

7.5.1 Physics-Informed Neural Networks (PINNs)

PINNs are a newly developed class of neural networks that utilize the underlying governing physical relations of a simulation domain to regularize the training data to produce robust physical function approximators [29]. As they model the functional mapping between structure and fields for a single device and source, they are distinct from the operator mapping concepts for general classes of problems introduced in Section 7.4. As illustrated in Fig. 7.8, PINNs take the governing partial

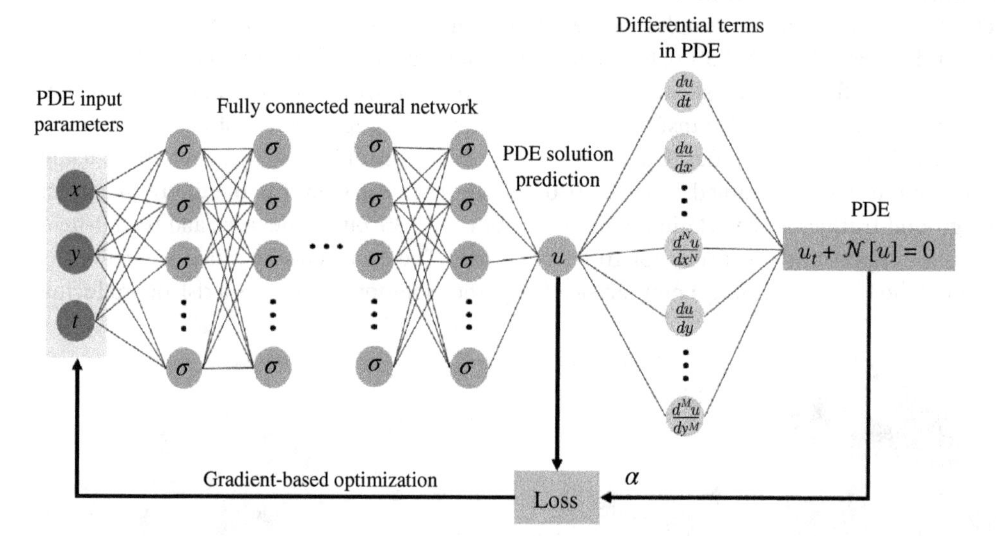

Figure 7.8 The general PINN training framework. A fully connected neural network (FCNN) is trained to predict the solution to a PDE, or more generally a set of PDEs, at the location in space-time specified by the inputted PDE parameters. The differential terms appearing in the PDE being solved are calculated using automatic differentiation. The loss is then constructed as a combination of data loss, which is known for a limited set of points in the domain, and the physics loss calculated across the domain using the PDE. A special case of PINN operation allows it to function without the use of any data and instead relies purely on the PDE itself to find the solutions.

differential equation (PDE) parameters as inputs, including space and time coordinates that specify a location in the solution space, and they output the corresponding solutions to the PDE. These networks therefore produce field values as a continuous function of space-time coordinates, exceeding the resolution limits set by discretized meshes and cubic scattered grids utilized by conventional CEM techniques (Section 7.2). PINNs can be employed to either deduce solutions to PDE problems or to discover the parameters that best satisfy a general PDE form based on data. The former use case, for which the governing PDEs are Maxwell's equations, will be the focus of this section.

The key to encoding the physics of a governing PDE into a PINN is to specify a training loss that balances data error (i.e., deviations between predicted and ground-truth values) with physical error (i.e., self-consistency of the predicted values and its derivatives with the governing PDEs). Consider a general one-dimensional PDE of the form:

$$\frac{\partial u}{\partial t} + \mathcal{N}[u] = 0, \; x \in \Omega, \; t \in [0, T] \tag{7.11}$$

where $u(t, x)$ is the underlying solution of the PDE within the domain of interest, Ω, and $\mathcal{N}$ is a nonlinear differential operator. The loss is:

$$MSE = MSE_d + \alpha MSE_f \tag{7.12}$$

where MSE_d is the data error and MSE_f is physical error. The latter can be further decomposed into volumetric and boundary terms, MSE_u and MSE_b, respectively, resulting in a total loss of:

$$MSE = MSE_d + \alpha(MSE_u + \gamma MSE_b); \; \alpha, \gamma > 0 \tag{7.13}$$

In this generalized formulation of the PINN loss, hyperparameters α and γ are used to tune the relative contributions from the data-driven, volumetric physical, and boundary physical errors to the overall loss. It is necessary to carefully adjust these relative weights in order to obtain a highly accurate converged solution for non-trivial problems [30]. Initial condition terms can also be added to the loss function for enforcement as needed.

Fully connected architectures are generally chosen for PINNs. To train the networks, predicted values are used to calculate the loss function, which is backpropagated back into the network using automatic differentiation (AD) to update the network weights and reduce prediction loss. AD is also used to calculate the derivative of predicted values with respect to the input space and time variables, making the evaluation of derivative-based PDE loss terms straightforward. Backpropagation can also be exploited by PINNs to optimize an objective functional and perform inverse design of devices, as explored in Section 7.5.2.

While PINNs introduce a qualitatively new way of solving PDEs with machine learning, they exhibit shortcomings that limit their direct application to solving broad classes of electromagnetics problems. As reviewed in Section 7.4, the efficacy of deep networks to learn full wave physics can strongly benefit from specialized network architectures. PINNs, as presented in this section, utilize generic, fully connected architectures that limit the complexity of problems and size of domains to which they are applicable. Furthermore, PINNs require the physical system (i.e., the geometry and dielectric distribution in the domain) to be fixed, limiting the method from being applied to general classes of electromagnetics problems.

7.5.2 Physics-Informed Neural Networks with Hard Constraints (hPINNs)

A broadly useful application of Maxwell solvers is photonic device optimization, which involves the identification of a dielectric distribution that maximizes a desired objective function within a solution space that is constrained by Maxwell's equations. For this purpose, a variant of PINNs with hard constraints, hPINNs, has been developed, for which network training is facilitated during a device optimization process without training data. The loss function in hPINNs is specified to satisfy the governing set of PDEs and simultaneously maximize the objective functional [31]. hPINNs use a loss function formulation different from the one employed by PINNs. PINNs cannot be used directly for inverse optimization because the gradient of the network's loss, generated from the set of governing PDEs, is not generally consistent with the gradient of the objective function. This means that a naive optimization implementation using PINNs would generally lead to a solution that does not satisfy the governing PDEs, which is a hard constraint of the problem. hPINNs introduce hard constraints using either the *penalty method* or the *augmented Lagrangian method*. Furthermore, hPINNs are trained using only physics information from the governing PDEs, without training data.

hPINNs directly impose boundary conditions in the formulation of the optimization problem. Formulations for Dirichlet and periodic boundary conditions have been developed, which can be straightforwardly extended to all other types of boundary conditions, such as Neumann and Robin. For Dirichlet boundary conditions, the neural network output is constructed as a sum of the analytical boundary condition and a scaled PINN output, which is scaled to zero where boundary conditions apply. The total field solution, $\hat{u}$, can therefore be expressed as:

$$\hat{u}(\mathbf{x}; \boldsymbol{\theta}_\mathbf{u}) = g(\mathbf{x}) + \ell(\mathbf{x})\mathcal{N}(\mathbf{x}; \boldsymbol{\theta}_\mathbf{u}) \tag{7.14}$$

where $\boldsymbol{\theta}_\mathbf{u}$ are the trainable weights of the network, $g(\mathbf{x})$ is the boundary conditions function imposed for the specified domain coordinates, $\mathbf{x}$, and $\mathcal{N}(\mathbf{x}; \boldsymbol{\theta}_\mathbf{u})$ is the neural network output that is scaled to zero by ℓ where boundary conditions apply. For non-trivial domains, spline functions are used to approximate $\ell(\mathbf{x})$. For periodic systems, periodic boundary conditions are implemented by expressing the periodic direction coordinate in $\mathbf{x}$ as a Fourier basis expansion, which imposes periodicity due to the periodicity of the basis functions.

The *penalty method* further enforces the hPINN's solution to adhere to PDE constraints. Consider the objective function $\mathcal{J}(\mathbf{u}; \gamma)$, which is dependent on the PDE solution, $\mathbf{u}(\mathbf{x})$, and the quantity of interest (i.e., the optimized device design), $\gamma(\mathbf{x})$. The design optimization problem is expressed as the following unconstrained problem:

$$\min_{\theta_u, \theta_\gamma} \mathcal{L}(\theta_u, \theta_\gamma) = \mathcal{J} + \mu_\mathcal{F} \mathcal{L}_\mathcal{F} \tag{7.15}$$

where $\mathcal{F}$ represents the N governing PDEs, $\mathcal{F} = \{\mathcal{F}_1, \mathcal{F}_2, \dots, \mathcal{F}_N\}$, and $\mu_\mathcal{F}$ is a fixed penalty coefficient. To prevent either the ill-conditioning or insufficient PDE constraint forcing of the gradient-based minimization of this equation, the design optimization problem is reframed as a sequence of unconstrained problems, with $\mu_\mathcal{F}$ incremented with each iteration by a constant multiplicative factor that is greater than 1. This thus yields,

$$\min_{\theta_u, \theta_\gamma} \mathcal{L}^k(\theta_u, \theta_\gamma) = \mathcal{J} + \mu_\mathcal{F}^k \mathcal{L}_\mathcal{F} \tag{7.16}$$

for the kth iteration. This optimization process yields two neural networks, $\hat{\mathbf{u}}(\mathbf{x}; \theta_u)$ and $\hat{\gamma}(\mathbf{x}; \theta_\gamma)$, the latter of which provides the structure of the optimized device. Note that this process trains the hPINNs without data, relying instead solely on the physics from the governing PDEs. While the penalty method is capable of optimization, it still suffers from potential ill-conditioning and also problem-dependent poor convergence rates. These are alleviated by the augmented Lagrangian method formulation of the optimization problem, which adds a Lagrangian term to the minimization expression, resulting in:

$$\min_{\theta_u, \theta_\gamma} \mathcal{L}^k(\theta_u, \theta_\gamma) = \mathcal{J} + \mu_\mathcal{F}^k \mathcal{L}_\mathcal{F} + \frac{1}{MN} \sum_{j=1}^{M} \sum_{i=1}^{N} \lambda_{i,j}^k \mathcal{F}_i[\hat{\mathbf{u}}(\mathbf{x}_\mathbf{j}); \hat{\gamma}(\mathbf{x}_\mathbf{j})] \tag{7.17}$$

Equation (7.17) is then optimized using a method inspired by the Lagrangian multiplier optimization method [32].

The penalty and Lagrangian methods were utilized to design a 2-D transmissive dielectric metamaterial structure capable of a desired holographic wavefront task. The neural network was constructed to impose periodic boundary conditions along one axis and Dirichlet boundary conditions along the other. The penalty method outperforms the soft constraints-optimized objective by about 2%, though it still suffers from the ill-conditioning problem in some hyperparameter regimes. The difficulty with convergence is overcome by the Lagrangian method. Both the well-converged penalty method and the Lagrangian method achieve average final solution PDE losses of about 10^{-5}, as defined by Eq. (7.18). Such strong convergence to physically valid solutions indicates that hPINNs are capable of serving as both full-wave solvers and gradient-based optimizers for photonics systems.

hPINNs are readily treated as surrogate electromagnetic solvers by dropping the objective function term in the loss function. In this formulation, the loss function simply becomes the mean squared sum of the real and imaginary components of the governing PDEs at each coordinate:

$$\mathcal{L}_{\mathcal{F}} = \frac{1}{2M}\sum_{j=1}^{M}((\Re[\mathcal{F}[\mathbf{x}_j]])^2 + (\Im[\mathcal{F}][\mathbf{x}_j])^2) \tag{7.18}$$

As a simple proof-of-concept demonstration, the hPINN solver, with the same boundary conditions as above, was employed to solve for the fields propagating through a domain with a constant permittivity. As illustrated in Figure 7.9, these resulting field profiles were qualitatively consistent with those solved using a conventional FDFD algorithm. Quantitatively, the converged solution for this test problem achieved an L^2 relative E-field error of 0.12% for both the real and imaginary components and an average PDE-informed loss, as defined by Eq. (7.18), of 0.0001% across the entire simulation domain. For the non-trivial case of an optimized dielectric distribution slab, both the penalty and Lagrangian methods achieved PDE losses of approximately 0.0012%.

7.5.3 WaveY-Net

WaveY-Net is a deep learning paradigm that combines operator learning concepts from Section 7.4 with physical regularization to produce a high-fidelity surrogate Maxwell solver for broad classes of electromagnetics systems [27]. The training overview and loss formulation for the generation of the field maps of periodic silicon nanostructure arrays from normally incident TM-polarized illumination is illustrated in Figure 7.10. The input to WaveY-Net is an image of the simulation domain and it outputs two channels, the real and imaginary magnetic field components of the solution. Only the magnetic fields, as opposed to all electromagnetic fields, are predicted to reduce the load on the network's learning capacity. The electric fields are subsequently computed by applying Ampere's law to the network-predicted magnetic field maps. In a manner consistent with PINNs, the WaveY-Nets loss function is a combination of a hyperparameter-scaled physical residue term, $L_{Maxwell}$, and a data loss term, L_{data}:

$$L_{total} = L_{data} + \alpha L_{Maxwell} \tag{7.19}$$

The Maxwell residue is calculated using the Helmholtz relation derived for the magnetic field:

$$\nabla \times \left(\frac{1}{\varepsilon}\nabla \times \hat{H}\right) - \omega^2\mu_0\hat{H} = 0 \tag{7.20}$$

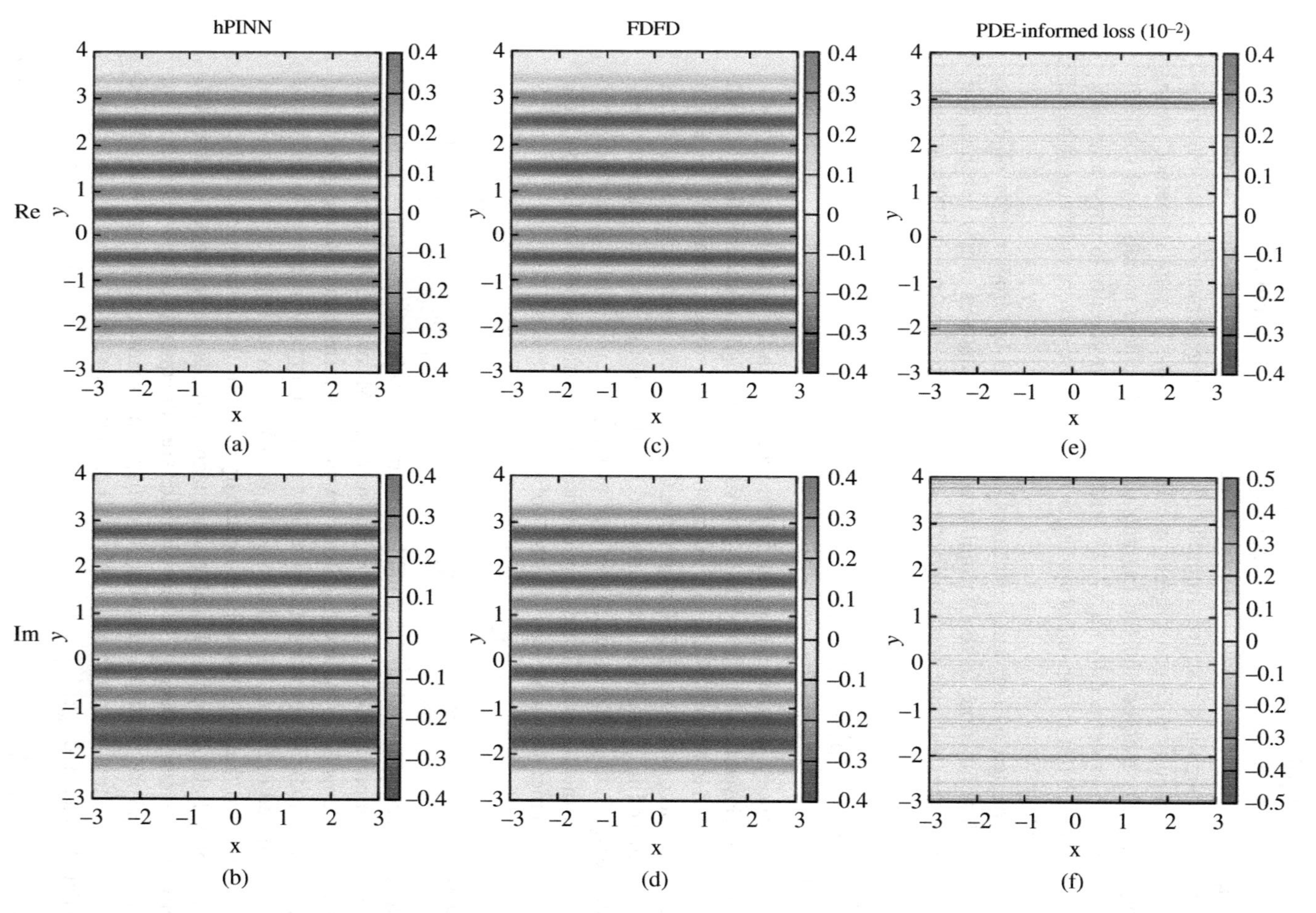

Figure 7.9 Benchmarking the hPINN in "forward" mode against FDFD, which is equivalent to training a PINN with physics information only. This demonstration was carried out on the trivial case of a domain with unity permittivity, but constrained by periodic and PEC boundary conditions. The hPINN (A) real and (B) imaginary component solutions are plotted across the entire domain alongside the FDFD (C) real and (D) imaginary component solutions. The PDE-informed loss, as defined in Eq. (7.18), of the forward-mode hPINN after convergence is plotted for the real (E) and imaginary (F) components. Source: Adapted from [33].

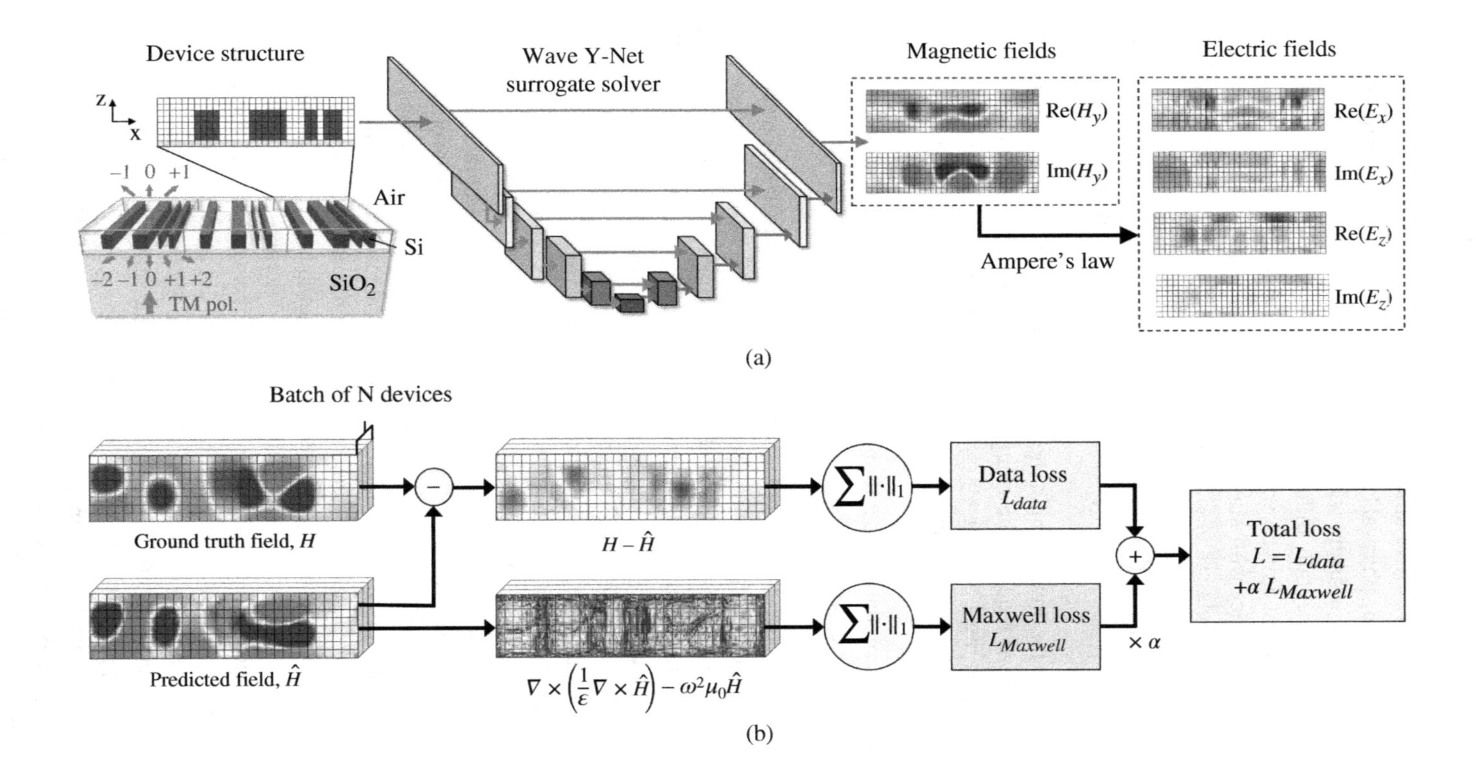

Figure 7.10 Overview of the WaveY-Net architecture and training technique, as presented in [27]. (a) A U-Net is trained to learn the electromagnetic response of silicon-based metagratings upon normally incident TM-polarized illumination from the substrate. Given the single-channel input consisting of the dielectric distribution of the input device, WaveY-Net outputs the real and imaginary components of the resulting magnetic field, which is then used to calculate the remaining electric field components using Ampere's Law. (b) The total loss used to train the WaveY-Net using Adam optimization is formulated as a combination of data loss and physics-based "Maxwell loss." The data loss is defined as the mean absolute error (MAE) between the batch of predicted magnetic fields and their corresponding ground truth, and the Maxwell loss is determined from the network-predicted field using the Helmholtz relation for the magnetic field. A hyperparameter, α, is used to tune the contribution of the Maxwell loss to the total loss. Source: Reproduced from [27]. Licensed under CC BY 4.0.

where $\hat{H}$ is the magnetic field predicted by the neural network. WaveY-Net is able to successfully implement robust physical regularization using Maxwell's equations by employing the Yee grid formalism to calculate L_{Maxwell}.

The high-level working principle of PINNs and WaveY-Net appear similar: both inject prior knowledge assumptions about governing physical laws explicitly into the loss function in a manner that makes the outputted fields more consistent with Maxwell's equations. In spite of this, PINNs and WaveY-Net are ultimately fundamentally different: PINNs learn the solution to a governing PDE over a continuous spatio-temporal domain with fixed geometries and boundary conditions, whereas WaveY-Net enforces consistency with Maxwell's equations using the Yee grid formalism to assist in the data training of entire classes of electromagnetic structures. Furthermore, unlike PINNs, Wavey-Net is not performance-bound to any specific type of neural network architecture, meaning that the framework can be applied to training the best-performing image segmentation neural networks, as outlined in Section 7.4, or to purpose-built and other high-performing future network architectures.

Calculating the physical residue for every point in the domain and calculating the electric fields from the magnetic field predictions requires the computation of spatial derivatives along each spatial axis at all points. This task must be done efficiently to manage the computational cost of the procedure. WaveY-Net casts Yee grid-discretized differentiation as a convolutional operation in the deep learning framework, enabling highly optimized parallelized computing for these calculations. Specifically, the differentiation operations are discretized using first-order central difference approximations and then executed using GPU-accelerated convolutional operations, as introduced in Section 7.3.1.

It has been demonstrated that WaveY-Net is more efficient than a data-only U-Net tasked with full E and H field prediction, and it performs equally as well as an orders of magnitude more computationally expensive FDFD solver for the task of adjoint optimization. When tasked with predicting the H field of a periodic nanostructure illuminated with TM-polarized light, the WaveY-Net demonstrated an average of 8.33% accuracy improvement for H field prediction and a 60.99% average accuracy improvement in the analytically calculated E fields over a data-only trained U-Net with the same structure, number of trained weights, and optimization process. The consequential improvement in the E field calculation is a direct result of the problem's physical underpinnings being encoded into the neural network by the WaveY-Net training scheme.

The hybridization of data and physics training enables exceptionally accurate prediction of full-wave electromagnetic fields that can be directly used to perform high level scientific computing tasks, such as the computation of performance gradients for gradient-based device optimization. The adjoint variables optimization method is one such algorithm and calculates the *adjoint gradient* of a particular device in order to iteratively improve its performance with respect to a metric. The computational graph of the adjoint optimization method for metagratings using forward and adjoint WaveY-Net surrogate solvers is illustrated in Figure 7.11a. For diffractive metagratings, the performance metric is the efficiency for which incident light is routed to a pre-determined angle. Gradients are calculated by evaluating the real dot product of electric fields from forward and adjoint simulations. As demonstrated in Figure 7.11b, WaveY-Net matches the adjoint gradient calculated via FDFD for a representative metagrating, whereas the purely data-trained U-Net is incapable of predicting the device's electromagnetic responses to a degree of accuracy high enough to faithfully calculate adjoint gradients. Compared to variations of the U-Net trained solely with data and no Maxwell loss, WaveY-Net consistently produces higher-fidelity field predictions, as demonstrated in Figure 7.12a–c. The normalized mean absolute error (MAE) of WaveY-Net's full field prediction for a set of test structures unseen during training is plotted

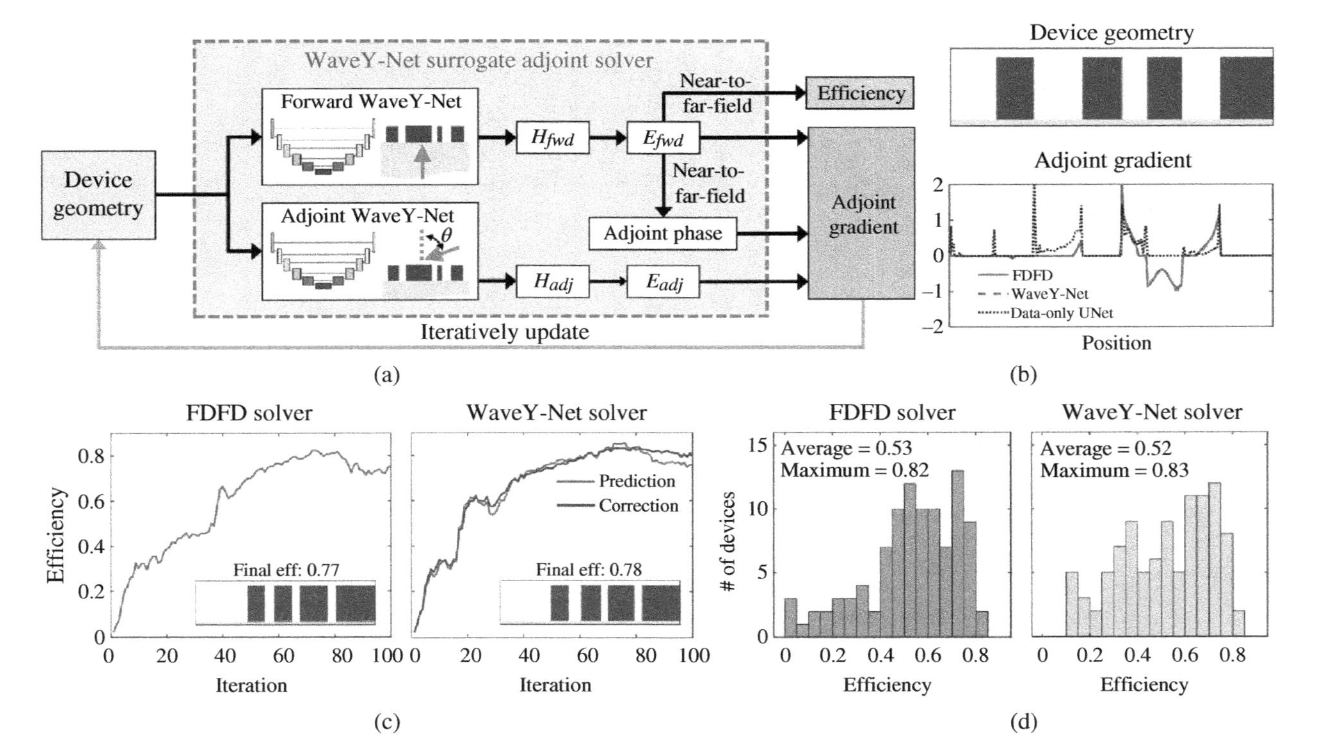

Figure 7.11 Local freeform metagrating optimization employing WaveY-Net as a surrogate adjoint solver. (A) Computational graph demonstrating the use of WaveY-Nets trained to predict the forward and adjoint responses of metagratings as surrogate solvers in the iterative adjoint optimization method. The field predictions from the separately trained WaveY-Nets are directly used to calculate the diffraction efficiency and the adjoint gradient, where the latter quantity drives the optimization process. (B) A randomly sampled representative metagrating device layout (top) and its corresponding adjoint gradients (bottom) from the process of local freeform metagrating optimization using WaveY-Net as a surrogate solver. FDFD is used to generate the ground-truth gradient values, which are nearly identical to those obtained using WaveY-Net. The data-only U-Net is incapable of capturing the smoothness of the wavelike field solution accurately enough to produce accurate gradients during the adjoint optimization process. (C) The efficiency trajectory of a single metagrating device, comparing the performance of utilizing an FDFD solver in the adjoint optimization process to that obtained using a WaveY-Net surrogate solver. The "prediction" efficiency curve of the WaveY-Net solver plot is directly predicted by WaveY-Net, whereas the "correction" efficiency curve is determined using an FDFD solver for the device optimized solely with the WaveY-Net surrogate solver. (D) Histograms depicting the efficiencies of 100 locally optimized devices, comparing the performance of an FDFD-based optimizer and a WaveY-Net surrogate solver optimizer. Source: Reproduced from [27]. Licensed under CC BY 4.0.

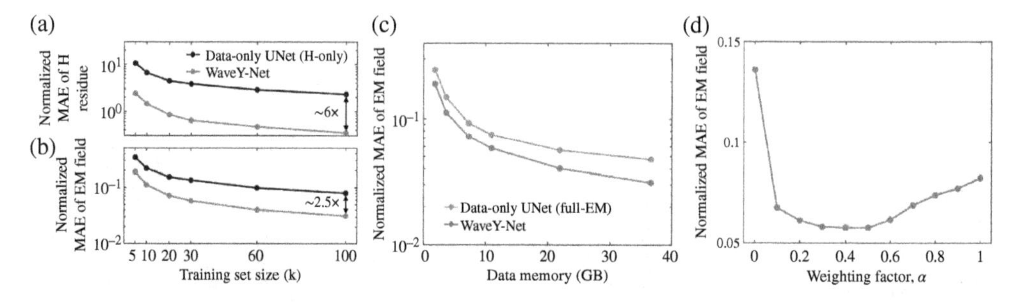

Figure 7.12 Numerical experiments benchmarking WaveY-Net field prediction performance. (a) Plot contrasting the predicted magnetic field MAE to that of a data-only U-Net that only predicts magnetic fields, for increasing training set size. (b) Plot contrasting the predicted electromagnetic field MAE to that of a data-only U-Net that only predicts magnetic fields, for increasing training set size. The electric field components are analytically calculated from the predicted magnetic fields using Maxwell's equations for both networks. (c) Plot contrasting the MAE of all electromagnetic field components outputted by a full-EM data-only U-Net and a WaveY-Net, plotted for increasing training set size. The data-only U-Net directly predicts all electromagnetic field components, whereas the WaveY-Net predicts the magnetic fields and the electric fields are subsequently calculated using Maxwell's equations. (d) Plot of a trained WaveY-Net's MAE of the predicted magnetic field with respect to the FDFD-generated test set ground truth versus the Maxwell loss coefficient α. The MAE is plotted as a sum of normalized $L_{maxwell}$ and L_{data} loss terms. Source: Reproduced from [27]. Licensed under CC BY 4.0.

as a function of the Maxwell loss coefficient, α from Eq. (7.19), in Figure 7.12d. The result is that the U-Net trained solely with data, when $\alpha = 0$, does not produce electromagnetic response simulations with sufficiently high fidelity to Maxwell's equations for the optimizer to be able to converge to an adequately efficient solution, In contrast, WaveY-Net, with an optimal α coefficient, produces devices with final efficiencies that are equally as high as those produced using FDFD, as demonstrated in Figure 7.11c–d.

Compared to conventional CEM methods, a trained WaveY-Net can solve for electromagnetic fields with speeds faster by orders of magnitude (Figure 7.13). This speed advantage is particularly amplified when the fields of multiple devices require evaluation, which can be performed in parallel

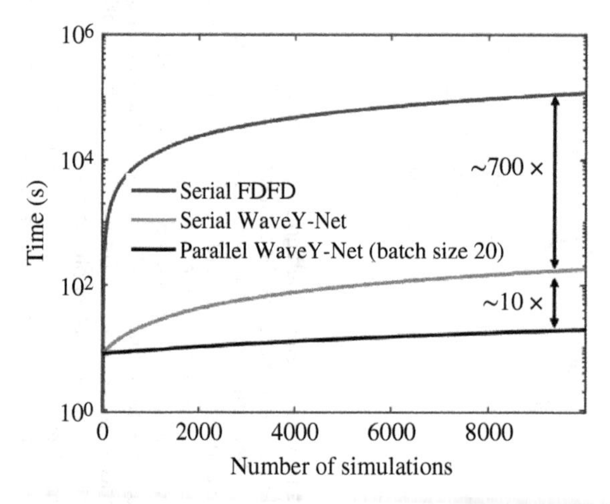

Figure 7.13 Comparison plot of computation time versus number of simulations for WaveY-Net utilized in serial and parallel modes of operation, benchmarked against serial FDFD. The parallel WaveY-Net processes batches of 20 devices simultaneously. Source: Reproduced from [27]. Licensed under CC BY 4.0.

using WaveY-Net due to its use of GPU hardware. WaveY-Net is thus particularly advantageous for applications requiring multiple batches of simulations, such as population-based optimization algorithms [34, 35]. The computationally efficient, high throughput, physically robust, and general nature of the WaveY-Net platform redefines the quantity of EM simulations and domain sizes that are considered computationally feasible by conventional CEM standards.

7.6 Conclusions and Perspectives

The rapid growth of machine learning in the last decade paved the way for a new class of CEM solvers that will provide practitioners the opportunity to operate beyond the confines of conventionally formulated solvers. Coupled with a concerted effort within the computational electromagnetics research community to share data and open-source code [36], the conditions are in place for the rapid development of high-performance, quasi-general machine learning-based EM solvers. In this chapter, we explored illustrative studies of purely machine learning-based surrogate solvers and hybrid conventional CEM-machine learning solvers that can operate orders of magnitude faster than conventional CEM solvers. These speed ups promise to transform CEM: by reducing the simulation and optimization time for electromagnetic devices from hours to seconds, the design and evaluation process is dramatically accelerated. High-speed full-wave simulators will enable new classes of devices and systems to be evaluated. For example, hybrid metamaterial-refractive elements have the potential to exhibit new wavefront shaping capabilities, but there is a computational mismatch between conventional wave-based simulators, which are slow and difficult to scale to large domains, and ray optics simulators, which are fast and operate on large scales. High-speed surrogate full-wave solvers could eliminate this mismatch and provide a practical route to simulating and optimizing these systems.

The concepts introduced in this chapter represent a relatively nascent, few-year effort, and demonstrations involving, thus far, relatively simple model systems. While promising, the discussed architectures and network training schemes are limiting in ways that make their extension to non-trivial problems a challenge. For example, the state-of-the-art image segmentation networks adapted to CEM are limited to rectilinearly organized pixels and convolutional operators, which are not directly compatible with irregular meshes. Furthermore, although the loss functions of the segmentation networks can be engineered to incorporate physical intuition into the learned function approximators, the neural network architectures themselves do not explicitly incorporate rigorous physical or mathematical structure tailored for CEM. Much more further research is required to develop purpose-built neural network architectures that can push the boundaries of surrogate solver accuracy, generalizability, and robustness.

For systems involving irregular meshes, such as those in FEM, graph neural networks (GNNs) are a newly developed class of neural networks that promise to be an active area for machine learning-based CEM research [33, 37]. GNNs are designed to process data that are represented on graphs, which are most generally composed of sets of vertices called *nodes* that are connected by a set of directed *edges*. Graphs therefore lend themselves to naturally represent meshes, which can either store electromagnetic field values directly or a set of basis functions that are subsequently used to determine the fields of interest. GNNs support the framework to process and learn regressive functions for data on graph structures, through operations that propagate information from the nodes, along the edges, in a spatially invariant manner (i.e. in a way that is applicable to any graph structure, with any number of neighboring nodes). This is similar to CNNs, which apply the convolutional operation to images composed of pixels, except that the CNN framework requires a fixed number of neighboring pixels for each pixel targeted by a convolutional operation. The GNN framework has already demonstrated its success in learning to make accurate predictions for incompressible fluid dynamics simulations from data only [38], which indicates that they are also likely to be able to learn physics problems that are governed by Maxwell's equations. GNNs thus unlock a new machine learning paradigm for mesh-based CEM that, at the time of writing, has not been substantively explored.

Fourier neural operators (FNOs) offer a novel formulation of neural networks that is capable of learning the mapping between the parameter space of a PDE directly to its solution space, rather than learning a discretized relationship for a single instance of a PDE [39]. The FNO is mesh-independent, as it learns a continuous function rather than the discretized weight matrices of conventional deep learning methods, making it a more flexible tool than discretizing conventional CEM methods, including FEM and FD techniques. This unlocks the ability to train the model using data on arbitrarily connected meshes and evaluate it on a completely different mesh type. The main working principle of the FNO is that it learns sets of global sinusoidal kernel functions that are integrated over the kernel's entire input domain, in contrast to the CNN's method of learning local kernels that are spatially convolved across the domain. The integration of the learned global kernel function across its entire input domain is performed using a Fourier transform, resulting in a very quick and efficient algorithm. Given that the inputs and outputs of PDEs are continuous, the FNO technique is ideal for learning the solution space of PDEs. The FNO has been successfully applied to a variety of incompressible fluid problems constrained by sets of PDEs [39], suggesting that the method is extendable to Maxwell's equations. The FNO is a sensible contender to succeed the state-of-the-art segmentation networks utilizing spatial convolutional operators currently powering most machine learning electromagnetics solvers.

Acknowledgments

We thank Jiaqi Jiang and Der-Han Huang for insightful discussions. Robert Lupoiu is supported by a graduate fellowship award from Knight-Hennessy Scholars at Stanford University. The authors acknowledge support from the National Science Foundation under award no. 2103301 and ARPA-E with award no. DE-AR0001212.

References

1 Kurrer, K.E. (2009). The history of the theory of structures: from arch analysis to computational mechanics. *International Journal of Space Structures* 23 (3). https://doi.org/10.1260/026635108786261018.

2 Jaluria, Y. and Atluri, S.N. (1994). Computational heat transfer. *Computational Mechanics* 14: 385–386.

3 Igel, H. (2017). *Computational Seismology*. Oxford: Oxford University Press.

4 Rylander, T., Inglestrom, P., and Bondeson, A. (2013). *Computational Electromagnetics*, 2e. Springer.

5 Hubing, T., Su, C., Zeng, H., and Ke, H. (2009). Survey of Current Computational Electromagnetics Techniques and Software.

6 Sheng, X.Q. and Song, W. (2012). *Essentials of Computational Electromagnetics*. Wiley.

7 Jin, J.-M. (2014). *The Finite Element Method in Electromagnetics*, 3e. Journal of Chemical Information and Modeling.

8 Yee, K.S. (1996). Numerical solution of initial boundary value problems involving Maxwell's equations in isotropic media. *IEEE Transactions on Antennas and Propagation* 14 (3): 302–307.

9 Courant, R., Friedrichs, K., and Lewy, H. (1967). On the partial difference equations of mathematical physics. *IBM Journal of Research and Development* 11 (2): 215–234.

10 Saad, Y. and Schultz, M.H. (1986). GMRES: a generalized minimal residual algorithm for solving nonsymmetric linear systems. *SIAM Journal on Scientific and Statistical Computing* 7 (3): https://doi.org/10.1137/0907058.

11 Liesen, J. and Strakos, Z. (2013). *Krylov Subspace Methods: Principles and Analysis.* Oxford University Press. ISBN 9780199655410.

12 Gedney, S.D. (1996). An anisotropic perfectly matched layer-absorbing medium for the truncation of FDTD lattices. *IEEE Transactions on Antennas and Propagation* 44 (12): 1630–1639.

13 Katz, D.S. and Taflove, A. (1994). Validation and extension to three dimensions of the Berenger PML absorbing boundary condition for FD-TD meshes. *IEEE Microwave and Guided Wave Letters* 4 (8): 268–270.

14 Guo, L., Li, M., Xu, S., and Yang, F. (2019). Study on a recurrent convolutional neural network based FDTD method. *2019 International Applied Computational Electromagnetics Society Symposium - China (ACES).* Nanjing, China: IEEE.

15 Yao, H.M. and Jiang, L.J. (2018). Machine learning based neural network solving methods for the FDTD method. *2018 International Symposium on Antennas and Propagation & USNC/URSI National Radio Science Meeting.* Boston, MA, USA: IEEE.

16 Yao, H.M. and Jiang, L. (2018). Machine-learning-based PML for the FDTD method. *IEEE Antennas and Wireless Propagation Letters* 18 (1): 192–196.

17 Key, C. and Notaros, B.M. (2020). Data-enabled advancement of computation in engineering: a robust machine learning approach to accelerating variational methods in electromagnetics and other disciplines. *IEEE Antennas and Wireless Propagation Letters* 19 (4): 626–630.

18 Key, C. and Notaros, B.M. (2021). Predicting macro basis functions for method of moments scattering problems using deep neural networks. *IEEE Antennas and Wireless Propagation Letters* 20 (7): 1200–1204.

19 Voulodimos, A., Doulamis, N., Doulamis, A., and Protopapadakis, E. (2018). Deep learning for computer vision: a brief review. *Computational Intelligence and Neuroscience* 2018: 7068349.

20 Liu, H., Liu, F., Fan, X., and Huang, D. (2021). Polarized self-attention: towards high-quality pixel-wise regression. *ArXiv.*

21 Long, J., Shelhamer, E., and Darrell, T. (2015). Fully convolutional networks for semantic segmentation. *Proceedings of the IEEE Conference on Computer Vision and Pattern Recognition (CVPR)*, (07–12 June 2015), pp. 3431–3440.

22 Tang, W., Shan, T., Dang, X. et al. (2018). Study on a 3D Possion's equation slover based on deep learning technique. *2018 IEEE International Conference on Computational Electromagnetics (ICCEM)*, January 2018, pp. 1–3. IEEE.

23 Kingma, D.P. and Ba, J. (2014). Adam: a method for stochastic optimization.

24 Ronneberger, O., Fischer, P., and Brox, T. (2015). U-Net: convolutional networks for biomedical image segmentation. In: *Medical Image Computing and Computer-Assisted Intervention - MICCAI 2015. MICCAI 2015, Lecture Notes in Computer Science*, vol. 9351 (ed. N. Navab, J. Hornegger, W. Wells, and A. Frangi), 234–241. Cham: Springer.

25 Pascanu, R., Mikolov, T., and Bengio, Y. (2013). On the difficulty of training recurrent neural networks. *Proceedings of the 30th International Conference on Machine Learning, PMLR*, Volume 28 (3), pp. 1310–1318.

26 Wiecha, P.R. and Muskens, O.L. (2020). Deep learning meets nanophotonics: a generalized accurate predictor for near fields and far fields of arbitrary 3d nanostructures. *Nano Letters* 20 (1): 329–338.

27 Chen, M., Lupoiu, R., Mao, C. et al. (2022). High speed simulation and freeform optimization of nanophotonic devices with physics-augmented deep learning. *ACS Photonics*. 9 (9): 3110–3123.

28 Qi, S., Wang, Y., Li, Y. et al. (2020). Two-dimensional electromagnetic solver based on deep learning technique. *IEEE Journal on Multiscale and Multiphysics Computational Techniques* 5: 83–88.

29 Raissi, M., Perdikaris, P., and Karniadakis, G.E. (2019). Physics-informed neural networks: a deep learning framework for solving forward and inverse problems involving nonlinear partial differential equations. *Journal of Computational Physics* 378: 686–707.

30 Zhu, Y., Zabaras, N., Koutsourelakis, P.S., and Perdikaris, P. (2019). Physics-constrained deep learning for high-dimensional surrogate modeling and uncertainty quantification without labeled data. *Journal of Computational Physics* 394: 56–81.

31 Lu, L., Pestourie, R., Yao, W. et al. (2021). Physics-informed neural networks with hard constraints for inverse design. *Figure license*. https://creativecommons.org/licenses/by-nc-nd/4.0/ (accessed 4 April 2023).

32 Bertsekas, D.P. (2014). *Constrained Optimization and Lagrange Multiplier Methods*. Academic Press.

33 Wu, Z., Pan, S., Chen, F. et al. (2021). A comprehensive survey on graph neural networks. *IEEE Transactions on Neural Networks and Learning Systems* 32 (1): 4–24.

34 Jiang, J. and Fan, J.A. (2019). Global optimization of dielectric metasurfaces using a physics-driven neural network. *Nano Letters* 19 (8): 5366–5372. PMID: 31294997.

35 Jiang, J. and Fan, J.A. (2019). Simulator-based training of generative neural networks for the inverse design of metasurfaces. *Nanophotonics* 9 (5): 1059–1069.

36 Jiang, J., Lupoiu, R., Wang, E.W. et al. (2020). MetaNet: a new paradigm for data sharing in photonics research. *Optics Express* 28 (9): 13670–13681.

37 Zhou, J., Cui, G., Hu, S. et al. (2020). Graph neural networks: a review of methods and applications. *AI Open* 1: 57–81.

38 Pfaff, T., Fortunato, M., Sanchez-Gonzalez, A., and Battaglia, P. (2021). Learning mesh-based simulation with graph networks. *International Conference on Learning Representations*.

39 Li, Z., Kovachki, N., Azizzadenesheli, K. et al. (2020). Fourier neural operator for parametric partial differential equations.

8

Design of Nanofabrication-Robust Metasurfaces Through Deep Learning-Augmented Multiobjective Optimization

Ronald P. Jenkins, Sawyer D. Campbell, and Douglas H. Werner

The Pennsylvania State University, University Park, PA, USA

8.1 Introduction

8.1.1 Metasurfaces

Over the past few decades, metasurfaces have been a source of much excitement to the engineering and optics communities due to their promise in achieving significant size reduction for optical systems, as well as the realization of previously impossible optical performance. What makes metasurfaces unique is that they provide a way to specify the boundary conditions between two spaces with subwavelength granularity. Relying on the generalized form of Snell's Law (Eq. (8.1)), metasurfaces can be used to introduce alternative phase gradients onto a surface, enabling anomalous reflection and refraction among other forms of wavefront engineering [1, 2].

$$n_t \sin(\theta_t) = n_i \sin(\theta_i) + \frac{\lambda_0}{2\pi}\frac{d\Phi}{dx} \tag{8.1}$$

As we can see from this equation, the angle of refraction changes when the phase gradient $\frac{d\Phi}{dx}$ changes. Anomalous refraction alone is a major point of interest as it is the backbone of so-called metalenses, which use metasurface augmentation to, for example, focus light from a flat surface [3–5]. Other applications and possibilities abound like the correction of various aberrations even in a broadband manner. Metasurfaces achieve their properties by relying on subwavelength structures, called unit cells, to interact with incident waves over a given bandwidth. At optical and near-infrared wavelengths, these unit cells are nanoscale structures and thus their physical realization relies on advanced nanofabrication methods.

8.1.2 Fabrication State-of-the-Art

Nanofabrication with E-beam lithography is a great resource for researchers interested in producing relatively small volume products with high precision [6]. Lithographic methods of fabrication rely on the exposure of a resist material to some incident field or particles (in the case of E-beam lithography, an electron beam). Once exposed by some dose of radiation, the parts of the resist material which have been exposed can be etched away with a chemical. E-beam lithography has been of particular interest to researchers, as it can be carried out for comparatively less cost than other methods of lithography. With enough care, nanoscale unit cells can typically be fabricated with the necessary precision – feature sizes as small as 2–4 nm have been produced using certain resists like polymethyl methacrylate (PMMA) or hydrogen silsesquioxane (HSQ) [7].

Advances in Electromagnetics Empowered by Artificial Intelligence and Deep Learning, First Edition.
Edited by Sawyer D. Campbell and Douglas H. Werner.

8.1.3 Fabrication Challenges

Although in small volumes these structures can be fabricated with the requisite precision, doing so requires very tight controls on the fabrication process. Thus, structures whose performance are sensitive to geometrical variations will only tolerate a very narrow range of possible processes (also known as the process window). The process window is very impactful for product throughput and quality control, and so is of critical concern when moving from research fabrication to industrial applications [8]. This presents a design problem for nanophotonic engineers to grapple with – namely, to design unit cells which tolerate probable geometrical variations and still operate optimally. The method discussed in this chapter primarily seeks to address the computational challenges that arise when designing and optimizing nanophotonic structures in this context.

8.1.3.1 Fabrication Defects

Many different kinds of fabrication defects (Figure 8.1) can contribute to unit cell performance degradation. These include such factors as erosion and dilation, side-wall angle, permittivity variations, and surface roughness among many others. Perhaps foremost among these to receive attention in the design community is erosion and dilation, also known as edge deviation. Edge deviation is a geometrical defect where the walls of a pattern are added to or removed from. In lithography,

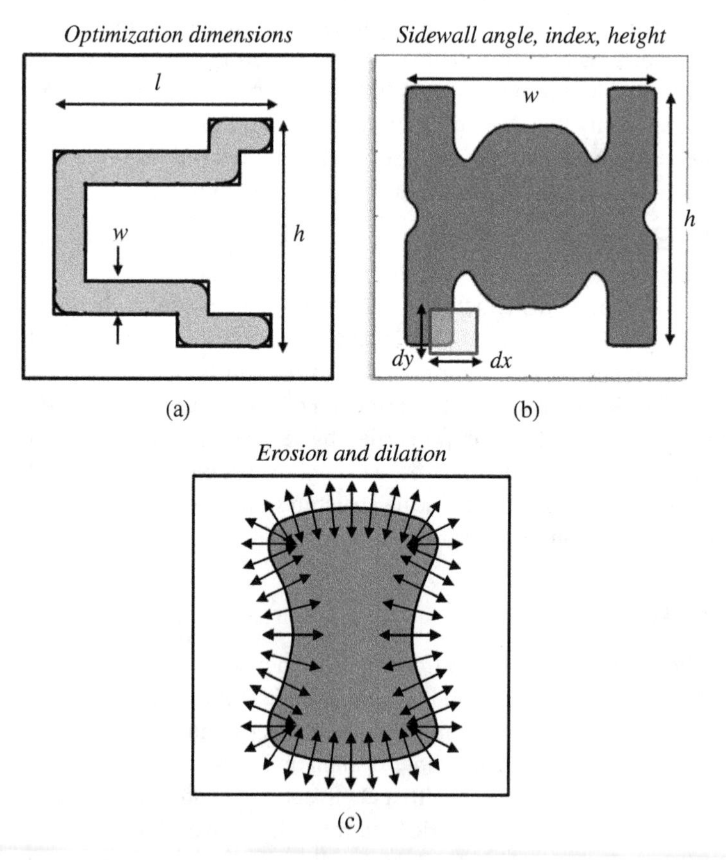

Figure 8.1 Fabrication defect varieties. Changes in optimization dimensions (a) as well as geometric parameters (b) will affect device performance. Of particular interest is the effect of erosion and dilation (c) on structures caused by imprecise dosing and/or etching.

this is caused by over- and under-dosing or etching – a phenomenon which can be controlled very effectively on the fabrication side with a sufficiently tight process window using techniques like the iso-focal dose method [9]. But very tight process windows have an associated cost. Scaling fabrication up to larger volumes of product while maintaining a miniscule process window may become financially problematic, and so the question arises: what can be done about fabrication defects?

8.1.4 Overcoming Fabrication Limitations

Successful and cost-effective realizations of larger-scale nanophotonic devices and metasurface-augmented optical systems at optical wavelengths will require improvements on the fabrication side, design side, or both. On the fabrication side, the focus is on reducing the so-called critical dimension (CD) or minimum feature size (MFS) possible at scale. These metrics describe the minimum length scale achievable for nanostructures with a given fabrication process type. For example, some recent work has produced wafer-scale Si metasurfaces with a MFS less than 100 nm using deep UV immersion lithography [10]. Others have developed computational models for lithographic fabrication which can characterize defects realistically, as well as methods for optimizing fabrication parameters to achieve a target structure [11–16]. On the design side, improvements to metasurface robustness produce a commensurate increase in computational load. In Section 8.2, detailed attention will be given to the existing methods for optimizing metasurfaces in a robustness context, specifically in the area of topology optimization (TO). However, as will be shown, designing for truly robust structures using prior methods under modest-to-large ranges of geometrical failure is intractable. This is the primary motivating factor that led to our work [17] wherein we augment an optimization process with deep learning (DL) to overcome the unavoidable computational overhead of thorough tolerance analysis.

8.2 Related Work

Many different kinds of unit cell structures have been used successfully in the past for metasurface design. Designs relying on concise parameterizations, like boxes, disks, I's and H's, are relatively easy to work with and optimize. In general though, more flexible geometrical definitions raise the performance ceiling when combined with optimization. Of preeminent status among the more complex metasurface parameterizations have been those designs optimized using the TO method.

8.2.1 Robustness Topology Optimization

TO is an exciting area of nanophotonic design that has been a focus of much interest for many years now. Viewed broadly, TO is a method that can be applied to many kinds of design problems from mechanical systems to electromagnetic ones. What sets TO apart from other optimization methods is the extreme flexibility it has to produce freeform structures. This unprecedented level of geometric specification has pushed the performance upper bound for metasurfaces much higher, as freeform shapes have the potential to be significantly more performant than canonical shapes (like disks, boxes, etc.) [18–25]. Common to all these is an assumed discretization for the structure – essentially structures are broken down into pixels (or voxels) and changes are made on a per-pixel basis. When applied to metasurface design, TO is typically used to create a 2D binary mask which defines exposed and unexposed areas for lithographic fabrication. TO works by computing a material gradient for each pixel and then iteratively updating the material distribution and recomputing the gradient until convergence.

The gradient itself is computed by comparing the real electric fields within the current structure with a target "adjoint" electric field. For more than two decades now, TO researchers have been developing methods for applying fabrication constraints within the TO paradigm [26]. Indeed, we use the morphological filter method for establishing a fixed MFS described in [26] as part of our own metasurface geometric parameterization.

It is true that structures which obey fabrication constraints will tend to be more robust. However, it has long been known that such robustness cannot be assumed, and so more recent TO work from the past decade has focused on developing modifications of TO which directly optimize for robustness. It was shown in [20] more than a decade ago that design sensitivity could be improved by incorporating the TO gradients from multiple designs together at once – not only the nominal structure but also the structure under positive and negative edge deviation. Variations on this same method have persisted to today with great success in improving the robustness of freeform TO metasurfaces [24]. These methods come with several advantages. First, they maintain the same geometric flexibility for which TO is so loved while simultaneously mitigating its tendency to produce highly sensitive structures. Second, they preserve the existing algorithmic structure of TO, integrating well into established inverse-design loops. However, they also have some disadvantages. Foremost, being a local method, it is inherently challenging for TO to locate globally optimal solutions. Additionally, the best multiobjective optimization (MO) methods are global and population-based rather than local like TO. This is relevant when it comes to robustness, as there are inherent tradeoffs between nominal performance and robustness which we will show later on. Finally, exhaustive robustness studies are simply intractable when a sufficient number of variations are tested to make strong claims about performance bounds. Any metasurface optimization relies on a full-wave solver to evaluate a structure's performance. Even with fast solvers like rigorous coupled-wave analysis (RCWA), the large numbers of design variation simulations required for exhaustive robustness analysis become computationally infeasible.

In order to overcome the limitations just listed, we chose to move away from TO and instead focus on multiobjective global evolutionary optimization. To make this change, we parameterized our metasurfaces using a different free-form design scheme than TO which will be discussed later. However, the most fundamental issue for exhaustive metasurface robustness optimization is the intractability of using full-wave solves for all edge deviation tests. To address this, we turned to DL as an avenue for cutting the total computational cost significantly.

8.2.2 Deep Learning in Nanophotonics

Recent years have seen a surge in interest in DL within the metasurface design and nanophotonics community more broadly [27–30]. Metasurface design specifically has been a major beneficiary of DL-augmentation in multiple ways. Some methods have sought to perform generic fast E-field predictions for use in inverse-design [31], whereas others have integrated DL more tightly into the inverse-design process in the form of generative models [32–35]. Concretely, DL has been used for the design of absorbers [36], diffraction gratings [37, 38], chiral metasurfaces [39], and all-dielectric metasurfaces [40–43]. This surge of interest has been driven by the successful application of DL in many other fields: image and natural language processing, biomedical applications, as well as other forms of engineering. DL applications to nanophotonics have been significantly inspired by existing applications, but researchers in this area have developed their own unique methods and goals. Deep neural networks (DNNs) act as trainable universal function approximators, which has led many to use it for optimization and inverse design. In fact, common to many relevant use cases of DNN is their application as a surrogate modeling technique for various kinds of optimization problems. By learning an approximation of the design problem's cost function, a DL model can stand-in for the full cost function, which will inevitably be more computationally expensive to run than a DL model. Of course, surrogate modeling methods have been used in electromagnetic inverse-design for many years [44–46]. Owing to the backpropagation method, DNNs are a particularly exciting inclusion in this mature area due to their remarkable ability to handle learning problems at scale. Whether against large numbers of inputs, outputs, or complexity of problem (network depth), DNNs have for years been the gold standard for scalable learning.

8.3 DL-Augmented Multiobjective Robustness Optimization

The aforementioned surrogate modeling method is the same route we charted for the following results using DNNs. By creating an alternative high-efficiency computing vehicle for testing edge deviations, it is possible to alleviate the computational load of full-wave simulations required for robust metasurface optimization.

8.3.1 Supercells

As was discussed before, the core of any metasurface design is its set of constituent unit cells. A library of unit cell designs can be used to form groups of unit cells called supercells. The resulting bulk behavior of the supercells can produce the kinds of phase profiles which make metasurfaces so interesting and useful for optical design. As we learn from Snell's law, due to the physics of wave propagation, transmission through a periodic grating will confine the scattered radiation to discrete angles called diffraction orders. The number of these angles depends on the periodicity of the grating compared to the wavelength of the light. Due to the conservation of energy, power delivered into the orders will sum to match the input power. Diffraction efficiency (DE) is a measurement of what fraction of the input power is delivered into a given order, and is critically important in supercell design. For supercells designed to diffract into a specific order, say the +1 order, we therefore care about maximizing the DE into the +1 order (DE_{+1}). A perfect version of such a supercell would reach a DE_{+1} of 1, leaving all the other available orders with a DE of 0.

8.3.1.1 Parameterization of Freeform Meta-Atoms

Of course, there are many possible ways of parameterizing a unit cell's mask geometry. These can range from simple structures like disks or boxes to freeform shapes like those produced through TO. One attractive freeform parameterization called shape-optimization works by using a level-set on a spline surface. This method has been used successfully in the past for designing many different metasurfaces and unit cells [47]. Figure 8.2 demonstrates typical shape optimization. A set of control points is laid out on a regular x/y grid, each with some user-selected height. The spline surface interpolates between the set of control points smoothly to form an arbitrary function in x, y. To form a final mask from this surface, a level-set function compares the height of the surface with a reference height. Those regions of the surface which are above the reference will be "filled in" in the mask, whereas those below will be voids. Optimization is done by tuning the heights of the control points. This approach covers a wide variety of possible freeform structures while nevertheless making the problem suitable for global optimization methods.

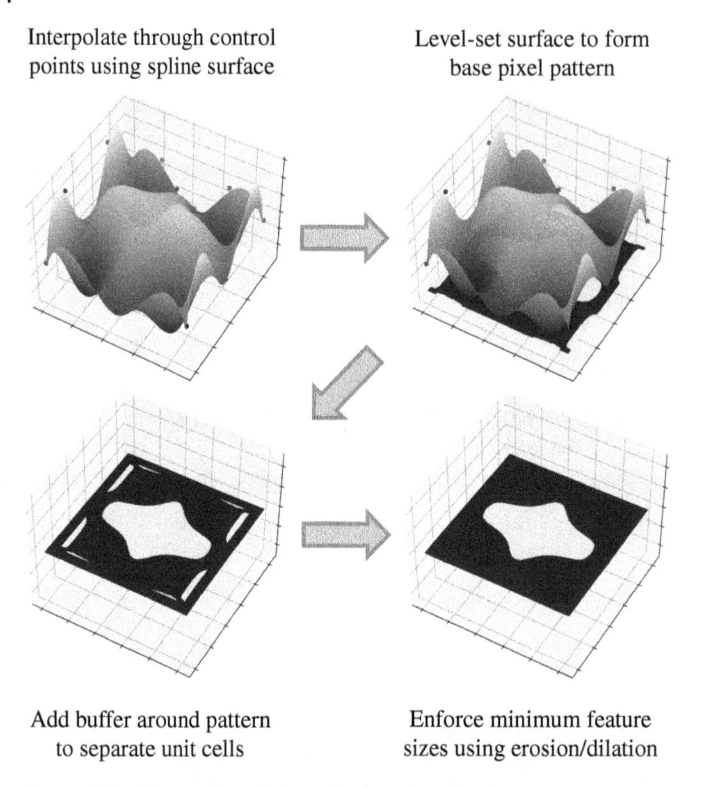

Figure 8.2 Illustration of the spline level-set method for freeform mask shapes.

Rather than constructing supercells from individually optimized unit cells, our study combines them together and optimizes the supercells directly (Figure 8.3). While this is not strictly necessary from a pure metasurface design standpoint, it is necessary to avoid a spurious source of performance variation. Since the purpose of robustness optimization is to characterize performance losses arising from geometric variations (edge deviation specifically here), it is necessary to rule out other potential sources of performance change first. A major contributor to performance degradation when going from unit cells to supercells is inter-cell coupling. When unit cells are electrically close, their near-fields will couple, altering their respective resonant modes. Thus, coupling between unit cells is problematic for teasing out the contribution to performance degradation from geometric variations. It has been shown by An et al. [48] that it is possible to characterize the coupling between adjacent unit cells in a supercell; however, this is a relatively recent development and so incorporating this technique may be an avenue for future progress in metasurface robustness optimization. Despite the direct use of supercells, in this study unit cells are still separated by gaps in order to preserve the analogy with truly library-based metasurface construction. The unit cell dimensions were chosen for an operational wavelength λ of 1.55 μm to be $\lambda/3 = 516.7$ nm on a side. When constructing each unit cell using the spline level-set method, 3×3 control points were initially defined in one corner before applying 4-fold symmetry to form a full 5×5 set of control points. For supercells composed of four unit cells each, this yielded a grating period of 2067 nm. This choice of supercell size permitted three diffraction orders, −1, 0, and +1 at −48.6°, 0°, 48.6°, respectively. The pattern for each unit cell was composed of Si embedded in air resting on a substrate of SiO_2, with incident fields impinging on the structure from the substrate side. The height of each cell was $\lambda/2 = 775$ nm. While this method is applied to the generation of 2D unit-cell structures, in general it could be extended to contiguous supercells or even into 3D structures which can achieve more performance stability over wide fields-of-view [49].

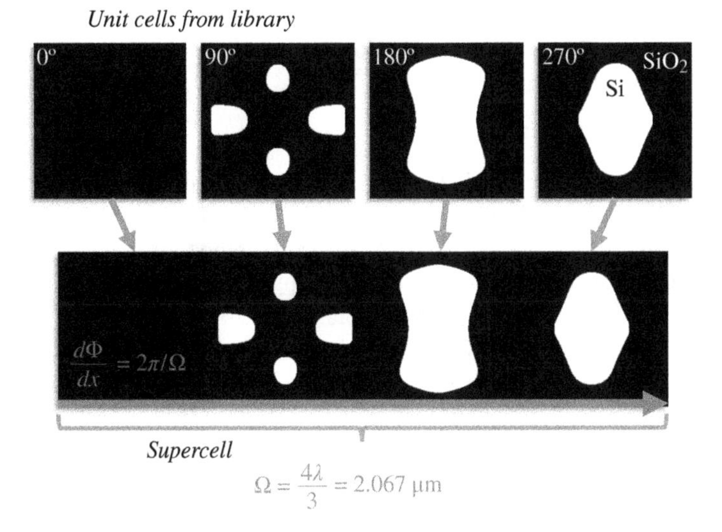

Figure 8.3 Supercell as a composite of unit cells from a phase-engineered library. A unit cell is electrically small (side length $< \lambda/2$), whereas a supercell is typically on the order of a wavelength. Here we compose four unit cells, each defined with an individual set of control points, into a single supercell for full-wave simulation or DNN prediction.

8.3.2 Robustness Estimation Method

8.3.2.1 Simulating Defects

As previously mentioned, we used the morphological filtering method described in [26] in order to simulate the effect of erosion and dilation on a supercell structure. This approach uses a standard image processing structuring element to affect changes to the boundaries between filled and empty regions of the supercell masks. For example, erosion of four pixels could be achieved by applying a 1 px "empty" structuring element against the mask four times to cause the boundary to recede by four pixels total. This method worked very well to provide high fidelity changes to the structure which were nevertheless physically meaningful and measurable.

8.3.2.2 Existing Estimation Methods

Estimating robustness as it relates to edge deviation is something which has been attempted previously [20, 24]. These methods represent the state-of-the-art in robustness estimation for TO. The general method that has been used is to evaluate extremes of variation of a structure, along with the nominal version of a given design. If the design under consideration might undergo up to ± 20 nm of variation, these altered geometries are tested along with the base geometry. In line with standard TO operation, the gradients for these alternate structures are also computed before being combined with the gradient of the nominal structure through a weighted sum.

8.3.2.3 Limitations of Existing Methods

Of course, this approach leaves the range between the nominal and each extreme open for potential performance losses, and as we will see later this can indeed be an issue for these methods. Thus, such methods cannot be considered to be "exhaustive," but rather they direct the TO away from many of the most irreparably sensitive solutions in the design space. However, a more fundamental issue is that this kind of method cannot tractably scale up to test larger numbers of variations. First, testing finer structural changes requires a more finely resolved mesh or grid representing the structure. For sufficiently small variations, this can cause a significant slowdown as the solver runtime for any individual test increases. Second, the additional runtime that comes from more full-wave solves per design may make large-scale optimization largely intractable. For these reasons, we wish to find an alternate method; one which can handle the high resolution and variation count which are required for exhaustive edge deviation tolerance analysis.

8.3.2.4 Solver Choice

In principle, any solver which can simulate periodic dielectric structures in 3D can be used for supercell robustness optimization. However in practice some options are more efficient than others, and so have gained widespread use in this area. RCWA is a powerful method for simulating periodic multi-layer dielectric structures which is probably the most popular choice for this kind of problem at the present [50]. This method is best geared toward pillar-based structures, like those most easily created through lithographic fabrication. RCWA works by assuming periodicity in the x and y directions, and then converting those components of Maxwell's equations to Fourier space. Structures in RCWA are represented through a stack of z layers, with each layer being a pixel pattern of permittivities and permeabilities. When it comes to estimating edge deviation robustness, we found that the x and y resolution of these layers becomes very important. Moreover, it should be noted that RCWA is one of the fastest full-wave solvers available for this class of problems. Were the following study to be conducted with another full-wave paradigm like the finite element method or the finite difference time domain method, the speedup results would be even more favorable to the inclusion of DL in the optimization.

8.3.3 Deep Learning Augmentation

Overcoming the limitations of existing methods is the primary motivation for introducing DL into the robust metasurface design process. The applicability of DL within electromagnetics is of course broad, but in this case we focus on the use of DNNs as a surrogate model as shown in Figure 8.4. A surrogate model is a mathematical model trained on some cost function which is a stand-in or "surrogate" for the cost function during optimization. Because full-wave electromagnetic solvers

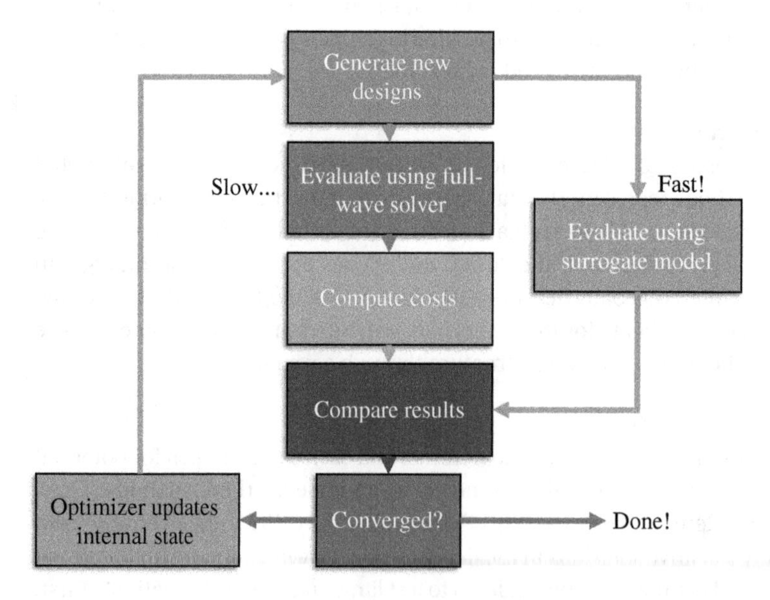

Figure 8.4 Operation of a typical surrogate-assisted optimization. DNNs can act as the surrogate, bypassing computationally expensive full-wave evaluations.

tend to have substantial runtime, surrogate modeling has been used widely in the field for many years. In the past, other modeling methods have been the go-to for engineers: radial basis functions or Gaussian process regression for example. DNNs are uniquely attractive among these methods as their ability to fit a target function scales very well with increases to input and output dimensionality, as well as complexity of the function. Thus, when it comes to modeling the behavior of freeform metasurface elements with many degrees of freedom, DNNs are an excellent fit. Surrogate modeling in optimization as a practice assumes that it's more efficient to spend time testing designs (which may or may not be part of an optimization) and training a model than just running a standard optimizer for longer. This may or may not be true depending on the problem under consideration. In general, the more use a surrogate model gets compared to full-wave function evaluations, the more worthwhile it is. This explains why surrogate modeling is so effective for robustness optimization since a surrogate model designed to understand metasurface performance under varying edge deviation can be tested many more times per design than the full-wave evaluation. This implies that upfront investment in training a DNN may be worth it, so long as the model can be trained with a sufficiently small training set. We will show how this can be done.

8.3.3.1 Challenges

While DNNs potentially offer a significant opportunity for improved optimization performance, this does not come without its challenges. Foremost is the high-resolution requirement arising from the reality that meta-elements can be highly sensitive to geometrical variation, as shown in Figure 8.5. Indeed, we found that for a supercell structure of more than 2 μm in length, significant changes in DE could be observed for as little as 2 nm of edge deviation. Ultimately this reality necessitates the use of a significantly higher resolution in RCWA than is typically used for DL of electromagnetic structures.

8.3.3.2 Method

Architecture A wide variety of network topologies have been used for DL in computational electromagnetics. Common to many are the use of convolutional neural networks (CNNs), which are particularly effective for learning image processing tasks [51]. For this application, we wish to compute

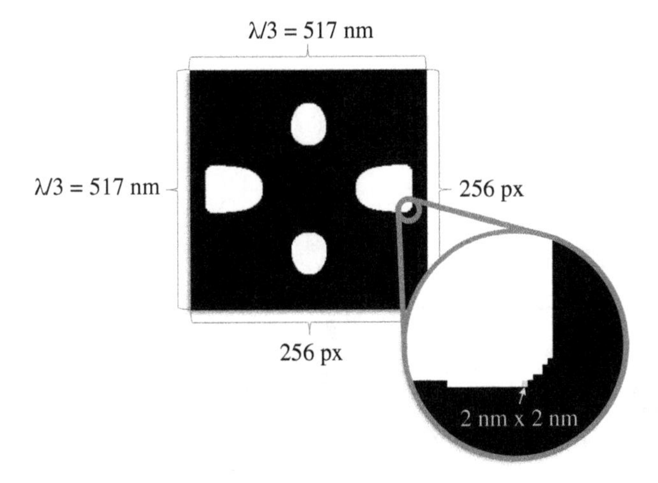

Figure 8.5 Erosion and dilation (edge deviation) pixel resolution requirement. Relevant changes to performance (as much as several percent in diffraction efficiency) can occur even for changes as small as 2 nm.

the DE of a supercell based on its mask. Once the mask has been computed from a design's control points at a sufficient resolution (in this case 1024 × 256), we wish to use it as an input to a DNN and the associated diffraction coefficient should be the output. The architecture we settled on, shown in Figure 8.6, uses a pair of networks each tasked with a different responsibility in the computation. The first network (Si → E) uses a U-Net style architecture [52] to compute internal E-fields for the

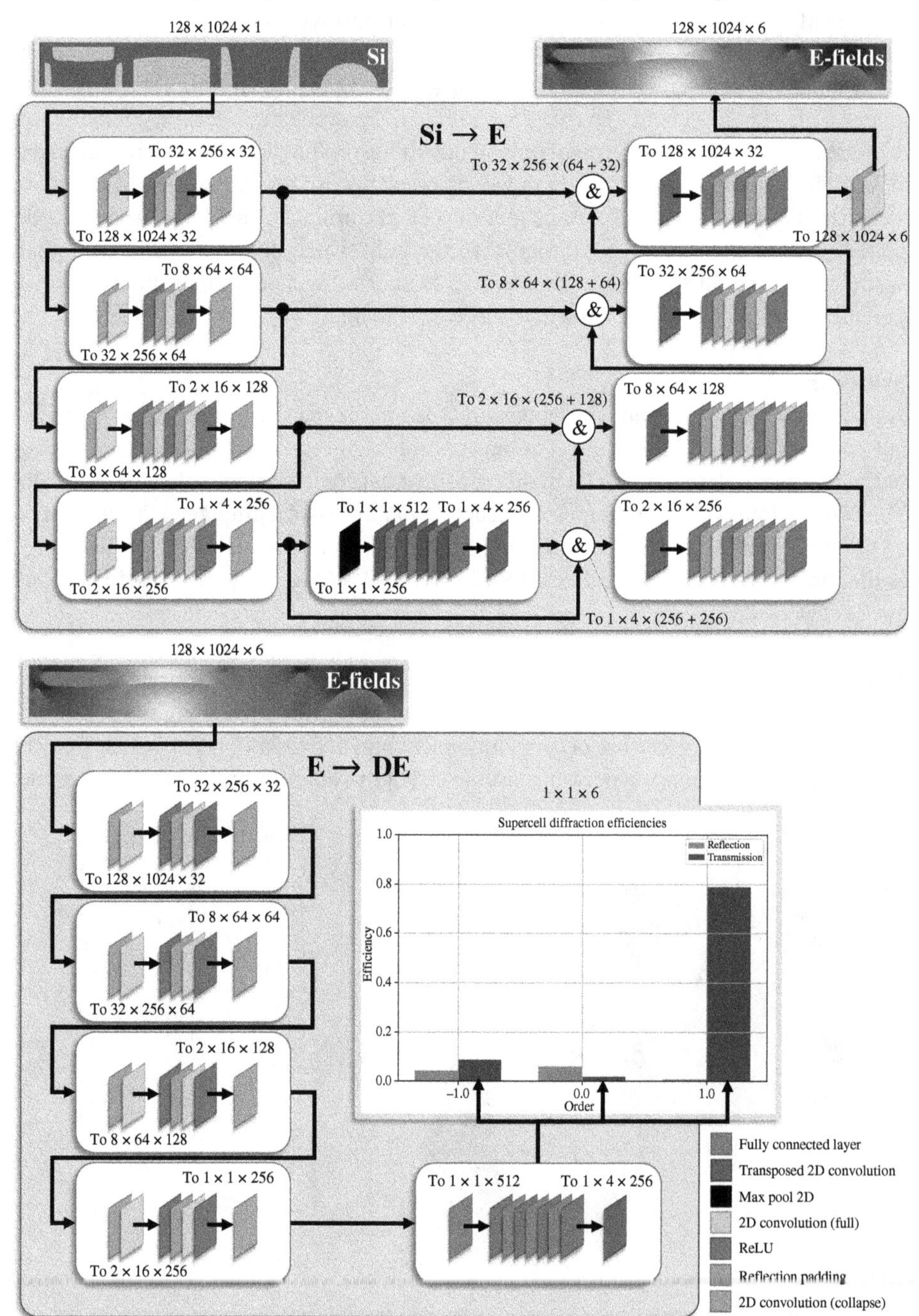

Figure 8.6 Network architecture visual (U-Net and CNN pair).

supercell using just the supercell mask. We found an increase in accuracy when training with an intermediate internal E-field stage as part of the prediction. It is also worth noting that because the unit cells all have 4-fold symmetry, any supercell will therefore have mirror-symmetry along the y axis. This results in symmetric (or antisymmetric) fields, and so the entire mask and E-field pair can be cut in half to save memory usage in the DNNs. Thus, our actual input size was 1024×128, an important consideration for memory economy. Like encoder topologies, the U-Net uses a series of progressively compressed convolutional layer groups to perform layered feature extraction on the input image before reversing the process back to the full size of the original image. What sets U-Nets apart from traditional encoders is the introduction of cross-link connections which concatenate computations across the compression region in the middle of the network. We found these connections were extremely helpful in improving the accuracy of the Si $\rightarrow$ E network as they aided in preserving large scale structure and fidelity in resulting E-field prediction. The E-fields produced by the U-Net were similarly 1024×128 pixels in x and y sizes, respectively, but additionally had six total features for the real and imaginary part of each component (x,y,z) of the E-field vector. The second stage of the network (E $\rightarrow$ DE) is a standard CNN which accepts E-fields in and produces diffraction coefficients for each possible diffraction order (these are complex numbers) as a result. The CNN architecture we selected was very similar to just the first half of the Si $\rightarrow$ E network. Blocks of convolutional layers processed and compressed the E-field image down several steps before arriving at a final prediction of diffraction coefficients for each order (-1, 0, $+1$). To compute final diffraction efficiencies (specifically for the $+1$ order of interest to this design problem), we can use Eq. (8.2):

$$\mathrm{DE}_{+1} = \frac{n_2}{n_1}\cos(\theta_t)|\mathrm{DO}_{+1}|^2 \tag{8.2}$$

Multiple different possible topologies and input/output schemes were tested during the development of this architecture, but this network pair was the one with which we had the most success.

Training Process Each neural network was trained independently, but both were able to draw from the same training set. While no hard and fast rule exists for how much data is needed to train a given network, for nanophotonic DL most work has converged on a sample count in the range of around 50,000 designs [32, 36, 37, 40, 42, 43]. Using the previously described supercell formation, a data set of 81,408 samples were collected recording input masks, intermediate E-fields, and final diffraction coefficients for each order. To fairly judge the accuracy of a network during convergence, it is best practice to withhold part of the training data as a validation set for tracking progress during training. Whereas the training set is used for updating the network during backpropagation, the validation set serves as a way for a researcher to watch training convergence and judge when to stop. Oftentimes, the training error can continue to decrease while the validation error stagnates, and this is an indicator that it's time to stop training. We chose a common division of 80% training set (65,024 designs), 20% validation set (16,384 designs). Neural network training can of course be configured with many external terms, all of which can have significant impacts on the training outcomes. These external factors are called "hyperparameters." Training for both networks used similar hyperparameters which are shown in Table 8.1.

In addition to the training and validation set sizes, initial learning rate, learning rate decay, and learning rate period were very important in guiding training. The learning rate is a scalar which multiplies against the weight gradient during network updates. We found that training converged better when this learning rate was decreased gradually throughout training. Thus, we used another scalar, the learning rate decay, to multiply against the learning rate with a period set by the learning rate period hyperparameter. There is no definitive guide on the best way to choose each of

Table 8.1 Training hyperparameters.

Hyperparameter	Value
Training set size	65,024
Validation set size	16,384
Initial learning rate	1×10^{-4}
Learning rate decay	0.9925
Learning rate period	200 iterations
Minibatch size	16

the parameters above, so ultimately trial-and-error was necessary to settle on the right choices. However, some guidelines did help direct us on some hyperparameter choices. For example, in [53], researchers found that networks trained using smaller mini-batch sizes (the set of designs analyzed between iterations of backpropagation) generalized better than those trained using large mini-batch sizes. Figure 8.7 shows training convergence curves for both the training and validation sets for both networks. Mean squared error (MSE) is a common metric for measuring the accuracy of the networks throughout training. MSE reports the difference on average between truth and prediction made by the network. At regular intervals, the training and validation sets are tested by the current network, and their MSE is calculated. When the validation MSE stops decreasing, it indicates that the network has converged and that training can stop. The validation sets for both networks converged after a significant number of iterations, although the training error continued to drop. Both networks were developed with the PyTorch library using a computer equipped with 4 NVidia RTX 2080ti graphics cards.

Model Error Estimates Many possible metrics have been proposed and used for evaluating DNN accuracy, and different ones apply better depending on the circumstance. Relative percent error (RPE) (Eq. (8.3)) normalizes the difference between guess and truth by the relative magnitude of the true value. This can be helpful in gauging the accuracy of unbounded values, as it provides a natural way to compare accuracies across different scales.

$$\mathrm{RPE}(x, x') = \frac{100}{N} \sum_{i=1}^{N} \left| \frac{x'_i - x_i}{x_i} \right| \tag{8.3}$$

where x_i is the true value, and x'_i is the prediction, and each are N dimensional vectors. For values where the set is bounded, however, a relative metric may not be as useful. In the case of diffraction efficiencies, for example, an absolute percent error (APE) (Eq. (8.4)) is more relevant to designers.

$$\mathrm{APE}(x, x') = \frac{100}{N} \sum_{i=1}^{N} |x'_i - x_i| \tag{8.4}$$

RPE can also complicate analysis of values when small amplitudes are not as relevant. This is important in the case of E-field prediction, where distinctions between small field values are simply unimportant compared to differences at larger values. Nevertheless, E-field values are unbounded which rules out an absolute error metric as can be used for DEs. The normalized cross correlation (NCC) (Eq. (8.5)) is a solution to this problem, focusing not on impact of amplitudes but rather on

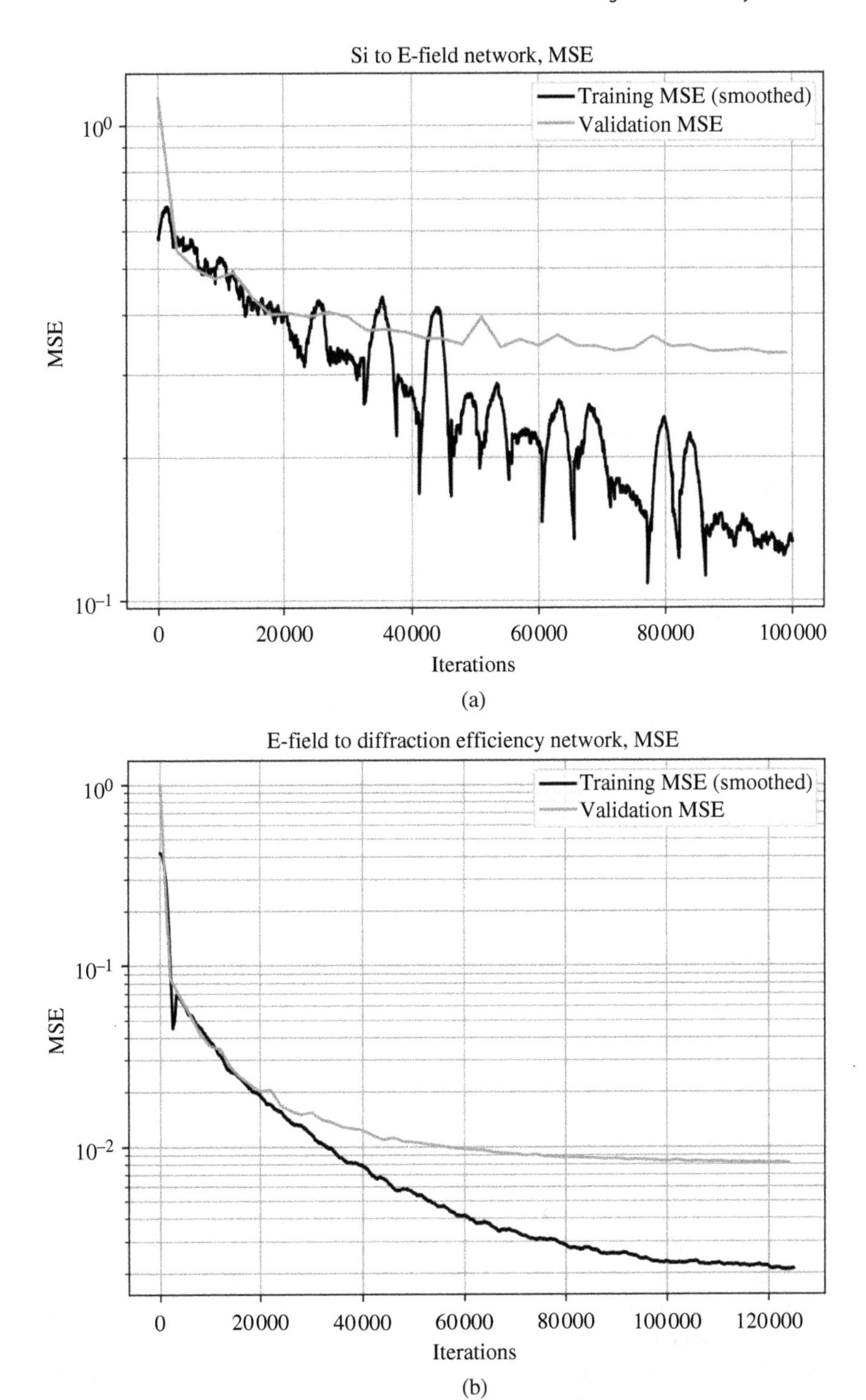

Figure 8.7 Training MSE curves for the Si to E-field network (a) and E-field to diffraction efficiency network (b).

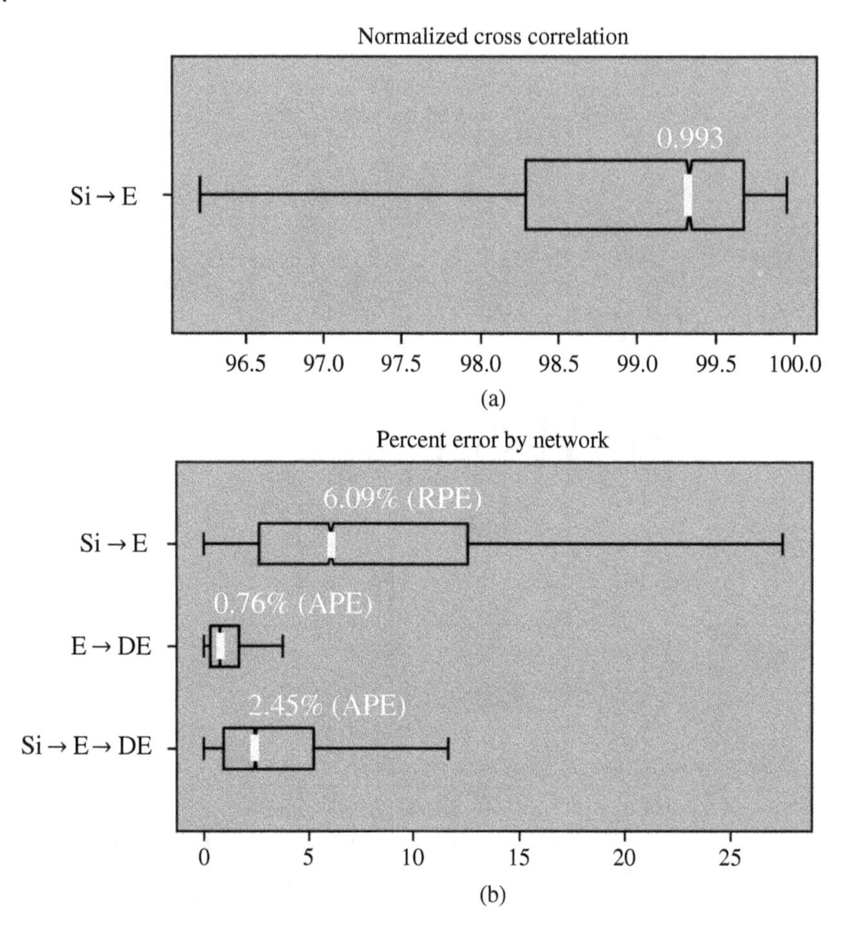

Figure 8.8 Error estimates for the two networks separately and together. (a) Normalized cross correlation of E-field predictions (higher is better). (b) Percent errors for the networks (lower is better). The final APE reached by the composite network indicates that the training reached a very high level of accuracy with half of predictions having less than 2.45% error.

relative values between dimensions. NCCs range from −1 to 1, with predictions reaching a NCC close to 1 being extremely accurate.

$$\mathrm{NCC}(x, x') = \frac{x \cdot x'}{\|x\|\|x'\|} \tag{8.5}$$

Figure 8.8 shows how each of the networks, as well as their operation together can be thoroughly analyzed using these metrics. Each network was tested by evaluating the validation set (16,384 designs) and computing the median error metric for the entire set. After training was complete, the median NCC of the Si → E network was found to be 0.993. Being well above 0.8 which is a common threshold for "good" prediction and very close to the maximum possible value of 1, this result indicates very good prediction on the part of the Si → E network. NCC indicates similarity of structure between predictions and true values, but an additional relevant measure is the maximum value of E-field predicted. Using the RPE metric, comparisons of max(E) for the predictions and true values yielded a median RPE of around 6.08%. The second network, E → DE, was tested alone using APE, with attention given only to its prediction of +1 DE as this is the prediction which is relevant for optimization. The median APE for just this network was 0.76% error, which is very good. Of course, as the networks are intended to operate serially, we also evaluated accuracy of the composite predictor Si → E → DE, which compounds inaccuracies from both networks. Fortunately, we found that the median APE for this composite network was good as well, reaching 2.45%. For any DL application, achieving a low predictive error is of course a critical piece of the puzzle. An extra motivator for this application, however, comes from the DNN's intended use as part of a global optimization procedure. In this context, the stakes are raised for network accuracy. Moreover, as we will see, good outcomes depend on carefully managing the interaction between the optimizer and DNN.

8.3.4 Multiobjective Global Optimization

For the past few decades, a wide variety of optimization methods have been used successfully in nanophotonics and electromagnetic design more generally [54–56]. When it comes to freeform metasurface design, the pool of suitable optimization methods shrinks. For one, being parameterized through a large number of design variables, the number of degrees of freedom for any freeform structure is huge. This is still true despite the structural simplifications offered through the previously described spline-based shape-optimization method. But what really makes supercell optimization challenging is the resulting complexity of the relevant cost functions.

8.3.4.1 Single Objective Cost Functions

Any constrained optimization problem can be understood as the minimization of a cost function f as shown in Eq. (8.6):

$$\min_{x} \{f(x)\}, \quad x \in \Omega \tag{8.6}$$

for some domain Ω. For supercell optimization, the cost function f will be something like Eq. (8.7):

$$f_1(x) = -\mathrm{DE}_{+1}(x) \tag{8.7}$$

Which is to say, the negative of the +1 order DE for a given input (in our case, x defines a set of control point heights for the supercell). The negative is used to provide compatibility with typical optimizers which assume minimization. What makes this optimization problem difficult is not necessarily that x is a high-dimensional vector (though that is true of this design problem). Rather, the function $\mathrm{DE}_{+1}(x)$ will be multimodal (i.e. having many minima), which is a great challenge for any state-of-the-art optimizer to deal with. So in addition to discovering how to improve a design, these optimization problems demand exploration mechanisms as well. From this point of view, it is easy to understand why TO may not be the final word on metasurface design. Indeed, because freeform metasurface costs are so multimodal and TO is a local method, it must be run many, many times in order to have a chance to find the global minimum of the cost function. These motivating factors led us to choose a global optimization method for supercell design rather than local.

8.3.4.2 Dominance Relationships

If our objective was just to optimize a supercell for +1 DE, then a single objective optimizer would do. However, attempting to optimize for robustness, a second objective complicates this matter. Simply put, there is no guarantee that a supercell designed for high nominal performance will be robust. In fact we will show later that these two goals, nominal performance and robustness, are actually in conflict. A simple method for optimizing two objectives simultaneously would be to add their costs together (perhaps with some weighting factor on each) and apply a single objective optimizer. This technique is called the "weighted-sum" approach and can achieve some level of

MO. However it fails to capture the full richness of what a MO truly is. For more discussion on why this is as well as a complete treatment of MO, we refer the reader to [57]. To put the problem succinctly, the weighted sum approach assumes that there is a single "solution" to the MO problem when this simply isn't the case for general multiobjective problems. This most certainly can't be true where competing objectives are concerned, as solutions which are optimal for one objective will be suboptimal for the other, and vice versa!

Non-dominated Sorting and the Pareto Front The core of true MO relies on a different understanding of how to compare two different cost function samples which is called non-dominated sorting (NDS). The dominance relationship shown in Eq. (8.8):

$$x_1 \prec x_2 => f_j(x_1) < f_j(x_2), \quad \forall j \in [1, 2, \dots, D] \tag{8.8}$$

states that design x_1 dominates design x_2 if each cost f_j is better for x_1 than for x_2 across all D costs. To illustrate this relationship, consider three different designs (A, B, and C) in two objectives (f_1 and f_2), as shown in Figure 8.9. The dominance relationship tells us that design A dominates design B because $f_1(A) < f_1(B)$ and $f_2(A) < f_2(B)$. On the other hand, and this is really what makes the dominance relationship special, A does not dominate C. Neither does C dominate A. In fact, A and C can be said to be "mutually non-dominating," as neither is completely better than the other; instead they tradeoff between the two objectives f_1 and f_2. Extending this comparison mechanism to a larger set of designs can be done using NDS. Mutually non-dominating designs will form subsets within the set of designs, called ranks. The "rank 1" will be the best subset, dominating all the others but being mutually non-dominating within. For a pair of competing objectives, there will be an ideal set of designs P which collectively dominate all the other solutions in the space, but which are nevertheless mutually non-dominating among themselves. P is called the Pareto set, and approximating this set through the iterative improvement of rank 1 solutions is the multiobjective equivalent to finding the global minimum of a cost function in a single objective optimization.

Many different kinds of MO optimizers have been devised over the years, but one successful category of these is multiobjective evolutionary algorithms (MOEAs). MOEAs are population-based methods designed to emulate biological evolution, employing reinforcement of both exploration

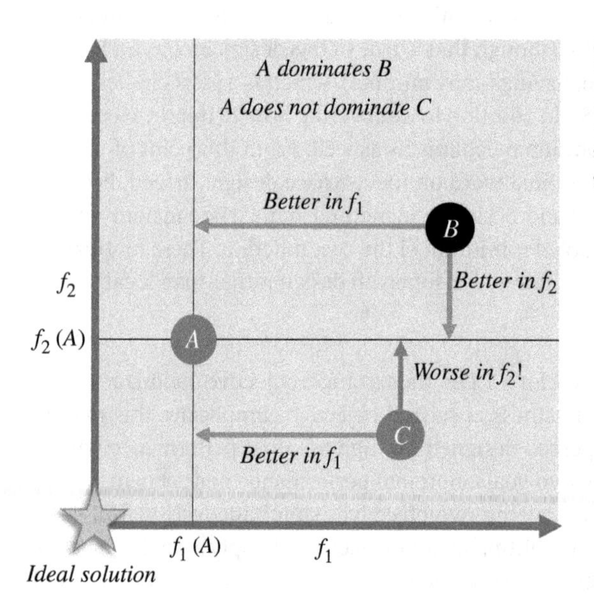

Figure 8.9 The dominances relationship between three points, A, B, and C. Design A dominates design B, but does not dominate design C. Designs A and C tradeoff between the two objectives f_1 and f_2 since neither is totally better than the other.

and exploitation mechanisms at a population level to drive optimization. As we will show later on, MOEAs can be used effectively for robustness optimization when the cost functions are properly designed.

8.3.4.3 A Robustness Objective

Back in the domain of robust supercell optimization, we can understand the Pareto set as the optimal tradeoff between nominal performance $\mathrm{DE}_{+1}(x)$ and robustness. But how do we measure robustness? We have previously discussed the need for exhaustive sampling of edge deviations in order to capture the dynamic DE variations which often accompany this defect. Within this framework, we can define a new objective which measures the change in performance from nominal arising from edge deviations. Specifically, Eq. (8.9) shows this additional cost which will be optimized alongside Eq. (8.7).

$$f_2(x) = \max_i |\mathrm{DE}_{+1}(x) - \mathrm{DE}_{+1}(x + \delta_i)| \tag{8.9}$$

where δ_i is the ith amount of edge deviation applied to nominal design x among the set of edge deviations tested. As this objective is minimized, solutions are increasingly stable. The computation of this objective is the primary place where the pretrained DNNs will provide a speedup over a traditional full-wave-only optimization.

8.3.4.4 Problems with Optimization and DL Models

The proposed robustness framework is nearly complete. However, one confounding issue remains which can have a major impact on any optimization method using DL (or another precomputed surrogate model). As an optimizer converges, it will seek out regions of the design space where the cost functions are as attractive (low) as possible. Additionally, cost functions which include DNNs will necessarily have some amount of error inherent to the network, since network accuracy is never guaranteed. Moreover DNNs which have a low median error for the validation set can nevertheless have very high error for certain unfortunate inputs. If this error leads to an artificially low (good) cost value in some places, this can be very problematic for an optimizer as it converges. Figure 8.10 illustrates this problem. If an optimizer begins converging in a region of the design space with this kind of artificially attractive cost value, the best solution's real cost can start to degrade.

8.3.4.5 Error-Tolerant Cost Functions

The way to resolve this issue is to modify the cost functions to penalize solutions which are excessively inaccurate. Equation (8.10) shows how this can be done:

$$f_{1,safe}(x) = -\mathrm{DE}_{+1}(x) + 1 \times 10^6 \cdot failure \tag{8.10}$$

where *failure* is computed through a comparison of the true $\mathrm{DE}_{+1}(x)$ and the neural network's prediction of this same nominal design's performance $\mathrm{DE}_{+1,test}$. If the absolute error between the two exceeds some difference (2% in our case) then *failure* = 1. Otherwise, *failure* = 0. By adding in a large value to f_1 where there is significant DNN inaccuracy, those regions of the design space are made unattractive to the optimizer protecting the integrity of the optimization results. A reality of this approach is that it does preclude the possibility of removing full-wave evaluation completely from the optimization. However, without this filtering method the optimization results cannot really be trusted, so it is a necessary addition.

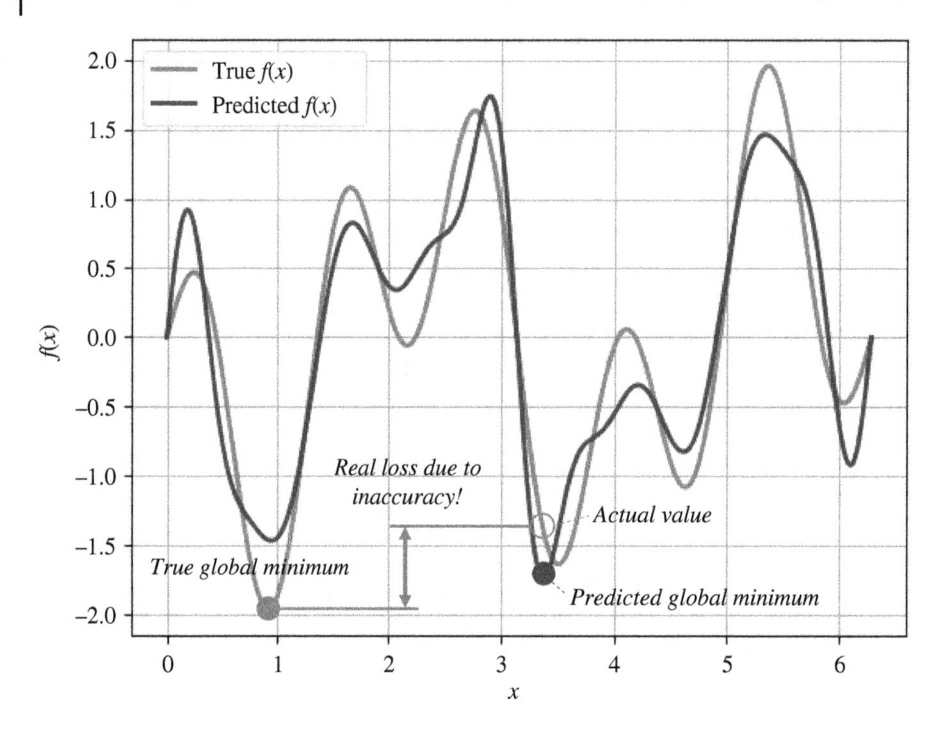

Figure 8.10 An "attractive but inaccurate" region of cost function arising from deep learning errors can lead an optimizer astray.

8.3.5 Robust Supercell Optimization

We chose to use the non-dominated sorting genetic algorithm II (NSGA-II) as the optimizer for this design problem to show how this method generalizes well to the class of global multiobjective optimizers which are commonly available to engineers [58]. NSGA-II is a very well-known and commonly used MOEA which has been influential enough to act as a kind of benchmark for comparing multiobjective optimizers in general. As a genetic algorithm, NSGA-II provides solid exploratory mechanisms through mutation and crossover which are critical to solving multimodal design problems. Moreover, as a multiobjective optimizer, NSGA-II lets us develop an approximation of the tradeoff between the two costs f_1 and f_2. As has been previously mentioned, the supercells are parameterized with 36 control point variables and 1 edge deviation variable. Thus, the optimization must operate on 36 input variables and 2 cost functions, a substantial design problem.

8.3.5.1 Pareto Front Results

The optimization results yielded important insights into the underlying tradeoffs between nominal efficiency and robustness of freeform supercells. Figure 8.11 shows the Pareto front which was found by NSGA-II within the space of the two objectives f_1 and f_2, although reexpressed in a way that is more readable. Rather than reporting the cost values directly, designs are plotted in analogous terms of nominal +1 DE and guaranteed +1 DE. These preserve the same multiobjective structure as the true cost functions while being easier to understand in the context of nanophotonic engineering. A couple of valuable pieces of information can be gleaned from these results. First, it's worth noting that an ideal design, which would have perfect nominal efficiency (100% on the y-axis) and also perfect guaranteed robustness (100% on the x-axis), is unreachable in the design space. Such a "utopia" point doesn't appear to exist in the set of feasible designs as the

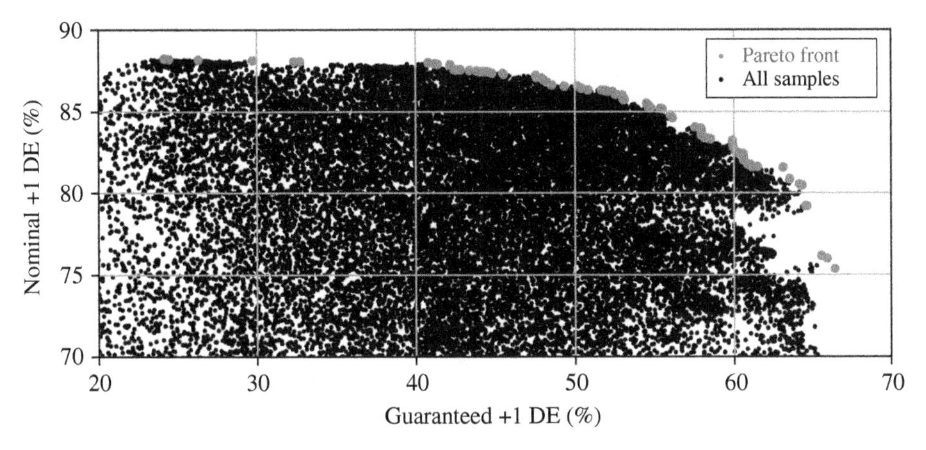

Figure 8.11 Results from NSGA-II optimization. The Pareto front (light gray) dominates all the other designs which were tested (black). Nominal performance and robustness are evidently in conflict in the Pareto front.

best-case tradeoff curve doesn't contain this solution. Instead, we see a clear exchange taking place – designs with high nominal efficiency are less robust while those with high robustness take a hit in nominal efficiency. Perhaps the most important outcome from this study comes from the observed slope of the tradeoff curve at various locations along the Pareto front. At the left of the plot where the Pareto front has maximized nominal performance, the curve is nearly flat. So for a very small sacrifice in nominal performance (say, as little as 2% absolute efficiency), it's possible to gain 10s of percents of guaranteed performance! This illustrates a common phenomenon in MO: the best performers for each objective will often do so at great expense to other objectives. The extreme ends of the Pareto front here have sharp changes which yield diminishing returns for one preferred objective. Let's examine some designs from the optimization results in more detail.

8.3.5.2 Examples from the Pareto Front

Looking closer at solutions from Figure 8.11, we can develop more valuable insight into the mechanisms underlying supercell robustness. Figure 8.12 shows three designs, A, B, and C, which have been picked from the set of designs tested by NSGA-II. Each of the three designs was plotted as a function of edge deviation, showing both the DNN prediction of their DE_{+1} as well as the true DE_{+1} computed through full-wave simulation at each edge deviation value. Designs A and B come from either end of the Pareto front and demonstrate the exchange made possible through this DL-augmented multiobjective analysis. Design A is the best solution found in terms of just nominal performance, reaching 86.3% absolute DE into the +1 order. However, as can be seen in the edge deviation curve, it suffers significant failure at +10 nm of edge deviation, dropping its guaranteed DE_{+1} to just 30.8%. On the other side of the Pareto front, design B accepts some loss in nominal performance (4.41% down from design A), but achieves a remarkable improvement in guaranteed efficiency of 35.9% more than design A. Any intermediate of this exchange can be chosen by the designer depending on their application, all made possible because of multiobjective analysis. In this case, it's easy to favor robustness because of such an extreme improvement for a comparatively small price in nominal performance. Had the problem been optimized only for nominal performance, however, robust solutions like B would never have been evident.

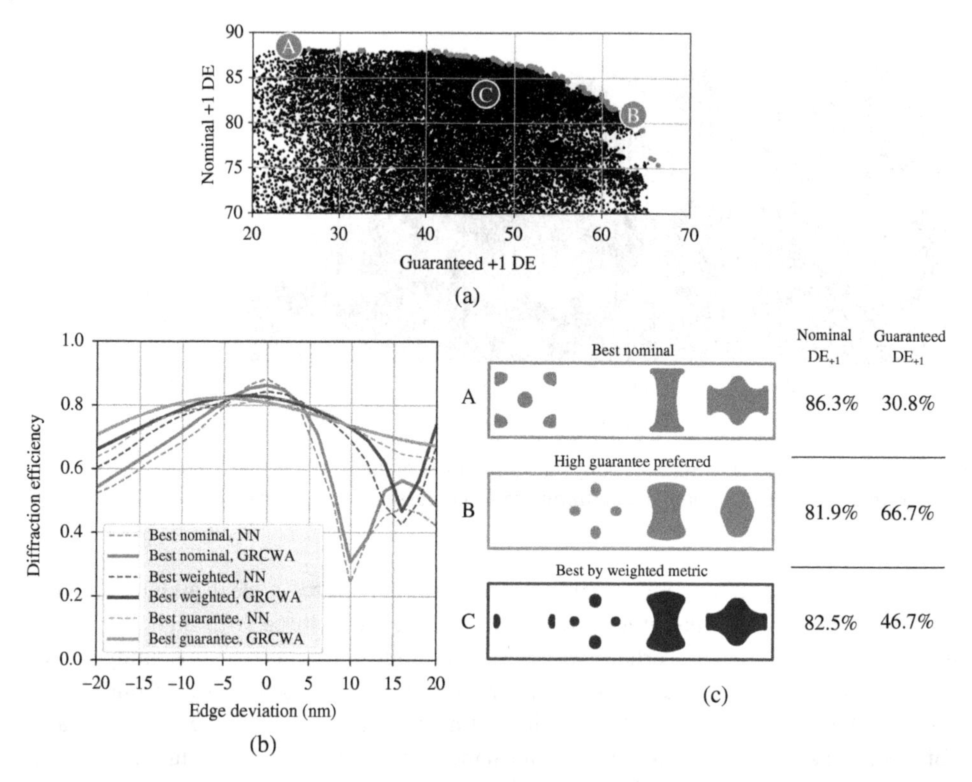

Figure 8.12 Comparison of three different designs tested by NSGA-II. (a) The dataset from Figure 8.11 where designs A, B, and C have been called out. (b) Edge deviation curves demonstrate how each design behaves under erosion and dilation defects. (c) The masks for each of the three designs are show along with the exact values computed with RCWA for nominal and guaranteed DE_{+1}.

8.3.5.3 The Value of Exhaustive Sampling

The resolution of the x-axis in the above edge deviation curves further illustrates how important it is for reliable robustness optimization to use exhaustive sampling. This stands in contrast to other existing methods which only consider the extreme values of edge deviation when computing a cost (a method necessitated by the unavoidable slowdown of running full-wave solves for many variations on each design). Figure 8.13 shows how robustness optimization using only extreme values of edge deviation is likely to lead to problematic solutions. This figure shows a transformation of the data collected from the original NSGA-II robustness optimization into a new space. The x-axis remains the same, but the y-axis has been exchanged. Instead of nominal performance, it provides an example weighted-sum metric which might be used during a weighted-sum single objective robustness optimization. Equation (8.11) expresses this candidate-weighted metric which combines nominal performance with performance at the extremes of edge deviation:

$$DE_{weighted}(x) = 0.5DE_{+1}(x) + 0.25DE_{+1}(x + \delta_{+20\,nm}) + 0.25DE_{+1}(x + \delta_{-20\,nm}) \tag{8.11}$$

where $\delta_{+20\,nm}$ and $\delta_{-20\,nm}$ represent the changes to the input variable x resulting from +20 nm or −20 nm edge deviation, respectively. The exact choices of weights in Eq. (8.11) are not particularly important and merely represent one possible prioritization among the different values. Whereas design C is clearly not Pareto optimal in the original Figure 8.11, we see that this design appears to be so in the transformed domain of Figure 8.13. It's easy to understand why this is the case when reviewing the edge deviation curve of design C in Figure 8.13. Design C is unique among the three

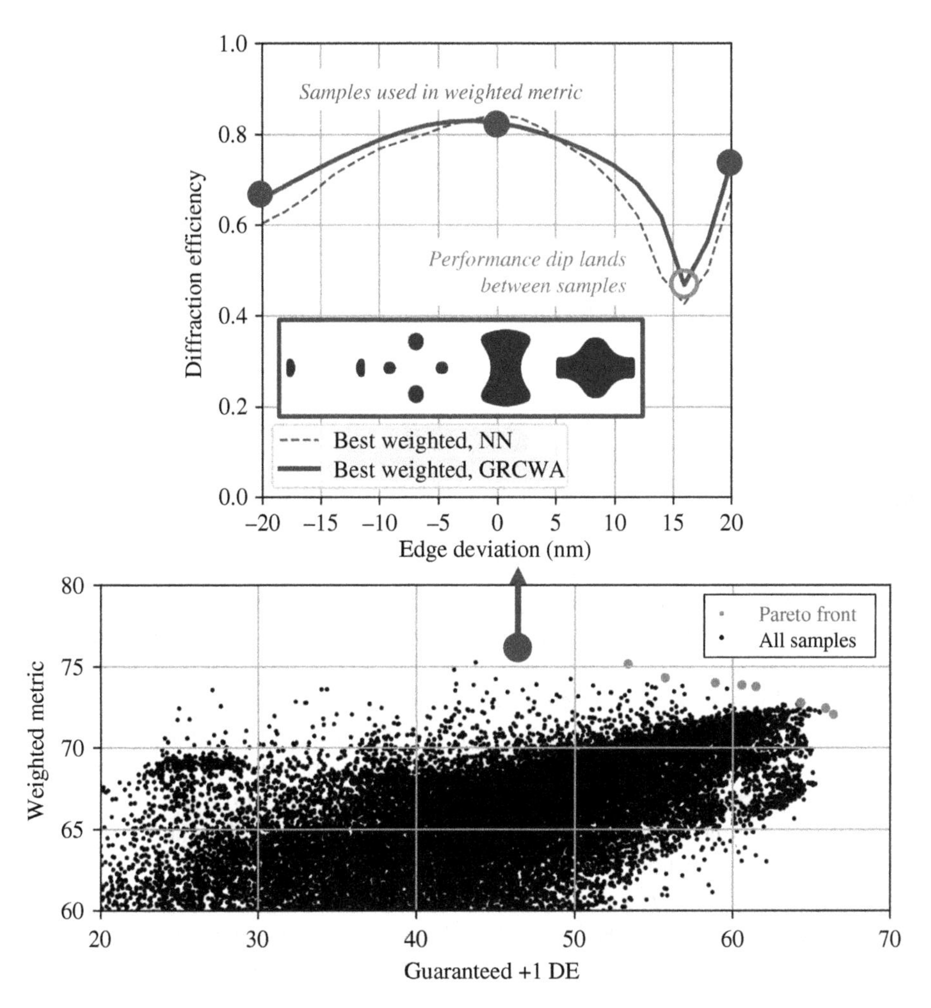

Figure 8.13 Transformation of the designs shown in Figure 8.12 into a weighted-sum space. In this new coordinate space, design B appears to be Pareto optimal. However, closer examination of its edge deviation curve (top) shows a performance dip between the coarse erosion and dilation samples.

in that it presents solid DEs at the nominal and extremes of edge deviation (0 nm and $\pm$20 nm), but has a significant fault between them. In fact, an optimizer using this weighted sum metric would be incentivized to place faults in just this way, avoiding the coarsely spaced test points. Without an exhaustive sampling approach (facilitated here by DL), weighted-sum based methods are liable to result in these kinds of solutions.

8.3.5.4 Speedup Analysis

Although we've shown that DL-augmented multiobjective robustness optimization enables exciting new ways of designing nanophotonic devices, it's necessary also to establish that this method offers a true performance advantage over competing techniques. In fact, the advantage of this method (and many DL-augmented design methods) hinges on its efficiency, since this is the primary way in which DL is assisting the optimization process. Many proposed DL-augmented design methods in the literature forgo this analysis, but we believe it is critical to establishing that our approach is truly useful to the field. Table 8.2 provides a comprehensive analysis of the

Table 8.2 Time analysis showing the speedup achieved against a purely full-wave version of the study.

Field description	RCWA-only	DL-augmented	Speedup
Single sample time (serial)	1.61 s	0.034 s	47.4
Data collection sample time (with internal fields)	N/A	3.35 s	N/A
Data collection total time	N/A	3.16 d	N/A
Training time	N/A	3.15 d	N/A
Full supercell evaluation (including edge deviations)	33.9 s	2.29 s	14.8
Single optimization (>100,000 samples)	5.60 wk	2.65 d	14.8
Total time (1 optimization)	5.60 wk	8.96 d	4.37
Total time (6 optimizations)	>8 mo	>3 wk	10.60

time taken by our method versus the time that would be taken by an equivalent full-wave-only optimization. The critical difference between using only RCWA as opposed to including DNN evaluations depends on the relationship between the upfront cost of collecting training data and training the networks versus the speedup gained back during optimization by replacing full-wave simulations with DNN predictions. Full robustness evaluation of a design using RCWA alone takes approximately 33 seconds. This appears to be fast, but in the context of a MO over 36 variables, it is painfully slow. Because such an optimization can be expected to take at least 100,000 function evaluations to converge, the resulting total time for a single optimization is more than 5.6 weeks! On the other hand, a full robustness evaluation using RCWA and DNN together takes approximately 2.29 seconds. Compared with an RCWA-only optimization, this would yield an excellent runtime of 2.65 days, an apparent 14.8 times speedup. In reality, the DNN's data collection time as well as training time must also be included which increases the total time for one optimization to a 8.96 days, a more modest speedup of 4.37 times. The performance advantage of DNNs as a precomputed surrogate model arises therefore as its use outstrips the startup time. Indeed, as Figure 8.14 shows, the speedup advantage will grow with additional optimization time

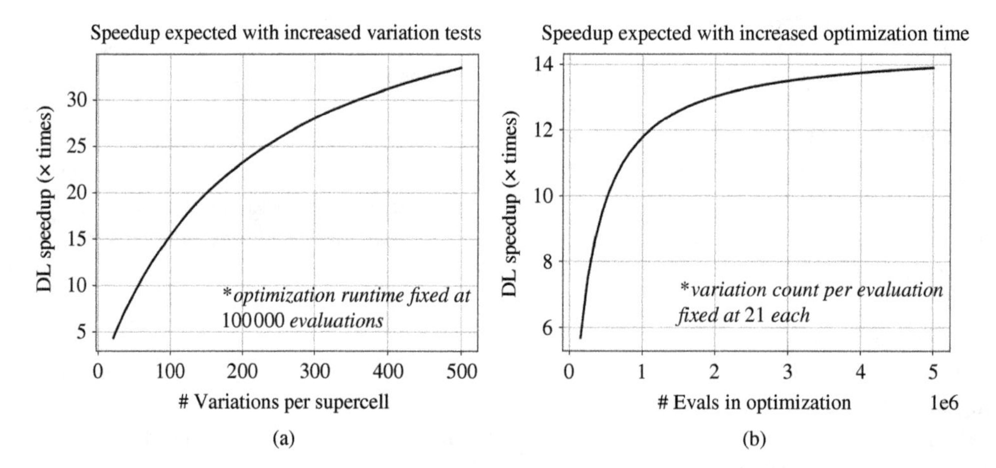

Figure 8.14 Speedup curves which demonstrate how DL-augmented optimizations provide a significant advantage in time versus equivalent full-wave-only optimizations. (a) Speedup with increased variations per supercell. (b) Speedup with network usage in optimizations. The more the pretrained DNN is used, the more the time investment in data collection and training becomes worthwhile.

or robustness evaluation granularity (say 1 nm edge deviation resolution rather than 2 nm). These speedup curves show that although more speedup is achieved the more the DNN is used, such a speedup is not unbounded. Diminishing returns in the expected speedup arise predictably as the total runtime involved for both RCWA-only and DL-augmented methods is dominated by RCWA evaluations. This is true even for the DL-augmented method due to the inclusion of RCWA evaluations which address the filtering concerns discussed in Section 8.2. As is often the case with optimization studies, despite our best efforts the development of the above supercell results wasn't done with a single run of the optimizer. Rather we had to conduct the above study over the course of no fewer than six optimization attempts as the exact parameters of optimization and filtering were tuned. Table 8.2 concludes that the total speedup we achieved with our method given full scope of the study was more than 10.6 times!

8.4 Conclusion

The proposed method for DL-augmented multiobjective robustness optimization provides new insights into the robustness characteristics of metasurface supercell designs. We demonstrated that it was possible to train a network pair to be extremely accurate in predicting DEs, using E-fields as an intermediate representation. The calculated values of DE were sufficiently close to truth as to enable the use of a benchmark optimizer, NSGA-II. Moreover, the cost functions employed in the multiobjective optimizer were tailored to suit the accuracy of the network, resulting in a stable and reliable design methodology. We also found that granular edge deviation tests were critical to capturing the full dynamic behavior of these meta-devices. Indeed, a simple transformation of the results demonstrated that only testing the extremes of variation was problematic, opening the door to solutions which were artificially attractive having carefully placed failures "in the gaps" between edge deviation samples. The Pareto front extracted from NSGA-II provided insight into the rate of exchange (for dielectric supercells) between nominal performance and robustness. The robust solution we presented yielded a remarkable increase of at least 35% more guaranteed DE into the +1 order compared to a design with just optimized nominal performance for a modest loss of <4.5% in nominal efficiency! Moreover, this method provided a significant speedup of more than 10 times what would have been achievable through an equivalent exclusively full-wave optimization. Indeed, we estimated that DL-augmentation saved multiple months of wait-time to complete the full set of optimizations for the study. Until the state-of-the-art in nanofabrication accuracy develops below the nanoscale, methods like the one presented in this chapter will be a critical piece of the puzzle for realizing wafer-scale metasurface designs. DL-augmented robustness is a step forward toward bringing wavefront engineering out of the lab and into the foundry.

8.4.1 Future Directions

Multiple avenues for additional work remain in this area. The proposed method could be extended to alternate formulations for supercells or unit cells to further develop our understanding of the robustness relationships between different types of structures. As previously mentioned, the issue of coupling necessitated that this analysis be performed at the level of supercells. But if coupling between unit cells could be differentiated from the impact of defects as part of the robustness analysis, this would open the door to the development of full unit-cell libraries with robustness characteristics, an exciting possibility. Finally, edge deviation is not the only relevant fabrication defect which could be considered. Future work with this technique may see its extension to additional kinds of fabrication error (like side-wall angle, surface roughness, and material index).

Acknowledgments

This research was supported in part by DARPA EXTREME (contract HR00111720032).

References

1 Chen, H.-T., Taylor, A.J., and Yu, N. (2016). A review of metasurfaces: physics and applications. *Reports on Progress in Physics* 79 (7): 076401.

2 Werner, D.H., Campbell, S.D., and Kang, L. (2020). *Nanoantennas and Plasmonics: Modelling, Design and Fabrication*. SciTech Publishing.

3 Khorasaninejad, M., Shi, Z., Zhu, A.Y. et al. (2017). Achromatic metalens over 60 nm bandwidth in the visible and metalens with reverse chromatic dispersion. *Nano Letters* 17 (3): 1819–1824.

4 Chen, W.T., Zhu, A.Y., Sanjeev, V. et al. (2018). A broadband achromatic metalens for focusing and imaging in the visible. *Nature Nanotechnology* 13 (3): 220–226.

5 Nagar, J., Campbell, S., and Werner, D. (2018). Apochromatic singlets enabled by metasurface-augmented GRIN lenses. *Optica* 5 (2): 99–102.

6 Cui, Z. (2008). Nanofabrication. *ECE Course Notes* 730: 91–144.

7 Chen, Y. (2015). Nanofabrication by electron beam lithography and its applications: a review. *Microelectronic Engineering* 135: 57–72.

8 Bossung, J.W. (1977). Projection printing characterization. *Proceedings of SPIE 0100, Developments in Semiconductor Microlithography II*, Volume 100, pp. 80–85. International Society for Optics and Photonics.

9 Keil, K., Choi, K.-H., Hohle, C. et al. (2008). Determination of best focus and optimum dose for variable shaped e-beam systems by applying the isofocal dose method. *Microelectronic Engineering* 85 (5–6): 778–781.

10 Hu, T., Tseng, C.-K., Fu, Y.H. et al. (2018). Demonstration of color display metasurfaces via immersion lithography on a 12-inch silicon wafer. *Optics Express* 26 (15): 19548–19554.

11 Hawryluk, R. (1981). Exposure and development models used in electron beam lithography. *Journal of Vacuum Science and Technology* 19 (1): 1–17.

12 Hudek, P. and Beyer, D. (2006). Exposure optimization in high-resolution e-beam lithography. *Microelectronic Engineering* 83 (4–9): 780–783.

13 Pinge, S., Qiu, Y., Monreal, V. et al. (2020). Three-dimensional line edge roughness in pre- and post-dry etch line and space patterns of block copolymer lithography. *Physical Chemistry Chemical Physics* 22 (2): 478–488.

14 Azumagawa, K. and Kozawa, T. (2021). Application of machine learning to stochastic effect analysis of chemically amplified resists used for extreme ultraviolet lithography. *Japanese Journal of Applied Physics* 60 (SC): SCCC02.

15 Mu, X., Chen, Z., Cheng, L. et al. (2021). Effects of fabrication deviations and fiber misalignments on a fork-shape edge coupler based on subwavelength gratings. *Optics Communications* 482: 126562.

16 Eissa, M., Mitarai, T., Amemiya, T. et al. (2020). Fabrication of Si photonic waveguides by electron beam lithography using improved proximity effect correction. *Japanese Journal of Applied Physics* 59 (12): 126502.

17 Jenkins, R.P., Campbell, S.D., and Werner, D.H. (2021). Establishing exhaustive metasurface robustness against fabrication uncertainties through deep learning. *Nanophotonics* 10 (18): 4497–4509.

18 Chen, Y., Zhou, S., and Li, Q. (2010). Multiobjective topology optimization for finite periodic structures. *Computers and Structures* 88 (11–12): 806–811.

19 Diaz, A.R. and Sigmund, O. (2010). A topology optimization method for design of negative permeability metamaterials. *Structural and Multidisciplinary Optimization* 41 (2): 163–177.

20 Wang, F., Jensen, J.S., and Sigmund, O. (2011). Robust topology optimization of photonic crystal waveguides with tailored dispersion properties. *JOSA B* 28 (3): 387–397.

21 Dong, H.-W., Wang, Y.-S., Ma, T.-X., and Su, X.-X. (2014). Topology optimization of simultaneous photonic and phononic bandgaps and highly effective phoxonic cavity. *JOSA B* 31 (12): 2946–2955.

22 Zhou, M., Lazarov, B.S., and Sigmund, O. (2014). Topology optimization for optical projection lithography with manufacturing uncertainties. *Applied Optics* 53 (12): 2720–2729.

23 Yi, G. and Youn, B.D. (2016). A comprehensive survey on topology optimization of phononic crystals. *Structural and Multidisciplinary Optimization* 54 (5): 1315–1344.

24 Wang, E.W., Sell, D., Phan, T., and Fan, J.A. (2019). Robust design of topology-optimized metasurfaces. *Optical Materials Express* 9 (2): 469–482.

25 Fan, J.A. (2020). Freeform metasurface design based on topology optimization. *MRS Bulletin* 45 (3): 196–201.

26 Sigmund, O. (2007). Morphology-based black and white filters for topology optimization. *Structural and Multidisciplinary Optimization* 33 (4): 401–424.

27 Massa, A., Marcantonio, D., Chen, X. et al. (2019). DNNs as applied to electromagnetics, antennas, and propagation-a review. *IEEE Antennas and Wireless Propagation Letters* 18 (11): 2225–2229.

28 Campbell, S.D., Jenkins, R.P., O'Connor, P.J., and Werner, D. (2021). The explosion of artificial intelligence in antennas and propagation: how deep learning is advancing our state of the art. *IEEE Antennas and Propagation Magazine* 63 (3): 16–27.

29 Khatib, O., Ren, S., Malof, J., and Padilla, W.J. (2021). Deep learning the electromagnetic properties of metamaterials a comprehensive review. *Advanced Functional Materials* 31 (31): 2101748.

30 Ma, W., Liu, Z., Kudyshev, Z.A. et al. (2021). Deep learning for the design of photonic structures. *Nature Photonics* 15 (22): 77–90.

31 Wiecha, P.R. and Muskens, O.L. (2020). Deep learning meets nanophotonics: a generalized accurate predictor for near fields and far fields of arbitrary 3D nanostructures. *Nano Letters* 20 (1): 329–338. PMID: 31825227.

32 Ma, W., Cheng, F., Xu, Y. et al. (2019). Probabilistic representation and inverse design of metamaterials based on a deep generative model with semi-supervised learning strategy. *Advanced Materials* 31 (35): 1901111.

33 Kudyshev, Z.A., Kildishev, A.V., Shalaev, V.M., and Boltasseva, A. (2020). Machine-learning-assisted metasurface design for high-efficiency thermal emitter optimization. *Applied Physics Reviews* 7 (2): 021407.

34 Zhu, D., Liu, Z., Raju, L. et al. (2021). Building multifunctional metasystems via algorithmic construction. *ACS Nano* 15 (2): 2318–2326. PMID: 33416319.

35 Ma, W. and Liu, Y. (2020). A data-efficient self-supervised deep learning model for design and characterization of nanophotonic structures. *Science China Physics, Mechanics & Astronomy* 63 (8): 1–8.

36 Qiu, T., Shi, X., Wang, J. et al. (2019). Deep learning: a rapid and efficient route to automatic metasurface design. *Advanced Science* 6 (12): 1900128.

37 Inampudi, S. and Mosallaei, H. (2018). Neural network based design of metagratings. *Applied Physics Letters* 112 (24): 241102.

38 Jiang, J. and Fan, J.A. (2019). Global optimization of dielectric metasurfaces using a physics-driven neural network. *Nano Letters* 19 (8): 5366–5372.

39 Ma, W., Cheng, F., and Liu, Y. (2018). Deep-learning-enabled on-demand design of chiral metamaterials. *ACS Nano* 12 (6): 6326–6334.

40 An, S., Fowler, C., Zheng, B. et al. (2019). A deep learning approach for objective-driven all-dielectric metasurface design. *ACS Photonics* 6 (12): 3196–3207.

41 Nadell, C.C., Huang, B., Malof, J.M., and Padilla, W.J. (2019). Deep learning for accelerated all-dielectric metasurface design. *Optics Express* 27 (20): 27523–27535.

42 An, S., Zheng, B., Shalaginov, M.Y. et al. (2020). A freeform dielectric metasurface modeling approach based on deep neural networks. *arXiv preprint arXiv:2001.00121.*

43 An, S., Zheng, B., Shalaginov, M.Y. et al. (2020). Deep learning modeling approach for metasurfaces with high degrees of freedom. *Optics Express* 28 (21): 31932–31942.

44 Koziel, S. and Leifsson, L. (2013). *Surrogate-based Modeling and Optimization.* Springer.

45 Leifsson, L. and Koziel, S. (2016). Surrogate modelling and optimization using shape-preserving response prediction: a review. *Engineering Optimization* 48 (3): 476–496.

46 Easum, J.A., Nagar, J., Werner, P.L., and Werner, D.H. (2018). Efficient multiobjective antenna optimization with tolerance analysis through the use of surrogate models. *IEEE Transactions on Antennas and Propagation* 66 (12): 6706–6715.

47 Whiting, E.B., Campbell, S.D., Kang, L., and Werner, D.H. (2020). Meta-atom library generation via an efficient multi-objective shape optimization method. *Optics Express* 28 (16): 24229–24242.

48 An, S., Zheng, B., Shalaginov, M.Y. et al. (2021). Deep convolutional neural networks to predict mutual coupling effects in metasurfaces. *Advanced Optical Materials* 10 (3): 2102113.

49 Zhu, D.Z., Whiting, E.B., Campbell, S.D. et al. (2019). Optimal high efficiency 3D plasmonic metasurface elements revealed by lazy ants. *ACS Photonics* 6 (11): 2741–2748.

50 Jin, W., Li, W., Orenstein, M., and Fan, S. (2020). Inverse design of lightweight broadband reflector for relativistic lightsail propulsion. *ACS Photonics* 7 (9): 2350–2355.

51 Krizhevsky, A., Sutskever, I., and Hinton, G.E. (2012). ImageNet classification with deep convolutional neural networks. *Advances in Neural Information Processing Systems 25 (NIPS 2012)*, pp. 1097–1105.

52 Ronneberger, O., Fischer, P., and Brox, T. (2015). U-Net: convolutional networks for biomedical image segmentation. In: *Medical Image Computing and Computer-Assisted Intervention. International Conference on Medical Image Computing and Computer-assisted Intervention, Lecture Notes in Computer Science*, vol. 9351 (ed. N. Navab, J. Hornegger, W. Wells, and A. Frangi), 234–241. Cham: Springer.

53 Keskar, N.S., Mudigere, D., Nocedal, J. et al. (2016). On large-batch training for deep learning: generalization gap and sharp minima. *arXiv preprint arXiv:1609.04836.*

54 Nagar, J. and Werner, D.H. (2017). A comparison of three uniquely different state of the art and two classical multiobjective optimization algorithms as applied to electromagnetics. *IEEE Transactions on Antennas and Propagation* 65 (3): 1267–1280.

55 Nagar, J. and Werner, D.H. (2018). Multiobjective optimization for electromagnetics and optics: an introduction and tutorial based on real-world applications. *IEEE Antennas and Propagation Magazine* 60 (6): 58–71.

56 Campbell, S.D., Sell, D., Jenkins, R.P. et al. (2019). Review of numerical optimization techniques for meta-device design. *Optical Materials Express* 9 (4): 1842–1863.
57 Deb, K. (2014). *Multi-Objective Optimization Search Methodologies*, 403–449. Springer.
58 Deb, K., Pratap, A., Agarwal, S., and Meyarivan, T. (2002). A fast and elitist multiobjective genetic algorithm: NSGA-II. *IEEE Transactions on Evolutionary Computation* 6 (2): 182–197.

9

Machine Learning for Metasurfaces Design and Their Applications

Kumar Vijay Mishra[1], *Ahmet M. Elbir*[2], *and Amir I. Zaghloul*[3]

[1] *Computational and Information Sciences Directorate (CISD), United States DEVCOM Army Research Laboratory, Adelphi, MD, USA*
[2] *Interdisciplinary Centre for Security, Reliability and Trust (SnT), University of Luxembourg, Luxembourg*
[3] *Bradley Department of Electrical and Computer Engineering, Virginia Tech, Blacksburg, VA, USA*

9.1 Introduction

The emerging industrial use-cases of sixth-generation (6G) and beyond wireless networks are envisaged to include industrial automation, autonomous vehicles, and smart infrastructure. These applications require significant improvements in data capacity, system latency, and quality-of-service reliability over the current 5G networks. In this context, *reconfigurable intelligent surface* (RIS) has been identified as a key enabling technology to program the *smart radio environment* (SRE), increase link quality, and reduce the hardware complexity [1, 2]. The RIS is made up of a *metasurface* (MTS) – a two-dimensional (2-D) reconfigurable electromagnetic (EM) layer composed of a large periodic array of subwavelength scattering elements (meta-atoms) with specially designed spatial features [3, 4]. Compared to electrically large arrays, the nearly passive meta-atoms offer lower cost and power consumption. The radio-frequency (RF) MTS performs customized transformations, such as beamforming, on a reflected incident wave through modified surface boundary conditions using Huygens' principle. For example, the MTS shifts the reflected phase of incident signal by creating a field discontinuity at the boundary of the surface. The arrangement and subwavelength structure of each meta-atom and, in turn, the array of space- and time-varying meta-atoms determine MTS aperture field distribution and control the direction and strength of reflected signal [5].

In a conventional wireless communication systems, the network optimization has been limited to control at the transmitter and receiver. This paradigm assumes that the wireless fading channel is uncontrollable and is a significant factor limiting the performance because of random signal reflections, diffraction, and scattering in the wireless environment. The RIS overcomes many of the aforementioned fading channel limitations through the ability of MTS to manipulate waves, achieve arbitrary aperture beamforming, and perform real-time analog spatial signal processing. This has spawned novel MTS-based RF applications such as intelligent beamforming [6], anomalous refraction and reflection [7], frequency selective and high-impedance surfaces [8], scattering reduction [9], polarization conversion [10], leaky-wave antenna [11], surface wave control [12], beam focusing [13], transmit-array antennas [14], reflect-array antennas [15], and holographic imaging [16]. Initial applications of RIS were limited to wireless communications for interference suppression [6], joint wireless information and power transmission [17], physical

Advances in Electromagnetics Empowered by Artificial Intelligence and Deep Learning, First Edition.
Edited by Sawyer D. Campbell and Douglas H. Werner.

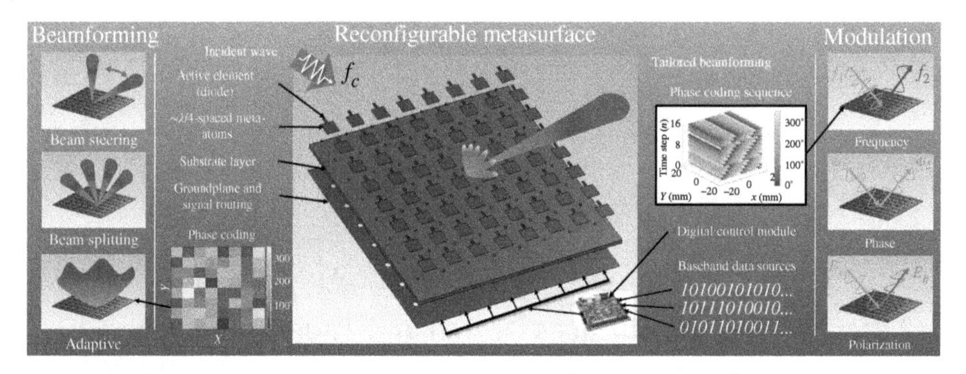

Figure 9.1 The RIS architecture (center) operating at carrier frequency f_c for wireless communication networks comprises meta-atoms located at below carrier wavelength (λ) spacing. It acts as both an endpoint transceiver and a relay. The RIS enables various beamforming functions (left column) including beam steering, splitting, and adaptive beamforming for customized radiation patterns by manipulating the phase coding of constituent meta-atoms. It is also capable of directly modulating (right column) the surface in frequency, phase, and polarization. Source: [25], Hodge et al., 2021/Cornell University.

layer security [18], and multi-beam design [19]. However, more recent works have introduced RIS to radar remote sensing [20–22] and joint radar-communications systems [23, 24].

In a wireless link, the RIS functions as either an electrically large antenna array at the endpoints or as an amplify-and-forward (AF) relay (Figure 9.1). By actively controlling and optimizing the amplitude/phase of each meta-atom across the aperture, the RIS maximizes the received signal-to-noise ratio (SNR) and provides adaptive beamforming to coherently focus the reflected signal on the receiver. Through joint optimization of the wireless channel and endpoints, RIS-assisted links are able to realize SRE.

Each scattering element typically includes an active tuning element, such as a varactor or PIN diode, whose bias voltage is software-controlled to change the EM response of the surface. The bias voltage for each meta-atom is pre-computed and modulated by a digital control module employing a field programmable gate array (FPGA) [26]. Each meta-atom is controlled by tuning its EM properties (susceptibility or impedance) which affects the spectral response of the reflected signal. This aids in producing tailored radiation patterns for diverse functions, such as beam steering, anomalous reflection, focusing, beam splitting, absorption, and direct modulation of the reflected signal.

There are several challenges in the design, fabrication, deployment, and processing of RIS. In applications such as radar and communications that have precise radiation pattern constraints, the RIS design often involves optimization of several complicated and irregular geometry parameters to meet the required resonant frequency, gain, polarization, bandwidth, and size constraints. The conventional design process could be very tedious. Further, post-deployment, the processing of RIS signals and optimized beamforming is also challenging because of high-dimensional nature arising from the use of several antennas. In this context, machine learning (ML) techniques have recently shown unprecedented performance in problems where it is challenging to develop an accurate mathematical model for feature representation. These methods are now also transforming the above-mentioned tedious approaches to design RIS and process its signals. In particular, as a class of ML techniques, deep learning (DL) methods have gained much interest recently for solving many challenging problems such as speech recognition, visual object recognition, rainfall estimation, and language processing [27–29]. These techniques offer advantages such as low computational complexity while solving optimization-based or combinatorial search problems as well as the ability to extrapolate new features from a limited set of features contained in a training set [27]. Recently, DL for MTS inverse design, wherein a meta-atom design if synthesized from a specific response, has become very popular. This has been applied for semi-automated inverse design of metamaterials [30], MTS [31, 32], and nanophotonic structures [33]. Note that the above ML/DL application to MTS/RIS design is different from using DL to perform signal processing functions in RIS-aided communications (see, e.g. [34] for a survey). In the following, we describe these aspects in detail.

9.1.1 ML/DL for RIS Design

The design and optimization of RIS hardware at the physical layer remains a formidable challenge. To date, RIS/MTS implementations remain quite limited. To realize the promise of RIS-assisted networks and SRE, more robust and automated MTS design techniques are required. Without capable RIS hardware, the benefit of RIS-assisted networks will be significantly reduced due to EM limitations. In general, canonical structures such as v-antennas, loaded-dipoles, and split-ring resonators are used to fabricate RIS. However, meta-atoms based on these geometries usually fall short of desired performance, particularly when anisotropic, broadband, and/or wide-angle responses are required. As a result, traditional MTS design approaches exhibit performance limitations, especially given the complexity of MTS hardware requirements and increasing functionality required for wireless nodes in next generation networks.

Designing a user-defined, arbitrary wave-front RIS or *metagrating* [35–37] is a challenging, labor-intensive, and long process. In general, a new MTS design entails numerous rounds of manual tuning and full-wave simulations that iteratively solve Maxwell's equations until a locally optimized design is achieved [35]. Initial designs are typically based on physical instincts and intuitive arguments. However, the final geometric structure and material characteristics are attained through iterative analyses.

The ML/DL approaches expend computational time and resources upfront as a fixed-cost to generate training data sets of device geometries and their associated spectral responses but are useful during the predication stage [38]. Deep neural networks (DNNs) are trained to map the nonlinear relationships between meta-atom geometry and spectral response. The power of DNNs comes from their multi-layered composition which allows them to learn the relationships between data with multiple levels of abstraction [27]. Once trained, a DNN efficiently produces the geometry of a meta-atom given a desired spectral response. The application of DL to the inverse design of MTS and nanophotonic structures is still in its early stages, and much more work is required to realize more generalized complex designs, reduce the amount of required training data, and result in increased efficacy. Nearly all of these works rely on supervised learning (SL) techniques for metamaterial performance predictions, which map known input–output pairs based on large training examples. In MTS design, applying such techniques does not result in new shapes different than the ones used in training. This severely limits the ability to generate customized MTS patterns. In [5, 37], we introduced the use of generative adversarial networks (GANs) to microwave MTS design that aids in discovering new shapes of meta-atoms.

9.1.2 ML/DL for RIS Applications

The next-generation millimeter wave (mm-Wave) massive multiple-input multiple-output (MIMO) systems require large antenna arrays with a dedicated RF chain for each antenna. This results in expensive and large system architectures which consume high power and processing resources.

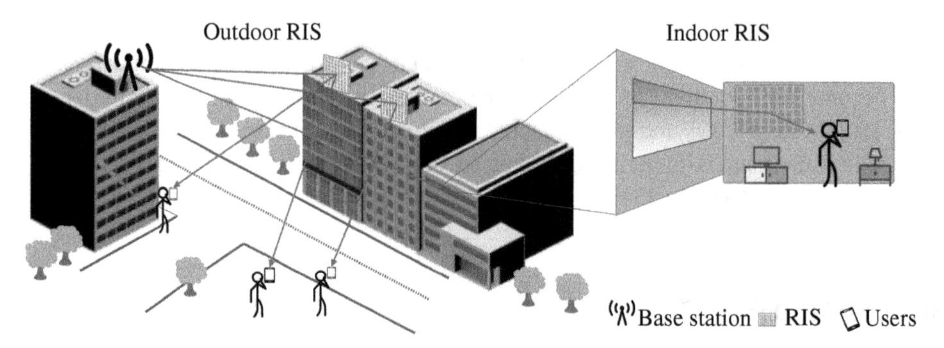

Figure 9.2 RIS-assisted wireless communications for outdoor and indoor deployments. A BS on top of the infrastructure (left) communicates with the users on ground through an intermediate RIS mounted on other buildings (center). The BS also serves users (right) inside the apartment building through an RIS placed on the wall of the room [34].

To reduce the number of RF chains while also maintaining sufficient beamforming gains, hybrid analog and digital beamforming architectures were introduced. However, the resulting cost and energy overheads using these systems remain a concern. Recently, RISs have emerged as a feasible solution [39] to implement low-cost and light-weight alternative to large arrays complexity in both outdoor and indoor applications, usually with separate operating frequencies or spectral bands (Figure 9.2).

The RISs reflect the incoming signal by introducing a pre-determined phase shift. This phase shift is controlled via external signals by the base station (BS) through a backhaul control link. As a result, the incoming signal from the BS can be manipulated in real-time, thereby, reflecting the received signal toward the users. Hence, the usage of RIS enhances the signal energy received by distant users and expands the coverage of the BS. It is, therefore, required to jointly design the beamformer parameters both at the RIS and BS. This achieves desired channel conditions, wherein the BS conveys the information to multiple users through the RIS [40]. Different from amplify-and-forward (AF) relay systems, an RIS can have both active and passive components, which can provide a flexible configuration; thus, it has less active transmit modules or totally reflects the received signal as a passive surface. Thus, the RIS is much more energy- and spectrum-efficient [41].

The accuracy of beamformer design strongly relies on the knowledge of the channel information. In fact, the RIS-assisted systems include multiple communications links, i.e. a direct channel from BS to users and a cascaded channel from BS to users through RIS. This makes the RIS scenario even more challenging than the conventional massive MIMO systems. Furthermore, the wireless channel is dynamic and uncertain because of changing RIS configurations. Consequently, there exists an inherent uncertainty stemming from the RIS configuration and the channel dynamics. These characteristics of RIS make the system design very challenging [40, 42]. To address the aforementioned uncertainties and non-linearities imposed by channel equalization, hardware impairments, and sub-optimality of high-dimensional problems, model-free techniques have become common in wireless communications [43]. In this context, DL is particularly powerful in extracting the features from the raw data and providing a "meaning" to the input by constructing a model-free data mapping with huge number of learnable parameters. Furthermore, DL is helpful when modeling the channel characteristics thanks to its data-driven structure. A learning model constructs a non-linear mapping between the raw input data and the desired output to approximate a problem from a model-free perspective [43]. Thus, its prediction performance is robust against the corruptions/imperfections in the wireless channel data. DL learns the feature patterns, which are easily updated for the new data and adapted to environmental changes. In the long run, this results in lower computational complexity than a model-based optimization [40]. DL-based solutions have significantly reduced run-times because of parallel processing capabilities. On the other hand, it is not straightforward to achieve parallel implementations of conventional optimization and signal processing algorithms [44]. The aforementioned advantages have led to DL superseding the optimization-based techniques in the RIS system design for physical layer of the wireless communications [43].

9.1.3 Organization

This chapter provides an overview of recent developments in using ML/DL for designing, deploying, and processing the physical layer of RIS. The rest of the chapter is organized as follows. In Section 9.2, we discuss various ML techniques for inverse RIS design. Then, we introduce various techniques DL for RIS design in Section 9.3 and provide a few case studies in Section 9.4. Then, we focus on DL-aided RIS applications for wireless systems in Section 9.5. including signal detection and channel estimation. For a more widely used application of RIS beamforming, we discuss various DL frameworks in Section 9.6. We also discuss current challenges in using ML/DL for RIS systems and highlight related future research directions in Section 9.7. We conclude in Section 9.8.

9.2 Inverse RIS Design

Communications-based analysis of RIS without physics-based EM-compliant models is a major limitation of current research. Until recently, prior works did not consider such realistic RIS implementations. As the parameter spaces of meta-atom geometry and constituent materials have grown, the conventional approaches to achieve the targeted EM response have become more tedious. In this context, learning models have demonstrated the ability to implicitly learn Maxwell's equations from training data within a constrained design space. The ML techniques have witnessed increased use in research to create surrogate models for MTS performance prediction, inverse design, and optimization. For an inverse MTS design problem, the input is an arbitrary design spectrum and the network finds or synthesizes a geometry to closely approximate the desired spectral response (Figure 9.3).

Major benefits of DL-based RIS design for wireless communications include:

- *EM-based surrogate models*: DL constructs a nonlinear mapping between the raw input data (meta-atom design) and the desired output to approximate the MTS response.
- *Inverse design*: Deep generative models (DGMs) are utilized to learn geometric features from training data and generate new meta-atom designs to achieve the spectral response.
- *Diverse EM surface representations*: DL-based MTS design admits flexible design representation. The input could be either vectors of discrete parameters describing the geometry, material, frequency, and angular design parameters or pixelated images to represent the geometry or phases of the meta-atom design. Whereas a fully connected neural network (NN) is well-suited to process the simple designs specified by the former representation, a convolutional network handles images appropriately to yield more complex MTS geometries.

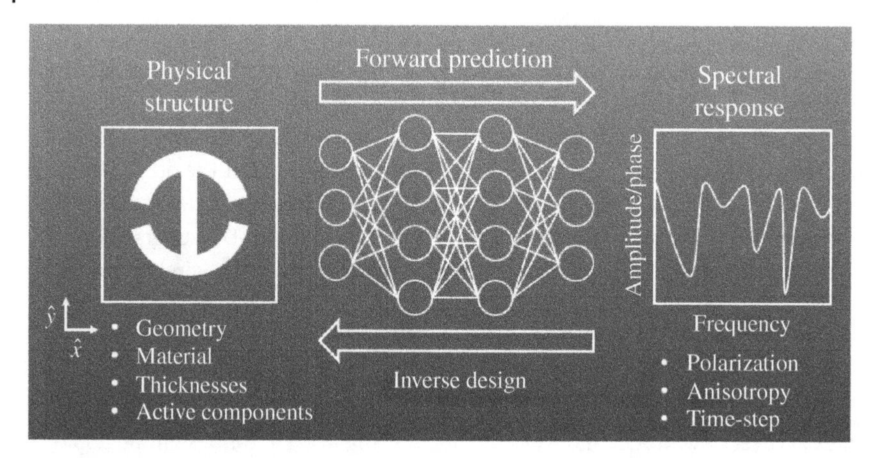

Figure 9.3 In inverse RIS design, ML algorithms learn and generalize complex EM relationships between the physical RIS structure (left column) and spectral response (right column) through training data. Source: [25] Hodge et al., 2021/Kumar Vijay Mishra.

Table 9.1 summarizes prior works on various techniques for RIS inverse design. The non-DL methods typically comprise several evolutionary optimization algorithms as listed below. The drawback of traditional optimization techniques is that they start from scratch with each new design. This often requires hundreds of additional full-wave simulations per design.

9.2.1 Genetic Algorithm (GA)

This is an iterative global optimization (GO) algorithm that has been used extensively in the design of pixelated coded MTS designs. Genetic algorithm (GA) is a nature-inspired algorithm that uses binary strings (chromosomes) to represent candidate designs [38]. During the optimization, the GA selects the best subset of design candidates from the previous generation to serve as starting points for mutation and crossover in the next design iteration. Recent GA applications include coding MTS [38] which demonstrates channel response modification, efficient polarization conversion, and phase-graded beam steering.

9.2.2 Particle Swarm Optimization (PSO)

A popular stochastic evolutionary computation technique, particle swarm optimization (PSO) is inspired by the movement and intelligence of swarms. Recently, it has been employed for shaping EM waves using pixelized coded MTSs [38]. The design procedure using PSO is tied to a full-wave EM solver and completely automatic. The software yields both microscopic meta-atom designs and the macroscopic aperture coding matrix. By changing the reflection phase difference between cells, this approach has produced designs of functional MTSs with circularly and elliptically shaped radiation beams and multi-beam patterns. This is useful for achieving customized radiation patterns to enhance link performance in the wireless communication channel. Similar efforts have used a simulated annealing algorithm for the design and optimization of a broadband diffusion MTS using anisotropic elements for scattering reduction. In [46], binary particle swarm optimization (BPSO) was used to automate the macroscopic layout of both passive and active apertures to realize user-defined dual-beam scattering radiation patterns. For example, this study used BPSO to realize a reflecting MTS with a left-handed circular polarization (LHCP) beam and a right-handed

Table 9.1 State-of-the-art on MTS inverse design.

Algorithm	Frequency	MTS layers	Data	Key features	Drawbacks
Evolutionary optimization techniques					
GA [38]	15–45 GHz	1	Parameter vector	Pixelized meta-atoms with discrete input design space when a contiguous structure is not required	Optimization from scratch for each design; output structures may be too complex to fabricate
PSO [38]	9.5–12 GHz	1	Binary matrix (2-D)	Swarm-based GO technique for pixelized meta-atom design; outperforms GA for various EM designs	Optimization from scratch for each design with parameter tuning
ACO [38]	1–4.5 GHz	3	Binary matrix (3-D)	MTS, including 3-D structures and wire grid arrays, with discrete design space and a contiguous structure	Optimization from scratch for each design; output structures may be too complex to fabricate
Learning methods					
ANN [33]	375–749 THz	1	Parameter vector	Performance prediction, inverse design, and optimization of nanophotonic particles	Limited design variables; applicable to only spherical dielectric nanoparticles
ANN [45]	193 THz	1	Parameter vector	Performance prediction and inverse design of metagratings	Limited set of parametric inputs; significant training overload
DNN [30]	30–80 THz	2	Parameter vector	Inverse design of chiral and multi-layer MTS	Design-specific architecture; limited design space
CNN [46]	10 GHz	1	Binary matrix (2-D)	Anisotropic digital coding MTS; PSO for beamforming	Significant training overload
CNN [47]	9.37 GHz	1	Binary matrix (2-D)	Hybrid CNN-GA for space-time modulation of programmable MTS; multi-beam steering	Binary phase coding limits beamforming performance; limited tunability
cDC-GAN [35]	170–600 THz	1	Image matrix (2-D)	Generative inverse design of transmission MTS	Significant training overload; limited to single layer designs and passive structures

(*Continued*)

Table 9.1 (Continued)

Algorithm	Frequency	MTS layers	Data	Key features	Drawbacks
cDC-GAN [37]	1–30 GHz	1	Image (2-D)	Reflective RF MTS; training set with published meta-atom structures to improve learning	Limited to single layer; post-processing required
cDC-GAN [5]	5–25 GHz	2	Image (3-D)	Multi-layer MTS; RGB-style matrix to represent multiple layers	No active elements; additional validation required
cDC-GAN [48]	5–25 GHz	3	Image (3-D)	Federated learning for multi-layer design	Significant training overload
cDC-VAE [49]	40–100 THz	1	Image (2-D)	Anisotropic MTS; encodes input into low-dimensional latent space	Significant training overload; post-processing required
TO-GAN [50]	231–600 THz	1	Image (2-D)	Free-form diffractive metagrating design for select wavelength-deflection angle pairs with topology refinement	Additional optimization required
GLOnet [51]	231–500 THz	1	Image (2-D)	Dielectric MTS design without training sets	Limited to single objective optimization; requires solving Maxwell's equations inside training loop

Source: [25] Hodge et al., 2021/Kumar Vijay Mishra.

circular polarization (RHCP) beam. Results of this study were experimentally verified. This digital coding approach has been applied to both passive and active R-MTS.

9.2.3 Ant Colony Optimization (ACO)

This is another swarm-based algorithm inspired by *stigmergy* in ant colonies in order to search for optimal solutions to graph-based problems [38]. Here, a number of *artificial ants* build solutions to an optimization problem and exchange information on their quality using a cooperation scheme similar to that utilized by real ants. In [38], inverse MTS design is performed based on multi-objective lazy ant colony optimization (MOLACO) to synthesize 3-D nano-antenna geometries with low-loss transmission performance and broad phase tunability. The ant colony optimization (ACO) is generally most useful for a discrete input design space and when a contiguous structure is required.

9.3 DL-Based Inverse Design and Optimization

The computational power and time required for evolutionary optimization algorithms grow exponentially with the number of design parameters. This is mitigated by DL-based inverse design for RIS. Prior works have employed a variety of network structures and algorithms based on the availability of data, RIS topology, and desired EM spectral response.

9.3.1 Artificial Neural Network (ANN)

The artificial neural networks (ANNs) were first used to approximate light scattering by multi-layer nanoparticles (meta-atoms) [33]. Similar to MTS, nanophotonic particles derive their frequency response from physical structure and the size constituent scatterers. Then, Inampudi and Mosallaei [45] used a similar technique for metagratings. Typical inverse design problems require optimization in high-dimensional space, which involves lengthy calculations and are typically solved using genetic algorithm or adjoint methods. However, the computational power and time required for GA optimization grows exponentially with the number of design parameters.

The primary application of ANNs in MTS design is performance approximation. The feedforward ANN is trained to be a high-fidelity surrogate model for performance prediction. Using training data consisting of meta-atom physical design parameters as inputs and frequency response as labels, the ANN is trained to approximate a complex physics simulation (such as finite-element method [FEM], method of moment [MoM], or finite-difference time-domain [FDTD] simulation). Through the training data, the ANN learns to map the scattering function of the meta-atom into a continuous, higher-order space where the derivative is found analytically through propagation. In [33], a trained ANN simulated spectral responses orders of magnitude faster than conventional full-wave simulations. This study used a fully connected ANN consisting of four layers with 250 neurons per layer resulting in 239,500 parameters. The inputs were the thickness of each meta-atom layer (the materials were fixed) and the outputs were the spectrum sampled at points between 400 and 800 nm. The results suggest that the ANN was not simply fitting the data, but rather discovered the underlying structure of input-to-output mapping to generalize the physics of the systems with the training set and solve problems not yet encountered.

A significant drawback of this approach is that the inputs are limited to the thicknesses of the meta-atom layers with fixed materials. This results in a lack of generalizability for the ANN that vastly limits the possible meta-atom design structures. While fixing the input parameters reduces the complexity of the ANN architecture, it limits the design space and optimal designs. Another drawback of this approach is that [33] required 50,000 examples using conventional simulation methods to generate training data. However, unlike evolutionary optimization methods such as GA or PSO, simulation of the training dataset is an upfront fixed cost because it only needs to be simulated once and is then leveraged for other designs. Additionally, the simulations for training data generation are highly parallelized unlike serial optimization techniques.

Once trained, Peurifoy et al. [33] shows that the ANN solves inverse design problems more quickly than its numerical counterparts because the gradient is found analytically, through back propagation, rather than numerically. Similar to inverse design, the ANN also optimizes for a desired property by altering the cost function used for the design without training the ANN. Their results show that the ANN performs inverse design and optimization more accurately than traditional numerical nonlinear optimization techniques.

9.3.1.1 Deep Neural Networks (DNN)

To model more complex meta-atom structures and increase performance prediction accuracy, DL has been applied to the on-demand design of chiral (a form of anisotropy) MTS [30]. Here, DNN – an ANN comprised many hidden layers to significantly expand learning and generalization ability – were employed to automatically design and optimize 3-D chiral MTS with strong anisotropic spectra at predetermined wavelengths. The network comprised two bidirectional networks that were constructed using partial stacking technique. This study limited the input design space (and hence the structures obtained) and predicted the reflection spectral response at 201 discrete frequency points for two orthogonal polarization and the cross-polarization coupling term resulting in a 3-by-201 spectral output vector. By fixing the inputs to be five specific design parameters, this DNN design approach is also limited in its generalizability to other physical structures in the design space. Full-wave simulation was used to generate the training data set for 30,000 example meta-atoms. The DNN achieved high efficiency and high accuracy for performance prediction and inverse design for anisotropic MTS, where the meta-atom design space is limited.

9.3.2 Convolutional Neural Networks (CNNs)

To improve on the lack of generalization and increase performance prediction accuracy, convolutional neural networks (CNNs) are used to design anisotropic digital coding MTSs. CNNs are a class of ANNs that use convolution functions to learn hierarchical patterns within data. These models learn generalized patterns across many spatial scales from their input data and are widely used on image data. In [46], a CNN predicted the reflection phase response of binary coded meta-atoms where each meta-atom contains 16-by-16 square sub-pixels and is mirrored with two-fold symmetry. The CNN used in this study is a 101-layer deep residual network, known as Resnet-101. The authors found that other networks with fewer layers resulted in less precise and robust performance predictions.

The results show an accuracy of 90.05% of phase responses with 2° error in the 360° phase. A drawback of this binary coding approach is that a 16-by-16 pixel meta-atom has 2^{16} potential design combinations. This study generated training data by simulating randomized pixel matrices. However, it was fundamentally inefficient in an analogous manner to GA because the training data is essentially random and does not contain the knowledge of canonical structures in the training data set. This likely results in significantly more required training data and greater network complexity.

Another drawback of this study is that it required full-wave simulation of 70,000 training examples 10,000 test examples to generate the training dataset.

A significant CNN advantage is that the meta-atom shape is directly input into the network rather than shape-specific design parameters. The convolutional filters allow the CNN to learn the physical structure that leads to given EM response, leading to a broader applicability of the model.

In [47], the element phases of a reconfigurable MTS were computed by a 11-layer CNN for multiple beam steering applications. The input was the parameter vector representing the target beam pattern and the output was a matrix that carried the 1-bit codes for a programmable 2304-element MTS. This technique to obtain the phase matrices reduced the time for producing almost similar beam patterns using conventional methods to a few milliseconds.

9.3.3 Deep Generative Models (DGMs)

Generative models are unsupervised or semi-SL models that infer a function to describe hidden structure from unlabeled data. Their functions include clustering, density estimation, feature learning, and dimension reduction. Whereas discriminative networks capture the relationship between meta-atom geometry and spectral response from a training set, DGMs focus on learning the properties of meta-atom geometry distributions [35, 37, 50, 51]. Major classes of DGMs (Figure 9.4) applied to MTS inverse design are as follows.

9.3.3.1 Generative Adversarial Networks (GANs)

In a GAN system, two ANNs compete to improve each of their models: the generative network learns to create inputs indistinguishable from the training data, while the discriminative network learns to identify true data from the output of the generative network. Training GANs involves jointly training a generator network and a discriminator network in a game theoretic approach to find a local Nash equilibrium. The goal of a generative model is to observe a collection of training examples and learn the underlying probability distribution that generates them. GANs are able to generate new samples from the estimated probability distribution. GANs were initially applied to generate photos, however, have been applied to many domains including speech and video generation. Very recently, GANs have been applied to generate new MTS hardware design including those not explicitly seen in the training dataset or current literature. At the end of a successful training process, GANs are able to produce realistic meta-atom designs, even for very complicated datasets and spectral responses.

In, [37], we introduce GANs to microwave MTS design. GANs are promising for low-cost MTS design with complex frequency and polarization dependent scattering responses. In [35], an input set of user-defined EM spectra is fed to GAN that generates candidate patterns to match the on-demand spectra with high fidelity. Here, DNNs are employed to approximate the spectra of the MTS and perform inverse design by generating meta-atom structures that yield user-defined input spectra. Once the model is trained, extensive parameter scans and trial-and-error procedures are bypassed. This conditional deep convolutional generative adversarial network (cDC-GAN) architecture uses three interconnected CNNs: generator, discriminator, and simulator. The simulator is a pretrained network that serves as a surrogate model for fast spectral performance prediction. In this study, S is a five-layer CNN with three fully connected layers at the output. The conditional generator networks accepts the desired spectral response and a latent noise vector to output potential meta-atom designs. The discriminator serves to train the generator by evaluating the distance of the distributions between the geometric patterns from training data and

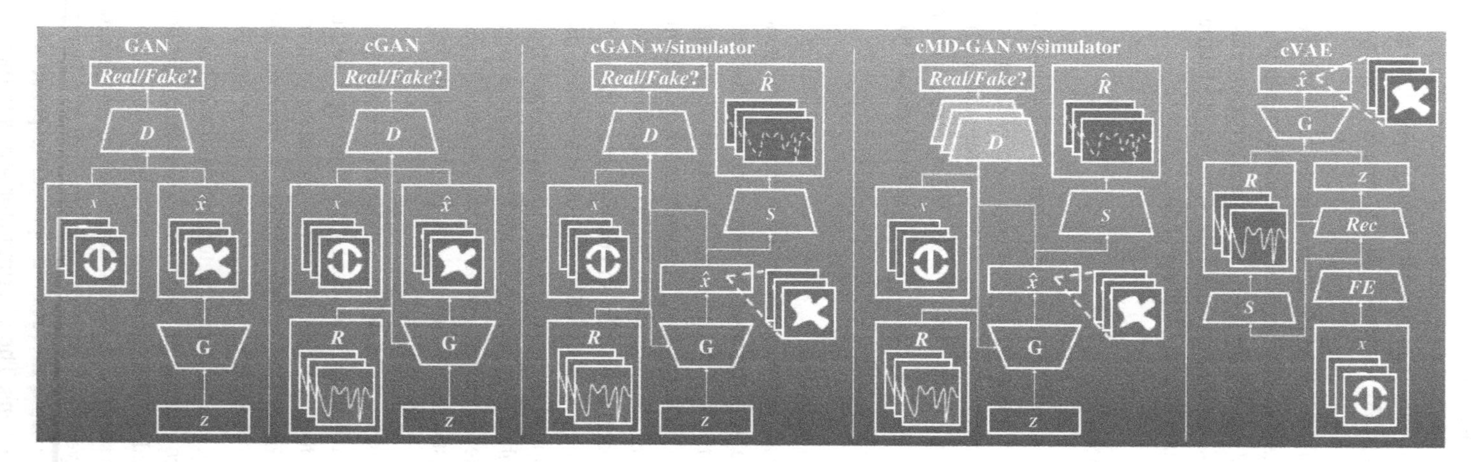

Figure 9.4 DGM architectures for RIS inverse design. The conventional GAN (left) lacks spectral information of the RIS structure x. The latent variable z is fed to the generator G to yield an estimated meta-atom structure $\hat{x}$. The discriminator D then makes a decision if $\hat{x}$ is a valid design. The cGAN (second from left), conditioned by the reflection spectra R, shows improved performance. A simulator neural network may also be added to cGAN (center) to accelerate training and also predict the performance $\hat{R}$ of generated meta-atoms. The cMD-GAN (second from right) comprises multiple discriminators, one for each layer. The cVAE (right) consists of an encoder–decoder network structure, where a feature extractor (FE) coupled with the recognition (Rec) network serves as the encoder to map the meta-atom structure to a lower-dimensional latent variable space. The generation model is a reconstruction (decoder) network. Source: [25] Hodge et al., 2021/Kumar Vijay Mishra.

generator. At the end of successful training, discriminator is unable to distinguish batches from generator and training set. This approach is shown to exhibit high accuracy in inverse design of meta-atoms.

In [37], a deep convolutional generative adversarial network (DC-GAN) is employed to generate anisotropic RF meta-atom designs. Using a small set of simulated spectra, the network learned the relationship between the physical structure of meta-atoms and their reflection spectra for vertical and horizontal polarizations. The DC-GANs generated meta-atom structures that resembled design features in the training data. To speed up training, the network was fed with parametric variations of twelve published meta-atom designs to a full-wave EM simulator. Starting out with parametric variations of canonical meta-atoms scatterers, the network picked up more efficiently than it would have from training with responses of randomized pixel data.

9.3.3.2 Conditional Variational Autoencoder (cVAE)

As an alternative to GAN approaches, Ma et al. [49] presents a probabilistic DGM that solves both forward and inverse problems at the same time. It is trained in an end-to-end manner and uses a conditional deep convolution VAE (cDC-VAE) architecture (Figure 9.4) comprising an encoder–decoder network structure. The encoder maps the meta-atom structure to a multivariate Gaussian distribution in the latent space, and the conditional decoder network inputs the reflection spectra and latent variable to generate meta-atom designs (Figure 9.4).

In [49], the RIS inverse design is modeled in a probabilistic generative manner to investigate the complex structure–performance relationship in an interpretable way and solve the one-to-many mapping issue that is intractable in deterministic models. It developed a semi-SL strategy that allows the model to utilize unlabeled data in addition to labeled data in an end-to-end training. The RIS design and spectral response are encoded into a low-dimensional latent space with a predefined prior distribution, from which the latent variables are sampled. The DGM, comprising prediction, recognition, and generation models, serves as a tool to accelerate the design, characterization, and even new discovery of MTS.

9.3.3.3 Global Topology Optimization Networks (GLOnets)

Recently, GANs utilized to learn structural features of topology-optimized (TO) metagratings for inverse design [50, 51]. TO is a method of optimizing a material layout or an array of pixels to maximize system performance given a set of constraints and boundary conditions. Unlike other approaches, simulation overload for TO does not increase with the number of RIS units. In [50], free-form diffractive metagratings were designed using TO-GAN. Here, DGMs were trained from images of periodic, TO metagratings to produce efficient scattering structures with the desired performance over a broad range of frequencies and angles. The network employed 5000 training examples for each angle. However, the performance of the best structures was not robust and additional refinement was needed to meet the desired performance. In [51], dielectric MTSs optimization was performed using a physics-informed cGAN. Global optimization-based generative networks (GLOnets) are able to search the design space for the global optimum design. Unlike other GAN approaches, GLOnets seek to fit a narrow-peaked function centered on the optimal solution without a training set. The GLOnet generates a distribution of meta-atoms to sample the global design space and then shifts the distribution toward a more optimal design. Training requires computing forward and adjoint EM simulations of output structures using backpropagation. In this work, GLOnets are shown to be successful and computationally efficient global TO for MTS and metagratings.

9.4 Case Studies

We perform two case studies for the design of single- and multi-layer RIS based on [37] and [5], respectively. The design approach in [5] introduced the cDC-GAN-based for jointly designing several layers of tensorial RIS. It represented three RIS layers with a $64 \times 64 \times 3$ red-green-blue (RGB) image matrix. The advantages of the cDC-GAN are that it trains classifiers in a semi-supervised manner and generates new free-form shapes not previously shown in the literature. However, GANs can be unstable and challenging to train. We validated the designs by simulating their spectrum using a full-wave EM solver and comparing the results to the desired spectrum. In this data representation, the top layer meta-atom design is represented as the first channel, the second layer meta-atom design is represented by the second channel, and a third layer is represented by the third channel using the conventional RGB image format.

9.4.1 MTS Characterization Model

Consider a two-dimensional (2-D) MTS lying in the x-y plane with z-axis being the direction of propagation. According to Huygens' principle, the EM fields created by arbitrary sources in an arbitrary volume V are found as the fields created by equivalent surface currents on the volume surface S [52]. Therefore, a known incident EM source, such as a plane wave, can be transformed into a desired transmitted or reflected wave using an MTS. The MTS creates the desired aperture field distribution or phase shift by modifying the effective boundary conditions of the EM surface.

The amplitude and phases of transmitted and reflected waves from MTS are functions of surface-averaged induced electric and magnetic current densities, $\boldsymbol{J}_e$ and $\boldsymbol{J}_m$, respectively. These effective surface current densities induced on MTS are described by average tangential electric ($\boldsymbol{E}_{t\pm}$) and magnetic ($\boldsymbol{H}_{t\pm}$) fields on each side of MTS as

$$\boldsymbol{J}_e = \boldsymbol{n} \times (\boldsymbol{H}_{t+} - \boldsymbol{H}_{t-}) \tag{9.1}$$

$$\boldsymbol{J}_m = -\boldsymbol{n} \times (\boldsymbol{E}_{t+} - \boldsymbol{E}_{t-}) \tag{9.2}$$

where $\boldsymbol{n}$ is the unit vector normal to the MTS. Examples of passive implementations of Huygens' metasurface (HMS) include reflectionless refraction, perfect anomalous reflection, and arbitrary antenna beamforming [7].

The induced surface currents $\boldsymbol{J}_e$ and $\boldsymbol{J}_m$ are related to their respective average tangential fields (applied on a thin slab of polarizable particles) by spatially varying electric surface impedance $\overline{\overline{Z}}_{se}$ and magnetic surface admittance $\overline{\overline{Y}}_{sm}$,

$$\boldsymbol{E}_{t,avg} = \overline{\overline{Z}}_{se} \cdot \boldsymbol{J}_e \tag{9.3}$$

$$\boldsymbol{H}_{t,avg} = \overline{\overline{Y}}_{sm} \cdot \boldsymbol{J}_m \tag{9.4}$$

Substituting (9.1) and (9.2) into (9.3) and (9.4), respectively, yield the following generalized sheet transition conditions (GTSCs) [3, 53] used for describing an MTS:

$$\boldsymbol{E}_{t,avg} = \overline{\overline{Z}}_{se} \cdot [\boldsymbol{n} \times (\boldsymbol{H}_{t+} - \boldsymbol{H}_{t-})] \tag{9.5}$$

$$\boldsymbol{H}_{t,avg} = \overline{\overline{Y}}_{sm} \cdot [-\boldsymbol{n} \times (\boldsymbol{E}_{t+} - \boldsymbol{E}_{t-})] \tag{9.6}$$

In case of single polarization, the tensor quantities above reduce to scalars [53]. Specifying the desired incident fields $\boldsymbol{E}_{t-}$ and $\boldsymbol{H}_{t-}$ and desired output fields $\boldsymbol{E}_{t+}$ and $\boldsymbol{H}_{t+}$ leads to computation of the required electric impedance and magnetic admittance at each (spatial) location of the MTS. The bianisotropy is included in these boundary conditions by introducing the tensor magneto-electric coupling coefficient $\overline{\overline{K}}_{em}$ as [7]

$$\boldsymbol{E}_{t,avg} = \overline{\overline{Z}}_{se} \cdot \boldsymbol{J}_e - \overline{\overline{K}}_{em} \cdot [\boldsymbol{n} \times \boldsymbol{J}_m] \tag{9.7}$$

$$\boldsymbol{H}_{t,avg} = \overline{\overline{Y}}_{sm} \cdot \boldsymbol{J}_m - \overline{\overline{K}}_{em} \cdot [\boldsymbol{n} \times \boldsymbol{J}_e] \tag{9.8}$$

The transmission and reflection spectral responses of MTS, described by vectors $\mathbf{T}$ and $\mathbf{R}$, respectively, are functions of the surface impedances at a particular incidence angle θ and frequency f (GHz). For instance, $\mathbf{T}(\theta, f) = h(\overline{\overline{Z}}_{se}, \overline{\overline{Y}}_{sm}, \overline{\overline{K}}_{em})$ [7, 54] where $h(\cdot)$ is a non-linear function. In this chapter, we fix $\theta = 0$ (broadside incidence) so that $\mathbf{T} = \mathbf{T}(0, f)$, where we have omitted the arguments for simplicity. From here on, we focus on only $\mathbf{T}$ because the design procedure using $\mathbf{R}$ is identical.

The EM wave is also characterized by its polarization. We consider two polarizations – "x" and "y" – wherein the electric field is parallel to the x- and y-directions, respectively. For an incident wave with a particular polarization, the MTS produces responses in both polarizations. For example, the response in x (y) polarization when the incident wave is also x-polarized (y-polarized) is the co-polar response $\mathbf{T}_{xx}$ ($\mathbf{T}_{yy}$). Similarly, cross-polar responses $\mathbf{T}_{xy}$ and $\mathbf{T}_{yx}$ are defined.

A multi-layer meta-atom consists of multiple layers of different shapes separated by dielectric spacers for structural support. Consider a 3-layer MTS (Figure 9.5) whose composite response

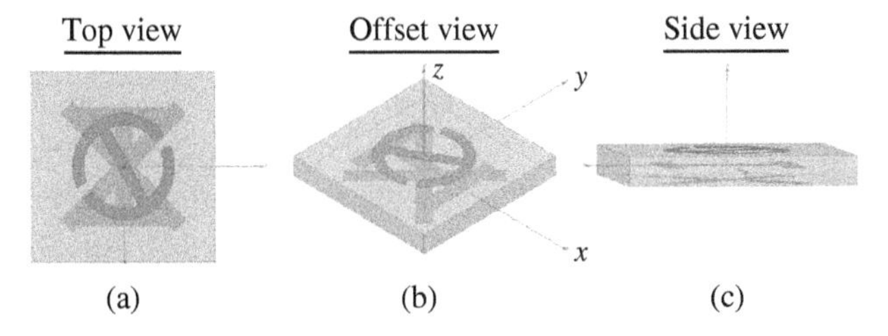

Figure 9.5 Illustration of a three-layer MTS unit-cell (meta-atom) in the three-dimensional x–y–z Cartesian space. The (a) top, (b) offset, and (c) side views were simulated in ANSYS HFSS software using periodic boundary conditions. The meta-atom occupies an 8 mm by 8 mm planar grid. The gold color rectangles represent copper traces on the top, middle, and bottom layers of meta-atom. The dielectric spacer is made up of 1.0 mm thick duroid material with relative permittivity $\epsilon_r = 2.2$. We used Floquet ports in HFSS to excite the meta-atom unit-cell by a wave traveling in the negative z-direction. Source: [48] Hodge et al., 2019/IEEE.

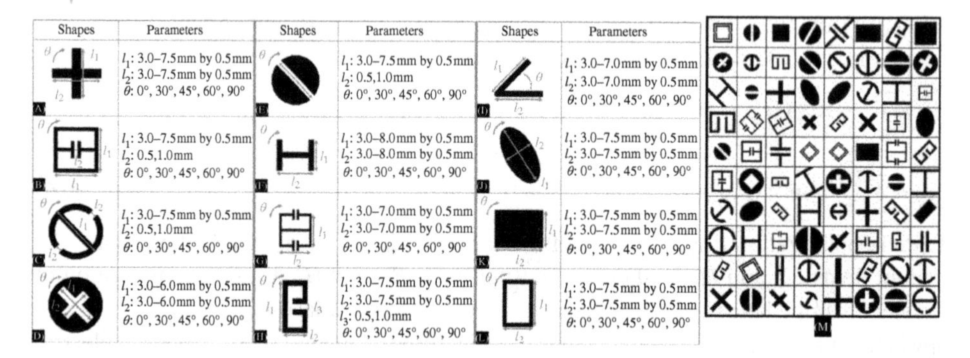

Figure 9.6 Meta-atom patterns from published literature. During training, several variations of these shapes are generated by changing parameter values as indicated. (M) 80 randomly selected training images of MTS unit cell designs from our design database used to train DC-GAN using the basic shapes (A)–(M). (A) Cross-resonator array. (B) Electric inductive-capacitive (ELC) resonator. (C) Symmetric split ring. (D) Asymmetric cross. (E) Circular patch with cuts inside. (F) H-shape. (G) ELC and non-bianisotropic split ring resonator. (H) Capacitively loaded loop (CLL). (I) V-shape. (J) Circular/elliptical patch. (K) Rectangular/square patch. (L) Square ring. Source: (A) Adapted from [55]; (B) Adapted from [56]; (C) Adapted from [57]; (D) Adapted from [58]; (E) Adapted from [11]; (F) Adapted from [59]; (G) Adapted from [60]; (H) Adapted from [61]; (I) Adapted from [62]; (J) Adapted from [63]; (K) Adapted from [64]; (L) Adapted from [9].

T is the superposition of the responses of individual layers. Our goal is to train the MD-GAN to implicitly learn physical quantities $\overline{\overline{Z}}_{se}$, $\overline{\overline{Y}}_{sm}$, and $\overline{\overline{K}}_{em}$ by mapping various design geometries to transmission spectra and produce new meta-atom designs for each layer to realize composite responses $\mathbf{T}_{xx}$, $\mathbf{T}_{yy}$, $\mathbf{T}_{xy}$, and $\mathbf{T}_{yx}$.

9.4.2 Training and Design

We evaluated proposed inverse design approach by implementing our distributed conditional multi-discriminator distributed generative adversarial network (cMD-GAN) architecture using PyTorch and performing simulations on an NVIDIA Tesla T4 GPU. During training, we included parametric variations of only those meta-atom shapes that have been extensively studied in the literature. Figure 9.6 lists these shapes and enumerates variations in the physical parameters to generate training data. The CNNs process matricized data, e.g. a color image composed of three matrices, each of which contains pixel intensities in RGB color channels. Rather than feeding a three-channel image matrix representing physical RGB colors as is conventionally done in image recognition, we exploit the three-channel matrix input into the CNN to represent spatial design of meta-atom scatterers in different layers of a multi-layer MTS. Prior works do not employ this innovative technique of representing multiple MTS layers as channels of an image matrix.

In the first case study, we generated single-layer meta-atom designs using cDC-GAN. The co-polarization and cross-polarization transmission responses of the resulting meta-atom designs (Figure 9.7) differed from EM simulators by less than a dB. One of the most exciting features of cDC-GAN is its ability to discover new geometries not previously found in the literature. This suggests that the model implicitly learned the physical relationships of Maxwell's equations rather than simply interpolating from past designs. We perform a second case study for multi-layer meta-atom design. Building on this technique, the federated learning (FL) approach in [48] employed a cMD-GAN (see Figure 9.4) for multi-layer RF MTS discovery (Figure 9.8). The results show the feasibility of GAN-based approaches for meta-atom discovery.

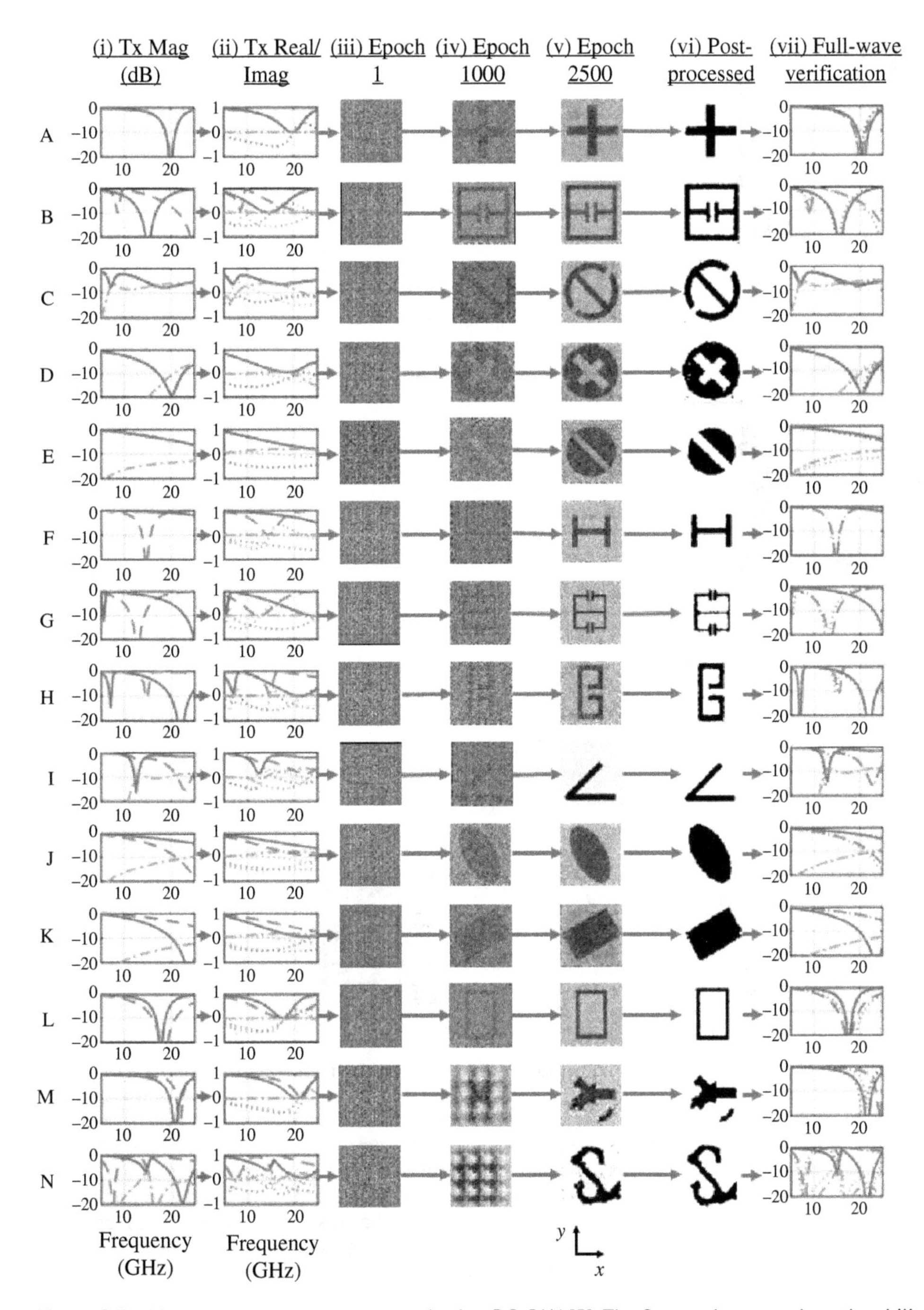

Figure 9.7 Meta-atom structures generated using DC-GAN [5]. The first twelve rows show the ability of the DC-GAN to regenerate canonical structures from the training data set. The last two rows show the ability of the DC-GAN to generate new meta-atom geometries, exhibiting spatial features similar to those in the training data set. Source: John A. Hodge, Kumar Vijay Mishra, and Amir I. Zaghloul. Deep inverse design of reconfigurable metasurfaces for future communications. arXiv preprint arXiv:2101.09131, 2021.

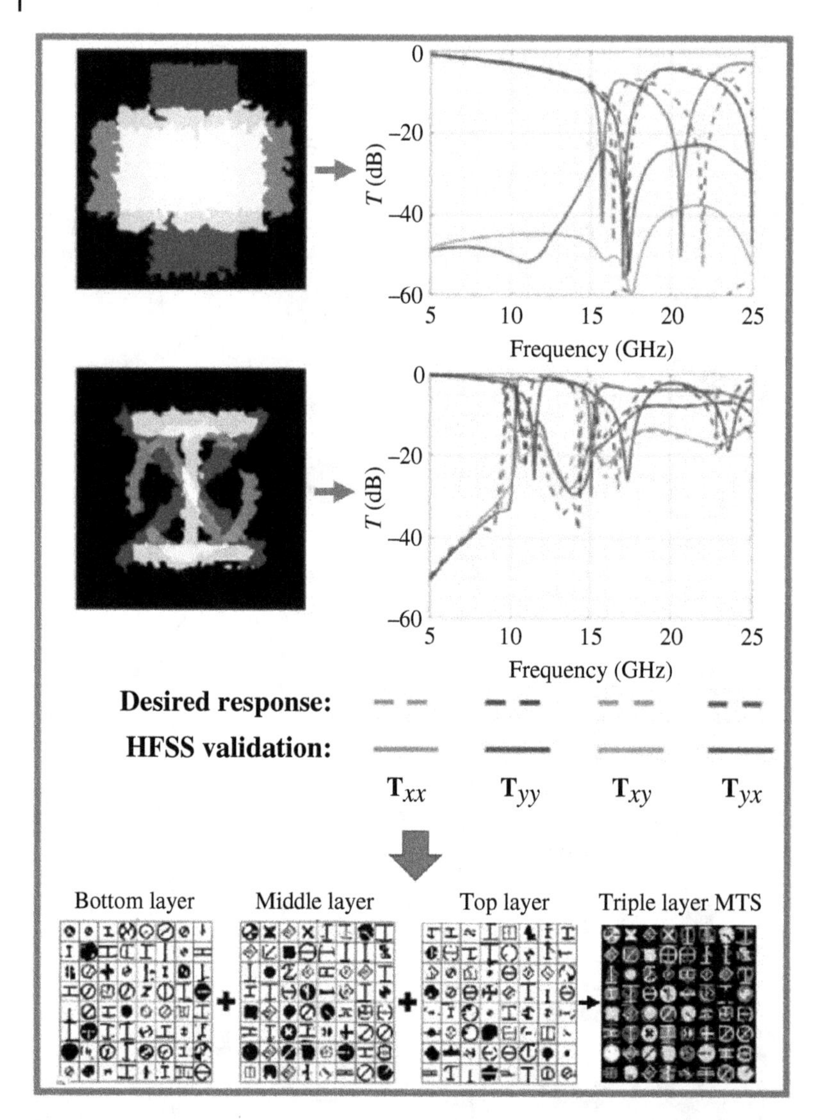

Figure 9.8 Two three-layer meta-atom designs (upper left) generated from cMD-GAN in [48]. The top (silver), middle (light gray), and bottom (dark gray) layers are metallic traces separated by dielectric spacers. The image matrices are post-processed to remove the background noise. The upper right shows desired input RF transmission response vectors (dashed lines) and the full-wave verification using the Ansys HFSS finite element method solver (solid lines) of generated design for 5–25 GHz when illuminated by a plane wave at normal incidence. The light gray and dark gray lines represent the respective x and y co-polarized transmission responses. Similarly, the silver and gray lines represent the cross-polarized responses for x and y polarized signals. The bottom row shows a composite of each layer of generated meta-atoms for a three-layer RIS. Source: [25] Hodge et al., 2021/Kumar Vijay Mishra.

9.5 Applications

Lately, the RIS-aided wireless systems have exploited DL to handle very challenging problems. For instance, signal detection in RIS requires development of end-to-end learning systems under the effect of channel and beamformers [65]. The channel needs to be estimated for multiple communication links, i.e. BS-user and BS-RIS-user [66]. Finally, beamformers are designed (by solving complex optimization problems) for phase shifters at both BS and passive elements of the RIS [67]. The DL-based techniques are able to handle the multidimensional, huge datasets in all these problems and may also be employed for channel modeling [39], where the conventional model-based approaches are not very useful. There have been recent surveys on applying DL [43] and RIS [39] individually to wireless communications. Here, we provide an overview of systems which jointly employ both approaches. In particular, we describe DL techniques (Table 9.2) for three important RIS problems: signal detection, channel estimation, and beamforming. Each of these requires different DL architectures, which have so far included supervised learning (SL), unsupervised learning (UL), reinforcement learning (RL) and federated learning (FL). The UL and RL do not require labeling; SL needs labeled dataset; and FL has distributed structure for model training. We provide a detailed synopsis of the advantages and shortcomings of each algorithm for these three applications in Section 14.5.1, 14.5.2, and 14.6.

In RIS-assisted scenario, wherein the BS with M antennas transmits K data symbols $s_k \in \mathbb{C}$ by using a baseband precoder $\mathbf{F} = [\mathbf{f}_1, \dots, \mathbf{f}_K] \in \mathbb{C}^{M \times K}$. Hence, the downlink $M \times 1$ transmitted signal becomes $\mathbf{s} = \sum_{k=1}^{K} \mathbf{f}_k s_k$. The transmitted signal is received from the k-user with two components, one of which is through the direct path from the BS and the another one is through the RIS. The received signal from the kth user can be given by

$$y_k = \left(\mathbf{h}_{D,k}^{\mathsf{H}} + \mathbf{h}_{A,k}^{\mathsf{H}} \mathbf{\Psi}^{\mathsf{H}} \mathbf{H}^{\mathsf{H}}\right) \mathbf{s} + n_k \tag{9.9}$$

where $n_k \sim \mathcal{CN}(o, \sigma_n^2)$ and $\mathbf{h}_{D,k} \in \mathbb{C}^M$ denote the direct channel between the BS and the kth user. The vector $\mathbf{h}_{A,k} \in \mathbb{C}^L$ expresses the RIS-assisted channel between the RIS and the kth user. $\mathbf{\Psi} \in \mathbb{C}^{L \times L}$ is a diagonal matrix, i.e. $\mathbf{\Psi} = \text{diag}\{\beta_1 \exp(j\phi_1), \dots, \beta_L \exp(j\phi_L)\}$. Here, $\beta_l \in \{0,1\}$ represents the on/off state of the RIS elements. In practice, the RIS elements cannot be perfectly turned on/off, Hence, β_l can be modeled as $\beta_l = \begin{cases} 1 - \epsilon_1 & \text{ON} \\ 0 + \epsilon_0 & \text{OFF} \end{cases}$ for $\epsilon_1, \epsilon_0 \geq 0$. $\phi_l \in [0,2\pi)$ is the phase shift of the reflective elements. Finally, the channel between the RIS and the BS is represented by $\mathbf{H} \in \mathbb{C}^{M \times L}$.

In mm-Wave transmission, the channel can be represented by the Saleh–Valenzuela (SV) model where a geometric channel model is adopted with limited scattering. Hence, we assume that the mm-Wave channels, i.e. $\mathbf{h}_{D,k}, \mathbf{h}_{A,k}$, and $\mathbf{H}$, include the contributions of N_D, N_A, and N_H paths, respectively. Thus, we can represent the channels $\mathbf{h}_{D,k}$ and $\mathbf{h}_{A,k}$ as $\mathbf{h}_{D,k} = \sqrt{\frac{M}{N_D}} \sum_{n_D=1}^{N_D} \alpha_{D,k}^{(n_D)} \mathbf{a}_D(\theta_{D,k}^{(n_D)})$, and $\mathbf{h}_{A,k} = \sqrt{\frac{L}{N_A}} \sum_{n_A=1}^{N_A} \alpha_{A,k}^{(n_A)} \mathbf{a}_A(\theta_{A,k}^{(n_A)})$, where $\{\alpha_{D,k}^{(n_D)}, \alpha_{A,k}^{(n_A)}\}$ and $\{\theta_{D,k}^{(n_D)}, \theta_{A,k}^{(n_A)}\}$ are the complex channel gains and received path angles for the corresponding channels, respectively. $\mathbf{a}_D(\theta)$ and $\mathbf{a}_A(\theta)$ are $M \times 1$ and $L \times 1$ steering vectors of the path angles as $\mathbf{a}_D(\theta) = \frac{1}{\sqrt{M}}[e^{j\omega_0}, \dots, e^{j\omega_{M-1}}]^T$, $\mathbf{a}_A(\theta) = \frac{1}{\sqrt{L}}[e^{j\omega_0}, \dots, e^{j\omega_{L-1}}]^T$ where $\omega_n = n\frac{2\pi d}{\lambda}\pi \sin(\theta)$ and $d = \lambda/2$ is the array spacing for the wavelength λ. Further, the mm-Wave channel between the BS and the RIS is given by

$$\mathbf{H} = \sqrt{\frac{ML}{N_H}} \sum_{n_H=1}^{N_H} \alpha^{(n_H)} \mathbf{a}_{BS}(\theta_{BS}^{(n_H)}) \mathbf{a}_{RIS}^{\mathsf{H}}(\theta_{RIS}^{(n_H)}) \tag{9.10}$$

where $\alpha^{(n_H)} \in \mathbb{C}$ denotes the complex gain and $\{\theta_{BS}^{(n_H)}, \theta_{RIS}^{(n_H)}\}$ are the angle-of-departure (AOD) and angle-of-arrival (AOA) angles of the paths, respectively. $\mathbf{a}_{BS}(\theta) \in \mathbb{C}^M$ and $\mathbf{a}_{RIS}(\theta) \in \mathbb{C}^L$ are the steering vectors. Let $\mathbf{G}_k \in \mathbb{C}^{M \times L}$ be the cascaded channel matrix between the BS and the kth user as $\mathbf{G}_k = \mathbf{H}\mathbf{\Gamma}_k$ where $\mathbf{\Gamma}_k = \text{diag}\{\mathbf{h}_{A,k}\}$. Then, we can write $\mathbf{H}\mathbf{\Psi}\mathbf{h}_{A,k} = \mathbf{G}_k \psi$, for which we have $\mathbf{\Psi} = \text{diag}\{\psi\}$.

Table 9.2 DL-based techniques for RIS-assisted wireless systems.

Learning scheme	NN architecture	Benefits	Drawbacks
Signal detection			
SL [65]	MLP with 3 layers	No need for channel estimation algorithm	Still needs to design beamformers and requires huge datasets and deeper NN architectures
Channel estimation			
SL [66]	Twin CNNs with 3 convolutional, 3 fully connected layers	Each user estimates its own channel with the trained model	Data collection requires channel training by turning on/off each RIS elements
FL [44]	A single CNN with 3 convolutional, 3 fully connected layers	Less transmission overhead for training, A single CNN estimates both cascaded and direct channels	Performance depends on the number of users and the diversity of the local datasets
SL [68]	DDNN with 15 convolutional layers	Leverages both compressed sensing (CS) and DL methods	Requires active RIS elements. High prediction complexity arising from CS algorithms
Beamforming			
SL [67]	MLP with 4 layers	Reduced pilot training overhead	Requires active RIS elements for channel training
UL [69]	MLP with 5 layers	Reduced complexity at the model training stage	Implicitly needs the reflect beamformers as labels
RL [70]	DQN with 4 layers	Provides standalone operation since RL does not require labels like SL	Longer training. Active RIS elements needed for channel acquisition

RL [40]	DDPG with 4-layered actor and critic networks	Better performance than DQN	Large number of NN parameters are involved
RL [71]	DDPG with actor and critic networks	Accelerated learning performance with the aid of optimization, shrinking the search space	Additional optimization tools needed
FL [72]	MLP with 6 layers	Less transmission overhead involved during model training	RIS must be connected to the PS
Secure beamforming			
RL [73]	DQN with 3 layers	Robust against eavesdropping	High model training complexity
Energy-efficient beamforming			
RL [42]	DQN	Energy-efficient and robust against channel uncertainty	RIS beamforming only
Indoor beamforming			
SL [41]	MLP with 5 layers	Reduces hardware complexity of multiple BSs and improves RSS for indoor environments	Learning model performance relies on room conditions

Source: [34] Elbir and Mishra, 2020/IEEE.

9.5.1 DL-Based Signal Detection in RIS

The signal detection comprises mapping the received symbols under the effect of channel and beamformers to transmit symbols (Figure 9.9). The signal detection problem can be formulated as

$$\hat{\mathbf{s}} = \underset{\mathbf{s}}{\text{argmin}} \left\| y_k - \left(\mathbf{h}_{D,k}^{\mathsf{H}} + \mathbf{h}_{A,k}^{\mathsf{H}} \mathbf{\Psi}^{\mathsf{H}} \mathbf{H}^{\mathsf{H}} \right) \mathbf{s} \right\|^2 \tag{9.11}$$

which requires the knowledge of the channel, i.e. $\mathbf{h}_{D,k}, \mathbf{h}_{A,k}$, and $\mathbf{H}$. Instead, DL-based model accepts the input data $\mathbf{y}_k = [y_{k,1}, \dots, y_{k,P_T}]^T$, where P_T is the number of collected observations. Then, the DL model is trained to construct a non-linear mapping between the corrupted data $\mathbf{y}_k$ and the clean symbols $\mathbf{s}$.

To leverage DL for signal detection, Khan et al. [65] devised a multi-layer perceptron (MLP) for mapping the channel and reflecting beamformer effected data symbols to the transmit symbols. The MLP is a feedforward neural network (NN) composed of multiple hidden layers. The framework in [65] uses three fully connected layers. Once the MLP is trained on a dataset composed of received-transmitted data symbols, each user feeds the learning model with the block of received symbols. These blocks account for the effect of channel and beamformers. Then, MLP yields the estimated transmit symbols.

A major advantage of this approach is its simplicity that the learning model estimates the data symbols directly, without a prior stage for channel estimation. Thus, this method is helpful reducing the cost of channel acquisition. In [65], a bit-error-rate (BER) analysis has shown that the DL-based RIS signal detection (DeepRIS) provides better BER than the minimum mean-squared-error (MMSE) and close performance to the maximum likelihood estimator.

However, a few challenges remain to achieve a reliable performance. The training data should be collected under several channel conditions and different beamformer configurations so that the trained model learns the environment well and reflects the accurate performance in different scenarios. This is a particularly challenging task because it requires collection of the training data

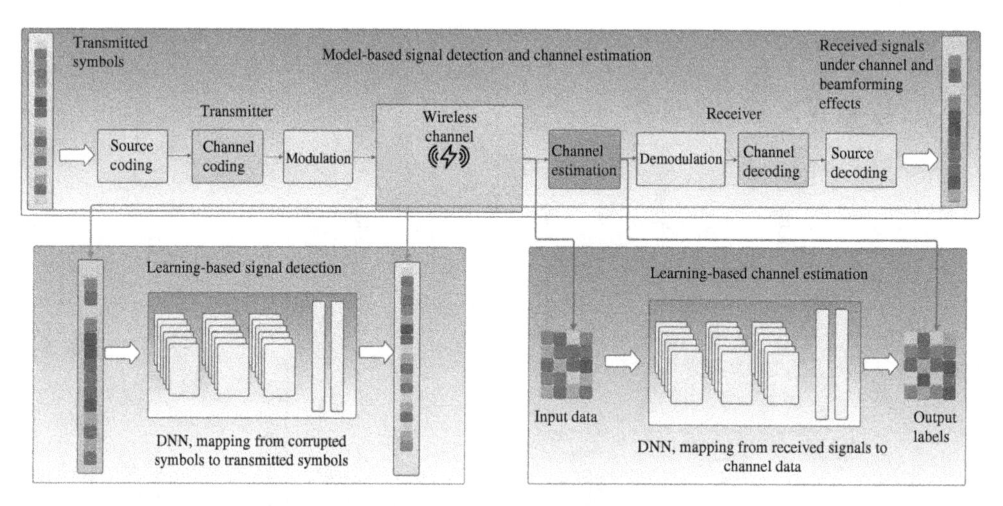

Figure 9.9 Model-based versus learning-based frameworks for signal detection and channel estimation. Model-based approach (top) comprises multiple subsystems to process the received signal. Learning-based signal detection (bottom, left) provides an end-to-end data mapping from the corrupted symbols under the channel effects at the receiver to the transmit symbols. Learning-based channel estimation (bottom, right) maps the input received signals to the channel estimate as output labels [34].

for different user locations. As a result, DL-based signal detection demands huge training dataset collected at different channel conditions.

9.5.2 DL-Based RIS Channel Estimation

The RIS is composed of a huge number of reflecting elements and, therefore, channel state acquisition is a major task in RIS-assisted wireless systems. A common approach is to turn on and off each individual RIS element one-by-one while also using orthogonal pilot signals to estimate the channel between the BS and the users through RIS. In particular, RIS channel estimation via DL involves constructing a mapping between the received input signals at the user and the channel information of direct and cascaded links (Figure 9.9). In this way, DL-based techniques reduce the pilot percentage and complexity in channel estimation stage [44].

The SL approach proposed in [66] estimates both direct and cascaded channels via twin CNNs. First, the received pilot signals at the user are collected by sequentially turning on the individual RIS elements. Then, the collected data are used to find the least squares (LS) estimate of the cascaded and the direct channels. Both CNNs are trained to map the LS channel estimates to the true channel data. The upshot is that each user estimates its own channels only once and feeds the received pilot data (LS estimate) to the trained CNN models. The CNNs have higher tolerance than MLP against the channel data uncertainties, imperfections (such as switching mismatch) of RIS elements.

When the model training is conducted at the user with huge datasets as in [66], the system may lack sufficient computational capability. This is overcome by FL-based training [44], where the learning model updates are computed at the devices (nodes) and aggregated at the BS (central server), thereby eliminating the transmission of raw data. FL significantly reduces the transmission overhead since the size of the datasets is usually larger than the size of the learning model, and its performance improves as the number of users increases [44, 72]. Furthermore, instead of using two CNNs as in [66], a single CNN in [44] jointly estimates both cascaded and direct channels.

Although FL reduces the transmission overhead during model training, its training performance is upper bounded by the centralized model training, i.e. training the model with the whole dataset at once. Therefore, the prediction performance of FL is usually poorer than the centralized learning (CL). As shown in Figure 9.10, CL and FL frameworks are compared with the MMSE and the LS estimation. We note that FL performs slightly poorer than CL in high SNR regimes. Despite this, FL significantly reduces the transmission overhead, e.g. approximately ten-fold reduction in the number transmitted symbols [44]. The performance of FL improves with the increase in the number of users or edge devices because this reduces the variance of the model updates aggregated at the BS. The diversity of the local dataset of the users also affects the training/prediction performance and better performance is obtained if the local datasets are close to uniformity.

Both SL- and FL-based channel estimation techniques suffer from high channel training overhead. In this context, compressive channel estimation with deep denoising neural networks (DDNNs) is very effective [68]. It employs a hybrid passive/active RIS architecture, where the active RIS elements are used for uplink pilot training and passive ones for reflecting the signal from the BS to the users. Once the BS collects the compressed received pilot measurements, complete channel matrix is recovered through sparse reconstruction algorithms such as orthogonal matching pursuit (OMP). Then, DDNN is used to improve the channel estimation accuracy by exploiting the correlation between the real and imaginary parts of the mm-Wave channel in angular-delay domain. During training, the input is the OMP-reconstructed channel matrix and the output is the noise, i.e. the difference between the OMP estimate and the ground truth channel data. This method leverages both compressed sensing (CS) and DL yielding a performance better than

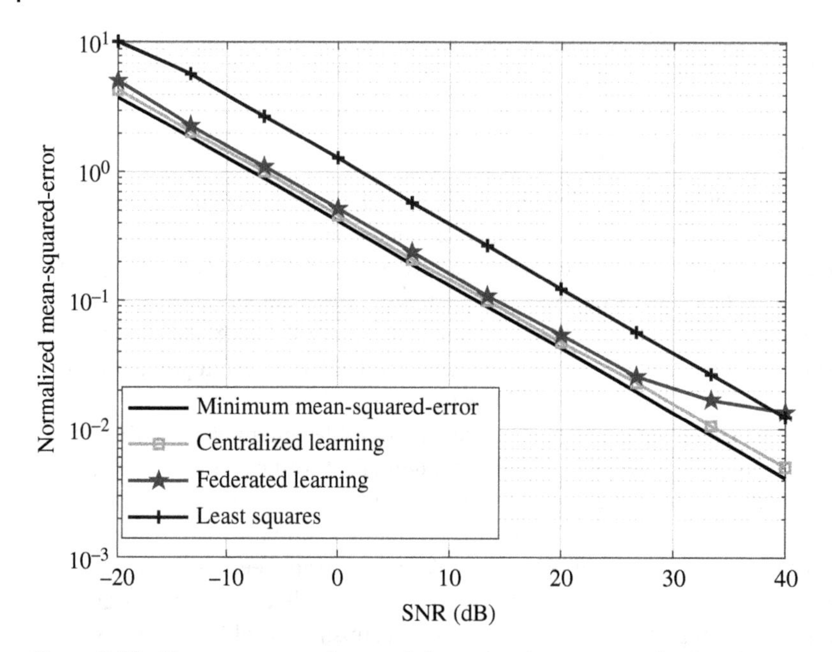

Figure 9.10 The mean-squared-error of channel estimates normalized against ground truth channel, obtained using CNN in centralized and HL frameworks, MMSE and LS. The BS consisted of 64 antennas and RIS employed 64 passive reflecting elements [34].

using these techniques individually. The major drawback is the additional hardware complexity introduced by the active RIS elements. Furthermore, OMP algorithm is used in place of the raw received pilot measurements for constructing the input. This requires repeated execution of the OMP algorithm thereby increasing the prediction complexity over the DL methods in [66] and [44].

Consider the downlink scenario where the BS transmits the orthogonal pilot signals $\mathbf{x}_p \in \mathbb{C}^M$, one at a single coherence time τ, with $p = 1, \dots, P$ and $P \geq M$. Hence, the total number of channel uses to estimate the direct channel is P. The received signal at the kth user can be given by

$$\mathbf{y}_k = \left(\mathbf{h}_{D,k}^{\mathsf{H}} + \boldsymbol{\psi}^{\mathsf{H}}\mathbf{G}_k^{\mathsf{H}}\right)\mathbf{X} + \mathbf{n}_k \tag{9.12}$$

where $\mathbf{X} = [\mathbf{x}_1, \dots, \mathbf{x}_P] \in \mathbb{C}^{M\times P}$ is the pilot signal matrix while $\mathbf{y}_k = [y_{k,1}, \dots, y_{k,P}]$ and $\mathbf{n}_k = [n_{k,1}, \dots, n_{k,P}]$ are $1 \times P$ row vectors and $\mathbf{n}_k \sim \mathcal{CN}(0, \sigma_n^2 \mathrm{I}_P)$. We assume that the pilot training has two phases: direct channel estimation (i.e. $\mathbf{h}_{D,k}$) and the cascaded channel estimation (i.e. $\mathbf{G}_k$). In phase I, we assume that all of the RIS elements are turned off, i.e. $\beta_l = 0, \forall l$, by using the BS backhaul link. We note here that by setting β_l as $\{1,0\}$ does not affect the direct and cascaded channels since they do not depend on the reflect beamformer $\boldsymbol{\Psi}$ as seen in (9.12). Then, the received baseband signal at the kth user becomes

$$\mathbf{y}_D^{(k)} = \mathbf{h}_{D,k}^{\mathsf{H}}\mathbf{X} + \mathbf{n}_{D,k} \tag{9.13}$$

Here, the direct channel $\mathbf{h}_{D,k}$ is selected as the label of the deep network with the corresponding input data of $\mathbf{y}_D^{(k)}$.

Once $\mathbf{h}_{D,k}$, being the estimated channel, is obtained, in the second phase of the training stage, the cascaded channel $\mathbf{G}_k$ can be estimated. This can be achieved via two approaches. In the first approach, $P = M$ pilot signals are transmitted when each of the RIS elements is turned on one by one. In this case, the BS sends a request to RIS via the micro-controller device in the backhaul link to turn on a single RIS element at a time. For the lth frame, the reflect beamforming vector becomes $\boldsymbol{\psi}^{(l)} = [0, \dots, 0, \psi_l, 0, \dots, 0]^T$ where $\beta_{\bar{l}} = \{0 : \bar{l} = 1, \dots, L, \bar{l} \neq l\}$ and the received signal from the cascaded channel at the kth user becomes

$$\mathbf{y}_C^{(k,l)} = \left(\mathbf{h}_{D,k}^{\mathsf{H}} + \mathbf{g}_{k,l}^{\mathsf{H}}\right)\mathbf{X} + \mathbf{n}_{k,l} \tag{9.14}$$

where $\mathbf{y}_C^{(k,l)} = [y_{C,1}^{(k,l)}, \dots, y_{C,P}^{(k,l)}]$ and $\mathbf{n}_{k,l} = [n_{k,1}^{(l)}, \dots, n_{k,P}^{(l)}]$ are $1 \times P$ row vectors. In (9.14), $\mathbf{g}_{k,l}$ represents the lth column of $\mathbf{G}_k$ as $\mathbf{g}_{k,l} = \mathbf{G}_k \boldsymbol{\psi}^{(l)}$. Then the LS estimate of $\mathbf{g}_{k,l}$ becomes

$$\hat{\mathbf{g}}_{k,l} = \left(\mathbf{y}_C^{(k,l)} \mathbf{X}^{\mathsf{H}} (\mathbf{X}\mathbf{X}^{\mathsf{H}})^{-1}\right)^{\mathsf{H}} - \mathbf{h}_{D,k} \tag{9.15}$$

By using $\hat{\mathbf{h}}_{D,k}$, (9.15) can be solved for $l = 1, \dots, L$. Then, we can construct the estimated cascaded matrix as $\hat{\mathbf{G}}_k = [\hat{\mathbf{g}}_{k,1}, \dots, \hat{\mathbf{g}}_{k,L}]$.

Then, the deep network accepts the received signals as input at the preamble stage. As a result, the input–output pairs become $\{\mathbf{y}_D^{(k)}, \mathbf{h}_{D,k}\}$ and $\{\mathbf{y}_C^{(k,l)}, \mathbf{g}_{k,l}\}$ for direct and cascaded channel estimation, respectively.

Now, let us consider model training via CL for channel estimation, wherein the training is performed by collecting the local datasets $\{\mathcal{D}_k\}_{k \in \mathcal{K}}$ from the users. Once the BS has collected the whole dataset $\mathcal{D} = \{\mathcal{D}_k\}_{k \in \mathcal{K}}$, the training is performed by solving the following problem

$$\begin{aligned} &\underset{\boldsymbol{\theta}}{\text{minimize}} \quad \mathcal{L}(\boldsymbol{\theta}) \\ &\text{subject to:} \quad f(\mathcal{X}^{(i)} | \boldsymbol{\theta}) = \mathcal{Y}^{(i)}, \quad i = 1, \dots, \mathsf{D} \end{aligned} \tag{9.16}$$

where $\mathsf{D} = |\mathcal{D}|$ is the number of training samples and $\mathcal{L}(\boldsymbol{\theta})$ denotes the loss function defined as

$$\mathcal{L}(\boldsymbol{\theta}) = \frac{1}{\mathsf{D}} \sum_{i=1}^{\mathsf{D}} \| f(\mathcal{X}^{(i)} | \boldsymbol{\theta}) - \mathcal{Y}^{(i)} \|_{\mathcal{F}}^2 \tag{9.17}$$

which is the mean squared error (MSE) between the label data $\mathcal{Y}^{(i)}$ and the prediction of the CNN, $f(\mathcal{X}^{(i)} | \boldsymbol{\theta})$.

On the other hand, in FL, the local datasets $\mathcal{D}_{k \in \mathcal{K}}$ are preserved at the users and not transmitted to the BS. Hence, FL-based model training is performed at the user side as

$$\begin{aligned} &\underset{\boldsymbol{\theta}}{\text{minimize}} \quad \overline{\mathcal{L}}(\boldsymbol{\theta}) = \frac{1}{K} \sum_{k=1}^{K} \mathcal{L}_k(\boldsymbol{\theta}) \\ &\text{subject to:} \quad f(\mathcal{X}_k^{(i)} | \boldsymbol{\theta}) = \mathcal{Y}_k^{(i)}, i = 1, \dots, \quad \mathsf{D}_k, \quad k \in \mathcal{K} \end{aligned} \tag{9.18}$$

where $\mathcal{L}_k(\boldsymbol{\theta}) = \frac{1}{\mathsf{D}_k} \sum_{i=1}^{\mathsf{D}_k} \| f(\mathcal{X}_k^{(i)} | \boldsymbol{\theta}) - \mathcal{Y}_k^{(i)} \|_{\mathcal{F}}^2$. Notice that the FL-based model training in (9.18) is solved at the user while the CL problem in (9.16) is handled at the BS. To efficiently solve (9.18) and (9.16), gradient descent (GD) is employed and the problems are solved iteratively. In CL, the gradient is computed over the whole dataset as $\mathbf{g}(\boldsymbol{\theta}_t) = \nabla \mathcal{L}(\boldsymbol{\theta}_t)$ and the parameter update is performed as

$$\boldsymbol{\theta}_{t+1} = \boldsymbol{\theta}_t - \eta \mathbf{g}(\boldsymbol{\theta}_t) \tag{9.19}$$

where η is the learning rate.

In FL, each user computes the gradients individually as $\mathbf{g}_k(\boldsymbol{\theta}_t) = \nabla \mathcal{L}_k(\boldsymbol{\theta}_t)$ to solve (9.18), then sends them to the BS, where the model parameters are updated as

$$\boldsymbol{\theta}_{t+1} = \boldsymbol{\theta}_t - \eta \frac{1}{K} \sum_{k=1}^{K} \mathbf{g}_k(\boldsymbol{\theta}_t) \tag{9.20}$$

Once the model is trained, each user can feed its received pilots signals to the CNN to predict its channel data.

9.6 DL-Aided Beamforming for RIS Applications

Beamforming in RIS-based communications has diverse applications such as RIS-only beamforming (passive), BS-RIS beamforming (active/passive), secure beamforming (eavesdroppers included), energy-efficient beamforming, and indoor RIS beamforming. There are specific DL challenges and solutions to each one of these problems.

In general, the beamforming design problem in RIS-assisted scenario maximizes the spectral efficiency of the system as

$$\begin{aligned} &\underset{\mathbf{f}_k,\forall k}{\text{maximize}} \sum_{k=1}^{K} \log_2(1+\mathsf{SINR}_k) \\ &\text{subject to: } |\beta_l| = 1, \phi_l \in S_\phi \\ &\sum_{k=1}^{K} \|\mathbf{f}_k\|^2 \leq P_t \end{aligned} \tag{9.21}$$

where P_t is the total transmit power and S_ϕ denotes the set of discrete phase-shifts. Also, we define the signal-to-interference-plus-noise ratio (SINR) as $\mathsf{SINR}_k = \frac{\mathbf{h}_{A,k}^{\mathsf{H}}\Psi\mathbf{f}_k s_k}{\sum_{n=1,n\neq k}^{K}\mathbf{h}_{A,k}^{\mathsf{H}}\Psi\mathbf{f}_n s_n+\sigma_n^2}$, wherein only RIS-reflected channel is assumed.

9.6.1 Beamforming at the RIS

The RIS beamforming requires passive elements continuously to reliably reflect the BS signal to the users. Here, the MLP architecture [67] is helpful in designing the reflect beamforming weights using active RIS elements [68]. These elements are randomly distributed through the RIS. They are used for pilot training, after which compressed channel estimation is carried out using OMP. During data collection, the reflect beamforming weights are optimized by using the estimated channel data. Finally, a training dataset is constructed with channel data and reflect beamformers as the input–output pairs for an SL framework. Note that the active RIS elements present similar shortcomings as in [68]. However, the method in [67] excels by leveraging DL for designing beamformers.

The labeling process in [67] demands solving an optimization problem for each channel instance in training data generation stage. One possible way to mitigate this is to use label-free techniques, such as UL. The UL approach in [69] for reflect beamforming design employs MLP with five fully connected layers. The network maps the vectorized cascaded and direct channel data input to the output comprising the phase values of the reflect beamformers. The loss function is selected as the negative of the norm of the channel vector, which may seem like an unsupervised approach because it does not minimize the error between the label and learning model prediction. However, this technique yields the phase information at the output uniquely for each training sample. Consequently, the beamformers implicitly behave like a label in the training process. In UL, the training data is clustered into smaller sets without a prior knowledge about the "meaning" of each clustered sets. However, in [69], the output of the NN is a design parameter, i.e. reflect beamformer phases, which have the complexity of beamformer optimization for each input.

In order to eliminate the expensive labeling process of the SL-based techniques, Taha et al. [70] employed RL to design the reflect beamformers for single-antenna users and BS. The RL is a promising approach which directly yields the output by optimizing the objective function of the learning model. First, the channel state is estimated by using two orthogonal pilot signals. An action vector is selected either by exploitation (using prior experience of the learning model) or exploration (using a predefined codebook). After computing the achievable rate based on the selected action vector from the environment, a reward or penalty is imposed by comparing with the achievable rate with a threshold. Upon reward calculation, a deep quality network (DQN) (Figure 9.12) updates the map from the input state (channel data) to the output action (action vector composed of reflect beamformer weights). This process is repeated for several input states until the learning model converges. While RL is not an RIS-specific technique, it is particularly useful in lowering the overhead of labeling process as compared to SL architectures deployed by recurrent neural network (RNN) or CNN models, which require labeled datasets. The RL algorithm learns reflect beamformer weights based on the optimization of the achievable rate. Thus, RL presents a solution for online learning schemes, where the model effectively adapts to the changes in the propagation environment. However, RL techniques have longer training times than the SL approaches because reward mechanism and discrete action spaces make it difficult to reach the global optimum. The label-free process implies that the RL usually has slightly poorer performance than the SL.

To accelerate the training stage by the use of continuous action spaces, a deep deterministic policy gradient (DDPG) (Figure 9.12) was introduced in [40]. Here, actor-critic network architectures are used to compute actions and target values, respectively. First, the learning stage is initialized by the use of input state excited by cascaded and direct channels. Given the state information, a deep policy network (DPN) (actor) constructs the actions (reflection beamformer phases). Here, the DPN provides a continuous action space that converges faster than the DQN architecture in [70]. The action vector is used by the critic network architecture to estimate the received SNR as objective. This SNR then yields the target beamformer vector under the learning policy. Using the gradient of DPN, the network parameters are updated and the next state is constructed as the combination of the received SNR and the reflecting beamformers. This process is repeated until it converges. An additional benefit of this approach is that it outperforms fixed-point iteration (FPI) algorithms used to solve reflect beamforming optimization. Moreover, the continuous action space representation with DPN in DDPG provides robustness of the learning model against changes in channel data. However, multiple NN architectures (actor and critic networks) increase the number of learning parameters and aggravate model update requirements for each architecture.

The model initialization in both DQN and DDPG may force the learning models to start far from the optimum point during the early stages of learning. This leads to a slow convergence and poor reward performance. In order to accelerate the learning process, Gong et al. [71] devised a joint learning and optimization technique. The key idea is to use DDPG to search for optimal action for each decision epoch during training. Then, a feasible beamformer vector is found via optimization in a convex-approximation setting. This reduces the search space of the DDPG algorithm and shortens training times.

Even if RL is a label-free approach that reduces the overhead during training data generation, training approaches in [40, 70, 71] demand expensive transmission overhead to be trained on huge datasets. This is mitigated in FL techniques (see Figure 9.11). The FL approach in [72] learns the RIS reflect beamformers by training an MLP by computing the model updates at each user with the local dataset. The model updates are aggregated in a parameter server (PS), which is connected to the

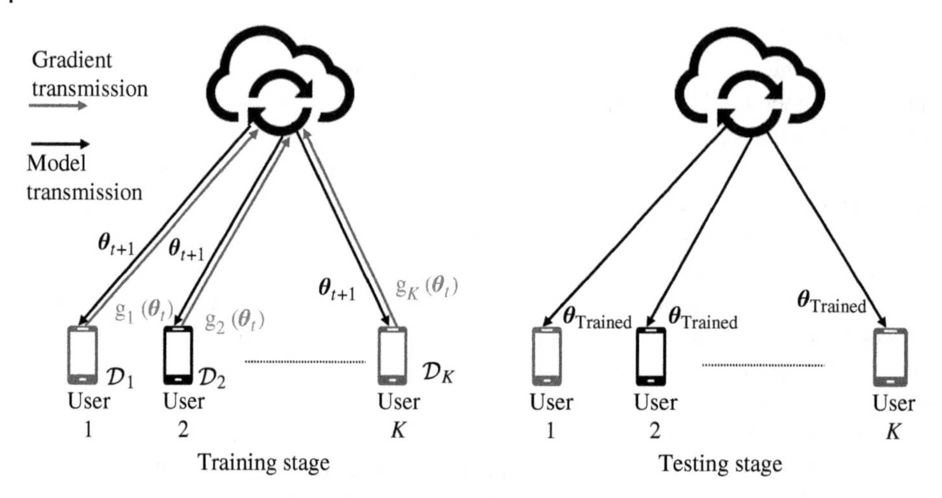

Figure 9.11 In the FL framework, each user processes its own local dataset $\mathcal{D}_k$, computes the model updates (gradients) $\mathbf{g}_k(\theta_t)$, and sends them to the PS. The server aggregates the collected model updates and the updated model parameters are sent back to the users as θ_{t+1}. Source: Adapted from [34].

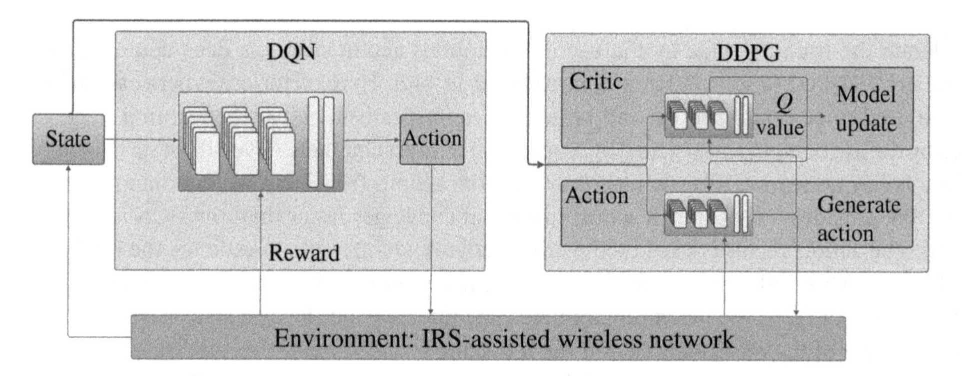

Figure 9.12 In RL, the DQN and DDPG architectures accept same state (channel data and received SNR) and environment data (beamformers to be evaluated). The DQN involves training a single neural network based on the reward determined from the environment. On the other hand, the DDPG has multiple neural networks, where actor-critic architectures are used to compute actions and target values, respectively [34].

RIS. The MLP input is the cascaded channel information and the output labels are RIS beamformer weights. The federated architecture lowers the transmission overhead during training. However, it is assumed that the PS is connected to the RIS. The simple architecture of the RIS could make this infeasible. It is more practical to access the PS via BS for model training.

9.6.2 Secure-Beamforming

Physical layer security in wireless systems is largely achieved through signal processing techniques, such as cooperative relaying and cooperative jamming. The hardware complexity is a major issue in these methods. The low-cost, less complex RIS-based systems have the potential to mitigate these problems. The RL-based secure beamforming [73] minimizes the secrecy rate by jointly designing the beamformers at the RIS and BS to serve multiple legitimate users in the presence of eavesdroppers. The RL algorithm accepts the states as the channel information of all users, secrecy rate, and transmission rate. Similar to [40], the action vectors are beamformers at the BS and RIS. The reward function is designed based on the secrecy rate of users. A DQN is trained to learn the beamformers by minimizing the secrecy rate while guaranteeing the quality-of-service requirements. The model training takes place at the BS, which is responsible for collecting the environment information (channel data) and making decisions for secure beamforming. This scheme is more realistic and reliable than that of [40, 70], which ignore the effect of eavesdroppers. The learning model includes high-dimensional state and action information, such as the channels of all users and beamformers of BS and RIS. This may necessitate more computing resources for training than non-secure RIS [40, 70] and conventional SL techniques [66, 67].

9.6.3 Energy-Efficient Beamforming

The RIS configuration dynamically changes depending on the network status. It is very demanding for the BS to optimize the transmit power every time when the on/off status of RIS elements is updated. This could be addressed by accounting energy-efficiency in the beamformer design problem. In [42], a self-powered RIS scenario maximizes the energy-efficiency by optimizing the transmit power and the RIS beamformer phases. In this DQN-based RL approach, the BS learns the outcome of the system performance while updating the model parameters. Thus, the BS makes decisions to allocate the radio resources by relying on only the estimated channel information. The RL framework has states selected as the estimated channels from users and the energy level of the RIS. Meanwhile, the action vector includes the transmit power, the RIS beamformer phases, and on/off status of the RIS elements. The learning policy is based on the reward which is selected as the energy-efficiency of the overall system. However, this work considers only RIS beamforming and ignores the same at the BS.

9.6.4 Beamforming for Indoor RIS

Different from the above scenarios, Huang et al. [41] addresses the RIS beamformer design problem in an indoor communications scenario to increase the received signal strength (RSS) (see Figure 9.2). This is particularly useful from the perspective of low hardware complexity because it eliminates deployment of multiple BSs to improve RSS. The MLP architecture in [41] accepts two-dimensional user position vector and yields the RIS beamformer phases at the output. Since the channel data is not employed as input, the network does not have to deal with severe environmental fluctuations. However, the learning model trains on specific room environments and may perform poorly for different room conditions or different obstacle distribution in the same room. This is mitigated in RL-based solutions which are highly adaptive to different environments [40, 70].

9.7 Challenges and Future Outlook

The techniques for RIS inverse design and processing are constantly evolving. Major challenges include reduction of training cost, gathering of labeled data, effective handling of system imperfections, and better data representations.

9.7.1 Design

New approaches are needed to increase the computational efficiency and reduce the amount of training required for DL-based RIS design. As mentioned below, reduction of design time and achieving full EM-compliance remains a major challenge.

9.7.1.1 Hybrid Physics-Based Models

Hybrid models, where training set is supplemented by physics-based analytical models, reduce the amount of required training data and increase learning efficiency. Analytical RF circuit-based models are available to predict the performance of several canonical meta-atom designs. To speed up the training data generation, these analytical circuit-based models could be used to supplement the training data set and reduce iterations of time-consuming full-wave EM simulations. It may also be feasible to create innovative DL design and optimization architectures that utilize physics-based analytical models within the ANN architecture. Another method to reduce the amount of required training data for multi-layer MTS designs is to use T-matrix data to analytically cascaded MTS designs from single-layer training data.

9.7.1.2 Other Learning Techniques

Transfer learning (TL) may also be used for expediting and improving the learning of a new task by using a previously trained neural network weights and bias as the initialization for the new ANN. Since all ANNs for meta-atom performance prediction and inverse design are implicitly learning Maxwell's equations, it is sensible that a network trained for one meta-atom design or frequency band is scaled and transferred to a related design. DQNs have also been studied to increase the efficiency of MTS holograms and automated multi-layer RIS design. Similar to evolutionary optimization techniques, RL is an area of ML concerned with how software agents ought to take actions in an environment in order to maximize the notion of cumulative reward through trial and error. Without the use of labeled training data, RL algorithms learn system dynamics through exploration to maximize a reward function. Here, deep reinforcement learning (DRL) algorithms, such as DQNs, have produced ML advances in a broad range of applications including robotics, strategy games, NLP, and computer vision. To date, DQNs have been studied to increase the efficiency of MTS holograms and automated multi-layer MTS design. However, research using D-RLs for MTS design and optimization is very limited and further research is needed to develop these techniques for MTS applications. D-RL-based networks hold the promise for automated self-learning of RIS in SRE that are able to adapt and optimize themselves for dynamic RF environments and modulations.

9.7.1.3 Improved Data Representation

More complex input data structures and representations are increasingly studied for DL-based RIS. While this chapter focused on discrete input parameters and image data structures are RIS design representations, graphical and sequential data structures have recently been proposed as alternatives. The graphical model has been used to represent EM systems with near-field coupling (as in coupled resonators). In this arrangement, graph nodes contain resonator attributes, such as material, geometry, and location, and graph nodes represent the near-field coupling factors. These graphical data structures are processed using graphical neural networks (GNNs). While yet to be extensively explored in the domain of MTS, GNNs have been applied to model a broad range of physical systems. GNNs have the potential to handle additional complexities to jointly optimize RIS design and operation in wireless communication networks. Additionally, sequential data structures are another data representation that is yet to be extensively explored in the context of MTS.

Similarly, sequential data structures are useful for representing time-sequence data in dynamic EM systems (as in RIS filters) and are learned using RNNs. In other domains, such as natural language processing (NLP), sequential data is often learned using RNNs, which are ANNs that use forward or backward connections to enable a memory of internal states between successive passes to the network. As dynamic operation of RIS becomes increasingly important in the development of wireless networks and SRE, it is likely that RNNs will become increasingly useful to model dynamic RIS.

9.7.2 Applications

Several challenges remain for DL architectures to reach their full potential in realizing significant performance gains and efficiency for RIS-assisted wireless systems. Given that it is an emerging technology, larger sets of real data are not yet available. Then, model training consumes much time and resources, including parallel processing and storage. Further, to achieve commercial viability of DL-based RIS-aided communications, dynamically adapting to changes in the environment is crucial. Finally, new RIS-specific implementation challenges have also been identified within emerging technologies such as terahertz communications, cell-free massive MIMO, drone operations, and open radio access network (RAN).

9.7.3 Channel Modeling

Channel Modeling is a challenging task in communication systems, especially with the large number of antennas due to the complexity of system architecture. In order to provide a reliable channel modeling performance, DL can be of help to construct a data-driven model based on the field measurements. In this case, SL schemes can be used to construct the channel model as a relationship between the input and output of a learning model [66]. Thus, DL-based methods are expected to become more frequently used in RIS-assisted wireless networks for channel modeling.

9.7.3.1 Data Collection

Massive data collection hampers successful performance of DL-based techniques for all wireless communications tasks: signal detection, channel estimation, and beamformer design. The signal detection requires collection and storage of transmit and receive data symbols for different channel conditions. The prerequisites for channel estimation and beamforming are even more tedious because of additional labeling process. This is difficult to overcome in, especially online scenarios. Apart from SL, the label-free structure in RL is particularly helpful but at the cost of training times. It is possible to relax the data collection requirements by realizing the propagation environment in a numerical EM simulation tool [44] and then using a more realistically simulated data. This is helpful in constructing the training dataset offline, but chances of failure remain in a real-world scenario. Very recently, public datasets for channel estimation problem in RIS-aided communications were made available in the 2021 IEEE SIgnal Processing Cup competition.

9.7.3.2 Model Training

The models are usually trained offline prior to their online deployment at a PS connected to the BS. In addition, the model training complexity increases with the number of RIS elements and number of RISs deployed between the users and the BS. This introduces huge transmission overhead for model training. The FL has potential to reduce this cost and enable a communication-efficient model training (see, e.g. Figure 9.10). Here, combining the label-free structure of RL and the communications efficiency of FL, i.e. federated reinforcement learning, could be the next step.

9.7.3.3 Environment Adaptation and Robustness

The behavior of the channel affects all DL-based tasks including channel estimation, beamforming, user scheduling, power allocation, and antenna selection/switching. Addressing the trade-off between the bias and the variance of the model output is essential for robust performance. This is usually achieved using a validation data so that the learning model does not either over-fit or under-fit the training data. Nonetheless, this does not generalize the learning performance to different environments. Moreover, the current DL architectures for wireless systems remain environment-specific because the input data space of their learning model is limited. As a result, the performance degrades significantly when the learning model is fed with the input from unlearned/uncovered data space. In order to cover larger data spaces and provide a robust performance against the changes in the environment, wider and deeper learning models are required. But the current DL architectures for wireless communications comprise less than a million neurons and are composed of only a few layers (Table 9.2) [44]. The giant learning models for image recognition or natural language processing consist of millions and billions of neurons, e.g. VGG (138 million), AlexNet (60 million), and GPT-3 (170 billion). Clearly, going wider and deeper in designing the learning models is of great interest for future DL-based RIS-aided systems.

9.8 Summary

We surveyed DL-based techniques for designing RIS hardware to be deployed for future wireless communications. When the design space and scale of the RIS arrays increases, learning-based architectures outperform evolutionary optimization techniques for both surrogate performance modeling and inverse design. The DL inverse design is flexible in admitting a variety of RIS unit structures. The DGMs are the most useful because of their ability to generate new designs not previously seen in the published literature. While active research and techniques in this area are still evolving, DL is a promising solution for the inverse design of RIS.

We also investigated DL architectures for RIS-assisted wireless systems for key applications of signal detection, channel estimation, and beamforming. We extensively discussed various learning schemes and model architectures, such as SL, UL, FL, and RL for RIS applications. The SL exhibits better performance than UL and RL because of label usage. The UL and RL are label-free schemes that provide less complexity during training data generation. However, UL still involves an optimization stage for each data instance. Among all, the RL is the most promising technique because of its standalone operation and the consequent ability to adapt to environmental changes at the cost of longer training times.

The FL reduces the transmission overhead significantly and can be integrated with the other learning methods. The combination of FL- and RL-based learning policies not only exhibits a communication-efficient model training but also provides environmental adaptation. Major research challenges include data collection, model training, and environment adaptation. These should be addressed simultaneously to provide a reliable DL architecture for the next generation RIS-assisted wireless systems. Specifically, the combination of FL and RL should be fed with the collection of huge datasets and massive neural networks so that a robust DL architecture is achieved.

Acknowledgments

The authors warmly acknowledge valuable contributions of Dr. John A. Hodge (Amazon) for the inverse design portion of this chapter, when he was a graduate student at Virginia Tech. K. V. M. acknowledges support from the National Academies of Sciences, Engineering, and Medicine via Army Research Laboratory Harry Diamond Distinguished Postdoctoral Fellowship.

References

1 Renzo, M.D., Debbah, M., Phan-Huy, D.-T. et al. (2019). Smart radio environments empowered by reconfigurable AI meta-surfaces: an idea whose time has come. *EURASIP Journal on Wireless Communications and Networking* 2019 (1): 1–20.

2 Hodge, J.A., Mishra, K.V., and Zaghloul, A.I. (2020). Intelligent time-varying metasurface transceiver for index modulation in 6G wireless networks. *IEEE Antennas and Wireless Propagation Letters* 19 (11): 1891–1895.

3 Holloway, C.L., Kuester, E.F., Gordon, J.A. et al. (2012). An overview of the theory and applications of metasurfaces: the two-dimensional equivalents of metamaterials. *IEEE Antennas and Propagation Magazine* 54 (2): 10–35.

4 Nguyen, Q., Mishra, K.V., and Zaghloul, A.I. (2019). Retrieval of polarizability matrix for metamaterials. *IEEE International Conference on Microwaves, Communications, Antennas and Electronic Systems*, pp. 1–5.

5 Hodge, J.A., Mishra, K.V., and Zaghloul, A.I. (2019). Joint multi-layer GAN-based design of tensorial RF metasurfaces. *IEEE International Workshop on Machine Learning for Signal Processing*, pp. 1–6.

6 Wu, Q. and Zhang, R. (2019). Towards smart and reconfigurable environment: intelligent reflecting surface aided wireless network. *IEEE Communications Magazine* 58 (1): 106–112.

7 Chen, M., Kim, M., Wong, A.M.H., and Eleftheriades, G.V. (2018). Huygens' metasurfaces from microwaves to optics: a review. *Nanophotonics* 7 (6): 1207–1231.

8 Sievenpiper, D., Zhang, L., Broas, R.F.J. et al. (1999). High-impedance electromagnetic surfaces with a forbidden frequency band. *IEEE Transactions on Microwave Theory and Techniques* 47 (11): 2059–2074.

9 Su, J., Lu, Y., Zhang, H. et al. (2017). Ultra-wideband, wide angle and polarization-insensitive specular reflection reduction by metasurface based on parameter-adjustable meta-atoms. *Scientific Reports* 7: 42283.

10 Zhu, H.L., Cheung, S.W., Chung, K.L., and Yuk, T.I. (2013). Linear-to-circular polarization conversion using metasurface. *IEEE Transactions on Antennas and Propagation* 61 (9): 4615–4623.

11 Minatti, G., Faenzi, M., Martini, E. et al. (2015). Modulated metasurface antennas for space: synthesis analysis and realizations. *IEEE Transactions on Antennas and Propagation* 63 (4): 1288–1300.

12 Maci, S., Minatti, G., Casaletti, M., and Bosiljevac, M. (2011). Metasurfing: addressing waves on impenetrable metasurfaces. *IEEE Antennas and Wireless Propagation Letters* 10: 1499–1502.

13 Mishra, K.V., Hodge, J.A., and Zaghloul, A.I. (2019). Reconfigurable metasurfaces for radar and communications systems. *URSI Asia-Pacific Radio Science Conference*, pp. 1–4.

14 Li, H., Wang, G., Xu, H.-X. et al. (2015). X-band phase-gradient metasurface for high-gain lens antenna application. *IEEE Transactions on Antennas and Propagation* 63 (11): 5144–5149.

15 Chen, H.-T., Taylor, A.J., and Yu, N. (2016). A review of metasurfaces: physics and applications. *Reports on Progress in Physics* 79 (7): 076401.

16 Glybovski, S.B., Tretyakov, S.A., Belov, P.A. et al. (2016). Metasurfaces: from microwaves to visible. *Physics Reports* 634: 1–72.

17 Kumar, C., Kashyap, S., Sarvendranath, R., and Sharma, S.K. (2022). On the feasibility of wireless energy transfer based on low complexity antenna selection and passive IRS beamforming. *IEEE Transactions on Communications* 70 (8): 5663–5678.

18 Mishra, K.V., Chattopadhyay, A., Acharjee, S.S., and Petropulu, A.P. (2022). OptM3Sec: optimizing multicast IRS-aided multiantenna DFRC secrecy channel with multiple eavesdroppers. *IEEE International Conference on Acoustics, Speech and Signal Processing*, pp. 9037–9041.

19 Torkzaban, N. and Khojastepour, M.A.A. (2021). Shaping mmWave wireless channel via multi-beam design using reconfigurable intelligent surfaces. *IEEE Global Communications Conference Workshops*, pp. 1–6.

20 Esmaeilbeig, Z., Mishra, K.V., and Soltanalian, M. (2022). IRS-aided radar: enhanced target parameter estimation via intelligent reflecting surfaces. *IEEE Sensor Array and Multichannel Signal Processing Workshop*, pp. 286–290.

21 Wang, Z., Mu, X., and Liu, Y. (2022). STARS enabled integrated sensing and communications. *arXiv preprint arXiv:2207.10748.*

22 Esmaeilbeig, Z., Mishra, K.V., Eamaz, A., and Soltanalian, M. (2022). Cramér–Rao lower bound optimization for hidden moving target sensing via multi-IRS-aided radar. *arXiv preprint arXiv:2210.05812.*

23 Wei, T., Wu, L., Mishra, K.V., and Bhavani Shankar, M.R. (2022). IRS-aided wideband dual-function radar-communications with quantized phase-shifts. *IEEE Sensor Array and Multichannel Signal Processing Workshop*, pp. 465–469.

24 Elbir, A.M., Mishra, K.V., Bhavani Shankar, M.R., and Chatzinotas, S. (2022). The rise of intelligent reflecting surfaces in integrated sensing and communications paradigms. *arXiv preprint arXiv:2204.07265.*

25 Hodge, J.A., Mishra, K.V., and Zaghloul, A.I. (2021). Deep inverse design of reconfigurable metasurfaces for future communications. *arXiv preprint arXiv:2101.09131.*

26 Hodge, J.A., Mishra, K.V., and Zaghloul, A.I. (2019). Reconfigurable metasurfaces for index modulation in 5G wireless communications. *IEEE International Applied Computational Electromagnetics Society Symposium*, pp. 1–2.

27 LeCun, Y., Bengio, Y., and Hinton, G. (2015). Deep learning. *Nature* 521 (7553): 436–444.

28 Bengio, Y., Courville, A., and Vincent, P. (2013). Representation learning: a review and new perspectives. *IEEE Transactions on Pattern Analysis and Machine Intelligence* 35 (8): 1798–1828.

29 Yu, D. and Deng, L. (2011). Deep learning and its applications to signal and information processing [exploratory DSP]. *IEEE Signal Processing Magazine* 28 (1): 145–154. https://doi.org/10.1109/MSP.2010.939038.

30 Ma, W., Cheng, F., and Liu, Y. (2018). Deep-learning-enabled on-demand design of chiral metamaterials. *ACS Nano* 12 (6): 6326–6334.

31 Zhang, Q., Liu, C., Wan, X. et al. (2018). Machine-learning designs of anisotropic digital coding metasurfaces. *Advanced Theory and Simulations* 2 (2): 1800132.

32 Qiu, T., Shi, X., Wang, J. et al. (2019). Deep learning: a rapid and efficient route to automatic metasurface design. *Advanced Science* 6 (12): 1900128.

33 Peurifoy, J., Shen, Y., Jing, L. et al. (2018). Nanophotonic particle simulation and inverse design using artificial neural networks. *Science Advances* 4 (6): eaar4206.

34 Elbir, A.M. and Mishra, K.V. (2020). A survey of deep learning architectures for intelligent reflecting surfaces. *arXiv preprint arXiv:2009.02540.*

35 Liu, Z., Zhu, D., Rodrigues, S.P. et al. (2018). Generative model for the inverse design of metasurfaces. *Nano Letters* 18 (10): 6570–6576.

36 Jiang, J., Sell, D., Hoyer, S. et al. (2018). Data-driven metasurface discovery. *arXiv preprint arXiv:1811.12436.*

37 Hodge, J.A., Mishra, K.V., and Zaghloul, A.I. (2019). RF metasurface array design using deep convolutional generative adversarial networks. *IEEE International Symposium on Phased Array Systems and Technology*, pp. 1–6.

38 Campbell, S.D., Sell, D., Jenkins, R.P. et al. (2019). Review of numerical optimization techniques for meta-device design. *Optical Materials Express* 9 (4): 1842–1863.

39 Gong, S., Lu, X., Hoang, D.T. et al. (2020). Towards smart wireless communications via intelligent reflecting surfaces: a contemporary survey. *IEEE Communications Surveys & Tutorials* 22 (4): 2283–2314.

40 Feng, K., Wang, Q., Li, X., and Wen, C.-K. (2020). Deep reinforcement learning based intelligent reflecting surface optimization for MISO communication systems. *IEEE Wireless Communications Letters* 9 (5): 745–749.

41 Huang, C., Alexandropoulos, G.C., Yuen, C., and Debbah, M. (2019). Indoor signal focusing with deep learning designed reconfigurable intelligent surfaces. *IEEE International Workshop on Signal Processing Advances in Wireless Communications*, pp. 1–5.

42 Lee, G., Jung, M., Kasgari, A.T.Z. et al. (2020). Deep reinforcement learning for energy-efficient networking with reconfigurable intelligent surfaces. *IEEE International Conference on Communications*, pp. 1–6.

43 Dai, L., Jiao, R., Adachi, F. et al. (2020). Deep learning for wireless communications: an emerging interdisciplinary paradigm. *IEEE Wireless Communications* 27 (4): 133–139.

44 Elbir, A.M. and Coleri, S. (2021). Federated learning for channel estimation in conventional and RIS-assisted massive MIMO. *IEEE Transactions on Wireless Communications* 21 (6): 4255–4268.

45 Inampudi, S. and Mosallaei, H. (2018). Neural network based design of metagratings. *Applied Physics Letters* 112 (24): 241102.

46 Zhang, Q., Liu, C., Wan, X. et al. (2019). Machine-learning designs of anisotropic digital coding metasurfaces. *Advanced Theory and Simulations* 2 (2): 1800132.

47 Shan, T., Pan, X., Li, M. et al. (2020). Coding programmable metasurfaces based on deep learning techniques. *IEEE Journal on Emerging and Selected Topics in Circuits and Systems* 10 (1): 114–125.

48 Hodge, J.A., Mishra, K.V., and Zaghloul, A.I. (2019). Multi-discriminator distributed generative model for multi-layer RF metasurface discovery. *IEEE Global Conference on Signal and Information Processing*, pp. 1–5.

49 Ma, W., Cheng, F., Xu, Y. et al. (2019). Probabilistic representation and inverse design of metamaterials based on a deep generative model with semi-supervised learning strategy. *Advanced Materials* 31 (35): 1901111.

50 Jiang, J., Sell, D., Hoyer, S. et al. (2019). Free-form diffractive metagrating design based on generative adversarial networks. *ACS Nano* 13 (8): 8872–8878.

51 Jiang, J. and Fan, J.A. (2019). Global optimization of dielectric metasurfaces using a physics-driven neural network. *Nano Letters* 19 (8): 5366–5372.

52 Tretyakov, S.A. (2015). Metasurfaces for general transformations of electromagnetic fields. *Philosophical Transactions of the Royal Society A: Mathematical, Physical and Engineering Sciences* 373 (2049): 20140362.

53 Epstein, A. and Eleftheriades, G.V. (2016). Huygens' metasurfaces via the equivalence principle: design and applications. *Journal of the Optical Society of America B* 33 (2): A31–A50.

54 Epstein, A. and Eleftheriades, G.V. (2016). Arbitrary power-conserving field transformations with passive lossless omega-type bianisotropic metasurfaces. *IEEE Transactions on Antennas and Propagation* 64 (9): 3880–3895.

55 Chen, H.-T. (2012). Interference theory of metamaterial perfect absorbers. *Optics Express* 20 (7): 7165–7172.

56 Schurig, D., Mock, J.J., and Smith, D.R. (2006). Electric-field-coupled resonators for negative permittivity metamaterials. *Applied Physics Letters* 88 (4): 041109.

57 Su, P., Zhao, Y., Jia, S. et al. (2016). An ultra-wideband and polarization-independent metasurface for RCS reduction. *Scientific Reports* 6: 20387.

58 Pereda, A.T., Caminita, F., Martini, E. et al. (2016). Dual circularly polarized broadside beam metasurface antenna. *IEEE Transactions on Antennas and Propagation* 64 (7): 2944–2953.

59 Sun, S., He, Q., Xiao, S. et al. (2012). Gradient-index meta-surfaces as a bridge linking propagating waves and surface waves. *Nature Materials* 11 (5): 426–431.

60 Nguyen, Q. and Zaghloul, A.I. (2018). Impedance matching metamaterials composed of ELC and NB-SRR. *IEEE Antennas and Propagation Society International Symposium*, pp. 1–2.

61 Hodge, J.A., Anthony, T., and Zaghloul, A.I. (2014). Enhancement of the dipole antenna using a capcitively loaded loop (CLL) structure. *IEEE International Symposium on Antennas and Propagation and USNC-URSI Radio Science Meeting*, pp. 1544–1545.

62 Yu, N., Genevet, P., Kats, M.A. et al. (2011). Light propagation with phase discontinuities: generalized laws of reflection and refraction. *Science* 334 (6054): 333–337. https://doi.org/10.1126/science.1210713.

63 Mencagli, M., Martini, E., and Maci, S. (2015). Surface wave dispersion for anisotropic metasurfaces constituted by elliptical patches. *IEEE Transactions on Antennas and Propagation* 63 (7): 2992–3003.

64 Fong, B.H., Colburn, J.S., Ottusch, J.J. et al. (2010). Scalar and tensor holographic artificial impedance surfaces. *IEEE Transactions on Antennas and Propagation* 58 (10): 3212–3221.

65 Khan, S., Durrani, S., and Zhou, X. (2021). Transfer learning based detection for intelligent reflecting surface aided communications. *IEEE Annual International Symposium on Personal, Indoor and Mobile Radio Communications*, pp. 13–16.

66 Elbir, A.M., Papazafeiropoulos, A., Kourtessis, P., and Chatzinotas, S. (2020). Deep channel learning for large intelligent surfaces aided mm-Wave massive MIMO systems. *IEEE Wireless Communications Letters* 9 (9): 1447–1451.

67 Taha, A., Alrabeiah, M., and Alkhateeb, A. (2021). Enabling large intelligent surfaces with compressive sensing and deep learning. *IEEE Access* 9: 44304–44321.

68 Liu, S., Gao, Z., Zhang, J. et al. (2020). Deep denoising neural network assisted compressive channel estimation for mmWave intelligent reflecting surfaces. *IEEE Transactions on Vehicular Technology* 69 (8): 9223–9228.

69 Gao, J., Zhong, C., Chen, X. et al. (2020). Unsupervised learning for passive beamforming. *IEEE Communications Letters* 24 (5): 1052–1056.

70 Taha, A., Zhang, Y., Mismar, F.B., and Alkhateeb, A. (2020). Deep reinforcement learning for intelligent reflecting surfaces: towards standalone operation. *IEEE International Workshop on Signal Processing Advances in Wireless Communications*, pp. 1–5.

71 Gong, S., Lin, J., Ding, B. et al. (2022). When optimization meets machine learning: the case of IRS-assisted wireless networks. *IEEE Network* 36 (2): 190–198.

72 Ma, D., Li, L., Ren, H. et al. (2020). Distributed rate optimization for intelligent reflecting surface with federated learning. *IEEE International Conference on Communications Workshops*, pp. 1–6.

73 Yang, H., Xiong, Z., Zhao, J. et al. (2021). Deep reinforcement learning based intelligent reflecting surface for secure wireless communications. *IEEE Transactions on Wireless Communications* 20 (1): 375–388.

Section IV

RF, Antenna, Inverse-Scattering, and Other EM Applications of Deep Learning

10

Deep Learning for Metasurfaces and Metasurfaces for Deep Learning

Clayton Fowler[1], *Sensong An*[1,2], *Bowen Zheng*[1], *and Hualiang Zhang*[1]

[1] *Department of Electrical & Computer Engineering, University of Massachusetts Lowell, Lowell, MA, USA*
[2] *Department of Materials Science & Engineering, Massachusetts Institute of Technology, Cambridge, MA, USA*

10.1 Introduction

The discovery that high-index dielectric nanostructures can support magnetic multipole responses in addition to electric ones [1] has led to a revolution in metasurface design, particularly with applications in optics. These additional magnetic resonances have enabled precise manipulation of light waves on a subwavelength scale [2] such that the phase and amplitude of an electromagnetic (EM) wave can be precisely tailored at the location of each nanostructure making up a metasurface [3]. This high degree of control over light has precipitated a variety of metasurface-based optical devices, including lenses [4–6], diffraction gratings [7, 8], spatial light modulators [9], and holograms [10]. By using metasurface-based designs, these devices can achieve superior performance over conventional components in terms of compactness, functionality, and fabrication.

However, finding nanostructures that meet the performance requirements for a particular device has historically been an inefficient matter of trial and error. It has so far proven implausible to establish simple guiding principles to predict a given structure's EM response based on its geometry, because small perturbations can sometimes change the response dramatically. As such, early metasurfaces were designed by conducting full wave simulations to sweep the parameter spaces of simple shapes (disks, crosses, etc.) to assemble a library of candidate structures from which the best ones could be chosen. While this approach has yielded some notable successes, it is practical only for simple structures which may not be physically capable of yielding a desired EM response, particularly for multi-functional devices. When employing more complex structures to overcome this barrier, the computational cost becomes prohibitively time-consuming, as the number of possible structures increases exponentially with the number of parameters used to define the structures. Therefore, computational methods that could predict EM responses more quickly (forward prediction) are highly desired. Additionally, the ability to specify an arbitrary EM response and generate an acceptable structure (inverse-design), is even more desirable. One powerful means of tackling both of these problems is the use of deep neural networks (DNNs).

DNNs, which were inspired by biological neural networks, are essentially highly complex functions with a large number of tunable parameters. By processing a large amount of training data that relates an input to an output, a neural network's parameters can be tuned to make accurate

Advances in Electromagnetics Empowered by Artificial Intelligence and Deep Learning, First Edition.
Edited by Sawyer D. Campbell and Douglas H. Werner.

predictions about new data points. In addition to the large number of degrees of freedom (DOF) ($\sim 10^6$ parameters or more) granted by neural networks, there are a variety of neural network architectures that can be chosen as an efficient means of making accurate predictions between the inputs and outputs of complex systems. In particular, neural networks have been demonstrated to be a highly effective means of relating the physical parameters of nanostructures with their EM responses.

In this chapter we present a variety of neural network architectures and methods that can be used to design metasurfaces for optical devices. We focus primarily on Huygens' metasurfaces, for which the goal is to find metasurface nanostructures that meet specified amplitude and phase requirements at one or more frequencies. Huygens' metasurfaces can be used for (but are not limited to) lenses, holograms, waveplates, birefringence, dichroism, polarizers, and beam steering. The versatility of Huygens' surfaces even allows for multiple functions to be incorporated into the same metasurface. We demonstrate architectures for fully connected neural networks, convolutional neural networks, generative adversarial networks, and recurrent neural networks (RNNs) in addressing both forward prediction and inverse design, and discuss conditions for when each is suitable. Lastly, we briefly describe the emerging field of neuromorphic photonics, for which metasurfaces can be employed as physical components of optical neural networks.

10.2 Forward-Predicting Networks

The goal of a forward neural network is to be able to input the physical properties of a structure (its geometry and materials) and predict its EM response. This is actually just a duplication of the full wave simulators used to generate the training data, but a properly trained forward neural network offers a tremendous computational speedup over full wave simulators (making predictions ~ms versus ~min), which makes a trained forward neural network a superior choice for implementation in conjunction with design optimization algorithms (genetic algorithms, particle swarm, inverse-design neural network, etc.). Training a forward neural network is an approachable problem in the sense that a well-constructed network with a sufficient amount of data should converge with good accuracy, because each structure has a unique response associated with it (contrast this with the inverse-design problem, where a given EM response can potentially be produced by multiple differing structures).

Generally, the EM response of a structure is characterized by using full wave simulation software to calculate its complex transmission and reflection coefficients with both phase and magnitude information. This enables quality structures to be selected for use in Huygens' metasurfaces. Additionally, the response can also be characterized by the induced EM field patterns in and around the structure. This field approach can be used to improve the overall efficiency of metasurfaces by accounting for coupling effects that occur between dissimilar structures and perturb the actual transmission and reflection coefficients away from their initially predicted values.

We consider several neural network architectures for predicting EM responses. First, we discuss fully connected neural networks, which are suitable for structures that can be characterized by a handful of parameters. They are versatile, simple, and straightforward to implement. Next, we look at 2D convolutional networks, which have been enormously successful for machine learning applications involving images. Their properties allow structures to be processed as images, which is particularly advantageous for making predictions about complex structures that can't be easily parameterized. Additionally, we will demonstrate how 2D convolutional networks can be used to correct for mutual coupling effects between dissimilar structures. Lastly, we will demonstrate sequential neural networks, which include RNNs and 1D convolutional networks. Sequential architectures are a natural choice for predicting EM spectra, since spectra are sequential with adjacent points being highly correlated.

10.2.1 FCNN (Fully Connected Neural Networks)

A fully connected neural network (FCNN) is a commonly seen fundamental DNN structure. A FCNN usually consists of several hidden layers, with all hidden neurons in each layer connected to each neuron in the next layer (Figure 10.1). The neurons in the hidden layers, along with their activation functions, yield a large trainable set of parameters which can be used to represent complicated functions according to the universal approximation theorem [11–13]. Therefore, it is possible to uncover hidden relations between variables, such as between nanophotonic structure geometries and their EM responses. Inspired by this idea, several DNNs that connect nanophotonic structures to their EM responses were constructed with FCNNs [14–23]. While some of these works use FCNNs only to predict the transmission/reflection spectrum of meta-atoms, in [14, 18] the authors constructed FCNNs that are capable of predicting both the transmission amplitude and phase. For a typical resonant type meta-structure, the amplitude and phase response of its transmission coefficient will abruptly change (especially the phase) around resonant frequencies, which increases the prediction difficulty for DNNs. A simple but effective method is to predict the real and imaginary parts of transmission coefficients and using those to calculate the amplitude and phase instead of finding them directly. This works because the real and imaginary parts remain smooth and non-singular around resonances (Figure 10.2).

During training, the parameters in the networks (hidden neurons in each layer) are optimized to minimize the differences between predicted results and ground truth. Once fully trained, the FCNNs are able to generate accurate EM responses as output given the physical parameters of meta-atoms as input. Since this forward prediction is based on a one-time calculation basis, it can be executed with almost no time cost, which makes it a perfect substitute for full wave simulation tools in optimization or inverse design tasks. On the other hand, the FCNNs with a large number of

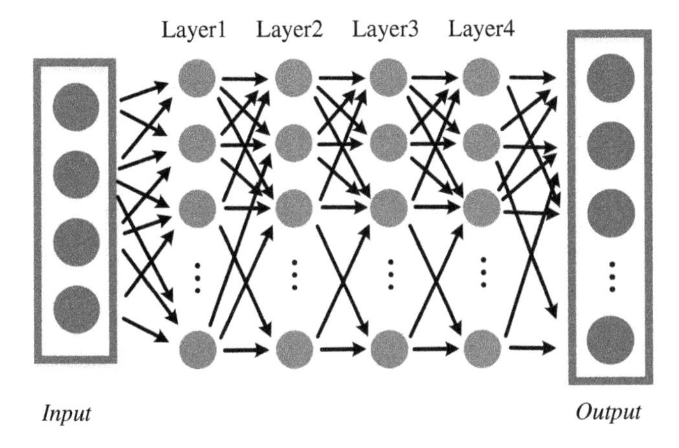

Figure 10.1 Schematic of a FCNN.

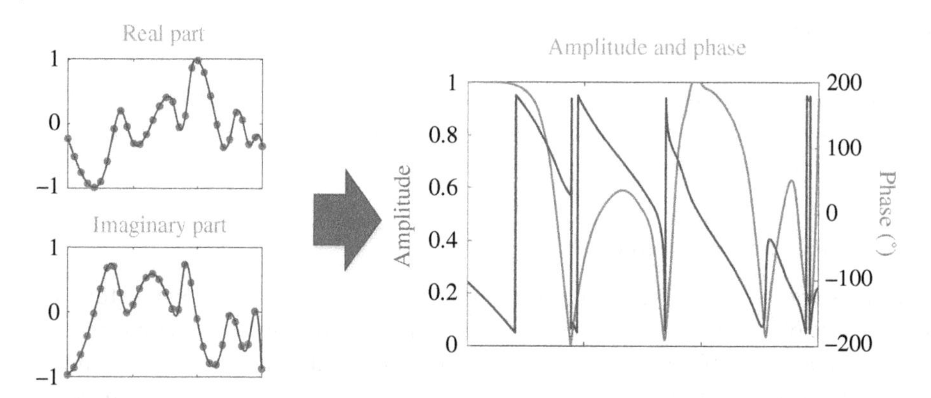

Figure 10.2 Spectrum responses of a meta-atom, complex transmission coefficients versus amplitude and phase.

hidden neurons are computationally intensive, which reduces its effectiveness when dealing with large-size inputs or outputs, especially with graphic inputs. We will elaborate more on this issue in Section 10.2.2.

10.2.2 CNN (Convolutional Neural Networks)

10.2.2.1 Nearly Free-Form Meta-Atoms

In previous works, the possibility of using FCNNs to predict the EM responses of meta-atoms in the shapes of cylinders [15, 17, 18, 24, 25], elliptic cylinders [26], spheres [22], and bars [20, 21] has been demonstrated. Compared with these simple structures, free-form meta-atoms have more design DOF and are thus more promising for multi-functional or high-efficiency designs. When constructing a DNN to predict the EM responses of simple-shaped meta-atoms, normal practice is to define the meta-atom structure with several parameters (e.g. the cylinders can be defined with their radius and height), and then assign these parameters as the input of an FCNN. However, the free-form patterns cannot be easily dealt with using the same approach because: (i) Those 2D patterns have to be pixelated and flattened into very large-dimensional 1D arrays before being fed into the dense layers, and FCNNs with a large number of hidden neurons quickly deplete the computational resources. (ii) Transforming the 2D patterns into 1D arrays means losing the relative positional information between each row (Figure 10.3), and this will lower the prediction accuracy, which has been validated in various computer vision tasks. As a result, a Convolution Neural Network has proven to be a superior to choice for predicting the EM performance of meta-atoms with free-form or nearly free-form patterns.

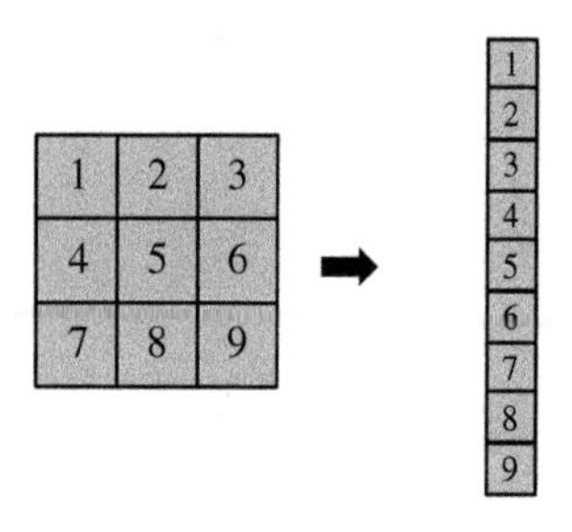

Figure 10.3 Example of a 2D pattern flattening into a 1D array.

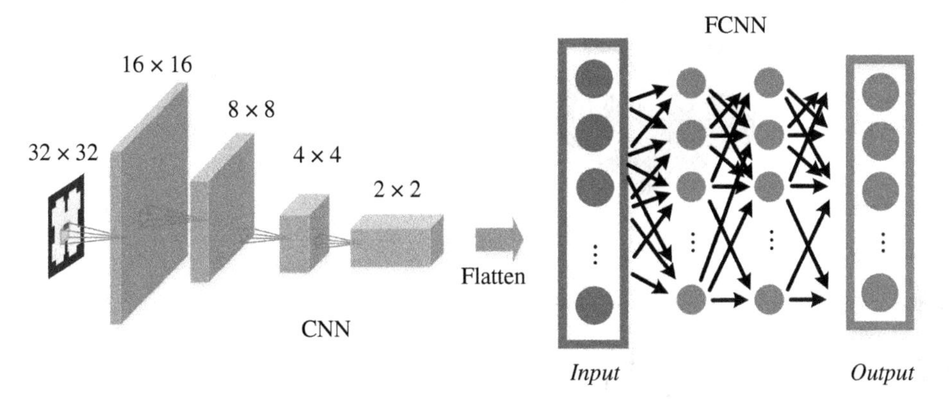

Figure 10.4 A Convolutional Neural Network that predicts the EM responses of meta-atoms [25].

The power in using convolutional neural networks comes from treating meta-atom unit cells as 2D images. As shown in Figure 10.4, normally the 2D cross-sections of meta-atoms were processed with several consecutive convolutional layers. The 2D dimensions of the patterns were reduced during this process due to the strided convolutions and pooling layers. Meanwhile, the depth (3rd dimension) of the tensors was enlarged because of the increasing number of filter kernels in each layer. After flattening into a 1D array, one or two dense layers (FCNNs) were used to translate the extracted information (Input) into Outputs that humans can understand. The dimensions of these dense layers are determined by specific functionalities: for computer vision tasks such as classification problems, the dimension of the final output is equal to the number of categories where the output represents a multi-nominal probability distribution; for EM prediction problems (which essentially are regression problems), the dimensions of output are the number of values being predicted. Compared with the DNNs that are composed of solely fully connected layers, the DNN based on CNN architecture (Figure 10.4) requires less computational resources, and is easier to converge while improving the final prediction accuracy.

Although we can learn from the successful stories of previous computer vision task solutions and directly adopt some well-constructed CNN architectures, for example, the AlexNet [27], the ResNet [28], and the VGGNet [29], there is still one fundamental difference between the meta-atoms and 2D images – meta-atoms have a third dimension (thickness) and are usually constructed with different materials, fitted to grids with different lattice sizes. These properties have a great impact on their EM responses and thus need to be considered as part of the input. However, there's no easy way to reconcile the dimensional mismatch between 2D images and 1D properties, since it's physically impossible to stack them together. In [25, 30] the authors proposed a way to solve this problem. The 1D properties were pre-processed using dense layers and then spatially tiled into 3D tensors. During the convolution operation, the intermediate tensors of 2D images and 1D properties were concatenated together and later processed with more convolutional layers and dense layers (Figure 10.5). This method increases the weight of the 1D properties, eases the dimensional mismatch, and relates the 1D and 2D inputs sufficiently, which further increases the training speed.

Since this CNN takes 2D patterns as input, it has integrated the functions of various FCNNs. When we are dealing with simple-shaped meta-atoms from different categories (e.g. cylinders and cubes), we used to have to collect multiple datasets and train distinct FCNNs. However, with the CNN showing in Figure 10.5, different patterns (e.g. circles and rings, Figure 10.6) can all be pixelated into 64 by 64 resolution images and assigned as the input of the CNN. In general, this CNN

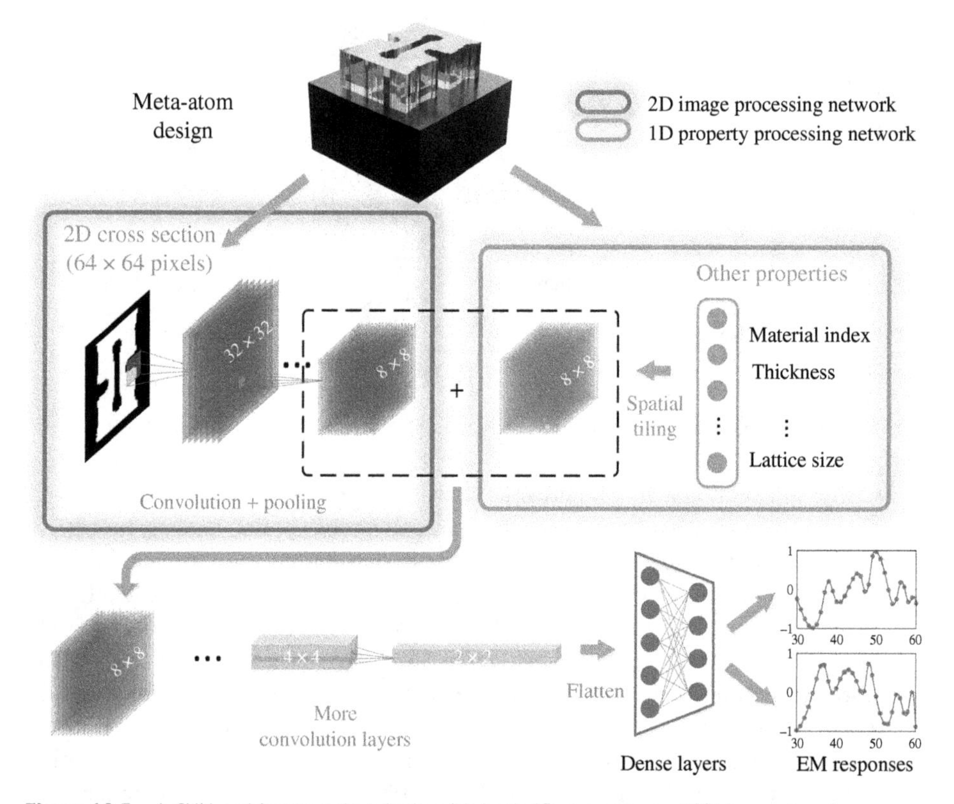

Figure 10.5 A CNN architecture that deals with both 1D-property and 2D-image inputs.

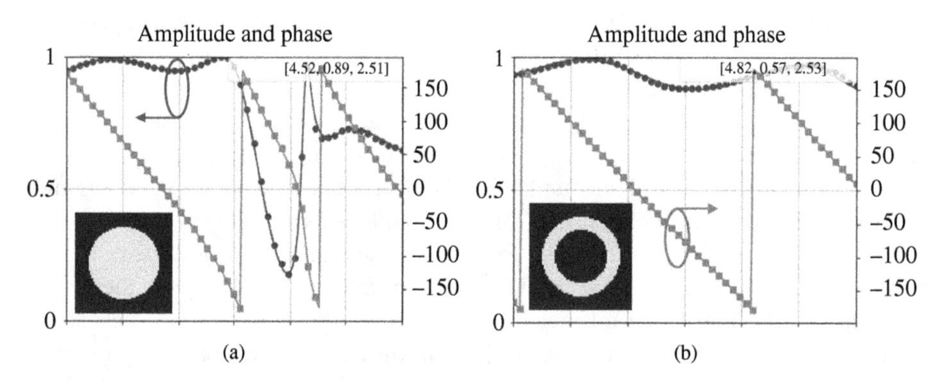

Figure 10.6 EM responses of (a) circle-shaped and (b) ring-shaped meta-atoms, predicted by CNN (dots) and ground truth (solid lines). Circle dotted line represents amplitude and square dotted line represents phase. 2D cross-sections of meta-atoms are included as insets.

can predict the accurate EM responses of meta-atoms from different categories. Only one dataset is needed for the training.

With the fully trained DNN that can predict the EM responses of different meta-atoms instantly, we can now easily validate our earlier statement, that "the free-form meta-atoms have more DOF and are thus more promising for multi-functional or high-efficiency designs." By comparing the performance between quasi-freeform shaped meta-atoms and those with simple shapes, such as circles, rectangles, and "H"s (Figure 10.7), we demonstrate how the added DOFs can be exploited

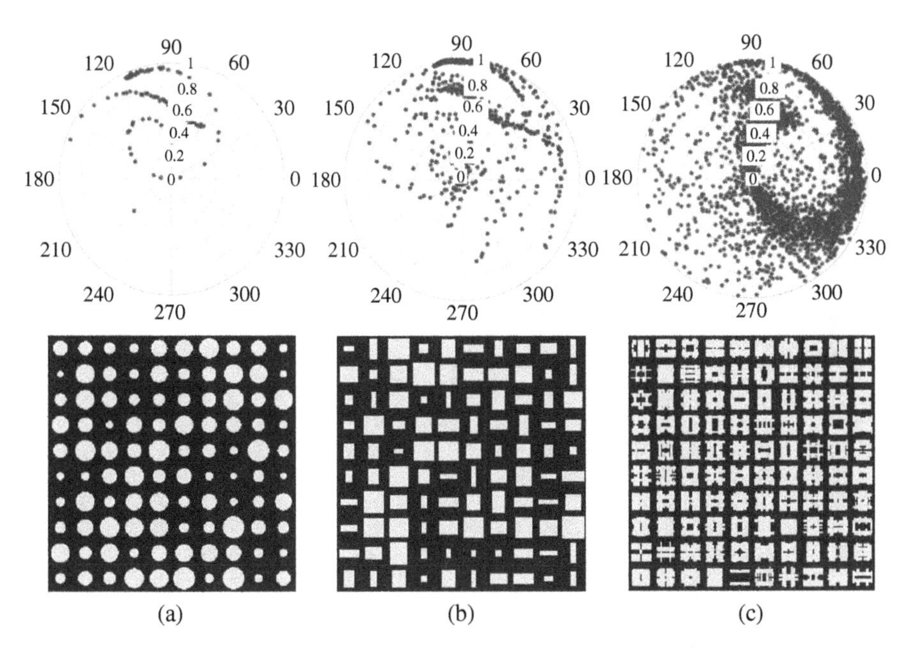

Figure 10.7 Comparing the performance of meta-atoms of (a) circle, (b) rectangle, and (c) freeform shapes using the well-trained CNN. Radial axes represent the meta-atoms' amplitude, angular axes represent phase.

to boost the overall efficiency and phase coverage of all-dielectric meta-atoms. With the same material, lattice size, thickness, and operating frequency, the meta-atoms with quasi-freeform cross-sections are superior to the meta-atoms from the other two categories (circles and rectangles) in terms of both efficiency and phase coverage. As we'll be showing in the following sections, this well-trained CNN can also be employed as a critic to examine the performance of meta-atoms generated by inverse design DNNs (Section 10.3) such as generative adversarial nets (GAN)s) (Section 10.3.2), and as a fast evaluation tool for DNN-based meta-device design optimizations (Sections 10.2.2.2, Figure 10.10).

10.2.2.2 Mutual Coupling Prediction

The traditional method for designing meta-atoms with high efficiency and accurate phase gradients has been considering structures with simple geometric shapes (such as circles [10, 31], rectangles [26, 32], H-shapes [33, 34], and plasmonic thin layers [35, 36]) or complicated free-form shapes [25, 37] and performing a parametric sweep over all dimensions to assemble a library (or train a neural network) covering the full design space. Then best-fit meta-atoms are selected from the library to approximate the ideal amplitude/phase map. During the simulations, unit cell or periodic boundary conditions were adopted, which assumes that each meta-atom structure under consideration is part of an infinite 2D array of identical structures. Thus, the amplitude and phase response calculations of the meta-atoms are based on the assumption that near-field coupling perturbations originate from identical neighbors (Figure 10.8a) and will be approximately the same when they are placed next to neighbors that are different when forming a metasurface (Figure 10.8b).

However, near-field coupling effects will differ from those used to calculate the original response. As a result, the phase and amplitude of each meta-atom will be deviated from their predicted values; thus, this method is accurate only when the mutual coupling effects between each meta-atom and its neighbors are reasonably small, which is generally not the case (note that this error will still

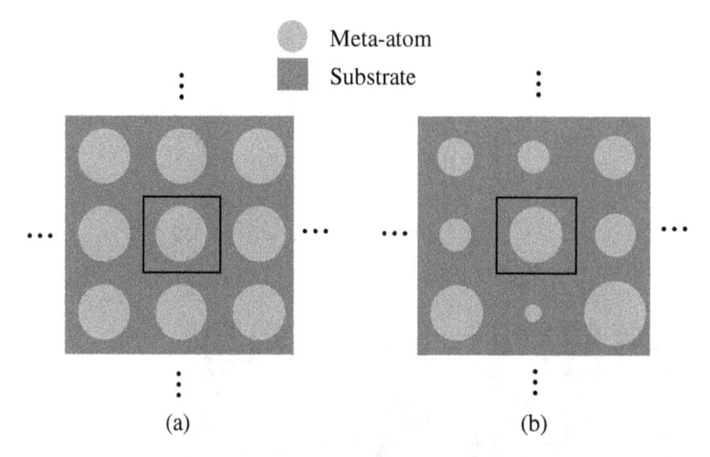

Figure 10.8 Meta-atom in simulation (a) and meta-atom in a practical metasurface device (b).

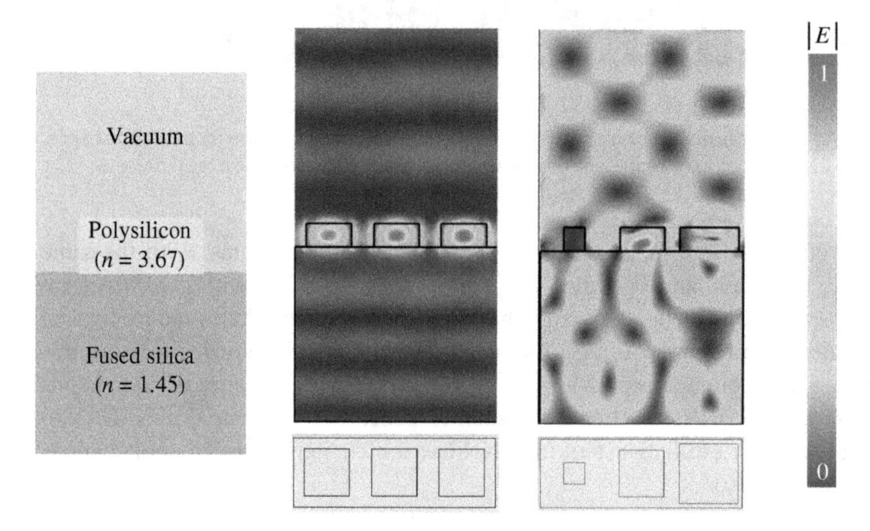

Figure 10.9 Simulated near-field electric field distribution of meta-atoms with identical and different neighbors.

manifest when using a neural network to generate structures as the underlying data will have been collected under the assumption that near-field coupling effects are small). For example, we examined and illustrated the near-field mutual coupling effect between meta-atoms constructed with a silicon-on-silica platform (Figure 10.9). The meta-atoms in the center of each three-meta-atom group are placed among identical neighbors (left) and different neighbors (right), respectively. It is clear that the perturbing effects of near-field coupling cause a significant change in the field distribution inside the central meta-atom.

In previous works, instances have been reported where the mutual coupling effect (either radiative coupling [38] or dipole coupling) plays such an important role in metasurfaces' overall performance that measures must be taken to minimize its effect. In [10] a periodic arrangement of 2 by 2 meta-atoms was adopted to decrease the coupling effects in a metasurface hologram, while in [39] the metasurface hologram was divided into several 17.35 by 17.35 μm^2 sub-arrays each constructed with identical meta-atoms. In [40] the height and diameters of the cylindrical meta-atoms that formed the metasurface beam deflector were slightly adjusted to achieve higher efficiency. In [41] a genetic algorithm (GA) was employed to find the meta-atoms' optimal dimensions and positions with strong coupling between neighbors taken into consideration. In some other works, the mutual coupling effects between adjacent meta-atoms were not only investigated, but also utilized to enhance the metasurfaces' performance. In [42] a strongly coupled resonator design was proposed and demonstrated in the terahertz and optical regimes, in which the coupling between neighboring resonators was tuned to enhance the effective refractive index. In [43] the near-field effects in high-index Mie-resonant nanoparticles were studied, and the distances between neighboring meta-atoms were tuned to realize continuous relative phase changes. In [44] a DNN was trained to predict the electric field of each 50 μm by 50 μm square area in a full-scale metasurface, which accounted for the interscatterer coupling effect. In [45] a numerical method (so-called local phase method [LPM]) was proposed to obtain the phase of each meta-atom within the metasurfaces while considering the mutual coupling effects. This approach quantifies the phase error of each element inside the metasurfaces, which enables the optimization of metasurfaces at an element level while accounting for near-field coupling effects. While it provides a way to measure the meta-atoms' accurate phase responses, the target meta-atom and all its neighbors need to be simulated as a whole in order to derive the performance of a single meta-atom, which is computationally intensive and time-consuming.

Alternatively, the actual EM responses of meta-atoms derived without periodic boundary condition assumption (hereinafter referred to as "local response") can be predicted with DNNs in a much faster way. Due to the DNNs' unique ability to uncover the hidden relationship between inputs and outputs, we can train a DNN that is capable of predicting a target meta-atom's local response given the dimensions of itself and all its neighbors. In [46] the authors randomly generated over 200,000 groups of meta-atoms placed among random neighbors and simulated all their local responses to assemble a training dataset. A CNN was then constructed (Figure 10.10), which took the dimensions of the target meta-atom and all its neighbors as input and generated the local responses of targets meta-atoms as output. With this well-trained DNN, a one-time matrix multiplication calculation (executed with almost no time cost) can generate the accurate local response prediction, which enables optimization of metasurface devices that suffer from performance deterioration caused by mutual coupling.

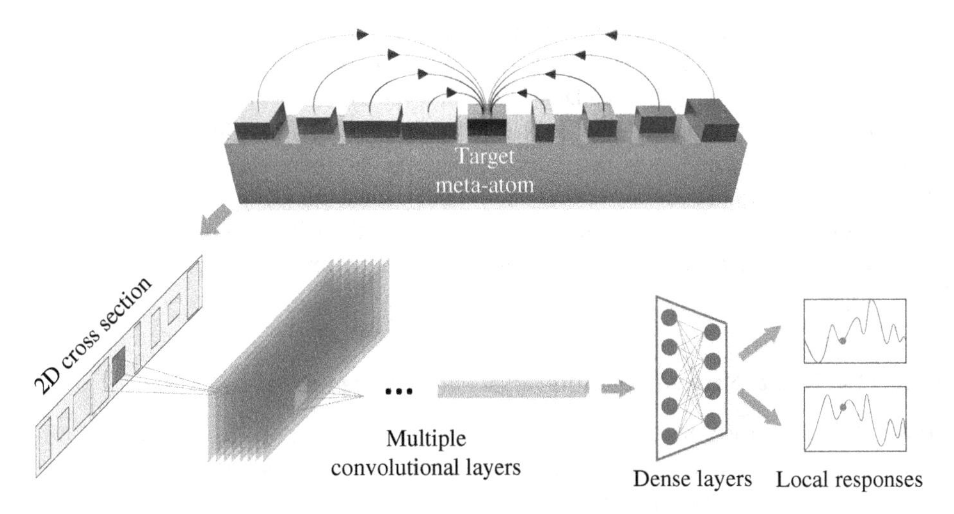

Figure 10.10 DNN that predicts the local responses of meta-atoms: network architecture.

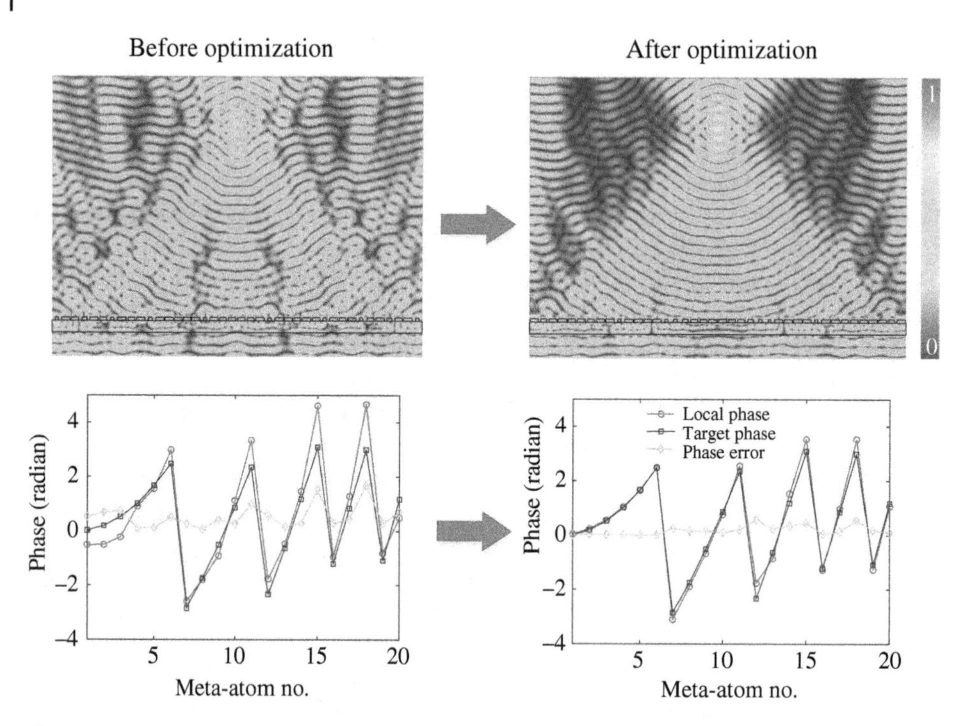

Figure 10.11 Optimization of a high numerical aperture metalens composed of 40 meta-atoms.

During metasurface device designs, we often find that the efficiency of the meta-device is lower than the theoretical calculations. Part of this is caused by fabrication defects. However, sometimes the full-wave simulated device efficiency is still lower than expected, and mutual coupling effect is usually part of the reason. For example, when we design a 1D cylindrical metalens composed of 40 meta-atoms using the conventional approach and then examine the local phase profiles of each individual meta-atom using the well-trained DNN (Figure 10.11), we can observe an obvious phase deviation (parallelogram dotted line) between the phase target (square dotted line) and the local phase profiles (circle dotted line), which also caused distorted phase fronts and significant scattering of the transmitted energy (as shown in the simulated E-field). Taking this metalens as the initial design, the dimensions of each meta-atom can be optimized to minimize the phase error, during which process the well-trained DNN was employed to evaluate the local phases of all meta-atoms after each optimization iteration. With a 20% enhancement in the focusing efficiency, it's clear that the transmitted energy is better focused due to the more accurate phase profile (after accounting for the mutual coupling effect). Importantly, since this DNN calculates the local responses of target meta-atoms on a one-time calculation basis, the whole optimization process only takes minutes – this indicates its potential for improvements in efficiency for large-scale metasurface device designs.

10.2.3 Sequential Neural Networks and Universal Forward Prediction

It is essentially always the case that the EM response of a metasurface over some range of frequencies (bandwidth) will need to be determined. Even for monochromatic applications, finding the response within a narrow bandwidth of the target frequency provides crucial information on the robustness of the design. With this in mind, it can be fruitful to think of the EM response as a sequence with adjacent points being highly correlated. The strong connections between neighboring points provide a large increase in information as compared to considering individual points on a spectrum, and thus a neural network that is designed to evaluate sequences is going to have an intrinsic advantage over one that looks only at individual data points.

There are two common types of sequential neural networks that can be utilized to predict EM responses, both of which benefit greatly from graphics processing unit (GPU) acceleration. The first type we consider is a RNN. These work by having sequential neurons pass feedback from their outputs, which is combined with their inputs (hence, "recurrent") and used to predict what the next points in the sequence will be. The second type of network is a 1D convolutional network. These networks work by finding sets of sequential patterns of arbitrary length, called filters, that appear most frequently in sequential data. By learning the best set of filters and corresponding weightings that describe a given set of training data, the network uses them to predict sequences accurately. Both of these types of networks can be used to make "universal" forward predicting networks, for which a single network can be trained to predict the EM responses of a given structure type for an arbitrary choice of structure and substrate materials and for any portion of the EM spectrum. The key to doing this is described in Sections 10.2.3.1–10.2.3.3.

10.2.3.1 Sequencing Input Data

Before we discuss the two kinds of networks, it is important to consider how input data will be processed. If the structure is relatively simple, such as a cylinder or a cross, then there will only be a few parameters needed to characterize the structure. For a simple case of a cylinder constructed with dispersionless and lossless materials, these are the structure height, cylinder radius, unit-cell size, index of refraction for the structures, and index of refraction for the substrate. Assuming that we measure the output transmission spectrum at 100 points to get good resolution, it is apparent that there is a very large parameter mismatch between the input and the output in that we are expecting the neural network to generate sequences of 100 points after seeing only a handful of input points. Intuitively, this is asking the network to generate a lot of information after seeing only a little, which is a difficult task, especially if we are interested in both phase and amplitude. Increasing the number of parameters that characterize the structure so as to provide more input data does not help as using more complex structures also increases the range of possible spectral outputs, and thus we are still trying to accurately predict a large number of data points from only a few input parameters. Furthermore, it is not clear how dispersion should be handled, in that there would be sequences of 100 points characterizing the optical material parameters, but only a few points specifying physical dimensions.

These problems can be resolved by normalizing the physical parameters of the structure with the electrical size, i.e. dividing them by λ/n [47]. This changes each physical parameter from a single point to a sequence which represents the normalized electrical size of the parameter and will be the same length as the output. Now the set of input data will typically be larger than the set of output data, which fixes the parameter mismatch, and we have a natural means to account for dispersion. There are several more advantages to this approach: (i) We can now construct our neural networks so as to gain information from both the connections between sequential spectrum data and also from direct connections between a structure's electrical size and its response at a particular frequency. (ii) We can forgo any additional normalization that would typically be performed when preprocessing neural network data, since the normalized electrical sizes will typically be ~ 1. (iii) Since the input data has been sequentialized, the neural networks can be constructed entirely using recurrent or convolutional layers that benefit from GPU acceleration. (iv) Most importantly, since all dimensions have been normalized by the wavelength, the data can be applied to any portion of the EM spectrum by taking advantage of the wavelength scalability of metasurfaces. This is how universal forward prediction can be achieved. However, care must be taken to ensure that an appropriate dispersion model is used to account for the broad range of material properties available across the EM spectrum. The 2nd-order Lorentz model fits low-loss materials very well.

$$\epsilon(\omega) = \epsilon_\infty + \frac{\beta_0 + i\omega\beta_1}{\alpha_0 + i\omega\alpha_1 - \omega^2} \tag{10.1}$$

10.2.3.2 Recurrent Neural Networks

RNNs are primarily used to make predictions about temporal data, but can be used for any kind of sequence. They tend to lose predictive power as sequences get longer, and so they perform better for narrowband applications. The original implementation was the simple RNN layer, but these are only effective for fairly short sequences (~10 points) and so have largely been supplanted by long-short term memory (LSTM) or gated recurrent unit (GRU) layers. LSTM and GRU layers are designed to detect both long-range and short-range behavior and thus increase the predictive range to ~100 data points. Consensus has not been reached as to whether LSTM or GRU are superior, and so it is largely up to the preferences of the designer. It is thought that LSTM cells will generally yield slightly better accuracy, whereas GRU cells are less complex and thus result in quicker network training and computation times. This suggests that one might consider using GRU cells while experimenting with the network architecture and then try switching over to LSTMs in the final design.

RNNs are assembled similarly as fully connected layers in that a series of LSTM (or GRU) layers can be stacked. However, since LSTM layers are more computationally intense than dense layers, typically fewer neurons per layer are used and the last layer must have only one neuron. It's important to set each layer to "return sequences" so that the network passes the entire sequence instead of just the last point. When predicting EM spectra [48, 49], we can make use of a technique for improving long range accuracy that is not available when predicting time-dependent data. Since there is no time dependency, it doesn't matter if we start the sequence from the highest frequency and go down or from the lowest frequency and go up. This means that we can create a branching network that calculates sequences starting from both the lowest and highest frequencies and then averages the two predictions. Each branch has its strongest prediction power where the other is weakest and thus should help compensate for the difficulties in predicting long sequences. Lastly, although a successful network can be constructed entirely from recurrent layers, it is possible to include fully connected layers at the end to improve long range prediction power. This is done by simply flattening the output of the last recurrent layer and then stacking as many fully connected layers as desired.

10.2.3.3 1D Convolutional Neural Networks

Although convolutional networks are primarily recognized for their tremendous successes in image processing and classification problems in their 2D form, there exists a 1D variant that is useful for handling sequences such as EM spectra [47, 50]. These networks search for common subsequences in the data, called filters, that can be weighted and aggregated to make predictions about the full spectra. Since filter lengths (kernel) can be directly specified in the construction of convolutional networks, they have a significant advantage over RNNs in predicting long sequences. Furthermore, they tend to be computationally less intensive than recurrent networks, and thus train more quickly.

Unlike with image processing, where convolutional networks take a 2D image and shrink it in its natural plane while stretching it in a 3rd dimension to extract information, EM spectra can be more easily handled by keeping the length and shape of the sequence constant. This makes it more simple to assemble networks, as there is no need to worry about matching dimensions between layers and there are fewer hyperparameters to tune, since for all layers the stride is set to one and "same" padding is used. The network can be constructed entirely from 1D convolutional layers. At this stage, it is up to the designer to choose the kernel size and number of filters used for each layer. The kernel size is essentially the length of patterns that the layer will try to find within each sequence, and the number of filters is the number of patterns that the layer will apply to the input sequence. The higher each of these numbers, then the more likely the network is going to be able to accurately fit (or possibly overfit) the data, but at the usual expense of required computing power and increased training time. It is generally efficient for the kernel size to be largest in the shallow layers and decrease in the deeper layers. The number of filters in the last layer should be equal to the number of output spectra, which for example, would be 4 to get transmission phase and magnitude for both x and y polarizations.

10.3 Inverse-Design Networks

Inverse design is inputting a desired EM response into a neural network and obtaining a structure(s) that will produce that response. This is the ultimate design goal in that only the performance specifications are needed and a design is produced without further need for optimization. Computationally it is a much more complex problem than forward neural network design because: (i) It's possible that no structures within the domain of the neural network can obtain the desired response. For such cases, a means of achieving a structure closest to the desired performance must be incorporated. (ii) A given response may be producible by multiple structures. This possibility can make it difficult for a neural network to converge during training. Furthermore, even if the neural network converges, it will typically only find one solution. This is undesirable, as fabrication restrictions may make one solution strongly preferable to another (i.e. for structures with identical responses, we prefer simple ones with robust tolerance to fabrication errors over complex ones).

We consider two types of neural network architecture for solving inverse-design problems. The first architecture uses direct neural networks and auto-encoding methods to help the network converge. This approach is relatively easy to implement but can produce at most one structure for a given EM response. The second method uses generative adversarial networks to find multiple structures with similar EM responses. These networks are trained through a competitive process wherein a generating network produces candidate structures and a discriminating network decides whether or not to accept the generated structures based on whether they produce the right response. The generating and discriminating networks are trained together to improve at their respective tasks until eventually the generator can produce structures that accurately yield the correct response. The method is very powerful, but it can be difficult for training to converge.

10.3.1 Tandem Network for Inverse Designs

Apart from meta-atom modeling and EM response prediction, the DNNs are also capable of inversely designing meta-atoms based on specific design goals. However, if we simply train a DNN

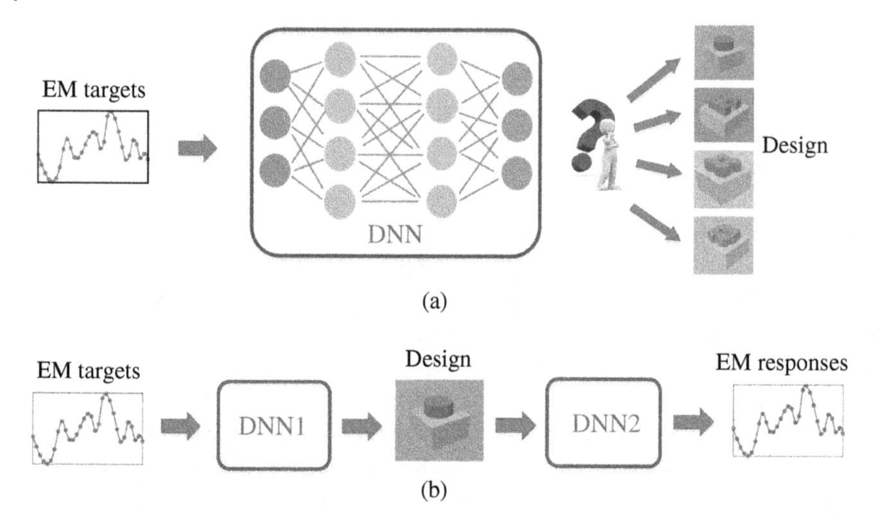

Figure 10.12 Schematic of a conventional meta-atom inverse design network (a) and a tandem network (b).

that takes the target EM response as input and generates meta-atom designs as output, the DNN will not converge with a good accuracy due to the multiple solution problems. Specifically for EM problems, it's common to have multiple structures with the same or similar EM responses. As a result, these one-to-many input–output data pairs in the training dataset contradict each other and confuse the optimizer, since choosing any direction to descend will simultaneously increase the error for all other data pairs (Figure 10.12). In [19] the authors proposed a tandem network structure to solve this non-unique solution problem. By cascading an inverse design DNN with a forward predicting DNN (Figure 10.12), the inverse designs generated by the first DNN are evaluated by the second DNN, which is fully trained and capable of predicting the accurate EM responses of meta-atom designs instantly. These two DNNs form a tandem structure (hence the name). Instead of using the differences between design parameters as the loss function, the loss function of a tandem network is the difference between EM targets and EM responses of generated designs, which eliminates the large training error caused by the non-unique solutions. With the tandem inverse design DNN, once the data set is created and the inverse DNN is trained, the design progress is non-recurring and the model generators are inquired only once per design target. Therefore, this approach is extremely time-efficient.

As a class of widely used devices in the EM field, Meta-filters or frequency-selective surfaces (FSS) represent a perfect application scenario to demonstrate the efficacy of the tandem inverse design network. The design objective of meta-filters is a pre-assigned target transmission spectrum that can be parameterized as a vector, suggesting that the design of meta-filters can be done with a tandem network [14]. As shown in Figure 10.13, without loss of generality, cylinder-shaped dielectric meta-atoms are investigated as the first example due to their robust shape and low fabrication complexity. During the design process, the meta-atoms are arranged in rectangular lattices, while their electrical permittivity, radius, height, and the gaps between adjacent meta-atoms are considered as design variables. The spectra of interest are set in the infrared regime from 30 to 60 THz (5–10 μm in wavelength). The meta-filter generator (DNN1 in Figure 10.12) employs the target spectrum responses as the input. In the output layer, an output vector is generated, which contains the parameters of the newly generated design. The design parameters are then designated as input for the consecutive predicting DNN (DNN2 in Figure 10.12), where the design's EM response is evaluated.

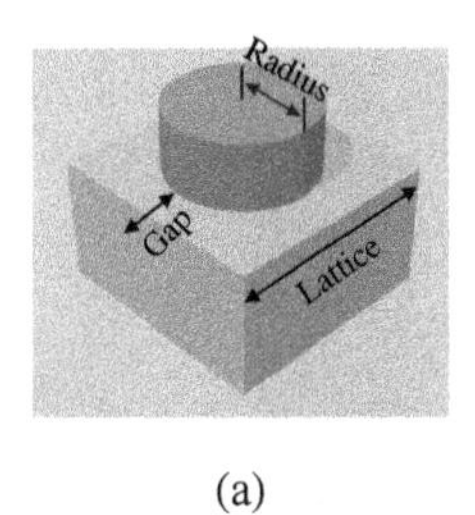

(a)

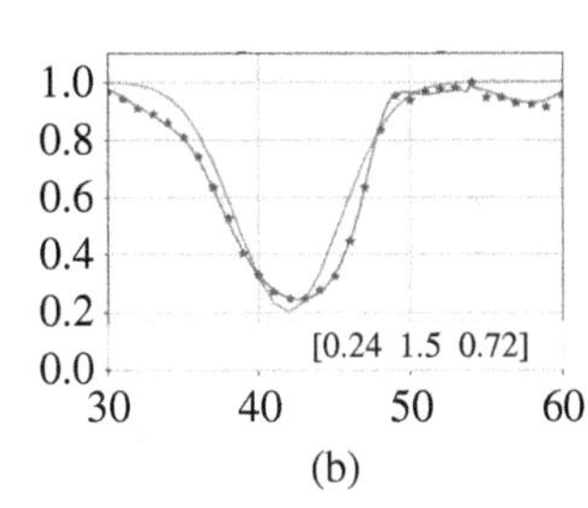

(b)

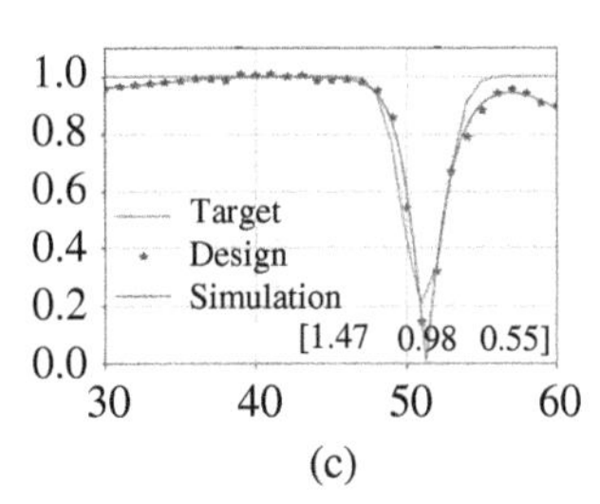

(c)

Figure 10.13 Cylinder meta-atom filter examples designed with a tandem DNN. Gap (μm), thickness (μm), and radius (μm) are given in the insets. (a) Meta-atom shapes and (b-c) Design targets and responses of meta-atoms generated by DNNs.

Finally, the transmission spectrum of the current design is compared to the target spectrum and the Euclidean distance between them is calculated. During training, the weights and biases in the hidden layers of the inverse design DNN (DNN1) are optimized to minimize this distance, while the values of hidden neurons in the modeling DNN (DNN2) remain unchanged. As a result, the inverse design DNN becomes "smarter" as training proceeds, eventually forming a network that can generate on-demand meta-filter designs on a one-time calculation basis. After being trained with 50,000 groups of randomly generated meta-atoms and their corresponding EM responses, an accurate predicting DNN is ready for use. With another 20,000 groups of human-defined spectrum targets as training datasets, the inverse design network can generate filter design parameters with EM responses closest to the human-defined input spectrum. Two examples were randomly selected and presented (Figure 10.13) for demonstration.

10.3.2 Generative Adversarial Nets (GANs)

Using tandem networks, we can realize the inverse design of simple-shaped meta-atoms that can be described using several parameters, for example, cylinders (radius and height), cubes (length, width, and height), H-shapes (length and width of each bar and height), etc. In Figure 10.7 we've also demonstrated how more DOFs can improve the meta-atoms' performance. Thus, it's natural to ask this question: could we build and train DNNs that can generate on-demand meta-atom designs with more complicated shapes?

The short answer is yes; however, this task cannot be easily done with tandem networks. For example, if we simply increase the size of the output tensor in tandem networks and make it large enough to contain a full 2D pattern (which represents the cross-section of a free-form meta-atom), and then train the tandem network, the training error will simply bounce up and down and refuse to converge to a local minimum. The reason is straightforward and intuitive – it may be easy to recognize the theme of a picture, but drawing high-quality images with a given theme requires years of practice. Similarly, for meta-atoms with free-form shapes, predicting the EM performances (small output) given the 2D patterns of the meta-atoms (large input) is easy, but generating on-demand patterns (large output) given the target EM performances (small input) requires composing and improvising, which further increase the training difficulty. Taking a free-form meta-atom with the cross-section composed of 16 by 16 pixels as an example (Figure 10.14). The meta-atom patterns are binarized, such that "1" represents high-index dielectric and "0" represents substrate. Each possible meta-atom design can thus be described by a 1 by 256 vector composed of "0"s and "1"s. With a certain design goal, the corresponding meta-atom designs that meet this design goal can be described by a certain binomial distribution. Considering the huge

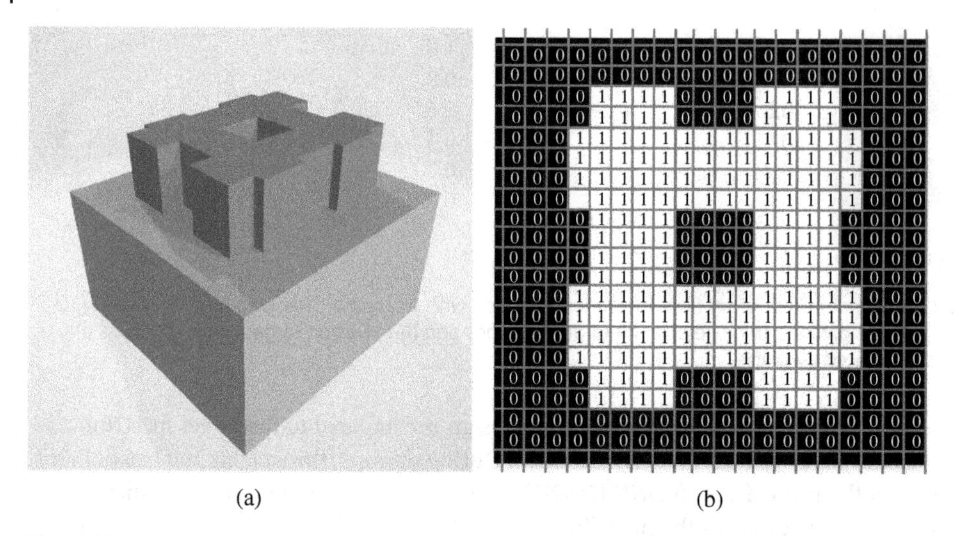

Figure 10.14 A nearly free-form meta-atom and the way to represent its 2D cross-section using a matrix. (a) Main view and (b) Top view.

size of this high-dimensional (256 dimensions in this example) discrete design space and the number of possible designs (2^{256}), finding a specific on-demand distribution is not an easy task (if not impossible).

On the other hand, GANs [51] have achieved great success in modeling distributions. Different from FCNNs, CNNs, and the tandem networks mentioned above, GANs have developed a unique approach to modeling specific binomial or multinomial distributions. GANs are usually composed of two main modules, a Generator and a Discriminator (Figure 10.15). The Generator (counterfeiter) learns to model the distribution of real data (authentic) and generates the fake data (counterfeit) to fool the Discriminator (expert). On the contrary, the Discriminator aims to detect generated samples as fake. In GANs, the Generators and the Discriminators are constructed with FCNNs [51] or CNNs [51, 52], and the trainings of the Generator and the Discriminator are processed in turns,

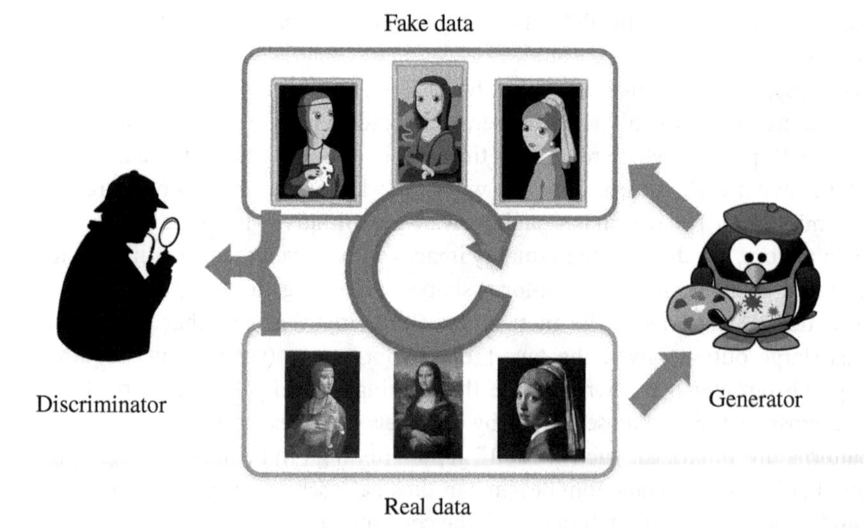

Figure 10.15 Schematic of GAN framework. Source: Leonardo da Vinci/Wikimedia Commons; Louvre Museum/Wikimedia Commons/Public Domain; Unknown Source/Wikimedia Commons/Public Domain.

aiming for opposite objectives. The key point of the training is to maintain the balance between the two sides: if the game starts with a Discriminator that can catch the fake every time, the Generator will not be able to progress. Sometimes certain measures have to be taken to maintain this balance, for example, the Generator often needs to be trained for several iterations before one iteration of Discriminator training. During this process, the Generator and the Discriminator work together to gradually narrow down the scope of the distribution of real data, which significantly lowers the training difficulty. In the end, the Discriminator is unable to differentiate the real and the fake, which indicates the Generator fully learns the distribution of the real data and is capable of generating more fake samples that are as good as the real ones from the very same distribution.

One major advantage of the GAN is that it learns the distribution of all data instead of memorizing one specific dataset. Theoretically, it's able to generate an unlimited number of outputs from the same distribution once it's fully trained. The Generator takes a latent vector as part of the input, which is composed of a group of randomly generated values that are normally distributed. The Generator converts this normal distribution into the distribution of the real dataset, which generates different outputs even with the same network with all parameters trained and fixed. This is also the major difference between the GAN and the tandem inverse design networks, which generates the output on a one-to-one basis.

One important variant of GAN is the conditional generative adversarial net (CGAN, [53], in which the Generator can be trained to generate on-demand images conditioned on inputs. In this case, the Discriminator is also conditioned, meaning it aims to tell whether the generated images are paired with the inputs. In this way, a conditional GAN can be used to generate designs with specific targets or from a given domain. With the target EM responses as conditions and the cross-sections of meta-atoms as the input/output images, we can train a CGAN that can generate on-demand meta-atom designs with given EM targets (Figure 10.16, [37]).

Without loss of generality, the dataset was assembled using meta-atoms with 1-μm-thick high-index dielectric material ($n = 5$) sitting on a dielectric substrate ($n = 1.4$). The cross-sections

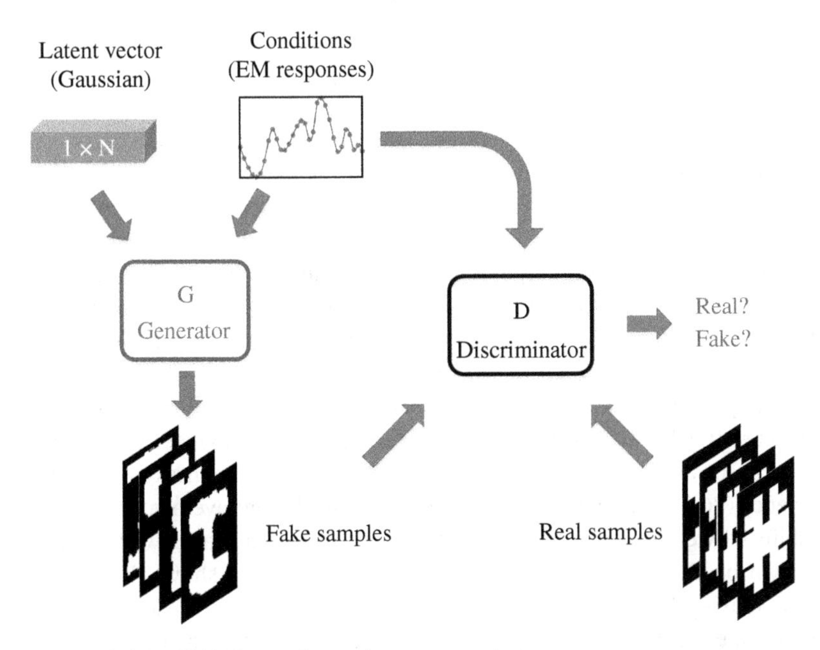

Figure 10.16 CGAN for on-demand meta-atom design.

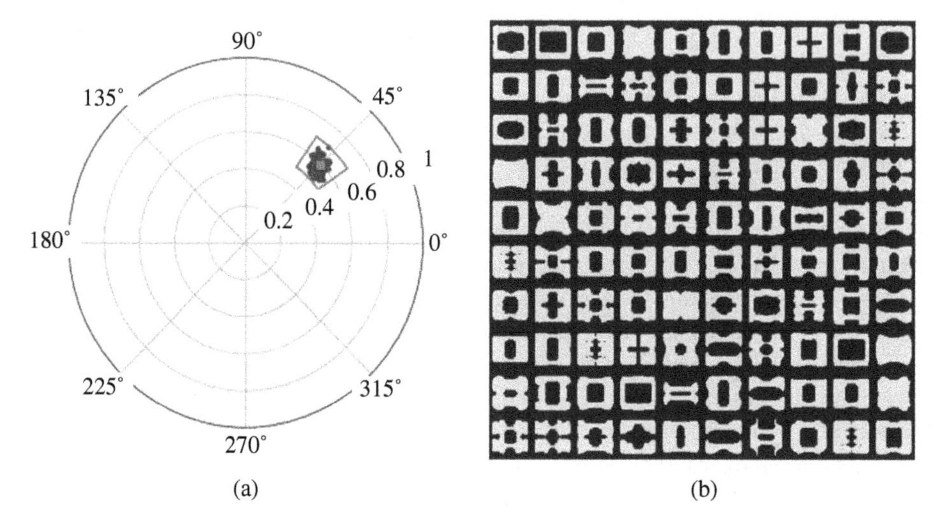

Figure 10.17 Meta-atom designs generated using a fully trained CGAN model. Radial axes represent the meta-atoms' amplitude, angular axes represent phase. Target is 0.6 (amplitude) 45° (phase) for this case. (a) Layout of designs generated by the CGAN and (b) Amplitude and phase of generated designs.

of meta-atoms were quasi-free-form patterns that are symmetric along both the x and y axes. After being trained with an adequate amount of data, the CGAN (Figure 10.16) can design various different-shaped meta-atoms with the same EM responses in a short time. The well-trained GAN was employed to consecutively generate 100 qualified designs to demonstrate its efficacy (Figure 10.17). It is worth noting that by examining the meta-atom structures generated by the CGAN, it may be possible to discover underlying physical characteristics pertaining to a particular EM response. Some of these characteristics are quite straightforward and even visible. For example, qualified patterns in Figure 10.17 can be roughly categorized into two categories: some have a similar appearance to letter "O," while the other ones resemble the letter "X." By processing the image through several convolutional layers, the neural network can uncover the common traits of these designs; these common traits can be used to categorize the designs into the same conditional distribution, which is highly non-intuitive.

Due to the one-to-many correlation between the targets and designs, the GANs provide an opportunity to visualize the mutual coupling effect existing in the meta-devices: for example, deflectors assembled with meta-atoms that have the same EM performances can be very different in efficiency. Here, 400 meta-atoms were generated with a fully trained GAN [37] that's capable of designing meta-atoms based on phase and amplitude targets. Among these 400 meta-atom designs, each group of 100 meta-atoms is created with the same amplitude target of 0.9 and different phase targets of 45°, 135°, 225°, and 315°, respectively (Figure 10.18). Subsequently, one meta-atom was randomly selected from each set of 100 geometries to assemble a beam deflector consisting of 4 meta-atoms. Two beam deflectors were created and simulated with a full-wave simulation tool to calculate their efficiencies and local responses. Although the periodic responses of all four meta-atoms in both deflectors are almost identical, their corresponding local responses can be very different. The high efficiency (86.3%) designs shown on the right have higher average amplitude and precise 90° phase shifts between adjacent meta atoms as compared with their low efficiency (51.6%) counterparts shown on the left. The big efficiency contrast not only shows the severe mutual coupling impact that existed in this beam deflector but also demonstrates the feasibility of using the GANs to assemble metasurfaces that are less prone to the mutual coupling effect.

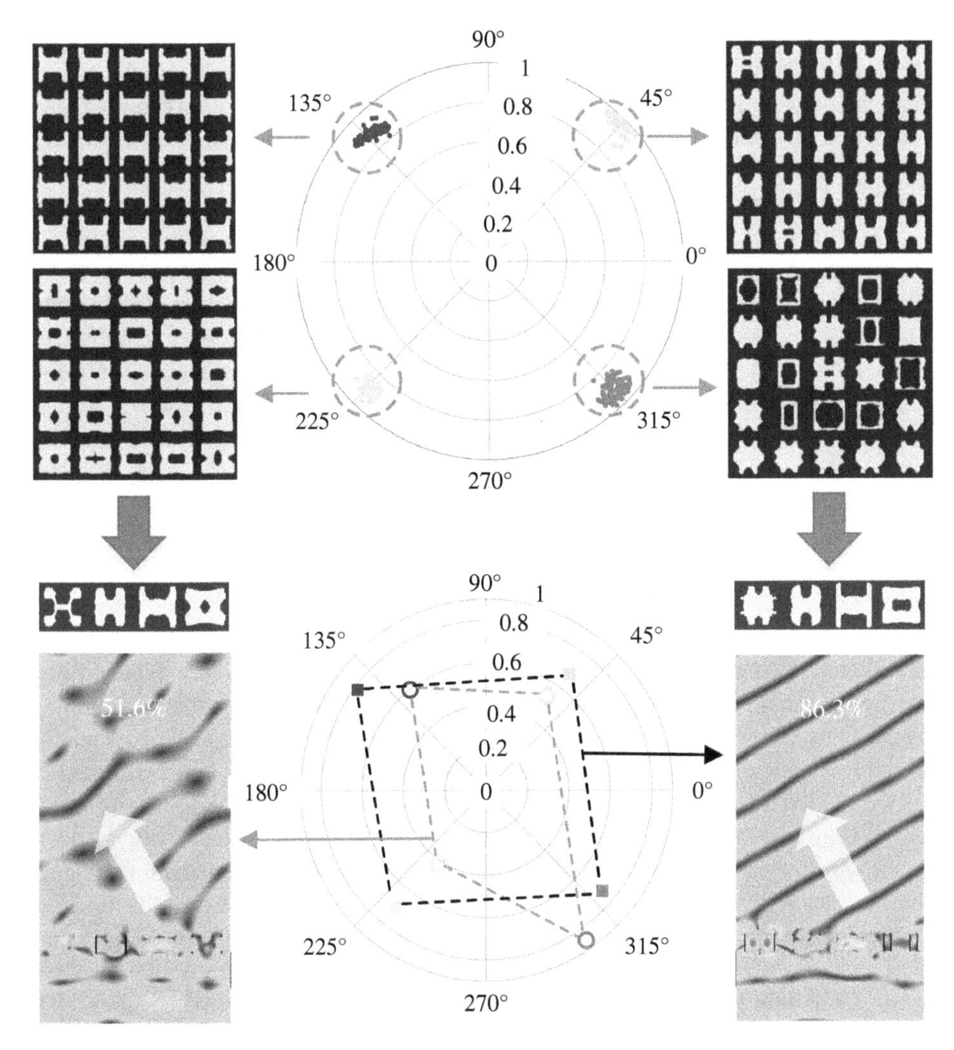

Figure 10.18 Mutual coupling in meta-deflectors.

10.4 Neuromorphic Photonics

Up to this point we have discussed how DNNs can be used to design metasurfaces. As a final topic, we briefly introduce an intriguing reversal of roles in which metasurfaces could be implemented to create neural networks. Whereas the neural networks we have considered up to this point have been implemented via software that instructs electrical circuits to perform calculations, the field of neuromorphic photonics seeks to create physical realizations of neural networks and use optical signals to perform computations. Using light along with physical versions of neurons instead of electronic ones offers high potential for faster computation times, less loss (and thus less waste heat), and reduced power consumption. The primary challenge of optical computing is in producing the same essential nonlinear effects for digital signal and information processing that are easily implemented with modern transistor technology. As such, current implementations of neuromorphic computing feature analog signals and/or hybrid architectures that convert optical signals to electrical ones for some parts of the processing before converting them back to optical signals. Additionally there are scalability challenges in matching the computing power of conventional integrated circuits with an optical chip of the same size. The ability to precisely manipulate light with metasurface nanostructures presents possible solutions to overcoming these challenges.

Metasurface nanostructures can replicate many of the functions needed in neuromorphic computing. Particularly, they can enable multiplexing by fanning light out(in) to separate(combine) signals sent at different wavelengths (analogous to a prism splitting white light into a rainbow) and introduce phase delays for interference effects that can be used to tune the magnitude of transmitted signals. Additionally, engineered nanostructures' ability to shape wavefronts and radiation patterns has potential to enable novel methods of forming optical connections and transmitting information. There have also been advances is using phase change materials to create reconfigurable structures that could be used to create dynamic and trainable neuromorphic networks. As the field is burgeoning, we defer presenting any implementations that will likely be made obsolete quickly, and refer the reader to the literature [54–59].

10.5 Summary and Outlook

Although we have described a variety of network architectures for finding meta-atoms suitable for use in metasurfaces, we have deliberately refrained from insisting on particular constructions, i.e. specifying the numbers of layers and neurons, etc. The reason being that the optimal network architecture is extremely situational. Even if we sampled the same structure twice to get two samples of the same size, we could likely find separate architectures that work best for each sample. As such, our approach has been to establish key concepts for the designer to consider when trying to meet particular design requirements, for it is up to them to determine what is most efficient in terms of computing power, data collection, fabrication ability, multifunctionality, and so forth. Is the task simple and a one-time design? Then a FCNN might be sufficient. Is a high-efficiency multifunctional design needed? Then a GAN might be the only realistic choice. Are multiple designs needed for devices operating at different portions of the electromagnetic spectrum? Then a near-universal type network could be the best.

There are a few additional concepts to keep in mind: (i) If in doubt, collect more data. Poorly constructed networks with a large amount of training data often outperform sophisticated networks with a low amount of training data. Sometimes there simply is no architecture that can compensate for under-sampling. (ii) Usual methods for regularization and improving accuracy, such as batch normalization and dropout, often work for metasurface design just as for other tasks, so those should usually be included. (iii) Complex networks take longer to perform calculations after being trained in addition to longer training. Also, they are more prone to overfitting, and so it is best to start relatively simple and then increase complexity as needed.

With the emergence of dielectric metasurfaces and their unprecedented ability to manipulate light nearly coinciding with the advent of highly powerful DNN computing methods, the optical devices used a few decades from now (or perhaps even sooner) may scarcely resemble those of today. Metasurfaces offer compactness, multifunctionality, and potential for novel devices that have yet to be conceived of that will be enabled via continuations of DNN approaches such as those presented here. The outlook of the field is exceptionally bright, with commercialization of some products having already begun and certainly more to follow. It is an exciting time to investigate the wealth of possibilities.

References

1 Kuznetsov, A.I., Miroshnichenko, A.E., Fu, Y.H. et al. (2012). Magnetic light. *Scientific Reports* 2 (1): 1–6.
2 Yu, N., Genevet, P., Kats, M.A. et al. (2011). Light propagation with phase discontinuities: generalized laws of reflection and refraction. *Science* 334 (6054): 333–337.
3 Decker, M., Staude, I., Falkner, M. et al. (2015). High-efficiency dielectric Huygens' surfaces. *Advanced Optical Materials* 3 (6): 813–820.
4 Yu, N. and Capasso, F. (2014). Flat optics with designer metasurfaces. *Nature Materials* 13 (2): 139–150.
5 Wang, S., Wu, P.C., Su, V.-C. et al. (2018). A broadband achromatic metalens in the visible. *Nature Nanotechnology* 13 (3): 227–232.
6 Chen, W.T., Zhu, A.Y., Sanjeev, V. et al. (2018). A broadband achromatic metalens for focusing and imaging in the visible. *Nature Nanotechnology* 13 (3): 220–226.
7 Ra'di, Y., Sounas, D.L., and Alù, A. (2017). Metagratings: beyond the limits of graded metasurfaces for wave front control. *Physical Review Letters* 119 (6): 067404.
8 Sell, D., Yang, J., Doshay, S. et al. (2017). Large-angle, multifunctional metagratings based on freeform multimode geometries. *Nano Letters* 17 (6): 3752–3757.
9 Li, S.-Q., Xu, X., Veetil, R.M. et al. (2019). Phase-only transmissive spatial light modulator based on tunable dielectric metasurface. *Science* 364 (6445): 1087–1090.
10 Zhao, W., Jiang, H., Liu, B. et al. (2016). Dielectric Huygens' metasurface for high-efficiency hologram operating in transmission mode. *Scientific Reports* 6 (1): 1–7.
11 Cybenko, G. (1989). Approximation by superpositions of a sigmoidal function. *Mathematics of Control, Signals and Systems* 2 (4): 303–314.
12 Hornik, K., Stinchcombe, M., and White, H. (1989). Multilayer feedforward networks are universal approximators. *Neural Networks* 2 (5): 359–366.
13 Hornik, K., Stinchcombe, M., and White, H. (1990). Universal approximation of an unknown mapping and its derivatives using multilayer feedforward networks. *Neural Networks* 3 (5): 551–560.
14 An, S., Fowler, C., Zheng, B. et al. (2019). A deep learning approach for objective-driven all-dielectric metasurface design. *ACS Photonics* 6 (12): 3196–3207.
15 Gao, L., Li, X., Liu, D. et al. (2019). A bidirectional deep neural network for accurate silicon color design. *Advanced Materials* 31 (51): 1905467.
16 Kiarashinejad, Y., Abdollahramezani, S., Zandehshahvar, M. et al. (2019). Deep learning reveals underlying physics of light–matter interactions in nanophotonic devices. *Advanced Theory and Simulations* 2 (9): 1900088.
17 Li, X., Shu, J., Gu, W., and Gao, L. (2019). Deep neural network for plasmonic sensor modeling. *Optical Materials Express* 9 (9): 3857–3862.
18 Lin, K.-F., Hsieh, C.-C., Hsin, S.-C., and Hsieh, W.-F. (2019). Achieving high numerical aperture near-infrared imaging based on an ultrathin cylinder dielectric metalens. *Applied Optics* 58 (32): 8914–8919.
19 Liu, D., Tan, Y., Khoram, E., and Yu, Z. (2018). Training deep neural networks for the inverse design of nanophotonic structures. *ACS Photonics* 5 (4): 1365–1369.
20 Ma, W., Cheng, F., and Liu, Y. (2018). Deep-learning-enabled on-demand design of chiral metamaterials. *ACS Nano* 12 (6): 6326–6334.
21 Malkiel, I., Mrejen, M., Nagler, A. et al. (2018). Plasmonic nanostructure design and characterization via deep learning. *Light: Science & Applications* 7 (1): 1–8.

22 Peurifoy, J., Shen, Y., Jing, L. et al. (2018). Nanophotonic particle simulation and inverse design using artificial neural networks. *Science Advances* 4 (6): eaar4206.

23 Tanriover, I., Hadibrata, W., and Aydin, K. (2020). Physics-based approach for a neural networks enabled design of all-dielectric metasurfaces. *ACS Photonics* 7 (8): 1957–1964.

24 Lin, R., Zhai, Y., Xiong, C., and Li, X. (2020). Inverse design of plasmonic metasurfaces by convolutional neural network. *Optics Letters* 45 (6): 1362–1365.

25 An, S., Zheng, B., Shalaginov, M.Y. et al. (2020). Deep learning modeling approach for metasurfaces with high degrees of freedom. *Optics Express* 28 (21): 31932–31942.

26 Kiarashinejad, Y., Zandehshahvar, M., Abdollahramezani, S. et al. (2020). Knowledge discovery in nanophotonics using geometric deep learning. *Advanced Intelligent Systems* 2 (2): 1900132.

27 Krizhevsky, A., Sutskever, I., and Hinton, G.E. (2012). ImageNet classification with deep convolutional neural networks. *Advances in Neural Information Processing Systems 25 (NIPS 2012)*, pp. 1097–1105.

28 He, K., Zhang, X., Ren, S., and Sun, J. (2016). Deep residual learning for image recognition. *Proceedings of the IEEE Conference on Computer Vision and Pattern Recognition*, pp. 770–778.

29 Simonyan, K. and Zisserman, A. (2014). Very deep convolutional networks for large-scale image recognition. *arXiv preprint arXiv:1409.1556.*

30 Levine, S., Pastor, P., Krizhevsky, A. et al. (2018). Learning hand-eye coordination for robotic grasping with deep learning and large-scale data collection. *The International Journal of Robotics Research* 37 (4–5): 421–436.

31 Arbabi, A., Arbabi, E., Kamali, S.M. et al. (2016). Miniature optical planar camera based on a wide-angle metasurface doublet corrected for monochromatic aberrations. *Nature Communications* 7 (1): 1–9.

32 Shalaev, M.I., Sun, J., Tsukernik, A. et al. (2015). High-efficiency all-dielectric metasurfaces for ultracompact beam manipulation in transmission mode. *Nano Letters* 15 (9): 6261–6266.

33 Zhang, L., Ding, J., Zheng, H. et al. (2018). Ultra-thin high-efficiency mid-infrared transmissive Huygens meta-optics. *Nature Communications* 9 (1): 1–9.

34 Shalaginov, M.Y., An, S., Zhang, Y. et al. (2021). Reconfigurable all-dielectric metalens with diffraction-limited performance. *Nature Communications* 12 (1): 1–8.

35 Ding, J., An, S., Zheng, B., and Zhang, H. (2017). Multiwavelength metasurfaces based on single-layer dual-wavelength meta-atoms: toward complete phase and amplitude modulations at two wavelengths. *Advanced Optical Materials* 5 (10): 1700079.

36 An, S., Ding, J., Zheng, B. et al. (2017). Quad-wavelength multi-focusing lenses with dual-wavelength meta-atoms. *CLEO: Science and Innovations*, pp. JW2A–104. Optical Society of America.

37 An, S., Zheng, B., Tang, H. et al. (2021). Multifunctional metasurface design with a generative adversarial network. *Advanced Optical Materials* 9 (5): 2001433.

38 Auguié, B. and Barnes, W.L. (2008). Collective resonances in gold nanoparticle arrays. *Physical Review Letters* 101 (14): 143902.

39 Chong, K.E., Wang, L., Staude, I. et al. (2016). Efficient polarization-insensitive complex wavefront control using Huygens' metasurfaces based on dielectric resonant meta-atoms. *ACS Photonics* 3 (4): 514–519.

40 Ollanik, A.J., Smith, J.A., Belue, M.J., and Escarra, M.D. (2018). High-efficiency all-dielectric Huygens metasurfaces from the ultraviolet to the infrared. *ACS Photonics* 5 (4): 1351–1358.

41 Cai, H., Srinivasan, S., Czaplewski, D. et al. (2019). Ultrathin metasurface for the visible light based on dielectric nanoresonators. *High Contrast Metastructures VIII*, Volume 10928, p. 109281M. International Society for Optics and Photonics.

42 Tan, S., Yan, F., Singh, L. et al. (2015). Terahertz metasurfaces with a high refractive index enhanced by the strong nearest neighbor coupling. *Optics Express* 23 (22): 29222–29230.
43 Lepeshov, S. and Kivshar, Y. (2018). Near-field coupling effects in Mie-resonant photonic structures and all-dielectric metasurfaces. *ACS Photonics* 5 (7): 2888–2894.
44 Zhelyeznyakov, M.V., Brunton, S., and Majumdar, A. (2021). Deep learning to accelerate scatterer-to-field mapping for inverse design of dielectric metasurfaces. *ACS Photonics* 8 (2): 481–488.
45 Hsu, L., Dupré, M., Ndao, A. et al. (2017). Local phase method for designing and optimizing metasurface devices. *Optics Express* 25 (21): 24974–24982.
46 An, S., Zheng, B., Shalaginov, M.Y. et al. (2021). Deep convolutional neural networks to predict mutual coupling effects in metasurfaces. *arXiv preprint arXiv:2102.01761*.
47 Fowler, C., An, S., Zheng, B. et al. (2021). A deep neural network near-universal dielectric meta-atom generator. *Flat Optics: Components to Systems*, pp. JW4D–4. Optical Society of America.
48 Yan, R., Wang, T., Jiang, X. et al. (2021). Efficient inverse design and spectrum prediction for nanophotonic devices based on deep recurrent neural networks. *Nanotechnology* 32 (33): 335201.
49 Noureen, S., Zubair, M., Ali, M., and Mehmood, M.Q. (2021). Deep learning based hybrid sequence modeling for optical response retrieval in metasurfaces for STPV applications. *Optical Materials Express* 11 (9): 3178–3193.
50 Deng, Y., Ren, S., Fan, K. et al. (2021). Neural-adjoint method for the inverse design of all-dielectric metasurfaces. *Optics Express* 29 (5): 7526–7534.
51 Goodfellow, I., Pouget-Abadie, J., Mirza, M. et al. (2014). Generative adversarial nets. *Advances in Neural Information Processing Systems 27 (NIPS 2014)*.
52 Radford, A., Metz, L., and Chintala, S. (2015). Unsupervised representation learning with deep convolutional generative adversarial networks. *arXiv preprint arXiv:1511.06434*.
53 Gauthier, J. (2014). Conditional generative adversarial nets for convolutional face generation. *Class Project for Stanford CS231N: Convolutional Neural Networks for Visual Recognition, Winter Semester* 2014 (5): 2.
54 Shastri, B.J., Tait, A.N., de Lima, T.F. et al. (2021). Photonics for artificial intelligence and neuromorphic computing. *Nature Photonics* 15 (2): 102–114.
55 Wu, C., Yu, H., Lee, S. et al. (2021). Programmable phase-change metasurfaces on waveguides for multimode photonic convolutional neural network. *Nature Communications* 12 (1): 1–8.
56 Khoram, E., Chen, A., Liu, D. et al. (2019). Nanophotonic media for artificial neural inference. *Photonics Research* 7 (8): 823–827.
57 Zhou, T., Lin, X., Wu, J. et al. (2021). Large-scale neuromorphic optoelectronic computing with a reconfigurable diffractive processing unit. *Nature Photonics* 15 (5): 367–373.
58 Wu, Z., Zhou, M., Khoram, E., Liu, B., and Yu, Z. (2020). Neuromorphic metasurface. *Photonics Research* 8 (1): 46–50.
59 Lin, X., Rivenson, Y., Yardimci, N.T. et al. (2018). All-optical machine learning using diffractive deep neural networks. *Science* 361 (6406): 1004–1008.

11

Forward and Inverse Design of Artificial Electromagnetic Materials

Jordan M. Malof, Simiao Ren, and Willie J. Padilla

Department of Electrical and Computer Engineering, Duke University, Durham, NC, USA

11.1 Introduction

In this chapter we overview the use of deep learning (DL) as a tool to study both the forward and inverse design of artificial electromagnetic materials (AEMs), depicted in Figure 11.1. Here we take AEMs to include: metamaterials, metasurfaces, plasmonics, photonic crystals, and frequency-selective surfaces. Although the principles underlying operation of these materials differ – as well as the range of free-space wavelength to characteristic feature size of the AEM – the forward and inverse deep learning methods we present here can be similarly applied to all.

Physical theories allow us to make predictions: given a complete description of a physical system, we can predict the outcome of measurements. This problem of predicting the result of measurements is called the modelization problem, the simulation problem, or the forward problem. Forward problems are usually well-posed, which means that: (a) a solution exists, (b) the solution is unique, (c) and the solution is insensitive to small changes in the initial values. As we will discuss, well-posed problems are generally easier to model than ill-posed problems.

In AEM disciplines the goal of the forward model is to make predictions of the electromagnetic scattering properties given a complete description of the physical system – the AEM geometry, desired frequency operational range, constituent materials, and embedding environment. Although historically experiments and experience were the basis of the scientific method, modern approaches to this goal include theory and numerical simulations. Even more recently, probabilistic models such as deep learning have been used for the formulation of hypotheses to explain observed phenomena. Forward model approaches are:

1. Theory
2. Numerical simulation
3. Experiment
4. Data-driven models (e.g. neural networks)

Although forward approaches (1)–(3) are well known, the data-driven model (4) is relatively new, and the focus of a portion of this chapter, which we briefly touch on next.

Data-driven models rely upon experimental data or other observations of a system to automatically build a model. The model parameters are then chosen so that the model provides the best explanation, or "fit," of the observed data according to some "loss function" (e.g. mean-squared error). Data-driven methods essentially automate the scientific method, by *automatically* finding

Advances in Electromagnetics Empowered by Artificial Intelligence and Deep Learning, First Edition.
Edited by Sawyer D. Campbell and Douglas H. Werner.

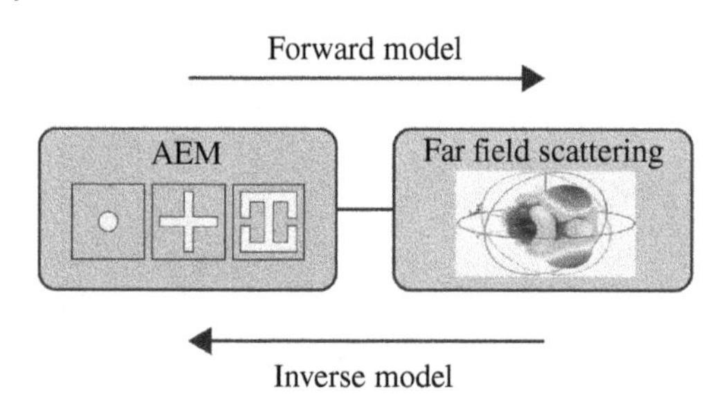

Figure 11.1 Schematic showing the forward and inverse models. The forward problem (top path), is, for example, to determine the electromagnetic scattering given an AEM geometry. The inverse problem (bottom path) is, given some electromagnetic scattering, to determine the AEM geometry which produced it.

some model that best explains the observed data. Deep learning (DL) models, and deep neural networks (DNNs) in particular, represent one class of data-driven models that have recently been shown capable of accurately modeling highly complex processes, including AEMs. In particular, DNNs have recently been shown capable of modeling the complex forward relationships between AEM structure and their electromagnetic properties.

Inverse problems are the reverse of forward problems, meaning that one is given the effect and the task is to recover the cause. Inverse problems are often ill-posed, meaning that one of the well-posed conditions (a)–(c) is violated. The AEM inverse problem, for example, may be to determine an AEM geometry which will provide a specific scattering response, shown as the bottom path of Figure 11.1. DNNs have also recently been shown capable of modeling challenging inverse relationships as well.

In this chapter we focus on DNN-based methods for both forward and inverse problems where the scattering is the frequency-dependent reflectivity coefficient $r(\omega)$ (or reflectance $R(\omega)$), transmissivity coefficient $t(\omega)$ (or transmittance $T(\omega)$), or the absorbance $A(\omega)$. We first describe the problem setting in more detail before specifying the forward problem and showing how DL can be used to find AEM geometries. We then discuss the significantly more challenging inverse problem including both well-posed and ill-posed variants. Finally we consider performance of various inverse approaches on benchmark datasets, highlighting advantages and disadvantages of included methods. We provide an outlook for DL forward and inverse design applied to the AEM system.

Forward and inverse problems are ubiquitous in many areas of science and engineering, and some of the methods we describe here may be applicable beyond AEM systems.

11.1.1 Problem Setting

AEMs are synthetic materials typically fashioned through placement of small repeating geometric patterns – termed unit-cells – onto a material substrate. These unit-cells, when filling a surface or volume, confer extraordinary properties to the AEM that can exceed those possible with the constituent building material. Although the unit-cells comprising AEMs can be quite simple, small changes in their geometry can have dramatic and unpredictable impacts on the electromagnetic scattering properties. Fortunately, AEMs with simple unit-cells and relatively well-understood constituent materials (e.g. metals and dielectrics) often admit closed-form mathematical expressions that relate the unit-cell geometry to its AEM properties. Therefore, searching for an AEM that exhibits particular properties can be achieved relatively quickly with trial-and-error, guided by physical insights or heuristics. This has led to AEMs that can achieve exotic properties such as invisibility cloaking and negative refractive index.

Despite such successes, it is conceivable that still-richer AEM property-profiles can be achieved if more complex geometries are employed, and/or if the AEMs are constructed with more challenging (i.e. less-understood) materials (e.g. newly discovered materials). However, physical understanding for such AEMs is poor – simple functional relationships, or even heuristic guidance, regarding unit-cell geometry and final AEM properties are often unavailable. The only contemporary means to estimate such AEM properties, given a unit-cell geometry, is computational electromagnetic simulation (CEMS), or fabrication.

In this chapter, our goal is to detail an alternative and more practical approach to design of AEMs, thereby removing the simulation bottleneck. We detail various DNN-based models which learn a closed-form mathematical relationship between the geometric parameters of the AEM-units and the resulting properties of the AEM – illustrated in Figure 11.1. DL models are computationally complex in the training stages (i.e. model inference), but notably are much faster to execute than CEMS, and a DNN-based system allows for design parameters to be obtained in a matter of seconds. The DNN approach for both forward and inverse design is a novel and computationally efficient means of achieving novel AEMs of unprecedented complexity.

11.1.2 Artificial Electromagnetic Materials

Deep learning has been used to study various classes of AEMs, and novel results have been demonstrated. We therefore define AEMs and give a brief description of working principles and the ranges over which they operate. An AEM is a synthetic material formed from (typically) periodic arrays of metal and/or dielectric inclusions in one, two (most common), or three dimensions. There are various classes of AEMs and these include: metamaterials, metasurfaces (conductive and all-dielectric metasurfaces [ADMs]), photonic crystals (PhC), frequency-selective surfaces (FSS), artificial dielectrics, and plasmonics. As mentioned, although many AEMs may be generally explored utilizing DL techniques, they operate in different regimes of free-space wavelength (λ_0) to periodicity (P), and therefore are described with different approaches. Here we primarily focus on AEMs which fall into two regimes of λ_0/P, although we briefly mention a third regime for completeness and comparison to other works [1–3].

The most common classes of AEMs are listed in Figure 11.2a, and may be categorized by the theories used to describe them, which are approximately correlated to the λ_0/P regime in which they operate. These theoretical approaches are denoted in Figure 11.2c and are, from left to right, Floquet–Bloch theory (FBT), resonant effective media (REM) theory, and various mixing theories – also termed effective media theory (EMT). A number of different AEM unit-cells are depicted in Figure 11.2b, and these are, from left to right: photonic crystal (and FSS), all-dielectric metasurfaces, metamaterials, plasmonic particles, metamaterials (electric ring resonators), and metamaterials (planar spirals). The boundary of each unit cell is shown as a black square and these unit-cells are used to tessellate a surface (2D) or – if a height is defined – may be used to fill a volume (3D). Although the second column (ADM cylinder and dielectric cube), as well as the fourth column (gold plasmonic sphere, plasmonic shell, and gold cube) shown in Figure 11.2b are depicted in 3D, most often only a 2D surface is formed, thereby constituting a metasurface. The geometrical dimensions most critical to operation of AEMs are shown in Figure 11.2b and are: the periodicity P, element length l, diameter of a plasmonic sphere D, and split-gap g of a metallic metamaterial.

We designate Regime 1 (see Figure 11.2c) to be where FBT [4, 5] is used to describe photonic crystals (also termed photonic band gap materials) and frequency-selective surfaces (FSSs). Here Regime 1 corresponds to $\lambda_0/P \approx 1$, and the electromagnetic properties are primarily governed by the periodicity P of the AEM. In Regime 2, resonant effective media theory is used to describe electromagnetic metamaterials and metasurfaces, and spans a huge swath of the electromagnetic spectrum – approximately given by $3 \leq \lambda_0/P \leq 1000$. AEMs which fall under Regime 2 are metal-based metamaterials and their electromagnetic properties are determined by their geometry – most notably g, l, and P, as shown in Figure 11.2. We make brief mention of

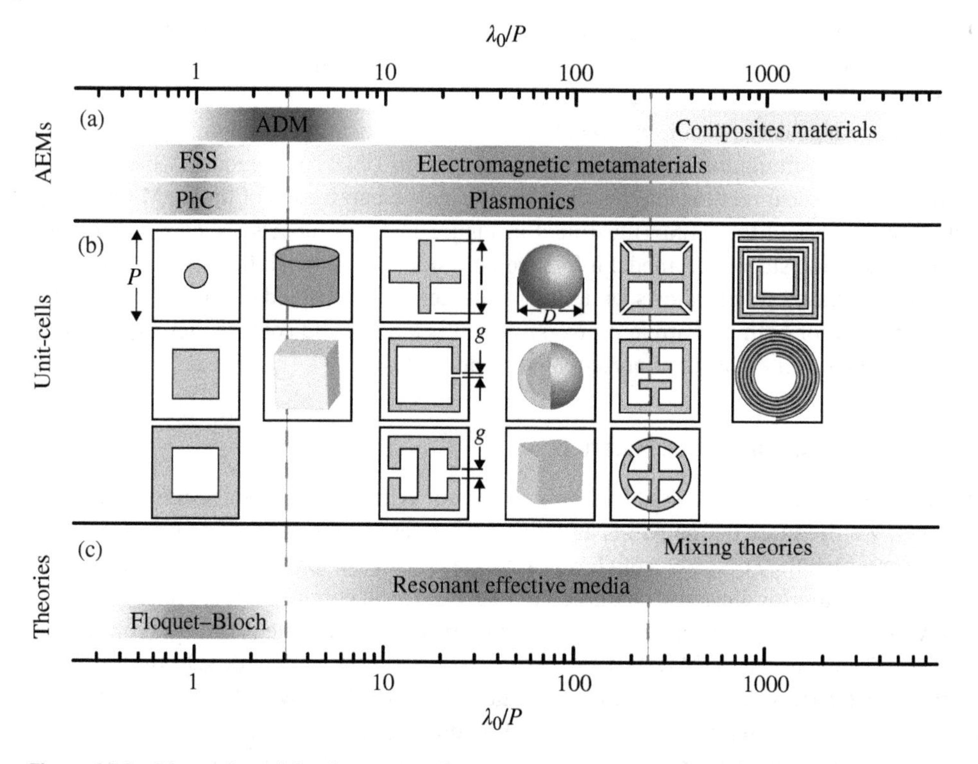

Figure 11.2 Illustration of the three operational ranges of AEMs versus free-space wavelength to periodicity λ_0/P. (a) The common names of the AEMs, (b) the unit-cells of some artificial electromagnetic materials, and (c) the theories applied in the different regimes.

Regime 3, where classical mixing theories (e.g. the Clausius–Mossotti relation, Maxwell–Garnett approximation, Bruggeman mixing model) [3] are applied, and generally termed effective media theory (EMT). The primary goal of EMT is to define averages – for example permittivity – such that these averaged quantities can be used in place of the microscopic details, while still accurately describing the scattering of electromagnetic waves in some range of the spectrum [3]. Regime 3 roughly spans $100 \leq \lambda_0/P \leq 4000$ [3]. Both Regimes 2 and 3 are effective medium theories; however, an important difference is that Regime 2 is focused on resonant AEMs, which yield a dispersive (frequency-dependent) electromagnetic response, whereas the only dispersion in Regime 3 – if it exists – directly arises from the constituent materials.

11.1.2.1 Regime 1: Floquet–Bloch

FBT is useful for the investigation of systems which possess periodic structure. The theory is named for both Floquet who studied partial differential equations with periodic coefficients in 1D [6], and Bloch who investigated the electron wavefunction in 3D matter [7]. There are two distinct categories of periodic systems to which FBT may be applied: (i) those with a continuous physical structure but where the electromagnetic properties (e.g. permittivity) vary periodically in the propagation direction, and (ii) structures with periodic boundary conditions [5]. The latter is more typical in AEM disciplines and the one that will be discussed here. In any case, all periodic structures have the common characteristic that they possess discrete passbands separated in frequency by stopbands. Here a passband (stopband) is a range of frequencies which have high transmittance (low transmittance).

Photonic Crystals Also called photonic band gap (PBG) or electromagnetic band gap (EBG) materials, photonic crystals (PhCs) are AEMs that consist of periodic structures that can be arranged to achieve a stopband or passband at a particular wavelength. For example, a change in the periodicity of the lattice – termed a defect – can be used to obtain a passband of electromagnetic (EM) radiation over a span of wavelengths [8, 9]. As mentioned, any periodic boundary condition will suffice, and thus PhC and FSSs may be implemented as metallic or dielectric inclusions embedded in air (or dielectric), or as voids in a dielectric or metallic medium – see Figure 11.2. Photonic crystals are easy to implement in 1D and 2D; however, a grand challenge of PhC is to achieve stopbands or passbands in 3D, i.e. for all incident angles and both polarizations [1]. Fundamentally, PhCs achieve pass and stop-band properties due to multiple reflections due to the periodicity, i.e. diffraction, and thus the periodicity of a PhC is of the order of the operational wavelength [10, 11].

Frequency-Selective Surfaces AEMs called frequency-selective surfaces (FSS) with features that are periodic in two dimensions and are designed to operate where the electromagnetic wave is normally incident to the surface. There are two major types of FSSs – those with unit-cells that have periodic apertures, and a complementary type consisting of periodic patches [12]. Design of FSS can be guided by circuit analog theory, where aperture-based FSSs have an inductive response yielding high-pass (for increasing frequency) or band-pass spectral features. On the other hand, when periodic patches are used the FSS has a capacitive response, and the "opposite" spectral response is obtained, i.e. low-pass or band-stop properties. A combination of multiple screens may thus be used to achieve both inductive and capacitive responses, thereby achieving more complex spectral features [13].

11.1.2.2 Regime 2: Resonant Effective Media

We discuss two types of REM: electromagnetic metamaterials and plasmonics. Both are fashioned from conductive metals and often gold, silver, aluminum, and copper are used due to their high electrical conductivities. Here we use the terms metamaterial and metasurface to refer to metallic-based AEMs, and the term ADM is reserved for the non-conductive counterpart. We note that the difference between a metamaterial and a metasurface is that a metamaterial may fill a volume, whereas a metasurface consists of a single functional layer of metamaterial, and here we use the terms metamaterial and metasurface interchangeably.

Electromagnetic Metamaterials Metamaterials are most often realized as periodic square arrays of sub-wavelength elements (see Figure 11.2) [14–16]. However, depending on the exact λ_0/P ratio achieved, metamaterials may be described by Lorentz oscillator forms of the material parameters $(\tilde{\varepsilon}, \tilde{\mu})$ with varying degrees of accuracy. In many cases, the electric and magnetic responses of metamaterials are well described using a modified Lorentz oscillator equation which includes spatial dispersion. For example, in CEMS, one may invert the complex scattering parameters $(\tilde{S}_{11}(\omega), \tilde{S}_{21}(\omega))$ obtained from metamaterial simulations, to obtain $\tilde{\varepsilon}_{eff}(\omega)$ and $\tilde{\mu}_{eff}(\omega)$. The effective permittivity and effective permeability obtained from simulation may be written as,

$$\begin{aligned} \tilde{\varepsilon}_{eff}(\omega, k) &= \zeta(k)\tilde{\varepsilon}(\omega) \\ \tilde{\mu}_{eff}(\omega, k) &= \zeta(k)\tilde{\mu}(\omega) \end{aligned} \tag{11.1}$$

where the term $\zeta(k) = \frac{1}{2}kd\cot(kd/2)$ is the spatially dependent term, and $k = \sqrt{\tilde{\varepsilon}_{eff}\tilde{\mu}_{eff}}k_0$ is the wavevector inside the metamaterial, $k_0 = \omega/c$ is the wavevector in free space, and d is the metamaterial thickness. The frequency-dependent Lorentz response is then given by,

$$\begin{aligned} \tilde{\varepsilon}(\omega) &= \zeta^{-1}\tilde{\varepsilon}_{eff}(\omega, k) \\ \tilde{\mu}(\omega) &= \zeta^{-1}\tilde{\mu}_{eff}(\omega, k) \end{aligned} \tag{11.2}$$

The frequency-dependent permittivity $\tilde{\epsilon}(\omega)$ and frequency-dependent permeability $\tilde{\mu}(\omega)$ obtained from Eq. (11.2) are well modeled by Lorentz oscillator equations of the form,

$$\begin{aligned} \tilde{\epsilon}(\omega) &= \epsilon_\infty + \frac{\omega_p^2}{\omega_0^2 - \omega^2 - i\omega\omega_s} \\ \tilde{\mu}(\omega) &= \mu_\infty + \frac{\omega_{p,m}^2}{\omega_{0,m}^2 - \omega^2 - i\omega\omega_{s,m}} \end{aligned} \tag{11.3}$$

where ϵ_∞ (μ_∞) specifies contributions to the permittivity (permeability) outside of the frequency range of consideration, ω_p^2 ($\omega_{p,m}^2$) are the plasma frequencies, ω_0 ($\omega_{0,m}$) the resonance frequencies, and ω_s ($\omega_{s,m}$) are the scattering frequencies. The metamaterial resonance frequency, $\omega_0 \approx (LC)^{-1/2}$, is governed by the capacitance (C), and inductance (L) given by the gap and wire self-inductance of the unit-cell, respectively. The "unwound" length ℓ (approximately the perimeter) of the metallic unit-cell element (see Figure 11.2) determines the resonance wavelength $\lambda_0 = 2\pi c/\omega_0$, where ω_0 is the resonance frequency and c is the speed of light in vacuum. The fundamental resonance exhibited by metamaterials is comparable to that of a half-wave dipole antenna, where $\ell = \lambda_0/2$. The oscillator strength defined as ω_p^2/ω_0^2 is proportional to the metal areal density, and determines peak and minimum values of $\mathrm{Re}\{\epsilon(\omega)\}$ near ω_0. The resonance width ω_s is governed by the loss in materials used to fashion the metamaterial as well as radiative loss.

Metamaterials have been demonstrated over a large portion of the electromagnetic spectrum including: radio frequency [17], microwave [18, 19], millimeter wave, terahertz [20], infrared [21, 22], and optical [23]. Although extremely sub-wavelength metal-based metamaterials have been shown which realize $\lambda_0/P = 1300$, which would place them in Regime 3, since they are resonant they are necessarily described by REM and are properly described by Regime 2. The capability of metamaterials to operate over nine orders of magnitude in frequency highlights their property of *electromagnetic similitude* [24, 25]. That is the electromagnetic scattering properties of metamaterials do not depend on their band structure or chemistry, but rather on their geometry, which may be scaled as detailed in the preceding paragraph.

11.1.2.3 All-Dielectric Metamaterials

The free-space wavelength to periodicity ratio of ADMs is approximately $2 < \lambda_0/P < 6$, as shown in Figure 11.2, and thus they operate in-between Regimes 1 and 2. ADMs exhibit a resonant response to electromagnetic waves, and may be characterized as either an electric or magnetic type. The characteristics of the resonant lineshape (oscillator strength, bandwidth, center frequency) may be modified by the geometry of the ADM, i.e. periodicity, filling fraction, material properties (complex permittivity), and particle shape. ADMs are classical and thus well described by Maxwell's equations, with two well-known approaches – the so-called "Mie theory" and waveguide theory.

The waveguide approach involves solving the homogeneous Helmholtz wave equation using the boundary conditions at the free-space dielectric particle interface. Various solutions are found and here we focus on hybrid electric (EH) or hybrid magnetic (HE) waveguide modes which are supported by various dielectric shapes, including spheres, cylinders, and cubes. There have been several terms used to describe hybrid modes in ADMs and for the fundamental hybrid electric these include: quasi-transverse electric (quasi-TE), even-mode, and electric dipole (ED). The fundamental hybrid magnetic modes have been termed: quasi-transverse magnetic (quasi-TM), odd-mode, and magnetic dipole (MD).

For the cylindrical ADM it is found that for the hybrid modes, the field along the cylindrical z-axis can be much larger than that in the transverse direction. For example for the electric (even) mode, $E_z \ll H_z$, which is the origin of the notation "EH" for this mode. The magnetic (odd) mode has $H_z \ll E_z$, i.e. "HE." Additionally, three indices are added to the hybrid modes to denote field variation in the azimuthal (n), radial (m), and cylindrical axis direction (l), i.e. the hybrid electric and magnetic modes are EH_{nml} and HE_{nml}. The fundamental hybrid electric and hybrid magnetic modes are thus EH_{111} and HE_{111}, respectively.

For cylindrical ADMs it is found that EH_{nml} and HE_{nml} modes are well described with Eq. (11.1), and modification of ω_p^2, ω_0, and ω_s may be achieved by changing the geometry of the particle, e.g. for cylinders the height primarily modifies the magnetic ω_0, and radius the electric ω_0 [26]. Since waveguide theory treats only the individual particle, it does not account for the periodicity or neighbor interaction. Empirically it is observed that the resonance wavelength ($\lambda_0 = 2\pi c/\omega_0$) of both the fundamental electric (EH_{111}) and magnetic (HE_{111}) modes of cylindrical ADMs scales as $\lambda_0/P \approx 0.58\sqrt{\epsilon_r}$. ADMs operate at approximately $\lambda_0/P \approx$ 2–5, and fall in-between Regimes 1 and 2. ADMs do indeed realize both periodic effects (Regime 1) and local resonances related to their geometry (Regime 2). Thus they may be designed to take advantage of both effects, thereby bringing about more complex responses than those possible with PhC or sub-wavelength metal-based metamaterials alone. Lastly we note that, unlike metal-based metamaterials, ADMs do not possess internal capacitive regions for field localization, but rather often exhibit significant evanescent field values outside of their volume. We have focus on waveguide theory and point to a reference on Mie theory for the interested reader [27].

Summary We have over-viewed a select number of AEMs, and have parsed these based on their free-space operational wavelength λ_0 to periodicity P ratio. Based on λ_0/P different theories are used to describe the underlying physics, as shown schematically in Figure 11.2. Although choice of constituent materials plays an important role, all AEMs utilize their geometry to primarily govern their scattering properties. However, despite the similar underlying geometry-based scattering design approach, AEMs operate in drastically difference regimes of λ_0/P. ADMs, however, are an alternative, as they naturally lie between Regimes 1 and 2.

As we will show AEM systems can benefit from a data-driven design approach. Although this is the case, conventional simulation and optimization methods are a more efficient design practice for some particular AEMs.

11.2 The Design Problem Formulation

The design problem assumes that we have some AEM of interest, and some electromagnetic or scattering properties of the AEM that we wish to manipulate, denoted $s \in S$. Typically, s comprises a numerical vector representing the AEM scattering properties of interest, and S refers to the domain of s (i.e. the values that s can take). Some common examples of s include the absorptance $A(\omega)$, reflectance $R(\omega)$, or transmittance $T(\omega)$ spectra of an AEM, sampled at specific frequencies (ω) of interest. In these examples each entry in s must be between zero and one – since $A(\omega)$, $R(\omega)$, and $T(\omega)$ are absolute value quantities – and therefore the domain of s is given by $S = [0, 1]^{N_s}$, where N_s refers to the number of frequency points sampled.

In the design problem it is also assumed that we can manipulate s by varying some controllable properties of the AEM, denoted $g \in G$, where g typically also comprises some vector of numerical values. Some common examples of g include the geometric structure of the AEM and/or the properties of its bulk construction material. Once again G represents the domain of g and represents the full set of g values that can be considered during the design process. The domain G is typically set by the designer based upon various factors. For example, G might be chosen to include designs that are likely to achieve a certain scattering based upon prior knowledge or modeling, or because it represents fabricable designs. G may also be chosen to limit the search space of designs since larger design domains generally require more computational resources (e.g. via simulation) to search.

Given the aforementioned definitions, the goal of the design problem is to choose g to achieve some desired properties of the AEM, denoted s^*. If we let $s(g)$ denote the properties associated with a design g, then our goal is to find g such that $s(g) = s^*$. In practice it is unlikely that a design exists that perfectly achieves s^*, and in many applications some error with respect to s^* can be tolerated. Therefore a more general framing of the design problem is to find some design, g, such that

$$L(s(g), s^*) < \varepsilon. \tag{11.4}$$

Here L refers to some measure of the difference, or error, between $s(g)$ and s^*. The variable ε refers to maximum tolerable error. Some commonly used measures of error include mean-squared error [28] or mean-absolute error [29] between $s(g)$ and s^*, although in general any measure that produces a scalar output is admissible.

The AEM design problem has been widely studied, and there are a variety of effective methods to solve it. Several papers and books can be found that summarize well-existing work on this topic [30]. Here we focus upon emerging methods based upon deep learning, especially DNNs, which have recently been found highly effective for AEM design [31–35].

AEM design with deep learning can be dichotomized into two general strategies, or paradigms: forward design and inverse design [31]. Historically forward design has been the most widely practiced due to its accessibility and ease-of-use compared to inverse design. Generally, inverse design is much more challenging to apply in practice, although in principle it offers several advantages over forward design. In this chapter we will discuss both paradigms and their relationship, as well as the role of DNNs within each. While deep learning has been demonstrated to dramatically enhance both forward and inverse design, it is inverse design where most of the recent advances, and therefore the most attention will be given.

11.3 Forward Design

In this section we discuss the forward design paradigm, which is illustrated in Figure 11.3. Forward design assumes that we have some model to compute the scattering properties s of a target AEM, given its design g. Mathematically, we can think of the forward model as some function, f, such that

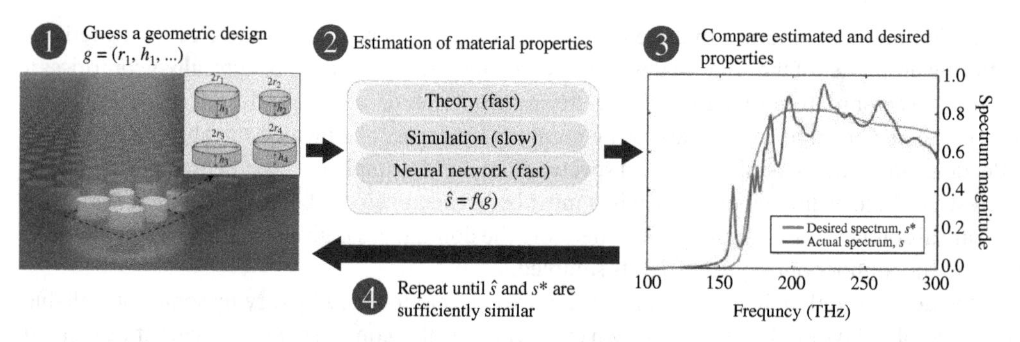

Figure 11.3 Workflow of forward design illustrated using a metasurface comprised repeating 2 × 2 supercells of cylindrical resonators. Each resonator in the supercell has an adjustable height and radius, constituting an 8-dimensional design vector, g. Our goal is to identify some g such that the AEM produces the absorption spectrum, denoted s^*, shown in the rightmost inset. In step (1) of the workflow, we identify some candidate design (or designs) and then in step (2) we evaluate their properties using a forward model of the metasurface.

$f(g) \approx s(g)$; or for simplicity, we will write $s = f(g)$. Using the forward model we can then repeatedly select candidate designs and evaluate their properties in search of a design that has sufficiently low error with respect to s^*. Mathematically, the goal of forward design is given by

$$L(f(g), s^*) < \varepsilon \tag{11.5}$$

which is identical to Eq. (11.4) except that we have replaced $s(g)$ with $f(g)$, the forward model.

The forward design paradigm is widely used and it can be highly effective; however, it imposes a fundamental trade-off between design time and solution quality. In general, evaluating more designs with the forward model will lead to higher-quality designs (lower $L(f(g), s^*)$), but at the cost of greater design time. Therefore, forward design problems can be characterized by two important properties: (i) search efficiency, which is the rate at which the design quality improves per design that is evaluated with the forward model; and (ii) evaluation time t, which is the time required to evaluate a single design using the forward model. These two properties then induce a curve of the performance achievable within a particular timeframe, denoted $\varepsilon(t)$, which can be useful for determining the feasibility and value proposition of a particular forward design problem – this is illustrated in Figure 11.4. We note that time can be converted to other useful metrics as well, such as monetary or computation costs.

In most cases, we will not actually know $\varepsilon(t)$ for a particular problem; however, there are a number of factors that influence search efficiency and evaluation time, making a given problem more or less feasible. The major role of DL methods in recent research has been to dramatically accelerate evaluation time; however, before discussing the role of DL we briefly summarize a few major factors that influence search efficiency and design time.

11.3.1 Search Efficiency

One important factor is the amount of *a priori* knowledge (via physics, or experience) that can be used to narrow the search for good designs. Such knowledge can be used to choose a relatively small search space, G, or to search more efficiently within a given G; both of these strategies will tend to increase the efficiency of the search process. Another similar approach is to use data-driven optimization methods to more efficiently search over G; these methods attempt to estimate the next

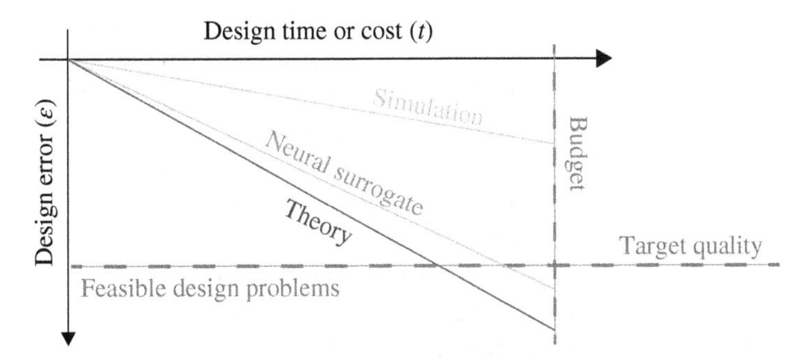

Figure 11.4 Illustration of the tradeoff between design quality, defined by ε, and design time, t that arise in forward design problems. Design error reduces with additional investment of time or computation in the design process; however, the rates of improvement vary depending upon whether Theory, Simulation, or Neural Surrogate Simulators (e.g. DNNs) are employed to evaluate designs. It is common that design problems are constrained by some computational budget of time or computation (vertical dashed line). If the desired quality of the design (horizontal dashed line) can be achieved within the budget, we call these feasible design problems.

best candidate designs based upon the performance of existing designs that have been evaluated. Some well-known examples include Nelder–Mead [36] or Bayesian optimization [37].

Another key characteristic of the forward design problem – as well as design problems in general – is the dimensionality of g, denoted $|g|$. In general, the volume of G, and therefore the total number of design solutions to consider, increases exponentially with $|g|$ – a problem sometimes called "the curse of dimensionality" [38]. To see this, let us discretize each design parameter so that there are Δ candidate settings of *each* design parameter. This implies that design candidates must be chosen from a $|g|$-dimensional grid with Δ total points along each axis. In a digital computer all numbers are technically discrete, and therefore even seemingly real-valued numbers in a digital computer are always sampled from a grid of this kind. Given this formulation, the total number of candidate designs is given by

$$K = \Delta^{|g|}, \tag{11.6}$$

which grows rapidly as either Δ or $|g|$ increase. This implies that increasing the number of free design parameters can quickly make a forward design problem intractable due to an explosion in the number of candidate designs. Another important consideration is the choice of Δ, which can sometimes be reduced substantially at little cost to design quality.

11.3.2 Evaluation Time

The time required to evaluate a given design depends primarily upon the mechanism used to evaluate the forward model, $f(g)$. Figure 11.3 lists the two most conventional mechanisms used to evaluate candidate solutions: (i) theory and (ii) computational simulation. Generally, the best mechanism is one based upon theory. A theoretical mechanism, as we define it here, has two properties: (a) it is an explicit mathematical expression for s in terms of g; and (b) the expression represents a parsimonious or succinct expression of the underlying relationship between s and g, leading to a computationally efficient evaluation of $f(g)$. These two properties together result in a computationally efficient mechanism to evaluate $f(g)$, and therefore they are also usually the fastest means to evaluate a forward model. Although we call these solutions "theory"-based, they can also be empirical models (e.g. models fit to data) as long as they satisfy criteria (a) and (b).

For many AEM problems, especially advanced problems, there may be no known explicit mathematical relationship between s and g. In these cases it is often still possible to evaluate $f(g)$ using CEMS. CEMS have the advantage that they can be used to evaluate the properties of many advanced AEM systems that are otherwise inaccessible. However, CEMS have the disadvantage that they often require significant time, expertise, and care to set up properly for a given AEM system. Additionally, CEMS are typically computationally slower than any theoretic solution: often by several orders of magnitude. This is a major disadvantage because it implies that, all other factors being equal, proportionally fewer designs can be evaluated as well. Generally a last resort to evaluate a forward model (not shown) is fabrication, which involves physically constructing the AEM and evaluating its properties experimentally. In most cases, however this strategy is impracticable due to the tremendous time and effort required to evaluate each candidate design.

11.3.3 Challenges with the Forward Design of Advanced AEMs

Tremendous success has been achieved using AEMs, resulting in materials with novel scattering properties. However, it is plausible that even more exotic and useful AEMs can be engineered if the design of more sophisticated AEMs is made possible. Advanced AEMs may include, for example, those with novel geometric structures, and/or new constituent materials, or more free design parameters. However, the physics for advanced AEMs of this kind is often limited, resulting in greater reliance upon CEMS for evaluation of the forward model, and less efficient design of – and search over – the design space G. Furthermore, the curse of dimensionality causes the design space to rapidly grow as more free design parameters are included. The result of these pressures is that the design of AEMs has quickly become intractable using conventional design methods, creating a major potential bottleneck in AEM research as a whole. In recent years deep learning has emerged as a potential solution to greatly alleviate this design bottleneck, and has enabled the forward design of much more advanced AEMs.

11.3.4 Deep Learning the Forward Model

The main role of DL in forward AEM design is to reduce – often dramatically – the computation time required to evaluate the forward model, bringing more complex AEMs within reach. The basic premise of DL-based approaches is to collect a dataset of design-property pairs, $D = \{(g_n, s_n)\}_{n=1}^{N}$, using some mechanism to evaluate the forward model (e.g. theory, CEMS). Then a DL model – usually a DNN – is "trained" to approximate the forward model using the data in D. The DNN is composed of parameters that influence how any input (i.e. g) is mapped to the model output (i.e. s), and during training these parameters are iteratively adjusted until the DNN makes accurate predictions for all of the data in D. Once trained, the DNN can then rapidly predict s for novel values of g, including those that were not present in the training dataset.

Trained DNNs are often several orders of magnitude faster than simulators (e.g. a factor of $\sim 9 \times 10^5$ [39]), making it possible to perform design for much more advanced AEMs than would otherwise be possible. In absolute terms, DNNs can usually evaluate hundreds to tens of thousands of candidate designs per second. This capability is not only useful for design, but also for general scientific inquiry, allowing researchers to rapidly evaluate the impact of specific AEM design changes. Due to these advantages, DNNs have now been employed for forward modeling in a variety of AEM applications. Table 11.1 lists some of the different types of AEMs where DL has been applied and some associated references.

11.3.4.1 When Does Deep Learning Make Sense?

Although DNNs dramatically accelerate the evaluation of candidate designs, they are not always a superior solution because of their need for training data. The number of simulations needed (N) depends upon (i) the desired accuracy of the DNN and (ii) the complexity of the forward function.

Table 11.1 Some published studies in AEM disciplines that apply DL.

AEM field	Literature applying DL
Metamaterials	[40–42]
Metasurfaces	[28, 39, 43]
Photonic crystals	[44–46]
Frequency-selective surfaces	[47–49]
Plasmonics	[50–52]

Source: Adapted from [7].

Furthermore, DNNs generally require more training data for a complex forward function than a simpler one to reach same level of accuracy. In recent studies, the number of simulations employed has been in the range of 10^2 to 10^6, with 10^4 being the most common scale [31]. DNNs are therefore most suitable when the acceleration provided by the DNN will clearly justify the initial simulation time required to obtain satisfactory model accuracy. This is often the case when there are a large number of candidate designs to consider, or the computation time to evaluate the forward model makes it intractable to effectively search the design space (e.g. as is often the case with CEMS).

11.3.4.2 Common Deep Learning Architectures

The DNNs employed for forward modeling in the AEM literature are mostly composed of feedforward fully connected networks, sometimes called multi-layer perceptrons [53]. Networks of this type are very general and make few assumptions about the properties of the input or output data that they are provided. Furthermore, theoretical results guarantee that fully connected networks with at least two layers can approximate any function (e.g. a forward function) to an arbitrary degree of accuracy [54], given a network of sufficient size (i.e. number of free parameters).

The size of DNNs is typically controlled by varying the number of layers in the network – often called its "depth" – or by varying the number of neurons in each layer of the network – often called its "width." In general, larger networks can approximate more complex forward functions and achieve higher accuracy; however, they also generally require greater quantities of training data to realize this advantage. Each AEM problem exhibits different forward modeling complexity and data collection budgets; consequentially, one major difference between DNNs in the AEM literature is their width and depth. Other common sources of variation among the DNN models in the literature involve their training procedures (e.g. learning rates, loss functions), or the inclusion of some special-purpose layers (e.g. convolutional or transposed convolutional layers [55]).

In principle, AEM problems are governed by similar underlying physics, and therefore some DNN-based approaches (e.g. DNN architecture, training procedures) may prove to be generally superior for modeling these processes. In the machine learning community it is conventional to perform benchmarking studies to uncover the relative performance of competing approaches; in such studies several methodologies are compared on multiple public datasets under fair and controlled conditions. At time of writing, this practice has not yet been widely adopted within the AEM community where machine learning is relatively new. One recent exception to this is the recent work by Deng et al. [56], which compared three recent DNN architectures on three public AEM datasets.

11.3.5 The Forward Design Bottleneck

Although DNNs can dramatically accelerate the evaluation of candidate designs, the forward design paradigm is fundamentally limited in its ability to scale to high-dimensional modeling problems. As given by Eq. (11.6), the number of candidate designs grows rapidly with the number of design parameters ($|g|$) or the step size of each parameter value (Δ), quickly making it impossible to evaluate even a small fraction of the candidate designs. For example, recent research has explored AEM problems with approximately 10^9 and 10^{12} design candidates, respectively [28, 39], making it difficult or impossible to evaluate all candidates. Even given a DNN-based model that can evaluate 10^5 solutions per second, it would still require approximately 115 days to search 10^{12} designs.

In principle, the design complexity of AEMs can grow substantially larger than the examples described above, and such complexity may even be necessary to realize the full potential of AEMs.

Although it is unnecessary to exhaustively examine the entire design space, the search process required for forward design imposes a tradeoff between design time and design quality that rapidly becomes unfavorable as complexity grows. Therefore forward design is ultimately limited in its ability to solve design problems as AEMs become more complex. In Section 11.4 we discuss inverse design, which provides a potential solution to the limitations of forward design, allowing design to scale to much higher-dimensional problems.

11.4 Inverse Design with Deep Learning

In this section we discuss the inverse design paradigm, which is illustrated in Figure 11.5. The inverse approach assumes that we have access to some model, termed the "inverse" model, that can estimate the structure of an AEM that will yield some specific electromagnetic properties, denoted $\hat{g} = f^{-1}(s)$. Therefore the inverse model takes as input an AEM's desired scattering properties and produces as output the design settings that will yield those properties. In principle, given an accurate inverse model, it is possible to directly infer the design setting needed to achieve any arbitrary s^*, rather than searching over large numbers of candidate designs, as is needed in the forward design paradigm. Consequently, inverse design scales substantially better with respect to AEM complexity than forward design.

As shown in Figure 11.5, to evaluate the effectiveness of an inverse model and its proposed designs, it is conventional to pass the proposed design solutions back through the forward model f and evaluate their properties, so that they can be compared to the desired value, s^*. This particular measure of error is referred to as "re-simulation" error of the inverse model [57]. Mathematically, the goal of inverse design is defined in terms of re-simulation error, and given by

$$L(f(f^{-1}(s^*)), s^*) < \varepsilon \tag{11.7}$$

which is identical to Eq. (11.4) except that we have replaced $s(g)$ with $f(f^{-1}(g)) = f(\hat{g})$, which is the estimated properties of the design proposed by the inverse model. Note that the evaluation of re-simulation error therefore requires evaluating the forward model. For advanced AEM problems, this will require passing the proposed design through a simulator (slow, but accurate), or using a

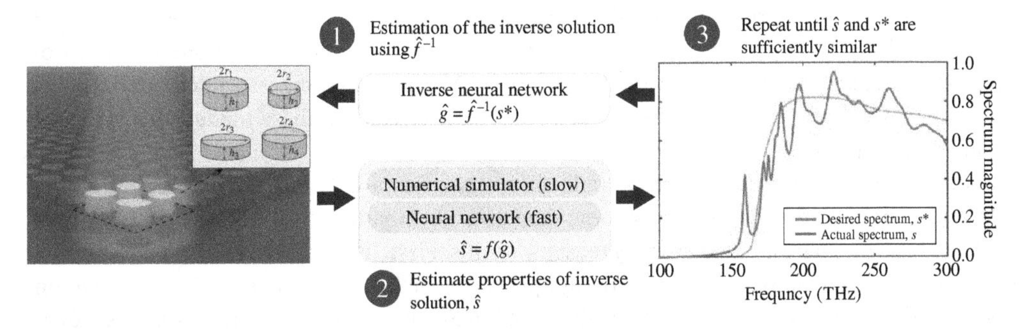

Figure 11.5 Workflow of inverse design illustrated using a metasurface comprised repeating 2×2 supercells of cylindrical resonators. Each resonator in the supercell has an adjustable geometric structure, denoted g. Our goal is to identify some g such that the AEM produces the absorption spectrum, denoted s^*, shown in the rightmost inset. In step (1) of the workflow, we use an inverse model, $\hat{g} = f^{-1}(s^*)$ to identify a candidate design setting, denoted $\hat{g}$, that will yield the desired properties. In step (2) we pass this proposed design through the forward model to estimate the properties associated with it, denoted $\hat{s}$. In step (3) we consider the similarity of $\hat{s}$ and s^* and then repeat these steps if they are not sufficiently similar.

Table 11.2 Inverse model types and their application to AEM problems.

Model	Applications to AEM problems
Conventional neural network (DNN)	[61–64]
Tandem (TD)	[65–68]
	[46, 52, 69, 70]
	[71–73]
	[45, 50, 74]
Variational auto-encoder (VAE)	[75–78]
	[79–82]
Mixture density network (MDN)	[83, 84]
Invertible neural network (INN)	[57]
Generative adversarial network (GAN)	[29, 77, 85]
	[86–88]
Evolutionary algorithm (EA)	[89–92]
Reinforcement learning (RL)	[51, 93, 94]
	[95, 96]
Neural adjoint (NA)	[28, 97–99]

Source: Adapted from [7].

neural network that has been trained to approximate the simulator (significantly faster, but less accurate).

While inverse design offers substantial speed and scalability advantages, these advantages depend upon the availability of an accurate inverse model, f^{-1}, which is often more challenging than deriving f. A large body of research has been conducted on general inverse modeling [58, 59]; however, we focus here on emerging methods involving DNN-based methods, which have recently been found highly effective for modeling f^{-1} [60], as well as solving AEM design problems [31–35]. Inverse problems exhibit unique challenges compared to forward modeling (described in Section 11.3), and a variety of specially designed DNN-based methods have been developed to overcome these challenges, termed deep inverse models (DIMs). Table 11.2 provides a list of major DIMs that have been applied to solve inverse AEM problems, and associated publications. In Section 11.4.2 we will discuss some of the major characteristics of these models that are most relevant to AEM problems.

Although inverse design can be substantially faster than forward design, the process of training DIMs, and their operation, does involve some time and costs that increase with the complexity of the AEM problem. There are two primary reasons for this. First, in similar fashion to conventional DNNs, DIMs also rely upon a dataset of design-property pairs, D, to learn a good approximation of the underlying inverse function. And as discussed in Section 11.3, more complex inverse functions also generally require greater quantities of (often simulated) training data to achieve the same level of accuracy, which contributes to the overall time and cost of design. A second limitation of inverse models is that modern DIMs can propose multiple different design solutions for a single target, s^*; and a relatively large number of proposals may need to be considered to find a good solution. Furthermore, to evaluate the re-simulation error of each proposed solution (Eq. (11.7)), the solution must be passed through the forward model, which requires additional computation time.

Although this computation time is relatively small (e.g. compared to generating D) it does contribute to a modest tradeoff between design time and design quality. As a result of these two dependencies, the time and costs associated with inverse modeling and DIMs increases with greater AEM complexity, and therefore cannot scale indefinitely with AEM complexity. In practice, however it still offers substantial speed and scalability advantages compared to forward modeling.

11.4.1 Why Inverse Problems Are Often Difficult

Learning the inverse model is essentially a regression problem where s comprises the independent variables, and g comprises the dependent variables that we wish to predict; this is identical to the forward modeling problem except that the roles of s and g are reversed. Therefore conventional regression models such as DNNs can be employed to approximate f^{-1}. As discussed in Section 11.3, a substantial body of work has focused on using DNNs to approximate the forward model, f, yielding impressive results [31, 32, 34, 100, 101]. This success is often attributed to the DNN's ability to approximate complex and highly non-linear functions. Despite this capability, and its associated success for forward modeling, DNNs often produce poor results when applied to inverse modeling tasks because DNNs are poorly suited to *ill-posed* problems [31, 102], and approximating f^{-1} is often ill-posed.

Ill-posed problems violate one of the three following Hadamard conditions [103]: (i) existence; (ii) uniqueness; and (iii) smoothness. In principle, inverse AEM problems can violate any of the three Hadamard conditions; however, most recent attention has been given to violations of condition (ii), which is sometimes called one-to-manyness, or non-uniqueness. In the context of AEM design, non-uniqueness arises when there exist multiple values of g that all yield similar AEM scattering (i.e. values of s), as illustrated in Figure 11.6. This is problematic for conventional DNNs because they can only produce a single output value for each input, making it impossible to simultaneously predict two (or more) g values for each input value of s. This can create significant problems during the DNN training process, yielding DNNs that make inaccurate predictions when trained on datasets that exhibit non-uniqueness.

In general, not every AEM problem will exhibit non-uniqueness [57]; however, it is difficult to determine in advance whether a particular problem is ill-posed. A recent study of three inverse

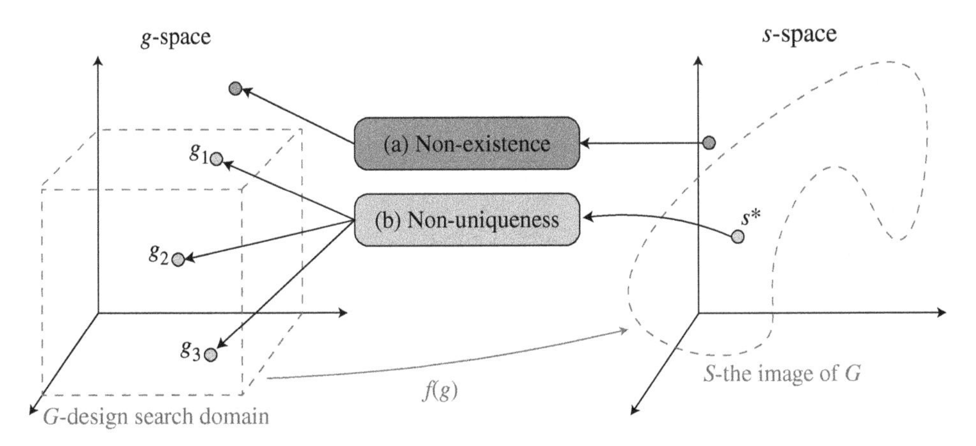

Figure 11.6 Illustration of two Hadamard conditions: (a) non-uniqueness and (b) non-existence. If these properties are violated, the function is said to be ill-posed. Ill-posed functions often require specialized deep neural networks, termed deep inverse models.

AEM problems found that two exhibited significant non-uniqueness, leading conventional DNNs to perform poorly. These considerations motivate the development and use of DIMs, which were also shown to identify highly accurate designs even when non-uniqueness was present [57]. The other two Hadamard conditions have received relatively little attention in the DIM literature. While condition (iii) is relatively uncommon in practice, in Section 11.5 we discuss condition (ii) further as an important consideration in future work. We refer the reader to other sources for a more thorough treatment of these other Hadamard conditions [31].

11.4.2 Deep Inverse Models

A variety of specialized deep learning models have been developed in recent years to overcome non-uniqueness, and they each do so in different ways. Given the diversity of DIMs, a discussion of their technical details is beyond the scope of this chapter, and here we instead focus upon describing important characteristics of modern DIMs that are especially relevant to AEM research and design. Table 11.3 lists each of the major DIMs explored in the AEM literature, along with three key properties that we describe below. We refer readers to other sources for a technical treatment of these methods [31, 104].

11.4.2.1 Does the Inverse Model Address Non-uniqueness?

The first important property of a DIM is whether it addresses the non-uniqueness that is often present in inverse AEM design problems. Most DIMs focus upon addressing non-uniqueness as indicated in Table 11.3, with the exception of conventional DNN models (e.g. those typically used for forward modeling). Although conventional DNNs are the only model that do not address non-uniqueness, it is not guaranteed that every inverse problem exhibits non-uniqueness, and in such cases conventional DNNs are often simpler to train and may achieve superior accuracy [57].

11.4.2.2 Multi-solution Versus Single-Solution Models

A second important property of DIMs is whether they can propose multiple solutions. Recent studies [60], including one recent AEM study [57], found that DIMs that are permitted to propose several solutions often lead to the discovery of substantially better solutions. This capability also

Table 11.3 Properties of deep inverse models.

Model	Non-uniqueness	Multiple solutions	Iterative
DNN	No	No	No
TD	Yes	No	No
VAE	Yes	Yes	No
MDN	Yes	Yes	No
INN	Yes	Yes	No
GAN	Yes	Yes	No
EA	Yes	Yes	Yes
RL	Yes	Yes	Yes
NA	Yes	Yes	Yes

Source: Adapted from [7].

makes it possible for the designer to consider several viable solutions that may have somewhat different scattering properties, or which may be more fabricable, and choose the one that is best-suited for their application. It is important to note that each additional solution requires an evaluation of the forward model, f, to evaluate its error, which imposes a (usually modest) trade-off between design quality and computation time.

11.4.2.3 Iterative Methods versus Direct Mappings

A third important property of some DIMs is that they rely on an iterative process for inferring each proposed inverse solution. Most DIMs attempt to learn a direct relationship from s to g, so that an estimate of g is essentially based upon a single calculation, making the inference of a solution computationally efficient. By contrast, iterative methods usually begin with an initial (often random) guess for the inverse solution, and then repeatedly make small changes to this initial guess to improve upon its quality based on a surrogate forward model (e.g. a trained DNN). This process is repeated for some fixed number of iterations, or until the quality of the solution (e.g. estimated re-simulation error) no longer improves. Recent work has found that iterative methods often achieve superior accuracy, although the computation required to infer solutions can be substantially larger than other non-iterative methods [57]. It is important to note however, that the additional computation time required for iterative methods is usually only a small fraction of the time for other steps in the design process, such as training the DIMs using D.

11.4.3 Which Inverse Models Perform Best?

The performance of DIMs can be evaluated using several criteria; however, we focus here upon their design accuracy, since this is of paramount importance to AEM researchers and designers. Despite the success and rapid growth of DIMs in AEM research, there has been little comparison of different DIMs under controlled settings where competing DIMs are examined on the same problems, and using the same training resources (e.g. quantities of training data). This type of testing – termed benchmarking – is common in the machine learning literature however it has not yet been widely adopted in the AEM community, making it difficult to compare existing DIMs. There have been two recent exceptions however [57, 105], which each compared several DIMs on multiple AEM problems.

Our discussion here relies most heavily upon Ren et al. [104] since they compared a much larger number of DIMs, and did so on more AEM design problems. In particular, Ren et al. [57] compared eight different state-of-the-art DIMs, which includes all of the approaches in Table 11.2, except for the reinforcement learning (RL) and generative adversarial network (GAN). The authors found that many DIMs provide excellent design accuracy; however, iterative methods tended to achieve the best overall accuracy. In particular, the Neural-Adjoint (NA) approach achieved consistently high levels of accuracy, and the best overall accuracy across different settings (e.g. number of solution proposals considered).

Although iterative methods tended to yield the best overall design accuracy, they also suffer from greater computational costs [57]. Computation time is another important factor for AEM researchers and designers. The experiments in Ren et al. [57] also considered the computation time required by each DIM for training and solution inference but found that, while there are significant differences in their average computation times, these differences were relatively small compared to the computational requirements shared by all of the DIMs (e.g. training data collection and the evaluation of proposed solutions).

11.5 Conclusions and Perspectives

Deep learning methods have led to a significant acceleration in both the forward and inverse design of AEMs. In the case of forward design, data-driven models based upon DNNs have been shown capable of accurately approximating the underlying physical relationships between the geometric structure of an AEM, g, and its electromagnetic properties, s. DNNs have exhibited this capability even for complex and poorly understood AEMs where the only alternative evaluation mechanism is CEMS. Once DNNs are trained to approximate these relationships, they have the capability of evaluating the properties AEMs within a small fraction of a second, dramatically accelerating both scientific investigation and forward design of advanced AEMs. Existing mechanisms to evaluate advanced AEMs are slow, severely limiting the complexity of AEMs (e.g. number of free design parameters) that can be effectively manipulated and designed. DNNs dramatically reduce evaluation time, bringing far more complex AEMs within the reach of designers.

Despite these advances, forward design does not scale well with increasing design complexity, and quickly becomes intractable as the number of free design parameters grows (i.e. dimensionality of g). One path to overcome this computational bottleneck is inverse design, where the goal is to *directly* predict the AEM design that would yield any desired AEM scattering properties. Unfortunately however, this approach requires the derivation of an inverse model, which is often much more challenging than learning a forward model. In recent years however a variety of specially designed DNNs, termed deep inverse models, have shown the potential to learn highly accurate inverse models. This approach is especially promising due to its ability to scale well with increasing problem complexity, making it a potential long-term solution for advanced AEM design as complexity continues to grow.

Despite the promise of deep learning for design, DNNs still suffer from several major limitations, greatly undermining their potential benefits. Next we summarize some of the biggest limitations of DNN-based approaches, and some opportunities to address them.

11.5.1 Reducing the Need for Training Data

Perhaps the biggest existing limitation of DNNs is their dependency upon large quantities of training data which, in the context of AEM design, must be collected using (relatively) slow computational simulation. Collecting these simulations is time-consuming, and as discussed in Section 11.3, the quantity of training data needed to train DNNs tends to increase as the complexity of the design problem increases (e.g. dimensionality of g), undermining the scalability of DNN-based design. We next briefly discuss several promising techniques to mitigate this limitation of DNNs.

11.5.1.1 Transfer Learning

The goal of transfer learning is to leverage the similarities across different problems (e.g. AEM design problems) to improve DNN training on one, or all, of the problems. A simple example of transfer learning is to utilize the training data generated from one task to help train a DNN for another different, but related, task. Transfer learning has been widely studied in the machine learning literature [106] and more recently it has been tremendously effective for reducing the quantity of data needed for training DNNs [107], often reducing it by several orders of magnitude. Transfer learning has just begun to be explored for AEM design [108, 109], showing promising results.

11.5.1.2 Active Learning

The premise of active learning is that less training data is necessary if it is chosen more intelligently. Active learning methods then attempt to estimate the most "informative" data to use for training machine learning models for a particular task [110, 111]. In the context of AEM design, active learning approaches can be used to choose which geometric settings, g, to simulate (i.e. to get s), in order to maximize the accuracy of a DNN being trained upon them. In principle, this approach will make it possible to use less training data to achieve the same level of DNN accuracy. Some recent research has begun to explore active learning techniques for AEM modeling [112], with promising results; however, a large number of methods have been developed [111], including for DNNs (e.g. [113]), offering substantial potential for additional exploration.

11.5.1.3 Physics-Informed Learning

The aim of this approach is to leverage prior knowledge about AEMs from physics to increase the effectiveness of DNNs. This can be accomplished by integrating prior physical knowledge into the structure of DNNs, their training procedures, or the collection of training data. Recent work has demonstrated the potential of physics-informed learning in AEM research [41, 114, 115], with impressive results for both the modeling of AEMs and design.

11.5.2 Inverse Modeling for Non-existent Solutions

Inverse design is a promising long-term paradigm for effectively designing more complex AEMs. As discussed, recent DIMs are highly effective at overcoming the non-uniqueness condition that is often present in inverse problems. However, another major challenge of inverse problems is the non-existence of solutions; this is a violation of Hadamard's second criterion for well-posed problems (see Section 11.4.1). Non-existence arises when the inverse model is given some s as input that cannot be realized by any design within the current space of designs being considered, G – this phenomenon is illustrated in Figure 11.6(b). In general, for an arbitrary value of s it is highly unlikely that there will exist some $g \in G$ such that $s = f(g)$, and therefore non-existence is an important problem in inverse modeling.

Despite its importance however, there is little discussion of non-existence in the literature on DIMs. Specifically most existing work on DIMs within the AEM community assumes that the input s value corresponds to some design within G. Some recent work has begun to explore non-existence in limited settings [28, 29, 63, 69, 88] by exploring values of s that are not *guaranteed* to exist, but without also ensuring that they don't exist, or controlling how different they are compared to the properties of realizable design solutions. An important question regarding inverse models is how they perform when a non-existent value of s^* is provided as input. For example, does the inverse model return the most accurate $g \in G$, given s^*, or does the model completely fail and provide pseudo-random design solutions?

Another important question regards the expansion of G so that it is more likely to include designs that can realize s^*. As mentioned in Section 11.5.1, the cost of collecting simulations over some input space G for training DNNs is one of the biggest limitations of DNN-based methods. When it is discovered that the designs present in G cannot yield the target scattering properties, this simulation process must be repeated again over some new, or extended, input space – an extremely costly outcome. Furthermore, the potential of this outcome makes AEM design with DNNs much more risky, because the total time and cost required to find a satisfying design becomes highly uncertain. This problem was recently termed "meta-inverse design" [31] and it is an important open problem in AEM design.

11.5.3 Benchmarking, Replication, and Sharing Resources

There has been tremendous growth in the use of DNN-based methods within the AEM community [31], yielding a wide variety of methods. Despite this proliferation of methodologies it is difficult to determine how these existing methods compare in performance (e.g. modeling accuracy, computation time) or whether any particular models tend to perform best. These difficulties arise because there is relatively little work comparing DNN-based methods under controlled experimental conditions – a practice often called "benchmarking" in the machine learning literature, where it is widely used. For DNN models, benchmarking typically involves comparing proposed methods against multiple existing state-of-the-art methods under controlled experimental conditions, whereby the methods are compared on the same (usually public) datasets, and all methods are provided with similar resources (e.g. quantities of training data, model sizes, or other factors). Benchmarking makes it possible to understand the relative advantages and disadvantages of existing methods, and generally discern whether methodological progress is being made over time.

Benchmarking is not yet widely adopted within the AEM community; however, it will be crucial to advance DNN research within the AEM community. As part of this effort, it will be important for the AEM community to cultivate practices of sharing software, datasets, and the trained DNN models generated as part of their research. Some benchmarking has been done very recently in the AEM community. The work by Deng et al. [56] benchmarked DNN-based methods for forward modeling on three different AEM problems, and published the datasets and models. The work by Ren et al. [57] and Ma et al. [105] compared multiple DIMs on several inverse problems.

Acknowledgments

We acknowledge funding from the Department of Energy under U.S. Department of Energy (DOE) (DESC0014372).

References

1 Vanbésien, O. (2012). *Artificial Materials*. London, Hoboken, NJ: ISTE Wiley. ISBN 978-1-84821-335-7.

2 Brener, I. (2020). *Dielectric Metamaterials: Fundamentals, Designs and Applications*. Duxford: Woodhead Publishing. ISBN 978-0-08-102403-4.

3 Choy, T. (2016). *Effective Medium Theory: Principles and Applications*. Oxford: Oxford University Press. ISBN 9780198705093.

4 Brillouin, L. (1953). *Wave Propagation in Periodic Structures; Electric Filters and Crystal Lattices*. New York: Dover Publications. ISBN 978-0486600345.

5 Collin, R. (1991). *Field Theory of Guided Waves*. New York: IEEE Press. ISBN 9780879422370.

6 Floquet, G. (1883). On linear differential equations with periodic coefficients. *Annales scientifiques de l'École Normale Supérieure, Serie 2* 12: 47–88. https://doi.org/10.24033/asens.220.

7 Bloch, F. (1929). Über die quantenmechanik der elektronen in kristallgittern. *Zeitschrift für Physik* 52 (7–8): 555–600. https://doi.org/10.1007/bf01339455.

8 Soukoulis, C.M. (1993). *Photonic Band Gaps and Localization*. New York: Springer Science+Business Media. ISBN 978-1-4899-1608-2.

9 Sakoda, K. (2005). *Optical Properties of Photonic Crystals*. Berlin, New York: Springer. ISBN 3-540-20682-5.

10 Yang, F. (2009). *Electromagnetic Band Gap Structures in Antenna Engineering*. Cambridge, New York: Cambridge University Press. ISBN 978-0-521-88991-9.

11 Joannopoulos, J.D. (2008). *Photonic Crystals: Molding the Flow of Light*. Princeton, NJ: Princeton University Press. ISBN 978-0691124568.

12 Munk, B.A. (2000). *Frequency Selective Surfaces*. Nashville, TN: Wiley.

13 Wang, D.S., Qu, S.-W., and Chan, C.H. (2016). Frequency selective surfaces. In: *Handbook of Antenna Technologies* (ed. Z.N. Chen, D. Liu, H. Nakano et al.), 471–525. Singapore: Springer.

14 Capolino, F. (2009). *Theory and Phenomena of Metamaterials*. Boca Raton, FL: CRC Press/Taylor & Francis. ISBN 978-1420054255.

15 Simovski, C. (2020). *An Introduction to Metamaterials and Nanophotonics*. New York: Cambridge University Press. ISBN 978-1108492645.

16 Engheta, N. (2006). *Metamaterials: Physics and Engineering Explorations*. Hoboken, NJ: Wiley-Interscience. ISBN 978-0471761020.

17 Wiltshire, M.C.K. (2001). Microstructured magnetic materials for RF flux guides in magnetic resonance imaging. *Science* 291 (5505): 849–851. https://doi.org/10.1126/science.291.5505.849.

18 Smith, D.R., Padilla, W.J., Vier, D.C. et al. (2000). Composite medium with simultaneously negative permeability and permittivity. *Physical Review Letters* 84 (18): 4184–4187. https://doi.org/10.1103/physrevlett.84.4184.

19 Shelby, R.A. (2001). Experimental verification of a negative index of refraction. *Science* 292 (5514): 77–79. https://doi.org/10.1126/science.1058847.

20 Yen, T.J. (2004). Terahertz magnetic response from artificial materials. *Science* 303 (5663): 1494–1496. https://doi.org/10.1126/science.1094025.

21 Linden, S. (2004). Magnetic response of metamaterials at 100 terahertz. *Science* 306 (5700): 1351–1353. https://doi.org/10.1126/science.1105371.

22 Enkrich, C., Wegener, M., Linden, S. et al. (2005). Magnetic metamaterials at telecommunication and visible frequencies. *Physical Review Letters* 95 (20). https://doi.org/10.1103/physrevlett.95.203901.

23 Fang, N. (2005). Sub-diffraction-limited optical imaging with a silver superlens. *Science* 308 (5721): 534–537. https://doi.org/10.1126/science.1108759.

24 Stratton, J. (2007). *Electromagnetic Theory*. Piscataway, NJ: IEEE Press. ISBN 978-0470131534.

25 Mo, T.C., Papas, C.H., and Baum, C.E. (1973). General scaling method for electromagnetic fields with application to a matching problem. *Journal of Mathematical Physics* 14 (4): 479–483. https://doi.org/10.1063/1.1666341.

26 Ming, X., Liu, X., Sun, L., and Padilla, W.J. (2017). Degenerate critical coupling in all-dielectric metasurface absorbers. *Optics Express* 25 (20): 24658. https://doi.org/10.1364/oe.25.024658.

27 Hergert, W and Wriedt, T (eds.) (2012). *The Mie Theory, Springer Series in Optical Sciences*, 2012e. Berlin, Germany: Springer-Verlag.

28 Deng, Y., Ren, S., Fan, K. et al. (2021). Neural-adjoint method for the inverse design of all-dielectric metasurfaces. *Optics Express* 29 (5): 7526. https://doi.org/10.1364/oe.419138.

29 So, S. and Rho, J. (2019). Designing nanophotonic structures using conditional deep convolutional generative adversarial networks. *Nanophotonics* 8 (7): 1255–1261. https://doi.org/10.1515/nanoph-2019-0117.

30 Herskovits, J. (1995). *Advances in Structural Optimization*. Netherlands, Dordrecht: Springer. ISBN 978-94-010-4203-1.

31 Khatib, O., Ren, S., Malof, J., and Padilla, W.J. (2021). Deep learning the electromagnetic properties of metamaterials-a comprehensive review. *Advanced Functional Materials* 31 (31): 2101748.

32 Huang, L., Xu, L., and Miroshnichenko, A.E. (2020). Deep learning enabled nanophotonics. In: *Advances and Applications in Deep Learning*, 65. IntechOpen. https://doi.org/10.5772/intechopen.93289.

33 Wiecha, P.R., Arbouet, A., Girard, C., and Muskens, O.L. (2021). Deep learning in nano-photonics: inverse design and beyond. *Photonics Research* 9 (5): B182–B200.

34 Jiang, J., Chen, M., and Fan, J.A. (2021). Deep neural networks for the evaluation and design of photonic devices. *Nature Reviews Materials* 6: 679–700.

35 Ma, W., Liu, Z., Kudyshev, Z.A. et al. (2021). Deep learning for the design of photonic structures. *Nature Photonics* 15 (2): 77–90.

36 Nelder, J.A. and Mead, R. (1965). A simplex method for function minimization. *The Computer Journal* 7 (4): 308–313.

37 Frazier, P.I. (2018). A tutorial on Bayesian optimization. *arXiv preprint arXiv:1807.02811*.

38 Bishop, C.M. (2006). *Pattern Recognition and Machine Learning*. Springer.

39 Nadell, C.C., Huang, B., Malof, J.M., and Padilla, W.J. (2019). Deep learning for accelerated all-dielectric metasurface design. *Optics Express* 27 (20): 27523. https://doi.org/10.1364/oe.27.027523.

40 Moon, G., Choi, J., Lee, C. et al. (2020). Machine learning-based design of meta-plasmonic biosensors with negative index metamaterials. *Biosensors and Bioelectronics* 164. https://doi.org/10.1016/j.bios.2020.112335.

41 Chen, Y., Lu, L., Karniadakis, G.E., and Dal Negro, L. (2020). Physics-informed neural networks for inverse problems in nano-optics and metamaterials. *Optics Express* 28 (8): 11618. https://doi.org/10.1364/oe.384875.

42 Tao, Z., You, J., Zhang, J. et al. (2020). Optical circular dichroism engineering in chiral metamaterials utilizing a deep learning network. *Optics Letters* 45 (6): 1403. https://doi.org/10.1364/ol.386980.

43 Jiang, J. and Fan, J.A. (2019). Global optimization of dielectric metasurfaces using a physics-driven neural network. *Nano Letters* 19 (8): 5366–5372. https://doi.org/10.1021/acs.nanolett.9b01857.

44 Chugh, S., Gulistan, A., Ghosh, S., and Rahman, B.M.A. (2019). Machine learning approach for computing optical properties of a photonic crystal fiber. *Optics Express* 27 (25): 36414–36425.

45 Singh, R., Agarwal, A., and Anthony, B.W. (2020). Mapping the design space of photonic topological states via deep learning. *Optics Express* 28 (19): 27893–27902. https://doi.org/10.1364/oe.398926.

46 Long, Y., Ren, J., Li, Y., and Chen, H. (2019). Inverse design of photonic topological state via machine learning. *Applied Physics Letters* 114 (18). https://doi.org/10.1063/1.5094838.

47 da Silva, M.R., Nóbrega, C.L., Silva, P.H.F., and D'Assunç ao, A.G. (2014). Optimization of FSS with Sierpinski island fractal elements using population-based search algorithms and MLP neural network. *Microwave and Optical Technology Letters* 56 (4): 827–831.

48 Nakmouche, M.F., Allam, A.M., Fawzy, D.E., and Lin, D.-B. (2021). Development of a high gain FSS reflector backed monopole antenna using machine learning for 5G applications. *Progress In Electromagnetics Research M* 105: 183–194.

49 Chan, C.H., Hwang, J.N., and Davis, D.T. (1992). Multilayered frequency selective surface design using artificial neural networks. *IEEE Antennas and Propagation Society International Symposium 1992 Digest*, pp. 1400–vol. IEEE.

50 Malkiel, I., Mrejen, M., Nagler, A. et al. (2018). Plasmonic nanostructure design and characterization via deep learning. *Light: Science and Applications* 7 (1). https://doi.org/10.1038/s41377-018-0060-7.

51 Baxter, J., Calà Lesina, A., Guay, J.M. et al. (2019). Plasmonic colours predicted by deep learning. *Scientific Reports* 9 (1): 1–9. https://doi.org/10.1038/s41598-019-44522-7.

52 He, J., He, C., Zheng, C. et al. (2019). Plasmonic nanoparticle simulations and inverse design using machine learning. *Nanoscale* 11 (37): 17444–17459. https://doi.org/10.1039/c9nr03450a.

53 da Silva Ferreira, A., da Silva Santos, C.H., Gonçalves, M.S., and Hernández Figueroa, H.E. (2018). Towards an integrated evolutionary strategy and artificial neural network computational tool for designing photonic coupler devices. *Applied Soft Computing* 65: 1–11.

54 Hornik, K., Stinchcombe, M., and White, H. (1989). Multilayer feedforward networks are universal approximators. *Neural Networks* 2 (5): 359–366.

55 LeCun, Y., Boser, B., Denker, J.S. et al. (1989). Backpropagation applied to handwritten zip code recognition. *Neural Computation* 1 (4): 541–551.

56 Deng, Y., Dong, J., Ren, S. et al. (2021). Benchmarking data-driven surrogate simulators for artificial electromagnetic materials. *35th Conference on Neural Information Processing Systems Datasets and Benchmarks Track (Round 2).*

57 Ren, S., Mahendra, A., Khatib, O. et al. (2022). Inverse deep learning methods and benchmarks for artificial electromagnetic material design. *Nanoscale* 14: 3958–3969.

58 Tarantola, A. (2005). *Inverse Problem Theory and Methods for Model Parameter Estimation.* Philadelphia, PA: Society for Industrial and Applied Mathematics. ISBN 0-89871-572-5.

59 Nakamura, G. and Potthast, R. (2015). *Inverse modeling.* IOP Publishing.

60 Ren, S., Padilla, W.J., and Malof, J.M. (2020). Benchmarking deep inverse models over time, and the neural-adjoint method. In: *Advances in Neural Information Processing Systems 33: Annual Conference on Neural Information Processing Systems 2020, NeurIPS 2020 (6–12 December 2020), Virtual* (ed. H. Larochelle, M.A. Ranzato, R. Hadsell et al.). https://proceedings.neurips.cc/paper/2020/hash/007ff380ee5ac49ffc34442f5c2a2b86-Abstract.html.

61 Chen, Y., Zhu, J., Xie, Y. et al. (2019). Smart inverse design of graphene-based photonic metamaterials by an adaptive artificial neural network. *Nanoscale* 11 (19): 9749–9755. https://doi.org/10.1039/c9nr01315f.

62 Tahersima, M.H., Kojima, K., Koike-Akino, T. et al. (2019). Deep neural network inverse design of integrated photonic power splitters. *Scientific Reports* 9 (1): 1–9. https://doi.org/10.1038/s41598-018-37952-2.

63 Zhang, T., Wang, J., Liu, Q. et al. (2019). Efficient spectrum prediction and inverse design for plasmonic waveguide systems based on artificial neural networks. *Photonics Research* 7 (3): 368. https://doi.org/10.1364/prj.7.000368.

64 Akashi, N., Toma, M., and Kajikawa, K. (2020). Design by neural network of concentric multilayered cylindrical metamaterials. *Applied Physics Express* 13 (4): 4. https://doi.org/10.35848/1882-0786/ab7cf1.

65 Liu, D., Tan, Y., Khoram, E., and Yu, Z. (2018). Training deep neural networks for the inverse design of nanophotonic structures. *ACS Photonics* 5 (4): 1365–1369. https://doi.org/10.1021/acsphotonics.7b01377.

66 Ma, W., Cheng, F., and Liu, Y. (2018). Deep-learning-enabled on-demand design of chiral metamaterials. *ACS Nano* 12 (6): 6326–6334. https://doi.org/10.1021/acsnano.8b03569.

67 Gao, L., Li, X., Liu, D. et al. (2019). A bidirectional deep neural network for accurate silicon color design. *Advanced Materials* 31 (51): 1905467. https://doi.org/10.1002/adma.201905467.

68 Hou, Z., Tang, T., Shen, J. et al. (2020). Prediction network of metamaterial with split ring resonator based on deep learning. *Nanoscale Research Letters* 15 (1). https://doi.org/10.1186/s11671-020-03319-8.

69 So, S., Mun, J., and Rho, J. (2019). Simultaneous inverse design of materials and structures via deep learning: demonstration of dipole resonance engineering using core-shell nanoparticles. *ACS Applied Materials and Interfaces* 11 (27): 24264–24268. https://doi.org/10.1021/acsami.9b05857.

70 Xu, L., Rahmani, M., Ma, Y. et al. (2020). Enhanced light-matter interactions in dielectric nanostructures via machine-learning approach. *Advanced Photonics* 2 (02). https://doi.org/10.1117/1.ap.2.2.026003.

71 Ashalley, E., Acheampong, K., Vázquez, L. et al. (2020). Multitask deep-learning-based design of chiral plasmonic metamaterials. *Photonics Research* 5. https://doi.org/10.1364/prj.388253.

72 Mall, A., Patil, A., Tamboli, D. et al. (2020). Fast design of plasmonic metasurfaces enabled by deep learning. *Journal of Physics D: Applied Physics* 53 (49): 49LT01. https://doi.org/10.1088/1361-6463/abb33c.

73 Pilozzi, L., Farrelly, F.A., Marcucci, G., and Conti, C. (2018). Machine learning inverse problem for topological photonics. *Communications Physics* 1 (1): 1–7. https://doi.org/10.1038/s42005-018-0058-8.

74 Phan, A.D., Nguyen, C.V., Linh, P.T. et al. (2020). Deep learning for the inverse design of mid-infrared graphene plasmons. *Crystals* 10 (2): 125. https://doi.org/10.3390/cryst10020125.

75 Ma, W., Cheng, F., Xu, Y. et al. (2019). Probabilistic representation and inverse design of metamaterials based on a deep generative model with semi-supervised learning strategy. *Advanced Materials* 31 (35): 1901111. https://doi.org/10.1002/adma.201901111.

76 Ma, W. and Liu, Y. (2020). A data-efficient self-supervised deep learning model for design and characterization of nanophotonic structures. *Science China: Physics, Mechanics and Astronomy* 63 (8). https://doi.org/10.1007/s11433-020-1575-2.

77 Kudyshev, Z.A., Kildishev, A.V., Shalaev, V.M., and Boltasseva, A. (2020). Machine-learning-assisted metasurface design for high-efficiency thermal emitter optimization. *Applied Physics Reviews* 7 (2). https://doi.org/10.1063/1.5134792.

78 Kudyshev, Z.A., Kildishev, A.V., Shalaev, V.M., and Boltasseva, A. (2020). Machine learning–assisted global optimization of photonic devices. *Nanophotonics* 1: 371–383.

79 Shi, X., Qiu, T., Wang, J. et al. (2020). Metasurface inverse design using machine learning approaches. *Journal of Physics D: Applied Physics* 53 (27): 275105.

80 Liu, Z., Raju, L., Zhu, D., and Cai, W. (2020). A hybrid strategy for the discovery and design of photonic structures. *IEEE Journal on Emerging and Selected Topics in Circuits and Systems* 10 (1): 126–135. https://doi.org/10.1109/JETCAS.2020.2970080.

81 Kiarashinejad, Y., Abdollahramezani, S., and Adibi, A. (2020). Deep learning approach based on dimensionality reduction for designing electromagnetic nanostructures. *npj Computational Materials* 6 (1): 1–12. https://doi.org/10.1038/s41524-020-0276-y.

82 Qiu, T., Shi, X., Wang, J. et al. (2019). Deep learning: a rapid and efficient route to automatic metasurface design. *Advanced Science* 6 (12). https://doi.org/10.1002/advs.201900128.

83 Unni, R., Yao, K., Han, X. et al. (2021). A mixture-density-based tandem optimization network for on-demand inverse design of thin-film high reflectors. *Nanophotonics* 10 (16). https://doi.org/10.1515/nanoph-2021-0392.

84 Unni, R., Yao, K., and Zheng, Y. (2020). Deep convolutional mixture density network for inverse design of layered photonic structures. *ACS Photonics* 7 (10): 2703–2712.

85 Jiang, J., Sell, D., Hoyer, S. et al. (2019). Free-form diffractive metagrating design based on generative adversarial networks. *ACS Nano* 13 (8): 8872–8878. https://doi.org/10.1021/acsnano.9b02371.

86 Christensen, T., Loh, C., Picek, S. et al. (2020). Predictive and generative machine learning models for photonic crystals. *Nanophotonics* 9 (13): 4183–4192. https://doi.org/10.1515/nanoph-2020-0197.

87 An, S., Zheng, B., Tang, H. et al. (2019). Generative multi-functional meta-atom and metasurface design networks. *arXiv preprint arXiv:1908.04851*.

88 Liu, Z., Zhu, D., Rodrigues, S.P. et al. (2018). Generative model for the inverse design of metasurfaces. *Nano Letters* 18 (10): 6570–6576. https://doi.org/10.1021/acs.nanolett.8b03171.

89 Zhang, T., Liu, Q., Dan, Y. et al. (2020). Machine learning and evolutionary algorithm studies of graphene metamaterials for optimized plasmon-induced transparency. *Optics Express* 28 (13): 18899. https://doi.org/10.1364/oe.389231.

90 Johnson, J.M. and Rahmat-Samii, V. (1997). Genetic algorithms in engineering electromagnetics. *IEEE Antennas and Propagation Magazine* 39 (4): 7–21. https://doi.org/10.1109/74.632992.

91 Forestiere, C., Pasquale, A.J., Capretti, A. et al. (2012). Genetically engineered plasmonic nanoarrays. *Nano Letters* 12 (4): 2037–2044. https://doi.org/10.1021/nl300140g.

92 Li, C.-J., Fang, Y.-C., and Cheng, M.-C. (2009). Study of optimization of an LCD light guide plate with neural network and genetic algorithm. *Optics Express* 17 (12): 10177–10188.

93 Wang, H., Zheng, Z., Ji, C., and Guo, L.J. (2020). Automated multi-layer optical design via deep reinforcement learning. *Machine Learning: Science and Technology* 2 (2): 025013.

94 Badloe, T., Kim, I., and Rho, J. (2020). Biomimetic ultra-broadband perfect absorbers optimised with reinforcement learning. *Physical Chemistry Chemical Physics* 22 (4): 2337–2342. https://doi.org/10.1039/c9cp05621a.

95 Sajedian, I., Badloe, T., and Rho, J. (2019). Optimisation of colour generation from dielectric nanostructures using reinforcement learning. *Optics Express* 27 (4): 5874. https://doi.org/10.1364/oe.27.005874.

96 Huang, Z., Liu, X., and Zang, J. (2019). The inverse design of structural color using machine learning. *Nanoscale* 11 (45): 21748–21758. https://doi.org/10.1039/c9nr06127d.

97 Peurifoy, J., Shen, Y., Jing, L. et al. (2018). Nanophotonic particle simulation and inverse design using artificial neural networks. *Science Advances* 4 (6): 1–8. https://doi.org/10.1126/sciadv.aar4206.

98 Asano, T. and Noda, S. (2018). Optimization of photonic crystal nanocavities based on deep learning. *Optics Express* 26 (25): 32704. https://doi.org/10.1364/oe.26.032704.

99 Miyatake, Y., Sekine, N., Toprasertpong, K. et al. (2020). Computational design of efficient grating couplers using artificial intelligence. *Japanese Journal of Applied Physics* 59 (SG): SGGE09.

100 Ma, W., Liu, Z., Kudyshev, Z.A. et al. (2020). Deep learning for the design of photonic structures. *Nature Photonics* 1–14. https://doi.org/10.1038/s41566-020-0685-y.

101 So, S., Badloe, T., Noh, J. et al. (2020). Deep learning enabled inverse design in nanophotonics. *Nanophotonics* 9 (5): 1041–1057. https://doi.org/10.1515/nanoph-2019-0474.

102 Mueller, J. (2012). *Linear and Nonlinear Inverse Problems with Practical Applications*. Philadelphia, PA: Society for Industrial and Applied Mathematics. ISBN 978-1-611972-33-7.

103 Hadamard, J. (1902). Sur les problèmes aux dérivées partielles et leur signification physique (on the problems with the derivative partial and their physical significance). *Princeton University Bulletin* 13: 49–52.

104 Ren, S., Padilla, W.J., and Malof, J.M. (2020). Benchmarking deep inverse models over time, and the neural-adjoint method. In: *Advances in Neural Information Processing Systems 33: Annual Conference on Neural Information Processing Systems 2020, NeurIPS 2020, December 6-12, 2020, virtual* (ed. H. Larochelle, M.A. Ranzato, R. Hadsell et al.). https://proceedings.neurips.cc/paper/2020/hash/007ff380ee5ac49ffc34442f5c2a2b86-Abstract.html.

105 Ma, T., Tobah, M., Wang, H., and Guo, L.J. (2022). Benchmarking deep learning-based models on nanophotonic inverse design problems. *Opto-Electronic Science* 1 (1): 210012–1.

106 Pan, S.J. and Yang, Q. (2009). A survey on transfer learning. *IEEE Transactions on Knowledge and Data Engineering* 22 (10): 1345–1359.

107 Tan, C., Sun, F., Kong, T. et al. (2018). A survey on deep transfer learning. In: *Artificial Neural Networks and Machine Learning. International Conference on Artificial Neural Networks, Lecture Notes in Computer Science*, vol. 11141 (ed. V. Køurková, Y. Manolopoulos, B. Hammer et al.), 270–279. Springer.

108 Qu, Y., Jing, L., Shen, Y. et al. (2019). Migrating knowledge between physical scenarios based on artificial neural networks. *ACS Photonics* 6 (5): 1168–1174. https://doi.org/10.1021/acsphotonics.8b01526.

109 Zhu, R., Qiu, T., Wang, J. et al. (2021). Phase-to-pattern inverse design paradigm for fast realization of functional metasurfaces via transfer learning. *Nature Communications* 12 (1): 1–10.

110 Cohn, D.A., Ghahramani, Z., and Jordan, M.I. (1996). Active learning with statistical models. *Journal of Artificial Intelligence Research* 4: 129–145.

111 Settles, B. (2012). Active learning. *Synthesis Lectures on Artificial Intelligence and Machine Learning* 6 (1): 1–114.

112 Pestourie, R., Mroueh, Y., Nguyen, T.V. et al. (2020). Active learning of deep surrogates for PDEs: application to metasurface design. *npj Computational Materials* 6 (1): 1–7.

113 Gal, Y., Islam, R., and Ghahramani, Z. (2017). Deep Bayesian active learning with image data. *International Conference on Machine Learning*, pp. 1183–1192. PMLR.

114 Fang, Z. and Zhan, J. (2019). Deep physical informed neural networks for metamaterial design. *IEEE Access* 8: 24506–24513. https://doi.org/10.1109/access.2019.2963375.

115 Raissi, M., Perdikaris, P., and Karniadakis, G.E. (2019). Physics-informed neural networks: a deep learning framework for solving forward and inverse problems involving nonlinear partial differential equations. *Journal of Computational Physics* 378: 686–707. https://doi.org/10.1016/j.jcp.2018.10.045.

12

Machine Learning-Assisted Optimization and Its Application to Antenna and Array Designs

Qi Wu[1,2], Haiming Wang[1,2], and Wei Hong[1,2]

[1] *State Key Laboratory of Millimeter Waves, School of Information Science and Engineering, Southeast University, Nanjing, Jiangsu Province, China*
[2] *Department of New Communications, Purple Mountain Laboratories, Nanjing, Jiangsu Province, China*

12.1 Introduction

With the explosive development of modern wireless communications and radars, the antennas, antenna arrays, and their associated systems are also experiencing evolutionary progression [1–3]. Traditionally, antenna and antenna array are designed based on knowledge-based analytical methodologies. Antenna and array theories provide insightful understanding of electromagnetic (EM) interactions within antenna structures and between array elements, including their radiation, reflection, and mutual coupling (MC) behaviors. Design guidelines, formulas, and equivalent circuits are then generated to provide designers with computationally inexpensive analytical models, therefore helps to build antenna and array prototypes with calculated dimensions, allocations, and excitations. While lots of efforts have been made to build relatively accurate analytical models for various antenna and array designs, in most cases, these knowledge-based methodologies can only provide an approximate estimation of the optimum designs. Moreover, when dealing with modern communication and radar systems with high requirements of package density, wind load and cooling performance along with radiation performance, the much complex EM environment brings more limitations on the applications of traditional knowledge-based methods.

In most cases, numerical-technique-based full-wave EM simulations are some of the only options in both validation and optimization procedures of antenna and array designs, due to their high evaluation accuracy. While the development of computers has extensively contributed to these processes, and has made the antenna and array models comprising millions of mesh cells evaluated in a reasonable time, these processes still demand significant computational cost. When applying EM simulations to solve tasks involving a large number of analyses, including optimization, sensitivity analysis or robust design, the computational burden are becoming unbearable. In these cases, global optimization algorithms, such as metaheuristic algorithms, are often necessary, due to the inherent characteristics of multi-modal with multiple local optima antenna and antenna array design problems. These algorithms typically require hundreds, thousands, or even tens of thousands evaluation times to find a solution. One of the most popular and successful applications of the metaheuristic algorithms in EM designs is the array factor optimization. However, when directly applying these algorithms

Advances in Electromagnetics Empowered by Artificial Intelligence and Deep Learning, First Edition.
Edited by Sawyer D. Campbell and Douglas H. Werner.

with practical EM-simulation-driven designs of either antenna element, feeding network, and antenna array with complex EM environment, such as installation fixture, connector, and many other active and passive system components, the corresponding computational costs would be tremendous.

One of the most promising techniques to solve this problem is to build computational cheap surrogate models to replace computational expensive full-wave EM simulations, which can be sorted as data-driven approaches. These methods use sampled training data (which are acquired using full-wave EM simulations mostly) to build explicit analytical formulas that are computationally cheap to evaluate. Various machine-learning (ML) methods, including Gaussian process regression (GPR) [4, 5], support vector regression (SVR), [6] and artificial neural network (ANN) [7], have been introduced to build surrogate models for many different design objectives in antenna and antenna array designs. By integrating ML methods into classical or newly proposed optimization algorithms, various kinds of machine learning-assisted optimization (MLAO) methods have been proposed to solve problems in antenna and antenna array designs including but not limited to single-objective antenna performance optimization, multiple-objective antenna performance optimization, sensitivity and robust antenna design, and antenna array pattern synthesis under MC and platform effects. The central topic of this chapter is the MLAO algorithm and its application in antenna and antenna array designs. In Section 12.2, the fundamental framework of MLAO algorithm is introduced. The applications of MLAO for antenna and array design are introduced in Section 12.3. Finally, Section 12.4 provides a conclusive discussion on MLAO for antenna and antenna array designs.

12.2 Machine Learning-Assisted Optimization Framework

This section first formulates the optimization task of antenna and antenna array, then explains the fundamental framework of MLAO algorithm. General antenna and antenna array design problems can be seen as multi-objective optimization problems, which can be formulated as:

$$\begin{aligned} &\underset{\mathbf{x}}{\text{Minimize}} : \mathbf{u}(\mathbf{x}) = [\mathbf{u}_1(\mathbf{x}), \mathbf{u}_2(\mathbf{x}), \dots, \mathbf{u}_k(\mathbf{x})]^T \\ &\text{subject to} : g_j(\mathbf{x}) \le 0; \quad j = 1,2,\dots,m \end{aligned} \tag{12.1}$$

where k is the number of objective functions, m is the number of inequality constraints, $\mathbf{x} \in E^n$ is a vector of design variables, and $\mathbf{f}(\mathbf{x}) \in E^k$ is a vector of objective functions $\mathbf{u}_i(\mathbf{x}) : E^n \to E^1$. The design space is defined as $\mathbf{X} = \mathbf{x}|\text{subject to} : g_j(\mathbf{x}) \le 0; j = 1,2,\dots,m$, and the criterion space is defined as $\mathbf{U} = \mathbf{u}(\mathbf{x})|\mathbf{x} \in \mathbf{X}$. In antenna and antenna array design, $\mathbf{x}$ is the vector of any designable parameters of the system, including but not limited to antenna geometries, element excitations and allocations (relative and absolute), platform geometries. The $\mathbf{u}(\mathbf{x}) = F(\mathbf{x}, \mathbf{y})$ is a scalar merit function that quantifies the quality of the design $\mathbf{x}$ for a given objective vector $\mathbf{y} = [y_1, \dots, y_k]^T$. These design objectives could be figures of interest, such as the reflection coefficient, gain, directivity, radiation patterns at concerned angles, etc. The function F is defined so that smaller values of $F(\mathbf{x}, \mathbf{y})$ correspond to better designs. Methods such as minimax formulation or L-square formulation are typically used for situations that target response is unknown or known, respectively. Because there may not exist a single point that minimizes all objectives simultaneously for (12.1), the idea of Pareto optimality is used to describe solutions for multi-objective optimization problems. A design point is defined as Pareto optimal if it is not possible to move from that point

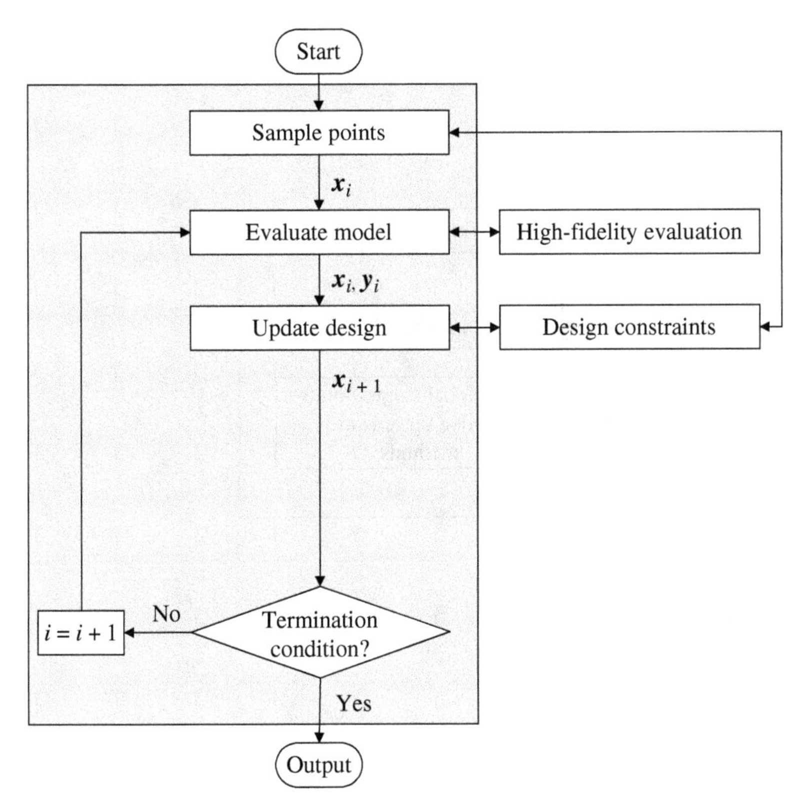

Figure 12.1 Traditional simulation-driven optimization flowchart.

and improve at least one objective function without detriment to any other objective functions. Another approach is to introduce a weighted sum of the merit functions and the constraints to derive a penalty function, which transforms the multi-objective optimization to the single objective optimization.

High-fidelity evaluations in antenna and antenna array design, such as full-wave EM simulations, are inherently computationally expensive. To solve the above-mentioned multi-/single-objective optimization problems for antenna and antenna array design, a massive amount of such evaluations are required by either global or local optimization methods, which lead to unaffordable computational burden. The generic flowchart of these direct simulation-driven optimizations is shown in Figure 12.1. The design points generated by both sampled and updated design points are evaluated through high-fidelity evaluations. The objective function is then calculated based on given requirements and constraints. By applying different optimization algorithms, the designs are updated based on the evaluation results to search for better design performance. Given different optimization algorithms, the update is implemented based on model responses and/or their derivatives. In practical cases, the entire optimization process could be very time-consuming.

The MLAO is one of the most promising approaches to achieve computational burden reduction [8]. The generic flowchart of MLAO is given in Figure 12.2. As a kind of surrogate model-based optimization method, MLAO manages to use data-driven methods aside from conventional analytical or equivalent circuit methods to build surrogate models $\mathbf{R}(\mathbf{x})$. Therefore, the models

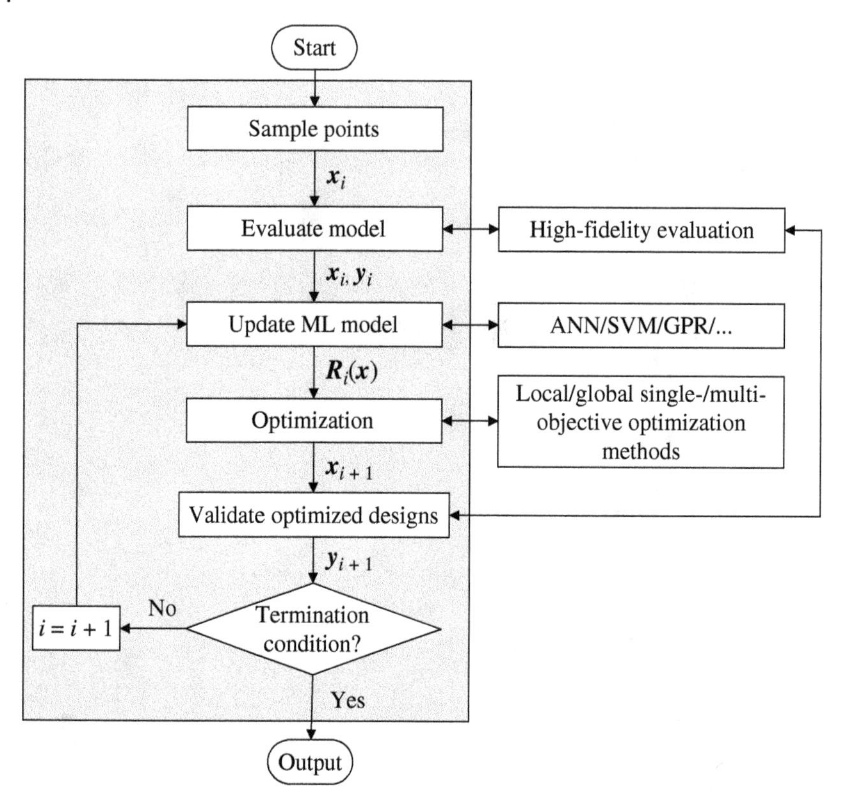

Figure 12.2 MLAO flowchart.

constructed using MLAO need no problem-specific knowledge and are easy to apply to various kinds of problems. Moreover, they are very cheap to evaluate, which is very important when applying optimization algorithms, especially those that need massive amounts of function evaluation calls. Various kinds of ML methods, such as ANN, support vector machine (SVM) and GPR, have been introduced to build surrogate models for different applications. As one of the most famous ML methods, ANNs have been introduced to the EM engineering in 1990s, and have found applications in radar, antenna and array design, circuit design, remote sensing, and many other fields. ANNs are more popular for application scenes with large data set. Compared with ANNs, the SVM owns the merit of improved generalization capability and is more suitable in cases where the size of the training sets is limited. Moreover, SVM requires fewer training patterns to obtain accurate results, which accelerates the training procedure. In contrast to the other two ML methods, GPR is able to provide not only the prediction but also the uncertainty of the new design point, which helps designers to explore global optima when few training points are given. By replacing original computational expensive high-fidelity evaluation with a fast yet reasonably accurate surrogate model built by ML in the optimization procedure, an approximate solution sequence $\mathbf{x}_i, i = 0,1, \ldots$ is obtained during an iterative process. These solutions are then validated using high-fidelity evaluations to see if terminal conditions are fulfilled. The surrogate model $\mathbf{R}(\mathbf{x})$ is updated at each iteration using obtained high-fidelity responses. The most important advantage of MLAO over conventional optimization methods is computational savings, due to the relatively much faster evaluation time of ML-assisted surrogate models compared with high-fidelity evaluations, and the substantially smaller number of iterations required by MLAO than that of conventional optimization methods. In Section 12.3, several most concerned perspectives in MLAO-based antenna and array design problems are addressed in detail.

12.3 Machine Learning-Assisted Optimization for Antenna and Array Designs

The most popular design case for antenna and array designs is to optimize antenna and array geometries to achieve better predefined performance including but not limited to gain, directivity, sidelobe level, reflection coefficient, physical size, and shaped radiation pattern. Based on the classic MLAO scheme, a great amount of researches have been done to further accelerate the optimization process in practical design cases with large number of variables, large parameter variable ranges, high computational cost of high-fidelity evaluations, requirement for robust designs and simultaneously achievement of multiple design goals. These methods emphasize one or some procedures of MLAO, and are described in Sections 12.3.1–12.3.5 with several typical examples shown. Three important techniques including design space reduction, variable-fidelity evaluations, and hybrid optimization methods are discussed. Moreover, two important applications of MLAO-based methods including robust design and antenna array synthesis are discussed.

12.3.1 Design Space Reduction

The number of training data points necessary to construct a reliable ML model grows rapidly with both design space dimensionality and parameter ranges for all kinds of ML methods. Unfortunately, practical antenna design and optimization tasks normally have design parameters >10 and large parameter ranges. Larger number of training data points need more times of high-fidelity evaluations. Moreover, the computational cost of both the training and prediction procedure will grow rapidly. Recently, many design space reduction methods have been proposed to alleviate this dilemma, based on the assumptions that a majority of design space domains are unhelpful for performance figures concerned. And computational effort could be saved if the parameter space regions that contain good designs can be identified before or within the MLAO process.

In [9], a constrained modeling method has been proposed by restricting the model domain to a vicinity of a manifold spanned by a set of reference designs. The method has been generalized for an arbitrary number of performance figures and an arbitrary allocation of the based designs in [10]. In [11], as shown in Figure 12.3, a performance-based nested surrogate modeling method is proposed, in which two levels of surrogates both realized using kriging are constructed. The method is demonstrated using a dual-band uniplanar dipole antenna and a ring slot antenna, in which accurate models are constructed for a wide range of operating conditions and geometry parameters using small training sets.

12.3.2 Variable-Fidelity Evaluation

The high cost of high-fidelity training data acquisition comes from using full-wave EM simulations with sufficient mesh cells in most MLAO design cases. By exploring the connections between variable-fidelity EM simulations, low-fidelity (LF) but time-saving data points of sufficient number

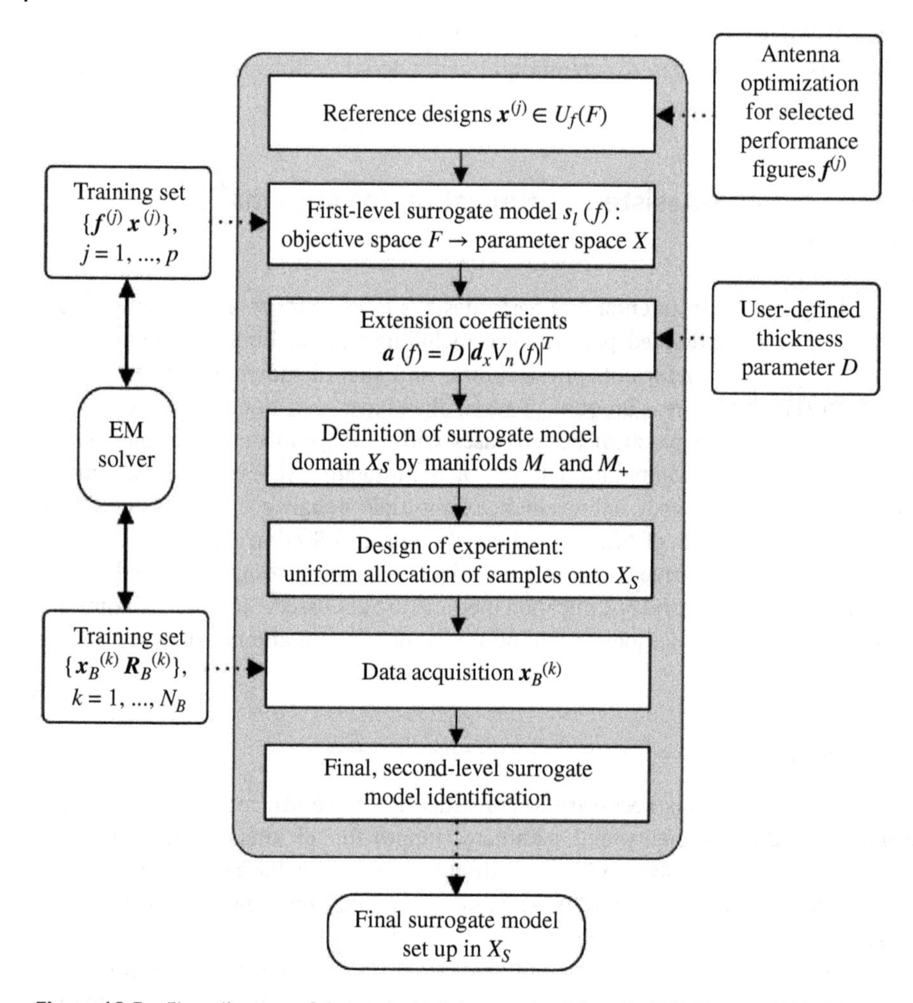

Figure 12.3 Flow diagram of the nested kriging methodology in [11]. Source: [11]/IEEE.

can be utilized to support small quantities of high fidelity (HF) but time-consuming data points to make accurate predictions. In [12], a two-stage framework for efficient antenna modeling is proposed by using the GPR to both establish the mapping between variable-fidelity databases and make predictions. In [13, 14], co-kriging was introduced to establish accurate antenna model using LF data and sparsely sampled HF data. In [5], the output space mapping is recommended to correct the discrepancy between the LF and HF data sets.

An efficient multi-stage collaborative machine learning (MS-CoML) method for multi-objective antenna optimization has been proposed in [15], with the modeling procedure shown in Figure 12.4. By combining the single-output Gaussian process regression (SOGPR) as well as symmetric and asymmetric multi-output Gaussian process regressions (MOGPRs) in a three-stage framework, the mappings between the EM models with variable fidelity and between different objectives are established. By exploring the correlations between different design objectives, the auxiliary databases for every design objective can be accurately established using the MS-CoML even if no LF responses for some objectives are given. The ML models for every design objective are then trained based on the obtained auxiliary databases.

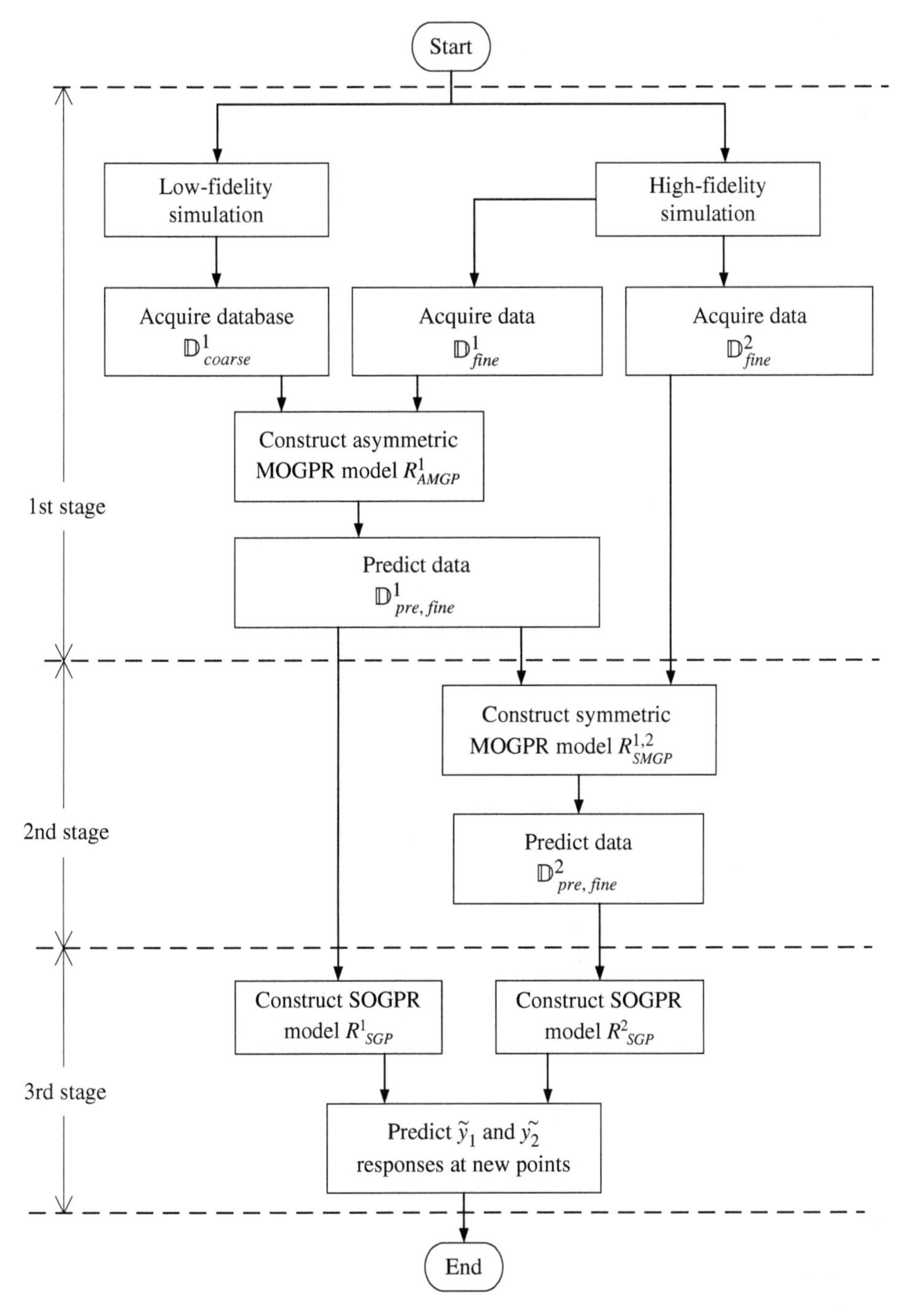

Figure 12.4 Flow diagram of MS-CoML in [15]. AMGP (asymmetric multiple-output Gaussian process), SMGP (symmetric multiple-output Gaussian process), SGP (single-output Gaussian process). Source: Adapted from [15].

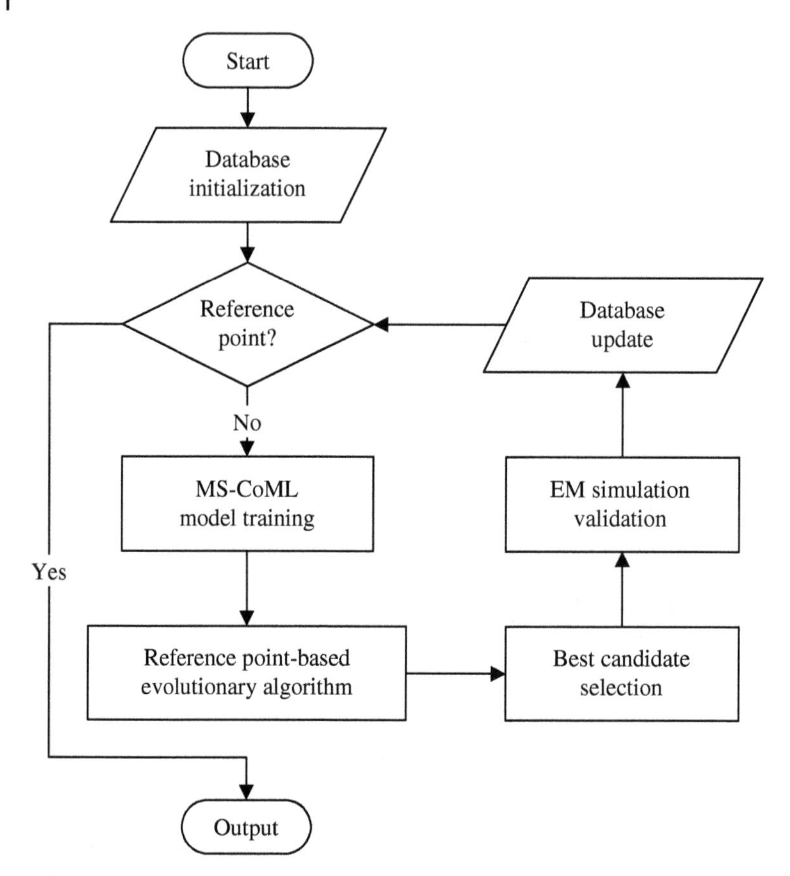

Figure 12.5 Flow diagram of the reference-point-based iterative multi-objective antenna optimization algorithm in [15]. Source: Adapted from [15].

Moreover, an iterative reference-point-based multi-objective optimization method based on the MS-CoML method has been proposed, with its flow diagram shown in Figure 12.5. The algorithm is based on classic MLAO scheme, with its modeling procedure updated using MS-CoML. Three antenna examples, i.e. single-band, broadband and multi-band designs operating at the Q, V, and S/C bands, have been chosen to verify the proposed method. Satisfactory results can be obtained by using data sets containing a very small proportion of HF results. Simulation results have shown that using the proposed MS-CoML method can greatly reduce the total optimization time without sacrificing accuracy.

12.3.3 Hybrid Optimization Algorithm

The MLAO algorithm is able to cooperate with various kinds of optimization methods for different applications. In [16], a method named surrogate-model-assisted differential evolution for antenna synthesis (SADEA) is proposed using Gaussian Process surrogate model and surrogate model-aware differential evolution (DE) optimization. Three antenna examples and two mathematical benchmark problems are shown using the proposed method. About 3–7 times speed enhancement is achieved when compared with traditional DE and particle swarm optimization method. In [17], a self-adaptive GPR modeling method-assisted DE optimization combined with radial basis function-assisted sequential quadratic programming local search is proposed for complex antenna optimization tasks. The method shows great improvement compared with conventional SADEA. In [18], the trust-region gradient search combined with the parameter space exploration realized using local kriging metamodels are proposed.

12.3.4 Robust Design

In practical communication and radar systems, uncertainties of various types may affect the antenna and array performance in an undesirable manner. Robust design, tolerance-aware design, or yield-driven design in modern antenna and array design aims to find a balance between the robustness and the performance of the final design, and provide guidelines for the integration and manufacturing process. However, robust design is a computationally expensive process because it normally involves multiple statistical analyzes. When using direct full-wave EM simulation-based stochastic optimization, the computational burden could be unbearable. The MLAO offers a promising solution to alleviate the computational cost, by helping to build surrogate models with high predictive ability to replace the EM simulations. In [19], the MLAO is introduced to find the worst-case performance (WCP) for EM devices. In [20], a novel sampling strategy is proposed to achieve a direct solution to search for the quasi-optimal shape of the maximum input tolerance hypervolume (MITH) of antenna cases based on trained surrogate model.

In [21], an efficient multilayer MLAO (ML-MLAO)-based robust design method is proposed for antenna and array applications. As shown in Figure 12.6, machine learning methods are introduced into multiple layers of the robust design, including worst-case analysis (WCA), MITH searching, and robust optimization to build surrogate models, considerably accelerating the whole robust design process. Based on the established surrogate model $\boldsymbol{R}_{s1}$ between the design parameters and antenna responses, the WCP values of different design points are calculated and utilized to build a surrogate model $\boldsymbol{R}_{s2}$ between the design parameters and the WCP. By using the two surrogate models established above, a training set consisting of antenna parameters and MITHs is constructed to

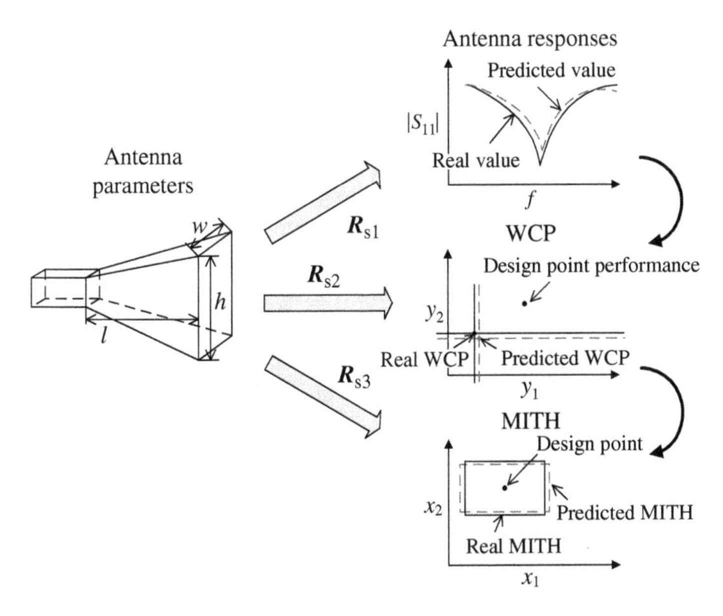

Figure 12.6 Surrogate models for antenna responses, WCP points, and MITHs in [21].
Source: [21]/IEEE/Public Domain CC BY 4.0.

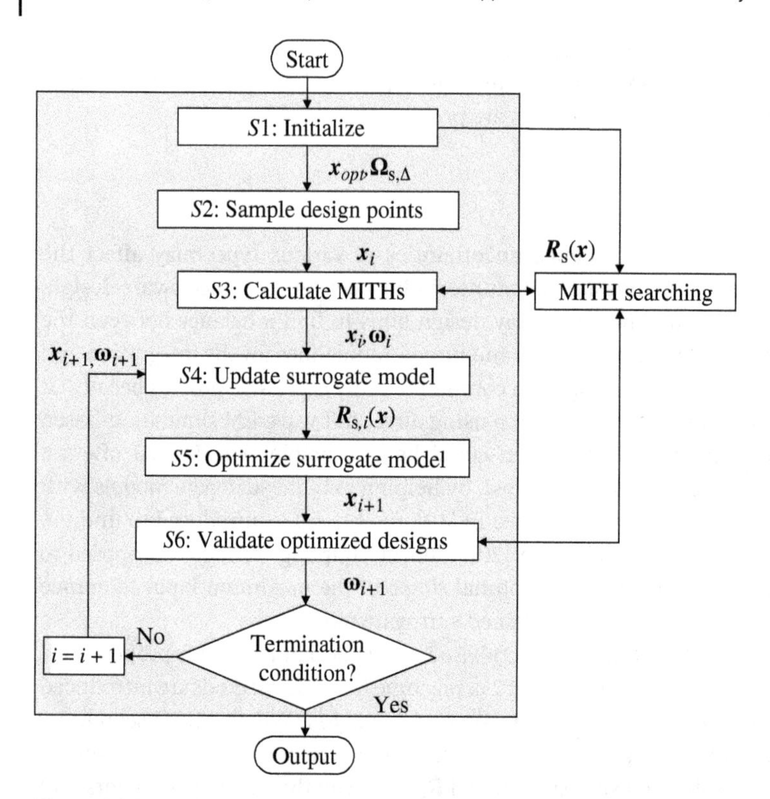

Figure 12.7 Flow diagram of the ML-MLAO algorithm for robust optimization in [21]. Source: [21]/IEEE/Public Domain CC BY 4.0.

build a surrogate model $\boldsymbol{R}_{s3}$. The flow diagram of the ML-MLAO algorithm is shown in Figure 12.7, which offers an efficient solution to build surrogate models, make predictions of new design points, and update surrogate models online, thereby efficiently searching for robust design points or Pareto fronts consisting of robust information.

12.3.5 Antenna Array Synthesis

A great number of excellent methods for finding optimized antenna arrays with proper element allocations and excitations have been proposed in the past several decades, including analytical methods [22, 23], stochastic optimization methods [24–26], compressive sensing (CS) [27], and many other hybrid methods. Most methods deal with isotropic elements without consideration of MC or platform effects, and inevitably lead to beam quality degradation for practical design tasks due to MC and platform effects. On the other hand, the direct combination of computational electromagnetic method (CEM) methods and optimization algorithms offers accurate but time-consuming choices for practical array design problems.

The MLAO serves as a great candidate to solve array design problems under MC and platform effects. In [28], models with variable fidelity have been introduced to optimize for both sidelobe level (SLL) and $|S_{11}|$ for a practical antenna array, which show great reduction of computational burden. While the initial intention of introduction of the ML method is to model the whole array performance under MC [29], recent studies have switched to building surrogate models for active element patterns (AEPs) and letting the conventional array synthesis method deal with the rest [30–32], which is apparently more practical considering the large computational costs for numbers of full-wave antenna array simulations. One great improvement of these MLAO-AEP-based algorithms is that they can easily deal with antenna arrays with free element positions; in other words, they offer another degree of freedom in antenna array design. In [31], an ANN is introduced to build very accurate surrogate models for AEPs under variable element location distributions, which greatly helps to optimize single beam, square-cosecant beam and flat top beam patterns of microstrip antenna arrays. In [33], an efficient machine learning-assisted array synthesis (MLAAS) method is introduced using efficient active base element modeling (ABEM), in which all possible element designs are accurately modeled into one active base element. This method is able to predict active element patterns for elements with arbitrary allocations and EM surroundings, therefore offering more degrees of freedom when compared with conventional ML-based array design methods.

12.4 Conclusion

In this chapter, MLAO-based algorithms have been discussed under the general approaches to antenna and array designs to largely alleviate the computational burden of conventional optimization methods. In practical antenna and array design cases, MLAO is able to cooperate with techniques including design space reduction, variable-fidelity evaluations, and hybrid optimization algorithms to further improve the design performance. In addition to antenna geometry optimization, the MLAO is able to deal with both robust design and antenna array synthesis problems.

References

1. Guo, Y.J. and Ziolkowski, R.W. (2021). *Advanced Antenna Array Engineering for 6G and Beyond Wireless Communications*. Wiley.
2. Haupt, R.L. and Rahmat-Samii, Y. (2015). Antenna array developments: a perspective on the past, present and future. *IEEE Antennas and Propagation Magazine* 57 (1): 86–96.
3. Xiao, Z., Han, Z., Nallanathan, A. et al. (2021). Antenna array enabled space/air/ground communications and networking for 6G. *arXiv preprint arXiv:2110.12610*.
4. Liu, B., Aliakbarian, H., Ma, Z. et al. (2014). An efficient method for antenna design optimization based on evolutionary computation and machine learning techniques. *IEEE Transactions on Antennas and Propagation* 62 (1): 7–18.
5. Koziel, S. and Ogurtsov, S. (2013). Multi-objective design of antennas using variable-fidelity simulations and surrogate models. *IEEE Transactions on Antennas and Propagation* 61 (12): 5931–5939.
6. Prado, D.R., López-Fernández, J.A., Arrebola, M., and Goussetis, G. (2018). Support vector regression to accelerate design and crosspolar optimization of shaped-beam reflectarray antennas for space applications. *IEEE Transactions on Antennas and Propagation* 67 (3): 1659–1668.
7. Kim, Y., Keely, S., Ghosh, J., and Ling, H. (2007). Application of artificial neural networks to broadband antenna design based on a parametric frequency model. *IEEE Transactions on Antennas and Propagation* 55 (3): 669–674.

8 Wu, Q., Chen, W., Wang, H., and Hong, W. (2020). Machine learning-assisted tolerance analysis and its application to antennas. *2020 IEEE International Symposium on Antennas and Propagation (ISAP)*, pp. 1853–1854. IEEE.

9 Koziel, S. and Bekasiewicz, A. (2017). On reduced-cost design-oriented constrained surrogate modeling of antenna structures. *IEEE Antennas and Wireless Propagation Letters* 16: 1618–1621.

10 Koziel, S. and Sigurdsson, A.T. (2018). Triangulation-based constrained surrogate modeling of antennas. *IEEE Transactions on Antennas and Propagation* 66 (8): 4170–4179.

11 Koziel, S. and Pietrenko-Dabrowska, A. (2019). Performance-based nested surrogate modeling of antenna input characteristics. *IEEE Transactions on Antennas and Propagation* 67 (5): 2904–2912.

12 Jacobs, J.P. and Koziel, S. (2014). Two-stage framework for efficient Gaussian process modeling of antenna input characteristics. *IEEE Transactions on Antennas and Propagation* 62 (2): 706–713.

13 Koziel, S., Bekasiewicz, A., Couckuyt, I., and Dhaene, T. (2014). Efficient multi-objective simulation-driven antenna design using co-Kriging. *IEEE Transactions on Antennas and Propagation* 62 (11): 5900–5905.

14 Koziel, S., Ogurtsov, S., Couckuyt, I., and Dhaene, T. (2013). Variable-fidelity electromagnetic simulations and co-Kriging for accurate modeling of antennas. *IEEE Transactions on Antennas and Propagation* 61 (3): 1301–1308.

15 Wu, Q., Wang, H., and Hong, W. (2020). Multistage collaborative machine learning and its application to antenna modeling and optimization. *IEEE Transactions on Antennas and Propagation* 68 (5): 3397–3409.

16 Liu, B., Aliakbarian, H., Ma, Z. et al. (2013). An efficient method for antenna design optimization based on evolutionary computation and machine learning techniques. *IEEE Transactions on Antennas and Propagation* 62 (1): 7–18.

17 Liu, B., Akinsolu, M.O., Song, C. et al. (2021). An efficient method for complex antenna design based on a self adaptive surrogate model-assisted optimization technique. *IEEE Transactions on Antennas and Propagation* 69 (4): 2302–2315.

18 Tomasson, J.A., Koziel, S., and Pietrenko-Dabrowska, A. (2020). Quasi-global optimization of antenna structures using principal components and affine subspace-spanned surrogates. *IEEE Access* 8: 50078–50084.

19 Xia, B., Ren, Z., and Koh, C.-S. (2014). Utilizing Kriging surrogate models for multi-objective robust optimization of electromagnetic devices. *IEEE Transactions on Magnetics* 50 (2): 693–696.

20 Easum, J.A., Nagar, J., Werner, P.L., and Werner, D.H. (2018). Efficient multiobjective antenna optimization with tolerance analysis through the use of surrogate models. *IEEE Transactions on Antennas and Propagation* 66 (12): 6706–6715.

21 Wu, Q., Chen, W., Yu, C. et al. (2021). Multilayer machine learning-assisted optimization-based robust design and its applications to antennas and arrays. *IEEE Transactions on Antennas and Propagation* 69 (9): 6052–6057.

22 Bucci, O.M., D'Urso, M., Isernia, T. et al. (2010). Deterministic synthesis of uniform amplitude sparse arrays via new density taper techniques. *IEEE Transactions on Antennas and Propagation* 58 (6): 1949–1958.

23 Liu, Y., Liu, Q.H., and Nie, Z. (2013). Reducing the number of elements in multiple-pattern linear arrays by the extended matrix pencil methods. *IEEE Transactions on Antennas and Propagation* 62 (2): 652–660.

24 Goudos, S.K., Gotsis, K.A., Siakavara, K. et al. (2013). A multi-objective approach to subarrayed linear antenna arrays design based on memetic differential evolution. *IEEE Transactions on Antennas and Propagation* 61 (6): 3042–3052.
25 Khodier, M.M. and Christodoulou, C.G. (2005). Linear array geometry synthesis with minimum sidelobe level and null control using particle swarm optimization. *IEEE Transactions on Antennas and Propagation* 53 (8): 2674–2679.
26 Yang, S.-H. and Kiang, J.-F. (2015). Optimization of sparse linear arrays using harmony search algorithms. *IEEE Transactions on Antennas and Propagation* 63 (11): 4732–4738.
27 Oliveri, G., Carlin, M., and Massa, A. (2012). Complex-weight sparse linear array synthesis by Bayesian compressive sampling. *IEEE Transactions on Antennas and Propagation* 60 (5): 2309–2326.
28 Koziel, S. and Ogurtsov, S. (2015). Fast simulation-driven optimization of planar microstrip antenna arrays using surrogate superposition models. *International Journal of RF and Microwave Computer-Aided Engineering* 25 (5): 371–381.
29 Ayestaran, R.G., Las-Heras, F., and Herrán, L.F. (2007). Neural modeling of mutual coupling for antenna array synthesis. *IEEE Transactions on Antennas and Propagation* 55 (3): 832–840.
30 Gong, Y. and Xiao, S. (2019). Synthesis of sparse arrays in presence of coupling effects based on ANN and IWO. *2019 IEEE International Conference on Computational Electromagnetics (ICCEM)*, pp. 1–3. IEEE.
31 Gong, Y., Xiao, S., and Wang, B.-Z. (2020). An ANN-based synthesis method for nonuniform linear arrays including mutual coupling effects. *IEEE Access* 8: 144015–144026.
32 Chen, W., Niu, Z., and Gu, C. (2020). Parametric modeling of unequally spaced linear array based on artificial neural network. *2020 9th Asia-Pacific Conference on Antennas and Propagation (APCAP)*, pp. 1–2. IEEE.
33 Wu, Q., Chen, W., Yu, C. et al. (2021). Machine learning-assisted array synthesis using active base element modeling. *IEEE Transactions on Antennas and Propagation* 70 (7): 5054–5065.

13

Analysis of Uniform and Non-uniform Antenna Arrays Using Kernel Methods

Manel Martínez-Ramón[1], José Luis Rojo Álvarez[2], Arjun Gupta[3], and Christos Christodoulou[1]

[1]Department of Electrical and Computer Engineering, The University of New Mexico, Albuquerque, NM, USA
[2]Departamento de Teoría de la señal y Comunicaciones y Sistemas Telemáticos y Computación, Universidad rey Juan Carlos, Fuenlabrada, Madrid, Spain
[3]Facebook, Menlo Park, CA, USA

13.1 Introduction

Kernel methods are a set of techniques that extend linear algorithms by endowing them with nonlinear properties under nonrestrictive assumptions by virtue of the Representer Theorem [1]. Kernel methods have been applied in particular to the support vector machine (SVM) for classification (SVC) [2] and regression (SVR) [3], and to classical algorithms, as it is the case of the kernel ridge regression (KRR) algorithm [4, 5]. Another very relevant methodology that uses kernels is the Gaussian Processes (GP) for Machine Learning [6]. In particular, the SVM methods have advantages in terms of their improved generalization capabilities and their robustness to outliers, thanks to their optimization criterion, which combines a linear loss function with a direct regularization through the control of the expressive capabilities of the machine by minimizing the so called Vapnik–Chervonenkis dimension of the machine. Therefore, SVMs are good candidates for their use in communications, in particular in antenna array processing [7, 8]. They have also been used in classification problems with radar and remote sensing [9].

Direction of arrival (DOA) has been mainly developed for spatially uniform arrays in 1 or 2 dimensions, using maximum likelihood (ML) approaches as the Minimum Variance Distortionless Method (MVDM) [10], and subspace techniques as the Multiple Signal Classification (MUSIC) [11, 12] or the Estimation of Signal Parameters Via Rotational Invariance Techniques (ESPRIT) [13]. Subspace methods have proven to be computationally superior to ML methods with significantly higher accuracy [14]. Nevertheless, nonuniform arrays are often used to provide array geometry flexibility. Furthermore, it has been proven that nonuniform arrays can provide decisive advantages in the form of enhanced resolution and reduced sidelobe levels [14, 15]. In order to perform DOA estimation in nonuniform arrays, linear approaches have been presented [16, 17] that divide the field of view in small sectors to compute linear interpolation of the steering vectors to create virtual uniform arrays, from which one can apply subspace methods for DOA estimation. GP methods have also been used in signal processing, in particular in DOA with nonuniform arrays [18, 19]. This methodology is based on a Bayesian framework that assumes a Gaussian distribution for the estimation noise and treats the model linear parameters as a latent variable with a Gaussian prior distribution. Appropriate inference over the latent prediction function allows us

Advances in Electromagnetics Empowered by Artificial Intelligence and Deep Learning, First Edition.
Edited by Sawyer D. Campbell and Douglas H. Werner.

to obtain not only a prediction from a given predictor input, but also a model for its probabilistic distribution. This allows the user to obtain a confidence interval of the prediction. Besides, the Bayesian approach avoids the need for parameter validation, which in turn makes it possible the use of models with a higher number of parameters than the ones used, for example, in SVMs, where all the free parameters need to be cross-validated.

In this chapter we summarize several main approaches to nonlinear uniform array beamforming using SVMs and DOA estimation with GP for regression. The formulation of the SVM and of the GP is provided in the complex plane in order to accommodate the solutions to the applications in array processing, where the signals to be processed are represented as complex envelope of the signals received by the antennas. In the case of SVM, we solve the optimization as a standard constrained minimization problem with a double number of constraints that cover the complex plane, and we prove that the final solution is formally identical to the one in the real axis; hence, the standard SVM optimization packages can be used. The non-linear formulation of SVM is derived for both spatial reference (SVM-SR) and time reference (SVM-TR) processing scenarios. The complex GP problem is solved by factorizing the signal probability distributions through the Bayes rule. The resulting algorithm is used for nonlinear array interpolation, which in turn solves the nonuniform array DOA problem.

13.2 Antenna Array Processing

From a signal processing point of view, both antenna array beamforming and detection of angle of arrival can be seen as a spatial signal filtering. Assume a linear, uniform array with L elements, illuminated with $M < L$ planar waves of angles of arrival θ_m, $0 \leq m \leq m-1$ (Figure 13.1). A linear filter for this purpose has the expression

$$y_n = \mathbf{w}^H \mathbf{x}_n + \epsilon_n \tag{13.1}$$

where $\mathbf{w}$ is a set of filter parameters, and $\mathbf{x}_n$ is an antenna snapshot taken at time instant n, with the form

$$\mathbf{x}_n = [x_{0,n} \cdots x_{L-1,n}]^T \tag{13.2}$$

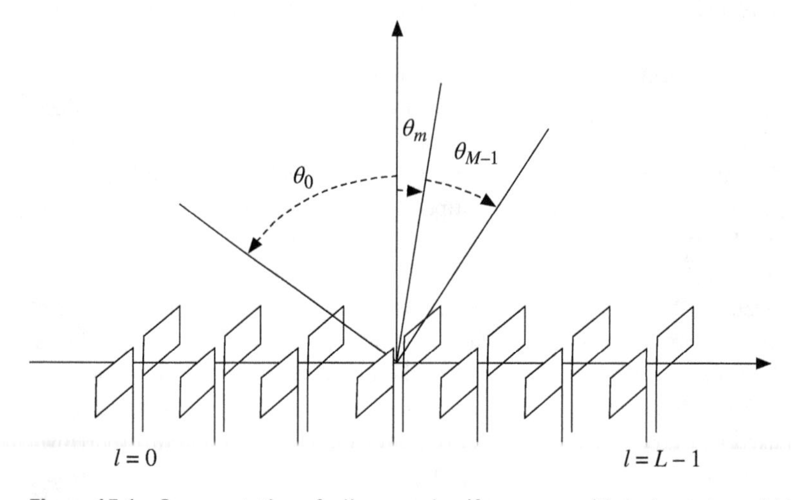

Figure 13.1 Representation of a linear and uniform array with L elements and M incoming signals with angles of arrival θ_m.

and it is defined as a column vector containing the samples $x_{l,n} \in \mathbb{C}$ of each one of the elements of the array, with $l = 0, \dots, L-1$. Variable ϵ_n denotes the model error and H stands for the Hermitian operator. This model can be used to construct either spatial reference and temporal reference beamforming, or detection of angle of arrival, which are summarized below.

13.2.1 Detection of Angle of Arrival

Detection of angle of arrival is usually done from the perspective of subspace methods, which rely on a principal component analysis of the autocorrelation matrix of the signal. These methods are called superresolution methods because of their low uncertainty, close to the Cramer–Rao bounds [20], which in particular holds for the MUSIC algorithm.

Assume first that a waveform of frequency f is illuminating an array of antennas spaced a distance d with angle of arrival θ. The phenomenon is illustrated in Figure 13.2, where continuous lines represent the frontwaves with phase $\phi = 0$ and the dashed lines represent the frontwaves with phase $\phi = \pi$. This wave produces a current in each one of the sensors with a phase equal to the phase of the wave at the position of each one of the antennas. The phase velocity of the wave is $v = \frac{c}{\sin(\theta)}$ and therefore, the travel time between sensors of a given frontwave is $T = \frac{d\sin(\theta)}{c}$. The phase difference between sensors is $\Omega = 2\pi f T = 2\pi f \frac{d\sin(\theta)}{c} = \frac{2\pi d \sin(\theta)}{\lambda}$, where λ is the wavelength. As a consequence, a spatial wave can be measured in the sensors whose sampled complex envelope $\mathbf{e}(\theta)$ can be expressed as

$$\mathbf{e}(\theta) = [1, e^{j2\pi\frac{d}{\lambda}\sin(\theta)}, \dots, e^{j2(M-1)\pi\frac{d}{\lambda}\sin(\theta)}]^T \tag{13.3}$$

This signal is usually called a steering vector, and the goal of DOA detection is to simultaneously estimate the frequencies $\Omega_k = \frac{2\pi d\sin(\theta_k)}{\lambda}$ of all incoming signals. The DOAS will be obtained by isolating them with the expression $\theta_m = \arcsin\left(\frac{\lambda\Omega_m}{2\pi d}\right)$.

We can derive the MUSIC methodology to estimate Ω_m according to a simple modification of the minimum variance distortionless response (MVDR) algorithm, which is a spatial filter $y_n(\Omega) = \mathbf{w}_\Omega^H \mathbf{x}_n$ that is optimized so that its response is minimized in variance while being unitary to a spatial signal $\mathbf{x}_n$, consisting of a complex exponential with frequency $\Omega = j2\pi\frac{d}{\lambda}\sin(\theta)$.

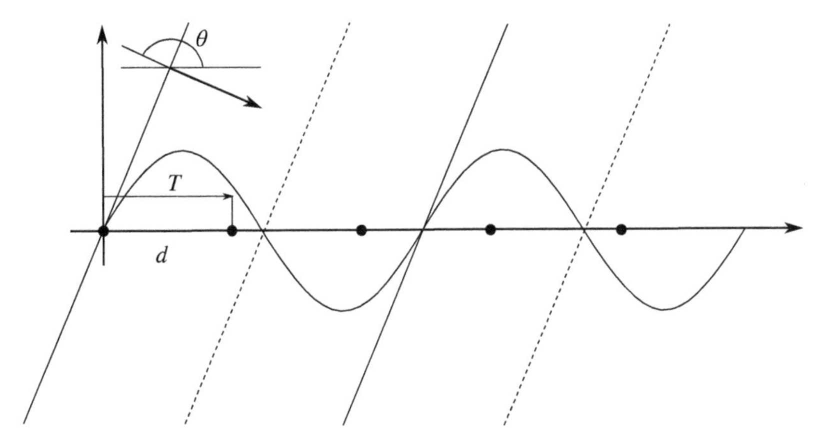

Figure 13.2 Electromagnetic wave traveling across an array of antennas with angle of arrival θ. The sinusoid represents the spatial wave observed in the direction of the array. The arrow represents the time $T = \frac{d\sin(\theta)}{c}$ needed for a frontwave to travel from a sensor to the next one.

According to this, we can write the problem as

$$\begin{aligned} & min \quad \mathbb{E}\left[\mathbf{w}_{\Omega}^{H}\mathbf{x}_{n}\right] \\ & \text{subject to: } \mathbf{w}_{\Omega}^{H}\mathbf{e}(\Omega) = 1 \end{aligned} \tag{13.4}$$

with

$$\mathbf{e}(\Omega) = \left[1, e^{j\Omega}, \dots, e^{j(L-1)\Omega}\right]^{T} \tag{13.5}$$

A Lagrange optimization for this filter has parameters given by

$$\mathbf{w}_{\Omega} = \frac{\mathbf{R}^{-1}\mathbf{e}_{\Omega}}{\mathbf{e}_{\Omega}^{H}\mathbf{R}^{-1}\mathbf{e}_{\Omega}} \tag{13.6}$$

and the variance of its output is

$$\mathbb{E}\left[\mathbf{w}_{\Omega}^{H}\mathbf{x}_{n}\right] = \frac{1}{\mathbf{e}_{\Omega}^{H}\mathbf{R}^{-1}\mathbf{e}_{\Omega}} \tag{13.7}$$

where $\mathbf{R} = \mathbb{E}\left[\mathbf{x}_n\mathbf{x}_n^H\right]$ is the snapshot autocorrelation matrix. Said matrix can be decomposed in eigenvectors and eigenvalues as $\mathbf{R} = \mathbf{Q}\mathbf{\Lambda}\mathbf{Q}^H$. If matrices $\mathbf{Q}_s$, $\mathbf{\Lambda}_s$, $\mathbf{Q}_n$, and $\mathbf{\Lambda}_n$ contain the signal and noise eigenvectors and eigenvalues, respectively, then $\mathbf{R}^{-1} = \mathbf{Q}_s\mathbf{\Lambda}_s^{-1}\mathbf{Q}_s^H + \mathbf{Q}_n\mathbf{\Lambda}_n^{-1}\mathbf{Q}_n^H$. If the noise eigenvalues are significantly smaller than the signal ones, therefore, we can take the approximation $\mathbf{R}^{-1} \approx \mathbf{Q}_n\mathbf{\Lambda}_n^{-1}\mathbf{Q}_n^H$. If all the noise eigenvalues are approximated by the noise variance σ_n^2, then the response of the filter can be approximated by

$$\mathbb{E}\left[\mathbf{w}_{\Omega}^{H}\mathbf{x}_{n}\right] = \frac{\sigma_n^2}{\mathbf{e}_{\Omega}^{H}\mathbf{Q}_n\mathbf{Q}_n^H\mathbf{e}_{\Omega}} \tag{13.8}$$

which leads to the MUSIC algorithm. Note that this response tends to infinity when Ω matches one of the spatial frequencies present in the signal, which makes the algorithm a superresolution one.

13.2.2 Optimum Linear Beamformers

Assume an L-element array antenna which is receiving M signals from DOAs θ_k, $0 \leq k \leq M-1$. The mathematical expression for the array snapshot can be formulated as

$$\mathbf{x}_n = \mathbf{A}\mathbf{s}_n + \mathbf{g}_n \tag{13.9}$$

where $\mathbf{s}_n$ is the vector of incoming signals, $\mathbf{g}_n$ models the observation additive white Gaussian noise (AWGN) noise, and $\mathbf{A}$ is the matrix of corresponding steering vectors, given by

$$\mathbf{A} = [\mathbf{e}(\theta_0), \mathbf{e}(\theta_1), \dots, \mathbf{e}(\theta_{M-1})] \tag{13.10}$$

The spatial correlation matrix of the received noisy signals is

$$\begin{aligned} \mathbf{R} &= \mathbb{E}[\mathbf{x}_n\mathbf{x}_n^H] = \mathbf{A}\mathbb{E}[\mathbf{s}_n\mathbf{s}_n^H]\mathbf{A}^H + \mathbb{E}[\mathbf{g}_n\mathbf{g}_n^H] \\ &= \mathbf{A}\mathbf{P}\mathbf{A}^H + \sigma^2\mathbf{I} = \sum_{k=1}^{M}\lambda_k\mathbf{v}_k\mathbf{v}_k^H \end{aligned} \tag{13.11}$$

where $\mathbf{P}$ is the autocorrelation matrix of the incoming signals, and it is assumed that the noise is uncorrelated and identically distributed with power σ^2. Scalars λ_k and vectors $\mathbf{v}_k$ are the eigenvalues and eigenvectors of $\mathbf{R}$, respectively; therefore, the highest eigenvalues are associated to eigenvectors $\mathbf{e}(\theta_l)$ corresponding to the impinging signals.

The beamformer is intended to optimally detect component $s_{d,n}$ in vector $\mathbf{s}_n$, and it can be expressed as a spatial filter with the form

$$y_n = \mathbf{w}^H \mathbf{x}_n + \epsilon_n \tag{13.12}$$

where y_n is the estimate of the desired symbol $s_{d,n}$ at instant n.

A temporal reference beamformer can be trained by simply transmitting a burst of known *training* signals by minimizing the squared error expectation, which leads to the Wiener solution

$$\mathbf{w} = \mathbf{R}^{-1}\mathbf{p} \tag{13.13}$$

where $\mathbf{p} = \mathbb{E}\left[s^*_{d,n}\mathbf{x}_n\right]$.

A spatial reference beamforming procedure consists of minimizing the estimator energy as in Eq. (13.4) while forcing the amplitude of an output corresponding to a unitary vector $\mathbf{e}(\theta_l)$ in the desired angle of arrival to be unitary. Using Lagrange optimization leads to the solution

$$\mathbf{w}_{\theta_d} = \frac{\mathbf{R}^{-1}\mathbf{e}(\theta_d)}{\mathbf{e}^H(\theta_d)\mathbf{R}^{-1}\mathbf{e}(\theta_d)} \tag{13.14}$$

The performance of these simple estimators can be improved by using Machine Learning in several aspects. The previously summarized algorithms, being linear in nature, are limited by the number of sensors in the array, plus the incoming signals must exhibit low correlations among them. In all the algorithms the optimization criterion is the minimization of the squared error, which may suffer from serious biases in the presence of signal outliers, mainly when the total number of training signals used to construct the correlation matrix is low. Linear and nonlinear approaches for DOA and beamforming can be used which can reduce these risks and can offer a reduced error rate, which are based on the well-known SVM and GP optimization criteria.

13.2.3 Direction of Arrival Detection with Random Arrays

The data model for a linear random array consists of M elements at nonuniform positions, where the average distance between elements is lower than $\lambda/2$. The array is illuminated by M signals with directions of arrival θ_m, $0 \leq m \leq M-1$. If the first array element is placed at the origin of the coordinate system, then the last one is placed at a distance $L\bar{d}$, where $\bar{d}$ is the average distance between elements. The $L-2$ remaining elements of the non-uniform array are randomly distributed along the axis where $0 < d_i < L \times \lambda/2$. Then, the model of the received signal is

$$\mathbf{x}_n = \mathbf{A}\mathbf{s}_n + \mathbf{g}_n \tag{13.15}$$

where $\mathbf{A}(\theta) = [\mathbf{a}(\theta_1), \mathbf{a}(\theta_2), \dots, \mathbf{a}(\theta_m)] \in \mathbb{C}^{L\times m}$ is the matrix containing nonuniform steering vectors

$$\mathbf{a}(\theta_m) = [1, e^{j2\pi\frac{d_1}{\lambda}\sin(\theta_m)}, \dots, e^{j2\pi\frac{d_{L-1}}{\lambda}\sin(\theta_m)}]^T \tag{13.16}$$

and d_i denotes the distances between elements.

The classical techniques for DOA with non-uniform arrays [14, 16] use an interpolation strategy to simulate a uniform array, from which a standard MUSIC algorithm can be applied. Since the strategy is linear, the signal field of view has to be divided into sectors that can be linearly interpolated with low error. Assuming that all the signals of interest are in the interval $[\theta_b, \theta_f]$, this one is divided into subintervals $\Delta\theta$. Each sector is represented by a nonuniform array manifold matrix

$\mathbf{A}_l(\theta)$ from which a uniform manifold $\tilde{\mathbf{A}}_l(\theta)$ is synthesized. The synthesis is done by a transformation matrix $\mathbf{T}$, according to the following expression,

$$\mathbf{T}\mathbf{A}_l(\theta) = \tilde{\mathbf{A}}_l(\theta) \tag{13.17}$$

Since both steering vectors can be computed, then

$$\mathbf{T} = \tilde{\mathbf{A}}_l(\theta)\mathbf{A}_l(\theta)^H[\mathbf{A}_l(\theta)\mathbf{A}_l(\theta)^H]^{-1} \tag{13.18}$$

Matrix $[\mathbf{A}_l(\theta)\mathbf{A}_l(\theta)^H]$ will be low rank, so the inversion is not possible unless a regularization term is added [21], and the final solution is

$$\mathbf{T} = \tilde{\mathbf{A}}_l(\theta)\mathbf{A}_l^H(\theta)[\mathbf{A}_l(\theta)\mathbf{A}_l^H(\theta) + \sigma_n^2\mathbf{I}]^{-1} \tag{13.19}$$

where $\sigma_n^2\mathbf{I}$ is the regularization term and σ_n^2 is the noise variance. This transformation is the used to compute an uniform version of the snapshot and, usually, the root-MUSIC algorithm [12] is applied to estimate the DOAs.

The solution requires a trade-off among sector size and error tolerance. Minimizing the sector size decreases the error but the sector has to be large enough to accommodate the signals of interest. A possible solution to this paradox proposed in [14] involves an initial estimation by virtue of Toeplitz completion and improving on the initial estimate through successive iterations, assuming that a viable initial estimate is available.

13.3 Support Vector Machines in the Complex Plane

13.3.1 The Support Vector Criterion for Robust Regression in the Complex Plane

The support vector regression (SVR) algorithm was introduced in [22] for regression in the real domain; thus, it cannot be directly applied to DOA or beamforming, but it can be easily modified for these applications by its extension to the complex plane. Besides, it is accepted that the observation noise in an antenna array application is mostly AWGN, but the received signal may suffer from correlated or uncorrelated non-Gaussian interferences in its sensors, which may lead to biases. The standard SVR cost function (known as ε-insensitive cost function) can be modified to make it robust to both classes of noise [23]. Assume the estimator of Eq. (13.12), but with a bias added to it, this is

$$y_n = \mathbf{w}^H\mathbf{x}_n + b + \epsilon_n \tag{13.20}$$

A complex version of the original SVR can be constructed as [24]

$$\begin{aligned} L_p = \tfrac{1}{2}||\mathbf{w}||^2 + C\sum_i \ell_\varepsilon(\xi_i, \xi_i') \\ + C\sum_i \ell_\varepsilon(\zeta_i, \zeta_i') \end{aligned} \tag{13.21}$$

subject to constraints

$$\begin{aligned} \mathbb{R}e\left(y_i - \mathbf{w}^T\mathbf{x}_i\right) &\leq \varepsilon + \xi_i \\ -\mathbb{R}e\left(y_i - \mathbf{w}^T\mathbf{x}_i\right) &\leq \varepsilon + \xi_i' \\ \mathbb{I}m\left(y_i - \mathbf{w}^T\mathbf{x}_i\right) &\leq \varepsilon + \zeta_i \\ -\mathbb{I}m\left(y_i - \mathbf{w}^T\mathbf{x}_i\right) &\leq \varepsilon + \zeta_i' \\ \xi_i, \xi_i', \zeta_i, \zeta_i' &\geq 0 \end{aligned} \tag{13.22}$$

Equation (13.21) defines the SVM criterion, which combines three terms. The first one, given by the minimization of the norm of the parameters, is related to the regularization of the problem in a Thikhonov sense [25], which was interpreted by Vapnik and Chervonenkis as the minimization of the structural risk of the estimator [26]. The second and third terms, weighted by a trade-off parameter C to be cross-validated, are the so-called empirical risk, where $\ell_\varepsilon(\cdot)$ is a convex cost function over the slack variables or losses ξ_i, ξ_i', ζ_i, ζ_i' corresponding to negative and positive errors of the real and imaginary parts of the signal, as they are defined in constraints in Eq. (13.22). Note that ε is an error tolerance term, which also needs to be cross-validated. Together with the positiveness constraints of the slack variables, the constraints on the real and imaginary errors impose that if the error is less than ϵ, then the slack variable is dropped to zero.

Cost function $\ell_\varepsilon(\cdot)$ over the slack variables is usually defined as simply the identity function in the original development of the SVR. In other words, for the case of a purely real error if $|\epsilon_i| < \varepsilon$, the cost function is zero, and otherwise its value is $||\epsilon_i|| - \varepsilon$, or a linear cost function over the errors. However, we define the function in three cases, one of them being a quadratic function over the errors, which accounts for those errors considered drawn from a Gaussian distribution. This section combines both with an ε-insensitive and a Huber [27] cost function. The combined cost function $\ell_R(\epsilon)$ has the expression given by [28]

$$\ell_R(\epsilon) = \begin{cases} 0 & |\epsilon| < \varepsilon \\ \frac{1}{2\gamma}(|\epsilon| - \varepsilon)^2 & \varepsilon \leq |\epsilon| \leq e_C \\ C\left(|\epsilon| - \varepsilon - \frac{\gamma C}{2}\right) & e_C \leq |\epsilon| \end{cases} \tag{13.23}$$

where $e_C = \varepsilon + \gamma C$ for the particular case where ϵ is real. Since in the present application the error is complex, this cost function is simply applied to each part of the error. Figure 13.3 illustrates the function and its derivative. A Lagrange functional can be constructed by adding the constraints (13.22), each of them multiplied by corresponding Lagrange variables $(\alpha_i, \alpha_i', \beta_i, \beta_i')$, to the functional (13.21). Computing the gradient of this functional with respect to the *variables*, i.e. with

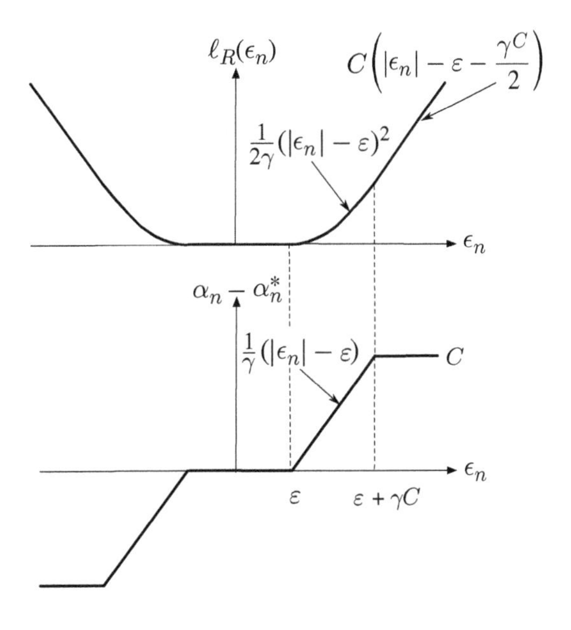

Figure 13.3 The upper panes shows the robust const function applied to the SVM criterion. The lower pane shows its derivative. In the optimum point of the optimization, this function determines the values of the dual coefficients $\alpha_n - \alpha_n^*$.

respect to $\mathbf{w}$, to b, and to slack variables $(\xi_i, \xi_i', \zeta_i, \zeta_i')$, leads to the following dual optimization problem when cost function is used in Eq. (13.23),

$$L_d = -\boldsymbol{\psi}^H (\mathbf{K} + \gamma \mathbf{I}) \boldsymbol{\psi} + \mathbb{R}e\left(\boldsymbol{\psi}^T \mathbf{d}^*\right) - \left(\boldsymbol{\alpha} + \boldsymbol{\alpha}' + \boldsymbol{\beta} + \boldsymbol{\beta}'\right) \mathbf{1}\varepsilon \tag{13.24}$$

where $\mathbf{K}$ is a Gram matrix of dot products between data, that is, whose entries are $\mathbf{K}_{i,j} = \mathbf{x}_i^H \mathbf{x}_j$; Ψ are complex Lagrange multipliers $\boldsymbol{\psi} = \boldsymbol{\alpha} - \boldsymbol{\alpha}' - j\left(\boldsymbol{\beta} - \boldsymbol{\beta}'\right)$; and $\boldsymbol{\alpha}, \boldsymbol{\alpha}', \boldsymbol{\beta}$ and $\boldsymbol{\beta}'$ are column vectors containing the real and imaginary parts of these multipliers, respectively. This dual functional has to be maximized with respect to them. The solution is given by

$$\mathbf{w} = \sum_{i=1}^{N} \psi_i \mathbf{x}_i \tag{13.25}$$

and the estimator for a new (test) sample $\mathbf{x}_*$ not previously seen in the training phase can be written as

$$y_* = \sum_{i=1}^{N} \psi_i \mathbf{x}_i^H \mathbf{x}_* + b \tag{13.26}$$

There are two interesting facts about Eq. (13.24). The first one is that it is just a complex version of the standard SVR dual functional to be optimized, where the modified cost function that has been applied simply adds a regularizing diagonal matrix $\gamma \mathbf{I}$ to matrix $\mathbf{K}$. The second one is related to the Lagrange multipliers that the cost function induces. For a sample that has a positive real component of the error, the corresponding Lagrange multiplier α_i is (and analogously for real negative and imaginary part of the error) fulfills

$$\alpha_i - \alpha_i' = \begin{cases} 0 & |\mathbb{R}e\left\{\epsilon_i\right\}| < \varepsilon \\ \frac{1}{\gamma}(\mathbb{R}e\left\{\epsilon_i\right\} - \varepsilon) & \varepsilon \leq |\mathbb{R}e\left\{\epsilon_i\right\}| \leq e_C \\ C & e_C \leq |\mathbb{R}e\left\{\epsilon_i\right\}| \end{cases} \tag{13.27}$$

The interest of this is that if the error magnitude is less than the tolerance ε, then it is ignored, and the sample does not contribute to the solution. If the error magnitude is between ε and e_C, then the cost function applied is the ML one for Gaussian noise (i.e. the cost function is quadratic) and then the contribution of the sample to the solution is proportional to its error (minus the tolerance). If the error magnitude is beyond the threshold e_C, then the contribution of the sample to the solution is saturated to C, which adds robustness against possible outliers, and this is one of the main differences between this solution and the minimum mean square error (MMSE).

In addition, the optimization for the dual functional in Eq. (13.24) can be implemented in any standard SVM package with two easy modifications. Indeed, the product $\boldsymbol{\psi}^H (\mathbf{K} + \gamma \mathbf{I}) \boldsymbol{\psi}$ is real and positive since $\mathbf{K}$ is positive definite. Therefore, all the imaginary parts of the product can be dropped and the following equality holds,

$$\boldsymbol{\psi}^H (\mathbf{K} + \gamma \mathbf{I}) \boldsymbol{\psi} = \begin{pmatrix} \boldsymbol{\alpha} \\ \boldsymbol{\beta} \end{pmatrix}^T \left[\begin{pmatrix} \mathbf{K}_{\mathbb{R}} & \mathbf{K}_{\mathbb{I}} \\ -\mathbf{K}_{\mathbb{I}} & \mathbf{K}_{\mathbb{R}} \end{pmatrix} + \gamma \mathbf{I} \right] \begin{pmatrix} \boldsymbol{\alpha} \\ \boldsymbol{\beta} \end{pmatrix} \tag{13.28}$$

where $\mathbf{K}_{\mathbb{R}}$ and $\mathbf{K}_{\mathbb{I}}$ contain the real and imaginary parts of the kernel matrix. This allows the use of the standard SVR optimizers in the implementation of the complex SVR in Eq. (13.24) using a quadratic programming procedure (e.g. see [29]) or an iterative recursive weighted least squares (IRWLS) procedure [30]. A more general theory about complex SVR can be found in [31].

13.3.2 The Mercer Theorem and the Nonlinear SVM

The SVR summarized above is linear in nature, but this limitation can be overcome by the use of the technique known as the Kernel trick. Mercer's theorem [32] justifies the use of positive definite functions (known as Mercer kernels) as dot products. The idea of the Kernel trick is that vectors $\mathbf{x}$ can be mapped to a higher (possibly infinite) dimension, complete and separable $\mathcal{H}$ endowed with a dot product (a Hilbert space), through a nonlinear transformation $\boldsymbol{\varphi}(\cdot)$. A linear approximating machine can be constructed in a higher dimensional space [2, 33] which will be nonlinear from the point of view of the input space.

Theorem 13.1 (Mercer [34]) There exists a function $\boldsymbol{\varphi} : \mathbb{R}^n \rightarrow \mathcal{H}$ and a dot product such that

$$K(\mathbf{x}_i, \mathbf{x}_k) = \boldsymbol{\varphi}(\mathbf{x}_i)^T \boldsymbol{\varphi}(\mathbf{x}_k) \tag{13.29}$$

if and only if $K(\cdot, \cdot)$ is a positive integral operator on a Hilbert space, i.e. if and only if for any function $g(\mathbf{x})$ for which

$$\int g(\mathbf{x}) d\mathbf{x} < \infty \tag{13.30}$$

the inequality

$$\int K(\mathbf{x}, \mathbf{y}) g(\mathbf{x}) g(\mathbf{y}) d\mathbf{x} d\mathbf{y} \geq 0 \tag{13.31}$$

holds [34].

Hilbert spaces provided with dot products that fit the Mercer's theorem are often called Reproducing Kernel Hilbert Spaces (RKHS).

Several functions that have been proven to be Mercer's kernels are the homogeneous polynomial kernel

$$K(\mathbf{x}_i, \mathbf{x}_k) = (\mathbf{x}_i^T \mathbf{x}_k)^p \tag{13.32}$$

and the inhomogeneous polynomial kernel

$$K(\mathbf{x}_i, \mathbf{x}_k) = (\mathbf{x}_i^T \mathbf{x}_k + 1)^p \tag{13.33}$$

Also, the square exponential kernel is very popular and widely used in many practical applications. Its expression is

$$K(\mathbf{x}_i, \mathbf{x}_k) = e^{-\frac{||\mathbf{x}_i - \mathbf{x}_k||^2}{2\sigma^2}} \tag{13.34}$$

It is straightforward to prove, from properties of the dot product, that its corresponding Hilbert space has infinite dimension, and the nonlinear transformation is not explicit. Note that this kernel is based on the Euclidean distance between vectors, resulting in a real dot product, this is $K_{\parallel} = \mathbf{0}$, which means that the data is transformed into (infinite dimension) real vectors inside the corresponding Hilbert space. Hilbert spaces with complex inner products exist. Refer to [31] for a more general view of this topic.

A nonlinear SVR is obtained by mapping the incoming data $\mathbf{x}[n]$ into a RKHS with a nonlinear transformation $\boldsymbol{\varphi}(\cdot)$. Nevertheless, the dot product of the corresponding spaces can be expressed as a function of the input vectors as in (13.29), and then matrix $\mathbf{K}$ containing all the dot products does not need the explicit knowledge of the transformation (which can be into an infinite dimension space), but only the dot product expressed as a function of the input data $\mathbf{x}$ is required.

The solution for the linear SVR is given by a linear combination of a subset of the training data mapped into the RKHS,

$$\mathbf{w} = \sum_{i=1}^{N} \psi_i \boldsymbol{\varphi}(\mathbf{x}_i) \tag{13.35}$$

And the expression of the regression test is now

$$y_* = \mathbf{w}^H \boldsymbol{\varphi}(\mathbf{x}_*) + b \tag{13.36}$$

where the bias b is added because the data is not necessarily centered around the origin in the Hilbert space. Plugging (13.35) into (13.36), yields

$$\begin{aligned} y_* &= \sum_{i=1}^{N} \psi_i \boldsymbol{\varphi}(\mathbf{x}_i)^H \boldsymbol{\varphi}(\mathbf{x}_*) + b \\ &= \sum_{i=1}^{N} \psi_i K(\mathbf{x}_i, \mathbf{x}_*) + b \end{aligned} \tag{13.37}$$

The resulting machine can now be expressed directly in terms of the Lagrange multipliers and the kernel dot products. In order to solve the dual functional which determines the Lagrange multipliers, the vectors are not required either, but only the Gram matrix $\mathbf{K}$ of the dot products between them. Here the kernel is used to compute this matrix, as in Eq. (13.29).

Once this matrix has been computed, solving for a nonlinear SVM is as easy as solving for a linear one. It can be shown that if the kernel fits the Mercer theorem, the matrix will be positive definite [32].

13.4 Support Vector Antenna Array Processing with Uniform Arrays

13.4.1 Kernel Array Processors with Temporal Reference

If a sequence of symbols is available for training purposes, a nonlinear array processor with temporal reference can be constructed from the model

$$y_n = \mathbf{w}^H \boldsymbol{\varphi}(\mathbf{x}_n) + b + \epsilon_n \tag{13.38}$$

A suitable solution can be obtained using the dual expression in Eq. (13.37) and with any suitable kernel dot product by solving the dual functional in Eq. (13.24) [7, 24]. This is the simplest possible array processor with temporal reference for quadrature amplitude modulation (QAM) signals with temporal reference.

13.4.1.1 Relationship with the Wiener Filter

The following property connects the beamformer constructed with the criterion applied to the SVM with the one where the optimization criterion is the MMSE.

Theorem 13.2 The SVR temporal reference processor approaches the Wiener (temporal reference) processor as $C \to \infty$, $\varepsilon = 0$ and $\gamma = 0$.

Proof: If $\varepsilon = 0$ and $C \to \infty$, the Lagrange multipliers of Eq. (13.27) become the estimation errors, i. e., $\psi_i = \frac{\epsilon_i^*}{\gamma}$. The dual functional (13.24) thus becomes

$$L_d = -\frac{1}{\gamma^2}\epsilon^H (\mathbf{K} + \gamma \mathbf{I}) \epsilon + \frac{1}{\gamma}\mathbb{R}e\left(\epsilon^H \mathbf{d}\right)$$

with

$$\mathbf{K} = \boldsymbol{\Phi}^T \boldsymbol{\Phi} \tag{13.39}$$

where $\boldsymbol{\Phi}$ is a matrix containing all column vectors $\boldsymbol{\varphi}(\mathbf{x}_n)$ corresponding to the training set and where ϵ is the vector of errors ϵ_i. Now, computing the gradient of this expression with respect to the errors and equaling it to zero results in the following expression,

$$-\frac{1}{\gamma^2}(\mathbf{K} + \gamma \mathbf{I})\,\epsilon + \frac{1}{\gamma}\mathbf{d} = 0 \tag{13.40}$$

The error is defined as the desired data minus the estimation, i.e., $\epsilon = \mathbf{d} - \boldsymbol{\Phi}\mathbf{w}$. Using this and the definition in Eq. (13.39), the weight vector $\mathbf{w}$ is

$$\mathbf{w} = (\mathbf{R} + \gamma \mathbf{I})^{-1}\mathbf{p} \tag{13.41}$$

with $\mathbf{R} = \frac{1}{N}\boldsymbol{\Phi}\boldsymbol{\Phi}^H$ and $\mathbf{p} = \frac{1}{N}\boldsymbol{\Phi}\mathbf{d}^*$.

By applying the Representer theorem, $\mathbf{w} = \boldsymbol{\Phi}\boldsymbol{\psi}$ and the solution for the multipliers in Eq. (13.37), we obtain

$$\boldsymbol{\psi} = (\mathbf{K} + \gamma I)^{-1}\mathbf{d}^* \tag{13.42}$$

This is a regularized version of the Wiener equations in a Hilbert space usually known as Kernel Ridge Regression (KRR) [4]. A Maximum A Posteriori solution is obtained by setting $\gamma = \sigma_n^2$. By setting $\gamma = 0$ in Eq. (13.41) or (13.42) we obtain the original Wiener equations.

13.4.2 Kernel Array Processor with Spatial Reference

The basic linear array processor with spatial reference in Eq. (13.14) uses the single constraint that the output for a unitary complex exponential produced by a continuous wave in the DOA is unitary. This single constraint cannot be used in nonlinear beamforming since the linearity properties simply do not hold. Therefore, we must establish constraints for exponentials with amplitudes distributed across the expected amplitude range of the signal. This is possible through the use of the SVR criterion and its constraints. But in order to accommodate the MVDR criterion to the one of the SVM, a principal component analysis in the Hilbert space is first needed, and it is explained below.

13.4.2.1 Eigenanalysis in a Hilbert Space

The following result, extracted from [35], is of importance to extend the spatial reference algorithms to its use with kernel methods. Here they are applied to spatial reference beamforming, and an application to kernel DOA can be found in [36].

Theorem 13.3 (Schoelkopf et al. [35]) Given the autocorrelation matrix of a set of N zero mean data samples (see Appendix A.13) mapped into a Hilbert space through mapping $\boldsymbol{\varphi}(\cdot)$, expressed as $\mathbf{R} = \frac{1}{N}\boldsymbol{\Phi}\boldsymbol{\Phi}^H$, which can be decomposed in eigenvectors and eigenvalues as

$$\mathbf{R} = \sum_{i=0}^{N} \mathbf{q}_i \lambda_i \mathbf{q}_i^H = \mathbf{Q}\Lambda\mathbf{Q}^H \tag{13.43}$$

where $\mathbf{\Lambda}$ is a diagonal matrix containing the eigenvalues, the corresponding kernel matrix $\mathbf{K} = \boldsymbol{\Phi}^H \boldsymbol{\Phi}$ can be decomposed with eigenvectors and eigenvalues as

$$\mathbf{K} = N \sum_{i=0}^{N} \mathbf{v}_i \lambda_i \mathbf{v}_i^H = N\mathbf{V}\mathbf{\Lambda}\mathbf{V}^H \tag{13.44}$$

where $q_i = \boldsymbol{\Phi}\mathbf{v}_i$.

Proof: By definition, eigenvalues $\mathbf{\Lambda}$ and eigenvectors $\mathbf{Q}$ of $\mathbf{R}$ satisfy

$$\mathbf{Q}\mathbf{\Lambda} = \mathbf{R}\mathbf{Q} \tag{13.45}$$

By virtue of the Generalized Representer Theorem [1], the eigenvectors can be expressed as a linear combination of the data set as

$$\mathbf{Q} = \boldsymbol{\Phi}\mathbf{V} \tag{13.46}$$

If Eqs. (13.45) and (13.46) are combined, then the following expression holds:

$$\boldsymbol{\Phi}\mathbf{V}\mathbf{\Lambda} = \frac{1}{N}\boldsymbol{\Phi}\boldsymbol{\Phi}^H\boldsymbol{\Phi}\mathbf{V} \tag{13.47}$$

Using the definition in Eq. (13.39) of kernel matrix and simplifying, we obtain

$$N\mathbf{V}\mathbf{\Lambda} = \mathbf{K}\mathbf{V} \tag{13.48}$$

Therefore, by assuming that $\mathbf{V}$ is an orthonormal set of vectors, we have that

$$\mathbf{K} = N\mathbf{V}\mathbf{\Lambda}\mathbf{V}^H \tag{13.49}$$

13.4.2.2 Formulation of the Processor

A simple solution to include the power minimization in a linear SVM beamformer has been introduced in [37]. Based on the same principle of that work, a direct spatial reference SVM beam processor can be constructed which minimizes the energy of the processor output and at the same time, and constraints are applied to the response of the processor to given signals. If the DOA of interest is known, an optimization functional can be formulated as

$$L_p = \frac{1}{2}\mathbf{w}^H\mathbf{R}\mathbf{w} + C\sum_i \ell_R(\xi_i + \xi_i') + C\sum_i \ell_R(\zeta_i + \zeta_i') \tag{13.50}$$

where the first element represents the expected energy of the output. It can be easily shown that $\mathbb{E}\left[|y_i|^2\right] = \mathbb{E}\left[\mathbf{w}^H\mathbf{x}\mathbf{x}^H\mathbf{w}\right] = \mathbf{w}^H\mathbb{E}\left[\mathbf{x}\mathbf{x}^H\right]\mathbf{w} = \mathbf{w}^H\mathbf{R}\mathbf{w}$, where $\mathbf{R} = \frac{1}{N}\boldsymbol{\Phi}\boldsymbol{\Phi}^H$. The second and third terms represent the application of cost function in Eq. (13.23) to slack variables or losses defined through constraints, this is,

$$\begin{aligned}
\mathbb{R}e\left(r_i - \mathbf{w}^H\boldsymbol{\varphi}(r_i\mathbf{e}_d) - b\right) &\leq \varepsilon + \xi_i \\
-\mathbb{R}e\left(r_i - \mathbf{w}^H\boldsymbol{\varphi}(r_i\mathbf{e}_d) - b\right) &\leq \varepsilon + \xi_i' \\
\mathbb{I}m\left(r_i - \mathbf{w}^H\boldsymbol{\varphi}(r_i\mathbf{e}_d) - b\right) &\leq \varepsilon + \zeta_i \\
-\mathbb{I}m\left(r_i - \mathbf{w}^H\boldsymbol{\varphi}(r_i\mathbf{e}_d) - b\right) &\leq \varepsilon + \zeta_i'
\end{aligned} \tag{13.51}$$

where $r_i, 1 \leq i \leq M$ are possible transmitted symbols with different amplitudes along a given amplitude range.

We apply a Lagrange optimization almost identical to the one of the standard complex SVR to the primal functional in Eq. (13.50) in order to obtain the primal solution,

$$\mathbf{w} = \mathbf{R}^{-1}\boldsymbol{\Phi}_d\boldsymbol{\psi} \tag{13.52}$$

with $\boldsymbol{\Phi}_d = [\varphi(r_1\mathbf{e}_d), \dots, \varphi(r_M\mathbf{e}_d)]^T$. This gives the dual

$$\begin{aligned} L_d = & -\tfrac{1}{2}\boldsymbol{\psi}^H\left[\boldsymbol{\Phi}_d^H\mathbf{R}^{-1}\boldsymbol{\Phi}_d + \gamma\mathbf{I}\right]\boldsymbol{\psi} + \mathbb{R}e(\boldsymbol{\psi}^T\mathbf{r}^*) \\ & -\varepsilon\mathbf{1}(\boldsymbol{\alpha} + \boldsymbol{\beta} + \boldsymbol{\alpha}' + \boldsymbol{\beta}') \end{aligned} \tag{13.53}$$

Autocorrelation matrix $\mathbf{R}$ cannot be computed if the corresponding Hilbert space has high or infinite dimensionality. Therefore, the results of Theorem 13.3 must be applied in order to change the inverse of the autocorrelation matrix by the kernel matrix, of dimensions $n \times N$. First, the matrix is expressed in terms of its principal components as $\mathbf{R}^{-1} = \mathbf{Q}\Lambda^{-1}\mathbf{Q}^H$, so that the dual is rewritten as

$$\begin{aligned} L_d = & -\tfrac{1}{2}\boldsymbol{\psi}^H\left[\boldsymbol{\Phi}_d^H\mathbf{Q}\Lambda^{-1}\mathbf{Q}^H\boldsymbol{\Phi}_d + \gamma\mathbf{I}\right]\boldsymbol{\psi} + \mathbb{R}e(\boldsymbol{\psi}^T\mathbf{r}^*) \\ & -\varepsilon\mathbf{1}(\boldsymbol{\alpha} + \boldsymbol{\beta} + \boldsymbol{\alpha}' + \boldsymbol{\beta}') \end{aligned} \tag{13.54}$$

The optimization of the dual (13.54) gives us the Lagrange multipliers $\boldsymbol{\psi}$ from which one can compute the optimal weight vector (13.52). Using Eq. (13.46) in (13.54) and using expression (13.49) gives the result

$$\begin{aligned} L_d = & \frac{1}{2}\boldsymbol{\psi}^H\left[N\mathbf{K}_d^H\mathbf{K}^{-1}\mathbf{K}_d + \gamma\mathbf{I}\right]\boldsymbol{\psi} \\ & -\mathbb{R}e(\boldsymbol{\psi}^T\mathbf{r}^*) + \varepsilon\mathbf{1}(\boldsymbol{\alpha} + \boldsymbol{\beta} + \boldsymbol{\alpha}' + \boldsymbol{\beta}') \end{aligned} \tag{13.55}$$

where $\mathbf{K}_d = \boldsymbol{\Phi}^H\boldsymbol{\Phi}_d$. This expression is formally identical to Eq. (13.24) and it can be optimized using the same standard SVM optimizers without modification.

The use of Eqs. (13.46) and (13.52) allows us to obtain the weight vector as a function of the dual parameters,

$$\begin{aligned} \mathbf{w} &= \mathbf{R}^{-1}\boldsymbol{\Phi}_d\boldsymbol{\psi} = \boldsymbol{\Phi}\mathbf{V}\mathbf{D}^{-1}\mathbf{V}^H\boldsymbol{\Phi}^H\boldsymbol{\Phi}_d\boldsymbol{\psi} \\ &= N\boldsymbol{\Phi}\mathbf{K}^{-1}\mathbf{K}_d\boldsymbol{\psi} \end{aligned} \tag{13.56}$$

which gives the output of the estimator as given by

$$y_n = \mathbf{w}^H\varphi(\mathbf{x}_n) + b = N\boldsymbol{\psi}^H\mathbf{K}_d\mathbf{K}^{-1}\mathbf{k}_n + b \tag{13.57}$$

where $\mathbf{k}_n = [K(\mathbf{x}_1, \mathbf{x}_n), \dots, K(\mathbf{x}_n, \mathbf{x}_n)]^T$ is the vector of dot products of the vector $\boldsymbol{\varphi}(\mathbf{x}_n)$ with all the training vectors $\boldsymbol{\varphi}(\mathbf{x}_i)$, $1 \le i \le N$.

13.4.2.3 Relationship with Nonlinear MVDM

The SVR applied to nonlinear array processing with spatial reference is related to a nonlinear version of the MVDM. The following result is useful to derive the expression of a nonlinear MVDR from the SVR.

Theorem 13.4 The SVR spatial reference array processor of Eq. (13.57) approaches the MVDR in the Hilbert Space as $C \to \infty$ and $\varepsilon = 0$

Proof: Following the same reasoning as in Theorem 13.2, if $\varepsilon = 0$ and $C \to \infty$, the Lagrange multipliers are proportional to the estimation errors, i.e. $\psi_i = \frac{\epsilon_i^*}{\gamma}$. The dual (13.53) is then

$$L = \frac{1}{2\gamma^2}\epsilon^H\left[\boldsymbol{\Phi}_d^H\mathbf{R}^{-1}\boldsymbol{\Phi}_d + \gamma\mathbf{I}\right]\epsilon - \frac{1}{\gamma}\mathbb{R}e(\epsilon^H\mathbf{r}) \tag{13.58}$$

Again, the optimization of this functional is done by computing its gradient with respect to $\mathbf{e}$ and nulling it, as follows,

$$\mathbf{0} = \frac{1}{\gamma_2}\boldsymbol{\Phi}_d^H\mathbf{R}^{-1}\boldsymbol{\Phi}_d\epsilon - \frac{1}{\gamma}\mathbf{r} + \frac{1}{\gamma}\epsilon \tag{13.59}$$

Knowing that $\epsilon = \mathbf{r} - \mathbf{w}^H \boldsymbol{\Phi}_d$, then the following result holds

$$\mathbf{w} = \mathbf{R}^{-1}\boldsymbol{\Phi}_d(\boldsymbol{\Phi}_d^H \mathbf{R}^{-1}\boldsymbol{\Phi}_d + \gamma \mathbf{I})^{-1}\mathbf{r} \tag{13.60}$$

which corresponds to a nonlinear (kernel) version of the MVDR processor in Eq. (13.6) except for a numerical regularization term.

Now using Eq. (13.52) in (13.60), the dual parameters can be solved as

$$\begin{aligned} \mathbf{R}^{-1}\boldsymbol{\Phi}_d \psi &= \mathbf{R}^{-1}\boldsymbol{\Phi}_d(\boldsymbol{\Phi}_d^H \mathbf{R}^{-1}\boldsymbol{\Phi}_d{}^{-1} + \gamma \mathbf{I})^{-1}\mathbf{r} \\ \psi &= (\boldsymbol{\Phi}_d^H \mathbf{R}^{-1}\boldsymbol{\Phi}_d + \gamma \mathbf{I})^{-1}\mathbf{r} \end{aligned} \tag{13.61}$$

which is an approximate solution of the optimization for $\varepsilon = 0$ and $C \rightarrow \infty$. This is also formally equivalent to a ridge regression (RR) algorithm, where component $\gamma \mathbf{I}$ plays the role of a regularization term.

13.4.3 Examples of Temporal and Spatial Kernel Beamforming

A simulation is here presented with four quadrature phase shift keying (QPSK) transmitters with unitary amplitude, where one of them is considered the signal of interest, with a DOA of 0° and where there is an interference coming from a DOA of −10°. The array consists of seven isotropic elements spaced a distance $d = \lambda/2$. For temporal reference beamforming, the training signal is a burst

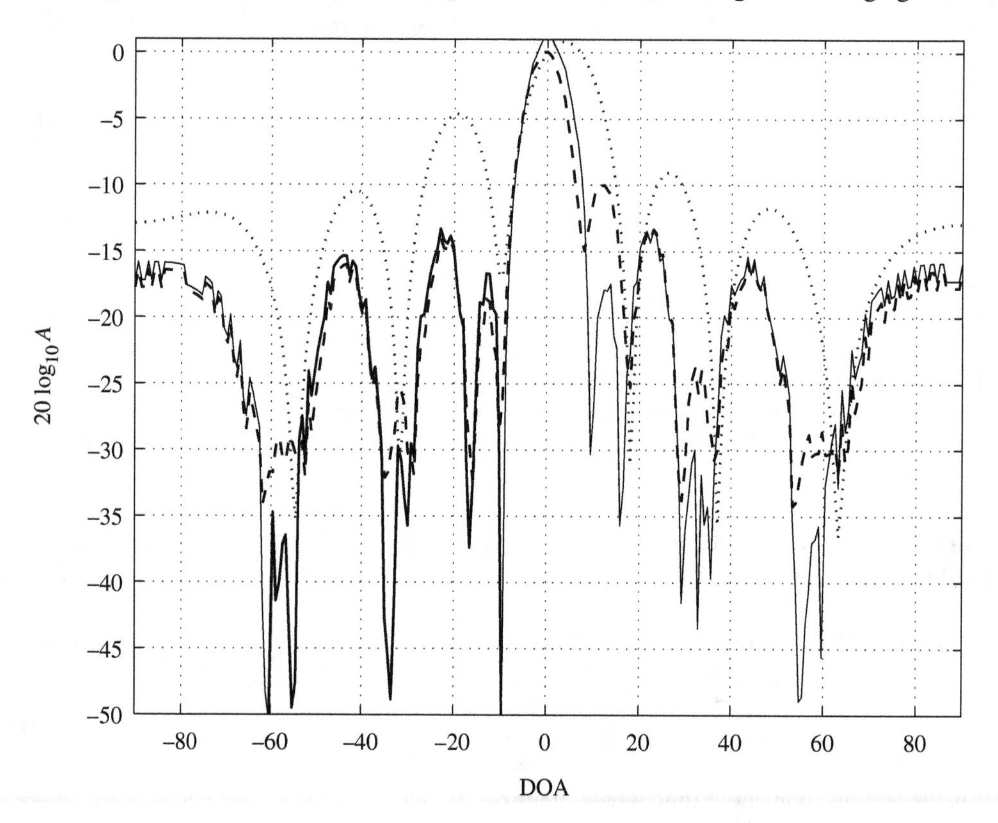

Figure 13.4 Radiation pattern of the SVM-TR (solid line) and the SVM-SR (dashed line) with Gaussian radial basis function (RBF) kernels in an array of seven elements and with one interference signal at a DOA of −10°. Dots correspond to the beam of the linear MVDM. Source: Reproduced from [38] with permission.

of 100 symbols, and the test is a sequence of 1000 new symbols. The experiments are reproduced from [38] and they consist of the comparison between the kernel temporal and spatial reference beamformers.

The chosen kernel for the experiment was the square exponential in Eq. (13.34) and the value used for parameter σ^2 was cross-validated in that work to show that for an array of 7 elements, the best validation results are obtained for values that range between $\sigma^2 = 3$ and $\sigma^2 = 30$. The range of acceptable values depends on the number of elements in the array. The optimal value of γ was also chosen by validation. In order to avoid cross validation in the rest of the SVM parameters (C and ε), they were heuristically fixed using the criteria in [39, 40]. Accordingly, parameters were fixed to $\sigma = 10, \gamma = 1e-6, C = 1, \varepsilon = 0$.

The approximate radiation patterns of the experiments are compared in Figure 13.4, which shows the results for the SVM estimator with temporal reference (SVM-TR) in continuous line, and with spatial reference (SVM-SR) in dashed line. The dots correspond to the performance of the linear MVDR. The SVM-TR beamformer corresponds to the model in Eq. (13.37) trained with the standard SVM optimizer formulated in Eq. (13.24). The SVM-SR is trained with Eq. (13.55). Figure 13.5 shows the experiment for the kernel RR temporal reference algorithm, whose coefficients are in Eq. (13.37) (Kernel-TR) and kernel RR spatial reference beamforming of Eq. (13.61) (Kernel-SR). It can be seen that the SVM methods exhibit the best rejection at −10° and that both kernel methods have a main lobe significantly thinner than the linear method.

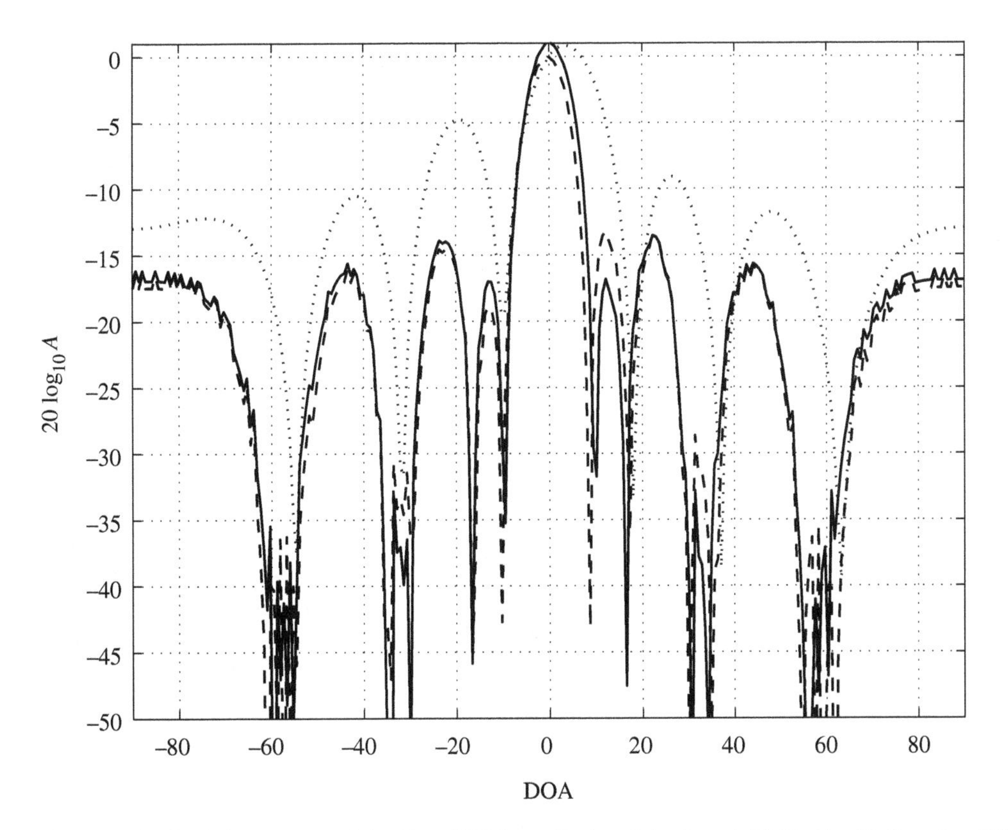

Figure 13.5 Radiation pattern of the Kernel-TR (cont. line) and the Kernel-SR (dash) with Gaussian RBF kernels in an array of seven elements and with one interference signal at a DOA of −10°. Dots correspond to the beam of the linear MVDM. Source: Reproduced from [38] with permission.

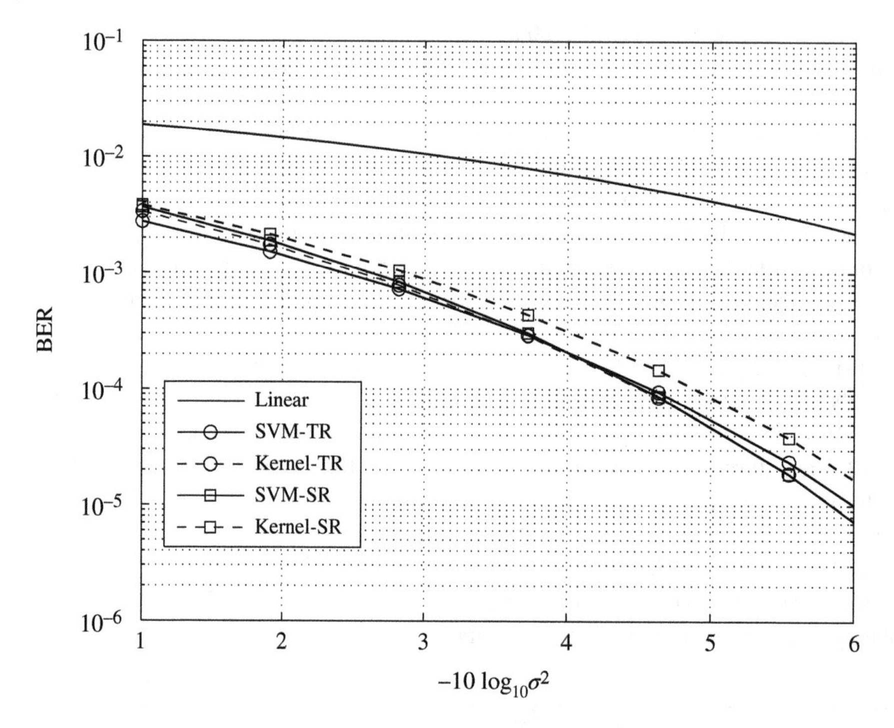

Figure 13.6 BER performance as a function of thermal noise power for linear algorithms, SVM-TR, SVM-SR, Kernel-TR, and Kernel-SR. Source: Reproduced from [38] with permission.

Finally, Figure 13.6 shows the bit error rate (BER) measured for the four processors with respect to the thermal noise. The continuous line shows the BER measured for the linear temporal and spatial reference beamforming algorithms. The array had seven elements and the interferences had DOAs of −10°, 10°, and 20°. The desired signal had a DOA of 0° and amplitudes were unitary. Kernel methods show similar performance, with an improvement of roughly 5 dB with respect to the linear approaches.

13.5 DOA in Random Arrays with Complex Gaussian Processes

In order to overcome the limitations of linear interpolation for random array processing, two different solutions were proposed in [18, 19]. Instead of interpolating the steering vector matrix divided in sectors, the methodologies interpolate the signal vector using stationary Mercer kernels as the interpolation basis functions, based on the works in [41–43]. The ideas in [19] are next summarized, which make use of a nonlinear complex GP for interpolation that also computes the confidence interval on the inference.

13.5.1 Snapshot Interpolation from Complex Gaussian Process

The standard GP for nonlinear regression and classification was introduced in [6]. By taking advantage of the Bayesian approach used in GP, we extend the idea to complex signals. A different nonlinear complex solution was proposed in [44]. We present here the approach adapted to nonuniform interpolation, following the models presented in [41–43]. Assume a nonuniform snapshot $\mathbf{x}_n$

at positions $0, \dots, d_l, \dots, d_{L-1}$. An interpolation function can be constructed by the use of basis interpolation functions $\phi(d)$, $0 \le d \le L\bar{d}$. The estimation has the form

$$x_n(d) = \mathbf{w}_n^H \boldsymbol{\varphi}(d) + \epsilon_n(d) \tag{13.62}$$

From a kernel trick standpoint, $\boldsymbol{\varphi}(d)$ is the transformation of the position of the element into a higher dimensional Hilbert space $\mathcal{H}$. Let us assume that this space has a kernel dot product $K(u, v) = \langle \boldsymbol{\varphi}(d_i), \boldsymbol{\varphi}(d_j) \rangle$ [4].

Vector $\mathbf{w}_n \in \mathcal{H}$ is the weight vector used to interpolate the snapshot at instant n. Error $\epsilon_n \in \mathbb{C}$ is modeled as an i.i.d. GP with probability distribution $p(\epsilon) \sim \mathcal{N}(\epsilon|0, \boldsymbol{\Sigma})$, and where $\boldsymbol{\Sigma} = \sigma_n^2 \mathbf{I}$ is the covariance between the real and the imaginary parts of the error, this is, Eq. (13.62) is a multivariate GP be expressed as the likelihood of the process $\mathbf{x}_n(d) = [x_{\mathbb{R},n}(d), x_{\mathbb{I},n}(d)]^T$, where subindexes $\mathbb{R}$, $\mathbb{I}$ represent the real and imaginary parts of the samples. Therefore, the known signal x_{n,d_l} likelihood function is

$$p(\mathbf{x}_{n,d_l}|\boldsymbol{\varphi}(d_l), \mathbf{w}) = \mathcal{N}(\mathbf{x}_{n,d_l}|\mathbf{w}_n^H \boldsymbol{\varphi}(d_l), \boldsymbol{\Sigma}_n) \tag{13.63}$$

By making use of the Bayes rule, this bivariate distribution can be decomposed into two univariate distributions with the form

$$p(\mathbf{x}_{n,d_l}|\boldsymbol{\varphi}(d_l), \mathbf{w}) = p\left(x_{\mathbb{R},n,d_l}|x_{\mathbb{I},n,d_l}, \boldsymbol{\varphi}(d), \mathbf{w}_{\mathbb{R}}\right) p\left(x_{\mathbb{I},n,d_l}|\varphi(d_l), \mathbf{w}_{\mathbb{I}}\right) \tag{13.64}$$

where $\mathbf{w}_{\mathbb{R}}, \mathbf{w}_{\mathbb{I}} \in \mathcal{H}$ are the linear parameters of the real and imaginary processes, respectively. Note that the first distribution at the right side of the expression depends on the imaginary part of the signal, but by virtue of the Bayes rule, an equivalent expression can be written by swapping the real and the imaginary parts of the signal.

Linear parameters $\mathbf{w}_{\mathbb{R}}$ and $\mathbf{w}_{\mathbb{I}}$ are assumed to be latent random variables with a Gaussian prior distribution given by

$$p(\mathbf{w}_{\mathbb{R}}) = \mathcal{N}(\mathbf{0}, \boldsymbol{\Sigma}_p) \tag{13.65}$$

Using the Bayes rule, the posterior distribution of the parameters is proportional to its prior times its likelihood. Since both are Gaussian distributions, the posterior distribution of the parameters is also Gaussian. Now, given all the known signals x_{n,d_l}, the interpolation of the signal at any arbitrary position d has a Gaussian predictive posterior distribution given by the means and variances,

$$\begin{aligned}
\mu_{\mathbb{I}}(d) &= \mathbf{x}_{\mathbb{I}}^T \left(\mathbf{K} + \sigma_n^2 \mathbf{I}\right)^{-1} \mathbf{k}(d) \\
\sigma_{\mathbb{I}}^2(d) &= k(d, d) - \mathbf{k}(d)^T \left(\mathbf{K} + \sigma_n^2 \mathbf{I}\right)^{-1} \mathbf{k}(d) \\
\mu_{\mathbb{R}|\mathbb{I}}(d) &= \mathbf{x}_{\mathbb{R}}^T \left(\mathbf{K} + \sigma_n^2 \mathbf{I}\right)^{-1} \left(\mathbf{k}(d) + w_{\mathbb{R}} \mu_{\mathbb{I}}(d)\right) \\
\sigma_{\mathbb{R}|\mathbb{I}}^2(d) &= \left(\mathbf{k}(d) + w_{\mathbb{I}} \mu_{\mathbb{I}}(d)\right)^T \left(\mathbf{K} + \sigma_n^2 \mathbf{I}\right)^{-1} \left(\mathbf{k}(d) + w_{\mathbb{R}} \mu_{\mathbb{I}}(d)\right)
\end{aligned} \tag{13.66}$$

where column vector $\mathbf{k}(d)$ contains dot products $k(d, d_l)$; $0 \le l \le L-1$, $\mathbf{K}$ is the matrix of kernel dot products $k(d_l, d_m)$ (with $0 \le l, m \le L-1$) between known distances; and $\mathbf{x}_{\mathbb{R}}, \mathbf{x}_{\mathbb{I}}$ contain the real and imaginary components, respectively, of snapshot $\mathbf{x}_n$. As it can be seen in Eq. (13.64), there is a linear dependency between $x_{\mathbb{R}}$ and $x_{\mathbb{I}}$, which is given by $w_{\mathbb{I}}$ in Eq. (13.66). This parameter is the maximum a posteriori value of the corresponding parameter inside vector $\mathbf{w}_{\mathbb{R}}$ in Eq. (13.64), which is easily computed as $w_{\mathbb{R}} = \mathbf{x}_{\mathbb{I}}^T \mathbf{x}_{\mathbb{R}} / \left(\mathbf{x}_{\mathbb{I}}^T \mathbf{x}_{\mathbb{I}} + \sigma_n^2\right)$

The predictive mean for the multivariate GP is then given by

$$\mu(d) = \begin{bmatrix} \mu_{\mathbb{R}|\mathbb{I}}(d) \\ \mu_{\mathbb{I}}(d) \end{bmatrix} \tag{13.67}$$

The predictive variance can be derived as

$$\boldsymbol{\Sigma} = \begin{bmatrix} \sigma^2_{\mathbb{R}|\mathbb{I}} + w_{\mathbb{I}}^2\sigma_{\mathbb{I}}^2 & w_{\mathbb{I}}\sigma_{\mathbb{I}}^2 \\ w_{\mathbb{I}}\sigma_{\mathbb{I}}^2 & \sigma_{\mathbb{I}}^2 \end{bmatrix} \tag{13.68}$$

Finally, the snapshot $\tilde{\mathbf{x}}_n$ interpolated at positions $d_k = k\overline{d}$ has elements

$$\tilde{x}_{n,d_k} = \mu_{\mathbb{R}|\mathbb{I}}(k\overline{d}) + j\mu_{\mathbb{I}}(k\overline{d}) \tag{13.69}$$

The autocorrelation of this signal can then be computed, and the directions of arrival can be estimated using a standard method as the root-MUSIC or the SVM approach introduced in [36].

13.5.2 Examples

The new GP algorithm is compared to the standard linear interpolation algorithm summarized in Section 13.2.3, labeled here as least squares (LS) algorithm. Both algorithms interpolate the signal and then the DOAs are estimated using the root-MUSIC algorithm. The standalone root-MUSIC algorithm with no interpolation is also tested, which is an optimal algorithm when the array is uniform, but as it can be seen, it fails when the array is non-uniformly distributed. The scenario contains two unitary QPSK signals with DOAs at 10° and 30°. The used kernel for the GP interpolator was a simple square exponential kernel with initial value of the width $\sigma = \overline{d}$. The final value of this parameter and the noise parameter σ_n^2 were inferred using ML through a gradient descent approach with respect to the parameters, i.e. by following the standard procedure for GP parameter optimization provided in [45].

The performance of the three algorithms is compared with respect to the number of elements and the signal-to-noise ratio (SNR). Regarding the LS algorithm, it was tested with a partition of six sectors of 30°, with four sectors of 45°, and with three sectors of 60°. The performance is measured as the mean square error (MSE) of the DOA estimation of each one of the algorithms. In order to compute the MSE error of the estimations with respect to the true values of the DOAs, we generated

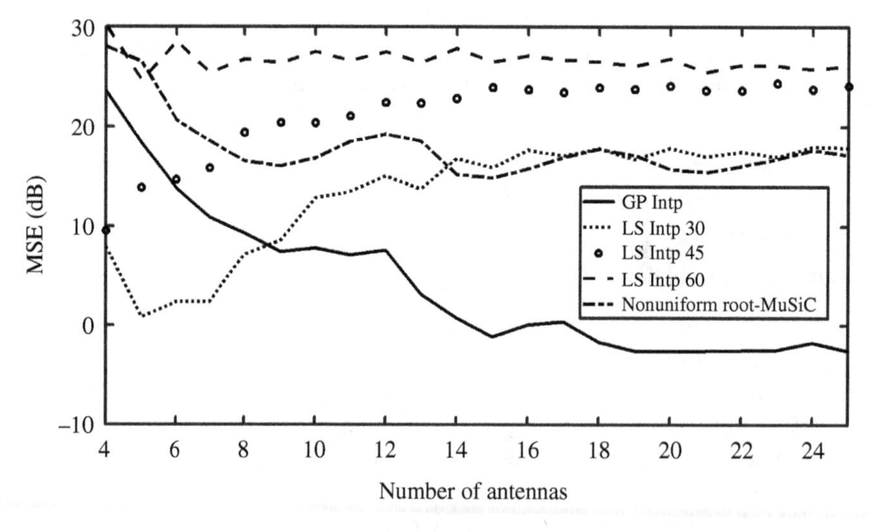

Figure 13.7 MSE (dB) versus an increasing number of elements for LS with 30°, 45°, 60° sectors, GP regression, and standalone root-MuSiC for two QPSK signals arriving from 10° and 30°, with an SNR of 3 dB. Source: Adapted from [19].

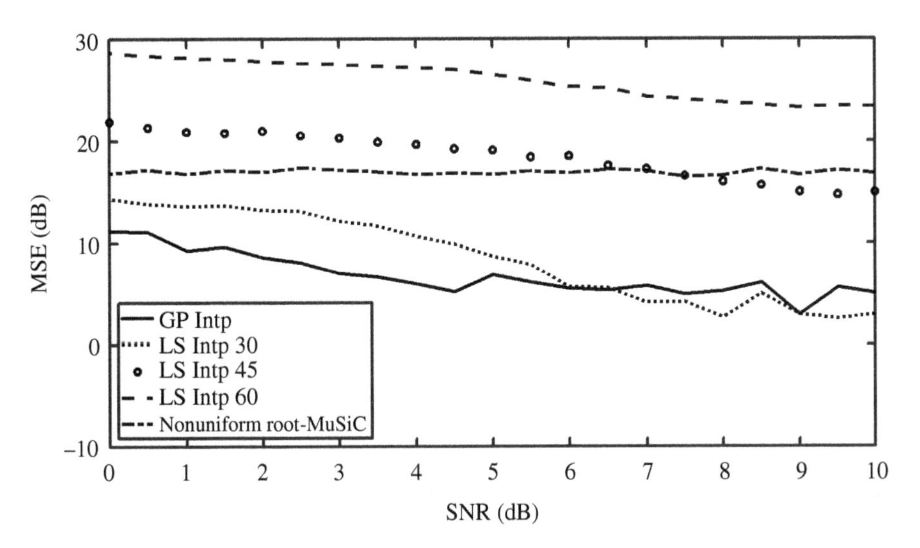

Figure 13.8 MSE (dB) plotted against an increasing SNR (dB) for LS with 30°, 45°, 60° sectors, GP, and standalone root-MUSIC for two QPSK signals arriving from 10° and 30°. Source: Adapted from [19].

100 simulations with random i.i.d. QPSK symbols. In each simulation, the array was randomly distributed. The first and last elements were positioned at $d_0 = 0$ and $d_{L-1} = (L-1)\lambda/2$, and the remaining elements were placed at positions d_l drawn from a uniform distribution centered at $l\lambda/2$ and distributed $\pm\lambda/4$ around the mean.

Figure 13.7 shows the performance of all the algorithms with respect to the number of sensors and with SNR of 3 dB, and Figure 13.8 shows their performance with respect to the SNR and with 10 sensors. As it can be seen in Figure 13.7, the performance of the LS algorithm is the best one when the number of sensors is low but higher than the number of sources, and the sectors are small (30°) and it decreases when the number of sensors increases. This is due to the fact that the interpolation is linear, and this approach is not applicable when the number of sensors is high. On the other hand, the GP approach has better performance when the number of sensors increases, which makes the signal more nonlinear, whereas for low number of sensors the approach tends to overfit. For 10 sensors, both approaches have similar performance and a reasonable robustness to low SNR, but the GP approach shows a performance about 3 dB better than the LS in low SNR. When the SNR is above 6 dB and the number of sensors is 10, both approaches have similar performance.

13.6 Conclusion

After describing the problems to be treated and their classical approaches, based on the MVDR and MUSIC algorithms, we introduced a complex version of the SVM regressor (SVR) suitable for use in antenna array processing, and its nonlinear counterpart through the kernel trick, together with a robust cost function to be used in conditions of AWGN and possibly nonlinear interference.

The nonlinear SVM approaches are related to nonlinear counterparts of Wiener and RR algorithms when the SVM parameters tend to certain limits. We introduced further algorithms for nonlinear array beamforming with temporal and spatial reference that have advantages over their linear counterparts. These advantages are clear when the algorithms are tested in signal detection, where the BER of the nonlinear algorithms are clearly lower than the linear ones, with advantages up to 5 dB in the presented simulations.

In order to perform DOA estimation of multiple signals in linear nonuniform arrays, we presented a methodology based on nonlinear interpolation of complex signals that uses a complex nonlinear GP. This approach has the advantage that it does not need to cross validate the model parameters and it offers a confidence interval of the prediction. As it can be seen in the simulations, it performs significantly better than the standard linear approach based on a LS squares interpolation of the nonuniform steering vectors.

A.13 Appendix: Subtracting the Mean in a Hilbert Space

In order to compute the covariance matrix from which to perform a principal component analysis, algorithm, in Theorem 13.3 it is assumed that the data has zero mean [35]. In order to ensure it, the mean has to be computed and subtracted as follows,

$$\tilde{\varphi}(\mathbf{x}[i]) = \varphi(\mathbf{x}[i]) - \frac{1}{N}\sum_{k}\varphi(\mathbf{x}[k]) \tag{A.13.1}$$

The kernel matrix of the centered data is

$$\tilde{\mathbf{K}} = \mathbf{K} - \mathbf{BK} - \mathbf{KB} + \mathbf{BKB} \tag{A.13.2}$$

where $\mathbf{B}$ is an $N \times N$ matrix whose elements are equal to $\frac{1}{N}$.

If a new vector is centered by using the mean computed with the previous data, the corresponding image in the feature space can be computed as

$$\tilde{\mathbf{k}}[n] = \mathbf{k}[n] - \mathbf{bK} - \mathbf{k}[n]\mathbf{B} + \mathbf{bKB} \tag{A.13.3}$$

where $\mathbf{b}$ is a row vector whose elements are equal to $1/N$.

Acknowledgments

This work has been partially supported by the King Felipe VI Endowed Chair of the University of New Mexico. This chapter is part of R&D project BigTheory (PID2019-106623RB-C41) funded by AEI (ERDF, A way of making Europe).

References

1 Schölkopf, B., Herbrich, R., and Smola, A.J. (2001). A generalized representer theorem. In: *Computational Learning Theory. International Conference on Computational Learning Theory, Lecture Notes in Computer Science*, vol. 2111 (ed. D. Helmbold and B. Williamson), 416–426. Berlin, Heidelberg: Springer-Verlag.

2 Burges, C. (1998). A tutorial on support vector machines for pattern recognition. *Data Mining and Knowledge Discovery* 2 (2): 1–32.

3 Smola, A., Schölkopf, B., and Müller, K.R. (1998). The connection between regularization operators and support vector kernels. *Neural Networks* 1 (1): 637–649.

4 Shawe-Taylor, J. and Cristianini, N. (2004). *Kernel Methods for Pattern Analysis*. Cambridge University Press.

5 Vovk, V. (2013). Kernel ridge regression. In: *Empirical Inference* (ed. Z. Luo and V. Vovk), 105–116. Berlin, Heidelberg: Springer-Verlag.
6 Rasmussen, C.E. and Williams, C.K.I. (2005). *Gaussian Processes for Machine Learning, Adaptive Computation and Machine Learning*. The MIT Press.
7 Martínez-Ramón, M. and Christodoulou, C.G. (2006). *Support Vector Machines for Antenna Array Processing and Electromagnetics, Synthesis Lectures on Computational Electromagnetics*. San Rafael, CA: Morgan & Claypool Publishers.
8 Martínez-Ramón, M., Gupta, A., Rojo-Álvarez, J.L., and Christodoulou, C.G. (2021). *Machine Learning Applications in Electromagnetics and Antenna Array Processing*. Artech House.
9 Camps-Valls, G. and Bruzzone, L. (2005). Kernel-based methods for hyperspectral image classification. *IEEE Transactions on Geoscience and Remote Sensing* 43 (6): 1351–1362.
10 Capon, J. (1969). High resolution frequency-wavenumber spectrum analysis. *Proceedings of the IEEE* 57 (8): 1408–1418.
11 Schmidt, R.O. (1979). Multiple emitter location and signal parameter estimation. *RADC Spectral Estimation Workshop*, Rome, NY, USA, pp. 243–258.
12 Barabell, A.J. (1983). Improving the resolution performance of eigenstructure-based direction-finding algorithms. *IEEE International Conference on Acoustics, Speech and Signal Processing - Proceedings*, Volume 8, pp. 336–339.
13 Roy, R., Paulraj, A., and Kailath, T. (1987). Comparative performance of esprit and music for direction-of-arrival estimation. *ICASSP'87. IEEE International Conference on Acoustics, Speech, and Signal Processing*, Volume 12, IEEE, pp. 2344–2347.
14 Tuncer, T.E. and Friedlander, B. (2009). *Classical and Modern Direction-of-Arrival Estimation*. Orlando, FL: Academic Press, Inc.
15 Oliveri, G. and Massa, A. (2011). Bayesian compressive sampling for pattern synthesis with maximally sparse non-uniform linear arrays. *IEEE Transactions on Antennas and Propagation* 59 (2): 467–481.
16 Friedlander, B. (1993). The root-music algorithm for direction finding with interpolated arrays. *Signal Processing* 30 (1): 15–29.
17 Bronez, T.P. (1988). Sector interpolation of non-uniform arrays for efficient high resolution bearing estimation. *ICASSP-88. IEEE International Conference on Acoustics, Speech, and Signal Processing*, Volume 5, pp. 2885–2888.
18 Gupta, A., Christodoulou, C.G., Martínez-Ramón, M., and Rojo-Álvarez, J.L. (2018). Kernel DOA estimation in nonuniform arrays. *2018 IEEE International Symposium on Antennas and Propagation USNC/URSI National Radio Science Meeting*, pp. 193–194.
19 Gupta, A., Christodoulou, C.G., Rojo-Álvarez, J.L., and Martínez-Ramón, M. (2019). Gaussian processes for direction-of-arrival estimation with random arrays. *IEEE Antennas and Wireless Propagation Letters* 18 (11): 2297–2300.
20 Van Trees, H.L. (2004). *Optimum Array Processing: Part IV of Detection, Estimation, and Modulation Theory*. Wiley.
21 Tuncer, T.E., Yasar, T.K., and Friedlander, B. (2007). Direction of arrival estimation for nonuniform linear arrays by using array interpolation. *Radio Science* 42 (04): 1–11.
22 Smola, A.J. and Schölkopf, B. (2004). A tutorial on support vector regression. *Statistics and Computing* 4 (3): 199–222.
23 Rojo-Álvarez, J.L., Camps-Valls, G., Martínez-Ramón, M. et al. (2005). Support vector machines framework for linear signal processing. *Signal Processing* 85 (12): 2316–2326.
24 Martínez-Ramón, M., Xu, N., and Christodoulou, C. (2005). Beamforming using support vector machines. *IEEE Antennas and Wireless Propagation Letters* 4: 439–442.

25 Tikhonov, A. and Arsenen, V. (1977). *Solution to ILL-Posed Problems.* V.H. Winston & Sons.

26 Vapnik, V.N. and Chervonenkis, A. (1971). On the uniform convergence of relative frequencies of events to their probabilities. *Theory of Probability and its Applications* 16: 264–280.

27 Huber, P.J. (1972). The 1972 wald lecture robust statistics: a review. *Annals of Mathematical Statistics* 43 (4): 1041–1067.

28 Rojo-Álvarez, J.L., Martínez-Ramón, M., de Prado-Cumplido, M. et al. (2004). Support vector method for robust ARMA system identification. *IEEE Transactions on Signal Processing* 52 (1): 155–164.

29 Platt, J. (1999). Advances in kernel methods: support vector learning. In: *Ch. Fast Training of Support Vector Machines Using Sequential Minimal Optimization* (ed. B. Scholkopf, C.J.C. Burgues, and A.J. Smola), 185–208. MIT Press.

30 Navia-Vázquez, A., Pérez-Cruz, F., Artés-Rodríguez, A., and Figueiras-Vidal, A. (2001). Weighted least squares training of support vector classifiers leading to compact and adaptive schemes. *IEEE Transactions on Neural Networks* 12 (5): 1047–1059.

31 Bouboulis, P., Theodoridis, S., Mavroforakis, C., and Evaggelatou-Dalla, L. (2014). Complex support vector machines for regression and quaternary classification. *IEEE Transactions on Neural Networks and Learning Systems* 26 (6): 1260–1274.

32 Aizerman, M.A., Braverman, E.M., and Rozoner, L. (1964). Theoretical foundations of the potential function method in pattern recognition learning. *Automation and Remote Control* 25: 821–837.

33 Vapnik, V. (1998). *Statistical Learning Theory, Adaptive and Learning Systems for Signal Processing, Communications, and Control.* Wiley.

34 Mercer, J. (1909). Functions of negative and positive type and their connection with the theory of integral equations. *Philosophical Transactions of the Royal Society of London, Series A: Mathematical, Physical and Engineering Sciences* 209: 415–446.

35 Schölkopf, B., Smola, A., and Müller, K.-R. (1996). Nonlinear Component Analysis as a Kernel Eigenvalue Problem. *Technical Report 44.* Tübingen, Germany: Max Planck Institut für biologische Kybernetik.

36 El Gonnouni, A., Martinez-Ramon, M., Rojo-Alvarez, J.L. et al. (2012). A support vector machine music algorithm. *IEEE Transactions on Antennas and Propagation* 60 (10): 4901–4910.

37 Gaudes, C.C., Santamaría, I., Via, J. et al. (2007). Robust array beamforming with sidelobe control using support vector machines. *IEEE Transactions on Signal Processing* 55 (2): 574–584.

38 Martínez-Ramón, M., Rojo-Álvarez, J.L., Camps-Valls, G., and Christodoulou, C.G. (2007). Kernel antenna array processing. *IEEE Transactions on Antennas and Propagation, Special Issue on Synthesis and Optimization Techniques in Electromagnetics and Antenna System Design* 55 (3): 642–650.

39 Kwok, J.T. and Tsang, I.W. (2003). Linear dependency between ε and the input noise in ε-support vector regression. *IEEE Transactions in Neural Networks* 14 (3): 544–553.

40 Cherkassky, V. and Ma, Y. (2004). Practical selection of SVM parameters and noise estimation for SVM regression. *Neural Networks* 17 (1): 113–126.

41 Rojo-Álvarez, J.L., Figuera-Pozuelo, C., Martínez-Cruz, C.E. et al. (2007). Nonuniform interpolation of noisy signals using support vector machines. *IEEE Transactions on Signal Processing* 55 (8): 4116–4126.

42 Figuera, C., Barquero-Pérez, O., Rojo-Álvarez, J.L. et al. (2014). Spectrally adapted mercer kernels for support vector nonuniform interpolation. *Signal Processing* 94: 421–433.

43 Rojo-Álvarez, J.L., Martínez-Ramón, M., Mu noz-Marí, J., and Camps-Valls, G. (2014). A unified SVM framework for signal estimation. *Digital Signal Processing* 26: 1–20.

44 Boloix-Tortosa, R., Arias-de-Reyna, E., Payan-Somet, F.J., and Murillo-Fuentes, J.J. (2018). Complex-valued Gaussian processes for regression: a widely non-linear approach. CoRR abs/1511.05710.

45 Sundararajan, S. and Keerthi, S.S. (2001). Predictive approaches for choosing hyperparameters in Gaussian processes. *Neural Computation* 13 (5): 1103–1118.

14

Knowledge-Based Globalized Optimization of High-Frequency Structures Using Inverse Surrogates

Anna Pietrenko-Dabrowska[1] and Slawomir Koziel[1,2]

[1] *Faculty of Electronics, Telecommunications and Informatics, Gdansk University of Technology, Narutowicza, Gdansk, Poland*
[2] *Engineering Optimization & Modeling Center, Reykjavik University, Menntavegur, Reykjavik, Iceland*

14.1 Introduction

Recent years have witnessed a significant increase of the topological complexity of passive high-frequency components [1, 2]. This mainly results from growing performance demands [3], functionality requirements (e.g. multi-band operation [4], spurious response suppression [5], unconventional phase responses [6], circular polarization [7], tunability [8]), and need for miniaturization [9]. The methods for size reduction of microwave circuits (transmission line [TL] folding [10], slow-wave phenomenon, e.g. replacing TLs by compact microstrip resonant cells, CMRCs [11]), and antennas (stubs [12], shorting pins [13], defected ground structures [14], stepped-impedance feed lines [15]), generally lead to geometrically intricate topologies featuring increased number of parameters [16, 17]. Circuit-theory-based approach (e.g. analytical or equivalent network models) is frequently incapable of tackling complexity of such structures, or simpler representations are not available at all in the case of some components, e.g. most of antenna structures. Furthermore, reliable evaluation of complex structures necessitates full-wave EM analysis, because of, e.g. occurrence of electromagnetic (EM) cross-coupling effects. Consequently, EM tools become indispensable, both at the initial stages of structure development and its design closure (final parameter tuning).

Due to intricate relationships between the component topology and its electrical and field characteristics, concurrent tuning of circuit dimensions using numerical procedures is nowadays indispensable for securing the best possible performance. As a matter of fact, unsupervised (automated) optimization is the only way to properly handle multiple objectives and constraints within parameter spaces of increased dimensionality. Still, its cost is sizeable even in the case of a local tuning. Moreover, for multimodal tasks (optimization of coding metasurfaces [18] or frequency selective surfaces [19], pattern synthesis [20]), multi-criterial design [21], or in the cases where a decent initial design is lacking (e.g. design of CMRC-based compact circuits [22], system re-design for different operating frequencies [23]), global search becomes a must, thereby making the optimization problem even more challenging.

Without a doubt, the most widely used global optimization procedures nowadays are population-based nature-inspired algorithms [24–26]. They originated in late 1960s (evolutionary strategies, ES [27]), although these techniques really took off in 1980s and early 1990s, to eventually dominate global search practice since 2000s (genetic algorithms [28], evolutionary

Advances in Electromagnetics Empowered by Artificial Intelligence and Deep Learning, First Edition.
Edited by Sawyer D. Campbell and Douglas H. Werner.

computation [29], ant systems [30], particle swarm optimization, PSO [31], differential evolution, DE [32]). Within the last decade, numerous variations of nature-inspired techniques have been devised (firefly algorithm [33], harmony search [34], gray wolf optimization [35], invasive weed optimization [36], spider monkey optimization [37]; the list goes on [33–41]). Nevertheless, majority of the recent routines are in a close resemblance of each other, so that only a small number of clearly distinct approaches can be identified. In population-based algorithms, information between each candidate solution (a population [42], a swarm [43], a pack [44]) is exchanged, in addition to producing new solutions with the use of exploitative operators (e.g. mutation [45]). Local minima are avoided by involving randomness of various types [46, 47]. Usually, implementation of nature-inspired algorithms is straightforward, the main obstacle remaining their poor computational efficiency: typically, an individual optimization run requires anything between a few hundreds and many thousands evaluations of the objective function. Clearly, this is a serious impediment if the response of a component under design needs to be simulated through full-wave EM analysis.

Practical applicability of nature-inspired algorithms in high-frequency design is restricted to the tasks for which the merit function is cheap to evaluate (such as array pattern synthesis using analytical array factor models [48]), EM analysis is rather fast (e.g. less than ten seconds), or possibility of parallelization exists. The latter requires sufficient resources both in terms of hardware and EM software licensing. Another approach consists in employing surrogate modeling techniques [49, 50]. A large variety of the available methods that are popular in high-frequency engineering include kriging [51], Gaussian process regression [52], several variations of neural networks [53–55], and, recently, polynomial chaos expansion [56]. For a practical case, the metamodel plays a role of a fast predictor, which undergoes further refinement with the use of the EM simulation data gathered throughout the optimization run. The infill criteria, determining the allocation of new training samples, may be aimed at space exploration (enhancement of global accuracy of the model) or its exploitation (optimum identification) [57]. Surrogates are frequently utilized in conjunction with machine learning techniques [58], or to pre-screen the design space, also in variable-fidelity regime [59].

The usage of data-driven metamodels for global optimization is—unfortunately—encumbered by the curse of dimensionality, and, even more importantly, by non-linearity of high-frequency circuit characteristics (scattering parameters versus frequency, antenna gain, axial ratio, etc.). In practice, only relatively simple devices can be handled, i.e. featuring a reduced number of parameters within rather narrow ranges [60, 61]. The range of applicability of metamodels may be significantly extended by employing the recently developed performance-driven modeling approach [62–65]. Therein, the surrogate is only built nearby a specific manifold, which corresponds to optimal or nearly optimal designs with regard to the relevant performance figures [62]. Confining the domain permits to construct reliable surrogates within broad ranges of geometry, material, and operating parameters of the circuit of interest. This has been achieved through a two-stage procedure, in which the design space is initially mapped onto a low-dimensional manifold with the use of a supplementary inverse surrogate. Next, the final metamodel is set up in the neighborhood of the aforementioned manifold. The volume of this region is significantly smaller than that of the original space. Applicability of the performance-driven approach has been also extended to variable-fidelity regime providing further computational savings [66, 67]. Other applications of performance-driven techniques include multi-objective design optimization [68, 69] or yield optimization [70, 71]. Yet another approach for expediting simulation-driven optimization and modeling procedures is the response feature technique [72], in which the design goals are formulated with regard to characteristic points of the component response [73]. As a consequence, significant savings can be achieved owing to a nearly linear relationship between the coordinates of the characteristic points and component dimensions, either in the form of expediting the search process [72] or limiting the number of training samples required to construct a reliable surrogate [74]. Incorporation of the response features into the performance-driven modeling techniques leads to a further, and significant reduction of the training data set sizes necessary to set up reliable surrogates [75].

This chapter discusses a recent technique for globalized parameter tuning of high-frequency devices [76, 77], in which encouraging regions of the design space are identified with the use of an inverse regression surrogate constructed using pre-selected random observables as training data. The said inverse model is established at the level of response features, which allows for handling innately non-linear system responses at a reasonable cost. The starting point produced by means of the inverse surrogate undergoes local tuning with the use of the trust-region gradient-based routine. The considered procedure has been numerically verified using several passive structures, including microwave and antenna components. The global search capability of the discussed algorithm has been achieved at a computational cost comparable to that of a local search. Moreover, the algorithm outperforms multiple-start gradient-based search and also particle swarm optimizer (employed as an example of population-based metaheuristics).

14.2 Globalized Optimization by Feature-Based Inverse Surrogates

This section introduces the optimization technique being the subject of the chapter. The presented procedure exploits inverse regression metamodels set up with the use of initially checked and accepted random observables, and the feature points of the system outputs evaluated using full-wave EM analysis. Close-to-linear relationship dependence of the said characteristic points on the design variables permits globalized search at a computational cost comparable to that of the local procedures. The inverse surrogate yields a decent initial design, which further undergoes a local refinement using a local gradient-based routine.

14.2.1 Design Task Formulation

Let us formulate the EM-driven design task as follows

$$\boldsymbol{x}^* = \arg\min_{\boldsymbol{x}} U(\boldsymbol{x}, \boldsymbol{F}_t) \tag{14.1}$$

with U being a scalar merit function, and $\boldsymbol{F}_t = [F_{t.1} \ldots F_{t.K}]^T$ representing a vector of target values of operating parameters. The quality of the design is quantified with the use of EM-simulated component characteristics, typically, scattering parameters $S_{ij}(\boldsymbol{x},f)$ (where i and j refer to the relevant ports of the structure at hand), gain $G(\boldsymbol{x},f)$, axial ratio $AR(\boldsymbol{x},f)$, etc. Therein, the vector of geometry parameters is denoted as $\boldsymbol{x}$, whereas f stands for the frequency.

Let us consider an example, in which a microwave coupler is to be optimized to fulfill the following requirements: (i) minimize matching and isolation, $|S_{11}|$ and $|S_{41}|$, at an operating frequency f_0, and (ii) ensure that the power split ratio $d_S(\boldsymbol{x},f_0) = |S_{21}(\boldsymbol{x},f_0)| - |S_{31}(\boldsymbol{x},f_0)|$ is equal to the intended value K_P (e.g. 0 dB when equal power split is of interest). Here, the operating parameter vector is $\boldsymbol{F}_t = [f_0\ K_P]^T$, with the possible definition of the objective function

$$U(\boldsymbol{x}, \boldsymbol{F}_t) = U\left(\boldsymbol{x}, [f_0\quad K_P]^T\right) = \max\{|S_{11}(\boldsymbol{x},f_0)|, |S_{41}(\boldsymbol{x},f_0)|\} + \beta[d_S(\boldsymbol{x},f_0) - K_P]^2 \tag{14.2}$$

In (14.2), the second component is a penalty term which ensures the target power split ratio, whereas β stands for the penalty factor which makes the contribution of the penalty term commensurable to that of the complete merit function.

The second example involves a dual-band coupler optimized for a specific substrate of relative permittivity ε_r. Moreover, the circuit is to minimize both $|S_{11}|$ and $|S_{41}|$ at the circuit operating frequencies $f_{0.1}$ and $f_{0.2}$, and, at the same time, enforce equal power split at $f_{0.1}$ and $f_{0.2}$. In this case, the operating parameter vector is $\boldsymbol{F}_t = [f_{0.1}\ f_{0.2}\ \varepsilon_r]^T$, whereas the merit function is formulated as

$$U(\boldsymbol{x}, \boldsymbol{F}_t) = U\left(\boldsymbol{x}, [f_{0.1}\ \ f_{0.2}\ \ \varepsilon_r]^T\right) = \max\left\{|S_{11}(\boldsymbol{x}, f_{0.1})|, \left|S_{41}\left(\boldsymbol{x}, f_{0.1}\right)\right|, \left|S_{11}\left(\boldsymbol{x}, f_{0.2}\right)\right|, \left|S_{41}\left(\boldsymbol{x}, f_{0.2}\right)\right|\right\} + \beta\left[d_S(\boldsymbol{x}, f_{0.1})^2 + d_S\left(\boldsymbol{x}, f_{0.2}\right)^2\right] \tag{14.3}$$

Finally, let us consider a multi-band antenna, which is supposed to be designed to improve its impedance matching at all operating frequencies, $f_{t.j}, j = 1, \ldots, K$. In this case, the operating parameter vector is $\boldsymbol{F}_t = [f_{t.1} \ldots f_{t.K}]^T$, whereas objective function is defined as

$$U(\boldsymbol{x}, \boldsymbol{f}_t) = \max_{\boldsymbol{x}}\left\{|S_{11}(\boldsymbol{x}, f_{t.1})|, \ldots, \left|S_{11}\left(\boldsymbol{x}, f_{t.K}\right)\right|\right\} \tag{14.4}$$

Clearly, it is possible to handle other conceivable design tasks in a similar manner.

14.2.2 Evaluating Design Quality with Response Features

Parameter tuning is an important stage of high-frequency design process, whose goal is to improve the performance parameters. As delineated in Section 14.2.1, this task may be formulated as an optimization problem, solving which, out of reliability, requires the involvement of EM simulations. In this chapter, we consider globalized optimization, which is frequently a must for various reasons, among them the absence of a reasonable initial design, or the existence of multiple local optima (which, in some cases, may fail to meet the intended targets). An example may be a design optimization of compact microwave passives, in which CMRC or similar unit cells replace conventional transmission lines [11]. In such cases, the inter-relations between design variables and cell responses are, as a rule, intricate [17]. This makes finding a decent starting point a daunting task. The same may pertain to the case when a given structure is to be re-designed, e.g. an antenna, for operating frequencies or substrates being significantly distant from that of the available design.

Nonlinearity of system responses both versus design variables and frequency, as well as dimensionality issues make global exploration of the design space troublesome. On the one hand, the cost of direct simulation-based global search involving nature-inspired algorithms more often than not turns out to be exorbitant. On the other hand, the aforementioned issues significantly impede the employment of surrogate models. This is because construction of reliable surrogates in high dimensional design spaces and, at the same time, within satisfactorily broad ranges of the design parameters is hard to accomplish.

Several exemplary cases, in which a local optimizer may not arrive at an acceptable solution as a result of the absence of a decent initial design and/or the need for re-designing a given component for operating parameters remote from that of the already available design are shown in Figure 14.1.

The aforementioned difficulties may be resolved through the exploitation of the problem-specific knowledge embedded in the EM-simulated system outputs in the form of the response features, i.e. by employing the feature-based optimization (FBO) technology [72]. Therein, the design task is formulated in terms of the said features (instead of the entire characteristics, as in the conventional approach). This allows for achieving a sizeable amount of information pertaining to the component of interest using only a small amount of EM-simulated data. The remarkable performance of FBO comes from the observation that the inter-relations between the feature point

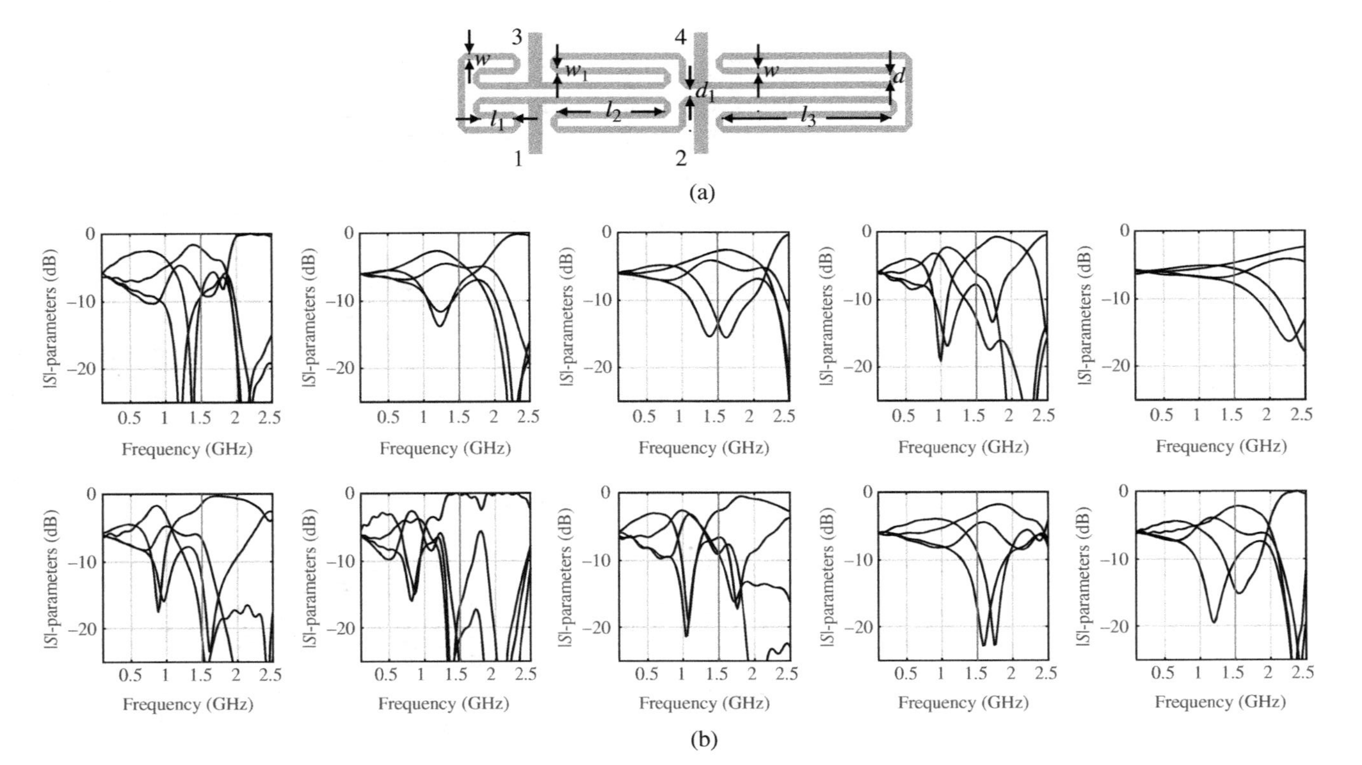

Figure 14.1 Scattering parameters of a compact microwave coupler versus frequency: (a) component geometry, (b) coupler responses at exemplary designs residing in the assumed design space; target operating frequency $f_0 = 1.6$ GHz is indicated using the vertical lines. Local optimizer launched from majority of the presented designs would not render satisfactory solution when executed with the use of the merit function in the form of (14.2), as a consequence of a considerable misalignment between the assumed and the required operating conditions.

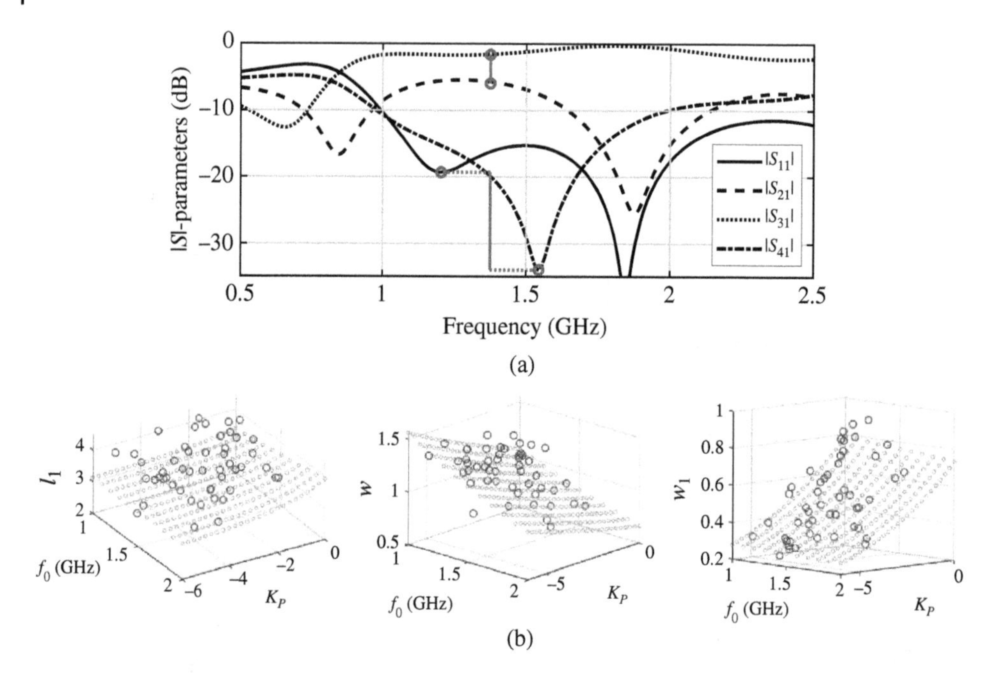

Figure 14.2 Compact rat-race coupler: (a) characteristic points (marked using circles) representing minima of matching $|S_{11}|$ and isolation $|S_{41}|$ characteristics, along with the power split ratio K_P evaluated at the approximate operating frequency f_0 of the coupler equal to the average frequency of the minima of $|S_{11}|$ and $|S_{41}|$ characteristics (indicated by thick vertical line); some feature points may not be discernible for some designs; (b) dependence of f_0 and K_P on the three selected geometry parameters, coupler designs are marked using circles; the gray smaller circles refer to the regression surrogate of the form $a_0 + a_1\exp(a_2 f_0 + a_3 K_P)$, which represents the trends between the circuit dimensions and its operating parameters.

coordinates and design variables are significantly less nonlinear that those of the system responses (see Figure 14.1b). This is presented in Figure 14.2 for a microwave coupler, and, in Figure 14.3, for an exemplary dual-band antenna.

The characteristic points definition relies on the actual shape of the circuit outputs, as well as the way in which the design task is formulated. The said points may simply refer to the frequency and level coordinates of the resonances [72], local minima and/or maxima of the return loss within the pass-band [74], or points delimiting an operational bandwidth or power split [23]. One of the fundamental components of the discussed optimization framework, particularly in its initial phase (Section 14.2.3), is the exploitation of response features. Yet, the features are mainly utilized for estimation of the current system operating parameters, and not directly (as in FBO [72]).

14.2.3 Globalized Search by Means of Inverse Regression Surrogates

As announced in Section 14.2.2, the efficiency of the methodology presented in this chapter is a result of a close-to-linear relationship between the design variables of the high-frequency component at hand and its operating conditions (frequency, power split, bandwidth), which has been employed for design space exploration. Some exemplary relationships of this type are shown in Figures 14.2b and 14.3b for a compact microstrip coupler and a dual-band antenna. These dependencies are estimated using the information gathered using random observables (parameter vectors). Clearly, the quality of some observables may be satisfactory from the point of view of the

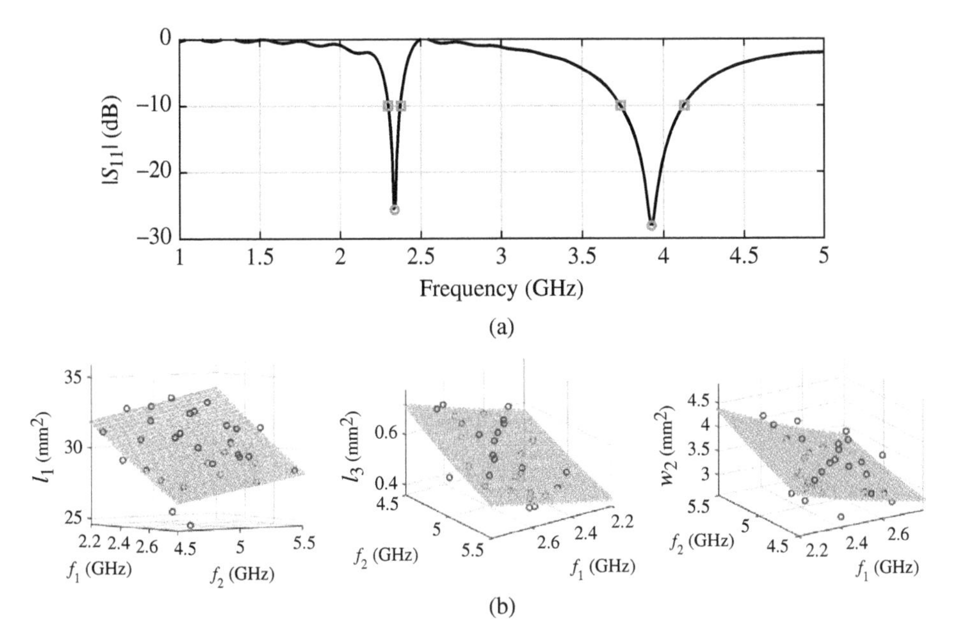

Figure 14.3 Dual-band dipole antenna: (a) characteristic points representing antenna resonances (marked using circles), along with −10 dB reflection levels (marked using squares); similarly as in Figure 14.2, some feature points may not be discernible for some designs; (b) dependence of the operating frequencies on the three selected geometry parameters, antenna designs are marked using circles; the gray smaller circles refer to the regression surrogate of the form $a_0 + a_1\exp(a_2 f_1 + a_3 f_2)$, which represents the trends between the antenna dimensions and its operating frequencies.

target performance requirements, whereas the quality of the remaining ones may be insufficient, and thus they will have to discarded. The quality of the random observables is assessed with the use of the characteristic points garnered from the simulated circuit response, through a comparison of the utility metrics with the assumed design goals. The subset of the observables of the highest quality is subsequently utilized for identification of an inverse regression surrogate. The purpose of the model is twofold: (i) finding the promising region of a design space region, and (ii) rendition of the infill points for the inverse model refinement. The global search is carried out in an iterative manner, by generating a single infill point per iteration, and replacing the worst observables by the ones that are closer to the assumed targets. The remainder of this section provides the details of the aforementioned procedure.

The following notation is used:

- $\boldsymbol{F}(\boldsymbol{x}) = [f_1(\boldsymbol{x}) \dots f_K(\boldsymbol{x})]^T$ – an operating parameter vector at the point $\boldsymbol{x}$ (e.g. operating frequency, bandwidth, power split ratio), derived from the EM simulated outputs. As explained above, the actual values of the operating parameters are assessed using the features (cf. Figures 14.2a and 14.3a); e.g. the coupler's operating frequency may be estimated as the mean value of the frequencies representing the minima of the matching and isolation characteristics. For the cases, in which it is impossible to extract certain parameters (e.g. if some characteristic points are indiscernible or do not reside within the frequency range of interest), we have $\boldsymbol{F}(\boldsymbol{x}) = [0 \dots 0]^T$;
- $\boldsymbol{L}(\boldsymbol{x}) = [l_1(\boldsymbol{x}) \dots l_K(\boldsymbol{x})]^T$ – an auxiliary vector comprising coefficients which reflect the design quality; the entries of the vector $\boldsymbol{L}(\boldsymbol{x})$ correspond to those of the vector $\boldsymbol{F}(\boldsymbol{x})$. Let us use an example: if the goal is to minimize the level of $|S_{11}|$ and $|S_{41}|$ at a certain operational frequency, the corresponding l_k may be defined as the average of the minimum levels of $|S_{11}|$ and $|S_{41}|$, with

the lower value indicating the observable of the better quality. Likewise, for a power split ratio: the corresponding l_k can be a difference between the target and estimated power split value. Another example may be the values of the antenna reflection at the resonant frequencies at a given design $\boldsymbol{x}$. In the cases, where certain elements of $\boldsymbol{L}(\boldsymbol{x})$ are impossible to be extracted, we use $\boldsymbol{L}(\boldsymbol{x}) = [0 \dots 0]^T$;

- $D(\boldsymbol{F},\boldsymbol{F}_t)$ – a function which quantifies the misalignment between the vector of operating parameters $\boldsymbol{F}$ and the target vector $\boldsymbol{F}_t$ (cf. Section 14.2.1) and; here, L_2-norm-based distance $D(\boldsymbol{F},\boldsymbol{f}_t) = ||\boldsymbol{F} - \boldsymbol{f}_t||$ is adopted;
- D_{accept} – control parameter (user-defined) utilized for termination of the global search stage. In particular, it is assumed that the current design is close enough to the assumed goal if $D(\boldsymbol{F},\boldsymbol{f}_t) \leq D_{accept}$.

Let us discuss the following outline to explicate the operation the specific steps along with their meaning:

1. *Observable generation*: Generate a set of designs $\boldsymbol{x}^{(j)}, j = 1, \dots, N$, allocated randomly within the parameter space X (typically, a box-constrained domain delimited the lower and upper parameter bounds), usually, according to a uniform probability distribution. The designs' rendition lasts until N designs fulfilling the condition $||\boldsymbol{F}(\boldsymbol{x}^{(j)})|| > 0, j = 1, \dots, N$ have been obtained.
2. *Inverse surrogate construction*: Identify an inverse regression model $r_I(\boldsymbol{F})$ with the values in X with the use of the set of triplets $\{\boldsymbol{F}(\boldsymbol{x}^{(j)}), \boldsymbol{L}(\boldsymbol{x}^{(j)}), \boldsymbol{x}^{(j)}\}_{j=1,\dots,N}$,; the surrogate serves for quantifying the relationships between the geometry and operating and parameters of the system under design. The details on analytical formulation of r_I will be provided in the remaining part of this section;
3. *Design prediction*: Identify a candidate design $\boldsymbol{x}_{tmp} = r_I(\boldsymbol{F}_t)$, where $\boldsymbol{F}_t$ denotes the target operating parameter vector (cf. Section 14.2.1) with the use of the inverse surrogate r_I to. If $||\boldsymbol{F}(\boldsymbol{x}_{tmp})|| > 0$ and $D(\boldsymbol{F}(\boldsymbol{x}_{tmp}),\boldsymbol{F}_t) < \max\{j = 1, \dots, N : D(\boldsymbol{F}(\boldsymbol{x}^{(j)}),\boldsymbol{F}_t)\}$, candidate vector $\boldsymbol{x}_{tmp}$ supersedes the design corresponding to the said maximum, and r_I is set anew.

Steps 2 and 3 are iterated in search of a design close enough to the design goals, i.e. the process is concluded if $D(\boldsymbol{F}(\boldsymbol{x}_{tmp}),\boldsymbol{f}_t) < D_{\max}$, with $D_{\max}$ being a user-defined acceptance threshold. This stage is followed by a local design refinement as described in Section 14.2.4. Observe that random designs are produced until an assumed number of the vectors featuring clearly discernible feature points have been gathered. The next stage is a construction of the inverse surrogate (a graphical illustration of this step is provided in Figures 14.2b and 14.3b). This model is subsequently utilized as a predictor which identifies the design of the operating parameters close as much as possible to the target $\boldsymbol{F}_t$. If the quality of the candidate design is deemed sufficient (as assessed by function $D(\boldsymbol{F},\boldsymbol{f}_t)$), the worst of the already-gathered observables is replaced by it.

As the procedure advances, the inverse surrogate will be more and more concentrated on the part of the design space encompassing parameter vectors featuring low values of the proximity function $D(\boldsymbol{F},\boldsymbol{f}_t)$. This is because this function governs the observable replacement in the entire dataset $\{\boldsymbol{x}^{(j)}\}$. As a consequence, the local predictive power of the model will be gradually enhanced. Figure 14.4 shows a graphical illustration of the Steps 1 through 3.

The inverse surrogate $r_I(\boldsymbol{F})$ is the key component of the considered quasi-global search. As previously explained, the said model is set up with the use of the triplets $\{\boldsymbol{F}(\boldsymbol{x}^{(j)}), \boldsymbol{L}(\boldsymbol{x}^{(j)}), \boldsymbol{x}^{(j)}\}_{j=1,\dots,N}$. Whereas its analytical form may remain fairly simple as the relationship between the operating conditions and design variables of the high-frequency component at hand is usually close-to-linear. Still, the surrogate needs to be sufficiently flexible to reflect the fact that the said relation may be

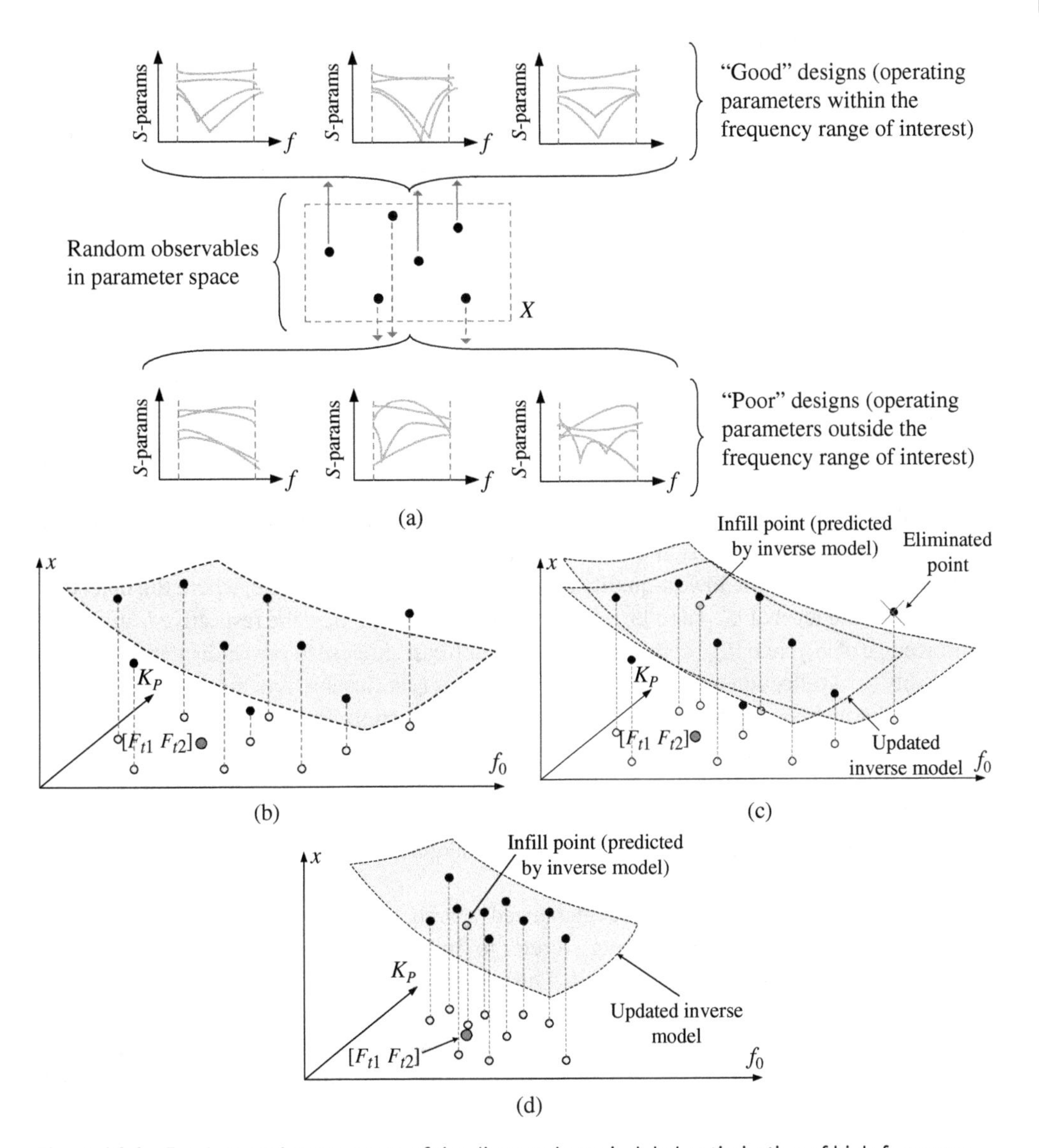

Figure 14.4 Fundamental components of the discussed quasi-global optimization of high-frequency components: (a) random observable acquisition: only the vectors featuring operating parameters (center frequency, power split ratio) from the simulation frequency range and the region of interest will be taken into account while building inverse model, (b) observables (•) in the two-dimensional operating parameter space f_0, K_P; also shown are the respective projections onto the f_0–K_P plane, the initial inverse surrogate (grey surface); dark grey circle marks the target operating parameters; (c) initial iteration of the globalized search: the infill point yielded by r_I (marked with a gray circle) is substituted for the worst design and r_I is refined; (d) final iteration: the observables are focused close to the targets and the updated surrogate renders the parameter vector which is next to the target; the procedure is terminated and followed by a local tuning (Section 14.2.4). In panels (b)–(d), a case of a single design variable x is considered; identical procedure is performed to for all variables.

nearly inversely proportional for some parameters. Hence, it takes the following form:

$$\boldsymbol{r}_I(\boldsymbol{F}) = \boldsymbol{r}_I\left([f_1 \ldots f_K]^T\right) = \begin{bmatrix} r_{I.1}(\boldsymbol{F}) \\ \cdots \\ r_{I.n}(\boldsymbol{F}) \end{bmatrix} = \begin{bmatrix} p_{1.0} + p_{1.1}\exp\left(\sum_{k=1}^{K} p_{1.k+1} f_k\right) \\ \cdots \\ p_{n.0} + p_{n.1}\exp\left(\sum_{k=1}^{K} p_{n.k+1} f_k\right) \end{bmatrix} \tag{14.5}$$

The model identification requires solving

$$[p_{j.0}\ p_{j.1} \cdots p_{j.K+1}] = \arg \min_{[b_0\ b_1 \ldots b_{K+1}]} \sum_{k=1}^{N} w_k \left[r_{I.j}(\boldsymbol{F}(\boldsymbol{x}^{(k)})) - x_j^{(k)} \right]^2, \quad j = 1, \ldots, n \tag{14.6}$$

where the following notation is used, $\boldsymbol{x}^{(j)} = [x_1^{(j)} \ldots x_n^{(j)}]^T$. The weights w_k are evaluated by taking into account the auxiliary vectors $\boldsymbol{L}(\boldsymbol{x}^{(j)})$

$$w_k = \left[1 - \max\left\{l_1(\boldsymbol{x}^{(j)}), \ldots, l_k(\boldsymbol{x}^{(j)})\right\}\right]^2, \qquad k = 1, \ldots, N \tag{14.7}$$

Additionally, the elements l_k undergo normalization, i.e. their values come from the interval [0, 1]: zero corresponds to the design of the highest-quality with regard to the kth operating condition, and one represents the lowest-quality design. In the previous example, where the objective was to minimize the level of $|S_{11}|$ and $|S_{41}|$ at the operational frequency, the respective l_k might be set as the average of $|S_{11}|$ and $|S_{41}|$ at their minima. In such case, l_k close to zero indicates the design of high-quality (low reflection and high isolation), whereas l_k is closer to one for a poorer design.

In general, the inverse surrogate r_I is basically a trend function approximating the set of the observables $\{\boldsymbol{x}^{(j)}\}$ in the weighted L-square sense (cf. (14.5)). The rationale behind introducing the weighting factors w_k is to differentiate the low-quality observables from the high-quality ones, so the impact of the latter on the model r_I would be greater. Figures 14.2b and 14.3b provide graphical illustrations of the inverse regression surrogate for the microstrip coupler and dual-band antenna of Figures 14.2a and 14.3a.

The operating flow of the considered globalized search procedure has been presented in Figure 14.5 as a pseudocode. Steps 1 through 4 refer to finding N observables featuring operating parameters being from the assumed ranges. This especially pertains to the frequency-related parameters which have to belong to the component simulation range. In Step 5, the inverse model r_I is built using these parameter sets. The subsequent steps involve: rendition of a candidate design $\boldsymbol{x}_{tmp}$ with the use of r_I, its assessment, and decision on its inclusion into the observable pool based on its quality. Subsequently, the inverse surrogate is reconstructed.

The termination condition is $D(\boldsymbol{F}(\boldsymbol{x}^{(0)}),\boldsymbol{F}_t) \le D_{accept}$, i.e. finding a design close enough to the assumed goal $\boldsymbol{F}_t$, which is subsequently utilized as a starting point $\boldsymbol{x}^{(0)}$ for local tuning (cf. Section 14.2.4). In the cases, where the procedure has been unable to find such a design, the process is terminated once computational budget has run out. Then, the best design rendered so far is returned.

14.2.4 Local Tuning Procedure

Globalized search procedure, delineated in Section 14.2.3, renders a vector $\boldsymbol{x}^{(0)}$ which satisfies the condition $D(\boldsymbol{F}(\boldsymbol{x}^{(0)}),\boldsymbol{F}_t) \le D_{accept}$. The threshold D_{accept} is set so as to enforce that the operating conditions at $\boldsymbol{x}^{(0)}$ are close enough to $\boldsymbol{F}_t$ to ensure that the target may be attained through a local search. Here, it is performed with the use of the trust-region (TR) gradient-based algorithm with numerical derivatives [78]. The TR routine renders a series of approximations $\boldsymbol{x}^{(i)}$, $i = 0, 1, \ldots$ to the optimal

1. Set $j = 1$;
2. Render a random vector $x^{(j)} \in X$;
3. **if** $||F(x^{(j)})|| > 0$
   Accept $x(j)$; set $j = j + 1$;
   end
4. **if** $j \leq N$ AND computational budget has not been exceeded
   Go to 2;
   else
   Go to 5;
   end
5. Set up the inverse surrogate $r_I(F)$ (cf. (5)-(7));
6. Obtain $x_{tmp} = r_I(F_t)$, where F_t are the target operating parameters (cf. Section 1.2.1);
7. **if** $||F(x_{tmp})|| > 0$ AND $D(F(x_{tmp}),F_t) < D_{\max} = \max\{j = 1, \ldots, N : D(F(x^{(j)}),F_t)\}$
   Supersede the vector featuring $D_{\max}$ in $\{x^{(j)}\}_{j=1,\ldots,N}$ by x_{tmp};
   else
   Produce random observables x_{tmp} until $D(F(x_{tmp}),F_t) < D_{\max}$ is met, then, the vector featuring $D_{\max}$ in $\{x^{(j)}\}_{j=1,\ldots,N}$ is superseded by x_{tmp}; if computational budget is exceeded go to 9;
   end
8. Reset the inverse surrogate $r_I(F)$ utilizing current set $\{x^{(j)}\}_{j=1,\ldots,N}$;
9. Find $x^{(0)} = x^{(jmin)}$, where $j_{\min} = \text{argmin}\{j = 1, \ldots, N : D(F(x^{(j)}),F_t)\}$;
10. **if** $D(F(x^{(0)}),F_t) \leq D_{accept}$ OR computational budget has been exceeded
    Go to 11
    else
    Go to 6
    end
11. Return $x^{(0)}$; END;

Figure 14.5 Pseudocode of the quasi-global optimization of high-frequency devices with inverse regression surrogate: the initial (global) optimization phase, followed by a local refinement (cf. Section 14.2.4).

solution $\boldsymbol{x}^*$. The upcoming iteration points are yielded by solving

$$\boldsymbol{x}^{(i+1)} = \arg \min_{\boldsymbol{x}; -\boldsymbol{d}^{(i)} \leq \boldsymbol{x} - \boldsymbol{x}^{(i)} \leq \boldsymbol{d}^{(i)}} U_L(\boldsymbol{x}, \boldsymbol{F}_t) \tag{14.8}$$

where the merit function U_L is of the same form as U (cf. (14.1)); yet, it is evaluated with the use of the linear model $\boldsymbol{G}^{(i)}(\boldsymbol{x},f)$ of the circuit outputs at $\boldsymbol{x}^{(i)}$. For the S-parameter S_{kl}, the said linear model is defined, as

$$\boldsymbol{G}^{(i)}(\boldsymbol{x},f) = S_{kl}(\boldsymbol{x}^{(i)},f) + \nabla_{Skl}(\boldsymbol{x}^{(i)},f)\cdot(\boldsymbol{x} - \boldsymbol{x}^{(i)}) \tag{14.9}$$

The sensitivities in (14.9) are evaluated through the finite differentiation. The trust region of (14.8) is defined as an interval $[\boldsymbol{x}^{(i)} - \boldsymbol{d}^{(i)}, \boldsymbol{x}^{(i)} + \boldsymbol{d}^{(i)}]$, where $\boldsymbol{d}^{(i)}$ represents the size vector adjusted in accordance with the conventional TR setup [78]. If, for the candidate solution $\boldsymbol{x}^{(i+1)}$, the merit function

Table 14.1 Global microwave design optimization: Control parameters.

Parameter rank	Notation	Interpretation	
Primary	N	Number of observables for constructing inverse surrogate	
	D_{accept}	Acceptance threshold for assessing designs yielded in the global search (cf. Section 14.2.3)	
Secondary	$N_{max.1}$	Computational budget: maximum number of full-wave simulations for:	Initial sampling
	$N_{max.2}$		Global search
	$N_{max.3}$		Local refinement
	ε	Termination threshold (for convergence in trust-region size and argument, cf. Section 14.2.4)	

value is reduced, i.e. if $U(\boldsymbol{x}^{(i+1)},\boldsymbol{F}_t) < U(\boldsymbol{x}^{(i)},\boldsymbol{F}_t)$, then $\boldsymbol{x}^{(i+1)}$ is accepted. Otherwise, it is discarded and the iteration is relaunched with a smaller $\boldsymbol{d}^{(i)}$.

The algorithm is terminated based on the following conditions: convergence in argument $||\boldsymbol{x}^{(i+1)}-\boldsymbol{x}^{(i)}|| < \varepsilon$, or reduction of the TR size, i.e. $||\boldsymbol{d}^{(i)}|| < \varepsilon$ (whichever comes first); in the numerical experiments, we set $\varepsilon = 10^{-3}$. The computational cost of the tuning process is decreased by replacing finite differentiation (which entails n EM simulations of the circuit in each iteration) by the Broyden formula [79] when the algorithm starts to converge, i.e. for $||\boldsymbol{x}^{(i+1)}-\boldsymbol{x}^{(i)}|| < 10\varepsilon$.

14.2.5 Global Optimization Algorithm

This section summarizes the operational flow of the quasi-global optimization framework considered in this chapter, whose main components, i.e. global search as well as local refinement procedures, have been delineated in Sections 14.2.3 and 14.2.4, respectively. The control parameters of the entire framework are provided in Table 14.1. Observe that only the first two parameters, N and D_{accept}, pertain specifically to the discussed framework, the other parameters are conventionally employed by numerical optimization procedures. Among these, the termination threshold determines the assumed resolution of the optimization process; $N_{max.k}$, $k = 1, 2, 3$, are adjusted with some margin so that, in practice, the optimization procedure will more likely terminate due to convergence, and not because the budget has been exceeded.

Regarding the parameter N, its value can be small, i.e. 10 or 20, since the inverse surrogate is constructed within low-dimensional space of operating parameters. The setup of D_{accept} is problem dependent. In particular, some engineering experience is required to estimate how far the operating parameters at the initial design may be from the target ones so that the target may be attained from $\boldsymbol{x}^{(0)}$ through a local optimization procedure. A rule of thumb for setting up D_{accept} is that the distances between the intended operating frequencies and the current ones at $\boldsymbol{x}^{(0)}$ are equal or smaller than the respective bandwidths of the structure under design.

To sum up, the operational flow of the overall optimization process consists of three main steps:

1. *Input arguments*:
   - Target operating vector $\boldsymbol{F}_t$,
   - Merit function U,
   - Design space X;
2. *Global search*: Find starting point $\boldsymbol{x}^{(0)}$ by executing the procedure of Section 14.2.3;
3. *Local refinement*: Render the ultimate optimal design $\boldsymbol{x}^*$ through the TR routine of Section 14.2.4.

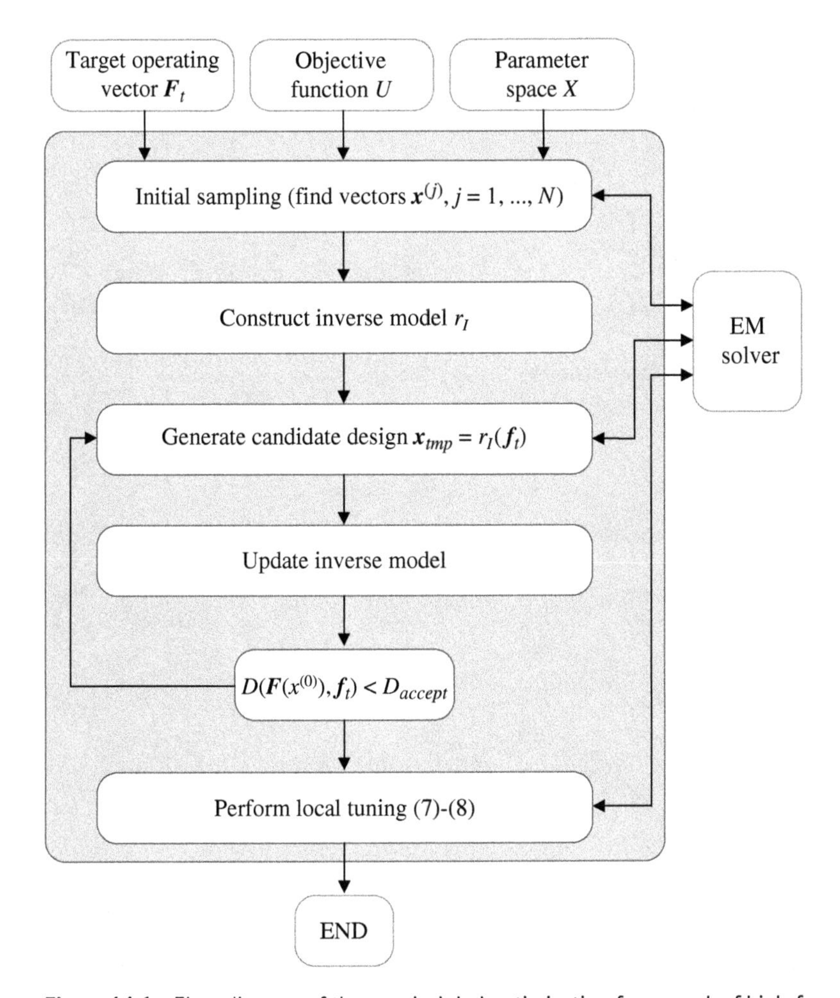

Figure 14.6 Flow diagram of the quasi-global optimization framework of high-frequency circuits.

The flowchart of the overall globalized search procedure is provided in Figure 14.6, with the global search stage represented by several components, and the local tuning shown as a single block.

14.3 Results

The performance of the optimization framework delineated in Section 14.2 is verified with the use of three microstrip components: a triple-band antenna, a miniaturized rat-race coupler (RRC), as well as a dual-band power divider. The numerical results obtained using the considered procedure are benchmarked against a local search starting from several random initial designs (to corroborate the necessity of performing global search) and a popular nature-inspired metaheuristic optimizer (to validate the efficiency of the discussed framework). Section 14.3.1 introduces the geometries of the verification structures and formulates the respective design problems. Whereas the numerical results yielded by the considered and the benchmark techniques are provided in Section 14.3.2. The discussion of the results, as well as their summary is given in Section 14.3.3.

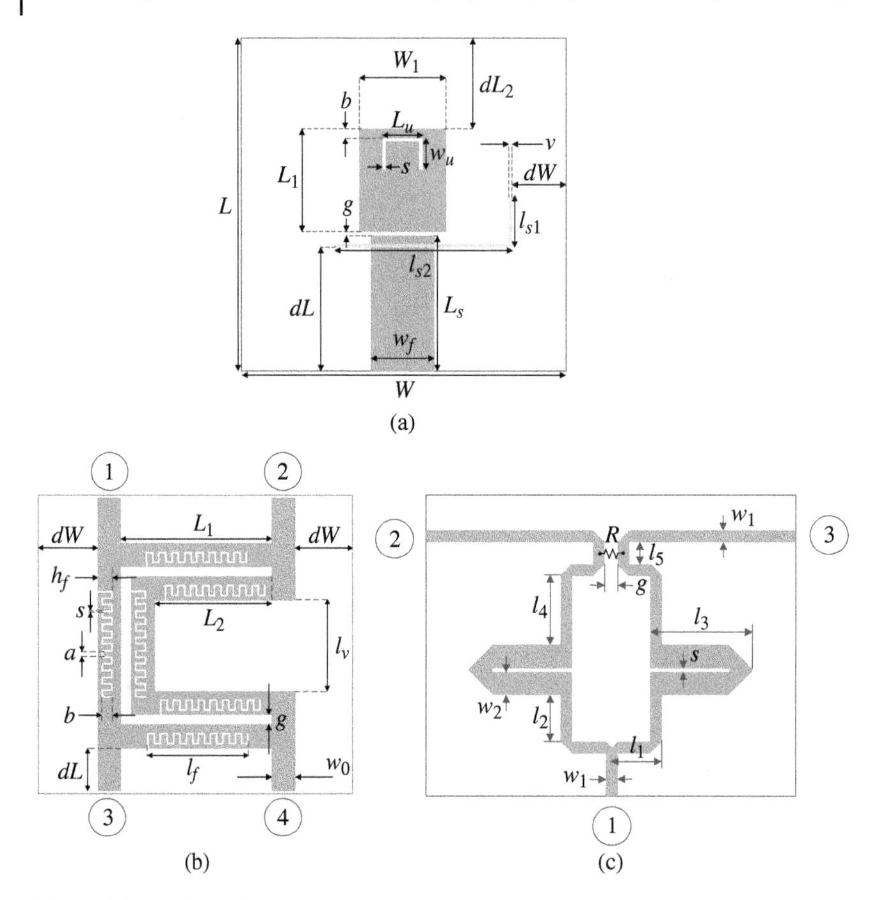

Figure 14.7 Microstrip devices used as verification case studies: (a) triple-band U-slotted patch antenna. Source: Adapted from Consul [80]. (b) Rat-race coupler with defected microstrip structure (RRC). Source: Adapted from Phani Kumar and Karthikeyan [81]. (c) dual-band power divider (PD), the lumped resistor denoted as *R*. Source: Adapted from [82].

14.3.1 Verification Structures

The discussed optimization procedure has been verified using the following three high-frequency components presented in Figure 14.7:

- Triple band U-slotted patch with L-slotted defected ground structure [80], implemented on a substrate of height equal to 3.064 mm (ε_r = 3.2). The vector of design variables is $\boldsymbol{x} = [L_1\ L_s\ L_{ur}\ W\ W_1\ dL_r\ dW_r\ g\ l_{s1r}\ l_{s2r}\ w_{ur}]^T$; the remaining parameters are fixed: $b = 1$, $w_f = 7.4$, $s = 0.5$, $w = 0.5$, $dL_2 = L_1$. Moreover, the following holds: $L = L_s + g + L_1 + dL_2$, $L_u = L_{ur}W_1$, $dL = dL_rL$, $dW = dW_rW$, $l_{s1} = l_{s1r}(L - dL)$, $l_{s2} = l_{s2r}(W - dW)$, $w_u = w_{ur}(L_1 - b - s)$. The design goal is to minimize antenna reflection at the operating frequencies $f_{t.1}$, $f_{t.2}$, and $f_{t.3}$, and the objective function is formulated similarly as in (14.4).
- Miniaturized RRC with a folded transmission line is implemented on a substrate of height 0.15 mm [81]. The independent geometry variables are $\boldsymbol{x} = [L_1\ b_r\ g\ h_{fr}\ s\ l_{fr}]^T$ (all dimensions in mm, except for those indicated with subscript *r* which are relative and unitless). We also have additional relationships: $L_2 = L_1 - g - w_0$, $a = (l_f - 17s)/16$, $b = (h_f - s)b_r$, $l_f = L_2\ l_{fr}$, $l_v = L_1 - 2g - 2w_0$, and $h_f = s + (w_0 - s)h_{fr}$; $dW = dL = 10$ mm. The width of input line w_0 is

adjusted to a specific substrate permittivity in order to keep 50 Ω input impedance. The design objective is to minimize the matching and isolation characteristics, $|S_{11}|$ and $|S_{41}|$, for an assumed substrate permittivity ε_r at the target operating frequency f_0, and, at the same time, to ensure equal power split. The merit function is formulated as in (14.2), yet, $K_P = 0$ dB.

- Dual-band equal-split power divider (PD) [82], is implemented on 0.81-mm-thick AD250 substrate of relative permittivity $\varepsilon_r = 2.5$. The designable parameters are $\boldsymbol{x} = [l_1\ l_2\ l_3\ l_4\ l_5\ s\ w_2]^T$ (expressed in mm); $w_1 = 2.2$ mm and $g = 1$ mm are fixed. The design objective is formulated as minimization of the input matching $|S_{11}|$, output matching $|S_{22}|$, $|S_{33}|$, and isolation $|S_{23}|$ at the target operating frequencies f_1 and f_2. The merit function is similar to (14.3), with the exception of handling the equal power split which is enforced by the structure symmetry.

The simulation models for all structures are evaluated using the time-domain solver of CST Microwave Studio. The specific design tasks (i.e. the intended values of the operating parameters), along with the lower and upper bounds on design variables, are provided in Table 14.2. Observe that parameter ranges are very wide (the average proportion between the upper and lower bounds equal 2.5, 4.6, and 10.3 for the triple-band antenna, the coupler and the power divider, respectively).

14.3.2 Results

The experimental setup of the quasi-global optimization framework considered in this chapter, as well as the benchmark techniques is provided in Table 14.3. The benchmark pool includes: the trust-region gradient-based procedure of Section 14.2.4 initiated from random starting points and particle swarm optimizer (a widely-used population-based metaheuristic). The aim of employing the TR algorithm is to prove that a local optimization algorithm often yields unsatisfactory results when the starting point is severely misaligned with the target, and, thus, it is insufficient for the discussed design task. Tables 14.4 through 14.6 gather the numerical results. Whereas Figures 14.8 through 14.10 present the optimized responses using the considered framework for the representative algorithm runs.

14.3.3 Discussion

The main points of the discussion of the results obtained using the considered global optimization framework with respect to both its efficacy and computational complexity, along with a comparison of its performance with the benchmark methods, are as follows:

Table 14.2 Design specifications and parameter spaces for verification structures.

	Design specification		Lower l and upper bounds u on design variables
Structure	Symbols	Target values	
I	$\boldsymbol{F}_t = [f_1\, f_2\, f_3]^T$	Case 1: $f_1 = 3.5$ GHz, $f_2 = 5.8$ GHz, $f_3 = 7.5$ GHz	$\boldsymbol{l} = [10\ 17\ 0.2\ 45\ 5\ 0.4\ 0.15\ 0.2\ 0.1\ 0.5\ 0.1]^T$ $\boldsymbol{u} = [16\ 25\ 0.6\ 55\ 15\ 0.5\ 0.3\ 0.8\ 0.4\ 0.65\ 0.5]^T$
II	$\boldsymbol{F}_t = [f_0]^T$	Case 1: $f_0 = 1.5$ GHz, $\varepsilon_r = 2.5$[a)] Case 2: $f_0 = 1.2$ GHz, $\varepsilon_r = 4.4$[a)]	$\boldsymbol{l} = [20.0\ 0.1\ 1.0\ 0.2\ 0.2\ 0.2]^T$ $\boldsymbol{u} = [40.0\ 0.95\ 5.0\ 0.95\ 0.5\ 0.8]^T$
III	$\boldsymbol{F}_t = [f_1\, f_2]^T$	Case 1: $f_1 = 3.0$ GHz, $f_2 = 4.8$ GHz Case 2: $f_1 = 2.0$ GHz, $f_2 = 3.3$ GHz	$\boldsymbol{l} = [10.0\ 1.0\ 10.0\ 0.5\ 1.0\ 0.1\ 1.5]^T$ $\boldsymbol{u} = [40.0\ 20.0\ 40.0\ 15.0\ 6.0\ 1.5\ 8.0]^T$

a) The component is to be optimized for a specific substrate of given relative permittivity ε_r.

Table 14.3 Setup of the numerical experiments.

Method	Control parameters	Termination condition	Comments
Globalized optimization algorithm (this chapter)	$N = 10$, $N_{max.1} = 100$, $N_{max.2} = 100$, $N_{max.3} = 500$, $D_{accept} = 0.2$	$\varepsilon = 10^{-3}$	cf. Section 14.2
PSO [83]	Population size 10 $\chi = 0.73$, $c_1 = c_2 = 2.05$	Maximum number of iterations (100)	Computational budget: 1000 EM simulations (to maintain reasonable computational cost of numerical experiments)
TR gradient-based algorithm	Conventional setup (e.g. [23])	$\varepsilon = 10^{-3}$	Gradients evaluated through finite differentiation; Algorithm termination: convergence in argument OR diminishing the TR size

Table 14.4 Triple-band antenna: optimization results.

	Optimization algorithm	Global optimization framework	PSO 50 iterations	PSO 100 iterations	TR gradient-based algorithm
Case 1 $f_1 = 3.5$ GHz,	Average objective function value (dB)	−26.4	−18.2	−19.3	−13.5
$f_2 = 5.8$ GHz,	Computational cost[a)]	144.7	500	1,000	84.2
$f_3 = 7.5$ GHz	Success rate[b)]	10/10	9/10	10/10	6/10

a) The cost expressed in terms of the number of EM simulations of the antenna structure under design.
b) Number of algorithms runs for which the operating frequencies fulfill the condition $D(\boldsymbol{F}(\boldsymbol{x}^*),\boldsymbol{f}_t) \le D_{accept}$.

Table 14.5 Rat-race coupler: optimization results.

	Optimization algorithm	Global optimization framework	PSO 50 iterations	PSO 100 iterations	TR gradient-based algorithm
Case 1 $f_0 = 1.5$ GHz,	Average objective function value (dB)	−18.6 dB	−17.6	−19.2	1.8 dB
$\varepsilon_r = 2.5$	Computational cost[a)]	85.5	500	1000	77.0
	Success rate[b)]	10/10	10/10	10/10	5/10
Case 2 $f_0 = 1.2$ GHz,	Average objective function value (dB)	−21.5 dB	−19.4	−22.5	7.6 dB
$\varepsilon_r = 4.4$	Computational cost[a)]	90.2	500	1000	83.8
	Success rate[b)]	10/10	9/10	10/10	5/10

a) The cost expressed in terms of the number of EM simulations of the coupler under design.
b) Number of algorithms runs for which the coupler's operating frequencies fulfill the condition $D(\boldsymbol{F}(\boldsymbol{x}^*),\boldsymbol{f}_t) \le D_{accept}$.

Table 14.6 Power divider: optimization results.

	Optimization algorithm	Global optimization framework	PSO 50 iterations	PSO 100 iterations	TR gradient-based algorithm
Case 1 $f_1 = 3.0$ GHz, $f_2 = 4.8$ GHz	Average objective function value (dB)	−33.9 dB	−19.6	−18.8	−12.3 dB
	Computational cost[a)]	99.1	500	1000	95.1
	Success rate[b)]	10/10	8/10	9/10	2/10
Case 2 $f_1 = 2.0$ GHz, $f_2 = 3.3$ GHz	Average objective function value (dB)	−23.6 dB	−18.8	−19.7	−20.6
	Computational cost[a)]	99.2	500	1000	93.8
	Success rate[b)]	10/10	8/10	9/10	7/10

a) The cost expressed in terms of the number of EM simulations of the power divider under design.
b) Number of algorithms runs for which the operating frequencies fulfill the condition $D(\boldsymbol{F}(\boldsymbol{x}^*), \boldsymbol{f}_t) \leq D_{accept}$.

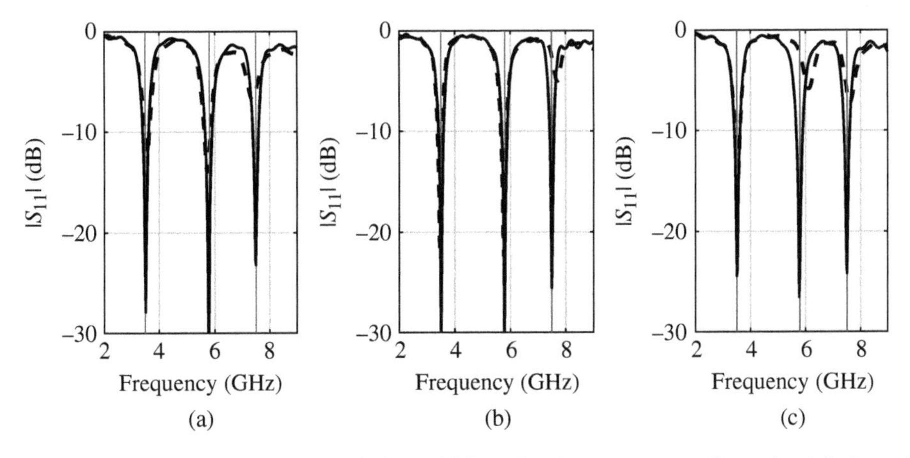

Figure 14.8 Triple-band antenna of Figure 14.7a: reflection response at the optimal designs yielded by the considered global optimization algorithm. The initial design $\boldsymbol{x}^{(0)}$ rendered in the global search stage is marked with dashed line, the optimized design is indicated with solid line. Target operating frequencies: (a) 3.5 GHz, (b) 5.8 GHz, and (c) 7.5 GHz are shown using vertical lines.

- Global search capability of the considered approach has been verified, since in all algorithm runs (ten per case), the designs satisfying the design specifications have been obtained. Moreover, as shown in Figures 14.8 through 14.10, the parameter vectors $\boldsymbol{x}^{(0)}$ yielded in the global search stage are of high quality (in the sense that their respective operating frequencies are close to the target); thus, local tuning proves to be sufficient. Whereas the local optimization algorithm fails in about half of the cases, with its performance being highly dependent on the starting point. As a consequence, the average value of the merit function at the optimal designs is significantly worse than it is in the case of the considered technique. PSO optimizer utilized as an example of a population-based metaheuristics performs considerably better; yet, the associated computational cost is high. Moreover, a noticeable difference of the design quality yielded after 50 and 100 PSO iterations can be discerned, which is an indicator that computational budget fixed at 500 EM simulations is far from sufficient for this algorithm.

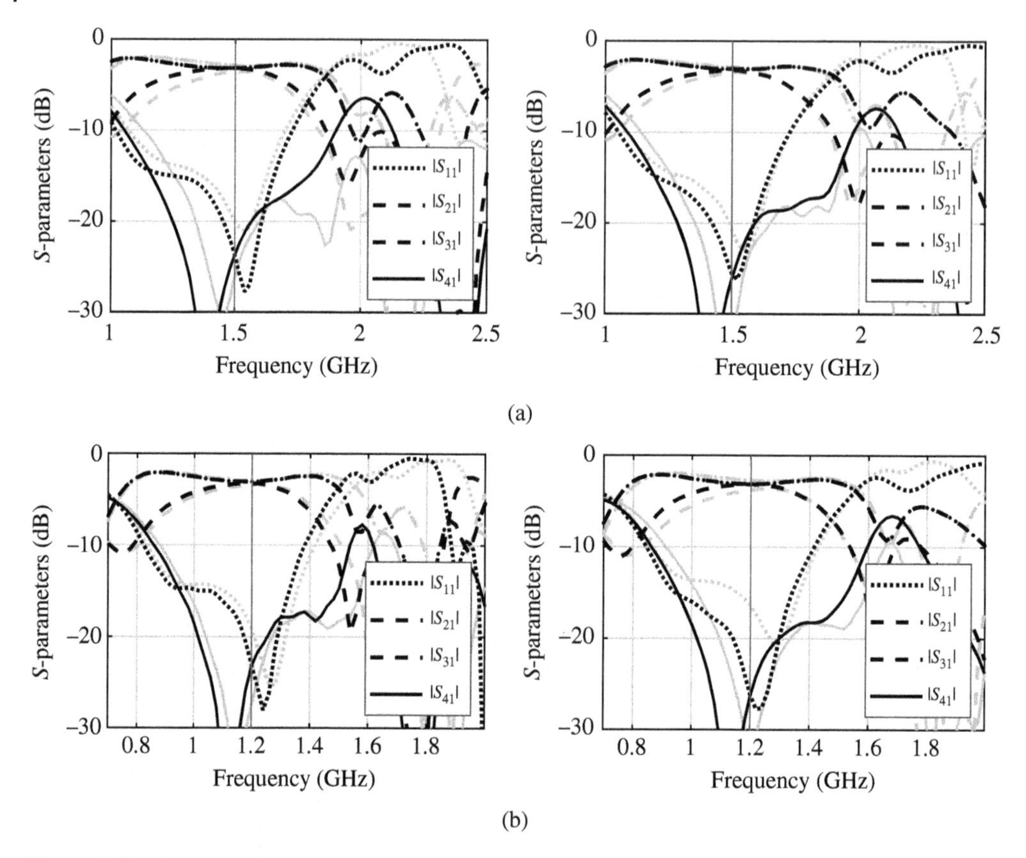

Figure 14.9 Rat-race coupler of Figure 14.7b: *S*-parameters at the optimal designs yielded by the considered global optimization algorithm for two representative algorithm runs: (a) Case 1, (b) Case 2. The initial design $\boldsymbol{x}^{(0)}$ rendered in the global search stage is marked with gray lines, the optimized design is indicated with black lines. Target operating frequency 1.2 GHz is shown using vertical lines.

- The design quality for the considered framework is significantly better than that of the local TR optimization algorithm and PSO. Despite the fact that the local search is capable of rendering designs of similar quality, this is only the case for the initial points close enough to the target. Whereas the discussed algorithm is virtually independent of the initial design, thereby making it considerably more robust.
- As far as computational cost is concerned, the considered framework is considerably cheaper than the population-based algorithm with the average optimization cost equal to only 145, 88, and 99 EM simulations for the triple-band antenna, RRC and power divider, respectively. Moreover, the discussed technique is only somewhat more expensive than local search (the cost is increased by 20% on the average across the set of the considered high-frequency structures). This stems from the fact that the expenses associated with the initial (global) search stage are low, only 36, 22, and 31 EM analyses on the average for the antenna, coupler, and power-divider, respectively. This degree of efficiency has been achieved due to a combination of the inverse modeling and response feature technology, in particular, construction of the surrogate within low-dimensional space of operating parameters. This is because it requires only a relatively moderate number of samples to represent interrelations between the operating parameters (e.g. operating frequency or power split) and design variables in a reliable manner.

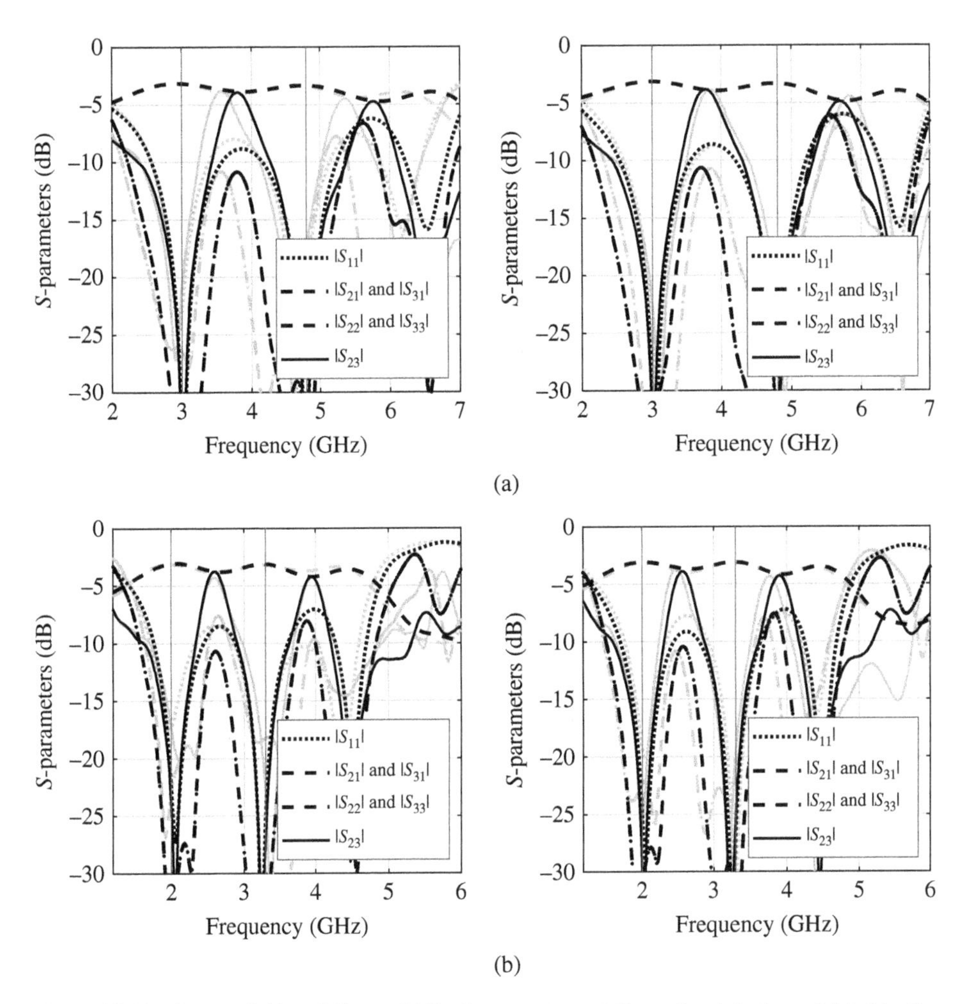

Figure 14.10 Power divider of Figure 14.7c: *S*-parameters at the optimal designs yielded by the considered global optimization algorithm for two representative algorithm runs: (a) Case 1, (b) Case 2. The initial design $\boldsymbol{x}^{(0)}$ rendered in the global search stage is marked with gray lines, the optimized design is indicated with black lines. Target operating frequencies are shown using vertical lines.

Therefore, the discussed framework is capable of performing a global search at a computational cost similar to that of the local optimizer. This makes it a cost-efficient alternative to the conventional global optimization algorithms, especially population-based metaheuristics. The one implicit assumption being that the design space for the problem is defined reasonably, meaning that it is not overly large; hence, the likelihood that responses of a randomly allocated design have a decent shape is reasonably high. As for any other optimization technique operating at the level of response features, a practical limitation is that the feature-based predictions of the operating parameters have to be extracted from the EM simulated characteristics of the high-frequency structure under design. This is performed on case-to-case basis; yet, the actual implementations may be to a certain degree transferred between the same shape of device responses (multi-band antenna, coupler, power divider, etc.). Automation of this procedure is a direction for the future work.

14.4 Conclusion

This chapter discussed a cost-efficient and reliable framework for quasi-global design optimization of high-frequency devices. The key concept of the procedure relies in an involvement of an inverse regression surrogate, which is built with the use of the information extracted from a pre-selected subset of randomly allocated parameter vectors and the corresponding EM-simulated responses of the high-frequency components. Construction of an inverse surrogate involves the response feature technology, which capitalizes on close-to-linear relationships between the operating and geometry parameters of the system under design. Numerical verification of the discussed procedure involved using three microstrip components, including a triple-band antenna, a miniaturized RRC, and a dual-band power divider. In each case, the efficacy of the technique, along with the solution repeatability has been validated based on ten independent algorithm runs. In all executions, the discussed algorithm yielded satisfactory designs, unlike the benchmark algorithms, especially multiple-start local optimizer, which failed in a considerable number of optimization runs. Moreover, the computational expenditures of the discussed framework are considerably lower than that of particle swarm optimizer utilized as a conventional global optimization algorithm. As a matter of fact, the cost of the considered framework is similar to the cost of local gradient-based tuning. The considered approach may be an attractive alternative to state-of-the-art global search algorithms, especially, nature-inspired algorithms, but also hybrid techniques involving forward surrogate modeling methods. The major advantages of the discussed framework are global search capability, low computational complexity, along with the enhanced immunity to parameter range and dimensionality issues.

Acknowledgment

The authors would like to thank Dassault Systemes, France, for making CST Microwave Studio available. This work is partially supported by the Icelandic Centre for Research (RANNIS) Grant 217771 and by National Science Centre of Poland Grant 2020/37/B/ST7/01448.

References

1 Ma, P., Wei, B., Hong, J. et al. (2018). A design method of multimode multiband bandpass filters. *IEEE Transactions on Microwave Theory and Techniques* 66 (6): 2791–2799.

2 Wen, S. and Dong, Y. (2020). A low-profile wideband antenna with monopole like radiation characteristics for 4G/5G indoor micro base station application. *IEEE Antennas and Wireless Propagation Letters* 19 (12): 2305–2309.

3 Hagag, M.F., Zhang, R., and Peroulis, D. (2019). High-performance tunable narrowband SIW cavity-based quadrature hybrid coupler. *IEEE Microwave and Wireless Components Letters* 29 (1): 41–43.

4 Gómez-García, R., Rosario-De Jesus, J., and Psychogiou, D. (2018). Multi-band bandpass and bandstop RF filtering couplers with dynamically-controlled bands. *IEEE Access* 6: 32321–32327.

5 Zhang, R. and Peroulis, D. (2018). Mixed lumped and distributed circuits in wideband bandpass filter application for spurious-response suppression. *IEEE Microwave and Wireless Components Letters* 28 (11): 978–980.

6 Liu, H., Fang, S., Wang, Z., and Fu, S. (2019). Design of arbitrary-phase-difference transdirectional coupler and its application to a flexible Butler matrix. *IEEE Transactions on Microwave Theory and Techniques* 67 (10): 4175–4185.

7 Bogdan, G., Bajurko, P., and Yashchyshyn, Y. (2020). Time-modulated antenna array with dual-circular polarization. *IEEE Antennas and Wireless Propagation Letters* 19 (11): 1872–1875.

8 Li, Q., Chen, X., Chi, P., and Yang, T. (2020). Tunable bandstop filter using distributed coupling microstrip resonators with capacitive terminal. *IEEE Microwave and Wireless Components Letters* 30 (1): 35–38.

9 Sheikhi, A., Alipour, A., and Mir, A. (2020). Design and fabrication of an ultra-wide stopband compact bandpass filter. *IEEE Transactions on Circuits and Systems II: Express Briefs* 67 (2): 265–269.

10 Firmansyah, T., Alaydrus, M., Wahyu, Y. et al. (2020). A highly independent multiband bandpass filter using a multi-coupled line stub-SIR with folding structure. *IEEE Access* 8: 83009–83026.

11 Chen, S., Guo, M., Xu, K. et al. (2018). A frequency synthesizer based microwave permittivity sensor using CMRC structure. *IEEE Access* 6: 8556–8563.

12 Hu, W., Yin, Y., Yang, X., and Fei, P. (2013). Compact multiresonator-loaded planar antenna for multiband operation. *IEEE Transactions on Antennas and Propagation* 61 (5): 2838–2841.

13 Ding, Z., Jin, R., Geng, J. et al. (2019). Varactor loaded pattern reconfigurable patch antenna with shorting pins. *IEEE Transactions on Antennas and Propagation* 67 (10): 6267–6277.

14 Zhu, S., Liu, H., Wen, P. et al. (2020). Vivaldi antenna array using defected ground structure for edge effect restraint and back radiation suppression. *IEEE Antennas and Wireless Propagation Letters* 19 (1): 84–88.

15 Wen, S., Xu, Y., and Dong, Y. (2021). A low-profile dual-polarized omnidirectional antenna for LTE base station applications. *IEEE Transactions on Antennas and Propagation* 69 (9): 5974–5979.

16 Chi, J.-G. and Kim, Y.-J. (2020). A compact wideband millimeter-wave quadrature hybrid coupler using artificial transmission lines on a glass substrate. *IEEE Microwave and Wireless Components Letters* 30 (11): 1037–1040.

17 Deng, J., Li, M., Sun, D. et al. (2020). Compact dual-band inverted-microstrip ridge gap waveguide bandpass filter. *IEEE Transactions on Microwave Theory and Techniques* 68 (7): 2625–2632.

18 Koziel, S. and Abdullah, M. (2021). Machine-learning-powered EM-based framework for efficient and reliable design of low scattering metasurfaces. *IEEE Transactions on Microwave Theory and Techniques* 69 (4): 2028–2041.

19 Li, Y., Ren, P., and Xiang, Z. (2019). A dual-passband frequency selective surface for 5G communication. *IEEE Antennas and Wireless Propagation Letters* 18 (12): 2597–2601.

20 Li, H., Jiang, Y., Ding, Y. et al. (2018). Low-sidelobe pattern synthesis for sparse conformal arrays based on PSO-SOCP optimization. *IEEE Access* 6: 77429–77439.

21 Rayas-Sanchez, J.E., Koziel, S., and Bandler, J.W. (2021). Advanced RF and microwave design optimization: a journey and a vision of future trends. *IEEE Journal of Microwaves* 1 (1): 481–493.

22 Jin, H., Zhou, Y., Huang, Y.M. et al. (2017). Miniaturized broadband coupler made of slow-wave half-mode substrate integrated waveguide. *IEEE Microwave and Wireless Components Letters* 27 (2): 132–134.

23 Pietrenko-Dabrowska, A. and Koziel, S. (2021). Fast design closure of compact microwave components by means of feature-based metamodels. *Electronics* 10 (1): 10.

24 Li, X. and Luk, K.M. (2020). The grey wolf optimizer and its applications in electromagnetics. *IEEE Transactions on Antennas and Propagation* 68 (3): 2186–2197.

25 Luo, X., Yang, B., and Qian, H.J. (2019). Adaptive synthesis for resonator-coupled filters based on particle swarm optimization. *IEEE Transactions on Microwave Theory and Techniques* 67 (2): 712–725.

26 Majumder, A., Chatterjee, S., Chatterjee, S. et al. (2017). Optimization of small-signal model of GaN HEMT by using evolutionary algorithms. *IEEE Microwave and Wireless Components Letters* 27 (4): 362–364.

27 Choi, K., Jang, D., Kang, S. et al. (2016). Hybrid algorithm combing genetic algorithm with evolution strategy for antenna design. *IEEE Transactions on Magnetics* 52 (3): 1–4, Article no. 7209004.

28 Ghorbaninejad, H. and Heydarian, R. (2016). New design of waveguide directional coupler using genetic algorithm. *IEEE Microwave and Wireless Components Letters* 26 (2): 86–88.

29 Ding, D., Zhang, Q., Xia, J. et al. (2018). Wiggly parallel-coupled line design by using multiobjective evolutionary algorithm. *IEEE Microwave and Wireless Components Letters* 28 (8): 648–650.

30 Zhu, D.Z., Werner, P.L., and Werner, D.H. (2017). Design and optimization of 3-D frequency-selective surfaces based on a multiobjective lazy ant colony optimization algorithm. *IEEE Transactions on Antennas and Propagation* 65 (12): 7137–7149.

31 Greda, L.A., Winterstein, A., Lemes, D.L., and Heckler, M.V.T. (2019). Beamsteering and beamshaping using a linear antenna array based on particle swarm optimization. *IEEE Access* 7: 141562–141573.

32 Cui, C., Jiao, Y., and Zhang, L. (2017). Synthesis of some low sidelobe linear arrays using hybrid differential evolution algorithm integrated with convex programming. *IEEE Antennas and Wireless Propagation Letters* 16: 2444–2448.

33 Baumgartner, P., Baurnfeind, T., Biro, O. et al. (2018). Multi-objective optimization of Yagi-Uda antenna applying enhanced firefly algorithm with adaptive cost function. *IEEE Transactions on Magnetics* 54 (3): 8000504.

34 Yang, S.H. and Kiang, J.F. (2015). Optimization of sparse linear arrays using harmony search algorithms. *IEEE Transactions on Antennas and Propagation* 63 (11): 4732–4738.

35 Li, X. and Guo, Y.-X. (2020). Multiobjective optimization design of aperture illuminations for microwave power transmission via multiobjective grey wolf optimizer. *IEEE Transactions on Antennas and Propagation* 68 (8): 6265–6276.

36 Zheng, T., Liu, Y., Sun, G. et al. (2020). IWORMLF: improved invasive weed optimization with random mutation and Lévy flight for beam pattern optimizations of linear and circular antenna arrays. *IEEE Access* 8: 19460–19478.

37 Al-Azza, A.A., Al-Jodah, A.A., and Harackiewicz, F.J. (2016). Spider monkey optimization: a novel technique for antenna optimization. *IEEE Antennas and Wireless Propagation Letters* 15: 1016–1019.

38 Liang, S., Fang, Z., Sun, G. et al. (2020). Sidelobe reductions of antenna arrays via an improved chicken swarm optimization approach. *IEEE Access* 8: 37664–37683.

39 Li, W., Zhang, Y., and Shi, X. (2019). Advanced fruit fly optimization algorithm and its application to irregular subarray phased array antenna synthesis. *IEEE Access* 7: 165583–165596.

40 Jiang, Z.J., Zhao, S., Chen, Y., and Cui, T.J. (2018). Beamforming optimization for time-modulated circular-aperture grid array with DE algorithm. *IEEE Antennas and Wireless Propagation Letters* 17 (12): 2434–2438.

41 Bayraktar, Z., Komurcu, M., Bossard, J.A., and Werner, D.H. (2013). The wind driven optimization technique and its application in electromagnetics. *IEEE Transactions on Antennas and Propagation* 61 (5): 2745–2757.

42 Rayno, J., Iskander, M.F., and Kobayashi, M.H. (2016). Hybrid genetic programming with accelerating genetic algorithm optimizer for 3-D metamaterial design. *IEEE Antennas and Wireless Propagation Letters* 15: 1743–1746.

43 Abdelhafiz, A., Behjat, L., and Ghannouchi, F.M. (2018). Generalized memory polynomial model dimension selection using particle swarm optimization. *IEEE Microwave and Wireless Components Letters* 28 (2): 96–98.

44 Goudos, S.K., Yioultsis, T.V., Boursianis, A.D. et al. (2019). Application of new hybrid Jaya grey wolf optimizer to antenna design for 5G communications systems. *IEEE Access* 7: 71061–71071.

45 Liu, F., Liu, Y., Han, F. et al. (2020). Synthesis of large unequally spaced planar arrays utilizing differential evolution with new encoding mechanism and Cauchy mutation. *IEEE Transactions on Antennas and Propagation* 68 (6): 4406–4416.

46 Karimkashi, S. and Kishk, A.A. (2010). Invasive weed optimization and its features in electromagnetics. *IEEE Transactions on Antennas and Propagation* 58 (4): 1269–1278.

47 Kovaleva, M., Bulger, D., and Esselle, K.P. (2020). Comparative study of optimization algorithms on the design of broadband antennas. *IEEE Journal of Multiscale Multiphysics Computational Technologies* 5: 89–98.

48 Bai, Y., Xiao, S., Liu, C., and Wang, B. (2013). A hybrid IWO/PSO algorithm for pattern synthesis of conformal phased arrays. *IEEE Transactions on Antennas and Propagation* 61 (4): 2328–2332.

49 Zhang, Z., Cheng, Q.S., Chen, H., and Jiang, F. (2020). An efficient hybrid sampling method for neural network-based microwave component modeling and optimization. *IEEE Microwave and Wireless Components Letters* 30 (7): 625–628.

50 Van Nechel, E., Ferranti, F., Rolain, Y., and Lataire, J. (2018). Model-driven design of microwave filters based on scalable circuit models. *IEEE Transactions on Microwave Theory and Techniques* 66 (10): 4390–4396.

51 Li, Y., Xiao, S., Rotaru, M., and Sykulski, J.K. (2016). A dual kriging approach with improved points selection algorithm for memory efficient surrogate optimization in electromagnetics. *IEEE Transactions on Magnetics* 52 (3): 1–4, Article no. 7000504.

52 Jacobs, J.P. (2016). Characterization by Gaussian processes of finite substrate size effects on gain patterns of microstrip antennas. *IET Microwaves, Antennas and Propagation* 10 (11): 1189–1195.

53 Ogut, M., Bosch-Lluis, X., and Reising, S.C. (2019). A deep learning approach for microwave and millimeter-wave radiometer calibration. *IEEE Transactions on Geoscience and Remote Sensing* 57 (8): 5344–5355.

54 Yu, X., Hu, X., Liu, Z. et al. (2021). A method to select optimal deep neural network model for power amplifiers. *IEEE Microwave and Wireless Components Letters* 31 (2): 145–148.

55 Na, W., Liu, W., Zhu, L. et al. (2018). Advanced extrapolation technique for neural-based microwave modeling and design. *IEEE Transactions on Microwave Theory and Techniques* 66 (10): 4397–4418.

56 Petrocchi, A., Kaintura, A., Avolio, G. et al. (2017). Measurement uncertainty propagation in transistor model parameters via polynomial chaos expansion. *IEEE Microwave and Wireless Components Letters* 27 (6): 572–574.

57 Couckuyt, I., Declercq, F., Dhaene, T. et al. (2010). Surrogate-based infill optimization applied to electromagnetic problems. *International Journal of RF and Microwave Computer-Aided Engineering* 20 (5): 492–501.

58 Torun, H.M. and Swaminathan, M. (2019). High-dimensional global optimization method for high-frequency electronic design. *IEEE Transactions on Microwave Theory and Techniques* 67 (6): 2128–2142.

59 Liu, B., Koziel, S., and Zhang, Q. (2016). A multi-fidelity surrogate-model-assisted evolutionary algorithm for computationally expensive optimization problems. *Journal of Computer Science* 12: 28–37.

60 Lim, D.K., Yi, K.P., Jung, S.Y. et al. (2015). Optimal design of an interior permanent magnet synchronous motor by using a new surrogate-assisted multi-objective optimization. *IEEE Transactions on Magnetics* 51 (11): 8207504.

61 Taran, N., Ionel, D.M., and Dorrell, D.G. (2018). Two-level surrogate-assisted differential evolution multi-objective optimization of electric machines using 3-D FEA. *IEEE Transactions on Magnetics* 54 (11): 8107605.

62 Koziel, S. and Pietrenko-Dabrowska, A. (2020). *Performance-Driven Surrogate Modeling of High-Frequency Structures*. New York: Springer.

63 Koziel, S. (2017). Low-cost data-driven surrogate modeling of antenna structures by constrained sampling. *IEEE Antennas and Wireless Propagation Letters* 16: 461–464.

64 Koziel, S. and Pietrenko-Dabrowska, A. (2019). Performance-based nested surrogate modeling of antenna input characteristics. *IEEE Transactions on Antennas and Propagation* 67 (5): 2904–2912.

65 Koziel, S. and Pietrenko-Dabrowska, A. (2020). Reduced-cost surrogate modelling of compact microwave components by two-level kriging interpolation. *Engineering Optimization* 52 (6): 960–972.

66 Pietrenko-Dabrowska, A. and Koziel, S. (2020). Antenna modeling using variable-fidelity EM simulations and constrained co-kriging. *IEEE Access* 8 (1): 91048–91056.

67 Pietrenko-Dabrowska, A. and Koziel, S. (2020). Surrogate modeling of impedance matching transformers by means of variable-fidelity EM simulations and nested co-kriging. *International Journal of RF and Microwave Computer-Aided Engineering* 30 (8): e22268.

68 Koziel, S. and Pietrenko-Dabrowska, A. (2019). Rapid multi-objective optimization of antennas using nested kriging surrogates and single-fidelity EM simulation models. *Engineering Computations* 37 (4): 1491–1512.

69 Pietrenko-Dabrowska, A. and Koziel, S. (2020). Accelerated multiobjective design of miniaturized microwave components by means of nested kriging surrogates. *International Journal of RF and Microwave Computer-Aided Engineering* 30: e22124.

70 Pietrenko-Dabrowska, A. (2020). Rapid tolerance-aware design of miniaturized microwave passives by means of confined-domain surrogates. *International Journal of Numerical Modelling* 33 (6): e2779.

71 Pietrenko-Dabrowska, A., Koziel, S., and Al-Hasan, M. (2020). Expedited yield optimization of narrow- and multi-band antennas using performance-driven surrogates. *IEEE Access* 8: 143104–143113.

72 Koziel, S. (2015). Fast simulation-driven antenna design using response-feature surrogates. *International Journal of RF & Microwave Computer* 25 (5): 394–402.

73 Koziel, S. and Pietrenko-Dabrowska, A. (2020). Expedited feature-based quasi-global optimization of multi-band antennas with Jacobian variability tracking. *IEEE Access* 8: 83907–83915.

74 Koziel, S. and Bandler, J.W. (2015). Reliable microwave modeling by means of variable-fidelity response features. *IEEE Transactions on Microwave Theory and Techniques* 63 (12): 4247–4254.

75 Koziel, S. and Pietrenko-Dabrowska, A. (2020). Design-oriented computationally-efficient feature-based surrogate modelling of multi-band antennas with nested kriging. *International Journal of Electronics and Communications* 120: 153202.

76 Pietrenko-Dabrowska, A. and Koziel, S. (2021). Globalized parametric optimization of microwave components by means of response features and inverse metamodels. *Scientific Reports* 11: 23718.

77 Koziel, S. and Pietrenko-Dabrowska, A. (2021). Global EM-driven optimization of multi-band antennas using knowledge-based inverse response-feature surrogates. *Knowledge-Based Systems* 227: 107189.

78 Conn, A.R., Gould, N.I.M., and Toint, P.L. (2000). *Trust Region Methods*, MPS-SIAM Series on Optimization. Philadelphia, PA: SIAM.

79 Koziel, S. and Pietrenko-Dabrowska, A. (2019). Expedited optimization of antenna input characteristics with adaptive Broyden updates. *Engineering Computations* 37 (3): 851–862.

80 Consul, P. (2015). Triple band gap coupled microstrip U-slotted patch antenna using L-slot DGS for wireless applications. In: *Communication, Control Intelligent Systems (CCIS)*, Mathura, India, 31–34.

81 Phani Kumar, K.V. and Karthikeyan, S.S. (2013). A novel design of ratrace coupler using defected microstrip structure and folding technique. In: *IEEE Applied Electromagnetics Conference (AEMC)*, Bhubaneswar, India, 1–2.

82 Lin, Z. and Chu, Q.-X. (2010). A novel approach to the design of dual-band power divider with variable power dividing ratio based on coupled-lines. *Progress In Electromagnetics Research* 103: 271–284.

83 Kennedy, J. and Eberhart, R.C. (2001). *Swarm Intelligence*. San Francisco, USA: Morgan Kaufmann.

15

Deep Learning for High Contrast Inverse Scattering of Electrically Large Structures

Qing Liu, Li-Ye Xiao, Rong-Han Hong, and Hao-Jie Hu

Institute of Electromagnetics and Acoustics, Xiamen University, Xiamen, China

15.1 Introduction

Electromagnetic (EM) inversion aims to solve an electromagnetic inverse scattering problem, i.e. to reconstruct model parameters such as permittivity, conductivity, and permeability of unknown objects located inside an inaccessible region by analyzing the scattered fields given the illumination of the domain of interest (DOI) [1, 2]. It has been widely applied in geophysical exploration [3], nondestructive evaluation [4], through-wall imaging [5], medical imaging [6, 7], remote sensing [8], and so on. Due to the intrinsic nonlinearity and the ill-posedness of the inverse scattering problem, it is a challenging task to find a reliable, accurate, and numerically efficient inversion method, especially for high contrast and electrically large problems [9].

Thus, this chapter discusses a series of deep learning-based methods we have developed in recent years for this purpose. Different from the previous deep learning-based methods, our approach can handle high contrast inverse scattering of electrically large structures. Meanwhile, our approach also aims to reduce the computational cost during the inversion process.

For the two-dimensional (2-D) inverse scattering problem, a dual-module nonlinear mapping module (NMM)-image-enhancing module (IEM) machine learning scheme is developed to perform reconstruction of inhomogeneous scatterers with high contrasts and large electrical dimensions. Compared with the previous works applying convolutional neural networks (CNNs) to EM inversion, the extreme learning machine (ELM) and a CNN with low training costs are used in the proposed modules to save training cost [10]. This high-contrast method has been developed for well-functioning sources and receivers as in other traditional machine learning inverse scattering methods. However, malfunctioning receivers are common in practical inversion environments, which cannot be treated by the traditional methods as there is damaged data. Thus, a new receiver approximation machine learning (RAML) method is proposed to repair the damaged data in the 2-D inverse scattering problem [11].

For the three-dimensional (3-D) inversion scattering problem, the direct application of the above NMM-ELM scheme is computationally expensive because of the large training number required for many 3-D unknown parameters. Thus, a semi-join extreme learning machine (SJ-ELM) model has been proposed to reduce the training sample set for super-resolution 3-D quantitative imaging of objects with high contrasts [12]. The semi-join strategy can decrease the inner matrix dimensions and improve the convergence performance of the model in super-resolution imaging. One important application of this 3-D inversion is the super-resolution 3-D microwave human

Advances in Electromagnetics Empowered by Artificial Intelligence and Deep Learning, First Edition.
Edited by Sawyer D. Campbell and Douglas H. Werner.

brain imaging; it is a typical high contrast electromagnetic inverse scattering problem with huge computational costs. Meanwhile, specific to the human brain imaging, it is particularly important to capture the feature of each tissue type to guarantee the imaging accuracy [13]. To solve this kind of inverse problem, a hybrid neural network electromagnetic inversion scheme (HNNEMIS) containing both shallow and deep neural networks is proposed, where both shallow and deep neural networks can capture the feature of each tissue to further enhance inversion accuracy.

15.2 General Strategy and Approach

15.2.1 Related Works by Others and Corresponding Analyses

Usually, EM inversion methods can be classified as noniterative and iterative inversion methods. Noniterative methods such as the backpropagation (BP) method [14], Born approximation (BA), the extended Born approximation methods [15, 16], and diagonal tensor approximation [17] can be used in relatively weak scattering scenarios, where the inverse problems can be formulated by linearized equations. Without iterations, such noniterative methods can provide reconstruction results rapidly, but they are not accurate for high contrast applications, although the improved methods such as those in [15-17] are more accurate than Born approximation for higher contrasts.

Based on the minimization procedure of the cost function, iterative inversion methods for high contrast applications can be further classified into deterministic inversion methods and stochastic inversion methods [18].

Deterministic inversion methods such as Born iterative method (BIM) [19, 20] and its variants, contrast source-type inversion method (CSI) [21], and subspace optimization method (SOM) [22, 23] can reconstruct the model parameters (such as the permittivity distribution) of the unknown scatterers through minimizing the misfit between the calculated and measured scattered field data iteratively under regularization. However, the major drawback of these iterative methods is that they are time-consuming and, thus, not suitable for the real-time reconstruction; furthermore, for high contrasts and electrically large problems, numerical convergence can be difficult.

Compared with the deterministic methods, the stochastic methods such as genetic algorithm (GA) [24–26], particle swarm optimization (PSO) [27–29], and memetic algorithm (MA) [30] can avoid being trapped into local minima of the cost function. However, for the stochastic methods, the dimension of unknown model parameters is severely restricted because of their high computational complexity, and thus they are especially challenged by 3-D voxel-based inversion problems where the dielectric parameters in all discretized cells need to be retrieved.

In recent years, machine learning has attracted increased attention in the areas of image processing and computer vision, such as image classification [31, 32] and segmentation [33, 34], and depth estimation [35]. Methodologies based on the artificial neural network have also been proposed and used to extract rather general information about the geometric and EM properties of the scatterers [36, 37]. Nevertheless, these methods used just a few parameters to represent the scatterers, such as their positions, sizes, shapes, and piecewise constant permittivity values. The applicability of this kind of parameterization is limited, since the scatterers in the DOI can be spatially inhomogeneous and their numbers can be arbitrary. A more versatile approach to representing the scatterers is the pixel and voxel bases for 2-D and 3-D problems, respectively, i.e. the values of the dielectric parameters in all the pixels/voxels are independent of each other. So far, based on the role of machine learning playing in the inverse scattering problem, ML-based inverse scattering methods can be classified into three types:

(a) **Iterative framework-based method**: A machine learning technique is inserted to the traditional iterative framework to map some components. As an example, in [38] the supervised descent learning module is embedded inside the iterative process, and the reconstruction of 2-D images was achieved through iterations based on the descent directions learned in the training stage.
(b) **Linearized approximation based method**: Combined with noniterative methods such as BP or BA, a machine learning technique is employed to learn the relationship between the inversion results obtained from linearized approximation and the corresponding ground truth to solve the inverse scattering problems [39-42]. In [39], a deep neural network architecture, termed DeepNIS, is proposed. DeepNIS consists of a cascade of three CNN modules, in which the inputs are the complex-valued images generated by the BP algorithm and the outputs are the super-resolution images of the dielectric parameter distribution of the unknown objects. The DeepNIS are shown to outperform conventional nonlinear inverse scattering methods in terms of both the reconstruction accuracy and computational time. In [40], three inversion schemes based on the U-Net CNN were proposed for inverse scattering problems. They are the direct inversion scheme, the BP scheme, and the dominant current scheme (DCS). In terms of the results of several representative tests, the DCS outperforms the other two schemes. In [41], several strategies to incorporate the physical expertise inspired by the traditional iterative algorithms such as CSI and SOM were adopted, and the inversion accuracy and efficiency were validated. In [42], an inversion method based on the BA and a 3-D U-Net is proposed for 3-D object inversion, where BA is employed to produce the preliminary 3-D images as the input of 3-D U-Net.
(c) **Direct inversion method**: The machine learning method is employed to directly learn the mapping relationship between the scattered field and the corresponding distribution of the model parameters. Yao et al. [43] developed a two-step deep learning approach that can reconstruct high-contrast objects using the cascade of a CNN and another complex-valued deep residual CNN. In [10], NMM-IEM is proposed, where the scattered field data is directly input into NMM to obtain the preliminary distribution of electrical parameter, then IEM is employed to further enhance the inversion quality. In [11], a new RAML method is proposed to repair the damaged data in the 2-D inverse scattering problem. For the 3-D inverse scattering problem, the direct application of the above NMM-ELM scheme is computationally expensive because of the large training number required. Thus, in [12], an SJ-ELM model is proposed to reduce the training sample set for super-resolution 3-D quantitative imaging of objects with high contrasts. The semi-join strategy can decrease the inner matrix dimensions and improve the convergence performance of the model in the performance of super-resolution imaging.

15.2.2 Motivation

As stated above, it is a challenging task for noniterative and iterative inversion methods to solve high contrast and large-scale inverse scattering problems. Meanwhile, for some machine learning-based methods, as the input of the traditional machine learning methods [such as DeepNIS, BP scheme (BPS), and DCS] is obtained from a linearized approximation, when the contrasts of objects are high, it is difficult to obtain a desired input with high accuracy from the linearized approximation.

Thus, inversion methods that can directly map the scattered field to the corresponding distribution of the model parameters without relying on a linearized approximation have been proposed. Usually, this kind of method contains two machine learning-based stages. The first stage is used to transfer the measured scattered electric field to the primary images. Then in the second stage, a deep learning technique is employed to further enhance the imaging quality of the output from the first stage. Meanwhile, to simultaneously reconstruct multiple electrical parameters such as relative permittivity and conductivity, machine learning methods with a multichannel structure have been also designed. Our recent works in these directions will be reviewed below.

15.3 Our Approach for High Contrast Inverse Scattering of Electrically Large Structures

In this section, the NMM-IEM machine learning scheme and RAML method are introduced for 2-D inverse scattering problem with high contrasts and large electrical dimension scatterers. For 3-D inverse problem, SJ-ELM model is firstly introduced for super-resolution 3-D quantitative imaging of objects with high contrasts. Then, for the important application of human brain imaging, an HNNEMIS method consisting of both shallow and deep neural networks is introduced to well capture the feature of each tissue.

15.3.1 The 2-D Inverse Scattering Problem with Electrically Large Structures

15.3.1.1 Dual-Module NMM-IEM Machine Learning Model

For the 2-D inversion, we have developed a dual-module machine learning scheme consisting of a NMM and an IEM [10]. There are two parts in this dual-module NMM-IEM machine learning model: NMM is employed to preliminarily convert the complex-valued scattered field data into the real-valued relative permittivity in the DOI, while IEM as a CNN is used to further enhance the reconstructed image from the NMM. Figure 15.1 shows the architecture of the proposed NMM-IEM method.

The scattered fields measured at the receiver arrays are complex-valued data, while the relative permittivity of the DOI is real. The NMM is to convert the complex-valued scattered field data into the real permittivity with their nonlinear relationship manifested by the ELM that has a simple architecture with low training cost. The NMM has four layers: the input layer, two hidden layers, and the preliminary imaging layer. The input layer nodes are filled with the column vector $\mathbf{x}_j = [x_{1j}, x_{2j}, \ldots, x_{Mj}]^T \in \mathbf{C}^M$, which contains the complex-valued E_z^{sct}, $\eta_0 H_x^{sct}$, and $\eta_0 H_y^{sct}$ for all the transmitter and receiver combinations, where η_0 is the intrinsic impedance of free space. The combination of E_z^{sct}, $\eta_0 H_x^{sct}$, and $\eta_0 H_y^{sct}$ could increase the diversity of the training data through the random matrix in the ELM to obtain more reliable reconstruction results. The subscript j denotes the jth training data set and M is the dimension of the input data set for each training. The NMM output column vector $\mathbf{o}_j = [o_{1j}, o_{2j}, \ldots, o_{Nj}]^T \in \mathbf{R}^N$ has real values, and its dimension N is the total pixel number in the DOI. It is evaluated by

$$\mathbf{o}_j = \overline{\overline{\alpha}}\, g_r(\overline{\overline{\mathbf{w}}}_r \mid \overline{\overline{\beta}}\, g_c(\overline{\overline{\mathbf{w}}}_c \mathbf{x}_j + \mathbf{b})| + \mathbf{p}) \tag{15.1}$$

where $j = 1, 2, \ldots, P$ if there are totally P sets of training data. $\overline{\overline{\mathbf{w}}}_c$ is an $L \times M$ complex-valued random weight matrix, and $\mathbf{b}$ is a column threshold vector including L complex-valued random numbers [44]. g_c is the nonlinear activation function, and its outputs are the L complex values in the neurons

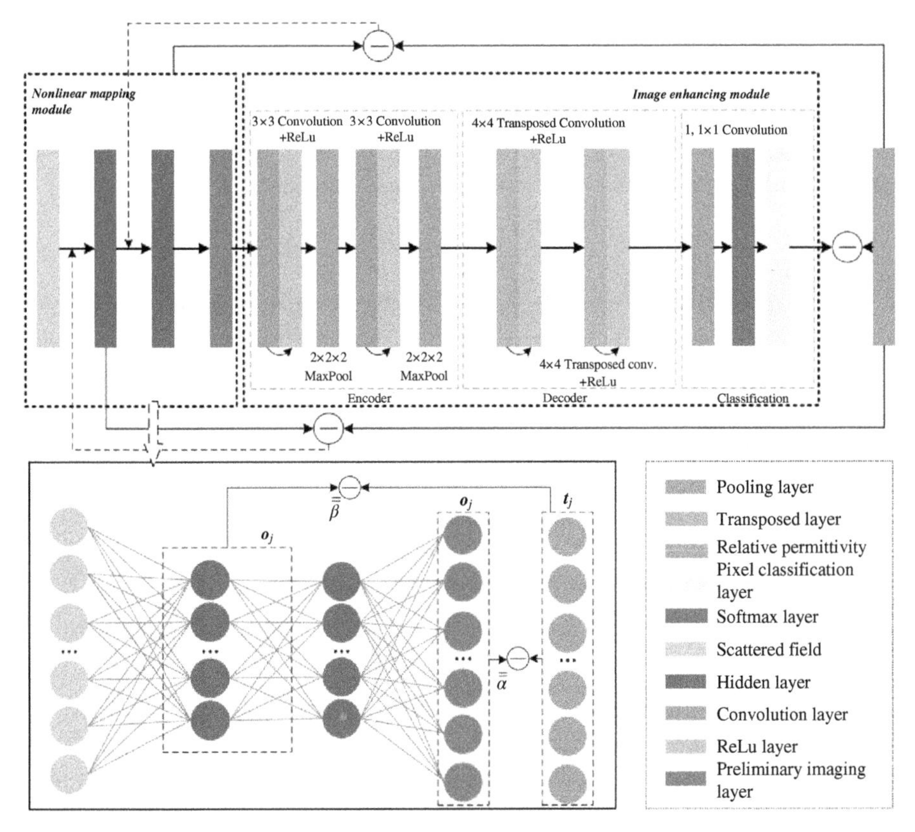

Figure 15.1 Architecture of the dual-module NMM-IEM machine learning scheme. It includes two parts: the NMM to preliminarily convert the complex-valued scattered field data into the real-valued relative permittivity in the DOI, and the IEM as a CNN to further enhance the reconstructed image from the NMM. Source: Xiao et al. [10]/IEEE.

of the first hidden layer. $\overline{\overline{\beta}}$ is a $K \times L$ complex-valued weight matrix connecting the neurons of the first hidden layer to the inputs of the second hidden layer and K is the neuron connection number to the output layer. It is determined by training. $\overline{\overline{\alpha}}$, g_r, $\overline{\overline{\mathbf{w}}}_r$, and $\mathbf{p}$ play the similar roles as $\overline{\overline{\beta}}$, g_c, $\overline{\overline{\mathbf{w}}}_c$, and $\mathbf{b}$, respectively, but work for the second hidden layer and have real values. $\overline{\overline{\alpha}}$ connects the neurons of the second hidden layer and the nodes in the preliminary imaging layer. The dimensions of $\overline{\overline{\alpha}}$, $\overline{\overline{\mathbf{w}}}_r$, and $\mathbf{p}$ are $N \times S$, $S \times K$, and S, respectively.

After many experiments, an inverse hyperbolic function expressed as

$$g_c = \text{arcsinh}(x) = \int_0^x \frac{dt}{\sqrt{1+t^2}} \tag{15.2}$$

is selected as the complex activation function g_c, and the sigmoid function expressed as

$$g_r = \text{sigmoid}(x) = \frac{1}{1+e^{-x}} \tag{15.3}$$

is selected as the real activation function g_r. It is assumed that the true relative permittivity values in the DOI are denoted as $\mathbf{t}_j = [t_{1j}, t_{2j}, \ldots, t_{Nj}]^T \in \mathbf{R}^N$. To obtain the optimum $\overline{\overline{\beta}}$, we minimize the mismatch between the inputs of the second hidden layer and the true relative permittivity values, i.e. we let $\sum_{j=1}^{P} \|\mathbf{o}_j - \mathbf{t}_j\| = 0$ and solve for the complex-valued weight matrix $\overline{\overline{\beta}}$. The real-valued weight $\overline{\overline{\alpha}}$ can be obtained following the similar procedure by minimizing the mismatch between the node values of the preliminary imaging and $\mathbf{t}_j$ for all the training data sets, i.e. we let $\sum_{j=1}^{P} \|\mathbf{o}_j - \mathbf{t}_j\| = 0$. Meanwhile, to match the dimensions for different vectors and matrices, we set $K = N$. Thus, the equation relating the weight matrix $\overline{\overline{\beta}}$ of the first hidden layer to the true relative permittivity vector $\mathbf{t}_j$ for all the training data sets can be compactly expressed as

$$\overline{\overline{\beta}} = \overline{\overline{\mathrm{T}}} \tag{15.4}$$

where $\overline{\overline{\mathbf{T}}} = [\mathbf{t}_1;\mathbf{t}_2;\ldots;\mathbf{t}_P]_{N\times P}$ and $\overline{\overline{\mathbf{G}}}_c$ is the combination of all the column vector output from the activation function g_c of the first hidden layer for all the training data sets $\mathbf{x}_j$ and has the dimensions of $L \times P$. Thus, the optimum complex-valued weight matrix $\overline{\overline{\beta}}$ can be computed by

$$\hat{\overline{\overline{\beta}}} = \overline{\overline{\mathbf{T}}}\,\overline{\overline{\mathbf{G}}}_c^{\dagger} \tag{15.5}$$

where the complex matrix $\overline{\overline{\mathbf{G}}}_c^{\dagger}$ is the Moore–Penrose generalized inverse of the complex matrix $\overline{\overline{\mathbf{G}}}_c$. The computation of the Moore–Penrose generalized inverse is referred to [45]. Following a similar procedure, we can obtain the optimum real matrix:

$$\hat{\overline{\overline{\alpha}}} = \overline{\overline{\mathbf{T}}}\,\overline{\overline{\mathbf{G}}}_r^{\dagger} \tag{15.6}$$

where the real-valued matrix $\overline{\overline{\mathbf{G}}}_r$ is the combination of all the column vector output from the activation function g_r of the second hidden layer and has the dimensions of $S \times P$. We can see that the ELM in the NMM can be trained at a low cost, since both unknown weight matrices of two hidden layers can be obtained by solving the matrix inverse only once.

The IEM is used to improve further the 2-D image output from the NMM. It is actually a CNN and consists of an encoder, a decoder, and a pixel classifier. The encoder is employed to analyze and contract the input data, and then the decoder synthesizes and expands the previously encoded feature. Finally, the decoded output is the input to the pixel classifier for a full-resolution segmentation. As shown in Figure 15.1, there are two layers in the encoder, two layers in the decoder, and one layer in the pixel classifier. In each layer of the encoder, a convolution kernel with the size of 3×3 and a rectified linear unit (ReLU) are followed by a 2×2 max pooling. In the decoder, each layer consists of a transposed convolution kernel with the size of 4×4 and a ReLu. Following the decoder, the convolution with the 1×1 kernel size produces the output with one channel. Then, a pixel classifier is employed to classify each pixel of the image for a full-resolution segmentation.

15.3.1.2 Receiver Approximation Machine Learning Method

The above dual-module NMM-IEM method assumes that all receivers are functioning properly. In a realistic inversion environment, however, it is common for some receivers to be damaged, hence the corresponding data become invalid. Thus, an RAML method is proposed for solving this problem with damaged data [11].

The RAML method has two channels for the real and imaginary parts of the approximated scattered field data, respectively. Each of these two channels is constructed based on the traditional ANN structure with double hidden layers. The specific structure of RAML is shown in Figure 15.2.

The input layer of RAML is filled with $\mathbf{p}_j = [\mathbf{p}_j^{real} + i\mathbf{p}_j^{imag}]^{\mathrm{T}} \in \mathbb{C}^M$, where $\mathbf{p}_j$ is the jth complex-valued training scattered data from other receivers around the damaged receiver.

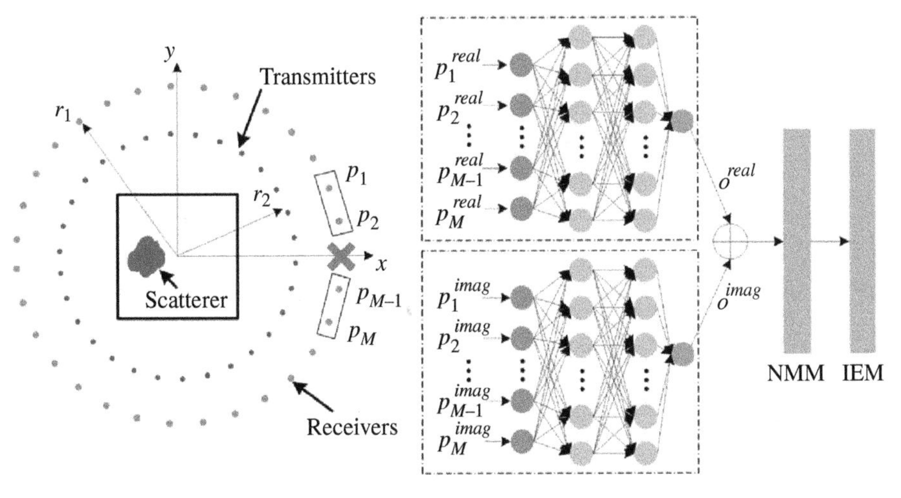

Figure 15.2 The architecture of RAML. The approximation process contains the real and imaginary parts. The input of RAML is other receivers within a wavelength around the damaged receiver, and its output is transferred to NMM-IEM. Source: Xiao et al. [12]/IEEE.

$\mathbf{p}_j^{real} = [p_{1j}^{real}, p_{2j}^{real}, \dots, p_{Mj}^{real}] \in \mathbf{R}^M$ and $\mathbf{p}_j^{imag} = [p_{1j}^{imag}, p_{2j}^{imag}, \dots, p_{Mj}^{imag}] \in \mathbf{R}^M$ are the real and imaginary parts of $\mathbf{p}_j$. The output of RAML is the approximated field data of the damaged receiver and is defined as

$$\mathbf{o}_j^{\mathrm{RAML}} = \mathbf{o}_j^{real} + i\mathbf{o}_j^{imag} \tag{15.7}$$

where $\mathbf{o}_j^{real}$ and $\mathbf{o}_j^{imag}$ are the real and imaginary parts of output, respectively. For the real part of $\mathbf{o}_j^{\mathrm{RAML}}$, it is evaluated by

$$\mathbf{o}_j^{real} = \overline{\overline{\boldsymbol{w}}}_3^{real} \left(g_r \left(\overline{\overline{\boldsymbol{w}}}_2^{real} \left(g_r \left(\overline{\overline{\boldsymbol{w}}}_1^{real} \mathbf{p}_j^{real} + \mathbf{b}_1^{real} \right) \right) + \mathbf{b}_2^{real} \right) \right) + \mathbf{b}_3^{real} \tag{15.8}$$

where $\mathbf{b}_1^{real}$, $\mathbf{b}_2^{real}$, and $\mathbf{b}_3^{real}$ are the bias of the first hidden layer, the second hidden layer, and the output layer, respectively, and their dimensions are equal to the number of nodes of the corresponding layer.

$\overline{\overline{\boldsymbol{w}}}_{31}^{real}$, $\overline{\overline{\boldsymbol{w}}}_2^{real}$, and $\overline{\overline{\boldsymbol{w}}}_3^{real}$ are the weight matrix connecting the input layer to the first hidden layer, the first hidden layer to the second hidden layer, and the second hidden layer to the output layer, respectively. Their values are optimized iteratively by the backpropagation algorithm. Meanwhile, the imaginary part of RAML has the same procedure as the real part. Then, $\mathbf{o}_j^{\mathrm{RAML}}$ is the input to NMM-IEM for further inversion. After many experiments, we found that using the data from the receivers within one wavelength around the damaged receiver as the input is the most beneficial and accurate for inversion. In other words, if the distance between the receivers is too large, for example more than one wavelength, the repair quality may be compromised.

15.3.2 Application for 3-D Inverse Scattering Problem with Electrically Large Structures

15.3.2.1 Semi-Join Extreme Learning Machine

For the 3-D inverse scattering problem, the number of voxels is usually much larger than the number of pixels in 2-D; thus, the direct application of the above NMM-ELM scheme for 2-D is computationally expensive for 3-D because of the large training number required. Hence, an

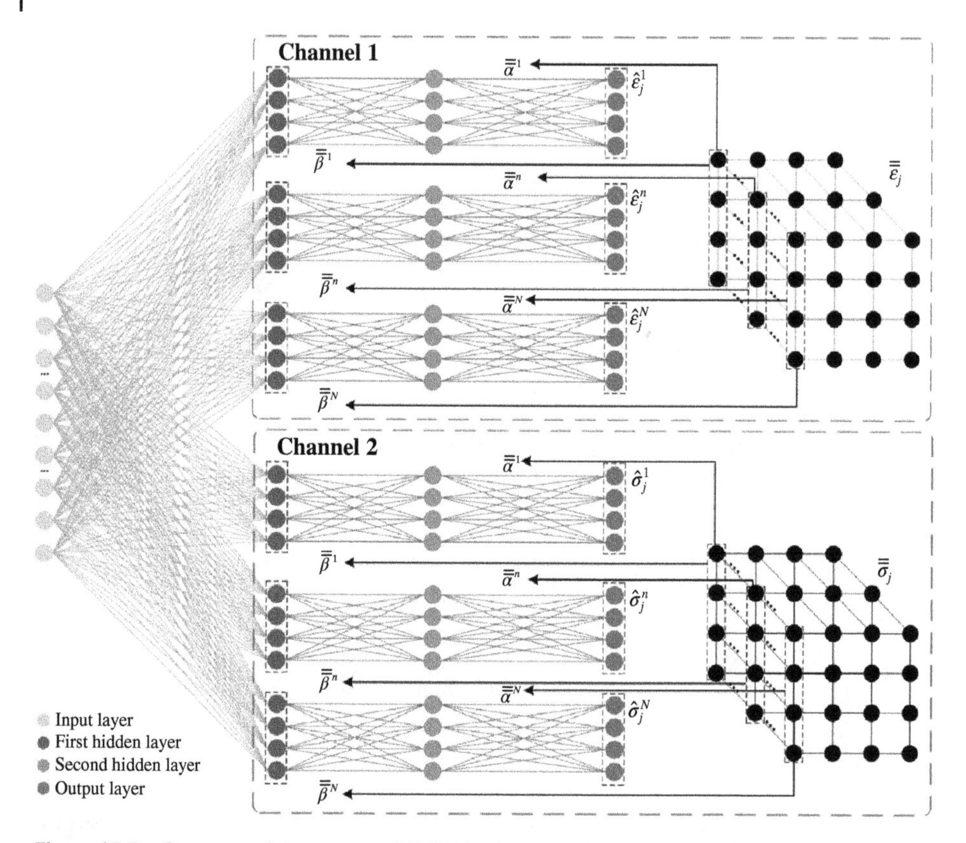

Figure 15.3 Structure of the proposed SJ-ELM scheme, which consists of two channels for mapping the relative permittivity and conductivity distributions of objects, respectively. Both channels have the same structure, and each channel consists of four layers, i.e. an input layer, two hidden layers, and an output layer. $\overline{\overline{\alpha}}^n$ and $\overline{\overline{\beta}}^n$ are the weight matrix of both hidden layers for the nth subset and they could be obtained from target as shown by arrows, respectively. Source: Xiao et al. [12]/IEEE.

SJ-ELM model has been proposed to reduce the training sample set for super-resolution 3-D quantitative imaging of objects with high contrasts [12]. Compared with the full-joint strategy in the traditional ELM [44], for the semi-joint strategy, the nodes in the hidden layer are not totally connected to the nodes in the output layer, so both the inner matrix dimensions and required memory storage of SJ-ELM are less and the convergence of the model is faster [12]. To relieve the convergence stress of the machine learning method during the inversion process, as shown in Figure 15.3, the permittivity and conductivity are reconstructed with two separate channels, respectively. The scattered fields obtained from the forward solver are complex valued data, while the relative permittivity and conductivity of each channel are real. Thus, each channel is employed to convert the complex scattered field data to the real permittivity and conductivity values with a simple structure having low training cost, respectively. However, the outputs of these two channels will be composed into the complex permittivity as the final inversion result of the proposed SJ-ELM.

Both channels have the same structure, and each channel consists of four layers, i.e. an input layer, two hidden layers, and an output layer. In each channel, each hidden layer receives the output of the previous layer in input and transmits it to the next layer. The input of jth training sample is a column vector $\mathbf{x}_j = [x_{1j}, x_{2j}, \cdots, x_{mj}]^T \in \mathbf{C}^m$ which contains the complex value E_x^{sct},

E_y^{sct}, and E_z^{sct} for all the transmitter and receiver combinations to the permittivity or conductivity distribution in 3-D space, where m is the input dimension of jth training sample, $j = 1, 2, \ldots, P$ and P presents the total set number of training data. The whole imaging space is discretized into N_1, N_2, and N_3 voxels in x, y, and z directions. Correspondingly, for the target of the jth training set, the corresponding jth output of the jth training sample could be expressed as a column matrix $\mathbf{o}_j = [o_{1j}, o_{2j}, \ldots, o_{N_1 \times N_2 \times N_3 j}]^T \in \mathbf{R}^{(N_1 \times N_2 \times N_3) \times 1}$, where $N_1 \times N_2 \times N_3$ is the total number of voxels in imaging domain. Usually in the 3-D super-resolution imaging, $N_1 \times N_2 \times N_3$ is a huge number, which will bring an unacceptable burden for computer memory. Thus, in the proposed semi-join strategy, the output matrix is uniformly divided into N subsets, which could be described as $\mathbf{o}_j = \mathbf{o}_j^1 \cup \mathbf{o}_j^n \cdots \cup \mathbf{o}_j^N$, where $o_j^n \in \mathbf{R}^{((N_1 \times N_2 \times N_3)/N) \times 1}$. For the convenient description, here we set $(N_1 \times N_2 \times N_3)/N = F$. After many numerical experiments and analyses, N is recommended to set as the maximum value of N_1, N_2, and N_3.

An input layer with m nodes, a complex-value hidden layer with L nodes, a real-value hidden layer with S nodes, and an output layer with F nodes are contained in SJ-ELM. The calculation of the nth subset can be expressed as:

$$\mathbf{o}_j^n = \overline{\overline{\alpha}}^n g_r\left(\overline{\overline{\boldsymbol{w}}}_r^n \mid \overline{\overline{\beta}}^n g_c\left(\overline{\overline{\boldsymbol{w}}}_c^n \mathbf{x}_j + \mathbf{b}^n\right) \mid + \mathbf{p}^n\right). \tag{15.9}$$

From the input layer to the first hidden layer, $\overline{\overline{\boldsymbol{w}}}_c^n$, $\mathbf{b}^n$, $\overline{\overline{\beta}}^n$, and g_c are involved. $\overline{\overline{\boldsymbol{w}}}_c^n$, which is an $L \times m$ complex-value random weight matrix, connects the input layer and the first hidden layer. $\mathbf{b}^n$, which is L complex-value random column vector, presents the threshold of the first hidden layer. $\overline{\overline{\beta}}^n$ is an $F \times L$ complex-value matrix which connects the complex-value hidden layer to the real-value hidden layer. Its elements will be determined in the training.

From the first hidden layer to the second hidden layer, $\overline{\overline{\boldsymbol{w}}}_r^n$, $\mathbf{p}^n$, $\overline{\overline{\alpha}}^n$, and g_r are involved. $\overline{\overline{\boldsymbol{w}}}_r^n$ is an $S \times F$ real-value random matrix as the weight matrix between the first hidden layer and second hidden layer. $\mathbf{p}^n$ is a vector with S real-value random numbers. $\overline{\overline{\alpha}}^n$ is an $F \times S$ real-value random matrix, which is the weights connecting the nodes in the second hidden layer and the output layer.

According to the ELM theory [45–48] the initial value of $\overline{\overline{\boldsymbol{w}}}_c^n$, $\mathbf{b}^n$, $\overline{\overline{\boldsymbol{w}}}_r^n$, and $\mathbf{p}^n$ are set randomly. Here "random" means that the elements in the weight matrices are randomly generated instead of being obtained in the training. The training cost is low since we only need to solve the matrices $\overline{\overline{\beta}}^n$ and $\overline{\overline{\alpha}}^n$.

Assume the corresponding true permittivity or conductivity values of the nth subset of the jth training sample is denoted as $\boldsymbol{t}_j^n$. Let $\sum_{j=1}^{P} \|\mathbf{o}_j^n - \boldsymbol{t}_j^n\| = 0$, and employ the layer-by-layer solution strategy to obtain the weight matrices $\overline{\overline{\alpha}}^n$ and $\overline{\overline{\beta}}^n$ of both hidden layers for the nth subset, respectively, where the second hidden layer is employed to make the model further converge to the target based on the results from the first hidden layer. The relationship between the weight matrix $\overline{\overline{\beta}}^n$ of the first hidden layer and $\boldsymbol{t}^n$ can be compactly expressed as

$$\overline{\overline{\beta}}^n g_c^n = \boldsymbol{t}^n, n = 1, \ldots, N \tag{15.10}$$

where the complex-value matrix $g_c^n = (\overline{\overline{\boldsymbol{w}}}_c^n \cdot \mathbf{x}_j + \mathbf{b}^n)_{L \times P}$ is the first hidden layer output matrix. Thus, the desired complex weight matrix of $\widehat{\beta}^n$ can be obtained by following equation,

$$\widehat{\beta}^n = \boldsymbol{t}^n g_c^{n\dagger}, n = 1, \ldots, N \tag{15.11}$$

where the complex matrix $g_c^{n\dagger}$ presents the Moore–Penrose generalized inverse of complex matrix g_c^n. The relationship between the weight matrix $\overline{\overline{\boldsymbol{\alpha}}}^n$ of the second hidden layer and $\boldsymbol{t}^n$ can be compactly expressed as

$$\overline{\overline{\boldsymbol{\alpha}}}^n g_r^n = \boldsymbol{t}^n, n = 1, \ldots, N \tag{15.12}$$

where the real value matrix $g_r^n = \left[g_r^n\left(\overline{\overline{\boldsymbol{w}}}_r^n|\overline{\overline{\boldsymbol{\beta}}}^n g_c| + \boldsymbol{p}^n\right)\right]_{K\times P}$ is the output matrix of the second hidden layer. Thus, the desired weight vector, which connects the ith hidden node and the output layer nodes, could also be computed by

$$\hat{\boldsymbol{\alpha}}^n = \boldsymbol{t}^n g_r^{n\dagger} \tag{15.13}$$

where complex matrix $g_r^{n\dagger}$ is the Moore–Penrose generalized inverse of complex matrix g_r^n.

The node number of the hidden layers is set on the basis of Hecht–Nelson method [49]: When the dimension of input data is n, the node number of the hidden layer is $2n+1$. Thus, the node numbers of the first and second hidden layers are $L = 2m+1$ and $S = 2F+1$, respectively, and the dimensions of $\overline{\overline{\boldsymbol{w}}}_c^n$, $\mathbf{b}^n$, $\overline{\overline{\boldsymbol{\beta}}}^n$, $\overline{\overline{\boldsymbol{w}}}_r^n$, $\mathbf{p}^n$, and $\overline{\overline{\boldsymbol{\alpha}}}^n$ could be further written as $(2m+1)\times m$, $(2m+1)\times 1$, $F\times(2m+1)$, $(2F+1)\times F$, $(2F+1)\times 1$, and $F\times(2F+1)$. For the total N subsets, the dimensions of $\overline{\overline{\boldsymbol{w}}}_c$, $\mathbf{b}$, $\overline{\overline{\boldsymbol{\beta}}}$, $\overline{\overline{\boldsymbol{w}}}_r$, $\mathbf{p}$, and $\overline{\overline{\boldsymbol{\alpha}}}$ could be written as $[(2m+1)\times N]\times m$, $[(2m+1)\times N]\times 1$, $F\times[(2m+1)\times N]$, $[(2F+1)\times N]\times F$, $[(2F+1)\times N]\times 1$, and $F\times[(2F+1)\times N]$, respectively. Based on the ELM theory, $\overline{\overline{\boldsymbol{\beta}}}$ and $\overline{\overline{\boldsymbol{\alpha}}}$ are the tuneable parameters, and their dimensions are $F\times[(2m+1)\times N]$ and $F\times[(2F+1)\times N]$.

In contrast, for the traditional full-join strategy, the dimensions of $\overline{\overline{\boldsymbol{w}}}_c$, $\mathbf{b}$, $\overline{\overline{\boldsymbol{\beta}}}$, $\overline{\overline{\boldsymbol{w}}}_r$, $\mathbf{p}$, and $\overline{\overline{\boldsymbol{\alpha}}}$ based on Hecht–Nelson method are $(2m+1)\times m$, $(2m+1)\times 1$, $(N\times F)\times(2m+1)$, $[2(N\times F)+1]\times(N\times F)$, $[2(N\times F)+1]\times 1$, and $(N\times F)\times[2(N\times F)+1]$, respectively. Thus, comparing the traditional full-join strategy and the semi-join strategy, the dimensions of $\overline{\overline{\boldsymbol{\beta}}}$ are the same; however, the dimension of $\overline{\overline{\boldsymbol{\alpha}}}$ based on the semi-join strategy is $(2F+1)/(2N\times F+1)$ of the ones based on the traditional full-join strategy, which could be obtained with the calculation of $\{F\times[(2F+1)\times N]\}/\{(N\times F)\times[2(N\times F)+1]\}$. Further, $(2F+1)/(2N\times F+1)$ will be less than 1 if N is larger than 1. Thus, it could be proven that the training burden based on the semi-join strategy is less than the ones based on the traditional full-join strategy, even though there are N sub-models in SJ-ELM.

Meanwhile, one should note that the sub-models are independent of each other, so there does not exist any information interchange between sub-models. Thus, the imaging process for a sub-model with the output dimensions of $(N_1\times N_2\times N_3)/N_{max}\times 1$ could be regarded as a 2-D imaging problem with respect to the output dimensions and structure, since the output of each sub-model is a 2-D slice of the imaging domain along the direction of dimension N_{max}, where N_{max} is the maximum value of N_1, N_2, and N_3. In other words, the proposed SJ-ELM divides the 3-D imaging problem into N_{max} sets of 2-D imaging slices, and the output of each sub-model is the corresponding slice of the 3-D imaging domain. Thus, different from other full-join methods which require a large number of training samples, the proposed model could solve a 3-D imaging problem with the cost of 2-D problems. Meanwhile, one should note that ELM randomly chooses hidden nodes and analytically determines the output weights for hidden layers of feedforward neural networks, which has much lower training costs than the traditional feedforward neural networks. Here "random" means that the elements in the weight matrices are randomly generated instead of being obtained in the training. This is the merit of the ELM. The training cost is low since we only need to solve the matrices $\overline{\overline{\boldsymbol{\alpha}}}$ and $\overline{\overline{\boldsymbol{\beta}}}$.

Furthermore, with the dimensions of output layer increasing which will lead to stronger nonlinearity, the node number of both hidden layers and/or number of training samples should also be largely increased to converge. Thus, the dimensions of $\overline{\overline{\beta}}$ and $\overline{\overline{\alpha}}$ in the full-join strategy are much larger than those in the semi-join strategy, which means the heavier computational burden and more difficult convergence in training. Therefore, compared with the full-join strategy, the semi-join strategy is more economical for the 3-D microwave imaging problem. However, one should note that more CPU time will be needed to train the several sub-models in parallel.

To illustrate the efficiency of SJ-ELM, we have added a performance evaluation to test the proposed SJ-ELM. Two approximation problems in complex domain used in [50] are employed to test the efficiency of the proposed model, and the fully complex extreme learning machine (CELM) [45] is selected as the comparison model. As given in the setting in [51], 10,000 training samples and 1000 testing samples are randomly drawn from the interval $[0 + i0, 1 + i]$. The first verification is based on a nonanalytic function

$$f(z) = f(x + \mathrm{i}y) = e^{\mathrm{i}y}(1 - x^2 - y^2) \tag{15.14}$$

and the second one is an analytic function given by

$$f(z) = f(x + \mathrm{i}y) = \sin(x)\cosh(y) + \mathrm{i}\cos(x)\sinh(y) \tag{15.15}$$

Considering the microwave imaging problem which maps the scattered field data to the permittivity and/or conductivity of objects, we modify the following function to map the complex domain to real domain as

$$f(z) = f(x + \mathrm{i}y) = |e^{\mathrm{i}y}(1 - x^2 - y^2)| \tag{15.16}$$

$$f(z) = f(x + \mathrm{i}y) = |\sin(x)\cosh(y) + \mathrm{i}\cos(x)\sinh(y)| \tag{15.17}$$

respectively for (15.17) and (15.18).

Here, a performance index is defined as the relative root-mean-squared error (RRMSE)

$$\mathrm{RRMSE} = \sqrt{\frac{\sum_{j=1}^{N} (y_j - \hat{y}_j)^2}{N}} \Bigg/ \sqrt{\frac{\sum_{j=1}^{N} (y_j)^2}{N}} \tag{15.18}$$

where y_j is the calculation result from (15.19) or (15.20) and $\hat{y}_j$ is the corresponding output from neural network. The average results of RRMSE and standard deviation (Dev) are shown in Table 15.1. The calculations are performed on an Intel i7-9700 3.0-GHz machine with 64-GB RAM. Here the node numbers of hidden layers are set as 50, 100, and 500, respectively, for CELM. The node number of both hidden layers from SJ-ELM is set to be the same value as the CELM. It could be seen that comparing with CELM, SJ-ELM has more accurate and stable performance, but has a little longer runtime due to its two hidden layers.

15.3.2.2 Hybrid Neural Network Electromagnetic Inversion Scheme

HNNEMIS consisting of a semi-joint back propagation neural network (SJ-BPNN) scheme and a U-Net is proposed to reconstruct super-resolution electrical properties distributions of the human brain. For the human brain imaging, it is more important to capture the feature of each tissue type to guarantee the imaging accuracy [13]. To solve this kind of inverse problem with such specific feature enhancement, a hybrid neural network electromagnetic inversion scheme (HNNEMIS) which contains SJ-BPNN and U-Net is proposed, where deep learning neural network can capture the feature of each tissue to further enhance inversion accuracy. In HNNEMIIS, SJ-BPNN is

Table 15.1 Performance Comparison of CELM and SJ-ELM.

	Node number of hidden layer			CELM	SJ-ELM
Nonanalytic function	50	CPU time (s)	Training	1.3252	1.5548
			Testing	0.1654	0.1949
		Training	RRMSE	0.0612	0.0097
			Dev	0.4826	0.0597
		Testing	RRMSE	0.0634	0.0114
			Dev	0.5103	0.0658
	100	CPU time (s)	Training	1.6952	1.8756
			Testing	0.2651	0.3215
		Training	RRMSE	0.0564	0.0092
			Dev	0.4327	0.0462
		Testing	RRMSE	0.0587	0.0126
			Dev	0.4912	0.0487
	500	CPU time (s)	Training	2.1635	2.6352
			Testing	0.1654	0.1949
		Training	RRMSE	0.0464	0.0083
			Dev	0.3621	0.0298
		Testing	RRMSE	0.0488	0.0096
			Dev	0.3916	0.0326
Analytic function	50	CPU time (s)	Training	1.2564	1.5485
			Testing	0.1649	0.1934
		Training	RRMSE	0.0497	0.0132
			Dev	0.5931	0.1108
		Testing	RRMSE	0.0536	0.0151
			Dev	0.6318	0.1236
	100	CPU time (s)	Training	1.6752	1.8542
			Testing	0.2632	0.3187
		Training	RRMSE	0.0461	0.0097
			Dev	0.4639	0.0312
		Testing	RRMSE	0.0494	0.0121
			Dev	0.4913	0.0367
	500	CPU time (s)	Training	2.2031	2.6217
			Testing	0.1713	0.2017
		Training	RRMSE	0.0431	0.0089
			Dev	0.3854	0.0216
		Testing	RRMSE	0.0458	0.0094
			Dev	0.4133	0.0227

Source: Xiao et al. [12]/IEEE.

proposed to transfer the measured scattered field into preliminary images of electrical properties, and U-Net is employed to enhance the quality of the preliminary images.

The first step of reconstruction with HNNEMIS is to restore electrical properties with the measured field. For the complex and nonlinear reconstruction of the human brain, the mature and robust back propagation neural network (BPNN) is adopted, which has the advantage of good self-learning, self-adapting, and generalization ability [52]. However, the full-join BPNN is very expensive if it is directly applied to learn the nonlinear relationship between the measured field and electrical properties for super-resolution with too many unknowns. Therefore, a SJ-BPNN scheme based on BPNN is proposed to handle this problem.

The SJ-BPNN scheme has three layers: an input layer, a hidden layer, and an output layer, as shown in Figure 15.4. For the ith training sample, the input of SJ-BPNN is a column vector $\mathbf{x}_j = [x_{1j}, x_{2j}, \ldots, x_{mj},]^T \in \mathbf{C}^{m\times 1}$ about the measured scattered field, and the output is also a column vector $\mathbf{o}_j = [o_{1j}, o_{2j}, \cdots, o_{N_1N_2N_3j},]^T \in \mathbf{R}^{N_1N_2N_3\times 1}$ about electrical properties distributions, where m is the dimension of the input data in the input layer, and $N_1N_2N_3$ are the dimensions of output.

Different from the traditional BPNN, SJ-BPNN divides the output into N subsets uniformly, and makes each subset of output only connect to a part of hidden layer network, as shown in Figure 15.4. In each subset, the number of output is $N_1N_2N_3/N = F$. The nth subset of output $\mathbf{o}_j^n$ can be written as

$$\mathbf{o}_j^n = g_l\left[\overline{\overline{\boldsymbol{w}}}_o^n \cdot g_s\left(\overline{\overline{\boldsymbol{w}}}_h^n \cdot \mathbf{x}_j + \mathbf{b}_h^n\right) + \mathbf{b}_o^n\right] \tag{15.19}$$

where $n = 1, 2, \ldots, N$, $j = 1, 2, \ldots, I$ and I is the total number of samples in the training dataset of SJ-BPNN. $\overline{\overline{\boldsymbol{w}}}_h^n$ and $\mathbf{b}_h^n$ are an $L\times m$ weight matrix and an $L\times 1$ threshold vector between the input layer and the hidden layer, respectively, where L is the number of nodes in the hidden layer of each subset. $\overline{\overline{\boldsymbol{w}}}_o^n$ and $\mathbf{b}_o^n$ are an $F\times L$ weight matrix and an $F\times 1$ threshold vector between the hidden layer and the output layer, respectively. g_l and g_s are the linear activation function of the output layer and the tan-sigmoid activation function of the hidden layer, respectively, and g_l can be expressed as

$$g_l(x) = x \tag{15.20}$$

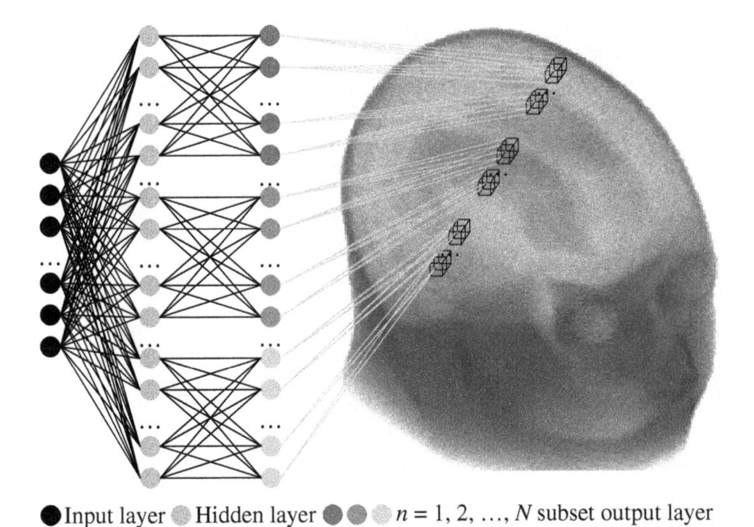

Figure 15.4 Architecture of SJ-BPNN including an input layer, a hidden layer, and an output layer.

For the proposed SJ-BPNN scheme, the dimensions of $\overline{\overline{\boldsymbol{w}}}_h^n$, $\mathbf{b}_h^n$, $\overline{\overline{\boldsymbol{w}}}_o^n$, and $\mathbf{b}_o^n$ to solve are $L \times m$, $L \times 1$, $F \times L$, and $F \times 1$ in each subset, respectively. Meanwhile, for the traditional full-join BPNN, the dimensions of $\overline{\overline{\boldsymbol{w}}}_h$, $\mathbf{b}_h$, $\overline{\overline{\boldsymbol{w}}}_o$, and $\mathbf{b}_o$ are $L' \times m$, $L' \times 1$, $N_1N_2N_3 \times L'$, and $N_1N_2N_3 \times 1$, respectively, where L' is the number of nodes in the hidden layer of the traditional BPNN and is always set to $L' = 2m+1$ based on the Hecht–Nelson method [49]. When the resolution of the inverse problem is ultra-high (such as $256 \times 256 \times 256$), $\overline{\overline{\boldsymbol{w}}}_o$ will be a large matrix requiring huge memory. In contrast, SJ-BPNN just needs to compute in parallel $\overline{\overline{\boldsymbol{w}}}_h^n$, $\mathbf{b}_h^n$, $\overline{\overline{\boldsymbol{w}}}_o^n$, and $\mathbf{b}_o^n$ of N subsets which are smaller than $\overline{\overline{\boldsymbol{w}}}_o$ in the traditional BPNN when $N > 1$, thus greatly reducing the computer memory to an acceptable level for a workstation. This is the main reason for adopting SJ-BPNN rather than the traditional BPNN to solve high-resolution impedance imaging of brain. Generally speaking, N needs to be a large enough number to make the network of each subset small enough so that the network can converge with small training samples and computer memory.

Just as the traditional BPNN, the convergence of SJ-BPNN is greatly affected by the selection of network parameters [53]. So, it is necessary to conduct simulation experiments for the selection of network parameters.

Usually, the quality of reconstruction results obtained by SJ-BPNN still has room for enhancement. Thus, the second step of reconstruction with SJ-BPNN is to improve the accuracy and robustness of the proposed inversion method.

To further enhance the imaging quality of output from SJ-BPNN, U-Net used in traditional image segmentation is employed in this work [34]. U-Net consists of a contracting path (down-sampling) and an expansive path (up-sampling), and its architecture is presented in Figure 15.5. The contracting path is a typical convolutional network. It has repeated applications of two 3×3 convolutions, two rectified linear units (ReLU), and a 2×2 max pooling. The typical convolutional network is frequently employed to image classification. However, if we want to classify each pixel or voxel of a biomedical image, the typical convolutional network is insufficient because it only outputs a single class label for image classification. Thus, the expansive path in U-Net is employed to increase the resolution of output. The expansive path is similar to the contracting path, but the max pooling in the contracting path is replaced by a 3×3 up-convolution in the expansive path.

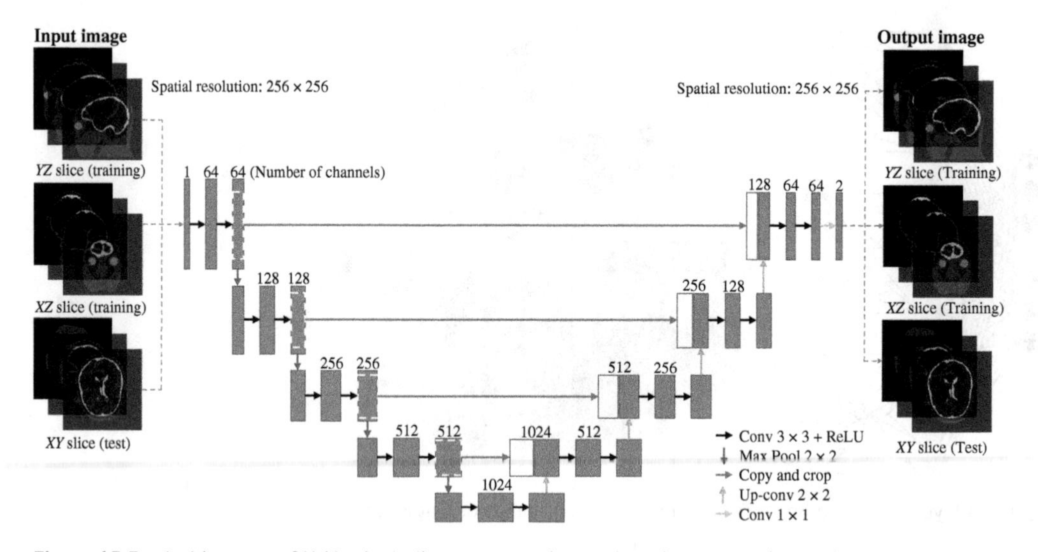

Figure 15.5 Architecture of U-Net including a contracting path and an expansive path.

The mature U-Net has been often adopted to image segmentation and enhancement for electromagnetic inverse scattering problems [54]. But most cases in literature are 2-D problems. If the input of U-Net is a 3-D image, the GPU memory consumed by U-Net is an unbearable burden for the most advanced GPUs. Therefore, we should modify the input of U-Net to suit high-resolution 3-D brain imaging. The 3-D images output from SJ-BPNN are cut into 2-D images by *XY*, *YZ*, and *XZ* slices, and these 2-D images are input into U-Net for learning and testing, where *YZ* and *XZ* slices are used to train U-Net and *XY* slices are used to test the trained U-Net, as shown in Figure 15.5. The effects of U-Net will be presented in experiments.

For a high-resolution 3-D brain imaging method using neural network, one of the biggest challenges is how to set up a reasonable and manageable training set to ensure training efficiency and reconstruction performance. If we employ the traditional training strategy which randomly sets some scatterers with random structures, sizes, positions and electrical properties to generate training samples for brain imaging, it will need at least thousands of samples to characterize the complex structures of the human brain and ensure reconstruction accuracy, resulting in a time-consuming and uneconomical training process. To address this problem, a brain imaging training strategy considering the *a priori* information of the human brain is proposed to build a dataset for learning. The schematic diagram of the brain imaging training strategy is presented in Figure 15.6.

The *a priori* information utilized here is that different human brains are similar in structure and electrical properties, which can be incorporated into the training process. Thus, we just need to consider the differences of the human brain when we set up the training samples.

First, based on the above principle, the human brain model is scaled with different factors in given ranges; the electrical properties of human brain tissues for different training samples are randomly set in given ranges according to different tissues, which makes the training strategy universal for different individuals.

Second, in consideration of possible individual differences in human brain tissues, we deform human brain tissues where the sizes of the deformation are based on a normal distribution.

Third, to reconstruct some abnormal situations such as tumors, bleeding, or lesions in the human brain, abnormal scatterers with different sizes and electrical properties are randomly distributed in the human brain.

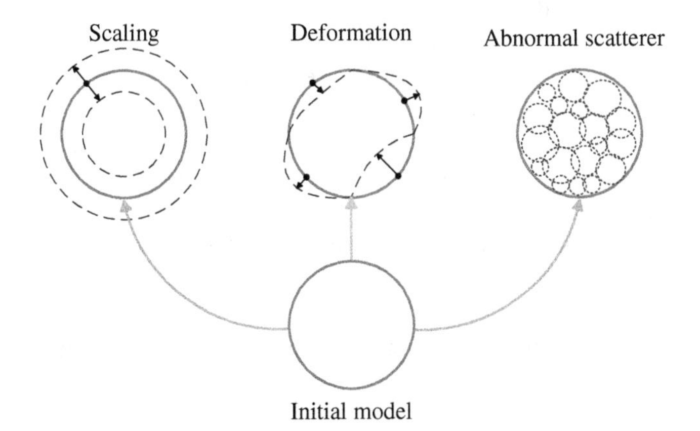

Figure 15.6 Schematic diagram of the brain imaging training strategy considering scaling and deformation of the human brain tissues and abnormalities.

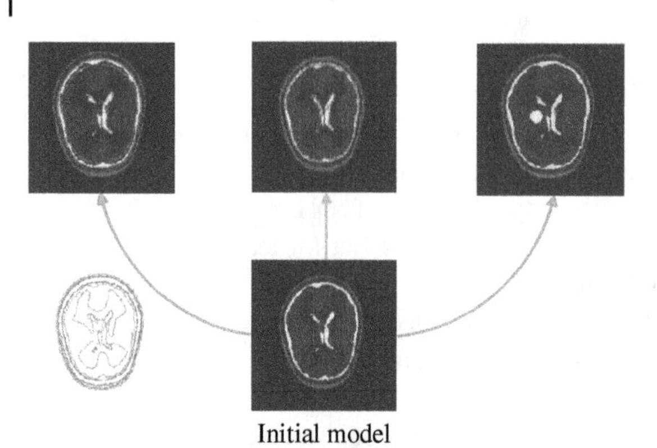

Figure 15.7 2-D slices of human brain models adopted with the brain imaging training strategy.

According to the above training strategy, the human brain models shown in Figure 15.7 are adopted. This brain imaging training strategy extracts the features of the human brain more easily and efficiently than the traditional strategy in training process, as demonstrated below.

15.4 Applications of Our Approach

15.4.1 Applications for 2-D Inverse Scattering Problem with Electrically Large Structures

15.4.1.1 Dual-Module NMM-IEM Machine Learning for Fast Electromagnetic Inversion of Inhomogeneous Scatterers with High Contrasts and Large Electrical Dimensions

In this section, two numerical examples are presented. The first one is used to validate the feasibility of the dual-module NMM-IEM machine learning scheme, and its implementation efficiency and accuracy are also compared with those of the conventional variational BIM (VBIM). In the second example, the reconstruction of the scatterers with high contrasts and/or large dimensions is performed and the anti-noise ability of the NMM-IEM is also evaluated. In both examples, the training model is the same. It is the variant of the "Austria" profile [5]. All simulations are performed on a personal computer with an Intel i7-9700 3.00 GHz CPU and 64 GB RAM. When the contrasts of the scatterers are too high, it is not feasible to use the BCGS-FFT presented in [55] to solve the integral equations to obtain the scattered fields. Therefore, the commercial software COMSOL using the finite-element method synthesizes the scattered field data that will be used for training and testing in the following.

In both numerical examples, the operating frequency is 300 MHz, corresponding to a wavelength of $\lambda_0 = 1$ m in free space. Following the Nyquist sampling theorem that the spatial distance between two receivers should roughly be half wavelength, 40 transmitters and 60 receivers are uniformly placed on two concentric circles with the radii of 4 and 4.5 m, respectively. The DOI shown in Figure 15.8 has the size of 5 m × 5 m and is discretized into 96 × 96 pixels with its center at the origin.

In order to quantitatively evaluate the reconstruction performance, the model misfit and data misfit under L_2 norm are defined as

$$\mathrm{Err}_{model} = \|\mathbf{m}_R - \mathbf{m}_T\| / \|\mathbf{m}_T\| \tag{15.21}$$

$$\mathrm{Err}_{data} = \|\mathbf{d}_R - \mathbf{d}_T\| / \|\mathbf{d}_T\| \tag{15.22}$$

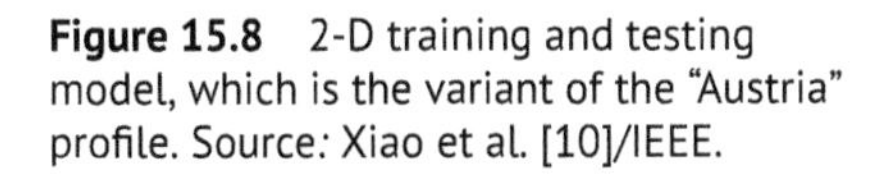

Figure 15.8 2-D training and testing model, which is the variant of the "Austria" profile. Source: Xiao et al. [10]/IEEE.

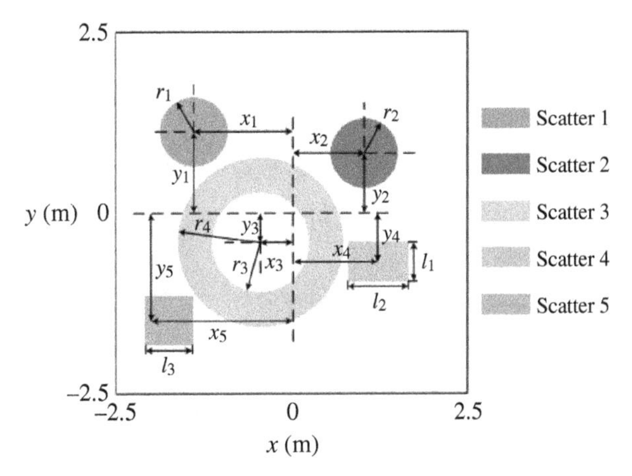

where $\mathbf{m}_T$ is the vector of the ground truth for the complex dielectric constant, with its element represents the true parameter value of the corresponding pixel. $\mathbf{m}_R$ is the vector of the reconstructed result, its element represents the reconstructed parameter value of the corresponding pixel. Similarly, $\mathbf{d}_T$ is the vector of measured scattered field data collected at all the receivers and $\mathbf{d}_R$ is the corresponding scattered field data vector of the reconstructed results.

The variant of the "Austria" profile is used to train the NMM and the IEM. As shown in Figure 15.8, the radii and centers of both disks, i.e. r_1, r_2, x_1, y_1, x_2, and y_2, and the inner and outer radii and centers of the ring, i.e. r_3, r_4, x_3, and y_3, are set as variables and assigned random values with their ranges listed in Table 15.2. Meanwhile, to enrich the training data sets, we add a rectangle and a square to the DOI. The centers of the rectangle and square, i.e. x_4, y_4, x_5, and y_5, and the side lengths of these two shapes, i.e. l_1, l_2, and l_3, are also set as variables and assigned random values. The relative permittivity values of the five scatterers are randomly set with different values in the range of [1, 8]. Hundred randomly generated training samples are employed to train the NMM. The Hecht–Nelson method is used to calculate the node number of the two hidden layers [49]. When there are n sets of training data, the node number of the hidden layer is empirically chosen as $2n+1$. Therefore, in the NMM, the dimensions of $\overline{\overline{\beta}}$ and $\overline{\overline{\alpha}}$ are 9216×201. Within one seconds, the values of $\hat{\overline{\overline{\beta}}}$ and $\hat{\overline{\overline{\alpha}}}$ could be obtained. The 100 sets of outputs from the NMM are used as the inputs to train and validate the IEM. Among them, 75 sets are used for training and 25 sets for validation.

Table 15.2 Parameter ranges for the scatterers in the inversion model shown in Figure 15.8.

	Parameter										
Range	r_1	x_1	y_2	r_2	x_2	y_2	r_3	r_4	x_3	y_3	l_1
Minimum	0.1	−2	−2	0.1	−2	−2	0.1	0.2	−2	−2	0.1
Maximum	1	2	2	1	2	2	1	2	2	2	2
Range	l_2	x_4	y_4	l_3	x_5	y_5	ε_1	ε_2	ε_3	ε_4	ε_5
Minimum	0.1	−2	−2	0.1	−2	−2	1	1	1	1	1
Maximum	2	2	2	2	2	2	8	8	8	8	8

Remark: the unit of length is meter ε_i is the relative permittivity value of ith scatter.
Source: Adapted from Xiao et al. [10].

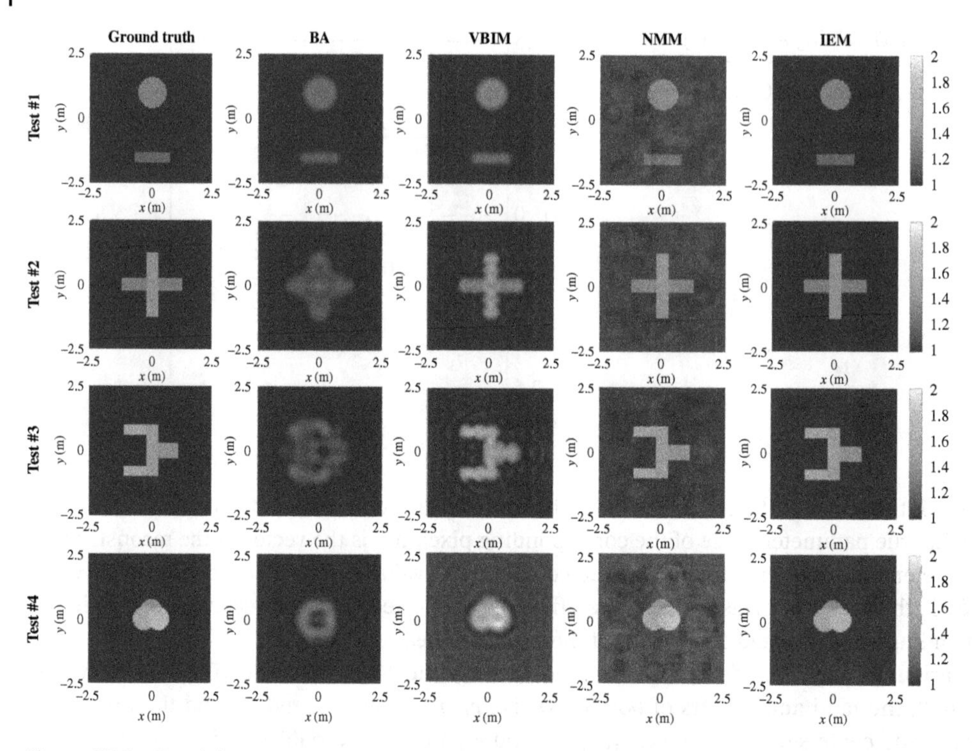

Figure 15.9 Four inhomogeneous scatterers are used to test the proposed dual-module machine learning scheme. From the first row to the fourth row, both the geometric shape complexity and contrasts increase. The first column is the ground truth, and from the second column to the fifth column, the inversion results from BA, VBIM, NMM, and IEM are shown, respectively. Source: Xiao et al. [10]/IEEE.

To validate the inversion accuracy and efficiency of the dual-module NMM-IEM machine learning scheme, we compare their inversion results with those from the conventional BA and VBIM. As shown in Figure 15.9, from Tests #1 to #4, the contrast of the scatterer gradually increases. Meanwhile, the geometric shape also becomes more and more complex. For the scatterers in the inversion domain including the simple disk and the rectangle with the low contrasts in Test #1, both BA and VBIM can reconstruct the scatterers well. As the contrast values increase, the inversion results by BA show larger and larger errors. In Test #4, even the iterative VBIM cannot reconstruct the three overlapped disks. However, the inversion results by the NMM and IEM are almost not affected by different contrast values and shape complexity, which is shown by the fourth and fifth columns of Figure 15.9. This is further quantitatively verified by the model misfits listed in Table 15.3. As can be seen, from Tests #1 to #4, the model misfits from BA and VBIM gradually increase. However, they almost remain unchanged for the NMM and the IEM. The IEM further improves the reconstructed images from the NMM by approximately reducing the model misfits by 2%. In addition, one should note that the discrepancies of model misfits between the VBIM and the IEM also gradually increase from Tests #1 to #4. This indicates that the proposed dual-module scheme has stronger adaptability. In other words, it is more competent to deal with the scatterers with high contrasts and complex geometric shapes compared with the conventional VBIM.

Figure 15.10 shows the variations of data misfits and model misfits for the inversion by VBIM. From Tests #1 to #4, it becomes more and more difficult for the VBIM to converge to a low data misfit. The final data misfit of Test #4 is the largest, while that of Test #1 is the smallest. This is quite different from the stable trend of data misfit values by the NMM-IEM mentioned above.

Table 15.3 Model misfits (%) of four tests shown in Figure 15.9.

Test	Method			
	BA	VBIM	NMM	IEM
#1	4.8230	2.0181	3.2615	1.0346
#2	8.2708	3.6295	3.3103	1.0951
#3	12.7513	5.5959	3.2916	1.0415
#4	13.0708	4.0328	3.2812	1.0864

Source: Xiao et al. [10]/IEEE.

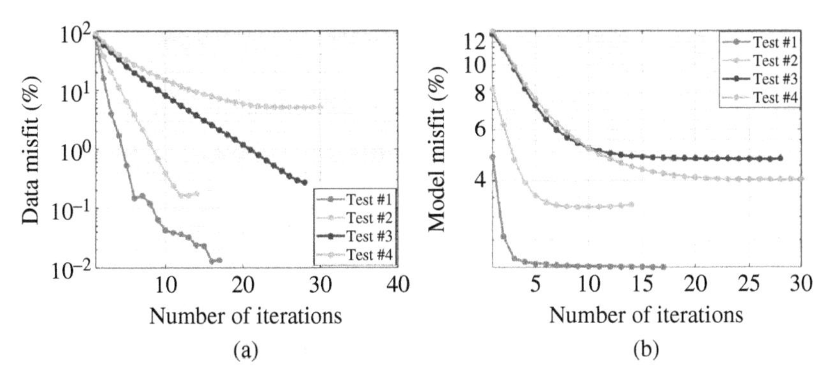

Figure 15.10 Variations of (a) data misfits and (b) model misfits in the VBIM iterations. Source: Xiao et al. [10]/IEEE.

For the model misfits shown in Figure 15.10b, Test #1 also has the smallest value. However, Test #3 has the larger final model misfit than Test #4. This is because the electrical size of the scatterer in Test #3 is larger than that of Test #4 although the contrast value used in Test #3 is smaller. Because each VBIM iteration takes about 46 seconds in Test #4, the total time cost of VBIM is much higher than that of the dual-module scheme that spends less than 1 seconds to obtain the final high-accuracy 2-D image. In addition, we also input the reconstructed 2-D images by the NMM in Test #4 into the VBIM solver as the initial values. Then, the VBIM takes around 983 s to obtain the final inversion results that have the model misfit as large as 4.017%. This is around four times of the model misfit by IEM. Obviously, it is inadvisable to replace the IEM module in the proposed scheme with a conventional iterative solver.

Then, three scatterers with high contrasts and large electrical dimensions, as shown in Figure 15.11, are used to test the proposed dual-module NMM-IEM scheme. From Tests #5 to #7, both the contrasts and dimensions gradually increase. Meanwhile, we add −20, −15, and −10 dB Gaussian white noise to evaluate further the anti-noise ability of the proposed method. Here, the noise level is defined by the signal-to-noise ratio (SNR) of power. Test #5 is designed with a tangency semicircle and a rectangle to evaluate the ability of NMM-IEM to distinguish different tangency scatterers, where the largest contrast is 3. The multiple scatterers in Test #6 are tangency, nested, and overlapped and have the largest contrast of 5. In Test #7, we select four concentric circles and a square overlapped together and the largest contrast is 7. As shown in Figure 15.11, as the noise increases, the reconstructed 2-D images become more and more blurred, which are clearly illustrated by the increasing model misfits listed in Table 15.4. However, the model

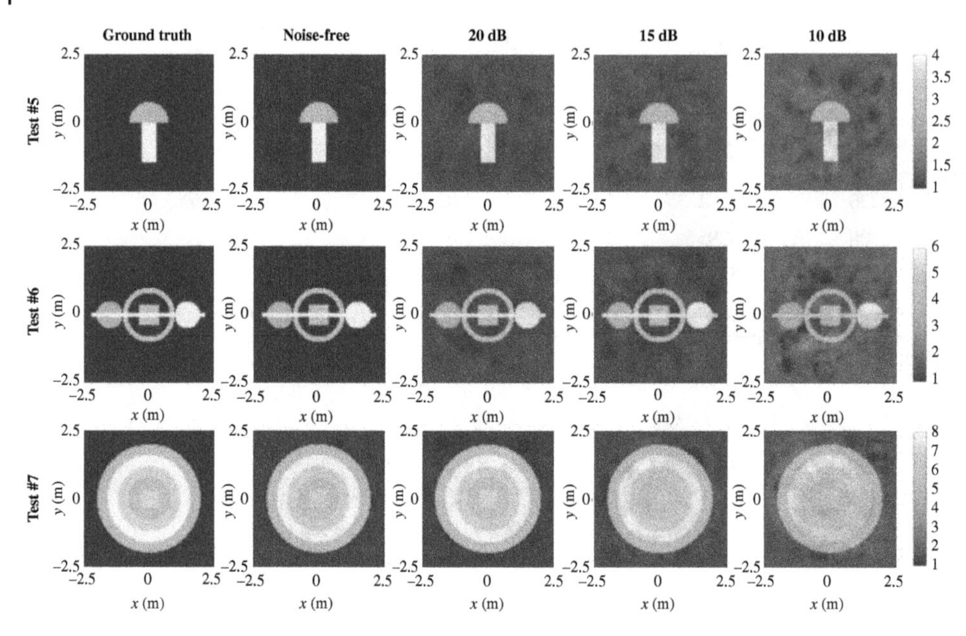

Figure 15.11 Three inhomogeneous scatterers with high contrasts and large electrical dimensions are used to test the proposed dual-module machine learning scheme. From the first row to the third row, both the complexity of geometric shapes and contrast increase. The first column is the ground truth, and from the second column to the fifth column, the inversion results of the NMM-IEM for noise-free, −20 dB noise, −15 dB noise, and −10 dB noise are shown, respectively. Source: Xiao et al. [10]/IEEE.

Table 15.4 Contrasts and model misfits (%) of three tests shown in Figure 15.11.

	Parameter				
Test	**Largest χ_ε**	**Noise-free**	**20 dB**	**15 dB**	**10 dB**
#5	3	1.2516	3.5433	7.2061	15.3315
#6	5	1.2814	3.5549	7.2123	15.3595
#7	7	1.3108	3.7316	7.2161	15.3634

Source: Xiao et al. [12]/IEEE.

misfits in three different tests almost keep the same levels for both noise-free and noisy scenarios, although both the electrical sizes and the largest contrast χ increase from Tests #5 to #7. The evaluated data misfits for the noise-free cases also show the similar stable trend. They are 1.1%, 1.34%, and 1.26% for Tests #5–#7, respectively. Note that the electrical dimensions of DOI here are $5\lambda_0 \times 5\lambda_0$, and in Test #7, the diameter of the largest circle is $4\lambda_0$. Numerical simulations show that the conventional solver such as VBIM does not converge for these three tests. It cannot tackle the scatterers with high contrasts and large electrical dimensions, but the proposed dual-module scheme can.

15.4.1.2 Nonlinear Electromagnetic Inversion of Damaged Experimental Data by a Receiver Approximation Machine Learning Method

In this section, the experimental data including TE and TM modes of four cases, i.e. "FoamDielInt," "FoamDielExt," "FoamTwinDiel," and "FoamMetExt," provided by Institute Fresnel [56], are

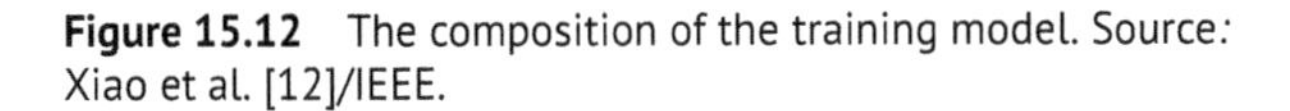

Figure 15.12 The composition of the training model. Source: Xiao et al. [12]/IEEE.

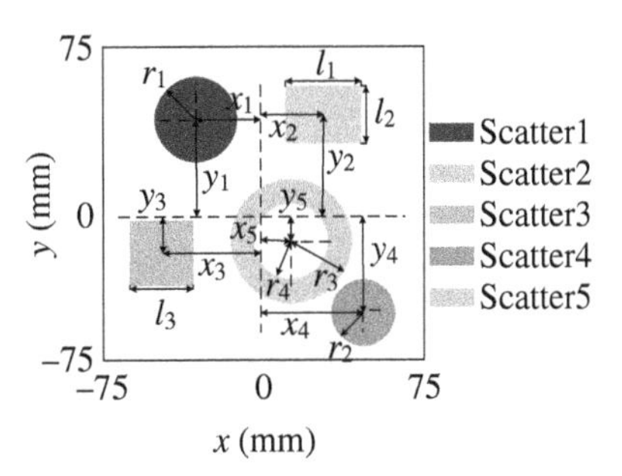

employed to evaluate the effectiveness of RAML. All simulations are performed on a personal computer with an Intel i7-8750H 2.20 GHz CPU and 16 GB RAM. The commercial software COMSOL with the full-wave EM simulation synthesizes the scattered field data used for training. Meanwhile, any other full-wave numerical simulation software can also be used to generate the training data.

The operating frequency for all cases is 5 GHz, and the corresponding wavelength is $\lambda_0 = 60$ mm in air. As shown in Figure 15.12, DOI with the size of 150 mm × 150 mm is discretized into 96 × 96 pixels and its center is at the origin. In [56], all the experiments were carried out in a microwave anechoic chamber. Both transmitters and receivers of all cases are placed on a circle with the radius of 1.67 m surrounding DOI. 241 receivers are placed with a step of 1° without any position closer than 60° from the source. The scatterers are placed at the center of the circle. The cross sections and the size of the scatterers are shown in the first column of Figure 15.13. Here, "FoamDielInt" and "FoamDielExt" are constructed with a foam cylinder and a plastic cylinder, "FoamTwinDiel" is constructed with a foam cylinder and two plastic cylinders, and "FoamMetExt" is constructed with a copper tube and a foam cylinder. Due to the large measurement error of the relative permittivity of metals at high frequencies, the relative permittivity of the copper tube in "FoamMetExt" is set to 1 and the conductivity is set to 5 S/m as in [56]. In "FoamDielInt" and "FoamDielExt," totally eight transmitters are placed with a step of 45°. In "FoamTwinDiel" and "FoamMetExt," totally 18 transmitters are placed with a step of 20°. To reconstruct the profile of scattering objects from experimental data to obtain simulation data, we use a simple calibration procedure [56] by using the complex ratio between the measured incident field and the calculated incident field at the same receiver location.

As shown in Figure 15.12, the training model consists of two circular cylinders, a ring, a rectangle, and a square, and their positions and sizes are randomly selected in a given range as shown in Table 15.5. r_1, r_2, x_1, y_1, x_4, and y_4 are the radii and the centers of both circular cylinders. l_1, l_2, x_2, and y_2 are set as the length, width, and center of rectangle. l_3, x_4, and y_4 are also set as the length and center of square. r_3, r_4, x_5, and y_5 are the inner and outer radii and center of the ring.

Meanwhile, the relative permittivity in the range of [1, 5] and the conductivity in the range of [0, 6 S/m] are randomly assigned to these five scatterers. We have used different numbers of samples as training data to compare the test results to determine the number of training samples. Finally, with tradeoff between the modeling cost and the model misfit, 200 training samples are generated and employed to train the proposed RAML.

In this section, 10 randomly selected receivers are set damaged, and the missing scattered field data are generated by RAML based on the receivers within one wavelength around the damaged one for the four cases. The ground truth together with the reconstructed relative permittivity and conductivity are shown in Figure 15.13. Meanwhile, to evaluate the performance of RAML, NMM-IEM

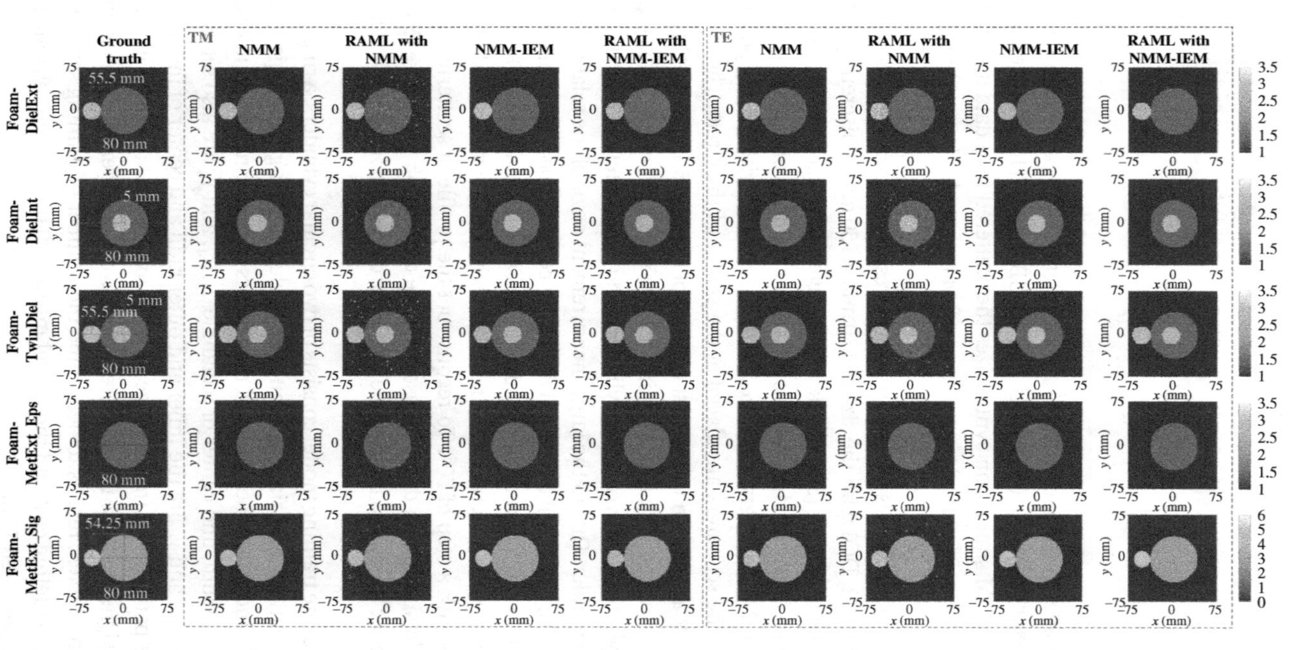

Figure 15.13 All four cases at 5 GHz in [56] are used to test the RAML. The 1st column is the ground truth, where the position and size of the scatterer are marked. From the 2nd column to the 5th column, the TM mode is used, while from the 6th column to the 9th column, the TE mode is used.The 2nd column, 4th column, 6th column, and 8th column are the results of the experimental data with no receiver damaged.The 3rd column, 5th column, 7th column, and 9th column are the results of the experimental data with ten receivers damaged. Source: Xiao et al. [12]/IEEE.

Table 15.5 Parameter Ranges for the scatterers in the Training Model Shown in Figure 5.12.

Parameter	r_1	x_1	y_1	l_1	l_2	x_2	y_2	l_3	x_3
Minimum	1	−70	−70	1	1	−70	−70	1	−70
Maximum	70	70	70	140	140	70	70	140	70
Parameter	y_3	r_2	x_4	y_4	r_3	r_4	x_5	y_5	
Minimum	−70	1	−70	−70	2	1	−70	−70	
Maximum	70	70	70	70	70	70	70	70	

Source: Xiao et al. [12]/IEEE.

Table 15.6 The Model Misfit and the Data Misfit of the Test Results of RAML with NMM-IEM Shown in Figure 15.13.

		TM				TE			
		NMM (%)	RAML with NMM (%)	NMM-IEM (%)	RAML with NMM-IEM (%)	NMM (%)	RAML with NMM (%)	NMM-IEM (%)	RAML with NMM-IEM (%)
Model misfit	FoamDielExt	2.41	15.64	1.05	4.23	3.23	9.63	1.24	4.86
	FoamDielInt	2.23	7.89	1.18	3.96	3.55	11.37	1.13	4.22
	FoamTwinDiel	1.71	15.98	1.05	5.84	1.87	10.43	1.17	5.63
	FoamMetExt_Eps	2.74	10.94	1.07	3.26	4.33	7.50	1.03	2.83
	FoamMetExt_Sig	3.50	19.86	1.26	13.87	3.58	15.33	1.07	9.61
Data misfit	FoamDielExt	7.56	16.32	3.06	4.95	9.63	10.61	3.56	5.31
	FoamDielInt	6.91	8.32	3.14	4.11	9.85	11.86	3.23	5.14
	FoamTwinDiel	4.12	16.59	2.87	6.32	4.32	11.63	2.98	4.13
	FoamMetExt	9.49	20.29	3.22	15.31	10.24	8.61	3.46	4.36

Source: Xiao et al. [12]/IEEE.

in is [10] selected as the comparison model. Due to the structure restriction, all the receivers for NMM-IEM should operate normally; thus, the whole measured data are input to NMM-IEM.

The inversion results of NMM, RAML with NMM, NMM-IEM, and RAML with NMM-IEM are shown in Figure 15.13. It is observed that when ten receivers are damaged, compared with NMM, the results of RAML with NMM have a bit of noisy points, but the profiles of the scatterers are basically complete. After denoising by IEM, the inversion result of RAML with NMM is enhanced, and the contour of the scatterer is almost the same as NMM-IEM with the complete experimental data. The data misfit and the model misfit of both models are also provided in Table 15.6. Although the model misfit of RAML with NMM-IEM with ten damaged receivers is higher than that from NMM-IEM without damaged receivers, the model misfit of RAML with NMM-IEM is still in an acceptable range; the reconstructed relative permittivity and the reconstructed conductivity have an error lower than 6% and 15%, respectively. Meanwhile, compared with NMM-IEM without any damaged measured data, the data misfit of RAML with NMM-IEM is also in an acceptable range smaller than 16%, even though ten receivers are damaged.

Meanwhile, we set the damaged data to zero to discuss whether the inversion results of NMM-IEM are still good enough when the receivers are damaged. As shown in Figure 15.14,

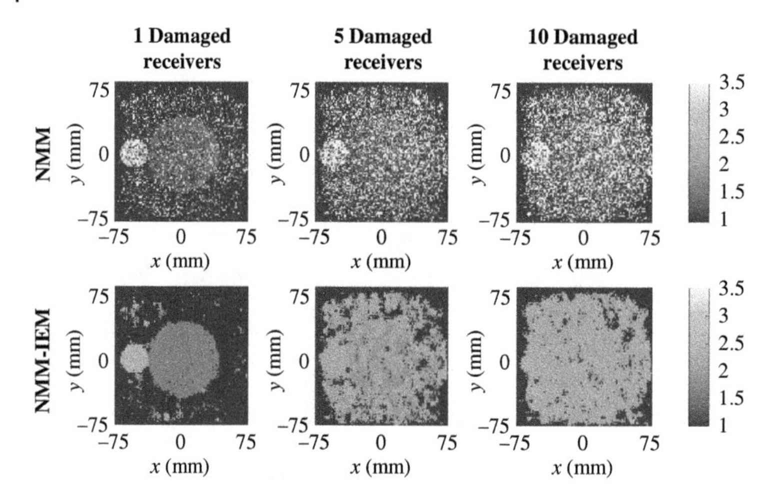

Figure 15.14 The results of zeroing damaged data input to NMM and NMM-IEM. Column 1 through column 3 are the cases where 1, 5, and 10 receivers are damaged, respectively. Source: Xiao et al. [12]/IEEE.

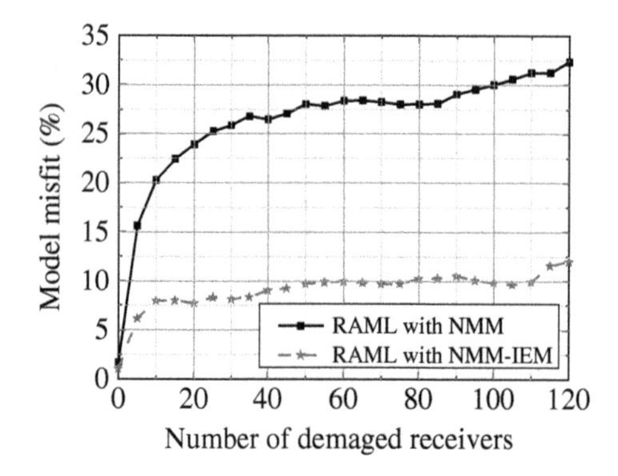

Figure 15.15 The model misfit versus the number of damaged receivers. Source: Xiao et al. [12]/IEEE.

if only one receiver is damaged, the NMM-IEM can obtain the approximate scatterer profile. As the number of damaged receivers increases, NMM-IEM cannot obtain effective inversion results.

To further evaluate the extreme ability of the proposed RAML, we set a sequence of damaged receivers where the number of damaged receivers is from 0 to 120, with the step of five. The most complicated case, i.e. "FoamTwinDiel," is selected as the numerical example. Figure 15.15 shows the model misfit of RAML with NMM and RAML with NMM-IEM versus the number of damaged receivers from 0 to 120.

Meanwhile, the reconstructed relative permittivity distribution with 30, 60, 90, and 120 damaged receivers are shown in Figure 15.16, respectively. As these results demonstrate, it is obvious that RAML with NMM can still maintain the approximate shape of the scatterers even though the input data are greatly damaged, and then after enhancement by IEM, the model misfit of RAML with NMM-IEM can still be kept below 15%. The results from the damaged data show that the proposed RAML with NMM- IEM can have good performance when half of the receivers are damaged, and the corresponding data are missing. As a nonlinear approximate method based on machine learning, RAML is recognized to have nonlinear mapping ability. It can nonlinearly approximate the

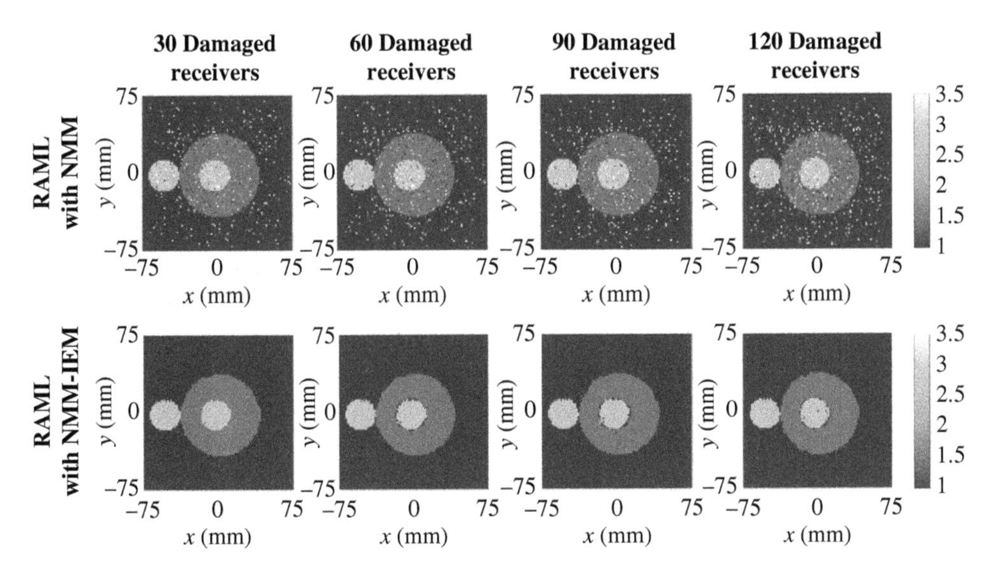

Figure 15.16 The most complicated case in [56], i.e. "'FoamTwinDiel," used to test the extreme of RAML with NMM-IEM. From 1st column to 4th column, the inversion results are the experimental data with 30, 60, 90, 120 receivers damaged, respectively. Source: Xiao et al. [12]/IEEE.

data from the damaged receivers based on other receivers around the damaged ones within one wavelength, and has good inversion performance.

15.4.2 Applications for 3-D Inverse Scattering Problem with Electrically Large Structures

15.4.2.1 Super-Resolution 3-D Microwave Imaging of Objects with High Contrasts by a Semi-Join Extreme Learning Machine

In this section, three numerical examples are presented to verify the validity of the proposed SJ-ELM for microwave imaging of 3-D objects. In the first example, the conventional VBIM is employed to compare and confirm the implementation efficiency and accuracy of SJ-ELM. The second example further compares the proposed model with VBIM to evaluate the super-resolution imaging ability of SJ-ELM. In the third example, the proposed model is evaluated by reconstructing the 3-D objects with high contrasts. All the inversions are performed on a workstation with 20-cores Xeon E2650 v3 2.3G CPU, 512 GB RAM. In the numerical examples, all the simulation settings, including excitation, receiving antennas, boundary conditions, etc., are the same in the forward models to generate the training and testing or synthetic "measured" data. Meanwhile, to avoid the "inverse crime", the meshes used in the forward modeling and inversion are different. The forward modeling mesh is smaller than that in the inversion, and they are not aligned. In fact, the forward modeling mesh can be curvilinear so that it captures the curved geometries, while the inversion mesh is the regular structured mesh with square elements, so there is a geometry approximation for curved objects.

In the three examples, the operating frequencies are 300, 500, and 600 MHz, respectively. The corresponding wavelengths of the three examples are $\lambda_0 = 1$ m, $\lambda_0 = 0.6$ m, and $\lambda_0 = 0.5$ m in air, respectively.

For the first example, totally 98 transmitters are uniformly located in two 2.4 m × 2.4 m planes at $z = -0.2$ m and $z = 1.2$ m, respectively. The scattered fields are collected by a 128-receiver array uniformly located in two 2.8 m × 2.8 m planes at $z = -0.1$ m and $z = 1.1$ m, respectively. The imaging domain D enclosing the objects has the dimensions of 0.8 m × 0.8 m × 0.8 m and is discretized into 40 × 40 × 40 voxels with its center at (0, 0, 0.5) m. The input of this example is the vector of scattered field, and each vector contains 98 (transmitters) × 128(receivers) elements. The corresponding outputs of two channels are the vectors of distribution of the relative permittivity and conductivity, respectively, and the vector has the dimension of (40 × 40 × 40) × 1. Based on the semi-join strategy, the output is discretized into 40 subsets.

The inversion domain D of the second example is with the dimensions of 1.4 m × 1.4 m × 0.4 m. The inversion domain D here is discretized into 70 × 70 × 20 voxels with its center at (0, 0, 0.3) m. Totally 128 transmitters are uniformly located in two 2.8 m × 2.8 m planes at $z = -0.1$ m and $z = 0.7$ m, respectively, and 162 receivers arrays uniformly located in two 3.2 m × 3.2 m planes at $z = -0.2$ m and $z = 0.8$ m, respectively. The imaging scenes, as shown in Figure 15.17, of the first and second examples are similar. The input of this example is the vector of scattered field, and each vector contains 128 (transmitters) × 162 (receivers) elements. The outputs of two channels are the vectors of distribution of the relative permittivity and conductivity, respectively, and the vector is with the dimension of (70 × 70 × 20) × 1. Based on the semi-join strategy, the output is discretized into 70 subsets.

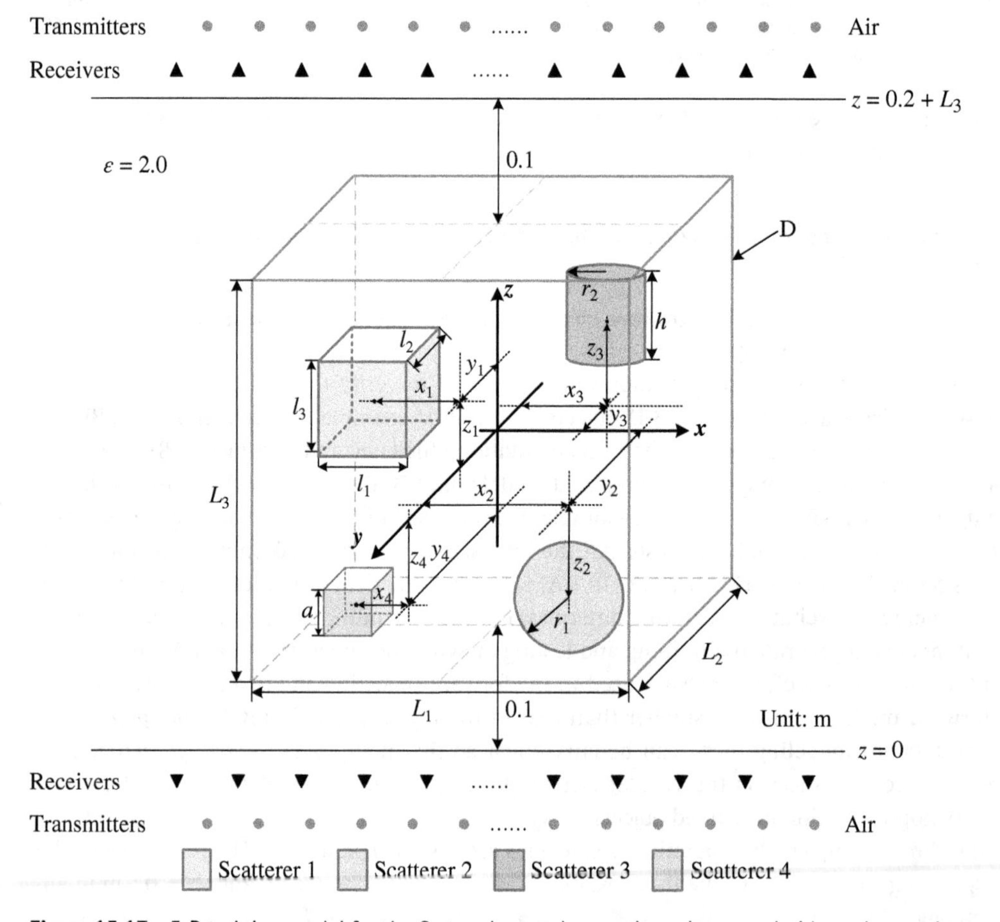

Figure 15.17 3-D training model for the first and second examples, where a cuboid, a cube, a sphere, and a cylinder as four non-overlapping scatterers in DOI are employed to train SJ-ELM, and the parameters of these scatterers are assigned random values with the ranges shown in Table 15.7. The transmitters and receivers are uniformly located in two planes on the upper and lower sides. Source: Xiao et al. [12]/IEEE.

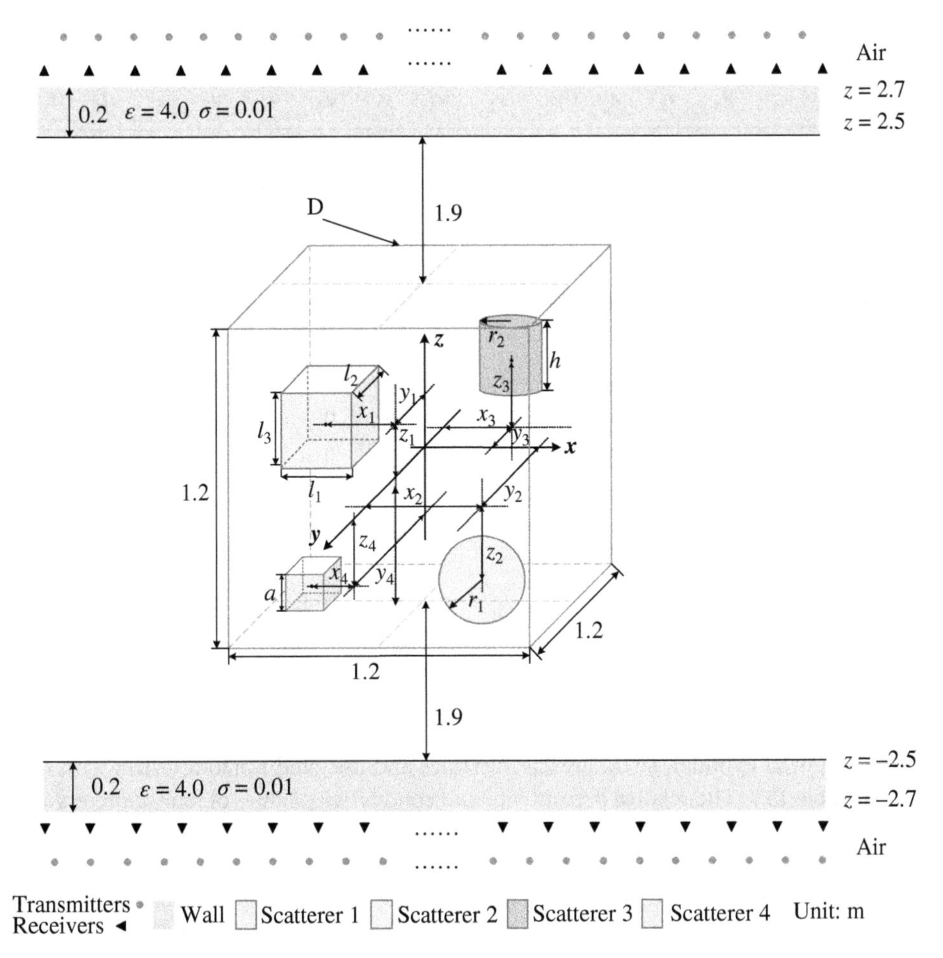

Figure 15.18 3-D training model for through-wall imaging as the third example where a cuboid, a cube, a sphere, and a cylinder as four non-overlapping scatterers in DOI are employed to train SJ-ELM, and the parameters of these scatterers are assigned random values with the ranges shown in Table II. The transmitters and receivers are uniformly located in two planes on the upper and lower sides. Source: Xiao et al. [12]/IEEE.

For the third example, another through-wall imaging circumstance, as shown in Figure 15.18, is considered to evaluate the proposed model. The center of inversion domain D is at (0, 0, 0) m and has the dimensions of $1.2\,\text{m} \times 1.2\,\text{m} \times 1.2\,\text{m}$. Here the inversion domain D is discretized into $60 \times 60 \times 60$ voxels. 162 transmitters are uniformly located in two $2\,\text{m} \times 2\,\text{m}$ planes at $z = -2.8$ m and $z = 2.8$ m, respectively. Meanwhile, the scattered fields are collected by a 242-receiver array uniformly located in two $2.5\,\text{m} \times 2.5\,\text{m}$ planes at $z = -2.9$ m and $z = 2.9$ m, respectively. Similar to the above two examples, the input of this example is the vector of scattered field, and each vector contains 162(transmitters) × 242(receivers) elements. The outputs of two channels are the vectors of distribution of the relative permittivity and conductivity, respectively, and the vector has the dimension of $(60 \times 60 \times 60) \times 1$. Based on the semi-join strategy, the output is discretized into 60 subsets.

In these three numerical examples, we use a cuboid, a cube, a sphere, and a cylinder as four non-overlapping scatterers in DOI to train SJ-ELM. As shown in Figures 15.17 and 15.18, the centers of the cuboid, sphere, cylinder, and cube, i.e. x_1, y_1, z_1, x_2, y_2, z_2, x_3, y_3, z_3, x_4, y_4, and z_4, the side lengths of cuboid and cube, i.e. l_1, l_2, l_3, and a, the radii of sphere and cylinder, i.e. r_1

Table 15.7 The parameter ranges for the objects shown in Figures 15.17 and 15.18 for Examples 1–3.

		x_1	y_1	z_1	x_2	y_2	z_2	x_3	y_3	z_3	x_4	y_4	z_4	l_1	l_2
Example 1	Minimum	−0.3	−0.3	0.2	−0.3	−0.3	0.2	−0.3	−0.3	0.2	−0.3	−0.3	0.2	0.02	0.02
	Maximum	0.3	0.3	0.8	0.3	0.3	0.8	0.3	0.3	0.8	0.3	0.3	0.8	0.2	0.2
Example 2	Minimum	−0.6	−0.6	0.2	−0.6	−0.6	0.2	−0.6	−0.6	0.2	−0.6	−0.6	0.2	0.02	0.02
	Maximum	−0.6	−0.6	0.4	−0.6	−0.6	0.4	−0.6	−0.6	0.4	−0.6	−0.6	0.4	0.1	0.1
Example 3	Minimum	−0.5	−0.5	−0.5	−0.5	−0.5	−0.5	−0.5	−0.5	−0.5	−0.5	−0.5	−0.5	0.02	0.02
	Maximum	0.5	0.5	0.5	0.5	0.5	0.5	0.5	0.5	0.5	0.5	0.5	0.5	0.4	0.4
		l_3	a	r_1	r_2	h	ε_1	σ_1	ε_2	σ_2	ε_3	σ_3	ε_4	σ_4	
Example 1	Minimum	0.02	0.02	0.02	0.02	0.02	2	0	2	0	2	0	2	0	
	Maximum	0.2	0.2	0.2	0.2	0.2	4	0.01	4	0.01	4	0.01	4	0.01	
Example 2	Minimum	0.02	0.02	0.02	0.02	0.02	2	0	2	0	2	0	2	0	
	Maximum	0.1	0.1	0.1	0.1	0.1	5	0.03	5	0.03	5	0.03	5	0.03	
Example 3	Minimum	0.02	0.02	0.02	0.02	0.02	1	0	1	0	1	0	1	0	
	Maximum	0.4	0.4	0.4	0.4	0.4	7	0.08	7	0.08	7	0.08	7	0.08	

Remark: the unit of length is meter. ε_i and σ_i are the relative permittivity and conductivity value of ith object, respectively.
Source: Xiao et al. [12]/IEEE.

and r_2, and the height of cylinder, i.e. h, are the variables and assigned random values with the ranges shown in Table 15.7. The relative permittivity and conductivity values of four scatterers are randomly set with different values in the range of [2, 4] and [0, 0.01] S/m for the first examples, [2, 5] and [0, 0.03] S/m for the second example and [1, 7] and [0, 0.08] S/m for the third example. The proposed SJ-ELM is trained with totally 200 randomly generated training samples for the three examples, respectively. The node numbers of two hidden layers are set based on the Hecht–Nelson method [49]. One should note that SJ-ELM should be trained for these three examples, respectively, due to the different imaging condition. Meanwhile, all the parameters in the trained model are settled. When the scattered field data is input to the trained model, the corresponding output is unique.

To evaluate the imaging accuracy and efficiency of the proposed SJ-ELM, its imaging results are compared with those from VBIM. As Figure 15.19 shown, Test #1 has a single object, Test #2 has two separated and different objects, while these two objects are tangent in Test #3. For Test #1, the radii of the outer and inner circles are 0.2 and 0.1 m (0.2λ and 0.1λ). For Test #2, the upper cuboid has the dimensions of $0.3\,\text{m} \times 0.2\,\text{m} \times 0.2\,\text{m}$ ($0.3\lambda \times 0.2\lambda \times 0.2\lambda$), and the other cuboid has the dimensions of $0.2\,\text{m} \times 0.2\,\text{m} \times 0.2\,\text{m}$ ($0.2\lambda \times 0.2\lambda \times 0.2\lambda$). For Test #3, the cuboid has the dimensions of $0.2\,\text{m} \times 0.2\,\text{m} \times 0.4\,\text{m}$ ($0.2\lambda \times 0.2\lambda \times 0.4\lambda$).

The contrasts of the three test samples are also gradually increased. As the contrast values increase and the geometric shape of objects becomes more complex, it is more and more difficult for VBIM to achieve a low data misfit and model misfit, which could be revealed in Table 15.8. In addition, one should note that the model misfits of VBIM are fluctuant for different test samples. On the contrary, the model misfits obtained from the proposed SJ-ELM model are always lower than VBIM as shown by Table 15.8.

Meanwhile, as shown by the 4th row of Figure 15.19, the imaging results obtained from the proposed SJ-ELM are not affected by different contrast values and geometric shape of objects

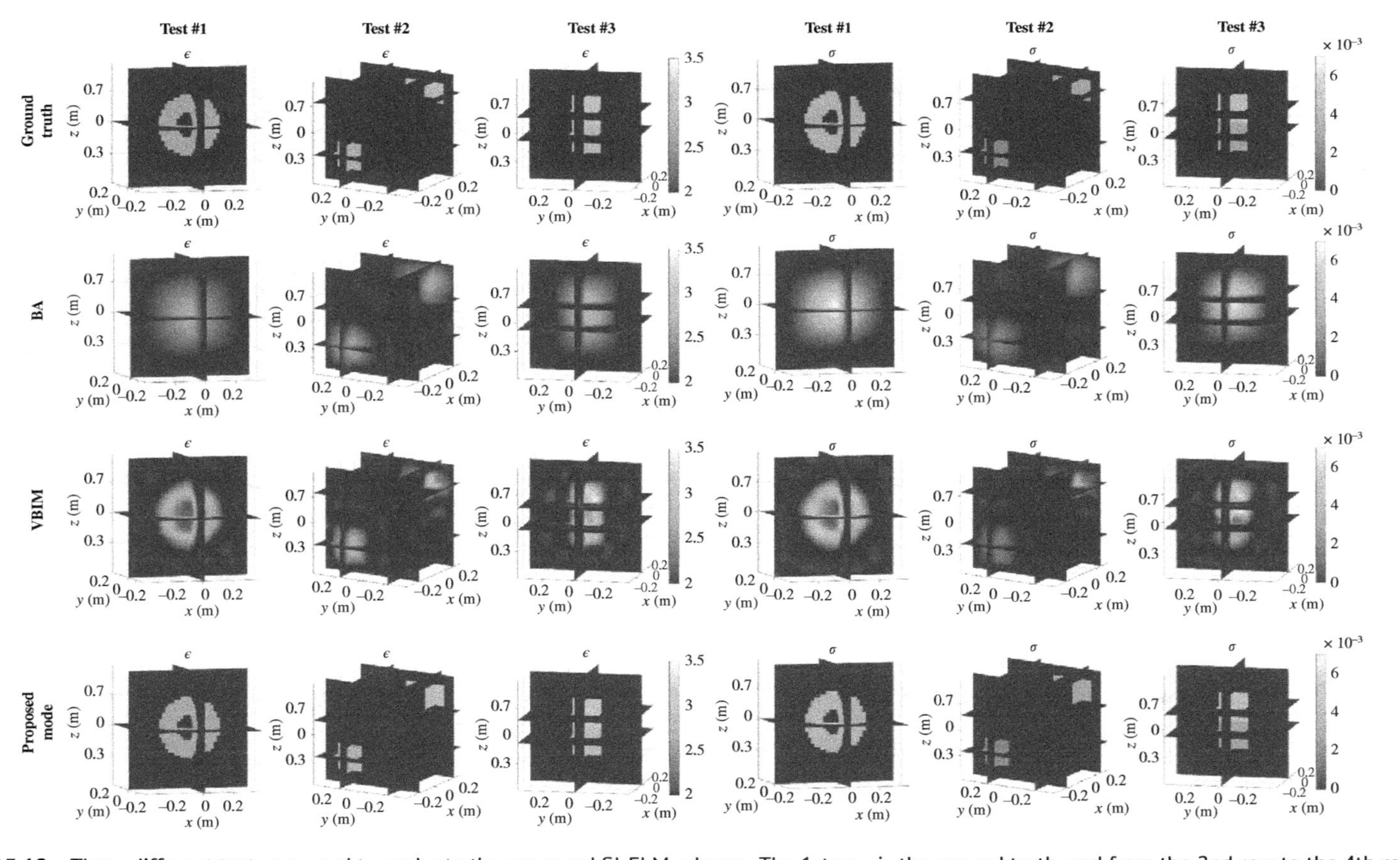

Figure 15.19 Three different tests are used to evaluate the proposed SJ-ELM scheme. The 1st row is the ground truth, and from the 2nd row to the 4th row, the imaging results from BA, VBIM, and the proposed model are shown, respectively. The 1st, 2nd, and 3rd columns show the distribution of relative permittivity, and 4th, 5th, and 6th columns show the distribution of conductivity. Source: Xiao et al. [12]/IEEE.

Table 15.8 The model misfits and data misfits of BA, VBIM, and the proposed model for tests #1–3.

	Model misfit (%)		
	BA	VBIM	Proposed model
Test #1	23.259	10.352	3.656
Test #2	30.231	13.346	3.764
Test #3	29.245	12.622	3.811
	Data misfit (%)		
	BA	VBIM	Proposed model
Test #1	17.322	0.725	2.517
Test #2	11.019	1.074	5.003
Test #3	18.254	0.985	4.156

Source: Xiao et al. [12]/IEEE.

and the model misfits, as shown in Table 15.8, obtained from the proposed model almost remain unchanged. This illustrates that the proposed SJ-ELM is more stable and competent to deal with microwave imaging compared with the conventional VBIM.

Furthermore, as an example, each VBIM iteration takes about 40 minutes in Test #3; thus, the total time cost in Tests #1-3 of VBIM is much higher than that of the trained SJ-ELM which spends less than 1 second to accurately image 3-D objects, even though about one-hour training time of SJ-ELM is added.

Thus, comparing with VBIM, the trained proposed SJ-ELM model can achieve more accurate and stable imaging performance and higher imaging efficiency.

Meanwhile, for the full-join strategy in this example, the dimensions of $\overline{\overline{\beta}}$ and $\overline{\overline{\alpha}}$ based on Hecht–Nelson method are $(40\times40\times40)\times(128\times98\times2+1)$ and $(40\times40\times40)\times(40\times40\times40\times2+1)$, i.e. $64{,}000\times25{,}089$ and $64{,}000\times12{,}8001$, respectively. However, for the semi-join strategy with parallel training in this example, the dimensions of $\overline{\overline{\beta}}$ and $\overline{\overline{\alpha}}$ based on Hecht–Nelson method are $(40\times40)\times[(2\times128\times98+1)\times40]$ and $(40\times40)\times[(2\times40\times40+1)\times40]$, i.e. $1600\times1{,}003{,}560$ and $1600\times128{,}040$. Thus, compared with full-join strategy, which requires a heavy burden for the memory and will lead to an enormous challenge for computational efficiency, the proposed semi-join strategy is easier to converge and more suitable for the 3-D MWI problem.

To test the super-resolution imaging ability of the proposed SJ-ELM, two parts are investigated. First, VBIM is employed as a comparison model to evaluate the proposed model with Test #4. Then, we design a series of test cases to further evaluate the super-resolution imaging ability of the proposed SJ-ELM in the noise-free and noisy environments, respectively. According to the definition in [57], as the image resolution is less than 0.25 wavelength, it can be called super-resolution. There are six objects in Test #4, as shown in Figure 15.20, and the adjacent spacings between them are $5/20\lambda$, $4/20\lambda$, $3/20\lambda$, $2/20\lambda$, and $1/20\lambda$, respectively; meanwhile, the relative permittivity and conductivity of them are gradually decreased, along the positive x direction. The height of each object is 0.1 m (or 0.167λ).

Figure 15.21 shows the imaging results of VBIM and the proposed model, respectively. It could be seen that the imaging results of VBIM cannot distinguish the fifth and sixth objects seperately; however, the proposed model could clearly image both relative permittivity and conductivity of the objects. The imaging performance of VBIM and the proposed model also could be revealed with

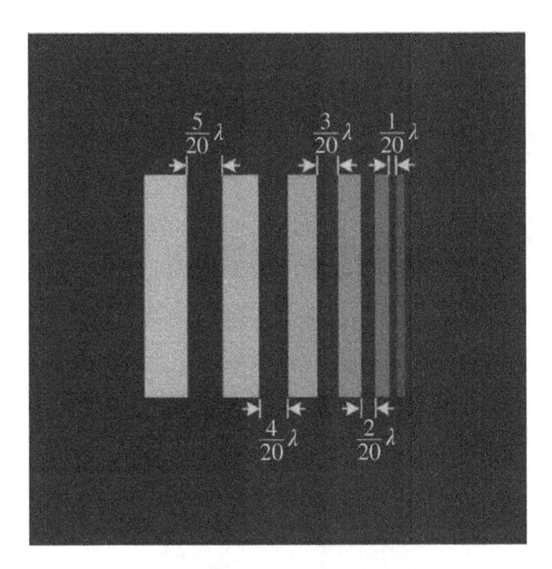

Figure 15.20 The top view of Test #4. The distances between objects are $5/20\lambda$, $4/20\lambda$, $3/20\lambda$, $2/20\lambda$, and $1/20\lambda$. Meanwhile, the relative permittivity and conductivity of the six objects are gradually decreased along the positive x direction, i.e. 3.2 to 2.2 with the step of −0.2 for the relative permittivity and 0.012 to 0.002 with the step of −0.002 for conductivity. Source: Xiao et al. [12]/IEEE.

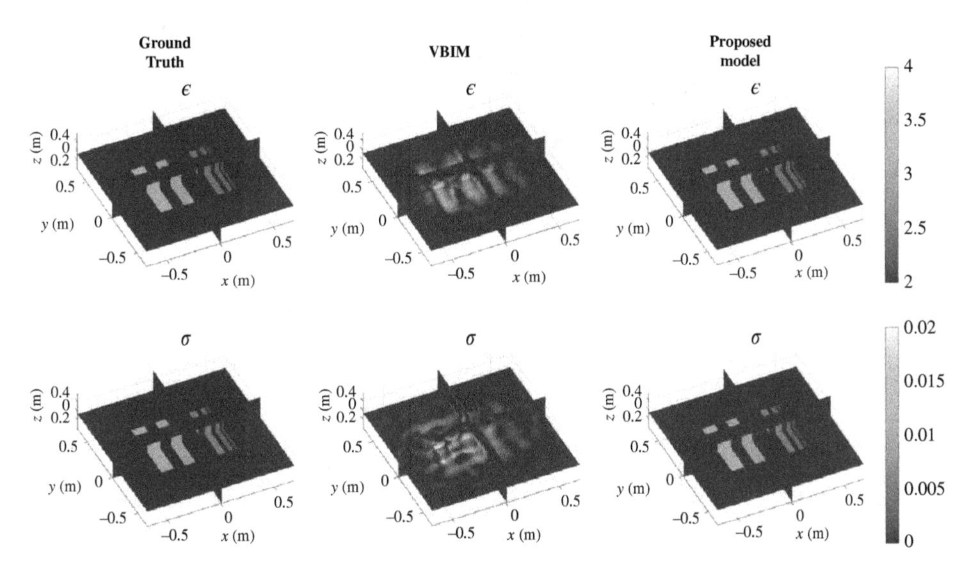

Figure 15.21 Test #4 is employed to evaluate the super-resolution imaging ability of the proposed scheme. The first column shows the ground truth of the relative permittivity and conductivity, respectively. The second and third columns are the imaging results of the relative permittivity and conductivity from VBIM and the proposed scheme, respectively. Source: Xiao et al. [12]/IEEE.

data misfits and model misfits which are 1.12% and 11.534% for VBIM and 2.592% and 3.215% for the proposed model, respectively. Thus, compared with VBIM in terms of imaging resolution, the proposed model could have better performance and achieve super-resolution imaging results.

Then, to further evaluate the imaging resolution of the proposed model, we design a series of tests as shown in Figure 15.22, where the adjacent spacings between objects are gradually decreased

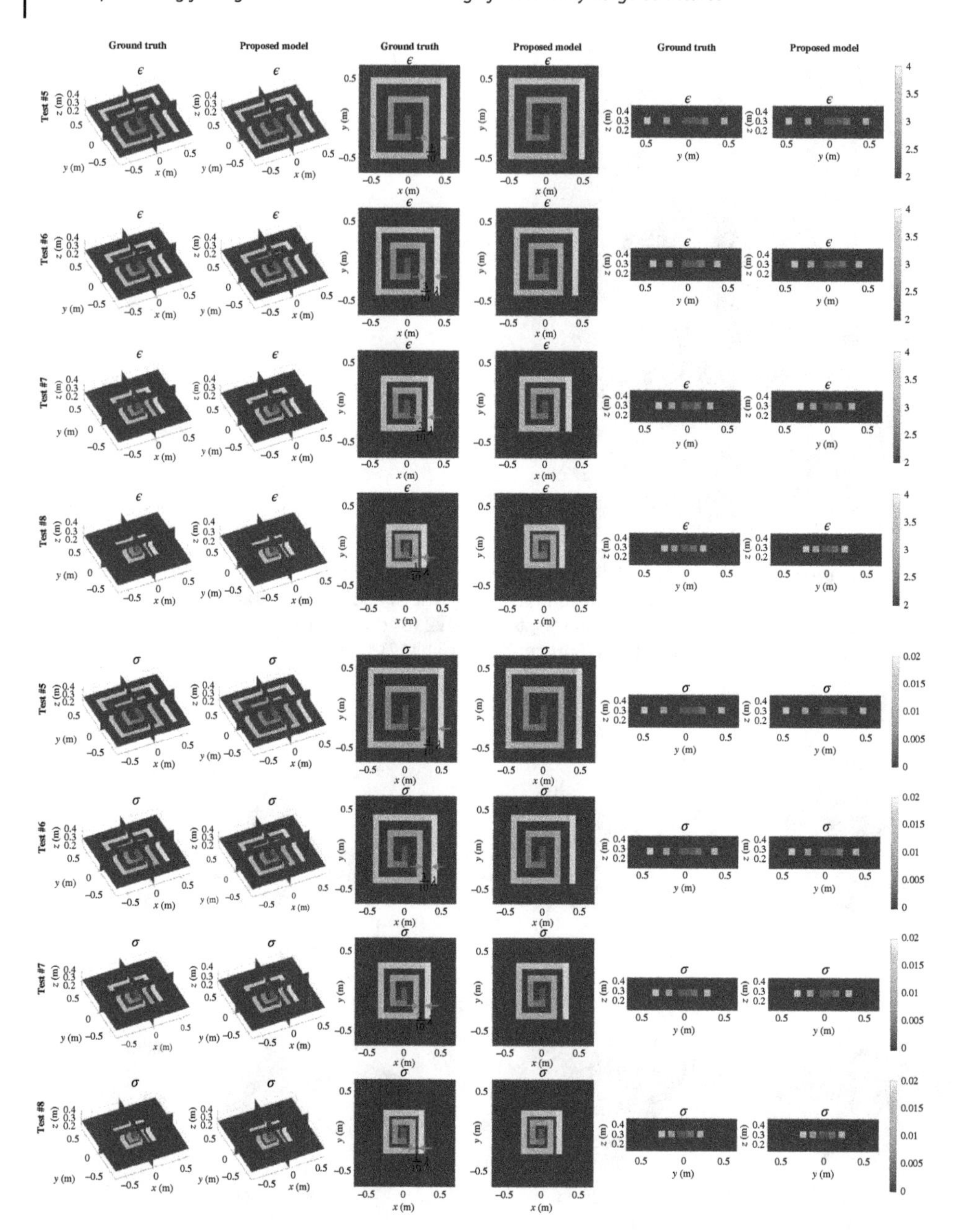

Figure 15.22 Tests #5–8 are employed to further evaluate the super-resolution imaging ability of the proposed scheme. The 1st, 3rd, and 5th columns are the 3-D perspective, *xy* slice at $z = 0.3$ m and *yz* slice at $x = 0$ m of ground truth, respectively. The 2nd, 4th, and 6th columns are the 3-D perspective, *xy* slice at $z = 0.3$ m and *yz* slice at $x = 0$ m of the imaging results from the proposed scheme. Source: Xiao et al. [12]/IEEE.

to $4/10\lambda$, $3/10\lambda$, $2/10\lambda$, and $1/10\lambda$ from Test #5 to Test #8, respectively. The height of each object is 0.1m (or 0.167λ). Meanwhile, the relative permittivity and conductivity of nine connected objects are gradually increased from the inside to the outside, i.e. 2.2 to 3.8 with the step of 0.2 for the relative permittivity and 0.002 to 0.018 with the step of 0.002 for conductivity. For these test cases, it is very difficult for VBIM to obtain a low model misfit. The evaluated data misfits of VBIM

Table 15.9 The model misfits of the proposed scheme for Tests # 5–8 in noise-free, 20 dB noise and 10 dB noise environments, respectively.

		Noise-free	20 dB	10 dB
Model misfit (%)	Test #5	3.585	6.103	16.703
	Test #6	3.624	6.232	16.654
	Test #7	3.611	6.098	16.695
	Test #8	3.632	6.112	16.709

Source: Xiao et al. [12]/IEEE.

are 0.716%, 0.712%, 0.705%, and 0.708% for Tests #5-8, respectively. Figures 15.22–15.24 are the imaging results obtained from the proposed model in the noise-free, −20 dB, and −10 dB Gaussian white noise environments (i.e. the SNR of power is infinity, 20 dB, and 10 dB, respectively). It could be seen the proposed model can image a stable imaging performance in the noise-free environment and have a stable trend of model misfits, even though the distances between objects changed, and the corresponding model misfits are small as shown in Table 15.9. The evaluated data misfits are 2.357%, 2.269%, 2.177%, and 2.634% for Tests #5-8, respectively, in the noise-free environment, which also reveal the similar stable trend. Meanwhile, the boundary between two connected objects can be clearly imaged, which is a challenging task to image two connected objects with close relative permittivity and conductivity by other traditional methods. Meanwhile, as shown in Figure 15.22, the xy slice at $z = 0.3$ m and yz slice at $x = 0$ m present the reconstructed results in a transverse plane and along the longitudinal axis, respectively. However, the anisotropy of the resolution is not observed from these two slices. This may be because our transmitter and receiver apertures are large enough to avoid obvious anisotropy in resolution. Then, as shown in Figures 15.23 and 15.24, as the noise increases, the imaging results become more and more blurred; however, the outline of objects also can be distinguished and the model misfits of four test cases, as shown in Table 15.9 almost remain the same levels for both 20 and 10 dB noise environments, although the distances between objects are closer. Thus, the proposed method has good anti-noise ability and super-resolution imaging ability in the noisy environment.

As shown in Figure 15.25, Test #9 consists of 27 spheres placed in three layers and each layer has three spheres. The center position, radius, relative permittivity, and conductivity of each sphere are listed in Table 15.10. The relative permittivity and conductivity of these spheres are gradually increased with the steps of 0.2 and 0.002 S/m from 1.2 and 0.002 S/m, respectively. The radius of each sphere is 0.32λ. Thus, the maxima of the relative permittivity and conductivity are 6.4 and 0.054 S/m, respectively.

Figure 15.25 shows the 3-D imaging results from the proposed model. Meanwhile, we select three slices located at $z = -0.32$ m, $z = 0$ m, and $z = -0.32$ m, as shown in Figure 15.26, to illustrate the imaging results clearly. It can be seen that both the permittivity and conductivity imaging results obtained from the proposed model can distinguish a good spatial resolution and provide accurate estimates of the contrast permittivity imaging and conductivity imaging, even though two adjacent spheres have close values of dielectric parameters or the object has a high contrast. The model misfit of this case is 3.88%, and the model misfits of relative permittivity and conductivity are 2.89% and 12.5%, respectively. The evaluated data misfit is 1.225%. However, for the conventional method such as VBIM, it is difficult to converge for this test which consists of objects with high contrasts, but the proposed SJ-ELM can obtain super-resolution imaging results in the application of through-wall imaging.

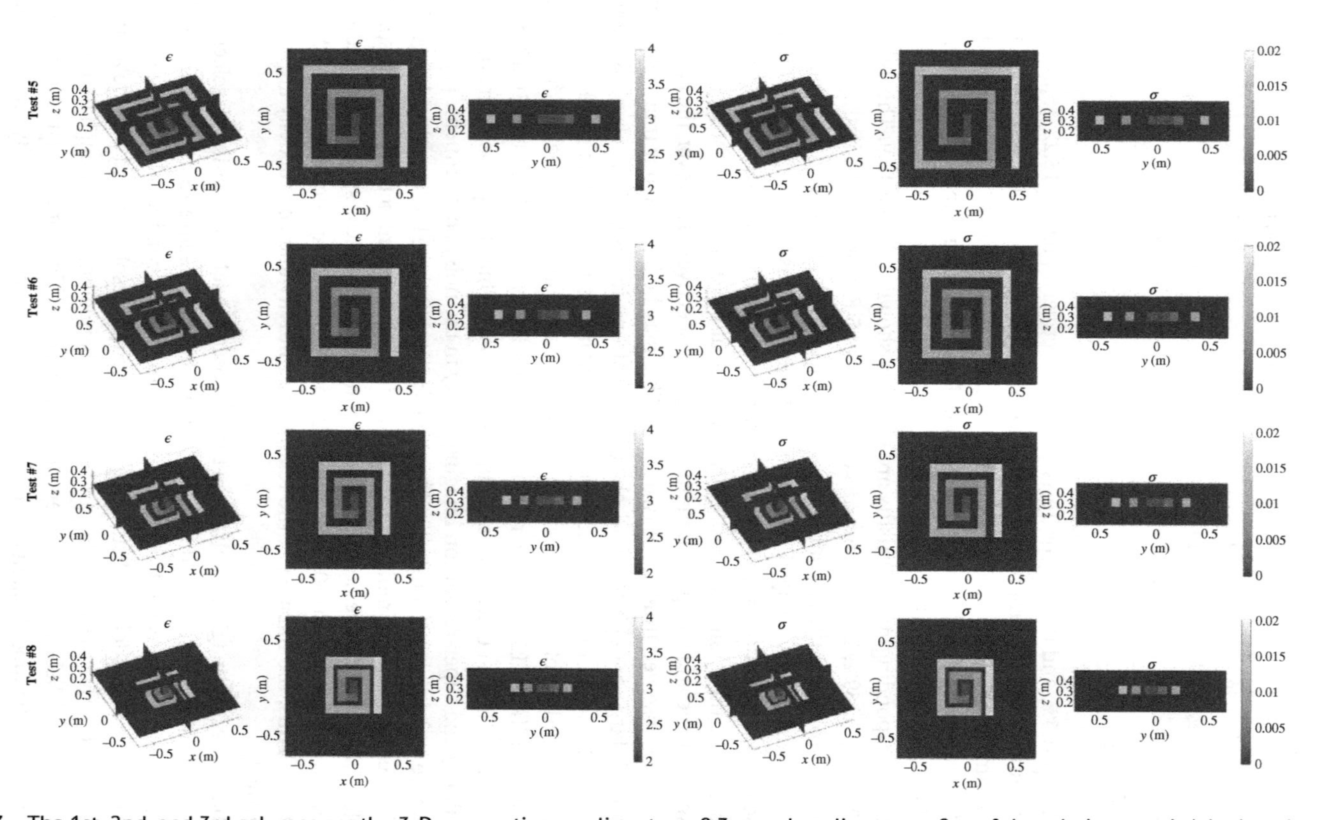

Figure 15.23 The 1st, 2nd, and 3rd columns are the 3-D perspective, *xy* slice at $z = 0.3$ m and *yz* slice at $x = 0$ m of the relative permittivity imaging results from the proposed scheme in the 20 dB noise environment. The 4th, 5th, and 6th columns are the 3-D perspective, *xy* slice at $z = 0.3$ m and *yz* slice at $x = 0$ m of the conductivity imaging results from the proposed scheme in the 20 dB noise environment. Source: Xiao et al. [12]/IEEE.

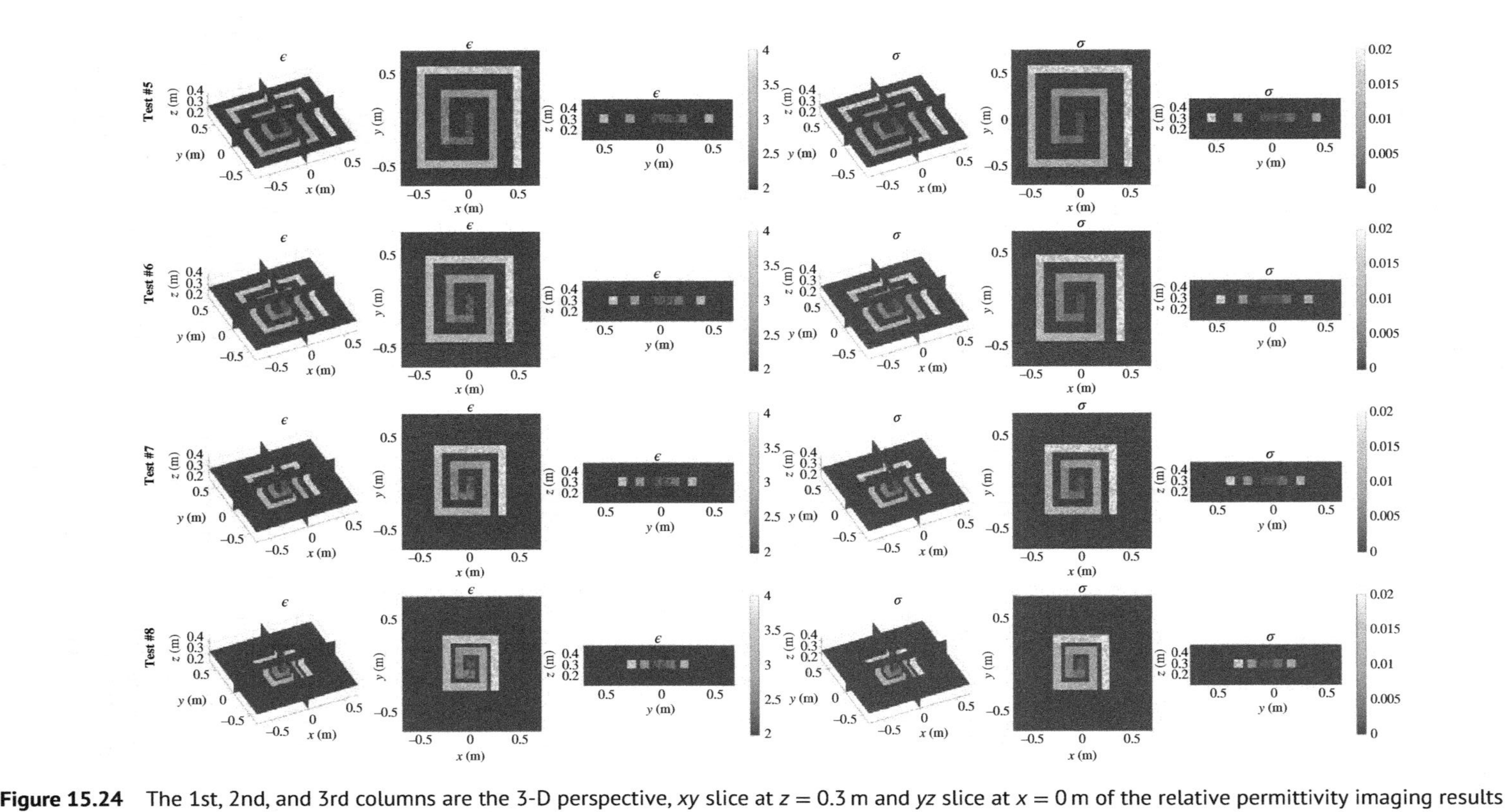

Figure 15.24 The 1st, 2nd, and 3rd columns are the 3-D perspective, xy slice at $z = 0.3$ m and yz slice at $x = 0$ m of the relative permittivity imaging results from the proposed scheme in the 10 dB noise environment. The 4th, 5th, and 6th columns are the 3-D perspective, xy slice at $z = 0.3$ m and yz slice at $x = 0$ m of the conductivity imaging results from the proposed scheme in the 10 dB noise environment. Source: Xiao et al. [12]/IEEE.

Table 15.10 The parameters of each sphere of Test #9.

No.	Center position (m)			Radius (m)	Relative permittivity	Conductivity (S/m)
	x	*y*	*z*			
1	−0.32	−0.32	−0.32	0.16	1.2	0.002
2	0	−0.32	−0.32	0.16	1.4	0.004
3	−0.32	0.32	−0.32	0.16	1.6	0.006
4	−0.32	0	−0.32	0.16	1.8	0.008
5	0	0	−0.32	0.16	2.0	0.010
6	0.32	0	−0.32	0.16	2.2	0.012
7	−0.32	0.32	−0.32	0.16	2.4	0.014
8	0	0.32	−0.32	0.16	2.6	0.016
9	0.32	0.32	−0.32	0.16	2.8	0.018
10	−0.32	−0.32	0	0.16	3.0	0.020
11	0	−0.32	0	0.16	3.2	0.022
12	−0.32	0.32	0	0.16	3.4	0.024
13	−0.32	0	0	0.16	3.6	0.026
14	0	0	0	0.16	3.8	0.028
15	0.32	0	0	0.16	4.0	0.030
16	−0.32	0.32	0	0.16	4.2	0.032
17	0	0.32	0	0.16	4.4	0.034
18	0.32	0.32	0	0.16	4.6	0.036
19	−0.32	−0.32	0.32	0.16	4.8	0.038
20	0	−0.32	0.32	0.16	5.0	0.040
21	−0.32	0.32	0.32	0.16	5.2	0.042
22	−0.32	0	0.32	0.16	5.4	0.044
23	0	0	0.32	0.16	5.6	0.046
24	0.32	0	0.32	0.16	5.8	0.048
25	−0.32	0.32	0.32	0.16	6.0	0.050
26	0	0.32	0.32	0.16	6.2	0.052
27	0.32	0.32	0.32	0.16	6.4	0.054

Source: Xiao et al. [13]/IEEE.

Throughout the model misfits and data misfits, which are obtained from the proposed SJ-ELM, from the first numerical example to the third numerical example, the contrasts of objects are gradually increased, and the model misfits remain at the same level, because the objective function of machine learning method is based on the model misfit. However, with the relative permittivity increasing, the corresponding data misfits are gradually decreased. This is because that with contrast increasing, the nonlinearity of the inversion problem is aggravated, and the caused multi-solution problem brings more solutions to satisfy the desired data misfit. The machine learning method could reconstruct the objects well, even though the objects are with high contrasts. Thus, with contrast increasing, the data misfit reveals a decreasing trend even though model misfits are at the same level. This is quite different from the conventional inverse

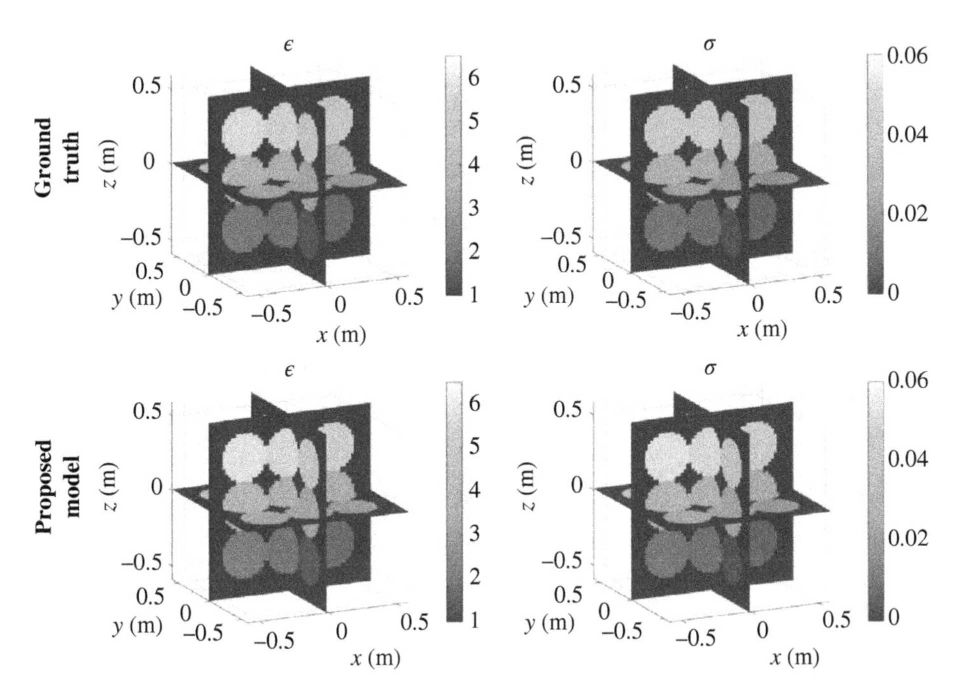

Figure 15.25 The relative permittivity and conductivity imaging results of Test #9 from the proposed scheme, where imaging objects are with high contrast. Source: Xiao et al. [12], © IEEE.

methods which still could have a very low data misfit when the objects are with high contrasts; however, the model misfit is also very large. Compared with conventional inverse methods, the advantage of machine learning-based method is that a stable model misfit could be obtained even though the objects have high contrasts.

To further evaluate the proposed model, the 3-D experimental data from the Fresnel database [58] are used in this test. Here, simulation data is used to train the model, and actual measured data is employed to test the trained model, where the simulation setting is the same as the experimental environment. A "CubeSpheres" target with co-polarized data, as shown in Figure 15.27, is adopted. Here we select two frequency points at 3 and 8 GHz for the evaluation. Each sphere has a diameter of 15.9 mm and a relative permittivity of 2.6. They are assembled in order to create a cube of side length 47.6 mm. In the measurement environment, the transmitter arrays are located on a sphere with the radius of 1.796 m surrounding the target, where the azimuthal angle θ_s is in the range from 20° to 340° with the step of 40° and the polar angle φ_s is in the range from 30° to 150° with the step of 15°. The receiver arrays are located on the azimuthal plane with the radius of 1.796 m, and the azimuthal angle θ_r is in the range from 0° to 350° with the step of 10°. The experimental scattered data are corrected for drift errors and calibrated, successively [59].

According to the experimental setup, the cubic volume imaging domain of dimensions $100 \times 100 \times 100\,\text{mm}^3$, centered at (0, 0, 0) of the coordinate system, is divided into $96 \times 96 \times 96$ voxels. Based on the measurement environment and the experimental setup, totally 200 training samples with frequency of 3 and 8 GHz are produced and employed to train the proposed model, respectively. The relative permittivities of four scatterers are randomly set with different values in the range of [1, 4]. The measured scattered field data at 3 and 8 GHz of "CubeSpheres" target are employed as the input of test, and the reconstructed relative permittivity distribution of both

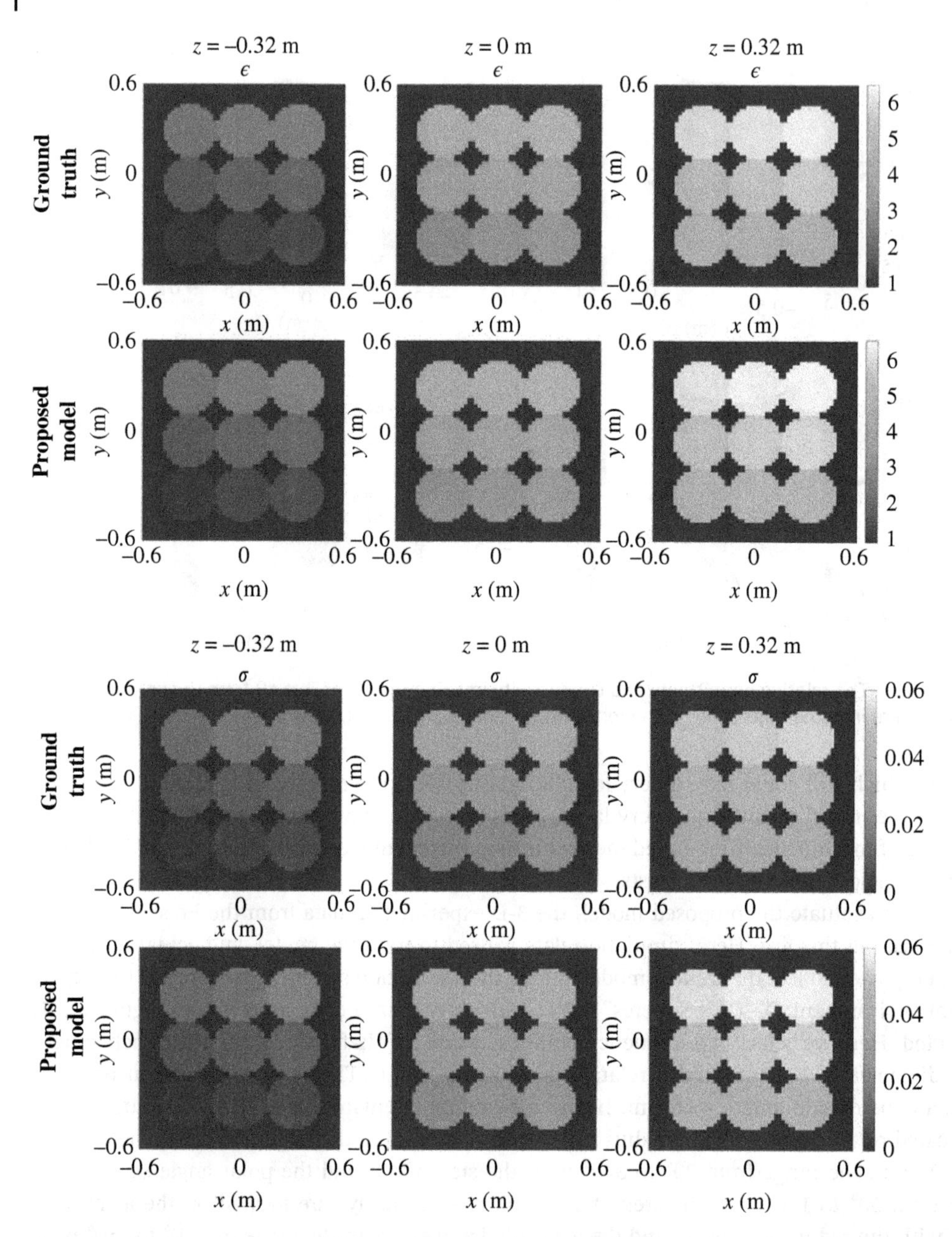

Figure 15.26 The relative permittivity and conductivity imaging results of *xy* slices at $z = -0.32$, 0, and 0.32 from the proposed scheme, respectively, where the relative permittivity and conductivity of each slice are gradually increased, and the model misfit of this case is 3.88%, and the model misfits of relative permittivity and conductivity are 2.89% and 12.5%, respectively. Source: Xiao et al. [12], © IEEE.

frequency points from the proposed model is obtained as shown in Figure 15.27. The model misfits of the reconstructed targets at 3 and 8 GHZ are 4.533% and 5.236%, respectively. Meanwhile, the data misfits for the reconstructed targets at 3 and 8 GHz are 4.692% and 6.132%, respectively. Because the nonlinearity at 8 GHz is stronger than that at 3 GHz, with the same number of the receiver arrays, the imaging errors, model misfit, and data misfit at 8 GHz are higher than those at

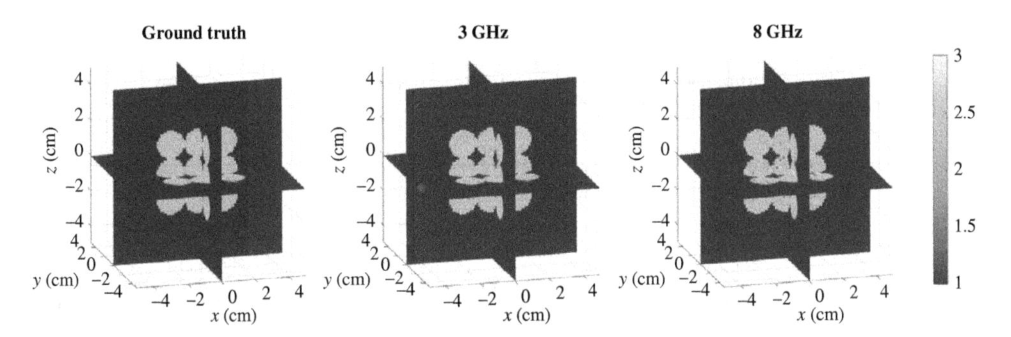

Figure 15.27 The ground truth and imaging results of relative permittivity profiles of "CubeSphere" at 3 GHz and 8 GHz by the proposed scheme, respectively. Source: Xiao et al. [12], © IEEE.

3 GHz. The results from the experimental data further verify the validity of the proposed model which has a good performance in the practical experimental environment for the microwave imaging problems.

15.4.2.2 A Hybrid Neural Network Electromagnetic Inversion Scheme (HNNEMIS) for Super-Resolution 3-Dimensional Microwave Human Brain Imaging

In this section, two numerical examples are presented to validate the proposed HNNEMIS for super-resolution 3-D microwave human brain imaging [13]. The first numerical example is employed to evaluate the HNNEMIS performance for the normal human brain. The second example is employed to evaluate the HNNEMIS performance for detecting human brain with a small abnormal scatterer. Both numerical examples are polluted by white Gaussian noise of different SNR to test the sensitivity to noise for the proposed method. All the forward and inverse problems are performed on a workstation with 20-cores CPU, 512 GB RAM, and NVIDIA Geforce RTX 3090 GPU.

The operating frequency of both examples is 300 MHz, so the corresponding wavelength is $\lambda_0 = 1$ m in air. The transmitting and receiving antennas (modeled as point electric dipoles) surround a human brain model at the positions listed in Table 15.11. There are 60 transmitters/receivers, so the signal transmitted from one transmitter will be received by other 59 receivers, namely $60 \times 59 \times 3 = 10{,}620$ measurements of scattered electric field in x, y, and z directions input into SJ-BPNN. In Figure 15.4, the outer shape of human brain model also can be observed. The virtual human brain model is constructed by NEVA electromagnetics [60].

As shown in Figure 15.28, the DOI for reconstruction is a $0.24 \times 0.24 \times 0.24\ \text{m}^3$ cubic box. For the first example, DOI is divided into $256 \times 256 \times 256$ voxels; thus, the number of unknowns (also the number of the output of SJ-BPNN and the input and output of U-Net) is 33,554,432. 200 training samples are employed to train HNNEMIS for the first example. For the second example, we challenge higher resolution by dividing the DOI into $512 \times 512 \times 512$ voxels, so the number of unknowns is 268,435,456, and the size of each voxel is $0.469 \times 0.469 \times 0.469\ \text{mm}^3$ (or $4.69 \times 10^{-4}\lambda \times 4.69 \times 10^{-4}\lambda \times 4.69 \times 10^{-4}\lambda$). For this highly ill-posed, nonlinear, and large-scale inverse problem, it is too prohibitively expensive to reconstruct with the traditional iterative methods. To the best of our knowledge, no human brain imaging based on the EM inversion with this resolution has been reported. Another 200 training samples are employed to train HNNEMIS for the second example.

The human brain model has 16 different tissues listed in Table 15.12 with the corresponding electrical properties at 300 MHz [61]. From Table 15.12, it can be seen that the electrical properties

Table 15.11 The position of each transmitter/receiver.

	Position (m)				Position (m)		
No.	x	y	z	No.	x	y	z
1	0.120	0.000	−0.060	31	0.111	−0.046	−0.060
2	0.118	0.023	−0.060	32	0.118	−0.023	−0.060
3	0.111	0.046	−0.060	33	0.115	0.000	0.000
4	0.100	0.067	−0.060	34	0.106	0.044	0.000
5	0.085	0.085	−0.060	35	0.081	0.081	0.000
6	0.067	0.100	−0.060	36	0.044	0.106	0.000
7	0.046	0.111	−0.060	37	0.000	0.115	0.000
8	0.023	0.118	−0.060	38	−0.044	0.106	0.000
9	0.000	0.120	−0.060	39	−0.081	0.081	0.000
10	−0.023	0.118	−0.060	40	−0.106	0.044	0.000
11	−0.046	0.111	−0.060	41	−0.115	0.000	0.000
12	−0.067	0.100	−0.060	42	−0.106	−0.044	0.000
13	−0.085	0.085	−0.060	43	−0.081	−0.081	0.000
14	−0.100	0.067	−0.060	44	−0.044	−0.106	0.000
15	−0.111	0.046	−0.060	45	0.000	−0.115	0.000
16	−0.118	0.023	−0.060	46	0.044	−0.106	0.000
17	−0.120	0.000	−0.060	47	0.081	−0.081	0.000
18	−0.118	−0.023	−0.060	48	0.106	−0.044	0.000
19	−0.111	−0.046	−0.060	49	0.100	0.000	0.060
20	−0.100	−0.067	−0.060	50	0.071	0.071	0.060
21	−0.085	−0.085	−0.060	51	0.000	0.100	0.060
22	−0.067	−0.100	−0.060	52	−0.071	0.071	0.060
23	−0.046	−0.111	−0.060	53	−0.100	0.000	0.060
24	−0.023	−0.118	−0.060	54	−0.071	−0.071	0.060
25	0.000	−0.120	−0.060	55	0.000	−0.100	0.060
26	0.023	−0.118	−0.060	56	0.071	−0.071	0.060
27	0.046	−0.111	−0.060	57	0.060	0.000	0.100
28	0.067	−0.100	−0.060	58	0.000	0.060	0.100
29	0.085	−0.085	−0.060	59	−0.060	0.000	0.100
30	0.100	−0.067	−0.060	60	0.000	−0.060	0.100

Source: Xiao et al. [13]/IEEE.

are very different and much larger than those for air, so human brain imaging is a high contrast EM inverse problem. In the training samples, the electrical properties of each human brain tissue are multiplied by the random multiplier between 0.8 and 1.2 (with the step of 0.1) based on the value shown in Table 15.12. The radii of the abnormal scatterers are randomly set to 4, 7, and 10 mm, and their relative permittivities are randomly set to 90, 120, and 150, and conductivities are randomly set to 1.038, 1.3840, and 1.73 S/m.

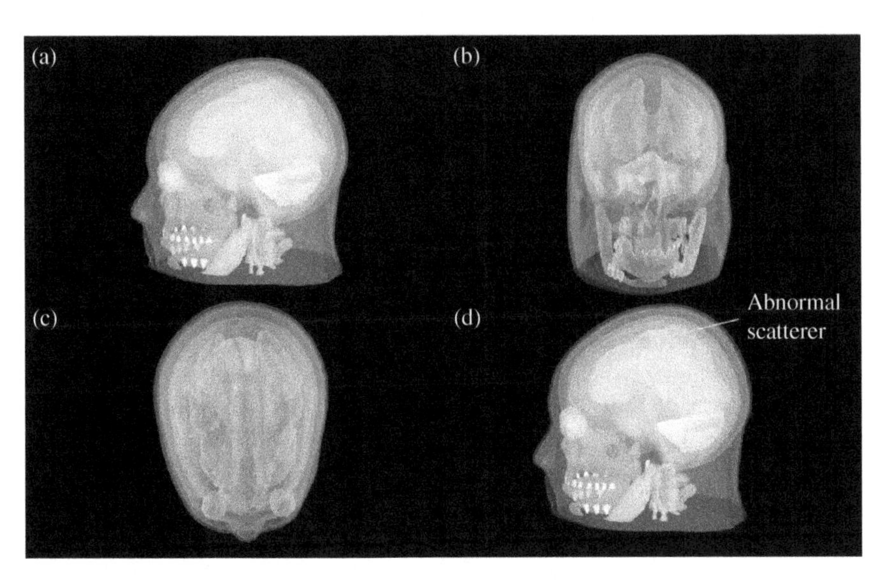

Figure 15.28 Two numerical examples used in this work: (a) side view of normal human brain used in Example I; (b) front view of normal human brain used in Example I; (c) top view of normal human brain used in Example I; and (d) side view of human brain with abnormal scatterer used in Example II. Source: Xiao et al. [13], © IEEE.

Table 15.12 Electrical properties of the human brain at 300 MHz.

Tissue	Relative permittivity	Conductivity (S/m)
Fat	11.7	0.0765
Muscle	58.2	0.771
Skull	23.2	0.216
CSF	72.7	2.22
Gray matter	60.0	0.692
White matter	43.8	0.413
Cerebellum	59.7	0.973
Eye	69.0	1.52
Mandible	13.4	0.0827
Tongue	58.9	0.745
Tooth	13.4	0.0827
Trachea	45.3	0.611
Blood vessel	48.3	0.537
Vertebrae	13.4	0.0827
Intervertebral disc.	47.2	0.911
Spinal cord	36.9	0.418

Source: Xiao et al. [13]/IEEE.

Table 15.13 Model misfits and data misfits of the proposed method for normal human brain example at different noise levels.

	10 dB	20 dB	30 dB	40 dB
Model misfit (%)	29.65	18.43	15.63	13.63
Data misfit (%)	27.25	11.37	5.83	3.68

Source: Xiao et al./IEEE.

To evaluate the inversion performance of the proposed HNNEMIS for the normal human brain, a normal human brain model which never appears in the training dataset is employed for testing. In this example, SJ-BPNN is accelerated with GPU, so that it takes about 89 GB computer memory and 8 hours in the training process, and in the testing process the memory and running time are reduced to 19 GB and 30 minutes, respectively. Meanwhile, the training time and testing time consumed by U-Net are about 60 minutes and 2 minutes, respectively.

The reconstructed 3-D slices of relative permittivity and conductivity distributions obtained from SJ-BPNN and the final results of HNNEMIS are shown in the first and second columns of Figure 15.29, respectively. The corresponding 2-D slices at $z = 0.02$ m plane are also provided in the bottom half of Figure 15.29. Preliminary imaging results obtained from SJ-BPNN can accurately image the shape, location, and size of each tissue of human brain, but there exist some noisy points at the boundary of tissues. Fortunately, U-Net in the second stage improves this issue. As shown in Figure 15.29, the final imaging results match well with the ground truth, even though human brain is a very complex organ. The model misfits between the reconstructed results and ground truth of SJ-BPNN and HNNEMIS are 18.62% and 12.52% and the corresponding data misfits are 8.36% and 0.73%. U-Net in the second phase can effectively reduce both model misfit and data misfit obtained from SJ-BPNN.

Then, to further evaluate HNNEMIS performance in noise environment, white Gaussian noises are added into the measure scattered electric field. With the relative noise level at −10, −20, −30, and −40 dB, the corresponding reconstructed permittivity and conductivity distributions including 3-D and 2-D slices obtained from SJ-BPNN and HNNEMIS are shown in Figure 15.30, respectively. As the inversion results of SJ-BPNN shown, with the relative noise level increasing, the boundary of each tissue is more and more blurry, and more and more noisy points appear in the brain domain. Then, the inversion results of SJ-BPNN are input to the trained U-Net to improve the imaging performance. It can be seen that the final inversion results have been improved from preliminary results. The corresponding model misfits and data misfits of the final results obtained from HNNEMIS are listed in Table 15.13. Thus, the proposed method has good anti-noise ability and super-resolution imaging ability for human brain imaging even in the noisy environment.

In this example, three cases, i.e. Tests # 2-4, containing an abnormal scatterer with random size, location and electrical parameters, are further employed to evaluate the inversion performance of the proposed HNNEMIS. These three cases never appear in the training dataset.

Meanwhile, one should note a higher resolution with $512 \times 512 \times 512$ voxels is required in this work. Compared with previous example, the number of unknowns is increased by eight times. This is a formidable challenge for the inversion method including traditional and machine

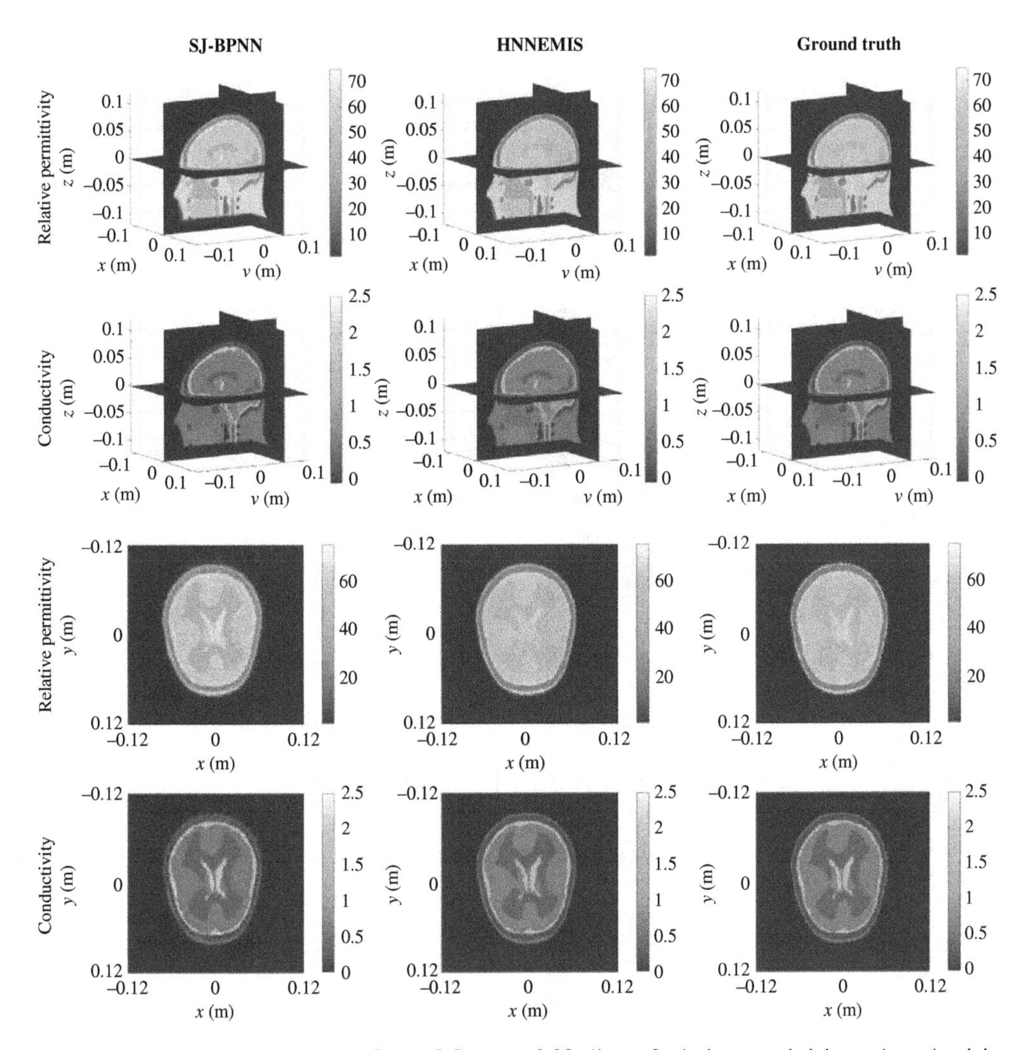

Figure 15.29 The reconstructed 3-D and 2-D at $z = 0.02$ slices of relative permittivity and conductivity distributions obtained from SJ-BPNN and HNNEMIS. Source: Xiao et al. [13], © IEEE.

learning-based methods. In this example, SJ-BPNN takes about 232 GB computer memory and 47 hours in the training process, and in the testing process the memory and running time are reduced to 19 GB and 110 minutes, respectively. U-Net takes about 60 and 2 minutes for the training and testing, respectively.

First, these three cases are tested in the noise-free environment, with the abnormal scatterers of Tests #2-4 at the planes of $z = 0.01$, 0.005, 0.02 m, respectively. As shown in Figure 15.31, the electrical parameters, shape, location, and size of each tissue of human brain can be well imaged. Meanwhile, relative permittivity, conductivity, shape, location, and size of each abnormal scatterer are also well imaged, even though some of these scatterers are very small, such as Test #3. The corresponding model misfits and data misfits of Test #2-4 are listed in Table 15.14. It can be seen, although the resolution of this example is much higher, both model misfits and data misfits remain at the same level as those in Example I in the noise-free environment.

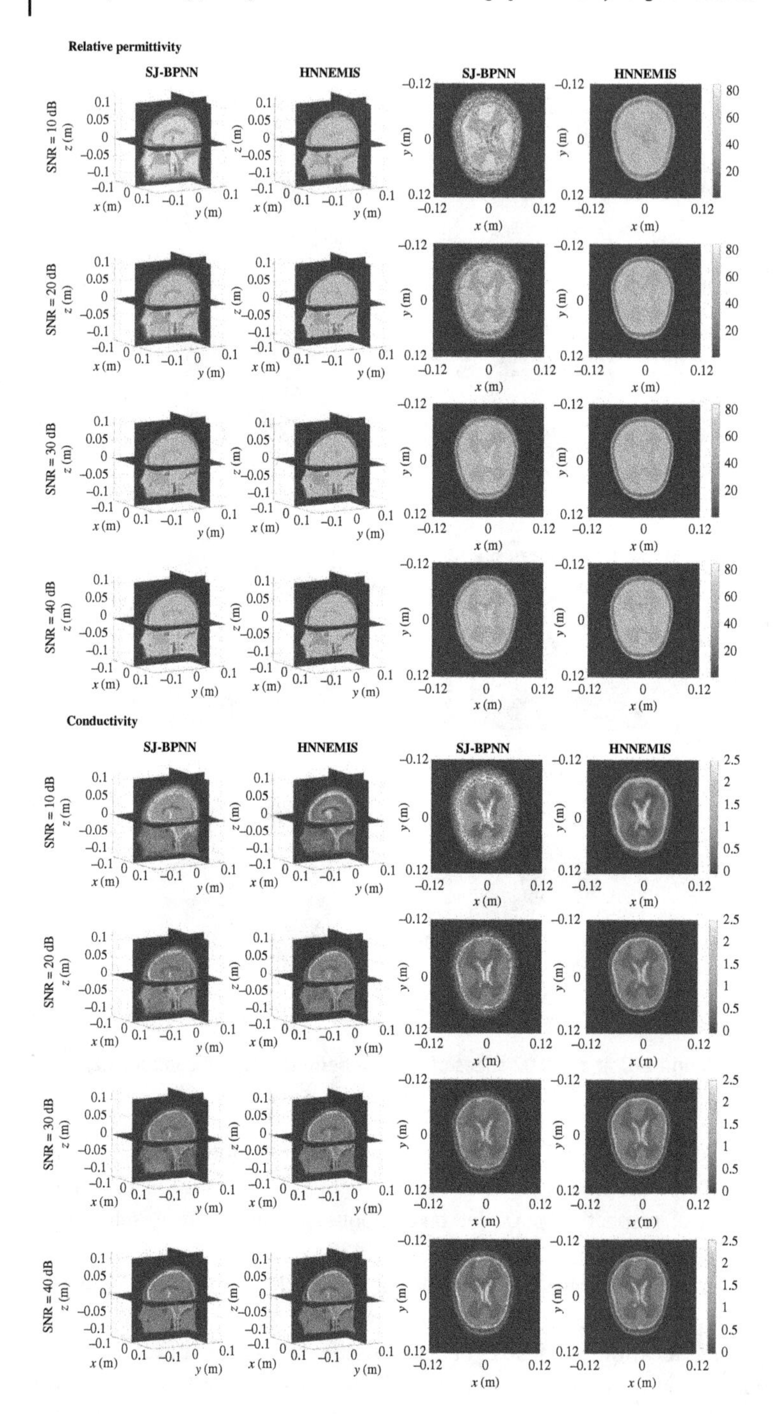

Figure 15.30 With the relative noise level at −10, −20, −30, and −40 dB, the reconstructed 3-D and 2-D at $z = 0.02$ slices of relative permittivity and conductivity distributions obtained from SJ-BPNN and HNNEMIS. Source: Xiao et al. [13], © IEEE.

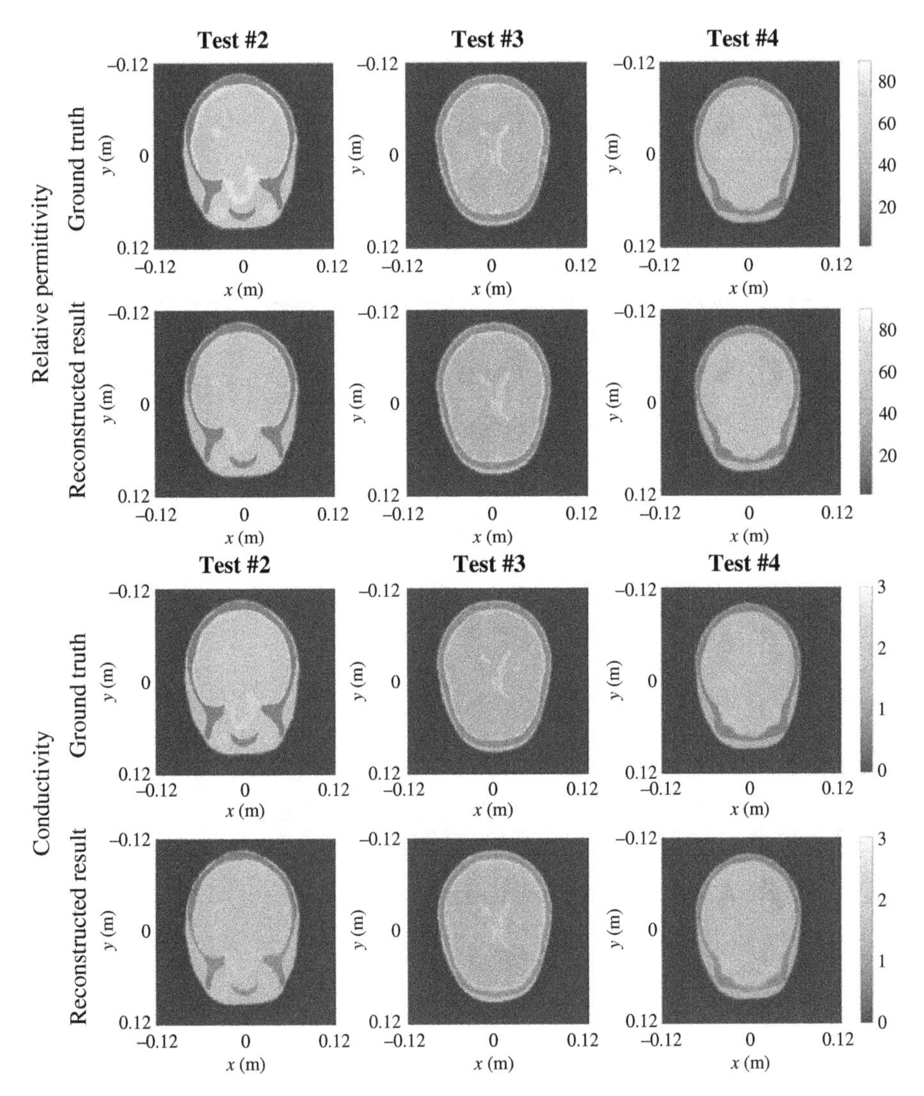

Figure 15.31 The reconstructed 2-D slices of relative permittivity and conductivity distributions obtained from HNNEMIS for the human brain with an abnormal scatterer. Source: Xiao et al. [13], © IEEE.

Then, white Gaussian noises with the relative noise level at −10, −20, −30, and −40 dB are added into the measured scattered electric field of Tests #2-4 to evaluate the proposed HNNEMIS performance in the noisy environment. As shown in Figure 15.32, for Tests #2 and 3, the abnormal scatterers with very small size can be detected at the approximate location at −10 dB noise; however, their shapes and sizes are hard to distinguish. With the relative noise level decreasing, this point has been improved that the imaging performance is getting better. For Test #4, due to the large size of the abnormal scatterer, it is easier to be detected than the other two cases. Thus, the size, location, and shape of the abnormal scatter can be detected with the relative noise level from −10 to −40 dB. The corresponding model misfits and data misfits of the reconstructed results are also listed in Table 15.14, where the inversion performance of the proposed HNNEMIS still can keep the same level as the performance in the last example in the noisy scenario.

Table 15.14 Model misfits and data misfits of the proposed method for human brain abnormal scatterer examples at different noise levels.

		Noise-free	10 dB	20 dB	30 dB	40 dB
Model misfit (%)	Test #2	7.63	21.78	15.43	11.18	9.64
	Test #3	9.22	22.04	16.59	12.96	10.70
	Test #4	10.15	22.51	17.57	13.87	11.83
Data misfit (%)	Test #2	0.79	26.28	11.46	5.54	3.64
	Test #3	0.83	27.44	12.69	6.35	4.26
	Test #4	0.94	29.37	13.19	7.47	4.93

Source: Xiao et al. [13]/IEEE.

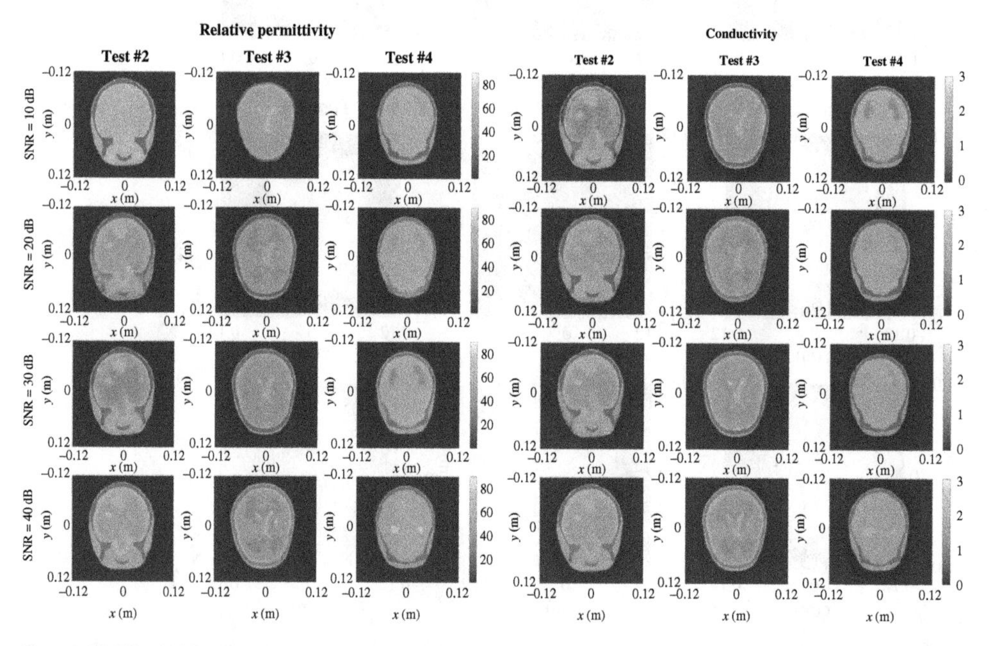

Figure 15.32 With the relative noise level at −10, −20, −30, and −40 dB, the reconstructed 2-D slices of relative permittivity and conductivity distributions obtained. Source: Xiao et al. [13], © IEEE.

15.5 Conclusion and Future work

15.5.1 Summary of Our Work

In this chapter, four machine learning-based inversion methods for high contrast inverse scattering of electrically large structures are summarized based on our recent works.

For the 2-D inversion problem, a dual-module NMM-IEM machine learning scheme is firstly introduced to deal with the reconstruction of inhomogeneous scatterers with high contrasts and large electrical dimensions. With the test of numerically simulated data and the experimental data measured in the laboratory, it is found that NMM-IEM has better inversion performance than the conventional methods, e.g. VBIM. Meanwhile, for the scatterers with high contrast and large electrical dimension, the effectiveness of NMM-IEM is also validated. Then, the RAML method is introduced to repair the damaged data due to malfunctioning receivers in the 2-D inverse scattering problem. RAML combined with NMM-IEM shows a great inversion performance when ten receivers are damaged for the measured data, for both TE and TM modes, provided by Institute Fresnel. In the extreme situation, it can have good inversion performance even though half of the receivers are damaged.

For the 3-D inversion problem, directly using the above 2-D method would be too expensive, thus the SJ-ELM model is firstly introduced for super-resolution 3-D quantitative imaging of objects with high contrasts. In SJ-ELM, two channels convert the complex-value scattered field data to the real permittivity and conductivity values of the imaging domain, respectively. Each channel is constructed with a shallow neural network structure based on the semi-join strategy. The semi-join strategy can decrease the inner matrix dimensions and improve the convergence performance of the model in the performance of super-resolution imaging, so it is employed to connect between the nodes of hidden layers and output layer to realize super-resolution imaging. Through comparison with a conventional method, e.g. VBIM, the proposed SJ-ELM model can achieve more accurate and stable imaging performance and higher imaging efficiency. In terms of imaging resolution, the proposed model can have better performance and achieve super-resolution imaging results. In the application of through-wall imaging, the proposed SJ-ELM can obtain super-resolution imaging results, even when the objects have high contrasts that are difficult for a conventional method such as VBIM to converge. The reconstruction of the laboratory measured scattered field data at 3 and 8 GHz for "CubeSphere" provided by Institute Fresnel demonstrates the adaptability of the proposed SJ-ELM for high-frequency experimental data. Meanwhile, for the human brain imaging, to capture the specific feature of each tissue type to guarantee the imaging accuracy, the HNNEMIS containing both shallow and deep neural networks is proposed for super-resolution 3-D microwave human brain imaging. The shallow neural network, termed SJ-BPNN, can nonlinearly transfer the measured scattered electric field into two output channels, namely the permittivity and conductivity of scatterers, respectively, using less computational costs in super-resolution 3-D imaging. The deep learning technique, i.e. U-Net, is employed to further enhance the imaging quality. Meanwhile, a novel training dataset construction strategy incorporating the characteristics of human brain is also proposed for human brain imaging to decrease the training cost and make neural networks fast convergent. Two numerical examples of normal human brain and the human brain with an abnormal scatterer are employed to validate the proposed HNNEMIS for super-resolution 3-D microwave human brain imaging, respectively.

15.5.1.1 Limitations and Potential Future Works

In spite of the successful application of machine-learning-based EM inverse scattering techniques, the major limitation of the above four inversion methods is that when the setups of frequency, transmitters, and receivers are changed, the trained model has to be re-trained. This will increase the training cost and decrease the generalization of the model. Thus, the future work will be focused on the following two aspects:

- The trained model can work in a wide band range. In the actual application, the frequency of the measured data does not remain a fixed value. To make the trained model more generalizable, the trained model should be effective at any frequency point in a wide band.
- The trained model can work when the setups of transmitters and receivers are changed, including the numbers and positions. In many applications, the numbers and positions of transmitters and receivers may not match the training setups. Thus, how to use the trained artificial neural network to perform such inverse scattering without retraining can be regarded as a potential future work.

References

1 Chew, W.C. (1995). *Waves and Fields in Inhomogeneous Media*. Wiley.
2 Colton, D. and Kress, R. (ed.) (1998). *Inverse Acoustic and Electromagnetic Scattering Theory*. New York: Springer.
3 Zhdanov, M.S. (ed.) (2004). *Geophysical Inverse Theory and Regularization Problems*. Amsterdam, The Netherlands: *Elsevier*.
4 Caorsi, S., Massa, A., Pastorino, M., and Donelli, M. (2004). Improved microwave imaging procedure for nondestructive evaluations of two-dimensional structures. *IEEE Trans. Antennas Propag.* 52 (6): 1386–1397.
5 Song, L.-P., Yu, C., and Liu, Q.H. (2005). Through-wall imaging (TWI) by radar: 2-D tomographic results and analyses. *IEEE Trans. Geosci. Remote Sens.* 43 (12): 2793–2798.
6 Abubakar, A., van den Berg, P.M., and Mallorqui, J.J. (2002). Imaging of biomedical data using a multiplicative regularized contrast source inversion method. *IEEE Trans. Microw. Theory Techn.* 50 (7): 1761–1771.
7 Meaney, P.M., Fanning, M.W., Li, S.P.P.D., and Paulsen, K.D. (2000). A clinical prototype for active microwave imaging of the breast. *IEEE Trans. Microw. Theory Techn.* 48 (11): 1841–1853.
8 Cui, T.J., Chew, W.C., Aydiner, A.A., and Chen, S. (2001). Inverse scattering of two-dimensional dielectric objects buried in a lossy Earth using the distorted born iterative method. *IEEE Trans. Geosci. Remote Sens.* 39 (2): 339–346.
9 Li, G. and Burkholder, R.J. (2015). Hybrid matching pursuit for distributed through-wall radar imaging. *IEEE Trans. Antennas Propag.* 63 (4): 1701–1711.
10 Xiao, L.Y., Li, J.W., Han, F. et al. (2020). Dual-Module NMM-IEM machining learning for fast electromagnetic inversion of inhomogeneous scatterers with high contrasts and large electrical dimensions. *IEEE Trans. Antennas and Propag.* 68 (8): 6245–6255.
11 Hu, H.J., Xiao, L.Y., Yi, J.N., and Liu, Q.H. (2021). Nonlinear electromagnetic inversion of damaged experimental data by a receiver approximation machine learning method. *IEEE Antennas Wireless Propag. Lett.* 20 (7): 1185–1189.
12 Xiao, L.Y., Li, J., Han, F. et al. (2021). Super-resolution 3-D microwave imaging of objects with high contrasts by a semijoin extreme learning machine. *IEEE Trans. Microw. Theory Techn.* 69 (11): 4840–4855.
13 Xiao, L.-Y., Hong, R., Zhao, L.-Y. et al. (2022). Hybrid neural network electromagnetic inversion scheme (HNNEMIS) for super-resolution 3-dimensional microwave human brain imaging. *IEEE Trans. Antennas Propag.* 70 (8): 6277–6286.
14 Solimene, R., Soldovieri, F., and Prisco, G. (2008). A multiarray tomographic approach for through-wall imaging. *IEEE Trans. Geosci. Remote Sens.* 46 (4): 1192–1199.
15 Lo Monte, L., Erricolo, D., Soldovieri, F., and Wicks, M.C. (2010). Radio frequency tomography for tunnel detection. *IEEE Trans. Geosci. Remote Sens.* 48 (3): 1128–1137.
16 Afsari, A., Abbosh, A.M., and Rahmat-Samii, Y. (2018). A rapid medical microwave tomography based on partial differential equations. *IEEE Trans. Antennas Propag.* 66 (10): 5521–5535.
17 Song, L.-P. and Liu, Q.H. (2005). A new approximation to three-dimensional electromagnetic scattering. *IEEE Geosci. Remote Sens. Lett.* 2 (2): 238–242.
18 Catapano, I., Randazzo, A., Slob, E., and Solimene, R. (2015). GPR imaging via qualitative and quantitative approaches. In: *Civil Engineering Applications of Ground Penetrating Radar* (ed. A. Benedetto and L. Pajewski), 239–280. Cham, Switzerland: Springer.
19 Wang, Y.M. and Chew, W.C. (1989). An iterative solution of the two dimensional electromagnetic inverse scattering problem. *Int. J. Imaging Syst. Technol.* 1 (1): 100–108.

20 Habashy, T.M., Oristaglio, M.L., and de Hoop, A.T. (1994). Simultaneous nonlinear reconstruction of two-dimensional permittivity and conductivity. *Radio Sci.* 29 (4): 1101–1118.

21 van den Berg, P.M. and Kleinman, R.E. (1997). A contrast source inversion method. *Inverse Probl.* 13 (6): 1607–1620.

22 Chen, X. (2010). Subspace-based optimization method for solving inverse scattering problems. *IEEE Trans. Geosci. Remote Sens.* 48 (1): 42–49.

23 Zhong, Y. and Chen, X. (2011). An FFT twofold subspace-based optimization method for solving electromagnetic inverse scattering problems. *IEEE Trans. Antennas Propag.* 59 (3): 914–927.

24 Pastorino, M., Massa, A., and Caorsi, S. (2000). A microwave inverse scattering technique for image reconstruction based on a genetic algorithm. *IEEE Trans. Instrum. Meas.* 49 (3): 573–578.

25 Qing, A., Lee, C.K., and Jen, L. (2001). Electromagnetic inverse scattering of two-dimensional perfectly conducting objects by real-coded genetic algorithm. *IEEE Trans. Geosci. Remote Sens.* 39 (3): 665–676.

26 Wildman, R.A. and Weile, D.S. (2007). Geometry reconstruction of conducting cylinders using genetic programming. *IEEE Trans. Antennas Propag.* 55 (3): 629–636.

27 Donelli, M., Franceschini, D., Rocca, P., and Massa, A. (2009). Three-dimensional microwave imaging problems solved through an efficient multiscaling particle swarm optimization. *IEEE Trans. Geosci. Remote Sens.* 47 (5): 1467–1481.

28 Salucci, M., Poli, L., Anselmi, N., and Massa, A. (2017). Multifrequency particle swarm optimization for enhanced multiresolution GPR microwave imaging. *IEEE Trans. Geosci. Remote Sens.* 55 (3): 1305–1317.

29 Caorsi, S., Donelli, M., Lommi, A., and Massa, A. (2004). Location and imaging of two-dimensional scatterers by using a particle swarm algorithm. *J. Electromagn. Waves Appl.* 18 (4): 481–494.

30 Caorsi, S., Massa, A., Pastorino, M. et al. (2003). Detection of buried inhomogeneous elliptic cylinders by a memetic algorithm. *IEEE Trans. Antennas Propag.* 51 (10): 2878–2884.

31 Krizhevsky, A., Sutskever, I., and Hinton, G.E. (2012). ImageNet classification with deep convolutional neural networks. In: *Proceedings of the 25th International Conference on Neural Information Processing Systems, Lake Tahoe Nevada, USA*, 1097–1105.

32 Russakovsky, O. et al. (2015). ImageNet large scale visual recognition challenge. *Int. J. Comput. Vis.* 115 (3): 211–252.

33 Girshick, R., Donahue, J., Darrell, T., and Malik, J. (2014). Rich feature hierarchies for accurate object detection and semantic segmentation. In: *Proceedings of the IEEE Conference Computing Vision Pattern Recognition, Columbus, OH, USA*, 580–587.

34 Ronneberger, O., Fischer, P., and Brox, T. (2015). U-net: Convolutional networks for biomedical image segmentation. In: *Proceedings of the18th International Conference on Medical Image Computing and Computer-Assistance Intervention, Munich, Germany*, 234–241.

35 Eigen, D., Puhrsch, C., and Fergus, R. (2014). Depth map prediction from a single image using a multi-scale deep network. In: *Proceedings of the Advanced Neural Information Processing Systems, Montreal, Canada2374*, 2366.

36 Caorsi, S. and Gamba, P. (1999). Electromagnetic detection of dielectric cylin ders by a neural network approach. *IEEE Trans. Geosci. Remote Sens.* 37 (2): 820–827.

37 Rekanos, I.T. (2002). Neural-network-based inverse-scattering technique for online microwave medical imaging. *IEEE Trans. Magn.* 38 (2): 1061–1064.

38 Guo, R., Song, X., Li, M. et al. (2019). Supervised descent learning technique for 2-D microwave imaging. *IEEE Trans. Antennas Propag.* 67 (5): 3550–3554.

39 Li, L., Wang, L.G., Teixeira, F.L. et al. (2019). DeepNIS: Deep neural network for nonlinear electromagnetic inverse scattering. *IEEE Trans. Antennas Propag.* 67 (3): 1819–1825.

40 Wei, Z. and Chen, X. (2019). Deep-learning schemes for full-wave nonlinear inverse scattering problems. *IEEE Trans. Geosci. Remote Sens.* 57 (4): 1849–1860.

41 Wei, Z. and Chen, X. (2019). Physics-inspired convolutional neural network for solving full-wave inverse scattering problems. *IEEE Trans. Antennas Propag.* 67 (9): 6138–6148.

42 Xiao, J., Li, J., Chen, Y. et al. (2020). Fast electromagnetic inversion of inhomogeneous scatterers embedded in layered media by Born approximation and 3-D U-Net. *IEEE Geosci. Remote Sens. Lett.* 17 (10): 1677–1681.

43 Yao, H.M., Sha, W.E.I., and Jiang, L. (2019). Two-step enhanced deep learning approach for electromagnetic inverse scattering problems. *IEEE Antennas Wireless Propag. Lett.* 18 (11): 2254–2258.

44 Huang, G.-B., Zhu, Q.-Y., and Siew, C.-K. (2006). Extreme learning machine: Theory and applications. *Neurocomputing* 70 (1–3): 489–501.

45 Li, M.-B., Huang, G.-B., Saratchandran, P., and Sundararajan, N. (2005). Fully complex extreme learning machine. *Neurocomputing* 68: 306–314.

46 Huang, G.-B., Zhu, Q.-Y., and Siew, C.-K. (2004). Extreme learning machine: a new learning scheme of feedforward neural networks. In: *Proceedings of the International Joint Conference on Neural Networks (IJCNN2004), Budapest, Hungary*, 25–29 July, 2004, Budapest, Hungary, 985–990.

47 Tamura, S. and Tateishi, M. (1997). Capabilities of a four-layered feedforward neural network: four layers versus three. *IEEE Trans. Neural Netw.* 8 (2): 251–255.

48 Huang, G.-B. (2003). Learning capability and storage capacity of two-hidden-layer feedforward networks. *IEEE Trans. Neural Netw.* 14 (2): 274–281.

49 Hecht-Nielsen, R. (1987). Kolmogorov's mapping neural network existence theorem. In: *Proceedings of the IEEE Conference on Neural Networks, New York, NY, USA*, New York, NY, USA, 11–13.

50 Arena, P., Fortuna, L., Re, R., and Xibilia, M.G. (1995). Multilayer perceptrons to approximate complex valued functions. *Int. J. Neural Syst.* 6 (4): 435–446.

51 Huang, G.-B., Li, M.-B., Chen, L., and Siew, C.-K. (2008). Incremental extreme learning machine with fully complex hidden nodes. *Neurocomputing* 71: 576–583.

52 Ding, S., Su, C., and Yu, J. (2011). An optimizing BP neural network algorithm based on genetic algorithm. *Artif. Intell. Rev.* 36 (2): 153–162.

53 Jin, W., Zhao, J.L., Luo, S.W., and Zhen, H. (2000). The improvements of BP neural network learning algorithm. In: *International Conference on Signal Processing, Beijing, China*, vol. 3, 1647–1649.

54 Hamilton, S.J. and Hauptmann, A. (2018). Deep D-Bar: Real-time electrical impedance tomography imaging with deep neural networks. *IEEE Trans. Med. Imag.* 37 (10): 2367–2377.

55 Lan, T., Liu, N., Liu, Y. et al. (2019). 2-D electromagnetic scattering and inverse scattering from magnetodielectric objects based on integral equation method. *IEEE Trans. Antennas Propag.* 67 (2): 1346–1351.

56 Geffrin, J.-M., Sabouroux, P., and Eyraud, C. (2005). Free space experimental scattering database continuation: Experimental set-up and measurement precision. *Inverse Problems* 21 (6): S117.

57 Cui, T.J., Chew, W.C., Yin, X.X., and Hong, W. (2004). Study of resolution and super resolution in electromagnetic imaging for half-space problems. *IEEE Trans. Antennas Propag.* 52 (6): 1398–1411.

58 Geffrin, J.M. and Sabouroux, P. (2009). Continuing with the Fresnel database: experimental setup and improvements in 3D scattering measurements. *Inverse Probl.* 25 (2): 1–18.
59 Yu, C., Yuan, M., and Liu, Q.H. (2009). Reconstruction of 3D objects from multi-frequency experimental data with a fast DBIM-BCGS method. *Inverse Probl.* 25 (2): 19–24.
60 Makarov, S.N. et al. (2017). Virtual human models for electromagnetic studies and their applications. *IEEE Rev. Biomed. Eng.* 10: 95–121.
61 Gabriel, C. (1996). *Compilation of the Dielectric Properties of Body Tissues at RF and Microwave Frequencies*. San Antonio, TX, USA: Brooks Air Force Base.

16

Radar Target Classification Using Deep Learning

Youngwook Kim

Electronic Engineering, Sogang University, Seoul, South Korea

16.1 Introduction

In operating radar, detecting and tracking a target is fundamental. That said, one of the critical advanced functions of radar is to recognize the type of a detected target or to classify the state/condition of the target. A technique for classifying an unknown target is called Non-Cooperative Target Recognition (NCTR) or Automatic Target Recognition (ATR) [1, 2]. NCTR is used to classify aerial targets from the ground or using airborne platform radar. In contrast, ATR means to classify ground targets from an airborne or satellite platform. Target classification results from analyzing the characteristics of the radar signal reflected from the target. The development of high-resolution radar has made it feasible to obtain radar data with improved resolution on the 1D range profile or on a 2D radar image such as SAR, range-Doppler diagram, and spectrograms. Moreover, high-resolution radar can further increase the classification probability. The classification of radar signals is used to recognize targets and provides the capability to determine the status of a target. In particular, when the target is a non-rigid body, the movement of parts of the target provides information pertaining to the target's motion. For instance, when radar illuminates human or animal movements, it can capture small as well as large motions of the subject, supporting many applications in defense, search and rescue, and health care.

In the past, NCTR or ATR relied on the experience of the skilled operator. However, the development of machine learning algorithms has sparked continuing research on automatic target feature extractions and target identification using the learned decision boundary. In particular, significant improvements in the performance of such machine learning algorithms as deep learning have attracted attention to trials applying the latest deep learning technology to radar [3, 4]. Machine learning, as the name suggests, refers to technology that trains a machine to replace human-defined tasks with software algorithms. Here, the machine means a mathematical model that has a high degree of freedom. Learning or training is intended to obtain the parameters of the mathematical model. A hyperparameter, which determines the structure and the degree of freedom of the mathematical model, is determined before training. In its current state, the performance of machine learning has exceeded the level of human classification capability, and many tasks are processed more efficiently than humans can manage. Furthermore, it has become possible to produce similar data by imitating existing data.

This section defines terms related to target classification, recognition, and identification. According to NATO AAP-6 Glossary of Terms and Definitions [5], classification refers to giving

Advances in Electromagnetics Empowered by Artificial Intelligence and Deep Learning, First Edition.
Edited by Sawyer D. Campbell and Douglas H. Werner.

the target a meta-class, while recognition involves giving the target a class, and identification means giving the target a sub-class. From a machine learning point of view, although these tasks—classification, recognition, and identification—represent the same kind of work, the terms are differentiated according to their classification purposes. Because the technical approaches are similar, the term in this chapter is unified as "classification."

This chapter describes the applications of deep learning to radar. Because radar works regardless of weather and light conditions, the use of this technology can be expanded to many fields. First, we will look into the topic of classifying the state of the target from micro-Doppler signatures. Examples of relevant studies include human activity classification, hand gesture classification, and drone detection. Classification using deep learning is well-known for working very successfully on two-dimensional (2D) images. Since a micro-Doppler signature is expressed as a 2D image, applying efficient deep convolutional neural networks (DCNN) for the purpose of micro-Doppler signature classification is a logical approach. Second, we will discuss synthetic aperture radar (SAR) image classification. SAR imaging is a technology that creates top-view images of terrain while scanning the ground using radar mounted on an airplane or artificial satellite. Detecting the movement of vehicles or of ships at sea in SAR images is a vital matter in the military operations and in border security. Since SAR is also a 2D image, DCNN can be expected to support target classification. Third, demand is increasing for solution that will classify targets in front of automobiles. Collision avoidance or autonomous driving of vehicles necessitates classifying the targets that automobile radar has detected on the road. Lastly, this chapter addresses the idiosyncrasies involved in using deep learning for radar. Although radar data is simpler than optical images from a camera, the number of data samples available is insufficient because of the high monetary and time costs required for radar measurement. Hence, the chapter will describe how the data deficiency problem can be addressed in terms of transfer learning and data augmentation. In addition, we will discuss continual learning, which can be useful for updating previously trained neural nets when a new type of data comes in.

16.2 Micro-Doppler Signature Classification

Researchers have extensively studied micro-Doppler signatures generated by the micro-motion of a target to identify the target or target motion detected by the radar. Even though the characteristics of the target (e.g. speed, length, range profile, and RCS) offer basic information, distinguishing a target using only these characteristics is not a simple matter. On the other hand, observing the micro-Doppler signatures generated by rotation, vibration, or movement of a non-rigid body as the target is moving can provide additional information about the target and its state.

Micro-Doppler signatures produced from micro-motion dynamics, such as mechanical vibrations or rotations, can be mathematically described. When a sinusoidal wave with a phase of zero is transmitted, the received signal from a target is modeled as follows [6]:

$$s(t) = A \cdot \exp(j(2\pi f_0 \cdot t + \phi(t))) \tag{16.1}$$

where A is the magnitude, f_0 is the carrier frequency of the transmitted signal, and $\phi(t)$ is the time-varying phase.

When a vibrating scatterer has an oscillation frequency of ω_γ and β is the vibration's magnitude, the time-varying phase becomes

$$\varphi(t) = \beta \cdot \sin(\omega_\gamma t) \tag{16.2}$$

Then, the received signal $s(t)$ becomes

$$s(t) = A \cdot \exp(j(2\pi f_o \cdot t + \beta \sin(\omega_\gamma t))) \tag{16.3}$$

Applying the Fourier transform, $s(t)$ can be written in terms of frequency components as

$$s(t) = A \sum_{n=-\infty}^{\infty} J_n(\beta) \cdot \exp[j(2\pi f_o + n \cdot w_v)t] \tag{16.4}$$

where $J_n(\beta)$ is the Bessel function. Observably, the vibration-induced micro-Doppler signature has harmonics of the vibrating frequency of ω_γ. Because the vibration feature is represented as modulated terms to the main body of a target, the radar return produced by the target body needs to be distinguished from that produced by its vibrating structure.

For a rotating scatterer, which illustrates another common micro-motion, the phase can be described as

$$\phi = \beta \cdot \sin(\Omega \cdot t + \theta_0) \tag{16.5}$$

where Ω is the rotation frequency and θ_0 is the initial phase. Therefore, the received signal becomes

$$s(t) = A \cdot \exp[j(2\pi f_o \cdot t + \beta \cdot \sin(\Omega \cdot t + \theta_0))] \tag{16.6}$$

Using the Fourier transform, the received signal becomes:

$$s(t) = A \cdot \sum_{k=0}^{N-1} J_k(\beta) \cdot \exp[j(2\pi f_o \cdot t + \beta \cdot \sin(\Omega \cdot t + k \cdot 2\pi/N))] \tag{16.7}$$

From the equation, it can be observed that the target rotation modulates the main frequency in an oscillatory manner. These micro-Doppler signatures are substantial characteristics in radar returns; thus, they can be further exploited for target detection and classification.

Since the micro-Doppler signature is a frequency-modulated component that changes with time, it is necessary to apply joint-time frequency analysis (JTFA) to identify its signatures. Using JTFA, the micro-Doppler signatures can be visualized as 2D monochrome images. One of the most popular JTFA approaches is the short-time Fourier transform (STFT), which performs a Fourier transform within a time window [6]. Sliding the time window allows the change in the frequency component over time to be obtained. Better resolution micro-Doppler images can be created by exploiting high-order methods, such as the Wigner–Ville distribution [7].

When the micro-Doppler signature is visualized as a two-dimensional image, the target can be classified through the distinctive features shown in the spectrogram, including bandwidth, periodicity, offset of Doppler frequency, power variance, and so on, which are human-determined [8]. However, when using the hand-crafted feature, a person's subjective judgment is involved. Furthermore, there is a chance that the performance of using hand-crafted features can be poor if the human fails to recognize crucial features that are hidden. To overcome this bottleneck, automatic feature extraction using DCNN, which is already widely used in image recognition, is being studied in conjunction with identifying a radar image.

The structure of a convolutional neural network (CNN) repeatedly combines a convolution filter and pooling, forming multiple layers, in order to extract features from images. In the last layers, the fully connected layer is connected to perform classification functionality. The structure of the CNN, which is hierarchical, is inspired by the human brain's visual cortex and its ability to recognize objects efficiently. CNN is one of the most successful deep learning structures for mapping between a 2D input data sample and the corresponding label provided by human annotators. A CNN structure that has a large number of layers, resulting in a deep layer, is called

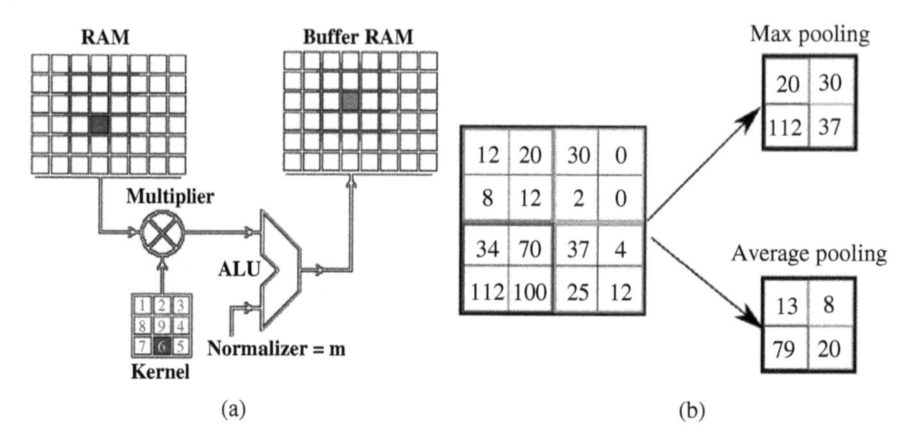

Figure 16.1 (a) Convolution filter, (b) Pooling operation. Source: Adapted from Kim and Moon [9].

DCNN. The convolution filter extracts the features of the image through the dot product with the partial image, as shown in Figure 16.1a. In the pooling layer, the data dimension is reduced for generalization purposes, as shown in Figure 16.1b. In order to prevent overfitting, a method, such as a dropout layer, can be used. Because the CNN has been explained many times in the literature [10], we will not describe it here in more detail.

In general, the number and size of convolution filters and the number of layers are determined empirically. The values and characteristics of the filters, which define the feature extraction, are found through the training of the given data. In the training process, approximately 80% of the data is used for training, while the remaining 20% of the data is used for verification and testing. Multiple recent studies have been conducted to classify radar targets based on micro-Doppler signatures using DCNN. Typical examples include investigations of such problems as recognizing human behaviors, hand gestures, animals, and drones.

16.2.1 Human Motion Classification

The first application of DCNN for recognizing micro-Doppler signatures entailed the problem of detecting and classifying human motion measured by radar [9]. Radar that can recognize the movement of a person with through-the-object capability makes it possible to analyze the actions of people in a building, as well as rescue people in a situation where visibility is not assured, as in the case of a fire. In addition, intrusion monitoring systems can be used in everyday life. An example of an application in the medical field is fall detection. In [9], the researchers investigated the feasibility of classifying human motion using DCNN. Seven human behaviors measured by 2.4 GHz continuous wave radar showed different micro-Doppler characteristics, an example of which is shown in Figure 16.2.

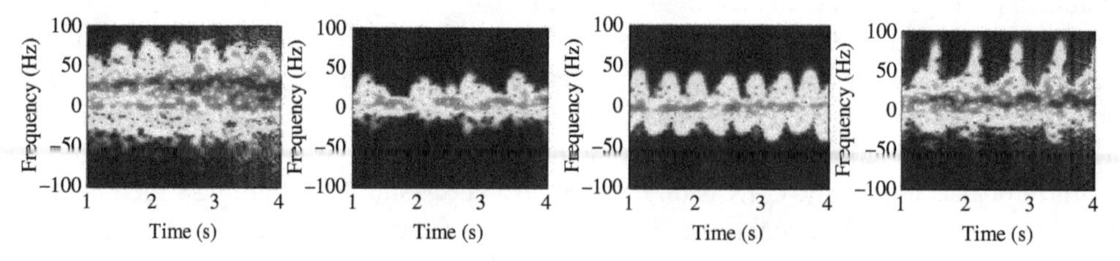

Figure 16.2 Examples of different micro-Doppler signatures. They are running, crawling, boxing still, and boxing forward from left to right [9].

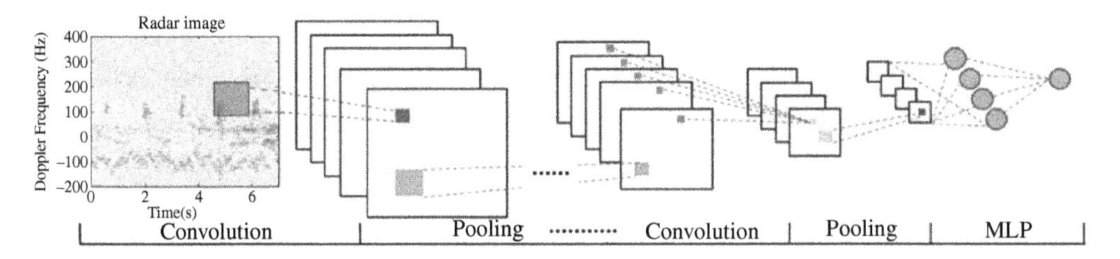

Figure 16.3 DCNN structure for micro-Doppler signature classification. Source: Kim and Moon [9].

A DCNN was designed for this problem. The study data included a total of 1008 images. Figure 16.3 illustrates the designed DCNN.

The number of convolution filters in each layer, along with the size of the convolution filter and the number of hidden nodes in the fully connected layer, was empirically optimized through four-fold cross-validation. The best model identified had three convolution layers, each with 5×5 20 filters, and two fully connected layers, the first of which featured 500 hidden nodes. A rectified linear unit (ReLU) served as the activation function [11]. The training process employed the stochastic gradient descent (SGD) method. The learning rate was 0.001, and the batch size was 84. Lastly, the average accuracy of classification was 90.9%, which is comparable to that found in [8]. While the micro-Doppler features appearing in the spectrogram were manually extracted and human activities could be classified using SVM in [8], the DCNN method had the advantage of automating the entire process without the subjective feature extraction process. The results demonstrated DCNN's potential for micro-Doppler-based target classification. Figure 16.4 presents the test-error curves with varying time window sizes in the spectrogram.

Even though feature extraction using DCNN is the key process in using deep neural networks for micro-Doppler classification, the effectiveness decreases when the data samples are insufficient. While convolution filters are commonly trained based on the data, an auto-encoder (AE) was suggested in [12] to construct the feature extraction network. Specifically, an AE consists of a feed-forward neural network that is intended to reconstruct the same input at the output. In other words, the AE tries to approximate $f(x) \approx x$ for a given input vector of x. Notably, when the number of available training samples was small, an unsupervised pre-training algorithm was highly effective in initializing the weights and biases of AE [13]. The AE was designed to offer unsupervised pre-training by first encoding the input and then decoding it to make the output the same as the input. This scheme facilitated extracting essential features from the input and reconstructing

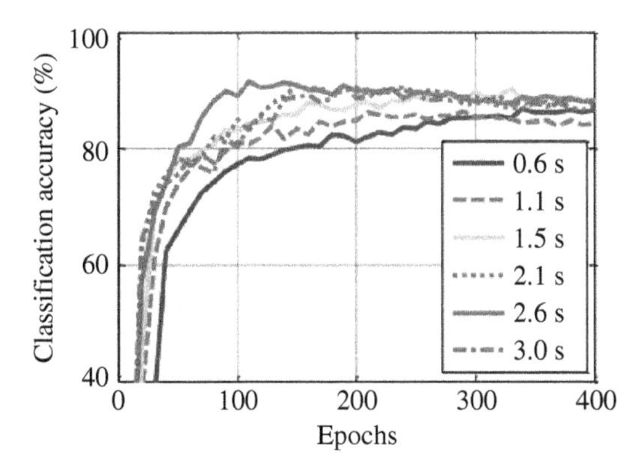

Figure 16.4 Test error curves with varying time windows in spectrogram. Source: Kim and Moon [9].

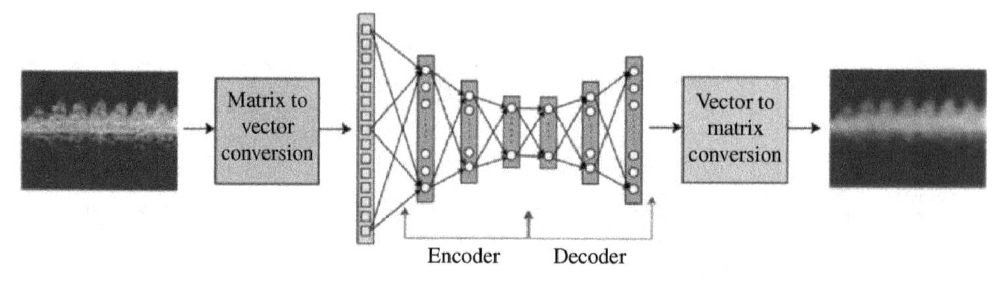

Figure 16.5 Structure of three-layer auto-encoder. Source: Seyfioğlu et al. [12].

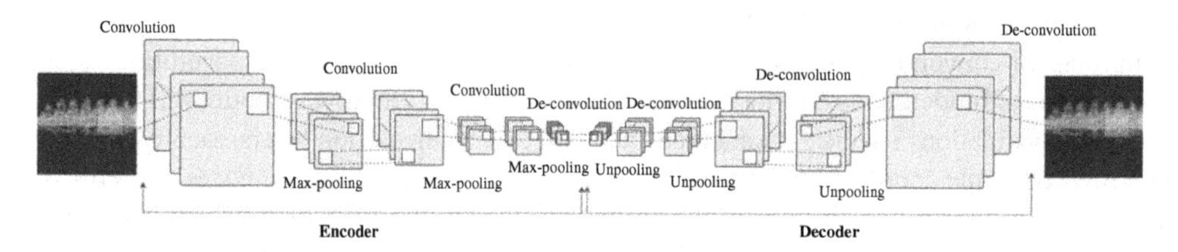

Figure 16.6 The structure of CAE. Source: Seyfioğlu et al. [12].

the input again using the extracted features, which is similar to a compression/decompression algorithm. Even though this process could lose information, the output could be approximated within a bounded error, as seen in Figure 16.5. In this case, the trained AE served as a feature extractor for the input images.

In [11], a convolutional autoencoder (CAE) was proposed for classifying micro-Doppler signatures. The structure consisted of convolutional filters and pooling layers, followed by unpooling layers and deconvolutional filters to reconstruct the original micro-Doppler signatures. After training the proposed convolutional AE, the decoder part was replaced with fully connected layers for classification purposes (Figure 16.6).

The effectiveness of CAE was verified by testing AE and SVM classifiers with ten-fold cross-validation for a dataset comprising 12 human motions, measured at 4 GHz using continuous-wave radar. Each deep learning structure was trained for 280 epochs. The validation data comprised 20% of the training samples. Figure 16.7 displays the accuracy after each epoch for the validation set. The methods that used unsupervised pre-training, such as AE and CAE, yielded high validation accuracies of 94.2% for discriminating 12 human activity classes. Because the activities involved aided and unaided motions, this outcome reveals the potential for radar-based health care systems.

In contrast to the above studies, which were based on monostatic radar, [14] examined the possibility of improving the accuracy of human activity classification by exploiting multistatic micro-Doppler signatures. The micro-Doppler signature is a function of the aspect angle. When the aspect angle approaches 90°, the micro-Doppler characteristic drastically changes. Moreover, the classification rate drops when the classifier is trained for the radar data with an aspect angle of 0 or 180°. In order to overcome this effect, a study was conducted to determine the movement of the human subject using radar images from various aspect angles through multistatic radar. In this study, data for a person walking were measured by FMCW radar operating at 2.4 GHz, using three linearly located radars located 50 m apart. At this time, data consisted of unarmed person and armed person (with metallic pole). A total of 200 data samples were created. The data measured by each radar were processed in parallel through three identical DCNN structures, and the final

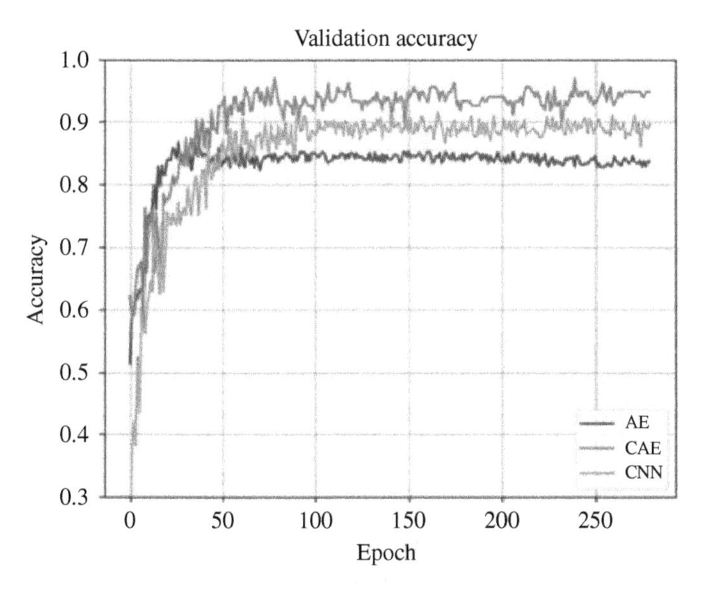

Figure 16.7 The result of AE, CNN, and CAE. Source: Seyfioğlu et al. [12].

decision was fused through binary voting (VMo-DCNN) or feature combination (Mul-DCNN). Notably, each DCNN used a VGG-f network. Because the VGG-f network was developed for color image processing, three inputs for RGB were necessary. However, the spectrogram with micro-Doppler information was a black-and-white image and had no RGB values, necessitating the inputting of more images. The research suggested the use of different temporal resolution images, which was inspired by multiresolution analysis, such as Wavelet. According to the results, the proposed Mul-DCNN structure outperformed VMo-DCNN in terms of gait classification and personal recognition, with an average of 99.63% (Figure 16.8 and Table 16.1).

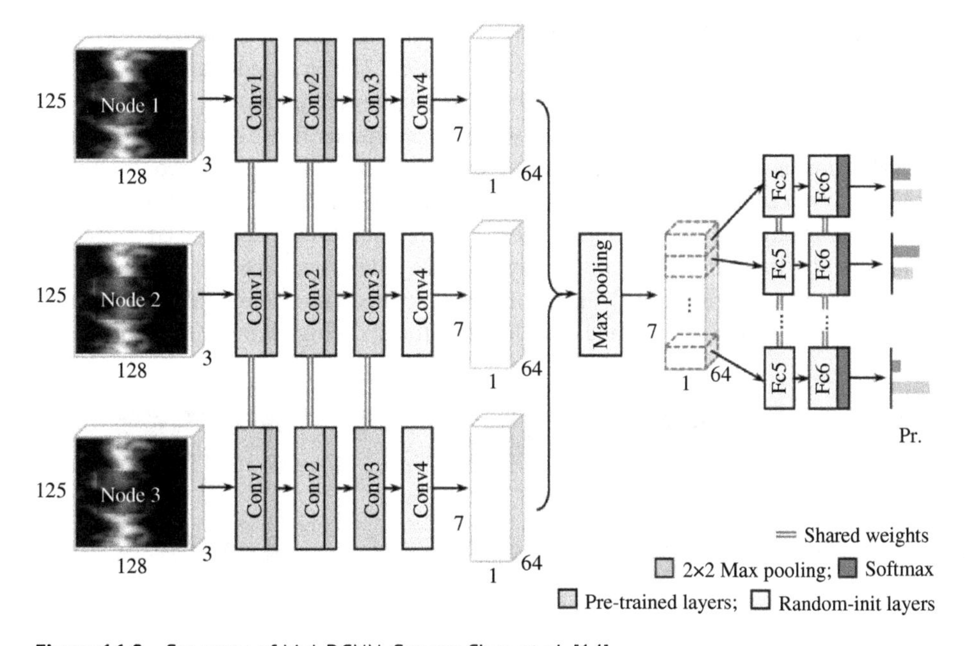

Figure 16.8 Structure of Mul-DCNN. Source: Chen et al. [14].

Table 16.1 The gait classification result.

		Rx node 1	Rx node 2	Rx node 3	Vmo-DCNN	Mul-DCNN
20% training	Minimum	98.75	95.62	94.37	98.12	98.75
	Maximum	100	98.75	100.00	100.00	100.00
	Average	99.50	97.17	98.00	99.33	99.63

Source: Chen et al. [14].

16.2.2 Human Hand Gesture Classification

Hand gesture recognition is another significant application of radar. Small radars that can recognize hand gestures play a significant role as a modality for a computer–human interface, biomechanical research, computer gaming, and defense. The use of radar has advantages in that it operates regardless of light conditions and without causing any privacy issues compared to optical cameras. In [15], the seven hand movements were measured with a Doppler radar operating at 5.8 GHz. During measurement, the radar was located 10 cm away from the hand, and 10 hand gestures were performed within the main lobe of its antenna. The gestures include (i) swiping from left to right, (ii) swiping from right to left, (iii) swiping from up to down, (iv) swiping from down to up, (v) rotating clockwise, (vi) rotating counterclockwise, (vii) pushing, (viii) holding, (ix) double pushing, and (x) double holding. STFT with a 256 ms time window was applied, yielding the spectrograms shown in Figure 16.9. Each hand gesture exhibits slightly different features in the spectrogram. Each gesture was measured 50 times, generating a total of 500 data samples. The spectrograms were cropped with a 2-seconds time window in generating the training dataset. Each cropped spectrogram was resized to 60×60. Regarding the DCNN structure, three layers having five, four, and two convolutional filters were employed, and each convolutional filter's size was 5×5. Pooling layers had a 2 : 1 reduction ratio. The hyperparameters, including the number of layers and filters, were empirically determined to maximize the accuracy. The number of epochs was 90 in the training process, and five-fold validation was used. The recorded accuracy was over 93%, an indication of the potential for using radar for hand gesture recognition.

A similar study was conducted to investigate dynamic continuous hand gesture recognition [16]. In practical terms, a system that would detect and recognize hand gestures continuously using unsegmented input streams was proposed to avoid any noticeable lag between performing a gesture and its classification. The diverse hand gesture data, measured using a 24 GHz FMCW radar, were classified by three methods: a 3D convolutional neural network (3D-CNN), a long short-term memory (LSTM) network [17], and a proposed connectionist temporal classification (CTC) algorithm in [18]. This study evaluated eight hand gestures: sliding a hand from right to left,

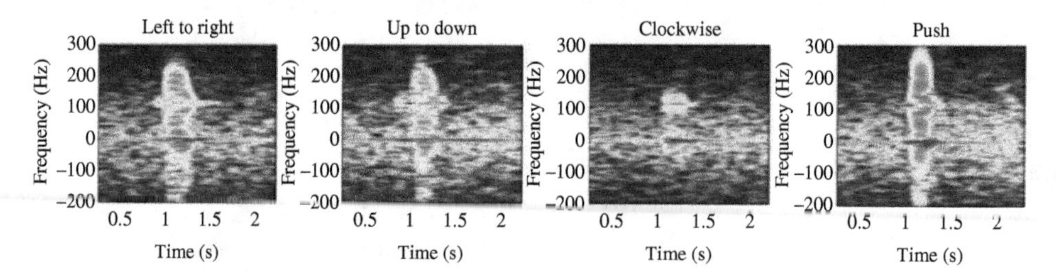

Figure 16.9 Examples of micro-Doppler signatures from diverse hand gestures [15].

sliding a hand from left to right, pulling, pushing, knocking, moving a hand up and down, waving a hand, and patting. The gestures involved moving the entire arm and hand, not individual fingers. Each gesture was measured for about 1.5 seconds and was performed for 100 repetitions with the hand at a distance of 1.5 m from the radar. A total of 3200 samples were collected. Spectrograms with time were stacked, and the 3D data served as the input data set.

This study was based on the ability of 3D-DCNN to classify 3D radar data. While 2D convolutional filters have been extensively used in many fields, these filters are not limited by the dimensions of the data. The study applied 3D convolutional filters to the 3D dataset, with the filter sliding in three directions to extract low-level feature representations. The output of the filter was another 3D matrix, which was a cube or cuboid. In comparison, deep recurrent neural networks (DRNNs) have shown a great capability for processing sequential data. The physics behind DRNNs is to recognize sequential information. A traditional multilayer perceptron neural network assumes that all inputs are instantaneous. However, many tasks require the inclusion of temporal information. A DRNN is intended to handle temporal data by designing a memory called a hidden state. This hidden state processes information from the previous memory as well as new inputs. Through this structure, a DRNN uses knowledge that exists in long sequences. However, this structure has a limitation in that temporal lookback length is practically constrained to a few steps because of the vanishing gradient problem. Hence, LSTM, a type of DRNN, was suggested to overcome this issue [17]. LSTM has a unique feature in the form of memory cells that determine which information to keep in the memory, including old information. Therefore, the output of LSTM is a function of previous memory, the current memory cell, and the current input. The use of the memory cell feature alleviates the vanishing gradient problem by storing the information even over a longer time.

Additionally, the CTC algorithm was suggested to recognize diverse continuous hand gestures without hand gesture pre-segmentation. For the purpose of better performance, training was performed by a CTC algorithm to predict class labels from gestures in unsegmented input streams. The experimental results reveal that the CTC algorithm could achieve a high recognition rate of 96%, higher than other gesture recognition methods (Figure 16.10).

16.2.3 Drone Detection

Applications for drones are drastically increasing, and monitoring them has become a significant issue because of the possible hazards posed by drones in crowded areas such as stadiums, parks, urban streets, or schools. In addition, the potential collision of a drone with a power line, radio transmitting tower, or any other critical facilities that might cause substantial damage to society should receive serious consideration. Avoiding these potential dangers necessitates detecting and tracking drones in advance to be able to control or even destroy them. In order to detect a drone at a far distance, the use of a radar system with a signal that propagates for a long distance with a low attenuation constant is desirable. Drone detection can be a critical application of deep learning in terms of analyzing the received radar signals.

Periodic micro-Doppler characteristics appear from the cyclically rotating blades of the drone. From this information, the number of motors and the characteristics of the drone can be distinguished. Even though the micro-Doppler characteristics of drones are revealing, the periodic signatures of the micro-Dopplers in the spectrogram are not separately distinguished, and the periodicity information can be difficult to utilize. Therefore, one study considered the use of a cadence-velocity diagram (CVD) resulting from the Fourier transform of each row of the spectrogram to take advantage of the periodic characteristics of the micro-Dopplers [19]. By applying

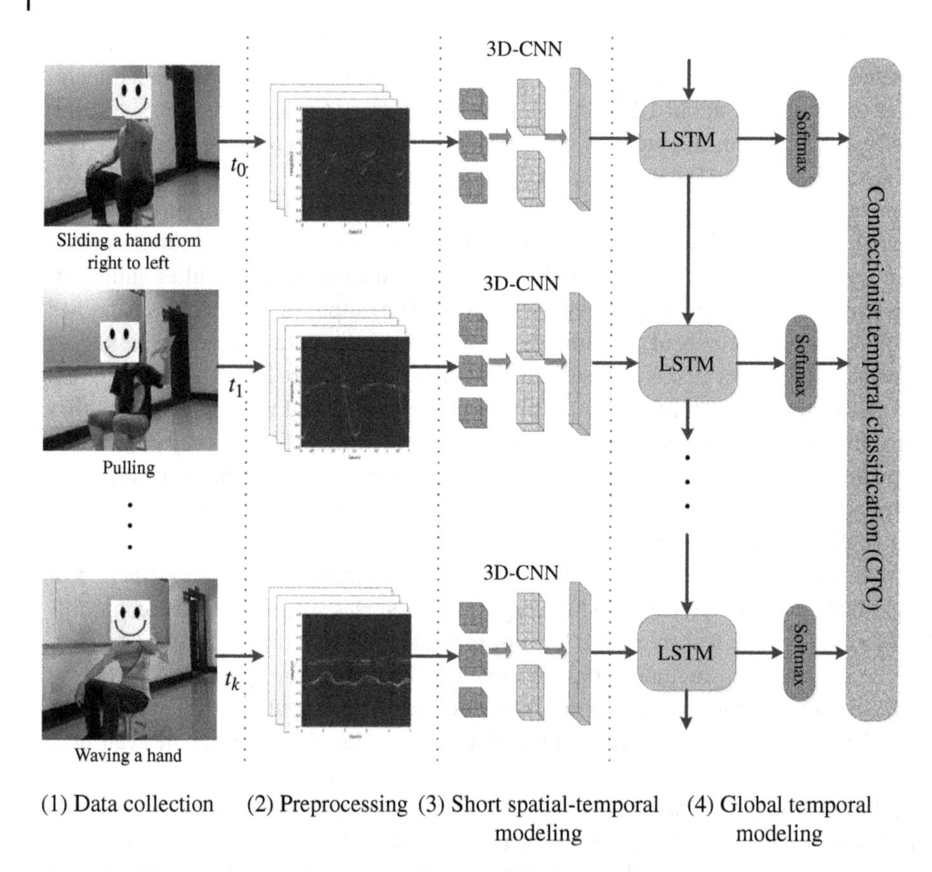

Figure 16.10 Network architecture for dynamic HGR. Source: Zhang et al. [16].

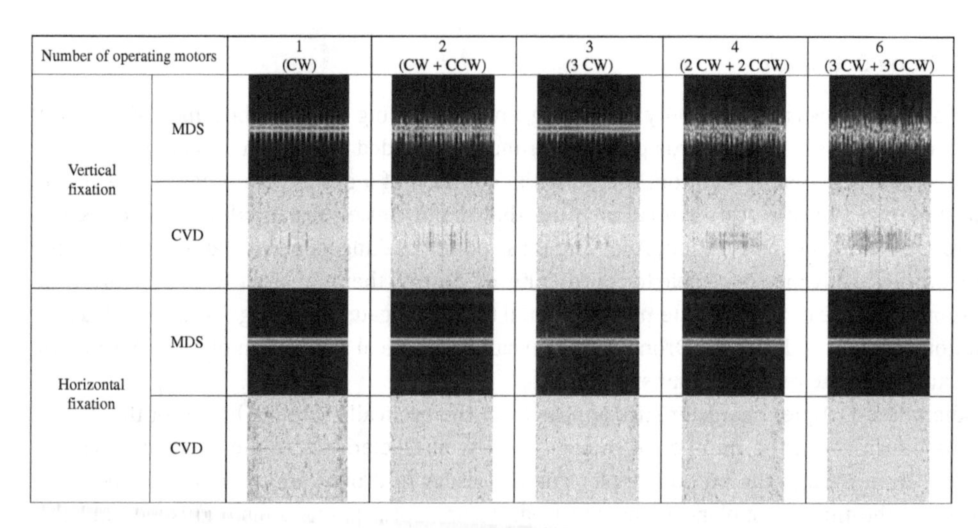

Figure 16.11 Examples of micro-Dopplers and CVD related to the rotor rotation direction. Source: Kim et al. [19].

DCNN to the combined image of the micro-Doppler spectrogram and CVD, the number of rotors could be found. Figure 16.11 provides an example of micro-Doppler signatures and CVD of a drone measured using Ku-band radar. Six types of drones based on the number of rotors were measured at different aspect angles. As a result of using the combination of the two images, the classification accuracy using GoogLeNet was 94.7%. In addition, quad-copter (Inspire 1) and hex-copter (F820) were classified with an accuracy of 100% due to their distinct features.

16.3 SAR Image Classification

Another substantial field of radar target classification is SAR image classification, which involves using SAR technology to visualize topographical and terrain surfaces. Recognizing targets based on SAR is a critical capability for military and geoscientific purposes. ATR using SAR images follows three steps: detection, then discrimination, and, lastly, classification, where the target identification takes place. Unlike an optical image, the target in a SAR image has a low signal-to-noise ratio (SNR); moreover, the shape of the object is distorted, making it generally difficult to identify with human eyes. In the past, a target of interest was identified using template matching. However, recent studies have sought to solve this problem through deep learning. Inspired by successful applications using optical images, the use of deep learning in remote sensing, including object detection in SAR, terrain surface classification, de-speckling [20], and SAR-optical data fusion [21], is now gaining considerable attention. Accordingly, the current study investigates vehicle detection on land and ship detection on the sea in SAR images.

16.3.1 Vehicle Detection

The first study that used deep learning for SAR target recognition can be found in [22], which was published in 2016. DCNN was trained using five types of SAR image patches, including buildings, road, vegetation, beach, and water, measured with TerraSAR-X. A total number of 2224 samples were divided into 1590 samples for training data and 634 for validation data. The proposed DCNN consisted of a total of seven layers, including two convolutional layers and a fully connected layer, and the image size was 100×100. Examples of targets are shown in Figure 16.12. The target was identified with an accuracy of 85.6%, which was an improvement over the result from using conventional feature extraction and SVM.

However, training SAR images with DCNN involves an overfitting problem because the degree of freedom of the DCNN model is, in general, much higher than the number of available SAR images. Thus, Chen et al. sought to overcome this issue by proposing new all-convolutional networks (A-ConvNets) in which only a few weights were connected [23]. The researchers used a publicly released dataset with 10 categories of ground targets, such as armored personnel carrier, tank, rocket launcher, air defense unit, truck, and bulldozer. Targets were measured using an

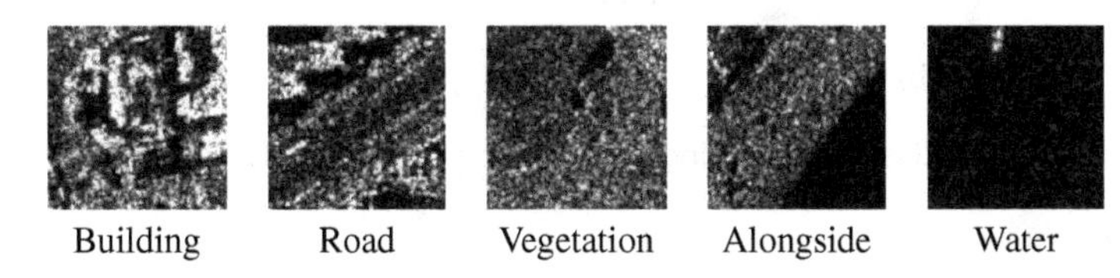

Figure 16.12 Five kinds of SAR Patch images classified by DCNN. Source: Zhao et al. [22].

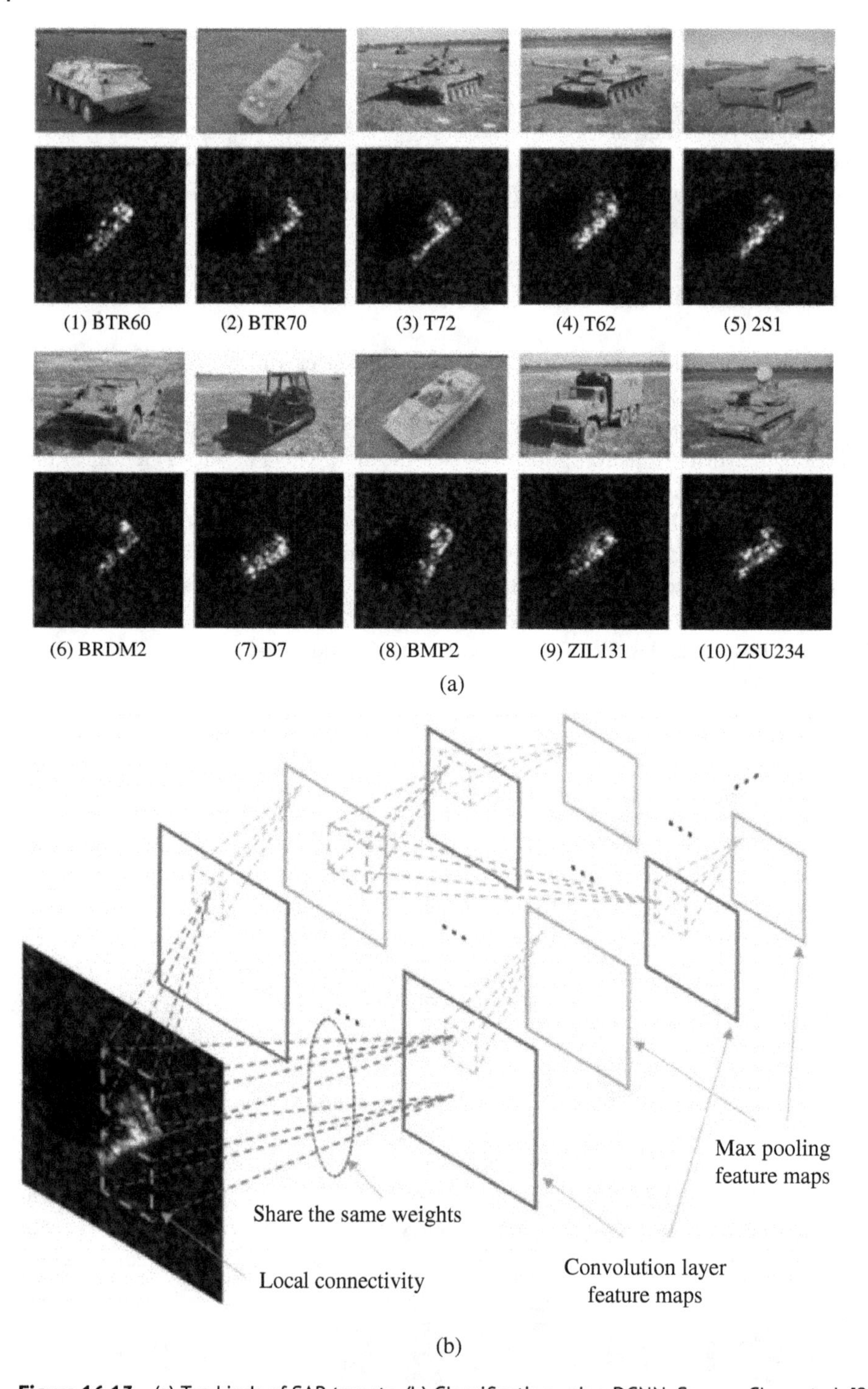

Figure 16.13 (a) Ten kinds of SAR targets, (b) Classification using DCNN. Source: Chen et al. [23].

X-band SAR sensor set to 1-ft resolution spotlight mode with 0° to 360° full aspect coverage. The total number of data samples was 2747. As shown in Figure 16.13, the study findings confirmed that a 99% identification accuracy could be obtained in the moving or stationary target recognition problem using this technique. Another recent study examined identifying a target from a SAR image using a single DCNN responsible for all three stages (as mentioned earlier, these are detection, discrimination, and classification) of SAR image processing [24].

16.3.2 Ship Detection

The problem of ship detection using SAR images can be applied in multiple areas, including port management, maritime rescue, cargo transportation, fishery surveillance, and national defense. Unlike satellite optical imagery, SAR imaging plays a particularly critical role in port management and maritime rescue in stormy areas, as 24/7 and all-weather monitoring services are able to acquire images in overcast areas as well as in clear circumstances. Problems encountered in SAR ship detection methods have been addressed in terms of a constant false alarm rate (CFAR) [25], super-pixel segmentation [26], visual saliency [27], and polarization decomposition [28]. Because these methods entail designing ship features manually, they have not been suitably robust in cases where SAR images have had different resolutions or in variable sea states involving different wind and wave conditions. However, advances in deep learning have allowed the ship detection problem to be converted into an image detection issue. In [29], an investigation was proposed to evaluate a unique SAR ship discrimination technique using a deep neural network classifier. Because the dataset available for the ship detection was relatively small, the use of deep highway networks was suggested to train the nets, which resulted in better performance compared to the conventional machine learning algorithm. A dataset was created using Sentinel-1 and RADARSAT-2 acquisition, resulting in a total of 42 dual and four single polarized images. The dataset was normalized to 21×21 sub-images that included ships (positives), ship-like areas (false positives), and ocean areas (negatives). The 1596 positive examples were identified by expert analysis, while 3192 false positive images were constructed, and 1596 sub-images were selected as negative ocean samples. This study applied a highway network that made use of a gating function. Efficiently training a deep network requires the network is able to transform or bypass signals adaptively. In the highway network, adaptive gating exists within each layer, which enables the transfer of information across multiple layers without losing information. According to the study results, most of the deep highway network configurations demonstrated high classification accuracy. The best performing classifier was a deep highway network with 20 hidden layers, yielding a mean accuracy of 96.67% (Figure 16.14).

Another investigation of ship detection using deep learning can be found in [30], in which the researchers built a large number of ship images cropped with multiresolution SAR. In this study, RetinaNet was adapted to address ship detection in multiresolution Gaofen-3 SAR imagery taken from the Gaofen-3 satellite launched in 2016. Eighty-six scenes at resolutions of 3, 5, 8, and 10 m were considered. The RetinaNet architecture has a backbone network for feature extraction and two subnetworks, one for classification and the other for box regression. RetinaNet uses the concept of transfer learning for the backbone network, such as ResNet [31], VGG-16 [14], or DenseNet [32] for feature extraction. Next, it uses a feature pyramid network (FPN) component to normalize the feature dimension for inputs with various scales. Specifically, FPN uses a pyramidal feature hierarchy of DCNN to represent multiresolution images. After normalization, these pyramidal features are inputted into two subnets for the classification and localization of objects. The experimental

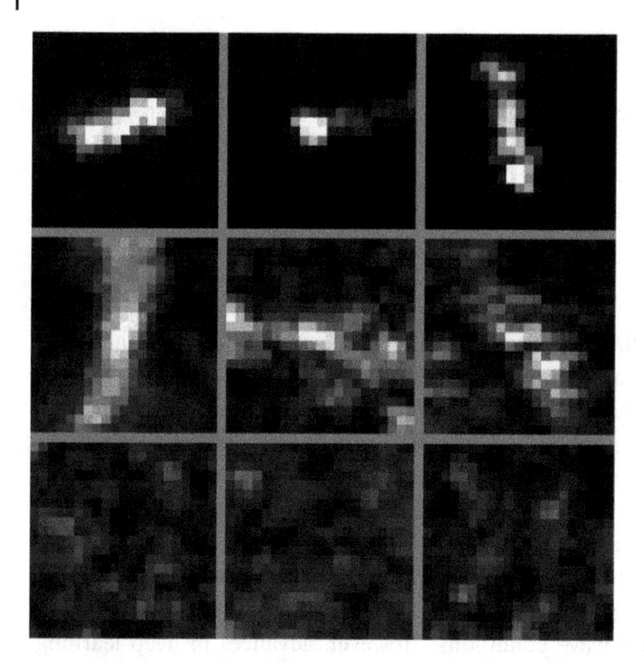

Figure 16.14 Nine SAR images. The first row consists of positives (ships), while the second row has false positives (ship-likes), and the third has negative examples (ocean). The first two columns are from Sentinel-1, and the last column contains images from RADARSAT-2. Source: Schwegmann et al. [29].

results revealed that RetinaNet could accomplish a high detection accuracy of over 97%. Furthermore, RetinaNet achieved the best performance in complex backgrounds in multiresolution SAR image, compared to other classifiers.

16.4 Target Classification in Automotive Radar

Demand has become strong for the development of autonomous vehicles, which are regarded as a new modality of transportation. Autonomous automobiles offer numerous advantages from many perspectives, including the improved safety while minimizing traffic congestion [33]. Even autonomous navigation involves the use of many sensors, and radar is an essential component because of its long-range detection capability in all weather and light conditions. Furthermore, compared to other sensors, such as cameras or LiDAR, radar can measure a target's range and speed with high accuracy at a relatively low cost.

Previous studies on automobile radar target classification have been based on the features in the range-Doppler domain of different targets [34]. However, performance has been shown to drastically degrade if targets are within the same range bin or Doppler bin. In [35], the investigators proposed a method that used RCS to classify targets regardless of target movement. This research was based on the fact that the intrinsic RCS behavior of a target is distinguishable and the statistical features of its RCS contain information for traffic target classification. By employing use of RCS statistics through simulation, a high classification accuracy of over 90% could be achieved for distant targets over a 50 m range for pedestrian detection. Using an RCS image with beam-steering capabilities yielded classification accuracy of more than 99% using an artificial neural network.

Meanwhile, an advanced approach that involved the range-Doppler domain was discussed in [36]. In the area of automotive radar, FMCW radar has been used in most cases owing to its simple hardware and high target detection capability. The most commonly used domain in detecting a target using radar has been the range–Doppler diagram. For the FMCW radar, it is possible to use 2D FFT, after stacking it for each frame when fast chirps are transmitted, to convert the received signal into a range–Doppler diagram in which each target shows different features according to its size and movement characteristics. Therefore, classifying the target using this domain is worth examining. This research addressed target classification in three ways when data were measured using automotive radar operating at 77 GHz.

The AWR1243 radar chip from Texas Instruments was used for the measurement campaign, setting the following parameters: chirp duration 54.34 μs, chirp slot 19.99 MHz/μs, chirps per frame 256, frame period 100 ms, frequency bandwidth 1086.15 MHz, and A/D sampling rate 5 MHz. Even though the chip provided MIMO capability, a single Rx channel was used in the measurement. Targets of interest consisted of humans, cars, cyclists, dogs, and road clutter. Each subject/object was located 3–30 m from the radar. Two to three subjects/objects per class were measured several times for 20 seconds. The researchers selected 1500 high-quality samples per class, making a total of 6000 data samples (Figure 16.15).

The first method involved classifying the target using 2D-CNN on the cropped image in the range–Doppler domain. Because of its powerful image classification capability, 2D-DCNN was employed to classify targets on a single range–Doppler diagram. After five-fold validation to determine a statistically meaningful result, the technique yielded classification accuracy of 95.23%.

In the second method, the researchers cropped the target on the range–Doppler, then made it a 3D matrix to find features that changed with time. The constructed 3D matrix could be classified using 3D CNN. In practice, the 3D convolutional filter has already seen use for event detection in videos, 3D medical images, and other applications; however, this technology has not been much used in conjunction with radar data. A feature of 3D convolutional filters is their ability to extract features in the time domain as well as the range–Doppler domain. With time as one of the axes of the 3D matrix, accuracy was calculated with 5 frames, 10 frames, 15 frames, and 20 frames for

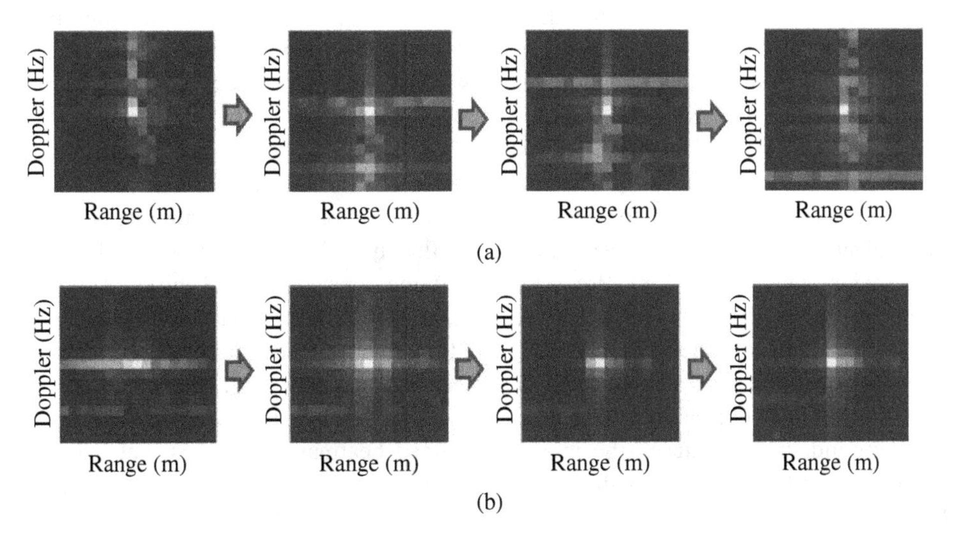

Figure 16.15 Time-varying signatures with a step at every five frames (500 ms) in the range-Doppler diagram: (a) human subject and (b) car. Source: Kim et al. [36].

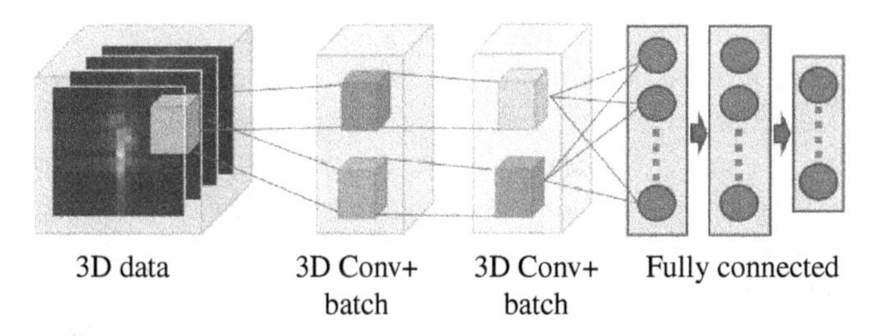

Figure 16.16 Structure of 3D-DCNNs. Source: Kim et al. [36].

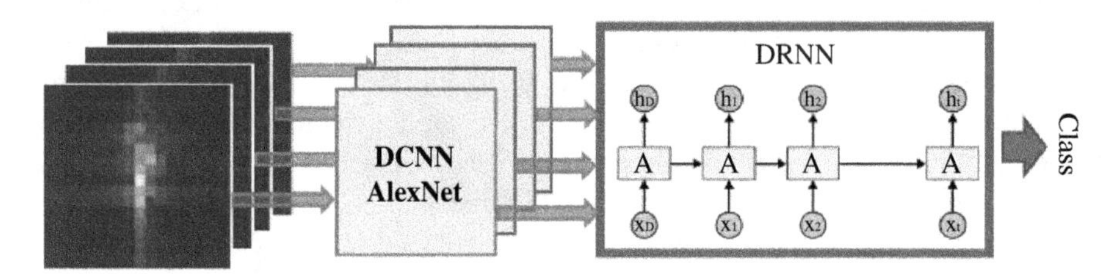

Figure 16.17 DRNN structure [36].

a duration of 100 ms/frame. Classification accuracies for different numbers of frames were calculated, revealing that the use of 20 frames produced 97.77% accuracy, which was better than that yielded by 2D-DCNN (Figure 16.16).

Lastly, in the third method, the investigators first selected a feature with a 2D CNN in order to distinguish features on the range–Doppler that changed with time and then used a recurrent neural network to classify the target using the features of the target over time. In this case, the experimental setup engaged recurrent neural networks to use memory to recognize the time-varying information. The use of DRNN along with DCNN was proposed to classify the 3D matrix in consideration of the fact that using DRNN is more effective than using 3D-DCNN for video classification [37]. In comparison to 3D-DCNN, DRNN was expected to result in higher accuracy. After processing a range–Doppler diagram in the AlexNet, the output was inputted into LSTM, as shown in Figure 16.17. According to the researchers' findings, the accuracy was over 98%, even for only 5 frames, exceeding the results achieved by 3D-DCNN. Moreover, using 20 frames yielded an accuracy of 99.68%.

In contrast, another study investigated the range-angle domain for target classification using TI's mmWave FMCW radar [38]. The researchers used the direction of arrival estimation as well as mechanical rotation to obtain angular information. The measurements were performed in diverse real-world scenarios with targets that included humans, a vehicle, and a drone. The radar operated at 77–81 GHz, and each frame consisted of 128 chirps. At each measurement, heatmaps of the target were saved as data samples. The You Only Look Once (Yolo) model, which comes from the family of convolutional neural networks for real-time fast object detection in camera-generated images, was used as a classifier [39]. The model takes advantage of the fact that objects in optical images have sharp edges that can separate them from the background. In the dataset, the images contained heatmaps of objects that had no distinct boundaries or featured shapes that made the objects more difficult to recognize. The proposed technique achieved an accuracy of 97.6% and 99.6% for classifying the UAV and humans, respectively, and an accuracy of 98.1% for classifying the car using the range-angle domain (Figure 16.18).

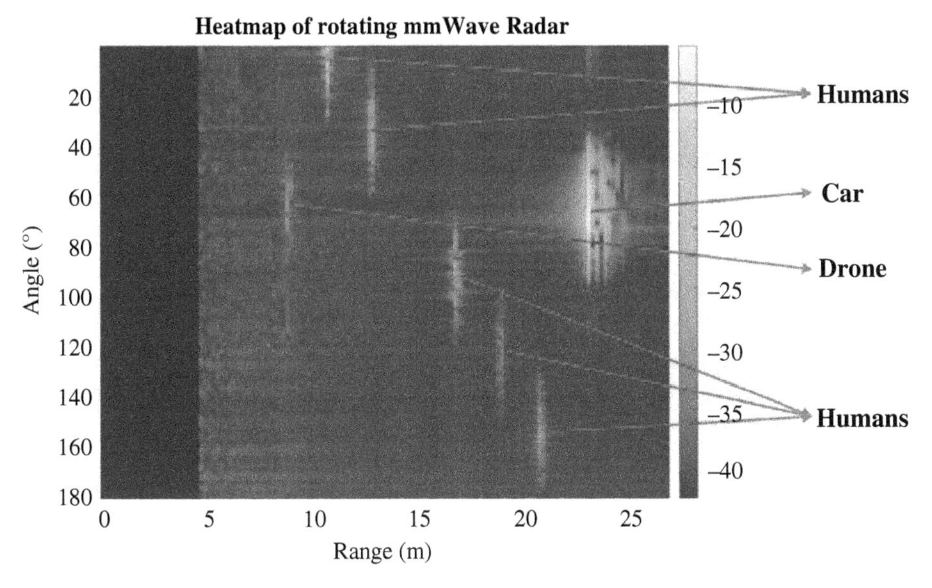

Figure 16.18 Radar target image in the range-angle domain. Source: Gupta et al. [38].

16.5 Advanced Deep Learning Algorithms for Radar Target Classification

This section presents the advanced deep learning algorithms that are useful for radar target classification. As shown in the example in Sections 16.2 and 16.3, using DCNN makes it possible to obtain higher accuracy classification performance for a 2D radar image through automatic feature extraction compared to previous methods. However, radar images have an intrinsic issue—a small number of samples, which differs from conventional optical camera images. In the case of a camera, images can be produced relatively easily, and the acquisition of millions of sufficiently different samples on the Internet is not difficult. However, most radar images are obtained via a radar measurement process that incurs substantial costs in terms of time and money. In addition, the measured radar data are rarely publicized. Therefore, realistically, the number of radar data samples that can be acquired ranges from a few hundred to a few thousand, at most, resulting in a data deficiency issue when using deep learning. Nevertheless, a deep learning model requires a corresponding number of data point to estimate a large number of parameters. Therefore, the small amount of available data imposes limits on exercising the spirit of the deep learning technique. Meanwhile, auto-encoders or all-convolutional networks can take place of deep learning machines when the available dataset is small. However, more fundamental solutions are possible. Three solutions can be offered to overcome this bottleneck. The first is to generate a radar image through simulation. Specifically, micro-Doppler images or SAR images can be generated through electromagnetic full-wave simulation or simplified models [40, 41]. Through simulation, a vast number of images can be formed, which can be used to train a machine learning model. That said, a new CAD model should be prepared to simulate whenever new radar images are required. In the case of simulating micro-Doppler signatures, the target motion should be mathematically modeled, or motion information should be provided; however, such information is occasionally not available.

The second approach concerns using transfer learning [42]. This method borrows a DCNN that has been trained for another application. In this case, the idea is to bring the trained model using optical images and apply it to radar image classification. The third approach entails using generative adversarial networks (GANs) [43] for data augmentation. This type of machine is trained to make a similar radar image based on an actual radar image and can synthesize as many images as desired. The following Sections 16.5.1 and 16.5.2 will examine how to use transfer learning and GANs for radar images, which can be used in general without being specific to the problem, in cases where the simulation method may not be easy to apply to various situations related to the target and environments.

Lastly, when classifying a detected target in actual radar operation, a new data set may be added if the characteristics of the target vary due to changes in the surrounding environment and climate or even in the case when a new class of target is measured. In this case, it is necessary to re-train the machine with the newly measured target characteristics or newly measured data from the new class. For this purpose, the discussion will turn to the application of continuous learning.

16.5.1 Transfer Learning

Transfer learning is a method of reusing successfully trained neural nets through a certain data set and applying the previous learning to a new data set. Effective transfer learning can take place if there is similarity between the two data sets or if the feature extraction method used in one data set also works in the other data set. In transfer learning, the input layer, the output layer, and the last fully connected layer that is connected to the output must be newly designed for a new task while the trained net for feature extraction is being reused. Therefore, re-training takes less learning time because only the last fully connected layer needs to be trained with new data.

In order to improve the performance of the classifier with a small number of radar images, nets trained with optical images can be borrowed. For example, in the annual ImageNet Large Scale Visual Recognition Challenge (ILSVRC), many groups compete for which algorithm has the highest success rate in classifying ImageNet using 1.5 million images of 1000 types. As shown in Figure 16.19, the performance has improved every year; in 2015, ResNet provided a more accurate performance than humans in classifying images. Numerous trained models are available, such as

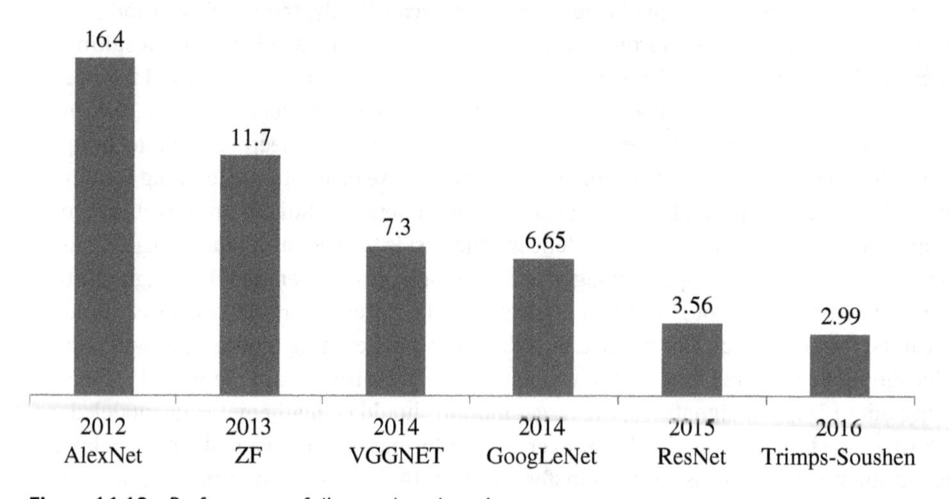

Figure 16.19 Performance of diverse deep learning.

AlexNet, VGG16, ResNet, GoogLeNet, and so on. Those models trained with ImageNet are publicly available. Applying these models, trained using optical camera images, to radar images has aroused much interest. In particular, AlexNet, VGG16, and ResNet have been popularly used to transfer learning for radar images.

The transfer learning technique was first applied to the micro-Doppler signature of human aquatic activities [44]. Because the acquisition of a data set regarding human swimming motion is not convenient, only some hundreds of radar images were available. Doppler radar operating at 7.25 GHz was used to measure human swimming, including such motions as freestyle, backstroke, breaststroke, pulling a boat, and rowing activities. The resulting 625 spectrograms from five human subjects have been used to re-train the fully connected layer after the convolutional neural networks of AlexNet and VGG16. Figure 16.20 displays examples of the output of convolution filters. This illustration demonstrates the different features that were extracted from the already-trained filters, such as shape, texture, and noises. The results of transfer learning reveal the improved performance; specifically, classification accuracy increased from 66.7% to 80.3%. In particular, it can be seen that the performance of VGG16, whose structure is deeper and more complex than that of AlexNet, yields higher accuracy.

A similar result was confirmed in [45]. The use of transfer learning was shown to enhance classification accuracy from 98.51% to 99.09% for 10-class object recognition in SAR images . As this discussion has shown, transfer learning is a potential candidate that can resolve the data deficiency problem in testing radar object classification. The reason why a network trained for optical images can work for radar images can be found in the fact that all images share the same characteristics regardless of the kind of image. For example, the shape of an object is defined by edges, color variation contains information, and geometric relationship matters.

Applying transfer learning to radar images involves two noteworthy idiosyncrasies. First, while optical images have RGB characteristics, radar images, in general, are monochrome in nature. Because the trained nets for ImageNet have three layers in the input for RGB, the same radar data can be inputted to the layers for RGB [44]. Inputting the same data does not take advantage of the three-layer capability. Thus, the use of multiresolution radar images has also been suggested to exploit the advantages of using three layers [14]. Second, not only is the resolution of a radar image less fine, in general, than that of an optical image, but the dimension of the radar image is also small. Therefore, the size of the convolutional filter should be small as well to capture the signatures.

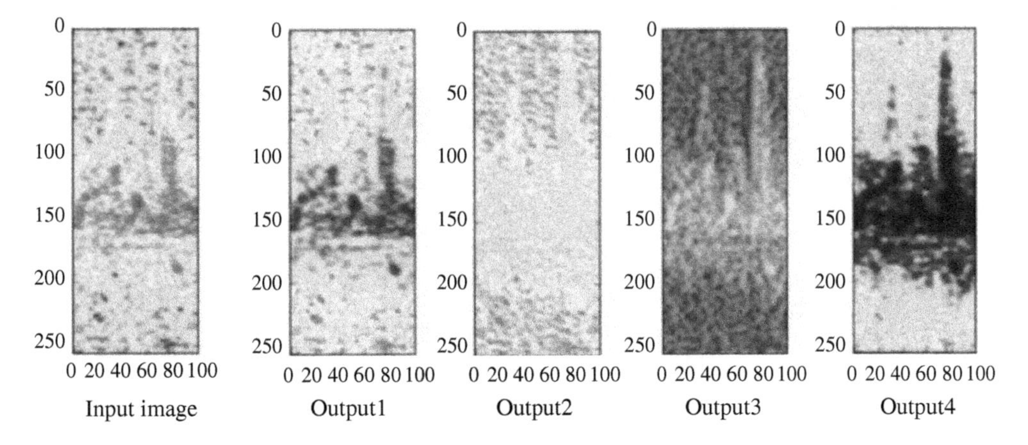

Figure 16.20 Output of convolutional filters of VGG16 [44].

16.5.2 Generative Adversarial Networks

The data deficiency issue under consideration can also be resolved by synthesizing radar data. The GAN model, which was proposed by Dr. Goodfellow in 2014, has been labeled as one of the most innovative models [43]. The purpose of GANs is to train a model that synthesizes data having a similar distribution to that of the original data. Research has shown that GANs can mimic even human creations, such as drawings and poems, as well as human behavior [46]. Data having a specific distribution can be created arbitrarily. If this technology is applied to a radar image, radar image data can be augmented by producing data that are similar to the original radar image.

To this end, GANs are composed of a generative network and a discriminative network and have a structure in which these two nets learn against each other through supervised learning. The generative network plays the role of forming a new image by receiving input from random noise. In contrast, a discriminator is a typical classifier that determines whether the input image is real or synthesized. In GANs, the discriminative network induces the most accurate judgment, while the generative network induces the discriminative network's judgment accuracy to decrease as much as possible. The fact that the image generated by the generative network causes poor classification accuracy in the discriminative network means that the generated synthesized image is analogous to the real image as to the level to which the discriminative network is confused. The general structure of GANs is shown in Figure 16.21.

This structure aims to maximize the following V while learning the discriminative network and the generative network alternately. The higher the V value, the better is the performance of the generative network.

$$V(D, G) = E_{x \sim Pdata(x)}[\log D(x) + E_{z \sim p(z)}[1 - \log D(G(z))] \tag{16.8}$$

Here, x is the sample from the real image, z is the input value of the generative network, $G(z)$ is the output of the generative network, and D is the classification accuracy of the discriminative network. In most cases, learning is performed through hundreds of repetitions, and the training process takes tens of hours or more, depending on the computer performance. However, once the GANs have been trained, it takes little time to generate a new image. Therefore, generating sufficient radar images using GANs can be an effective method to augment a data set to be used in the training process in DCNN. An example of this approach can be found in [47], in which micro-Doppler

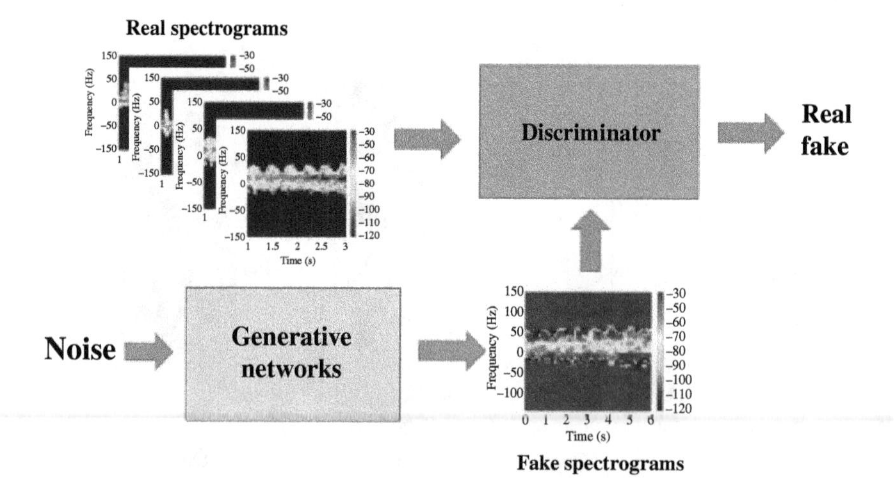

Figure 16.21 Structure of GANs. Source: Alnujaim et al. [47].

Generator neural network
Noise, size = 100
Fully connect
Leaky Rectification Unit
ConvTranspose2d, stride = 2, filters = 256, size = 4 × 4
Batch Normalization
Leaky Rectification Unit
Dropout (0.4)
ConvTranspose2d, stride = 2, filters = 64, size = 4 × 4
Active function

(a)

Discriminator neural network
Image, size = 64 × 64 × 1
Conv2d, stride = 2, filters = 64, size = 4 × 4
Batch Normalization
Leaky Rectification Unit
Dropout (0.4)
Conv2d, stride = 2, filters = 128, size = 4 × 4
Batch Normalization
Leaky Rectification Unit
Dropout (0.4)
Fully connect
Leaky Rectification Unit
Fully connect
Active function

(b)

Figure 16.22 Structure of GAN used in [47], (a) generative network, (b) discriminative network. Source: Source: Alnujaim et al. [47].

signatures from human motion were synthesized by GAN. A GAN was designed for each activity, and 144 data samples (12 humans × 4 iterations × 3 extractions) were used to train each GAN. The trained GANs produced 1,440 data samples for each activity. Figure 16.22 shows the structures used. Including the synthesized images in the training dataset improved the classification accuracy from 90.5% to 97.6%, which was higher than the use of transfer learning for the same data set (Figure 16.23).

Another study attempted to synthesize micro-Doppler signatures from different aspect angles using conditional GANs (cGANs) [48]. While general GANs receive noise as an input, cGANs receive an input vector and produce output based on it. Because micro-Doppler signatures from non-rigid body motions are a function of the radar's aspect angle, a large amount of data from diverse aspect angles is necessary for effective target classification based on machine learning algorithms. Circumventing the high cost of radar measurements, the synthesis of micro-Doppler signatures using deep learning can provide an alternate solution. In [49], training data were constructed by simulation. The micro-Doppler signatures of human motion were simulated from different angles, from 0° to 315°, with 45° increments. The cGANs were trained to synthesize the micro-Doppler signatures for each specific angle given micro-Doppler signatures from a reference angle, as shown in Figure 16.24. Examples of synthesized micro-Doppler signatures are presented in Figure 16.25. Observably, the synthesized images are very analogous to the true answers. The output from the cGANs was evaluated by mean-square errors and structural similarity indexes.

GAN can also be used for SAR image synthesis. Even though the demand for high-quality SAR images is increasing rapidly, this technology involves high monetary and labor costs due to the limitations of current SAR devices. Thus, a Dialectical GAN was proposed to generate high-quality SAR images in order to enhance the quality of SAR images while reducing the costs [49]. A dialectical GAN, based on a combination of conditional WGAN-GP (Wasserstein Generative Adversarial Network—Gradient Penalty) loss functions and spatial gram matrices under the rule

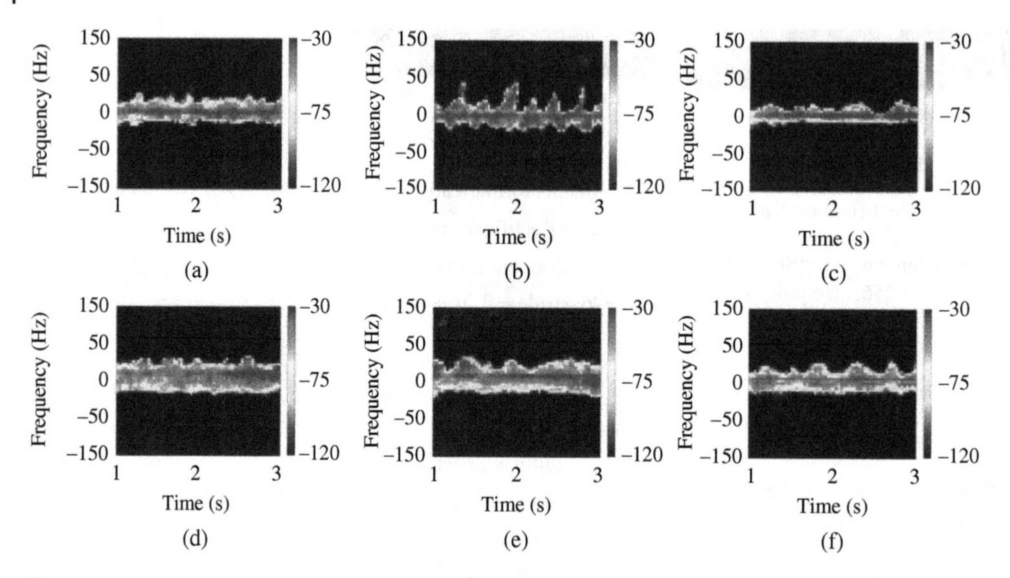

Figure 16.23 Synthesized micro-Doppler signatures from GANs. (a) Boxing while moving forward, (b) boxing while standing in place, (c) crawling, (d) running, (e) walking, and (f) walking hunched over while holding as tick. [47]

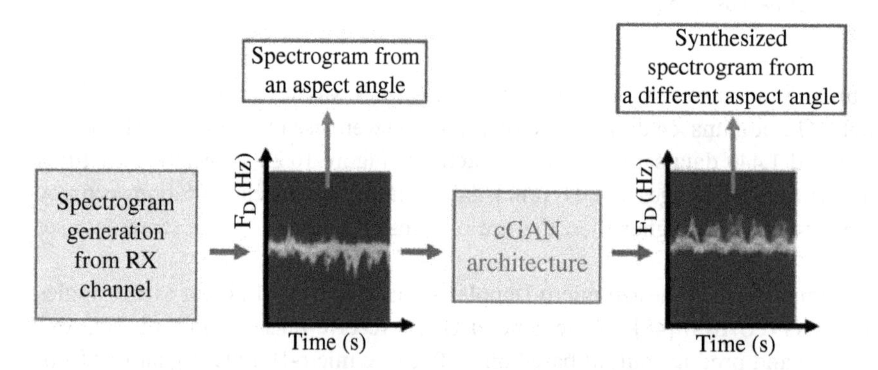

Figure 16.24 System processing steps to generate Pix2Pix output of one angle [48].

of dialectics, was used. According to the research report, a low-resolution SAR image with large ground coverage was successfully translated into a high-resolution SAR image. This research focused to translate Sentinel-1 images to TerraSAR-X image. Experimental results verified that the SAR image translation using WGAN-GP performed better than traditional methods. Since this field has not yet been sufficiently researched, and many applications remain undiscovered, additional studies are needed in the future (Figure 16.26).

16.5.3 Continual Learning

When identifying radar targets through machine learning algorithms, it is now very common to train a model with given training data. That is, the training data remain the same. However, even if a model with good performance is created, the data distribution of the input changes due to changes in propagation characteristics and target reflection characteristics according to climatic conditions over time, and the performance of the model may deteriorate accordingly. Furthermore,

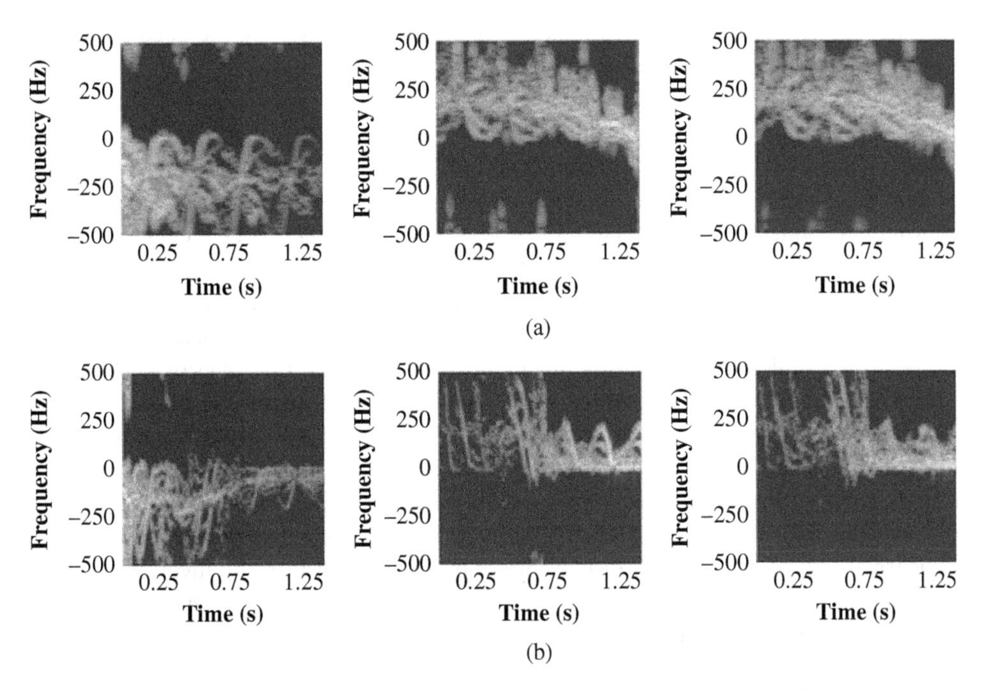

Figure 16.25 Example results from cGAN. Three spectrograms (left to right) are 0° input, ground truth at the desirable aspect angle, and output spectrogram of the desired angle from cGAN: (a) run at 135°, and (b) run to jump to walk at 180° [48].

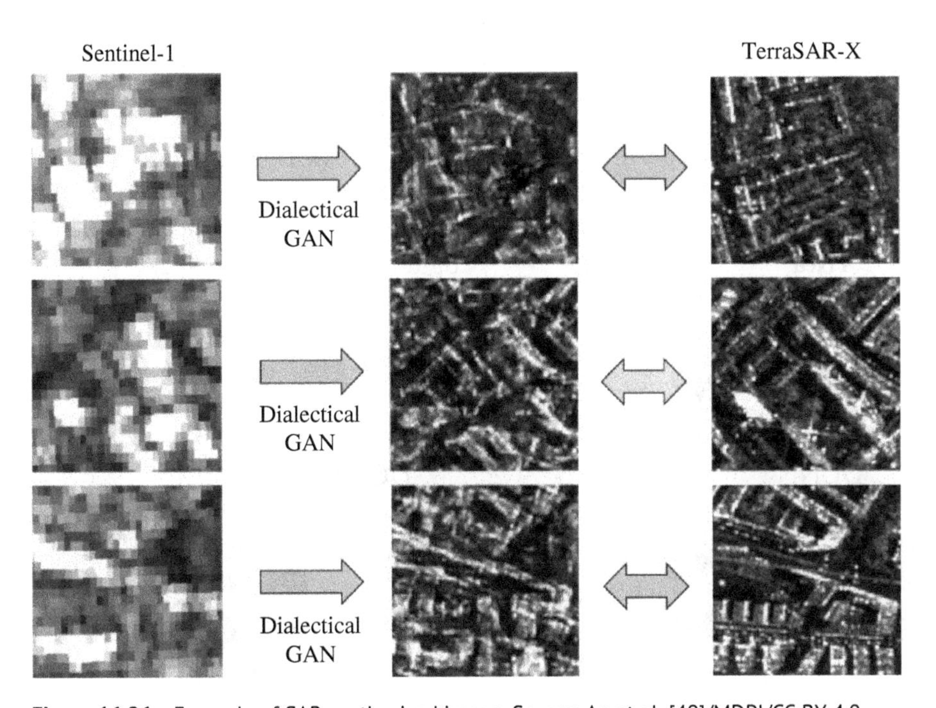

Figure 16.26 Example of SAR synthesized image. Source: Ao et al. [49]/MDPI/CC BY 4.0.

the appearance of a new, unexpected target and the addition of a class to be classified both lead to the need to re-train the model. In this case, it is natural to re-train the model from scratch by adding new data to the existing data. Although this method may represent the most basic and intuitive approach, learning all the weight values of the model for the entire dataset every time new data is collected takes a lot of training time; moreover, a large loss in terms of the consumption of computer resources, such as memory, is another problem. In addition, re-training is not desirable in that information about the previous dataset is overwritten, resulting in a new network that is focused on new data. This problem is called Catastrophic Forgetting.

Overcoming this issue entails using a continuous learning technique that learns from new data while not forgetting previously learned knowledge. Continual learning addresses the stability-plasticity dilemma. There are mainly two streams for this objective, involving regularization-based and exemplar-memory-based methods. Regularization-based continual learning is a method of updating an already-trained neural net using new coming data. After determining the important nets among the trained nets, either excluding the net or setting high strength regularization to update can be enforced to prevent the nets from changing significantly. Through this approach, new data can be learned without losing a significant amount of previously learned information. In contrast, exemplar-memory-based learning involves saving a few samples of the previous data in a limited memory buffer and reusing it with the new data. Thus, the catastrophic forgetting problem can be overcome by training the net with the combined data set.

The performance of continual learning in radar micro-Doppler classification has been reported in [50]. This research revisited the classification of seven kinds of human motions using 12 subjects via micro-Doppler. Two scenarios were tested using continual learning. In the first scenario, domain incremental learning, increased the number of people tested in the learning data; meanwhile, the human behavior to be distinguished remained the same. That is, the class was fixed while the learning data increased. In the second scenario, involving class incremental learning, a novel activity class was added to the whole data sample. In other words, the number of classes and the learning data both increased (Figure 16.27).

The domain incremental learning was tested by regularization-based continual learning and exemplar-memory-based learning. The regularization-based methods included elastic weight

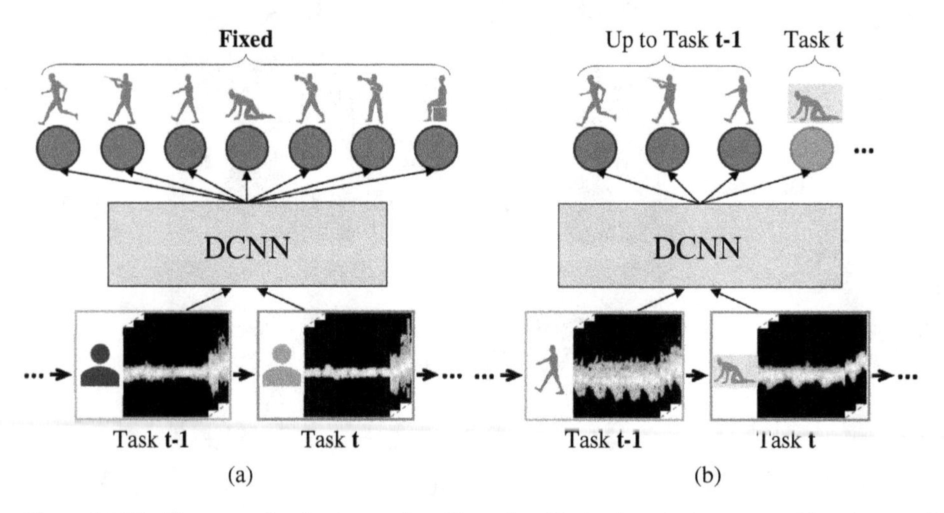

Figure 16.27 Two scenarios for the continual learning, (a) the domain incremental learning, and (b) class incremental learning. Source: Lee et al. [50].

Table 16.2 Results for domain incremental learning [51].

Methods		Accuracy
Fine tuning		80.30%
Regularization-based	EWC	85.08%
	SI	83.14%
	MAS	84.75%
Exemplar-memory-based	GEM	92.41%
	ER	92.06%

consolidation (EWC) [51], synaptic intelligence (SI) [52], and memory-aware synapses (MAS) [53], while the exemplar-memory-based methods included gradient episodic memory (GEM) [54] and experience replay (ER) [55]. The results showed that exemplar-memory-based learning produced a better performance as shown in Table 16.2. As far as the domain incremental learning was concerned, the incremental classifier and representation learning [56] yielded the highest accuracy.

16.6 Conclusion

This chapter discussed the various examples of deep learning algorithms to radar image classification. A new methodology using deep learning was presented in the area of radar target classification, and its performance was verified in the micro-Doppler and SAR fields. In particular, the use of DCNN to radar image classification has produced remarkable results, applying new algorithms developed from the machine learning field in a progression that is gradually overcoming the existing limitations. For example, the introduction of new models, including GANs, to radar images can be more actively applied to such fields as training data generation and de-noising. However, many topics still remain to be studied in the future, reflecting the need to go beyond applying machine learning that processes and recognizes various objects in various situations to the existing military radar, especially extending the topic to include radar technology for vehicle and health care applications in future investigations.

Unlike conventional optical images, radar images have a small data size and are black and white rather than featuring RGB colors. In particular, in the radar image, because the values of the matrix are continuous and change slowly, the boundary is not clear. Moreover, radar image has a lot of noise characteristics. Therefore, it is necessary to develop and study appropriate filters in deep learning algorithms for radar image processing. In addition, in order for machine learning to be put to practical use in radar, it will be necessary to develop a high-speed CPU and GPU and to optimize the learned net to reduce computational complexity for real-time processing.

References

1 Blacknell, D. and Griffiths, H. (2013). *Radar Automatic Target Recognition (ATR) and Non-Cooperative Target Recognition (NCTR)*. The Institution of Engineering and Technology.

2 Furukawa, H. (2018). Deep learning for end-to-end automatic target recognition, *IEICE Technical Report*, vol.117, arXiv:1801.08558.

3 Neumann, C. and Brosch, T. (2020). Deep learning approach for radar applications. In: *21st International Radar Symposium (IRS), Warsaw, Poland.*

4 Geng, Z., Yan, H., Zhang, J., and Zhu, D. (Oct 2021). Deep-learning for radar: a survey. *IEEE Access* 9: 141800–141818.

5 NATO (2013). AAP-06 Edition 2013. NATO Glossary of Terms and Definitions.

6 Chen, V.C. and Ling, H. (1999). Joint time-frequency analysis for radar signal and image processing. *IEEE Signal Processing Magazine* 16: 81–93.

7 Tan, C.M. and Lim, S.Y. (2002). Application of Wigner-Ville distribution in electromigration noise analysis. *IEEE Transactions on Device and Materials Reliability* 2 (2): 30–35.

8 Kim, Y. and Ling, H. (2009). Human activity classification based on micro-Doppler signatures using a support vector machine. *IEEE Transactions on Geoscience and Remote Sensing* 47: 1328–1337.

9 Kim, Y. and Moon, T. (2016). Human detection and activity classification based on micro-Dopplers using deep convolutional neural networks. *IEEE Geoscience and Remote Sensing Letters* 13: 2–8.

10 LeCun, Y., Bengio, Y., and Hinton, G. (2015). Deep learning. *Nature* 521: 436–444.

11 Glorot, X., Bordes, A., and Bengio, Y. (2011). Deep sparse rectifier neural networks. In: *Proceedings of the 14th International Conference on Artificial Intelligence Statistics, Fort Lauderdale, Florida*, 315–323.

12 Seyfioğlu, M., Özbayoğlu, A., and Gürbüz, S. (2018). Deep convolutional autoencoder for radar-based classification of similar aided and unaided human activities. *IEEE Transactions on Aerospace and Electronic Systems* 54: 1709–1723.

13 Hinton, G. and Salakhutdinov, R. (2006). Reducing the dimensionality of data with neural networks. *Science* 313 (5786): 504–507.

14 Chen, Z., Li, G., Fioranelli, F., and Griffiths, H. (2018). Personnel recognition and gait classification based on multistatic micro-Doppler signatures using deep convolutional neural networks. *IEEE Geoscience and Remote Sensing Letters* 15: 669–673.

15 Kim, Y. and Toomajian, B. (2016). Hand gesture recognition using micro-Doppler signatures with convolutional neural network. *IEEE Access* 4: 7125–7130.

16 Zhang, Z., Tian, Z., and Zhou, M. (2018). Latern: dynamic continuous hand gesture recognition using FMCW radar sensor. *IEEE Sensors Journal* 18: 3278–3289.

17 Hochreiter, S. and Schmidhuber, J. (1997). Long short-term memory. *Neural Computation* 9 (8): 1735–1780.

18 Arsalan, M. and Santra, A. (2019). Character recognition in air-writing based on network of radars for human-machine interface. *IEEE Sensors Journal* 19: 8855–8864.

19 Kim, B., Kang, H., and Park, S. (2016). Drone classification using convolutional neural networks with merged Doppler images. *IEEE Geoscience and Remote Sensing* 14: 38–42.

20 Wang, P., Zhang, H., and Patel, V. (2017). SAR image despeckling using a convolutional neural network. *IEEE Signal Processing Letters* 24 (12): 1763–1767.

21 Hughes, L., Schmitt, M., Mou, L. et al. (2018). Identifying corresponding patches in SAR and optical images with a pseudosiamese CNN. *IEEE Geoscience and Remote Sensing Letters* 15 (5): 784–788.

22 Zhao, J., Guo, W., Cui, S. et al. (2016). Convolutional neural network for SAR image classification at patch level. In: *IEEE International Geoscience and Remote Sensing Symposium (IGARSS), Beijing, China*.

23 Chen, S., Wang, H., Xu, F., and Jin, Y. (2016). Target classification using the deep convolutional networks for SAR images. *IEEE Transactions on Geoscience and Remote Sensing* 54: 4806–4817.

24 Srinivas, U., Monga, V., and Raj, R.G. (2014). SAR automatic target recognition using discriminative graphical models. *IEEE Transactions on Aerospace and Electronic Systems* 50: 591–606.

25 Gao, G. (2010). Statistical modeling of SAR images: a survey. *Sensors* 10: 775–795.

26 Li, M.D., Cui, X.C., and Chen, S.W. (2021). Adaptive superpixel-level CFAR detector for SAR inshore dense ship detection. *IEEE Geoscience and Remote Sensing Letters* 1–5.

27 Xu, L., Zhang, H., Wang, C. et al. (2016). Compact polarimetric SAR ship detection with m-δ decomposition using visual attention model. *Remote Sensing* 8.

28 Parikh, H., Patel, S., and Patel, V. (2020). Classification of SAR and PolSAR images using deep learning: a review. *International Journal of Image and Data Fusion* 11 (1): 1–32.

29 Schwegmann, C.P., Kleynhans, W., Salmon, B.P. et al. (2016). Very deep learning for ship discrimination in synthetic aperture radar imagery. In: *IEEE International Geoscience and Remote Sensing Symposium (IGARSS)*, 104–107. Beijing, China: IEEE.

30 Wang, Y., Wang, C., Zhang, H. et al. (2019). Automatic ship detection based on RetinaNet using multi-resolution Gaofen-3 imagery. *Remote Sensing* 11: 531.

31 He, K., Zhang, X., Ren, S., and Sun, J. (2015). Deep residual learning for image recognition, arXiv:1512.03385.

32 Huang, G., Liu, Z., Maaten, L., and Weinberger, K. (2018). Densely connected convolutional networks, arXiv:1608.06993.

33 Bilik, I., Longman, O., Villeval, S., and Tabrikian, J. (2019). The rise of radar for autonomous vehicles: signal processing solutions and future research directions. *IEEE Signal Processing Magazine* 36: 20–31.

34 Ng, W., Wang, G., Siddhartha, Z.L., and Dutta, B. (2020). Range-Doppler detection in automotive radar with deep learning. In: *2020 International Joint Conference on Neural Networks*, 1–8. Glasgow, Scotland, UK: IEEE.

35 Cal, X. and Sarabandi, K. (2019). A machine learning based 77 GHz radar target classification for autonomous vehicles. In: *IEEE International Symposium on Antennas and Propagation and USNC-URSI Radio Science Meeting*, 371–372. Atlanta, Georgia, USA: IEEE.

36 Kim, Y., Alnujaim, A., You, S., and Jeong, B. (2020). Human detection based on time-varying signature on range-Doppler diagram using deep neural networks. *IEEE Geoscience and Remote Sensing Letters* 18: 426–4430.

37 Ullah, A., Ahmad, J., Muhammad, K. et al. (2018). Action recognition in video sequences using deep bi-directional LSTM with CNN features. *IEEE Access* 6: 1155–1166.

38 Gupta, S., Rai, P., Kumar, A. et al. (2021). Target classification by mmWave FMCW radars using machine learning on range-angle images. *IEEE Sensors Journal* 21: 19993–20001.

39 Redmon, J., Divvala, S., Girshick, R., and Farhadi, A. (2016). You only look once: Unified, real-time object detection, arXiv:1506.02640.

40 Ram, S., Christianson, C., Kim, Y., and Ling, H. (2010). Simulation and analysis of human micro-Dopplers in through-wall environments. *IEEE Transactions on Geoscience and Remote Sensing* 48: 2015–2023.

41 Weijie, X., Hua, L., Fei, W. et al. (2014). SAR image simulation for urban structures based on SBR. In: *IEEE Radar Conference*, 0792–0795. Cincinnati, Ohio, USA: IEEE.

42 Pan, S. and Yang, Q. (2010). A survey on transfer learning. *IEEE Transactions on Knowledge and Data Engineering* 22: 1345–1359.

43 Goodfellow, I., Pouget-Abadie, J., Mirza, M. et al. (2014). Generative Adversarial Networks, arXiv:1406.2661.

44 Park, J., Rios, J., Moon, T., and Kim, Y. (2016). Micro-Doppler based classification of human activities on water via transfer learning of convolutional neural networks. *Sensors* 16: 1990.

45 Huang, Z., Pan, Z., and Lei, B. (2017). Transfer learning with deep convolutional neural network for SAR target classification with limited labeled data. *Remote Sensing* 9: 907.

46 Karthika, S. and Durgadevi, M. (2021). Generative adversarial network (GAN): a general review on different variants of GAN and applications. In: *6th International Conference on Communication and Electronics Systems (ICCES), Coimbatore, India*, 1–8. IEEE.

47 Alnujaim, I., Oh, D., and Kim, Y. (2019). Generative adversarial networks for classification of micro-Doppler signatures of human activity. In: *IEEE Geoscience and Remote Sensing Letters*, vol. 17, 396–400. IEEE.

48 Alnujaim, I., Ram, S., Oh, D., and Kim, Y. (2021). Synthesis of micro-Dopplers of human activities from different aspect angles using conditional generative adversarial networks. *IEEE Access* 9: 46422–46429.

49 Ao, D., Dumitru, C., Schwarz, G., and Datcu, M. (2018). Dialectical GAN for SAR image translation: from Sentinel-1 to TerraSAR-X. *Remote Sensing* 10: 1597.

50 Lee, D., Park, H., Moon, T., and Kim, Y. (2021). Continual learning of micro-Doppler signature-based human activity classification. *IEEE Geoscience and Remote Sensing Letters* 19: 1–5.

51 Jamesetal, K. (2017). Overcoming catastrophic forgetting in neural net- works. *Proceedings of the National Academy of Sciences of the United States of America* 114 (13): 3521–3526.

52 Zenke, F., Poole, B., and Ganguli, S. (2017). Continual learning through synaptic intelligence. In: *Proceedings of the 34th International Conference Machine Learning (JMLR), Sydney, Australia*, vol. 70, 3987–3995.

53 Aljundi, R., Babiloni, F., Elhoseiny, M. et al. (2018). Memory aware synapses: learning what (not) to forget. In: *Proceedings of the European Conference Computer Vision, Munich, Germany*, 139–154.

54 Lopez-Pazand, D. and Ranzato, M. (2017). Gradient episodic memory for continual learning. In: *Proceedings of the Advances Neural Information Processing Systems*, 6467–6476.

55 Chaudhryetal, A. (2019). On tiny episodic memories in continual learning, arXiv:1902.10486. http://arxiv.org/abs/1902.10486.

56 Rebuffi, S.-A., Kolesnikov, A., Sperl, G., and Lampert, C.H. (2017). ICaRL: incremental classifier and representation learning. In: *Proceedings of the IEEE Conference on Computer Vision Pattern Recognition (CVPR), Honolulu, Hawaii, USA*, 2001–2010.

17

Koopman Autoencoders for Reduced-Order Modeling of Kinetic Plasmas

Indranil Nayak[1], Mrinal Kumar[2], and Fernando L. Teixeira[1]

[1] *ElectroScience Laboratory and Department of Electrical and Computer Engineering, The Ohio State University, Columbus, OH, USA*
[2] *Department of Mechanical and Aerospace Engineering, The Ohio State University, Columbus, OH, USA*

17.1 Introduction

Often called the fourth state of matter, plasma is the most abundant form of matter found in the universe. A plasma can be defined as a quasi-neutral ionized mixture of neutral, positive-, and negatively-charged, mutually interacting particles. In addition to its obvious importance to astrophysics and ionosphere/magnetosphere phenomena, the study of plasmas is also of importance for many other applications as noted below. In order to understand the complex plasma behavior, computer simulations play a crucial role, as experimental verification is not always feasible. Broadly speaking, plasmas can be simulated by means of magnetohydrodynamics equations [1–3] (fluid model), or as a collection of charged particles with mutual interactions mediated by the electromagnetic fields [4–6] and governed by Maxwell–Vlasov equations (kinetic models). The magnetohydrodynamics or fluid model captures well the large-scale behavior of certain plasmas at length scales larger that the mean free-path. Kinetic models, on the other hand, are better suited for scenarios where the mean free-path is much larger than the characteristics dimensions of the problem. Electromagnetic particle-in-cell (EMPIC) plasma simulations play an important role in the analysis of such "collisionless" plasmas, which are present in high-power microwave sources, laser ignited devices, particle accelerators, vacuum electronic devices, and many other applications [4–11]. EMPIC plasma simulations will be the focus of this chapter. We will discuss some of the challenges associated with traditional EMPIC simulation algorithms and possible strategies to overcome those challenges using machine learning based reduced-order models (ROMs).

Historically, EMPIC algorithms have been a popular choice for simulating kinetic plasmas due to their ability to accurately capture complicated transient nonlinear phenomena [12–14]. EMPIC algorithms essentially simulate the motion of charged particles (constituent of plasma) under the influence of electromagnetic fields in a self-consistent manner. In general, EMPIC algorithms assume a coarse graining of the phase-space, whereby "superparticles" are employed to represent a large number of actual charged particles. EMPIC algorithms also invoke Debye shielding, i.e. the fact that direct (pairwise) Coulomb interactions are effectively screened between (super)particles and therefore can be neglected. These assumptions work very well for most applications mentioned earlier. Despite the accuracy of EMPIC simulations, they incur high very high computational costs.

Advances in Electromagnetics Empowered by Artificial Intelligence and Deep Learning, First Edition.
Edited by Sawyer D. Campbell and Douglas H. Werner.

This is a direct consequence of the large number of superparticles necessary to obtain accurate results. However, several recent studies [15–19] have indicated that most of the high-fidelity plasma systems can be effectively described through a lower dimensional feature space. This serves as the motivation to design appropriate ROMs based on EMPIC algorithms that can capture essential features of the high-fidelity plasma simulations through only a few characteristic "modes."

These observations dovetail with the recent surge in data-driven modeling in the scientific community [20, 21] due to advances in "big data" processing [22–24], hardware capabilities [25, 26], and data-driven (including machine learning based) techniques [27–29]. Part of the motivation for some of these developments comes from the fact that oftentimes there is only partial information (or no information) available regarding the underlying physics or guiding principles of the problem under study [30, 31]. In other cases, the dynamic system is so large and involves so many uncorrelated variables that modeling from first principles is simply not feasible (such as viral pandemics). In such cases, data-driven methods are the only option. Finally, in disciplines such as fluid dynamics and plasma physics, the underlying fundamental equations are well understood, however the majority of problems can only be solved numerically requiring extensive computational resources. The idea of a data-driven ROM is to process the data collected from a single or selected few number of high-fidelity simulations to generate a low-dimensional model that can capture the dominant system behavior. Such ROMs can then be used for low-cost model extrapolation (for example, to generate late-time data from early-time shortwindow simulations of a dynamic system) and/or model interpolation (for example, to generate data for a model across wide range of parameter(s) from the results of a small, sampled set of parameter values) of quantity of interest. Proper orthogonal decomposition (POD) [32, 33], dynamic mode decomposition (DMD) [34–36], and Koopman autoencoders (KAE) [37, 38] are some examples of such data-driven reduced-order techniques. These methods not only help to reduce the computational load of high-fidelity simulations, but also provide physical insights into complex problems by extracting relevant low-dimensional features. Linear ROMs can further be used as a linear system identification tool, facilitating control applications for nonlinear systems.

In this chapter, we explore the use of KAE as an effective reduced-order tool for modeling the nonlinear current density evolution in EMPIC simulations, with the aim of reducing the computational burden. We discuss how deep learning architectures provide a natural data-driven route for approximating finite-dimensional Koopman invariant subspaces. The present chapter is organized as follows: Section 17.2 provides a short tutorial on Maxwell–Vlasov equations governing the dynamics of collisionless kinetic plasmas. Section 17.3 covers the details of EMPIC algorithm used to simulate kinetic plasmas and discusses the inherent computational challenges associated with particle-in-cell algorithms. Section 17.4 provides a background on the Koopman operator theory and discusses the deep learning-based KAE architecture. Some illustrative results are presented to describe the effectiveness of KAE in modeling plasma dynamics. Section 17.5 discusses how physics information can be incorporated in the training of KAE geared towards kinetic plasma problems. Finally, Section 17.6 provides conclusions and future outlook.

17.2 Kinetic Plasma Models: Overview

As mentioned in the Introduction of this chapter, there are two primary types of models used to describe the plasma state: kinetic models [39, 40] and fluid models [1–3]. Kinetic models take into account the motion of the constitutive particles and their interaction with electromagnetic fields in order to predict the collective plasma behavior. Collisionless kinetic plasma systems can be described using the Maxwell-Vlasov system. In this system, Maxwell's equations (17.1) govern the spatiotemporal evolution of the electromagnetic field

$$\nabla \times \mathbf{E} = -\frac{\partial \mathbf{B}}{\partial t} \tag{17.1a}$$

$$\nabla \times \mathbf{B} = \mu_0 \epsilon_0 \frac{\partial \mathbf{E}}{\partial t} + \mu_0 \mathbf{J} \tag{17.1b}$$

where $\mathbf{E}, \mathbf{B}$ and $\mathbf{J}$, respectively, represents electric field intensity, magnetic flux density and current density. The vacuum permittivity and permeability are denoted as ϵ_0 and μ_0, respectively. The particle dynamics on the other hand is governed by the Vlasov equation (17.2):

$$\frac{\partial f}{\partial t} + \mathbf{v} \cdot \nabla f + \frac{q}{\gamma m_0} (\mathbf{E} + \mathbf{v} \times \mathbf{B}) \cdot \nabla_{\mathbf{v}} f = 0 \tag{17.2}$$

where $f(\mathbf{r}, \mathbf{v}, t)$ describes the particle number density, a function of particle position $\mathbf{r}$, velocity $\mathbf{v}$ and time t. For simplicity, we assume here only a single species of particles (e.g. electrons). The particle charge and rest mass are denoted as q and m_0, respectively. The relativistic factor γ is given by $\gamma^2 = (1 - v^2/c^2)^{-1}$, with c as the speed of light and v being the particle velocity magnitude. Note that the factor $q(\mathbf{E} + \mathbf{v} \times \mathbf{B})$ is associated to the Lorentz force acting on the charged particle.

The particle number density function $f(\mathbf{r}, \mathbf{v}, t)$ can be thought of as superposition of density functions of all the plasma constitutive particles. Given the very large number of charged particles in a typical problem, consideration of each individual particle separately makes the problem computationally intractable. In order to overcome this difficulty, a "coarse graining" of the phase space is typically performed in computer simulations whereby, instead of single electrons/ions, "superparticles" are considered. Simply stated, each superparticle represents a large collection of charged particles. A superparticle can be approximately represented as a point charge with charge $S_p q$ and rest mass $S_p m_0$, where S_p is the superparticle ratio. Under such point charge approximation, the number density function for the pth superparticle can be expressed as $f_s(\mathbf{r}, \mathbf{v}, t) \approx S_p \delta\left(\mathbf{r} - \mathbf{r}_p(t)\right) \delta\left(\mathbf{v} - \mathbf{v}_p(t)\right)$, where δ denotes the Dirac delta distribution. The particle number density function in (17.2) can then be written as

$$f_s(\mathbf{r}, \mathbf{v}, t) \approx \sum_p S_p \delta\left(\mathbf{r} - \mathbf{r}_p(t)\right) \delta\left(\mathbf{v} - \mathbf{v}_p(t)\right) \tag{17.3}$$

where the summation over p runs over all the N_p superparticles. From the above approximation, the number density function of superparticles can be updated by simply updating position and velocity of each superparticles following the Lorentz force and kinematic equations, i.e.

$$\frac{\partial}{\partial t} \mathbf{v}_p = \frac{q_p}{m_p} (\mathbf{E} + \mathbf{v}_p \times \mathbf{B}), \qquad \text{and} \qquad \frac{\partial}{\partial t} \mathbf{r}_p = \mathbf{v}_p \tag{17.4}$$

The EMPIC algorithm couples the above kinetic particle equations to Maxwell's field-update equations in order to self-consistently solve the Maxwell–Vlasov system of equations.

17.3 EMPIC Algorithm

17.3.1 Overview

EMPIC algorithms have been widely used for collisionless plasma problems for their ease of implementation and their ability to accurately capture transient and nonlinear effects. There are many EMPIC algorithm variants [13, 41–45]. However, at their core, any EMPIC algorithm implements

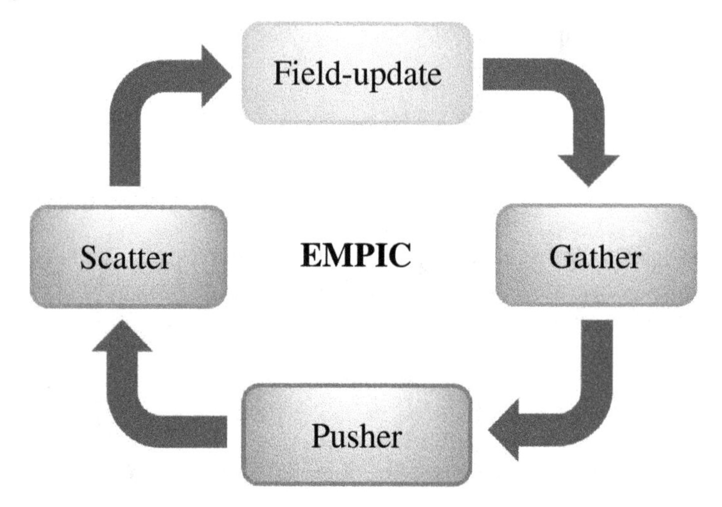

Figure 17.1 Cyclic stages of the EMPIC algorithm at each time-step.

a marching-on-time evolution wherein, at each time-step, four stages are involved (Figure 17.1): (i) field-update, (ii) gather, (iii) pusher, and (iv) scatter. These four stages are executed in a cyclic fashion for every time-step.

Within a EMPIC algorithm, the field-update solution of Maxwell's equations can be obtained by conventional partial-differential equation (PDE) solvers such as finite differences or finite elements. The former is typically used in conjunction with structured (regular) meshes and the latter with unstructured (irregular) meshes. The fields **E** and **B** and the current density **J** are defined on mesh elements (element edges and facets) whereas the (super)particles are defined on ambient space (i.e. they can move continuously anywhere across the various finite difference or finite element cells comprising the solution domain). This is where the "particle-in-cell" terminology comes from [46].

Before delving into the more detailed discussion of the particular EMPIC algorithm considered here, it is important to discuss the type of Maxwell field discretization scheme that we will be employing. Depending on the chosen field discretization, the details of the equations governing the four cyclic steps may vary.

The spatial domain of interest can be discretized by a structured (regular) mesh or by an unstructured (irregular) mesh. A common example of a structured mesh is a Cartesian one where each mesh element is a rectangle with identical dimensions. On the other hand, unstructured meshes are typically comprise triangular elements in two dimensions or tetrahedral elements in three dimensions. The shape of each triangle or tetrahedron can vary so as to conform to the specific problem geometry. In general, structured meshes have certain advantages over unstructured meshes such as yielding simpler discretization. However, unstructured meshes can better capture complex geometries and are free from staircase errors. Unstructured meshes can also provide better grid (mesh) dispersion performance in plasma problems and mitigate spurious numerical artifacts such as artificial Cherenkov radiation [47].

One of the critical aspects related to the spatial discretization of EMPIC algorithms based on unstructured meshes is charge conservation [48–50]. On the finite element mesh, the macroscopic charge density ρ is a variable based on mesh nodes and the macroscopic current density **J** is a variable based on mesh element edges. Such assignment choice is based on first-principle rules set by discrete exterior calculus, as discussed in [51]. Both these macroscopic quantities need to be computed numerically from the instantaneous positions and velocities of the charged (super)particles in ambient space. However, unless this calculation is performed in a consistent manner, charge conservation can be violated. Indeed, many attempts at developing unstructured mesh EMPIC algorithms over the years have led to violation of conservation laws and other spurious effects. Charge conservation has been a especially challenging problem for unstructured mesh EMPIC algorithms because of the irregular connectivity among the mesh elements. Ad hoc approaches such as pseudo-currents [49] or Poisson correction steps [48] have been utilized to in order to enforce charge conservation a posteriori; however, these approaches either subtly alter the physics of the problem or call for an additional, time-consuming linear solver at each time-step. Recently, the problem of charge conservation was successfully resolved from first-principles by employing Whitney forms as the correct finite-element basis functions to both represent the fields and macroscopic charges/currents as mesh variables, and to project the effects of the movement of the ambient point charges back onto the mesh variables during the scatter procedure [41]. Other variants of first-principle conservation strategies in EMPIC algorithms have been considered in [13, 44, 50]. In this chapter, we will follow references [14, 41, 52] and discuss the implementation of a finite-element-based charge-conservation EMPIC algorithm on unstructured meshes from first principles with a matrix-free field-update scheme. This implementation is used to generate the full-order numerical results presented later on in this chapter. The key feature of Whitney forms is that they represent consistent interpolants for discrete differential forms of various degrees representing the macroscopic variables ρ, $\mathbf{J}$, $\mathbf{E}$, and $\mathbf{B}$ on the mesh. In the language of the exterior calculus of differential forms, the electric field intensity $\mathbf{E}$ is considered as a differential form of degree one or 1-form, the magnetic flux density $\mathbf{B}$ as a 2-form, and the (Hodge dual of the) current density $\mathbf{J}$ as a 1-form, and the charge density ρ as a 0-form. A detailed discussion about the discrete exterior calculus of differential forms is out of scope for this chapter. The interested reader is referred to [41, 51, 53, 54] for more details on this topic. However, we will next discuss a few important aspects which are pertinent to explain the four cyclic procedures in the EMPIC algorithm and later on to their relation to the reduced-order approaches.

In a nutshell, the present charge-conserving EMPIC algorithm on unstructured meshes has been developed based on first principles. In each time-step, the electric field intensity values $\mathbf{E}$ (defined on element edge), and the magnetic flux density values $\mathbf{B}$ (defined on element facet) are solved using a finite element time-domain algorithm with a spatial discretization based on discrete exterior calculus approach as mentioned above. The discrete time-update is based on a second-order leap-frog algorithm [14, 52]. The $\mathbf{E}$ and $\mathbf{B}$ fields are then interpolated at each particle position (gather) using Whitney forms, and used to update the position and velocity of each particle via Lorentz force law and the kinematic equations of motion (pusher). Finally, the motion of charged particles is mapped back to the mesh variables representing the macroscopic current $\mathbf{J}$ and charge ρ densities (scatter). These steps are repeated again in the next time-step. The details of each of these steps are as follows:

17.3.2 Field Update Stage

In the present finite element setting, the electric field intensity $\mathbf{E}^{(n)}(\mathbf{r})$ and the magnetic flux density $\mathbf{B}^{(n+\frac{1}{2})}(\mathbf{r})$ at the nth (half-integer time-steps for $\mathbf{B}$) time-step can be represented using vector proxies of the Whitney forms [42, 55–57] as

$$\mathbf{E}^{(n)}(\mathbf{r}) = \sum_{i=1}^{N_1} e_i^{(n)} \mathbf{W}_i^1(\mathbf{r}) \tag{17.5a}$$

$$\mathbf{B}^{(n+\frac{1}{2})}(\mathbf{r}) = \sum_{i=1}^{N_2} b_i^{(n+\frac{1}{2})} \mathbf{W}_i^2(\mathbf{r}) \tag{17.5b}$$

where N_1 and N_2 denote, respectively, the total number mesh element edges and mesh element facets. The basis functions $\mathbf{W}_i^1(\mathbf{r})$ and $\mathbf{W}_i^2(\mathbf{r})$ are the vector proxies of Whitney 1- and 2-forms [55, 56]. These functions have a one-to-one correspondence with the element edges and element facets of the mesh. The e_i and b_i coefficients are the degrees of freedom (DoF) for the discretized **E** and **B** fields, respectively. For ease of discussion, with mild abuse of language, we will refer to the above vector proxies of Whitney forms simply as Whitney forms. Note that for a two-dimensional (2-D) unstructured mesh composed of triangular elements, the mesh element facets are simply the triangular elements themselves. One can similarly represent the electric current density and the charge density on the mesh as

$$\mathbf{J}(\mathbf{r})^{(n+\frac{1}{2})} = \sum_{i=1}^{N_1} j_i^{(n+\frac{1}{2})} \mathbf{W}_i^1(\mathbf{r}) \tag{17.6a}$$

$$Q(\mathbf{r})^{(n)} = \sum_{i=1}^{N_0} q_i^{(n)} W_i^0(\mathbf{r}) \tag{17.6b}$$

where N_0 is the total number of nodes on the mesh and W_i^0 is a set of Whitney 0-forms having one-to-one correspondence with the mesh nodes, with j_i and q_i being the DoFs for current density and charge density, respectively. From the above, it is clear that the Whitney forms in (17.5) and (17.6) acts as basis functions in the usual finite element context, and the unknown DoFs are simply the weights corresponding to each basis function. Using the leap-frog discretization for the time derivatives to construct the time update scheme, the following discrete Maxwell's equations are obtained [14, 41, 42],

$$\mathbf{b}^{(n+\frac{1}{2})} = \mathbf{b}^{(n-\frac{1}{2})} - \Delta t \mathbf{C} \cdot \mathbf{e}^{(n)} \tag{17.7a}$$

$$\mathbf{e}^{(n+1)} = \mathbf{e}^{(n)} + \Delta t \left[\star_{\epsilon}\right]^{-1} \cdot \left(\tilde{\mathbf{C}} \cdot \left[\star_{\mu^{-1}}\right] \cdot \mathbf{b}^{(n+\frac{1}{2})} - \mathbf{j}^{(n+\frac{1}{2})}\right) \tag{17.7b}$$

with constraints

$$\mathbf{S} \cdot \mathbf{b}^{(n+\frac{1}{2})} = 0 \tag{17.8a}$$

$$\tilde{\mathbf{S}} \cdot \left[\star_{\epsilon}\right] \cdot \mathbf{e}^{(n)} = \mathbf{q}^{(n)} \tag{17.8b}$$

where **e**, **b**, **j** and **q** represent column vectors of DoFs, i.e. $\mathbf{e} = [e_1\ e_2\ \dots\ e_{N_1}]^T$, $\mathbf{b} = [b_1\ b_2\ \dots\ b_{N_2}]^T$, $\mathbf{j} = [j_1\ j_2\ \dots\ j_{N_1}]^T$, and $\mathbf{q} = [q_1\ q_2\ \dots\ q_{N_0}]^T$, with "$T$" denoting the transpose. In addition, **C** and **S** refer to the incidence matrices encoding the discrete representation of the curl and divergence operators on the (primal) mesh while $\tilde{\mathbf{C}}$ and $\tilde{\mathbf{S}}$ represent their counterparts on the dual mesh [51, 54, 58, 59]. These matrices are related through $\tilde{\mathbf{C}} = \mathbf{C}^T$ and $\tilde{\mathbf{S}} = \mathbf{S}^T$ [58, 59]. The incidence operators are metric-free in the sense that their elements are equal to 1,0 or -1, depending on the connectivity of mesh elements and their assigned relative orientation. Since, **E**, **B** and **J** are directional in nature, e_i, b_i and j_i are signed scalar quantities. We should point out that **J** defined in (17.6a) and (17.7) is actually the Hodge dual of 2-form current density. In the language of differential forms, the current density is a 2-form defined on the dual mesh facets, where as the dual of current density is a 1-form defined on the primal mesh edges (from the Hodge duality, there is a one-to-one correspondence between dual mesh facets and primal mesh edges). However, to facilitate the discussion, we will again abuse the language slightly and refer to **J** as the current density. Note that if the initial conditions satisfy the pair of Eq. (17.8) (which correspond to the divergence constraints $\nabla \cdot \mathbf{B} = 0$ and $\nabla \cdot \epsilon_0 \mathbf{E} = \rho$) then these equations are automatically satisfied [14] at all subsequent times steps. The metric information is encoded in the symmetric positive definite matrices $\left[\star_{\mu^{-1}}\right]$

and $[\star_\epsilon]$ in (17.7b), representing the discrete Hodge star operators [42, 51, 56]. One of the features of (17.7) is the explicit nature of the field update, which obviates the need for a linear solver at each time-step. However, this comes at the one-time cost of calculating the approximate inverse $[\star_\epsilon]^{-1}$ in Eq. (17.7b). Instead of computing $[\star_\epsilon]^{-1}$ directly (which is computationally impractical for large problems), a sparse approximate inverse (SPAI) can be precomputed with tunable accuracy [14, 42].

17.3.3 Field Gather Stage

As explained before, the DoFs of the fields and currents are defined on the mesh; however, the particles can be anywhere in the solution domain. In order to model the effect of fields on the charged particles, it is necessary to interpolate the fields at the particle position. This is performed at the gather stage using the same Whitney forms as in Eqs. (17.5a) and (17.5b). The fields at p^{th} particle position $\mathbf{r}_p$ are simply given by

$$\mathbf{E}^{(n)}(\mathbf{r}_p) = \sum_{i=1}^{N_1} e_i^{(n)} \mathbf{W}_i^1(\mathbf{r}_p) \tag{17.9a}$$

$$\mathbf{B}^{(n+\frac{1}{2})}(\mathbf{r}_p) = \sum_{i=1}^{N_2} b_i^{(n+\frac{1}{2})} \mathbf{W}_i^2(\mathbf{r}_p) \tag{17.9b}$$

17.3.4 Particle Pusher Stage

Once the fields are interpolated at the particle position, the position and velocity of the particle can be updated using the Lorentz force equation and Newton's laws of motion, considering relativistic effects if necessary. The pth particle kinematic equations in the continuous time domain are given by

$$\gamma_p^2 = \frac{1}{1 - |\mathbf{v}_p|^2/c^2} \tag{17.10}$$

$$\frac{d\mathbf{r}_p}{dt} = \mathbf{v}_p = \frac{\mathbf{u}_p}{\gamma_p} \tag{17.11}$$

$$\frac{d\mathbf{u}_p}{dt} = \frac{q}{m_0}\left[\mathbf{E}\left(\mathbf{r}_p, t\right) + \mathbf{v}_p \times \mathbf{B}\left(\mathbf{r}_p, t\right)\right] \tag{17.12}$$

where the relativistic factor is given by γ_p. Using a standard finite-difference approximation for the time derivatives, the discrete-time version of (17.11) and (17.12) becomes

$$\frac{\mathbf{r}_p^{(n+1)} - \mathbf{r}_p^{(n)}}{\Delta t} = \frac{\mathbf{u}_p^{(n+\frac{1}{2})}}{\gamma_p^{(n+\frac{1}{2})}} \tag{17.13}$$

$$\frac{\mathbf{u}_p^{(n+\frac{1}{2})} - \mathbf{u}_p^{(n-\frac{1}{2})}}{\Delta t} = \frac{q}{m_0}\left(\mathbf{E}_p^{(n)} + \frac{\overline{\mathbf{u}}_p}{\overline{\gamma}_p} \times \mathbf{B}_p^{(n)}\right) \tag{17.14}$$

where $\overline{\mathbf{v}}_p$ denotes the mean particle velocity between the $(n \pm \frac{1}{2})^{th}$ time-steps, and given the mean $\mathbf{u}_p$ and mean γ_p, we set $\overline{\mathbf{u}}_p = \overline{\gamma}_p \overline{\mathbf{v}}_p$. The mean velocity for non-relativistic case ($\gamma_p \to 1$) is simply $\overline{\mathbf{v}}_p^n = \mathbf{v}_p^n = \left(\mathbf{v}_p^{n+\frac{1}{2}} + \mathbf{v}_p^{n-\frac{1}{2}}\right)/2$. In the relativistic regime, however, the mean velocity $\overline{\mathbf{v}}_p$ should be calculated carefully. The reader is referred to [52] for more details on the different types of relativistic pushers.

17.3.5 Current and Charge Scatter Stage

In the scatter stage, the effect of the collective movement of charged particles are mapped back to currents defined on the element edges of the mesh, and mapped to charges defined on the mesh nodes. This mapping is performed using the Whitney forms to ensure consistency, and in particular, charge conservation on the mesh. The charge density is assigned to the mesh nodes from the instantaneous positions of the charged particles using Whitney 0-forms, whereas Whitney 1-forms are used to map current density to mesh element edges from the trajectory traversed by charged particles during a single time-step [14, 41].

The total charge assigned to the ith node at the nth time-step is calculated as follows

$$q_i^{(n)} = \sum_p q W_i^0(\mathbf{r}_p^{(n)}) = \sum_p q\lambda_i(\mathbf{r}_p^{(n)}) \tag{17.15}$$

where the summation index p runs over all the N_p superparticles, q is the charge and $\mathbf{r}_p^{(n)}$ is the position of the pth superparticle at the nth time-step, W_i^0 represents the Whitney 0-form or equivalently, the barycentric coordinate $\lambda_i(\mathbf{r}_p)$ of the point $\mathbf{r}_p$ with respect to the ith node. Similarly, the current density generated at mesh element edges due to movement of charged particles are calculated using Whitney 1-forms. Let us consider the pth (super)particle moving from $\mathbf{r}_p^{(n)}$ to $\mathbf{r}_p^{(n+1)}$ during one time-step (comprising Δt time duration). The current density $j_{ik,p}^{(n+\frac{1}{2})}$ generated due to movement of the pth particle[1] at the ik^{th} edge (indexed by its two end nodes, i^{th} and k^{th} nodes) is given in terms of the line integral of the Whitney 1-form along the path from $\mathbf{r}_p^{(n)}$ to $\mathbf{r}_p^{(n+1)}$, i.e.

$$\begin{aligned} j_{ik,p}^{(n+\frac{1}{2})} &= \frac{q}{\Delta t}\int_{\mathbf{r}_p^{(n)}}^{\mathbf{r}_p^{(n+1)}} \mathbf{W}_{ik}^1(\mathbf{r}_p)\cdot d\mathbf{l} \\ &= \frac{q}{\Delta t}\left[\lambda_i(\mathbf{r}_p^{(n)})\lambda_k(\mathbf{r}_p^{(n+1)}) - \lambda_i(\mathbf{r}_p^{(n+1)})\lambda_k(\mathbf{r}_p^{(n)})\right] \end{aligned} \tag{17.16}$$

where $\mathbf{W}_{ik}^1(\mathbf{r}_p)$ is the Whitney-1 form with respect to ik^{th} edge calculated at $\mathbf{r}_p$. In the above, we have used the standard relation between Whitney 1-forms and Whitney 0-forms to evaluate the integral in closed form [41]. The total current density generated at the ik^{th} edge is obtained by summing the contributions from all superparticles:

$$j_{ik}^{(n+\frac{1}{2})} = \sum_p \frac{q}{\Delta t}\left[\lambda_i(\mathbf{r}_p^{(n)})\lambda_k(\mathbf{r}_p^{(n+1)}) - \lambda_i(\mathbf{r}_p^{(n+1)})\lambda_k(\mathbf{r}_p^{(n)})\right] \tag{17.17}$$

It can be shown that due to this intrinsic relation between Whitney-1 and 0 forms, the following discrete charge continuity equation (17.18) is always satisfied

$$\frac{\mathbf{q}^{(n+1)} - \mathbf{q}^{(n)}}{\Delta t} + \tilde{\mathbf{S}}\cdot\mathbf{j}^{(n+\frac{1}{2})} = 0 \tag{17.18}$$

For more details, the reader is referred to [41, 52].

17.3.6 Computational Challenges

High-fidelity EMPIC simulations in general suffer from high runtime and large memory requirements. Along with the potentially large number of mesh elements, the position and velocities of thousands or millions of superparticles need to be stored in memory to execute each time-step.

1 Note that the j_{ik} in (17.17) and j_i in (17.6a) represent the same quantity. The only difference is how we index the mesh element edge over which j is defined. In (17.17) we use the indices of the two end nodes to index an mesh element edge, while we directly use edge indices in (17.6a).

The computational complexity of a typical EMPIC algorithm for each time-step is given by $\mathcal{O}(N^s + N_p)$ [60], where N is the aggregate mesh size and N_p is the total number of superparticles. The specific dependence of computational complexity on the mesh size depends on the particular choice of field update scheme (17.7). Typically for implicit field updates, $s \geq 1.5$. However, for an explicit field-update scheme such as in (17.7), $s = 1$. Depending on the problem setup, especially if the solution domain is large, resulting in large N, the repeated solution of (17.7) for many time steps can pose significant computational challenges.

The dependence on N_p arises from the EMPIC stages involving the superparticles, namely, the *gather*, *pusher* and *scatter* stages. These stages require the EMPIC algorithm to perform computations across the entire set of superparticles at any given time-step. Although these computations are embarrassingly parallelizable, they can become a serious bottleneck unless a extremely large number of processors are available. Typically, the number of superparticles is far greater than the number of mesh elements ($N_p \gg N$), resulting in $\mathcal{O}(N + N_p) \approx \mathcal{O}(N_p)$. In other words, the cost of superparticles computations can dominate the computational burden of EMPIC simulations in serial computers. In Section 17.4, we shall discuss the Koopman operator theory, and how ROMs such as KAE (Section 17.4.3) can be used to overcome the computational challenges presented by EMPIC simulations.

17.4 Koopman Autoencoders Applied to EMPIC Simulations

17.4.1 Overview and Motivation

In order to understand the motivation behind adopting a Koopman ROM such as KAE for plasma problems, first and foremost we must formulate the problem we are trying to solve, and investigate how the Koopman approach can be beneficial in that endeavor. We are essentially trying to devise a ROM which will help us perform the time-update of dynamic variables such as **E**, **B** and **J** with a reduced memory requirement, lower time complexity, and sufficient accuracy by capturing the dominant physical features of a nonlinear kinetic plasma problem. As noted before, large memory/storage requirement is the direct result of the large number of mesh elements and particles that need to be stored for executing an EMPIC algorithm. A data-driven ROM helps us tackle this issue by extracting the dominant characteristic features (modes/eigenfunctions) from high-fidelity data and modeling the dynamics governed by such reduced number of "features" or "modes," and thus reducing the size of the original problem. Apart from memory usage, another aspect contributing to the high computational load is the long simulation time required by EMPIC simulations. As mentioned in Section 17.3.6, typical EMPIC algorithms have time complexity of $\mathcal{O}(N^s + N_p)$ at each time-step. In most of the scenarios, $N_p \gg N$, so the steps involving the (super)particles act as the primary bottleneck in expediting EMPIC simulations.

In the literature, there are several model order reduction techniques available for modeling plasma phenomena, such as POD [32, 33, 61], bi-orthogonal decomposition (BOD) [62, 63], principal component analysis (PCA) [64], among others. Although these methods are successful in reducing the size of the problem in certain cases, they fail to provide any analytical estimates for the time evolution of the system. On the other hand, reduced-order methods based on Koopman operator theory such as DMD [19, 34–36] and KAE [37, 38, 65] are able to extrapolate the solution for future times by using analytical expressions derived from linear systems theory.

We are primarily interested in the reduced-order modeling of time evolution of **j** since it directly involves the particles. Equations (17.7a) and (17.7b) are just the discretized version of

Maxwell's equations which are linear in nature. However, in the EMPIC setting, the time variation of current density DoF $\mathbf{j}^{(n+\frac{1}{2})}$ in (17.7b) is dependent on the fields as they dictate the particle motion. The gather, pusher, and the scatter stages takes into account this wave-particle interaction and governs the time update of $\mathbf{j}$. This makes the time evolution of $\mathbf{j}$ and the overall dynamics of the problem nonlinear, leading to nonlinear time variation of $\mathbf{e}$ and $\mathbf{b}$. The presence of nonlinearity in the dynamics is one of the motivations for using the Koopman approach, as we will discuss in Section 17.4.2. In some cases, such as for physics-informed learning, we might need to explicitly model $\mathbf{e}$ and $\mathbf{b}$ as well. We will briefly discuss those in Section 17.5.

In the next subsections, we will first discuss the basic Koopman operator theory (Section 17.4.2) and then address the construction of deep learning based KAE architecture (Section 17.4.3). The KAE architecture helps realize a finite-dimensional approximation of the infinite-dimensional Koopman operator.

17.4.2 Koopman Operator Theory

B.O. Koopman in the early 1930s demonstrated that Hamiltonian flows governed by nonlinear dynamics can be analyzed via an infinite-dimensional linear operator on the Hilbert space of observable functions [66]. The spectral analysis of this infinite-dimensional linear operator, or the so called "Koopman operator theory" provides an operator theoretic understanding of dynamical systems. The Koopman operator theory essentially tells us that the state space corresponding to a *nonlinear* dynamical system can be transformed into an infinite-dimensional observable space where the dynamics is *linear*.

In order to understand how this works, let us consider a discrete time nonlinear dynamical system evolving on a N-dimensional manifold $\mathcal{M}$ [67]. The discrete time evolution of state $\mathbf{x}$ is given by the flow map $F : \mathcal{M} \mapsto \mathcal{M}$, where $\mathcal{M}$ is the N-dimensional manifold ($\mathbf{x} = [x_1 \; x_2 \; \dots \; x_N]^T$)

$$\mathbf{x}^{(n+1)} = F(\mathbf{x}^{(n)}) \tag{17.19}$$

The discrete-time Koopman operator, denoted by $\mathcal{K}$, operates on the Hilbert space of observable functions $g(\mathbf{x})$ (observable functions take the state as input and produces a scalar valued output, $g : \mathcal{M} \mapsto \mathbb{C}$), such that

$$\mathcal{K}g(\mathbf{x}^{(n)}) = g(F(\mathbf{x}^{(n)})) = g(\mathbf{x}^{(n+1)}) \tag{17.20}$$

In other words, $\mathcal{K}g = g \circ F$, where "$\circ$" represents the composition of functions. The infinite dimensionality of the Koopman operator stems from the fact that the observable functions $g(\mathbf{x})$ which satisfy (17.20) form an infinite-dimensional function space, i.e. $g(\mathbf{x}) = \sum_{i=1}^{\infty} \alpha_i \zeta_i(\mathbf{x})$, where $\zeta_i(\mathbf{x})$ are the basis observable functions and $\alpha_i \in \mathbb{C}$. Naturally, the eigenvalues $\lambda_i \in \mathbb{C}$ and the eigenfunctions $\phi_i : \mathcal{M} \mapsto \mathbb{C}$ associated with $\mathcal{K}$ are infinite in number such that $\mathcal{K}\phi_i(\mathbf{x}) = \lambda_i \phi_i(\mathbf{x})$, $i = 1,2 \dots \infty$. Given that the Koopman eigenfunctions span the observable functions, one can write $g(\mathbf{x}) = \sum_{i=1}^{\infty} \phi_i(\mathbf{x}) v_i$, where $v_i \in \mathbb{C}$. In a similar fashion, a vector valued observable $\mathbf{g}(\mathbf{x}) = [g_1(\mathbf{x}) \; g_2(\mathbf{x}) \; \dots \; g_p(\mathbf{x})]^T$ can be expressed as $\mathbf{g}(\mathbf{x}) = \sum_{i=1}^{\infty} \phi_i(\mathbf{x}) \mathbf{v}_i$, where $\mathbf{v}_i = [v_{1i} \; v_{2i} \; \dots \; v_{pi}]^T$ are defined as Koopman modes. Using the initial state, $\mathbf{x}^{(0)}$ along with linearity of the Koopman operator, we have

$$\mathbf{g}(\mathbf{x}^{(n)}) = \mathcal{K}^n \mathbf{g}(\mathbf{x}^{(0)}) = \mathcal{K}^n \sum_{i=1}^{\infty} \phi_i(\mathbf{x}^{(0)}) \mathbf{v}_i = \sum_{i=1}^{\infty} \mathcal{K}^n \phi_i(\mathbf{x}^{(0)}) \mathbf{v}_i = \sum_{i=1}^{\infty} \lambda_i^n \phi_i(\mathbf{x}^{(0)}) \mathbf{v}_i \; . \tag{17.21}$$

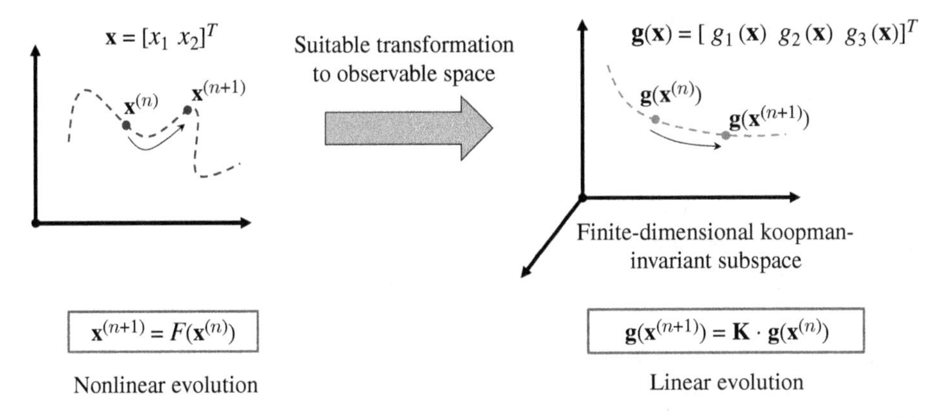

Figure 17.2 Schematic illustration of the Koopman approach for modeling the time evolution of nonlinear dynamical systems.

Although (17.21) helps us decompose the underlying dynamics in terms of Koopman modes $\mathbf{v}_i$ and corresponding eigenvalues λ_i, from a computational perspective it is not practical to work in an infinite-dimensional space. However, for a finite-dimensional Koopman invariant subspace, it is possible to find a finite-dimensional matrix representation $\mathbf{K}$ (Figure 17.2), of the infinite-dimensional Koopman operator $\mathcal{K}$ [68]. Consider a class of observable functions $g_j(\mathbf{x})$ belonging to the Koopman invariant subspace $\mathcal{S}_{\mathcal{K}}$ of dimension d. As the subspace is $\mathcal{K}$-invariant, $g_j(\mathbf{x}) \in \mathcal{S}_{\mathcal{K}} \implies \mathcal{K}g_j(\mathbf{x}) \in \mathcal{S}_{\mathcal{K}}$. Furthermore,

$$g_j(\mathbf{x}^{(n)}) = \sum_{i=1}^{d} \alpha_{ji} s_i(\mathbf{x}^{(n)}), \quad \mathcal{K}g_j(\mathbf{x}^{(n)}) = g_j(\mathbf{x}^{(n+1)}) = \sum_{i=1}^{d} \beta_{ji} s_i(\mathbf{x}^{(n)}) \tag{17.22}$$

s_i being the basis in $\mathcal{S}_{\mathcal{K}}$ and $\alpha_{ji}, \beta_{ji} \in \mathbb{C}$. Assuming the existence of $\mathbf{K}$ such that

$$\mathbf{g}(\mathbf{x}^{(n+1)}) = \mathbf{K} \cdot \mathbf{g}(\mathbf{x}^{(n)}) \tag{17.23}$$

$$\Rightarrow \mathbf{B} \cdot \mathbf{s}(\mathbf{x}^{(n)}) = \mathbf{K} \cdot \mathbf{A} \cdot \mathbf{s}(\mathbf{x}^{(n)}) \quad [\text{where } [\mathbf{B}]_{j,i} = \beta_{ji},\ [\mathbf{A}]_{j,i} = \alpha_{ji}, \ \text{ and } \mathbf{s} = [s_1\ s_2\ \dots\ s_d]^T]$$

$$\Rightarrow (\mathbf{B} - \mathbf{K} \cdot \mathbf{A}) \cdot \mathbf{s}(\mathbf{x}^{(n)}) = \mathbf{0} \quad [\text{Dimension of } \mathbf{B}, \mathbf{A} :\ p \times d; \text{ dimension of } \mathbf{K} :\ p \times p\,] \tag{17.24}$$

$$\Rightarrow \mathbf{B} = \mathbf{K} \cdot \mathbf{A} \quad [\text{Since } \mathbf{s} \text{ can be any set of basis.}]. \tag{17.25}$$

If (17.24) is to be satisfied for any $\mathbf{s}(\mathbf{x})$, then $\mathbf{B} = \mathbf{K} \cdot \mathbf{A}$ which results in a unique $\mathbf{K}$ if $\mathbf{A}$ is square and full rank. This requires the observables $g_1(\mathbf{x}), g_2(\mathbf{x}), \dots, g_p(\mathbf{x})$ to be linearly independent and span $\mathcal{S}_{\mathcal{K}}$ with $p = d$. Note that $\mathbf{K}$ is a $p \times p$ matrix, the finite-dimensional restriction of the infinite-dimensional Koopman operator. Clearly, a careful choice of the observable functions $g(\mathbf{x})$ is crucial for the effectiveness of the Koopman approach, and the development of algorithms or heuristics for identifying optimal observables is still an active area of research. The primary takeaway from (17.24) to (17.25) is that, with an appropriate choice of observables, we can bypass the infinite sum in (17.21) and make the problem tractable.

However, there is a last piece of the puzzle that needs to be addressed. The Koopman mode decomposition (17.21) consists of the eigenvalues λ_i, eigenfunctions $\phi_i(\mathbf{x})$, and Koopman modes

$\mathbf{v}_i$ corresponding to the infinite-dimensional $\mathcal{K}$. We need to establish how these quantities relate to the eigenvalues λ_{ki} and eigenvectors of the finite-dimensional $\mathbf{K}$. For the linear dynamical system,

$$\mathbf{g}(\mathbf{x}^{(n+1)}) = \mathbf{K} \cdot \mathbf{g}(\mathbf{x}^{(n)}) \tag{17.26}$$

$$\mathbf{K}\mathbf{u}_i = \lambda_{ki}\mathbf{u}_i \quad , \mathbf{K}^*\mathbf{w}_i = \overline{\lambda}_{ki}\mathbf{w}_i \quad , \quad [i = 1,2,\ldots,p] \tag{17.27}$$

where $\mathbf{u}_i$ and $\mathbf{w}_i$ are the right and left eigenvectors of $\mathbf{K}$, respectively. The "$*$" denotes complex conjugate transpose and the overbar denotes complex conjugate. The $\mathbf{w}_i$ are defined in such a way that $\langle \mathbf{u}_i, \mathbf{w}_j \rangle = \delta_{ij}$, where $\langle \cdot, \cdot \rangle$ denotes inner product in $\mathbb{C}^p$. Let us define a scalar valued observable $\psi_i(\mathbf{x})$ such that

$$\psi_i(\mathbf{x}) = \langle \mathbf{g}(\mathbf{x}), \mathbf{w}_i \rangle = \overline{w}_{i1}g_1(\mathbf{x}) + \overline{w}_{i2}g_2(\mathbf{x}) + \cdots + \overline{w}_{ip}g_p(\mathbf{x}) \quad , \quad [i = 1,2,\ldots,p] \tag{17.28}$$

where $\mathbf{w}_i = [w_{i1}\ w_{i2}\ \ldots\ w_{ip}]^T$. Note that $\psi_i(\mathbf{x})$ is a linear combination of Koopman observables. As a result, from linearity of the Koopman operator, $\psi_i(\mathbf{x})$ is also a Koopman observable. Thus, $\mathcal{K}$ upon operating on $\psi_i(\mathbf{x}^{(n)})$ results in

$$\begin{aligned}
\mathcal{K}\psi_i(\mathbf{x}^{(n)}) &= \psi_i(\mathbf{x}^{(n+1)}) = \langle \mathbf{g}(\mathbf{x}^{(n+1)}), \mathbf{w}_i \rangle = \langle \mathbf{K}\mathbf{g}(\mathbf{x}^{(n)}), \mathbf{w}_i \rangle \\
&= \langle \mathbf{g}(\mathbf{x}^{(n)}), \mathbf{K}^*\mathbf{w}_i \rangle = \langle \mathbf{g}(\mathbf{x}^{(n)}), \overline{\lambda}_{ki}\mathbf{w}_i \rangle \\
&\Rightarrow \mathcal{K}\psi_i(\mathbf{x}^{(n)}) = \lambda_{ki}\langle \mathbf{g}(\mathbf{x}^{(n)}), \mathbf{w}_i \rangle \\
&\Rightarrow \mathcal{K}\psi_i(\mathbf{x}^{(n)}) = \lambda_{ki}\psi_i(\mathbf{x}^{(n)})
\end{aligned} \tag{17.29}$$

It turns out that eigenvalues of $\mathbf{K}$ are also the eigenvalues of $\mathcal{K}$ ($\lambda_{ki} = \lambda_i$) and $\psi_i(\mathbf{x})$ are nothing but the eigenfunctions $\phi_i(\mathbf{x})$,

$$\phi_i(\mathbf{x}) = \psi_i(\mathbf{x}) = \langle \mathbf{g}(\mathbf{x}), \mathbf{w}_i \rangle = \overline{w}_{i1}g_1(\mathbf{x}) + \overline{w}_{i2}g_2(\mathbf{x}) + \cdots + \overline{w}_{ip}g_p(\mathbf{x}). \tag{17.30}$$

In other words, $\phi_i(\mathbf{x}) \in \text{span}\{g_j(\mathbf{x})\}, j = 1,2,\ldots,p$ [67]. However, unlike $\mathbf{K}$, $\mathcal{K}$ has infinite eigenvalues. Through simple algebraic manipulations we can show that λ_i^k ($k \in \mathbb{Z}^+$) is also an eigenvalue of $\mathcal{K}$ with eigenfunction $\phi_i^k(\mathbf{x})$ [69]. Up to this point, no assumptions have been made about $\mathbf{K}$, except that it exists. Let us assume that $\mathbf{K}$ has a full set of eigenvectors, i.e. $\mathbf{u}_i$ are linearly independent and form a basis. With this assumption, we can write the solution for a linear system as follows

$$\begin{aligned}
\mathbf{g}(\mathbf{x}^{(n)}) = \sum_{i=1}^{p} \langle \mathbf{g}(\mathbf{x}^{(n)}), \mathbf{w}_i \rangle \mathbf{u}_i &= \sum_{i=1}^{p} \langle \mathbf{K}^n \mathbf{g}(\mathbf{x}^{(0)}), \mathbf{w}_i \rangle \mathbf{u}_i \\
&= \sum_{i=1}^{p} \langle \mathbf{g}(\mathbf{x}^{(0)}), (\mathbf{K}^*)^n \mathbf{w}_i \rangle \mathbf{u}_i
\end{aligned} \tag{17.31}$$

$$\begin{aligned}
\Rightarrow\ \mathbf{g}(\mathbf{x}^{(n)}) &= \sum_{i=1}^{p} \langle \mathbf{g}(\mathbf{x}^{(0)}), (\mathbf{K}^*)^n \mathbf{w}_i \rangle \mathbf{u}_i = \sum_{i=1}^{p} \langle \mathbf{g}(\mathbf{x}^{(0)}), (\overline{\lambda_i})^n \mathbf{w}_i \rangle \mathbf{u}_i \\
&= \sum_{i=1}^{p} \lambda_i^n \langle \mathbf{g}(\mathbf{x}^{(0)}), \mathbf{w}_i \rangle \mathbf{u}_i \\
\Rightarrow\ \mathbf{g}(\mathbf{x}^{(n)}) &= \sum_{i=1}^{p} \lambda_i^n \phi_i(\mathbf{x}^{(0)}) \mathbf{u}_i, \quad \text{[Using 17.30]}
\end{aligned} \tag{17.32}$$

Eqs. (17.21) and (17.32) don't contradict each other if we consider $\mathbf{u}_i$, the right eigenvectors of $\mathbf{K}$, as the Koopman modes $\mathbf{v}_i$. In that case (17.32) is just the finite truncation of infinite summation in (17.21). Equation (17.32) may also be interpreted as the set of eigenfunctions $\phi_i(\mathbf{x})$, $[i = 1,2,\ldots,p]$ forming the finite-dimensional $\mathcal{K}$-invariant subspace in question. In summary, in a suitable observable space where the dynamics is linear with $\mathbf{g}(\mathbf{x}^{(n+1)}) = \mathbf{K} \cdot \mathbf{g}(\mathbf{x}^{(n)})$, the following points can be noted:

- The eigenvalues of $\mathbf{K}$ are the eigenvalues of the linear infinite-dimensional Koopman operator $\mathcal{K}$.
- If $\mathbf{K}$ has full set of eigenvectors, then the (right) eigenvectors represent Koopman modes associated with $\mathcal{K}$.
- If $\mathbf{K}$ does not have a full set of eigenvectors, then it remains an open question whether they represent Koopman modes [36].

17.4.3 Koopman Autoencoder (KAE)

In what follows, we assume that the reader is familiar with basic concepts of neural networks and we keep the associated discussion very brief. In a nutshell, given a collection of input $\mathbf{x}$ and output $\mathbf{y}$ dataset $\{(\mathbf{x}_i, \mathbf{y}_i)\}_{i=1}^{s}$, a neural network "learns" the mapping from input to output through an optimization process and predicts $\mathbf{y}_i$ for a given $\mathbf{x}_i$, outside the training dataset, i.e. for $i > s$. Autoencoders [70, 71] can be seen as a special type of feed-forward fully-connected neural networks which have essentially two separate components: an (i) encoder and a (ii) decoder. The encoder component maps (encodes) the input to a lower dimensional space and the decoder decodes it back to the original high-dimensional space. Autoencoders are primarily used in data compression to facilitate machine learning methods and to make efficient use of memory. As a result of its ability to compress data while retaining relevant information, autoencoders are a natural choice for nonlinear model order reduction.

In the context of Section 17.4.2, KAE seeks to obtain a suitable vector-valued observable $\mathbf{g}(\mathbf{x})$, such that $\mathbf{g}(\mathbf{x}^{(n+1)}) = \mathbf{K} \cdot \mathbf{g}(\mathbf{x}^{(n)})$ as in (17.26). KAE models have been successfully implemented in modeling of fluids, sea-surface temperature [37, 38, 72] and many other complex nonlinear phenomena. KAE models have also been recently applied to plasma systems, in particular to model the evolution of plasma currents in EMPIC simulations [65]. In what follows, we will describe the basic aspects of an KAE architecture and provide some illustrative results of its application to the reduced order modeling of kinetic plasma problems.

Similar to an generic autoencoder, a KAE has an encoding χ_e and a decoding χ_d layer. In between those two, there is an extra linear layer which approximates $\mathbf{K}$. In other words, the encoder mapping χ_e approximates the transformation $\mathbf{g}(\cdot)$ i.e. $\chi_e(\mathbf{x}) \approx \mathbf{g}(\mathbf{x})$.

Figure 17.3 shows the KAE architecture. It should be noted that, in typical KAE architecture, the linear layer in the middle only approximates the forward dynamics C ($\approx \mathbf{K}$) [38]. However, we will focus here on an improved type of KAE architecture [37] where the backward dynamics D is also taken into consideration, such that $C \cdot D \approx I$ (identity matrix). This consideration improves the stability of the solution and is referred to as consistent KAE [37] to distinguish it from the earlier, more traditional KAE. However, for brevity we will refer to it simply as KAE. The input and output layers (white) consist of $N_{in} = N_{out} = N$ neurons which are typically equal to the dimension of the state which we are attempting to model. Both the encoding and decoding layers have N_h neurons in the hidden (gray) layer. The "bottleneck" layer (or encoded layer, in black) has N_b neurons with $N_b \ll N$ (order-reduction). The linear layers are followed by a hyperbolic tangent activation layer except the bottleneck layer. The state at the nth time-step $\mathbf{x}^{(n)}$ is fed to the input layer, which is then encoded to $\mathbf{g}(\mathbf{x}^{(n)})$. The linear layer in the middle section of the network advances the dynamics in the forward (or backward) directions by k time-steps to generate $\mathbf{g}(\mathbf{x}^{(n\pm k)})$, which is then decoded back to the original state space giving us $\hat{\mathbf{x}}^{(n\pm k)}$:

$$\mathbf{x}^{(n+k)} \approx \hat{\mathbf{x}}^{(n+k)} = \chi_d \circ C^k \circ \chi_e(\mathbf{x}^{(n)}) \tag{17.33a}$$

$$\mathbf{x}^{(n-k)} \approx \hat{\mathbf{x}}^{(n-k)} = \chi_d \circ D^k \circ \chi_e(\mathbf{x}^{(n)}) \tag{17.33b}$$

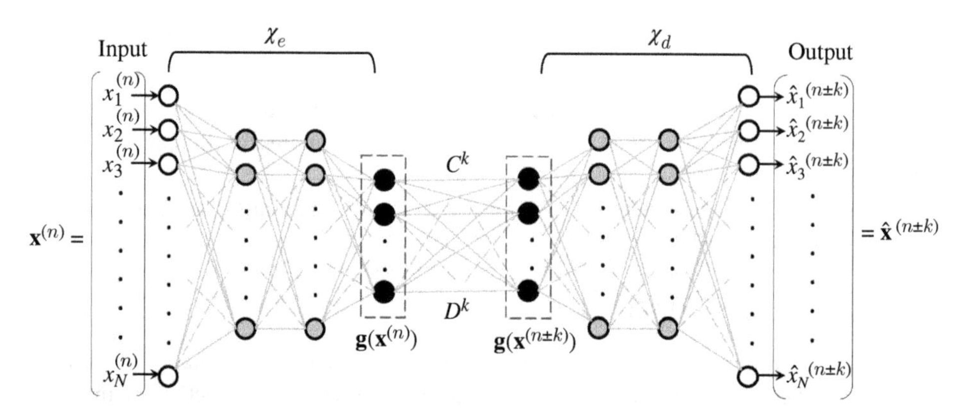

Figure 17.3 Schematic diagram of a consistent KAE architecture. The input and output layers are shown by the white neurons (nodes) whereas the hidden layers are shown by gray neurons. The black neurons denote the encoded layer where it learns the (reduced-order) linear dynamics.

Here, "$\circ$" denotes the composition of two operators. The χ_e and χ_d are not exactly symmetric because χ_d has an extra nonlinear activation layer at the end. Before feeding the autoencoder, the input data are scaled in the range $[-1,1]$. Since the hyperbolic tangent activation function also varies from -1 to 1, the final activation layer for χ_d ensures that the output is in the desired range. Note that typically, data are sampled regularly across certain time-steps Δn. In that case, C^k advances the dynamics by k time samples or $k\Delta n$ time-steps. In order to facilitate a concise explanation of the functionality of KAE, we assume $\Delta n = 1$.

The autoencoder is trained to minimize the total loss function L_{tot} which consists of four separate components denoted as (i) identity loss L_{id}, (ii) forward loss L_{fwd}, (iii) backward loss L_{bwd}, and (iv) consistency loss L_{con}:

$$L_{tot} = \gamma_{id}L_{id} + \gamma_{fwd}L_{fwd} + \gamma_{bwd}L_{bwd} + \gamma_{con}L_{con} \tag{17.34}$$

where $\gamma_{id}, \gamma_{fwd}, \gamma_{bwd}$, and γ_{con} are the user defined weights. The identity loss L_{id} measures the autoencoder's ability to reconstruct the state as it is at a particular time-step n by first encoding it to a lower dimensional subspace and then by decoding it back to the original state space. L_{id} can be defined as

$$L_{id} = \frac{1}{2l}\sum_{n=1}^{l} \|\hat{\mathbf{x}}^{(n)} - \mathbf{x}^{(n)}\|_2 \tag{17.35}$$

where l is number of time samples in the training data. The consistency loss L_{con} measures consistency of the matrices C and D, i.e. how closely they follow the relation $C \cdot D \approx I$. It can be expressed as

$$L_{con} = \sum_{i=1}^{N_b} \frac{1}{2i}\|D_{i\star}C_{\star i} - I_{N_b}\|_F + \frac{1}{2i}\|C_{\star i}D_{i\star} - I_{N_b}\|_F \tag{17.36}$$

where $D_{i\star}$ is the upper i rows of D, $C_{\star i}$ is the i left most columns of C, $\|\cdot\|_F$ denotes the Frobenius norm, and I_{N_b} is the identity matrix of dimension $N_b \times N_b$. The forward loss L_{fwd} measures the autoencoder's ability to reconstruct $\mathbf{x}^{(n+k)}$ by encoding $\mathbf{x}^{(n)}$, forwarding the dynamics by k time-steps (or time samples) in encoded space and then decoding it back to the original state space. Mathematically, L_{fwd} can be expressed as

$$L_{fwd} = \frac{1}{2k_m l}\sum_{k=1}^{k_m}\sum_{n=1}^{l} \|\hat{\mathbf{x}}^{(n+k)} - \mathbf{x}^{(n+k)}\|_2^2 \tag{17.37}$$

where k_m is the maximum number of time-steps utilized in the forward direction. Finally, the backward loss L_{bwd} plays a role similar to L_{fwd} except it deals with reconstruction in the backward direction, that is $\mathbf{x}^{(n-k)}$. As such, the backward loss is expressed as

$$L_{bwd} = \frac{1}{2k_m l} \sum_{k=1}^{k_m} \sum_{n=1}^{l} ||\hat{\mathbf{x}}^{(n-k)} - \mathbf{x}^{(n-k)}||_2^2 \tag{17.38}$$

where k_m now represents the maximum number of time-steps utilized in the backward direction. Note that for notational convenience we have explained the KAE operation in the above by assuming that each time-step data corresponds to one sample in training data. However, as mentioned earlier, the input data does not need to correspond to the sampling of every consecutive time-step. Instead, there might be regular or irregular intervals comprising several time-steps between consecutive input data samples. Apart from the usual neural network training hyperparameters such as learning rate, training epochs etc., the tunable parameters $\gamma_{id}, \gamma_{fwd}, \gamma_{bwd}, \gamma_{con}, N_h$, and N_b play a crucial role in determining the extrapolation accuracy of the KAE model.

17.4.3.1 Case Study I: Oscillating Electron Beam

In the following, we will illustrate the application of the ideas discussed above in the simulation of electron plasma beams. We will first revisit and address in more detail the problem first discussed in [65], which deals with a 2D oscillating electron beam. Consider a 2D square domain of size 1 × 1m as shown in Figure 17.4a. The solution domain is discretized by an unstructured mesh with $N_0 = 1647$ nodes, $N_1 = 4788$ edges, and $N_2 = 3142$ triangular cells for the EMPIC simulation. Perfect electric conductor (PEC) boundary conditions are assumed for the fields. Superparticles representing $S_p = 2 \times 10^5$ electrons, are injected at random points along the x coordinate in the bottom of the domain following a uniform distribution in the range [0.4 m, 0.6 m]. The injection rate is 10 superparticles per time-step ($\Delta_t = 0.01$ ns). The superparticles have an initial y directional

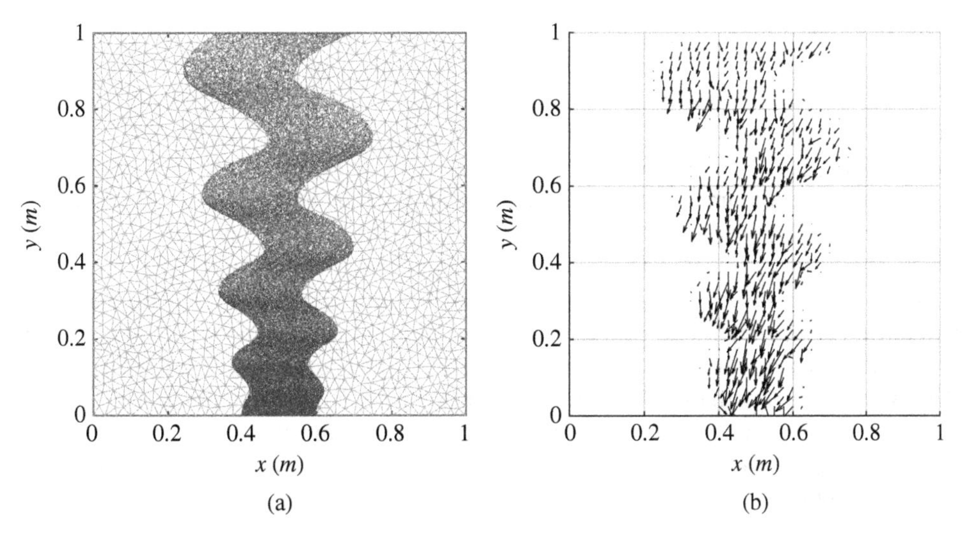

Figure 17.4 (a) Snapshot of an oscillating 2D electron beam propagating along the y direction at $n = 40{,}000$. The net charge transport occurs along the longitudinal direction, and the beam oscillation occurs along the transverse (lateral) direction. The dark gray dots represent the superparticles in the backdrop of the unstructured mesh shown by the light grey lines. Source: This figure is adapted from [65]. (b) The quiver plot of current density at $n = 40{,}000$. The black arrows indicate the direction of current density, with their relative size indicating their relative magnitude.

velocity of $+5 \times 10^6$ m/s. They are absorbed as they hit the opposite boundary. An external voltage bias of $V_b = 2$ kV is set to accelerate the electrons in positive y direction. An external, oscillating transverse magnetic flux density field $B_{ext} = B_0 \sin(2\pi t/T_b)\,\hat{\mathbf{z}}$ is applied where $B_0 = 10^{-3}$ T and $T_b = 20$ ns. The EMPIC simulation is run for $n = 240{,}000$ time steps and the current density DoFs ($\mathbf{j}^{(n+1/2)}$) are sampled from $n = 16{,}000$ to $n = 240{,}000$ with sampling interval $\Delta n = 40$, providing total 5601 time samples (or snapshots) for training and testing.

As discussed earlier, the various stages of the EMPIC algorithm involving particles act as computational bottlenecks, restricting our ability to speed up the simulations. The steps involving particles are necessary for the time update of the current density $\mathbf{j}^{(n+1/2)}$, which is an integral part of the full Maxwell's time update Eq. (17.7). If we can otherwise model and predict the time evolution of $\mathbf{j}$, we can bypass the time consuming steps involving particles in the EMPIC algorithm. Note, however, that the inherent nonlinearity of the problem makes it challenging to accurately model the evolution of $\mathbf{j}$. In addition, $\mathbf{j}$ is not guaranteed to have a smooth behavior in the spatial or in the temporal domain. As noted before, when dealing with nonlinear problems, it makes sense to exploit nonlinear transformations which are realizable by nonlinear activation layers in KAE [65].

As can be seen in Figure 17.4a, the charged particles are confined to a limited region of space along the horizontal direction centered around the midpoint $x = 0.5$ m. As a result, the corresponding current density (Figure 17.4b) is zero outside this region of space. So, a new vector $\mathbf{j}_{nz}^{(n+1/2)}$ is formed with the edges (i) where current density DoF is nonzero, i.e. $j_i^{(n+1/2)} \geq 10^{-20}$ A/m for all the training time samples. Note that in the context of Figure 17.3, $\mathbf{x}^{(n)} = \mathbf{j}_{nz}^{(n+1/2)}$, since we want to model the time-evolution of snapshots of current density (Figure 17.4b) using KAE. This results in $N_{in} = N_{out} = N = 2072$. The other hyperparameters are chosen as follows: $N_h = 256$ and $N_b = 32$ with $\gamma_{id} = 1$, $\gamma_{fwd} = 2$, $\gamma_{bwd} = 0.1$, and $\gamma_{con} = 0.01$. A total of 2000 training samples are used spanning from $n = 16{,}000$ to 95,960 with consecutive time samples sampled at 40 time-steps ($\Delta n = 40$) apart. The model is trained for 100 time samples ($k_m = 100$) in the forward and the backward directions, and tested on 3500 time samples in the extrapolation region. The key training parameters are summarized in Table 17.1.

The parameters N_{in} and N_{out} depend on the number of mesh elements over which we seek to model the current density. The actual N_h and N_b values can be decided based on trial and error. However, the physics of the problem provides some insight on how to choose N_b. For example, N_b essentially denotes the number of "modes" or "features" for modeling the dynamics in the transformed space. The Koopman eigenvalues (λ) are related to the frequencies ($\omega = \ln|\lambda|/\Delta t$, Δt being the time interval between training samples) under which the Koopman modes evolve in time. Since we are dealing with real-valued data, the Koopman eigenvalues exist in complex-conjugate pairs, except for those residing on the real axis. If we consider the original state space, there are two primary frequency components for this electron beam problem: the static component (representing the net, longitudinal particle transport from the bottom to the top of the domain) and the transversal oscillatory component with oscillation frequency matching that of the external magnetic flux. So, we can expect at the very minimum two modes (three eigenvalues) to roughly represent the dynamics in the original state space. However, in case of KAE we are dealing with a transformed space and we do not have insights regarding how these frequency components may be precisely

Table 17.1 Important KAE training parameters for oscillating beam.

Parameters	Δn	k_m	N_{in}, N_{out}	N_h	N_b	γ_{fwd}	γ_{id}	γ_{bwd}	γ_{con}
Value	40	100	2072	256	32	2	1	0.1	0.01

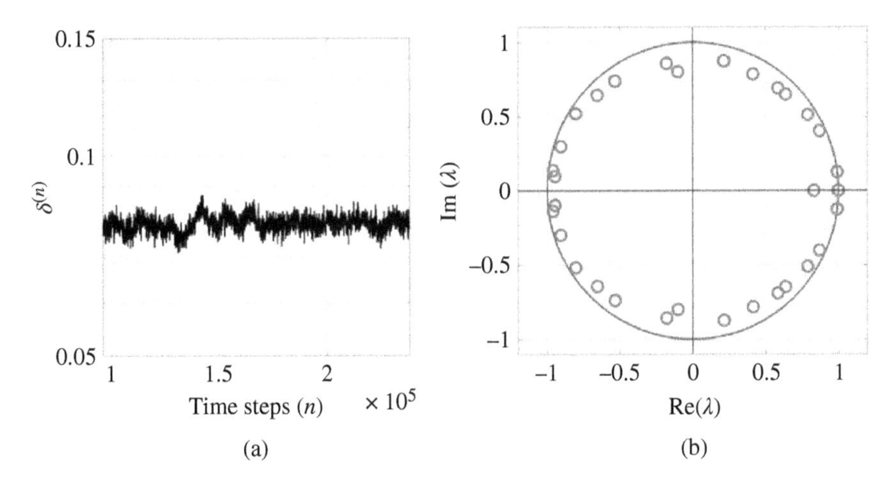

Figure 17.5 (a) Relative 2-norm error for $\mathbf{j}_{nz}^{(n+1/2)}$ in the extrapolation region. (b) The Koopman eigenvalues (small circles) corresponding to C, the forward dynamics. The large circle represents the unit circle in the complex plane. Source: This figure is adapted from [65].

transformed (through a nonlinear transformation) to the feature space. However, $N_b > 3$ acts as a good starting point. High values of N_h might result in a large number of trainable parameters which could lead to overfitting. On the other hand, while small N_h avoids overfitting, the model may not be able to capture all the intricacies of the dynamics. The weights for the loss functions also play a crucial role in determining the extrapolation accuracy of the KAE model. Intuitively, it makes more sense to assign larger weights to γ_{fwd} and γ_{id} as our primary interest is to predict (extrapolate) the solution in future time. After gaining a rough idea about the range of various weights, grid search can be implemented to determine their optimum values.

Note that in the context of this problem, (17.33a) can be re-written as

$$\mathbf{j}_{nz}^{(n+\frac{1}{2}+k\Delta n)} \approx \hat{\mathbf{j}}_{nz}^{(n+\frac{1}{2}+k\Delta n)} = \chi_d \circ C^k \circ \chi_e \left(\mathbf{j}_{nz}^{(n+\frac{1}{2})} \right) \tag{17.39a}$$

$$\mathbf{j}_{nz}^{(n+\frac{1}{2}-k\Delta n)} \approx \hat{\mathbf{j}}_{nz}^{(n+\frac{1}{2}-k\Delta n)} = \chi_d \circ D^k \circ \chi_e \left(\mathbf{j}_{nz}^{(n+\frac{1}{2})} \right) \tag{17.39b}$$

Figure 17.5a shows that the relative error in the predicted $\mathbf{j}_{nz}$ is around 8%. Note that the relative error is calculated over all the mesh element edges associated with $\mathbf{j}_{nz}$. It is important to mention that we can always go back to $\mathbf{j}$ (or $\hat{\mathbf{j}}$) from $\mathbf{j}_{nz}$ (or $\hat{\mathbf{j}}_{nz}$), simply by zero padding. The expression for the relative error at n^{th} time-step is given by

$$\delta^{(n)} = \frac{\|\hat{\mathbf{j}}_{nz}^{(n+\frac{1}{2})} - \mathbf{j}_{nz}^{(n+\frac{1}{2})}\|_2}{\|\mathbf{j}_{nz}^{(n+\frac{1}{2})}\|_2} \tag{17.40}$$

where hatted quantities represent the KAE reconstruction and $\|\cdot\|_2$ denotes the L^2 norm. The stability of the solution, even for long term predictions can be attributed to the Koopman eigenvalues, as depicted in Figure 17.5b, staying inside or on the unit circle in the complex plane. This can be better understood from the perspective of theory of linear systems. Let us go back to the (17.26), (17.32) which tells us that under suitable transformation $\mathbf{g}(\cdot)$, any dynamical system can be represented through linear dynamics:

$$\mathbf{g}(\mathbf{x}^{(n+1)}) = \mathbf{K} \cdot \mathbf{g}(\mathbf{x}^{(n)}) \tag{17.41}$$

$$\mathbf{g}(\mathbf{x}^{(n)}) = \sum_{i=1}^{p} \lambda_i^n \phi_i(\mathbf{x}^{(0)}) \mathbf{u}_i \tag{17.42}$$

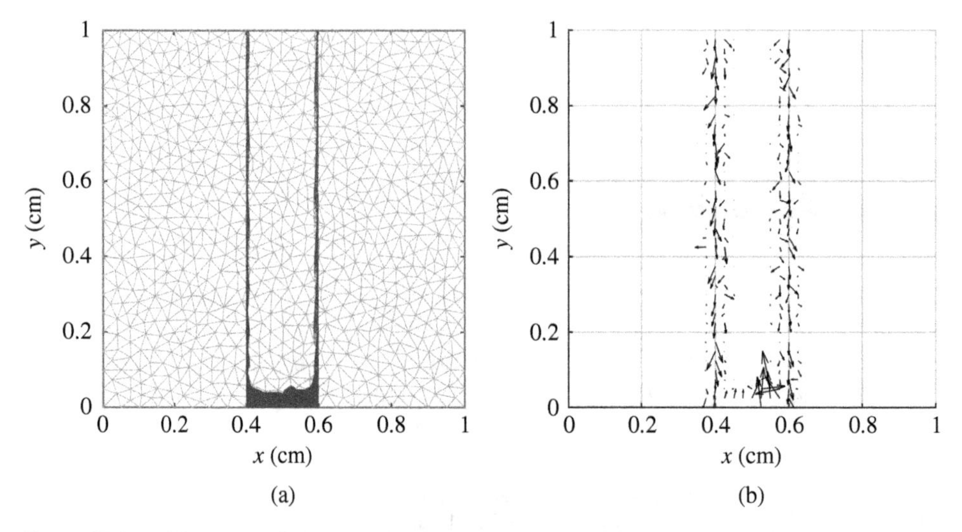

Figure 17.6 (a) Snapshot of virtual cathode formation at $n = 80{,}000$. The dark gray dots represent the superparticles, whereas the light gray lines indicate the unstructured mesh. The virtual cathode impedes most charge transport through the center beam region beyond the cathode. Two leakage streams are observed, corresponding to the electrons that are able to laterally circumvent the virtual cathode repulsion. (b) The quiver plot of current density at $n = 80{,}000$.

In our setting, $\mathbf{x} \equiv \mathbf{j}_{nz}$, $\mathbf{g} \equiv \chi_e$, and $\mathbf{K} \equiv C$. Equations (17.32) or (17.42) tells us that for a stable systems $|\lambda_i| \leq 1$, or in other words, eigenvalues of $\mathbf{K}$ should lie on or inside the unit circle, which is the case in Figure 17.5b.

17.4.3.2 Case Study II: Virtual Cathode Formation

Next, we test KAE for a different electron beam problem involving a virtual cathode formation. Figure 17.6a shows a 2D electron beam propagating along the positive y direction and forming the virtual cathode. The solution domain (xy plane) has dimensions 1 cm × 1 cm and is discretized by an unstructured mesh consisting of $N_0 = 844$ nodes, $N_1 = 2447$ edges, and $N_2 = 1604$ triangular cells. Superparticles are injected with initial velocity $\mathbf{v}_0 = 5 \times 10^6\ \hat{\mathbf{y}}$ m/s at the bottom of the cavity with uniform random distribution in the spatial interval [0.4, 0.6] cm along x. The injection number rate of superparticles is set to 30 per time step (with superparticle ratio of 5×10^5 and $\Delta_t = 0.2$ ps). A strong confining magnetic field, $B_{ext} = B_y\ \hat{\mathbf{y}}$ is applied in the y direction, with $B_y = 100$ T.

The current density is sampled at every 80th time-steps, generating a total of 2000 time samples. A virtual cathode formation with stable oscillations takes place after the transient regime is over. We are essentially trying to model these oscillations in the current density (Figure 17.6b) using the KAE. We consider a total of 800 time samples for training, starting from the 400th time sample onward. An approach similar to the one considered in the previous example is taken in order to model the current density; however, unlike the previous case where the oscillations are prominent due to external magnetic flux, the oscillations in this case are more subtle and localized. The oscillations are localized in the sense that they take place at the root of the beam, close to the virtual cathode, and a lingering effect can be observed at the edges due to the lateral beam leakage. The higher concentration of superparticles results in "smoother" time variation of the current density at the root of the beam compared to boundary of the beam (Figure 17.8). Such weak localized oscillations make the modeling task challenging which is reflected in relatively higher error in reconstruction in the extrapolation region.

The KAE training parameters in this case are chosen as follows (Table 17.2):

Figure 17.7 shows on average relative error of 15% in the predicted current density. Similar to the oscillating beam case, the Koopman eigenvalues residing inside or on the unit circle ensures stability in the reconstruction. However, relatively high error in reconstruction can be attributed to limitations of fully connected neural networks in learning high-frequency functions. Note that KAE is realized by fully connected layers and fully connected NNs are known to have spectral bias [73] also known as "F-principle" [74]. As mentioned earlier, the time signature of current density is not so "smooth" for this particular test case, especially at the mesh edges located at two sides of the beam. Figure 17.8 shows how smoothness of current density time signature varies based on the density of superparticles. The edge indexed 530 (indexed with respect to 2447 edges) is located at the root of the beam where superparticles are concentrated. Consequently the time variation of j is smooth and KAE reconstruction follows the original high-order solution closely as can be seen in Figure 17.8a. Same comments can not be made for the edge indexed 777, which encounters much lower density of superparticles resulting in non-smooth time variation in j (Figure 17.8b). This phenomena leads to higher relative error in the KAE reconstruction.

One of the primary reasons behind the non-smooth nature of the current variation is the point nature of the superparticles, i.e. representation of number density function for a superparticle through a delta distribution. Consideration of finite-size superparticles in EMPIC algorithms can mitigate this issue to some extent, but it poses additional implementation challenges. One straightforward approach would be to inject a much large number of superparticles while keeping the $N_p \times S_p$ same. But recall that time-step complexity of EMPIC algorithms is approximately $\mathcal{O}(N_p)$. therefore, a careful consideration should be made regarding the trade-off between smoothness and computational cost.

Table 17.2 Important KAE training parameters for virtual cathode.

Parameters	Δn	k_m	N_{in}, N_{out}	N_h	N_b	γ_{fwd}	γ_{id}	γ_{bwd}	γ_{con}
Value	80	100	459	256	32	2	1	0.1	0.01

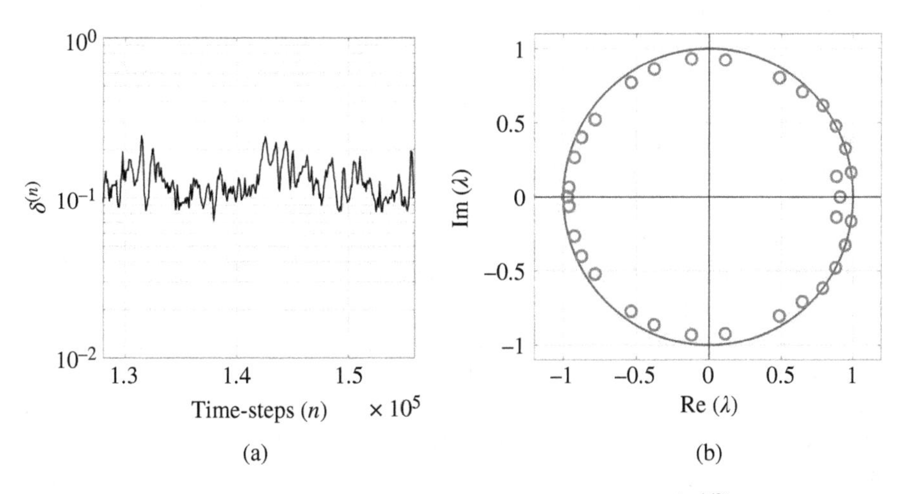

Figure 17.7 (a) Relative 2-norm error in the extrapolation region for $\mathbf{j}_{nz}^{(n+1/2)}$ (b) The relative position of the Koopman eigenvalues (corresponding to forward dynamics) w.r.t. the unit circle in the complex plane.

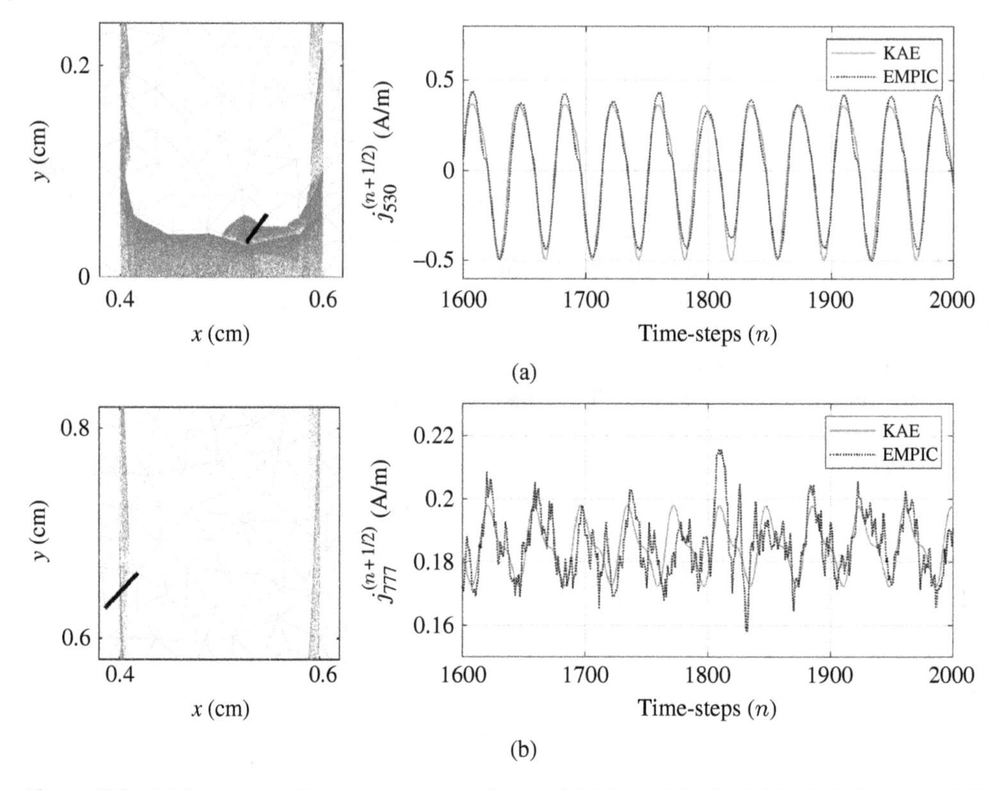

Figure 17.8 (a) Current density sampled at a mesh edge (highlighted by the thick black line, top left figure) located at the root of the beam with larger concentration of (super)particles (gray dots). The top right figure shows the corresponding reconstruction of current density DoF. (b) KAE reconstruction of the current density (bottom right) is observed at an edge experiencing leakage current (bottom left).

17.4.4 Computational Gain

As noted, one of the primary bottlenecks in expediting the EMPIC simulations stems from the large number of superparticles, resulting in the time complexity of $\mathcal{O}(N + N_p)$ at each time-step for an explicit field-update. The three EMPIC stages involving superparticles (N_p) are necessary for the time update of current density $\mathbf{j}$. The initial results shown here for the two test cases show good potential for the modeling of current density using KAE. Once the model is trained, the current density at any time-step can be queried by $\mathcal{O}(N_1)$ or equivalently $\mathcal{O}(N)$ operations. Recall that N_1 is the number of mesh element edges over which $\mathbf{j}$ evolves and N is the aggregate mesh dimension. Using KAE to model the time evolution current density, we can effectively reduce the time-step complexity of EMPIC simulations down to $\mathcal{O}(N)$. This is a huge improvement considering that $N \ll N_p$ in typical scenarios. Figure 17.9 illustrates this approach, which was explored in a recent work, but in the context of DMD [75].

However, there are some downsides of using data-driven extrapolation of current density at each time-step. The first and most obvious drawback is the loss of accuracy in the extrapolated current. We already discussed what makes modeling of current density challenging and how some scenarios can be especially difficult to model. In particular, any error in the extrapolated current density will introduce subsequent error in the electric field, which in turn will add an error in the magnetic flux, according to the field-update equations in (17.7). This error, typically larger than the ordinary discretization error due to EMPIC algorithm, is cumulative with each time-step. As a result,

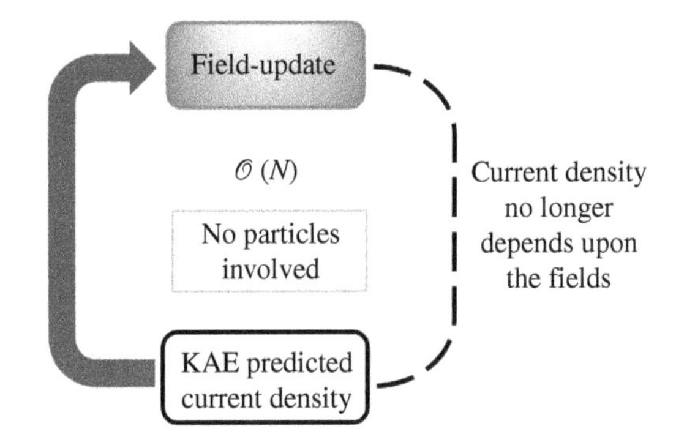

Figure 17.9 EMPIC with KAE predicted current density ($\hat{\mathbf{j}}$) only involvs field update with $\hat{\mathbf{j}}$, with time-step complexity of $\mathcal{O}(N)$.

the field values progressively deviate from the original solution with time. Error accumulation is a common issue in most types of time-series predictive algorithms. The second drawback comes from the potential loss of physical consistency in the solution in the extrapolation region. With KAE extrapolated currents, it is not always possible to ensure energy or momentum conservation, for example, simply due the fact that the typical KAE models are purely data-driven and do not enforce all physics constraints of the problem while training. One of the possible solutions to tackle the drawbacks mentioned above is using the physics constraints in training of KAE models, to be discussed in Section 17.5. Again, due to the cumulative nature of the effects, these drawbacks become more relevant for very long-term predictions. For short or moderate time extrapolation intervals, KAE models are expected to provide reasonable computational gain while providing acceptable accuracy.

17.5 Towards A Physics-Informed Approach

Motivated by the recent breakthroughs in "big data" processing [22, 23], hardware and memory architecture evolution [25, 26], and machine learning (ML) algorithms [29], there has been a surge in machine learning based data-driven modeling of physical systems across the scientific community. Deep learning methods provide us with tools for extracting features from multi-fidelity, multidimensional data, which are essential for characterizing complex systems [76]. However, modeling and forecasting multiscale multiphysics systems from a purely data-driven perspective has its own limitations. This is because machine learning models, and in particular deep learning models involving neural networks, tend to lack general interpretability. Models trained purely on data might fit the observations well in the training region, but when it comes to predicting beyond the observed dataset, these models typically fail in the presence of any new underlying physical regimes not present in the training dataset (for example, beyond a phase transition) [77]. Also, the accuracy of any data-driven model depends on the quality of the data used to train that particular model. Often, the available data are corrupted with noise. Therefore, it becomes important to have some *a* priori information regarding the governing physics model of the system to distinguish "physical data" from "non-physical data" (noise). Another motivation for including physics information into the training process stems from the data-hungry nature of ML models. In many cases, certain parts of a physical system might be inaccessible to experiments or a large amount of time might be necessary to collect/generate sufficient amount of data. In such data-limited scenarios, one needs to incorporate physics constraints to guide the ML model through "low-data" regime. In the past, one of the strategies for the inclusion of physics knowledge in the training of machine learning models included the so-called knowledge-based neural networks [78–81]. It is of note that neural networks have been used in the past to directly solve PDEs [82] and to emulate certain types of measurement data as well [83]. However, a number of recent advances in ML techniques have shed new light [84, 85] on the topic of physics-informed machine learning (PIML). The reader is referred to [77] for a recent account of ongoing developments in this area.

Depending on the available data, there can be broadly three categories of prior information levels encountered while trying to model a physical system. First, a limited amount of data and a comprehensive knowledge of the governing physics leads to traditional approaches where physics equations are used to describe a system. On the other extreme, an absolute lack of information regarding the underlying physics and the presence of a large dataset motivates a purely data-driven "black-box" approach. The more common scenario lies somewhere in the middle, where some physics-information regarding the system is available (for example some of the governing equations but not the full boundary conditions, energy conservation, expected underlying symmetries, etc.), and some data is available through measurements or partial simulations. A "physics-informed" or "physics-guided" machine learning approach is suitable for such scenarios.

We have so far demonstrated the effectiveness of KAE for modeling EMPIC simulations without involving the physics of the problem (other than using high-fidelity simulations to provide the underlying data for training). Therefore, this approach is more data-centric rather than physics-centric. As a result, there is no guarantee that reconstructed current density (or the electromagnetic field for that matter) will follow the physics relations discussed in (17.7). One way to make the model aware of the underlying physics is to introduce new loss term(s) incorporating physical constraints (laws and/or boundary conditions) so that the physics information can be included during the training process. In the context of EMPIC kinetic plasma simulations, the energy constrain or the contraints arising from the Maxwell's equations can be taken into account while training KAE. The new physics loss term will be simply equivalent to the equation error in the discretized Maxwell's equations (17.7), or energy equation. These involve the quantities $\mathbf{e}$, $\mathbf{b}$, $\mathbf{j}$, $\mathbf{C}$, $\tilde{\mathbf{C}}$ and the Hodge star matrices $\left[\star_{\mu^{-1}}\right]$ and $\left[\star_{\epsilon}\right]$. Recall that $\mathbf{C}$ and $\tilde{\mathbf{C}}$ depends only on the connectivity among mesh elements whereas $\left[\star_{\mu^{-1}}\right]$ and $\left[\star_{\epsilon}\right]$ contains the metric information from the computational mesh. In other words, for a fixed mesh, $\mathbf{C}$, $\tilde{\mathbf{C}}$, $\left[\star_{\mu^{-1}}\right]$ and $\left[\star_{\epsilon}\right]$ are fixed. For each of $\mathbf{e}$, $\mathbf{b}$, and $\mathbf{j}$ we can have separate KAE architectures, the outputs of which will be connected through the equation error from physical constraints. Note that there are several training challenges associated with physics-informed learning in general [86]. Also, it becomes difficult for the neural networks to learn the physics laws from loss functions originated by complicated equations such as (17.7). This is an area of ongoing research and further investigations need to be carried out to take full advantage of physics-informed approach.

17.6 Outlook

Data-driven machine learning models in general perform very well for interpolation in time or the parameter domain. However, when it comes to extrapolating beyond the scope of training dataset, the data-driven models typically do not fare well. A key strength of ML models is their ability to extract relevant information or features entirely from data, without the explicit knowledge of the underlying physics driving the system. While the reliance on data is a desirable strength of ML models, it can also become a limitation when it comes to predicting "beyond" the observed data. Since the data-driven models "learn" from the training data itself, it is not wise to expect these models to predict something out of the ordinary or fundamentally different behavior (such as phase transitions) not already covered by the training dataset. In the context of time extrapolation of certain plasma properties such as current density, electric field etc., it is ambitious to expect the ML models to predict equilibrium or limit cycle behavior, from very limited amount of transient data, or vice versa. KAE is also not immune to these. However, KAE has certain advantages over traditional methods for long-term forecasting of dynamical systems. Extrapolating periodic data such as limit cycle behavior might seem like a trivial task, but the traditional statistical tools such as auto regressive integrated moving average (ARIMA), and Fourier-based models are not well-suited for very long term predictions of high-dimensional complex nonlinear systems, especially when assumption of stationarity is violated. Neural network (NN) based models such as recurrent neural network (RNN) and long short-term memory (LSTM) can effectively handle nonlinearlity, and perform well for short-term predictions. However, for long-term forecasting of high-dimensional complex nonlinear dynamics, KAE has several advantages over RNN or LSTM, from better generalizability and interpretability to better handling of nonlinearity. Moreover, KAEs are designed to learn the reduced-order dynamics in a low-dimensional space, leading to better scalability, which might be crucial for modeling EMPIC simulations consisting of thousands and millions of mesh elements (edges, facets etc.). The high-dimensionality, inherent nonlinearities, and complex nature of plasma dynamics, make the Koopman based methods such as KAE an ideal choice for reduced-order modeling purpose.

The overdependency of data-driven models such as KAE on the data and thus their inability to accurately extrapolate beyond the training dataset in complex problems can be further addressed by including prior physics information in the model in the form of PIML. There are inherent challenges associated with physics-informed learning and it is not always straight forward to fuse physics with the data-driven models. However, if implemented successfully, physics-informed learning has the potential to overcome the barriers mentioned earlier. Further investigations need to be carried out to determine the efficacy of integrating physics information into the training KAE models for plasma-related problems.

Acknowledgments

This work was supported in part by the Defense Threat Reduction Agency (DTRA) grant No. HDTRA1-18-1-0050, in part by the US Department of Energy Grant DE-SC0022982 through the NSF/DOE Partnership in Basic Plasma Science and Engineering and in part by the Ohio Supercomputer Center grant No. PAS-0061. The content of the information does not necessarily reflect the position or the policy of the U.S. federal government, and no official endorsement should be inferred.

References

1 Priest, E. (2014). *Magnetohydrodynamics of the Sun*. Cambridge University Press.
2 Goldstein, M.L., Roberts, D.A., and Matthaeus, W.H. (1995). Magnetohydrodynamic turbulence in the solar wind. *Annual Review of Astronomy and Astrophysics* 33 (1): 283–325.
3 Sovinec, C.R., Glasser, A.H., Gianakon, T.A. et al. (2004). Nonlinear magnetohydrodynamics simulation using high-order finite elements. *Journal of Computational Physics* 195 (1): 355–386.
4 Gold, S.H. and Nusinovich, G.S. (1997). Review of high-power microwave source research. *Review of Scientific Instruments* 68 (11): 3945–3974.

5 Booske, J.H. (2008). Plasma physics and related challenges of millimeter-wave-to-terahertz and high power microwave generation. *Physics of Plasmas* 15 (5): 055502.

6 Benford, J., Swegle, J.A., and Schamiloglu, E. (2015). *High Power Microwaves*. CRC Press.

7 Lapenta, G., Brackbill, J.U., and Ricci, P. (2006). Kinetic approach to microscopic-macroscopic coupling in space and laboratory plasmas. *Physics of Plasmas* 13 (5): 055904.

8 Nayak, I., Na, D., Nicolini, J.L. et al. (2020). Progress in kinetic plasma modeling for high-power microwave devices: analysis of multipactor mitigation in coaxial cables. *IEEE Transactions on Microwave Theory and Techniques* 68 (2): 501–508.

9 Karimabadi, H., Loring, B., Vu, H.X. et al. (2011). Petascale kinetic simulation of the magnetosphere. *Proceedings of the 2011 TeraGrid Conference: Extreme Digital Discovery*, TG '11, 1–2. New York, NY, USA: Association for Computing Machinery. ISBN 9781450308885.

10 Chen, Y., Tóth, G., Hietala, H. et al. (2020). Magnetohydrodynamic with embedded particle-in-cell simulation of the geospace environment modeling dayside kinetic processes challenge event. *Earth and Space Science* 7 (11): e2020EA001331.

11 Zheng, R., Ohlckers, P., and Chen, X. (2011). Particle-in-cell simulation and optimization for a 220-GHz folded-waveguide traveling-wave tube. *IEEE Transactions on Electron Devices* 58 (7): 2164–2171.

12 Eppley, K. (1988). The Use of Electromagnetic Particle-in-Cell Codes in Accelerator Applications. *Technical Report SLAC-PUB-4812; CONF-881049-46 ON: DE89004968*. Menlo Park, CA, USA: Stanford Linear Accelerator Center.

13 Kraus, M., Kormann, K., Morrison, P.J., and Sonnendrücker, E. (2017). GEMPIC: geometric electromagnetic particle-in-cell methods. *Journal of Plasma Physics* 83 (4): 905830401.

14 Na, D., Moon, H., Omelchenko, Y.A., and Teixeira, F.L. (2016). Local, explicit, and charge-conserving electromagnetic particle-in-cell algorithm on unstructured grids. *IEEE Transactions on Plasma Science* 44 (8): 1353–1362.

15 Pandya, M. (2016). Low edge safety factor disruptions in the Compact Toroidal Hybrid: operation in the low-Q regime, passive disruption avoidance and the nature of MHD precursors. PhD thesis. Auburn University.

16 Van Milligen, B.P., Sánchez, E., Alonso, A. et al. (2014). The use of the biorthogonal decomposition for the identification of zonal flows at TJ-II. *Plasma Physics and Controlled Fusion* 57 (2): 025005.

17 Byrne, P.J. (2017). Study of external kink modes in shaped HBT-EP plasmas. PhD thesis. Columbia University.

18 Kaptanoglu, A.A., Morgan, K.D., Hansen, C.J., and Brunton, S.L. (2020). Characterizing magnetized plasmas with dynamic mode decomposition. *Physics of Plasmas* 27 (3): 032108.

19 Nayak, I., Kumar, M., and Teixeira, F.L. (2021). Detection and prediction of equilibrium states in kinetic plasma simulations via mode tracking using reduced-order dynamic mode decomposition. *Journal of Computational Physics* 447: 110671.

20 Brunton, S.L. and Kutz, J.N. (2019). *Data-Driven Science and Engineering: Machine Learning, Dynamical Systems, and Control*. Cambridge University Press.

21 Montáns, F.J., Chinesta, F., Gómez-Bombarelli, R., and Kutz, J.N. (2019). Data-driven modeling and learning in science and engineering. *Comptes Rendus Mécanique* 347 (11): 845–855.

22 Qin, S.J. (2014). Process data analytics in the era of big data. *AIChE Journal* 60 (9): 3092–3100.

23 Labrinidis, A. and Jagadish, H.V. (2012). Challenges and opportunities with big data. *Proceedings of the VLDB Endowment* 5 (12): 2032–2033.

24 Hajjaji, Y., Boulila, W., Farah, I.R. et al. (2021). Big data and IoT-based applications in smart environments: a systematic review. *Computer Science Review* 39: 100318.

25 Pan, W., Li, Z., Zhang, Y., and Weng, C. (2018). The new hardware development trend and the challenges in data management and analysis. *Data Science and Engineering* 3 (3): 263–276.
26 Sze, V., Chen, Y.-H., Emer, J. et al. (2017). Hardware for machine learning: challenges and opportunities. *2017 IEEE Custom Integrated Circuits Conference (CICC)*, 1–8. IEEE.
27 Schmidhuber, J. (2015). Deep learning in neural networks: an overview. *Neural Networks* 61: 85–117.
28 Iten, R., Metger, T., Wilming, H. et al. (2020). Discovering physical concepts with neural networks. *Physical Review Letters* 124 (1): 010508.
29 Liu, G.-H. and Theodorou, E.A. (2019). Deep learning theory review: an optimal control and dynamical systems perspective.
30 Scher, S. (2018). Toward data-driven weather and climate forecasting: approximating a simple general circulation model with deep learning. *Geophysical Research Letters* 45 (22): 12616.
31 Campbell, S.D. and Diebold, F.X. (2005). Weather forecasting for weather derivatives. *Journal of the American Statistical Association* 100 (469): 6–16.
32 Beyer, P., Benkadda, S., and Garbet, X. (2000). Proper orthogonal decomposition and Galerkin projection for a three-dimensional plasma dynamical system. *Physical Review E* 61 (1): 813.
33 Kaptanoglu, A.A., Morgan, K.D., Hansen, C.J., and Brunton, S.L. (2021). Physics-constrained, low-dimensional models for MHD: first-principles and data-driven approaches. *Physical Review E* 104: 015206.
34 Schmid, P.J. (2010). Dynamic mode decomposition of numerical and experimental data. *Journal of Fluid Mechanics* 656: 5–28.
35 Schmid, P.J., Li, L., Juniper, M.P., and Pust, O. (2011). Applications of the dynamic mode decomposition. *Theoretical and Computational Fluid Dynamics* 25 (1–4): 249–259.
36 Tu, J.H., Rowley, C.W., Luchtenburg, D.M. et al. (2014). On dynamic mode decomposition: theory and applications. *Journal of Computational Dynamics* 1 (2): 391–421.
37 Azencot, O., Erichson, N.B., Lin, V., and Mahoney, M. (2020). Forecasting sequential data using consistent Koopman autoencoders. *International Conference on Machine Learning*, 475–485. PMLR.
38 Lusch, B., Kutz, J.N., and Brunton, S.L. (2018). Deep learning for universal linear embeddings of nonlinear dynamics. *Nature Communications* 9 (1): 1–10.
39 Hockney, R.W. and Eastwood, J.W. (1988). *Computer Simulation Using Particles*. CRC Press.
40 Birdsall, C.K. and Langdon, A.B. (2004). *Plasma Physics Via Computer Simulation*. CRC Press.
41 Moon, H., Teixeira, F.L., and Omelchenko, Y.A. (2015). Exact charge-conserving scatter-gather algorithm for particle-in-cell simulations on unstructured grids: a geometric perspective. *Computer Physics Communications* 194: 43–53.
42 Kim, J. and Teixeira, F.L. (2011). Parallel and explicit finite-element time-domain method for Maxwell's equations. *IEEE Transactions on Antennas and Propagation* 59 (6): 2350–2356.
43 Evstatiev, E.G. and Shadwick, B.A. (2013). Variational formulation of particle algorithms for kinetic plasma simulations. *Journal of Computational Physics* 245: 376–398.
44 Squire, J., Qin, H., and Tang, W.M. (2012). Geometric integration of the Vlasov-Maxwell system with a variational particle-in-cell scheme. *Physics of Plasmas* 19 (8): 084501.
45 Jianyuan, X., Hong, Q., and Jian, L. (2018). Structure-preserving geometric particle-in-cell methods for Vlasov-Maxwell systems. *Plasma Science and Technology* 20 (11): 110501.
46 Meyer-Vernet, N. (1993). Aspects of Debye shielding. *American Journal of Physics* 61 (3): 249–257.

47 Na, D.-Y., Nicolini, J.L., Lee, R. et al. (2020). Diagnosing numerical Cherenkov instabilities in relativistic plasma simulations based on general meshes. *Journal of Computational Physics* 402: 108880.

48 Eastwood, J.W. (1991). The virtual particle electromagnetic particle-mesh method. *Computer Physics Communications* 64 (2): 252–266.

49 Marder, B. (1987). A method for incorporating Gauss' law into electromagnetic PIC codes. *Journal of Computational Physics* 68 (1): 48–55.

50 Pinto, M.C., Jund, S., Salmon, S., and Sonnendrücker, E. (2014). Charge-conserving FEM–PIC schemes on general grids. *Comptes Rendus Mecanique* 342 (10–11): 570–582.

51 Teixeira, F.L. (2014). Lattice Maxwell's equations. *Progress in Electromagnetics Research* 148: 113–128.

52 Na, D.-Y., Moon, H., Omelchenko, Y.A., and Teixeira, F.L. (2018). Relativistic extension of a charge-conservative finite element solver for time-dependent Maxwell-Vlasov equations. *Physics of Plasmas* 25 (1): 013109.

53 Warnick, K.F., Selfridge, R.H., and Arnold, D.V. (1997). Teaching electromagnetic field theory using differential forms. *IEEE Transactions on education* 40 (1): 53–68.

54 Teixeira, F.L. and Chew, W.C. (1999). Lattice electromagnetic theory from a topological viewpoint. *Journal of Mathematical Physics* 40 (1): 169–187.

55 Bossavit, A. (1988). Whitney forms: a class of finite elements for three-dimensional computations in electromagnetism. *IEE Proceedings A-Physical Science, Measurement and Instrumentation, Management and Education-Reviews* 135 (8): 493–500.

56 He, B. and Teixeira, F.L. (2006). Geometric finite element discretization of Maxwell equations in primal and dual spaces. *Physics Letters A* 349: 1–14.

57 He, B. and Teixeira, F.L. (2007). Differential forms, Galerkin duality, and sparse inverse approximations in finite element solutions of Maxwell equations. *IEEE Transactions on Antennas and Propagation* 55 (5): 1359–1368.

58 Clemens, M. and Weiland, T. (2001). Discrete electromagnetism with the finite integration technique. *Progress In Electromagnetics Research* 32: 65–87.

59 Schuhmann, R. and Weiland, T. (2001). Conservation of discrete energy and related laws in the finite integration technique. *Progress In Electromagnetics Research* 32: 301–316.

60 Wolf, E.M., Causley, M., Christlieb, A., and Bettencourt, M. (2016). A particle-in-cell method for the simulation of plasmas based on an unconditionally stable field solver. *Journal of Computational Physics* 326: 342–372.

61 Nicolini, J.L., Na, D., and Teixeira, F.L. (2019). Model order reduction of electromagnetic particle-in-cell kinetic plasma simulations via proper orthogonal decomposition. *IEEE Transactions on Plasma Science* 47 (12): 5239–5250.

62 de Witt, T.D. (1995). Enhancement of multichannel data in plasma physics by biorthogonal decomposition. *Plasma Physics and Controlled Fusion* 37 (2): 117.

63 Dudok de Wit, T., Pecquet, A.-L., Vallet, J.-C., and Lima, R. (1994). The biorthogonal decomposition as a tool for investigating fluctuations in plasmas. *Physics of Plasmas* 1 (10): 3288–3300.

64 Bellemans, A., Magin, T., Coussement, A., and Parente, A. (2017). Reduced-order kinetic plasma models using principal component analysis: model formulation and manifold sensitivity. *Physical Review Fluids* 2 (7): 073201.

65 Nayak, I., Teixeira, F.L., and Kumar, M. (2021). Koopman autoencoder architecture for current density modeling in kinetic plasma simulations. *2021 International Applied Computational Electromagnetics Society Symposium (ACES)*, 1–3.

66 Koopman, B.O. (1931). Hamiltonian systems and transformation in Hilbert space. *Proceedings of the National Academy of Sciences of the United States of America* 17 (5): 315.

67 Kutz, J.N., Brunton, S.L., Brunton, B.W., and Proctor, J.L. (2016). *Dynamic Mode Decomposition: Data-Driven Modeling of Complex Systems*. SIAM.

68 Brunton, S.L., Brunton, B.W., Proctor, J.L., and Kutz, J.N. (2016). Koopman invariant subspaces and finite linear representations of nonlinear dynamical systems for control. *PLoS ONE* 11 (2): e0150171.

69 Rowley, C.W., Mezić, I., Bagheri, S. et al. (2009). Spectral analysis of nonlinear flows. *Journal of Fluid Mechanics* 641: 115–127.

70 Baldi, P. (2012). Autoencoders, unsupervised learning, and deep architectures. *Proceedings of ICML Workshop on Unsupervised and Transfer Learning*, 37–49. JMLR Workshop and Conference Proceedings.

71 Le, Q.V. (2015). A tutorial on deep learning part 2: autoencoders, convolutional neural networks and recurrent neural networks. *Google Brain* 20: 1–20.

72 Rice, J., Xu, W., and August, A. (2020). Analyzing Koopman approaches to physics-informed machine learning for long-term sea-surface temperature forecasting. *arXiv preprint arXiv:2010.00399*.

73 Rahaman, N., Baratin, A., Arpit, D. et al. (2019). On the spectral bias of neural networks. *International Conference on Machine Learning*, 5301–5310. PMLR.

74 Xu, Z.-Q.J., Zhang, Y., Luo, T. et al. (2019). Frequency principle: Fourier analysis sheds light on deep neural networks. *arXiv preprint arXiv:1901.06523*.

75 Nayak, I., Teixeira, F.L., Na, D.-Y. et al. (2023). Accelerating particle-in-cell kinetic plasma simulations via reduced-order modeling of space-charge dynamics using dynamic mode decomposition. *arXiv preprint arXiv:2303.16286*.

76 Reichstein, M., Camps-Valls, G., Stevens, B. et al. (2019). Deep learning and process understanding for data-driven earth system science. *Nature* 566 (7743): 195–204.

77 Karniadakis, G.E., Kevrekidis, I.G., Lu, L. et al. (2021). Physics-informed machine learning. *Nature Reviews Physics* 3 (6): 422–440.

78 Towell, G.G. and Shavlik, J.W. (1993). Extracting refined rules from knowledge-based neural networks. *Machine Learning* 13 (1): 71–101.

79 Wang, F. and Zhang, Q.-J. (1997). Knowledge-based neural models for microwave design. *IEEE Transactions on Microwave Theory and Techniques* 45 (12): 2333–2343.

80 Ghaboussi, J., Garrett, J.H. Jr., and Wu, X. (1991). Knowledge-based modeling of material behavior with neural networks. *Journal of Engineering Mechanics* 117 (1): 132–153.

81 Noordewier, M.O., Towell, G.G., and Shavlik, J.W. (1991). Training knowledge-based neural networks to recognize genes in DNA sequences. *Advances in Neural Information Processing Systems 3 (NIPS 1990)*, 530–536.

82 Lagaris, I.E., Likas, A., and Fotiadis, D.I. (1998). Artificial neural networks for solving ordinary and partial differential equations. *IEEE Transactions on Neural Networks* 9 (5): 987–1000.

83 Marashdeh, Q., Warsito, W., Fan, L.-S., and Teixeira, F.L. (2006). Nonlinear forward problem solution for electrical capacitance tomography using feed-forward neural network. *IEEE Sensors Journal* 6 (2): 441–449.

84 Karpatne, A., Watkins, W., Read, J., and Kumar, V. (2017). Physics-guided neural networks (PGNN): an application in lake temperature modeling. *arXiv preprint arXiv:1710.11431*.

85 Raissi, M., Perdikaris, P., and Karniadakis, G.E. (2019). Physics-informed neural networks: a deep learning framework for solving forward and inverse problems involving nonlinear partial differential equations. *Journal of Computational Physics* 378: 686–707.

86 Wang, S., Yu, X., and Perdikaris, P. (2022). When and why PINNs fail to train: a neural tangent kernel perspective. *Journal of Computational Physics* 449: 110768.

Index

Advances in Electromagnetics Empowered by Artificial Intelligence and Deep Learning, First Edition.
Edited by Sawyer D. Campbell and Douglas H. Werner.

o

p

q

r

IEEE PRESS SERIES ON ELECTROMAGNETIC WAVE THEORY

Field Theory of GuidedWaves, Second Edition
Robert E. Collin

Field Computation by Moment Methods
Roger F. Harrington

Radiation and Scattering of Waves
Leopold B. Felsen, Nathan Marcuvitz

Methods in ElectromagneticWave Propagation, Second Edition
D. S. J. Jones

Mathematical Foundations for Electromagnetic Theory
Donald G. Dudley

The Transmission-Line Modeling Method: TLM
Christos Christopoulos

Methods for Electromagnetic Field Analysis
Ismo V. Lindell

Singular Electromagnetic Fields and Sources
Jean G. Van Bladel

The PlaneWave Spectrum Representation of Electromagnetic Fields, Revised Edition
P. C. Clemmow

General Vector and Dyadic Analysis: Applied Mathematics in Field Theory
Chen-To Tai

Computational Methods for Electromagnetics
Andrew F. Peterson

Finite Element Method Electromagnetics: Antennas, Microwave Circuits, and Scattering Applications
John L. Volakis, Arindam Chatterjee, Leo C. Kempel

Waves and Fields in Inhomogenous Media
Weng Cho Chew

Electromagnetics: History, Theory, and Applications
Robert S. Elliott

Plane-Wave Theory of Time-Domain Fields: Near-Field Scanning Applications
Thorkild B. Hansen

Foundations for Microwave Engineering, Second Edition
Robert Collin

Time-Harmonic Electromagnetic Fields
Robert F. Harrington

Antenna Theory & Design, Revised Edition
Robert S. Elliott

Differential Forms in Electromagnetics
Ismo V. Lindell

Conformal Array Antenna Theory and Design
Lars Josefsson, Patrik Persson

Multigrid Finite Element Methods for Electromagnetic Field Modeling
Yu Zhu, Andreas C. Cangellaris

Electromagnetic Theory
Julius Adams Stratton

Electromagnetic Fields, Second Edition
Jean G. Van Bladel

Electromagnetic Fields in Cavities: Deterministic and Statistical Theories
David A. Hill

Discontinuities in the Electromagnetic Field
M. Mithat Idemen

Understanding Geometric Algebra for Electromagnetic Theory
John W. Arthur

The Power and Beauty of Electromagnetic Theory
Frederic R. Morgenthaler

Electromagnetic Modeling and Simulation
Levent Sevgi

Multiforms, Dyadics, and Electromagnetic Media
Ismo V. Lindell

Low-Profile Natural and Metamaterial Antennas: Analysis Methods and Applications
Hisamatsu Nakano

From ER to E.T.: How Electromagnetic Technologies Are Changing Our Lives
Rajeev Bansal

ElectromagneticWave Propagation,Radiation, and Scattering: From Fundamentals to Applications, Second Edition
Akira Ishimaru

Time-Domain Electromagnetic Reciprocity in Antenna Modeling
Martin Štumpf

Boundary Conditions in Electromagnetics
Ismo V. Lindell, Ari Sihvola

Substrate-Integrated Millimeter-Wave Antennas for Next-Generation Communication and Radar Systems
Zhi Ning Chen, Xianming Qing

Electromagnetic Radiation, Scattering, and Diffraction
Prabhakar H. Pathak, Robert J. Burkholder

Electromagnetic Vortices:Wave Phenomena and Engineering Applications
Zhi Hao Jiang, Douglas H.Werner

Advances in Time-Domain Computational Electromagnetic Methods
Qiang Ren, Su Yan, Atef Z. Elsherbeni

Foundations of Antenna Radiation Theory: Eigenmode Analysis
Wen Geyi

Advances in Electromagnetics Empowered by Artificial Intelligence and Deep Learning
Sawyer D.Campbell, Douglas H. Werner

Printed in the USA
CPSIA information can be obtained
at www.ICGtesting.com
LVHW052007051023
759641LV00002B/1